The Nautical Gazetteer; or, Dictionary of maritime geography, etc.

Anonymous

The Nautical Gazetteer; or, Dictionary of maritime geography, etc.
Anonymous
British Library, Historical Print Editions
British Library
1847, 48
1-256 p. ; 8°.
1302.l.11.

The BiblioLife Network

This project was made possible in part by the BiblioLife Network (BLN), a project aimed at addressing some of the huge challenges facing book preservationists around the world. The BLN includes libraries, library networks, archives, subject matter experts, online communities and library service providers. We believe every book ever published should be available as a high-quality print reproduction; printed on- demand anywhere in the world. This insures the ongoing accessibility of the content and helps generate sustainable revenue for the libraries and organizations that work to preserve these important materials.

The following book is in the "public domain" and represents an authentic reproduction of the text as printed by the original publisher. While we have attempted to accurately maintain the integrity of the original work, there are sometimes problems with the original book or micro-film from which the books were digitized. This can result in minor errors in reproduction. Possible imperfections include missing and blurred pages, poor pictures, markings and other reproduction issues beyond our control. Because this work is culturally important, we have made it available as part of our commitment to protecting, preserving, and promoting the world's literature.

GUIDE TO FOLD-OUTS, MAPS and OVERSIZED IMAGES

In an online database, page images do not need to conform to the size restrictions found in a printed book. When converting these images back into a printed bound book, the page sizes are standardized in ways that maintain the detail of the original. For large images, such as fold-out maps, the original page image is split into two or more pages.

Guidelines used to determine the split of oversize pages:

• Some images are split vertically; large images require vertical and horizontal splits.
• For horizontal splits, the content is split left to right.
• For vertical splits, the content is split from top to bottom.
• For both vertical and horizontal splits, the image is processed from top left to bottom right.

PART I.   OCTOBER, 1842.

THE

NAUTICAL GAZETTEER;

OR

DICTIONARY

OF

MARITIME GEOGRAPHY.

UNDER THE   ESPECIAL SANCTION

OF   THE

LORDS COMMISSIONERS   OF THE ADMIRALTY

LONDON:

HENRY HURST, KING WILLIAM STREET, CHARING CROSS.

SOLD ALSO BY

R. B. BATE, CHART-AGENT TO THE ADMIRALTY, Poultry;

NORIE AND CO., CHART AND MAP-SELLERS TO THE ADMIRALTY, Leadenhall Street;

AND BY ALL BOOKSELLERS IN TOWN AND COUNTRY.

Price 2s. 6d.

ADDRESS.

It is not the custom to regard any effort to be useful with indifference, and the attempt to concentrate information of the present character from sources widely scattered must be admitted to come within the category of utility. Our national greatness reposes almost exclusively upon a maritime superiority, in which science is united with mechanical proficiency, or, in other words, cultivated intellect with manual skill and experience. Neither the dauntless enterprise and courage struggling against physical obstacles which marked the period of early British navigation, nor its historical details down to the close of the last century; abounding as they do in pictures of matchless perseverance and triumphant result, exhibited that union of intellectual with physical effort which advanced the science of navigation so greatly during the half century that has just run its course. Within the term of human memory the subjugation of the ocean has gone rapidly forward. Great progress has been made in acquiring a knowledge of the phenomena of the winds and waves, although in exploring their causes we may as yet have done no more than pick up a pebble here and there from the shore of the great ocean of truth. We are advancing from secondaries to primaries. We are busy in penetrating into existing mysteries through the collection of facts for data, and in establishing their correctness. We are exploring those great, perhaps simple because powerful agencies, which work out atmospheric and oceanic changes, ruling tides, directing the course of currents, and wielding the terrific energies of the tempest. More remarkable still, we are in the present day doing all this not with the exclusive view of avoiding what may be prejudicial to navigation, but with that of turning this very prejudicial action to account. This advancement of the intellectual part of the science has overcome many of its physical difficulties. The suffering experienced at sea by former navigators arising from disease, unwholesome atmosphere, and ill-found vessels, has been exchanged for the healthiest position in which the human body can be placed, and in well-found vessels, proportional security. The modern seaman incurs voyages of discovery with ease under which the amazing power of endurance of his forefathers must have sunk, and this facility he owes to scientific advancement. In voyages of exploration and in examining and surveying the most dangerous coasts, the patient persevering spirit of investigation is eminently awakened. Even the course and action of hurricanes are submitting to a careful investigation. In regard to these the theory of their progression, so recently advanced recently, may perhaps be found correct and made available for use of society. In the laborious and hazardous task of surveying, the names of Murice, Owen, Smyth, Vidal, Bayfield, Stokes—but it is invidious to particularize,—here with others contributed largely to expand the sphere of knowledge, and widen the bounds of maritime geography, securing to themselves a boundless debt of national gratitude. The Government of late years has not been wanting in lending assistance to this most important branch of the science, and it has recorded the results under the revision and superintendence of one of the most distinguished hydrographical officers this or any other country ever had the advantage of possessing.

It is under such circumstances and at such a period that the present compilation is submitted to general use. As yet but few points on the earth's surface are established in a perfect agreement of position, a disadvantage to which time and superior instruments will administer a remedy. Many, therefore, are drawn from former authorities which have already recorded them, others differ, and some are a mean of two or three observations by different individuals; a circumstance inseparable from the vast mass of materials collating. This very fact has been the cause of a degree of compression in the different articles not at first contemplated. The object of the concentration of so many localities will, however, be attained and a reference will always be at hand for the purpose of instruction and information, to gratify curiosity in relation to what is at present scattered far and wide over the oceanic surface. For the Australasian regions and those in the Pacific Ocean, where a vast extent of coast has been explored, named, and colonized, within human memory, a work of reference of the present character does not exist in any shape, and was therefore imperiously required to carry onward the march of improvement and extend the limits of the world.

THE
NAUTICAL GAZETTEER;
OR
DICTIONARY OF MARITIME GEOGRAPHY.

AA, a river in the N.E. part of France, having its source at Bourthes, in the department of the Pas de Calais, when, passing by Fauquembergues and St. Omer, where it becomes navigable, it flows to Watten and divides into two branches; one still named the Aa, falls into the sea at Gravelines, the other under the name of the Colme, passes to Bergues, where it joins the canal of Dunkirk. The Aa communicates with the Lys by the canal of Neuf-fosse, and with that of Calais and Bourbourg. Its mouth forming the port of Gravelines, lies E. of Calais; small vessels, at half-flood, pass up to the town, but boats only can go up to St. Omer. It is high water ¼ before 12. See *Gravelines.*

AA, THE, a river of Holland, at the port of Blockzyll. See *Blockzyll.*

AAG ISLAND or HOLM, on the coast of Norway, in lat. 58° N.

AAGADEER, or SANTA CRUZ, Africa, a town, 6 leagues S.E. of Cape Ghir, at the bottom of the bay of that name. It is the last port of Morocco in the Atlantic ocean. The town is on the summit of a hill, strongly fortified. The road is the best for shipping on the coast of Morocco. To anchor enter so far that the castle may bear N.N.E., and the storehouses E.N.E., which is to the S. of a rocky ledge. The best anchorage is in 6 or 7 fathoms, Cape Ghir bearing N. The anchor to be kept ready in case of storm, which is soon perceived by the action of the sea before it commences. Cape Ghir, or Ras Aferni, is in lat. 30° 37′ 30″ N., long. 9° 52′ W. High water 2ʰ 15ᵐ, rise 10 feet.

AAHUUS, AAHUYS, ACHUSEN, or HARHUS, North Jutland. A city having a good trade, the port standing on a river in lat. 56° 10′ N. and long. 9° 50′ E. The road is good, but the sea is tremendous with S.E. winds. A packet-boat sails daily from hence. Moles have been constructed for protecting vessels drawing 7 feet water. The harbour light is fixed on the pilot-house; there is a light also on the southern mole, visible 4 or 5 miles distance.

AALAND, or ALAND ISLANDS, a group in the Baltic, at the entrance of the Gulf of Bothnia, N.N.E. from Stockholm, in lat. 60° to 61° N., and long. 20° to 21° E. The passage into the gulf is on the W. side. The principal island is 20 miles long and 8 wide. Castleholm is the chief town.

AALBEC, a fishing town of Denmark, midway between the Scaw and Fladstrand. The fishermen here burn *two red lights* from September 1 to May 1, 15 ells distant from each other, and 8 feet above the sea, visible about 1¼ league.

AALBORG, the second city of Denmark, on the Lym Fiorden, or S. coast of Lymsint, so called from the number of eels taken there. An English

Vice-Consul resides here. There is a bar which prevents ships drawing more than 10 feet from going over. There are two buoys at the entrance, and two beacons on the S. shore. A reef at the entrance is avoided by bringing the two buoys into one, a little W. of the Muedburg mountains. Lat. 56° 50′ N., long. 9° 46′ E.

ABACO, GREAT and LITTLE. See *Bahamas.*

ABACOU POINT, Hayti, or St. Domingo, on the S. coast, composed of two reefs extending about one-third of a mile seaward. These may be passed without danger at the distance of 1½ mile in 40 fathoms. The town of Cayes lies 11 miles N. ⅜ E. from Abacou point, which is in lat. 18° 4′ N., long. 74° W.

ABAJO ISLET, Graciosa, off Point Carapacho, one of the Azores. About 2 cables' length distance, through this channel, if need be, the largest ships may sail to the anchorage of Praya.

ABB'S HEAD, a high promontory that, from the S. or N., appears first like an island. It is in lat. 55° 55′ 30″ N., long. 2° 8′ W. Close to the head is deep water.

ABBASCO, ABASCIA, or ABYAS, one of the countries bordering on the Black sea on the N.E., extending from 43° to 45° N. lat.

ABBEVILLE, France, department of the Somme, a city with a pop. of 18,000. It is seated on the Somme, several branches of which pass through it and the canal of Angoulême outside the walls. Ships of 100 and 150 tons come up with the tide to the city. There are manufactures of woollen cloth, of cotton, muslin, cordage, soap, tanning, and others established here. High water 10ʰ 30ᵐ. This city is 15 miles from the sea, in lat. 50° 6′ 55″ N., long. 1° 50′ 5″ E.

ABBOTSBURY, a town in Dorsetshire, situated in a valley, the inhabitants of which live principally by fishing. It stands at the N.W. end of the Chesil bank, which connects the Isle of Portland with the main. The ruins of St. Catherine's chapel form a conspicuous sea-mark between Abbotsbury and the shore. The town is 8 miles E. of Bridport. Pop. 880.

ABER, North Wales, E.N.E. of Bangor 6 miles, it stands at the mouth of the small river Gwyngregyn, and is a ferry-place to Anglesea and no more. It is sometimes called Abergwyngregin, from the cockles found on the coast. In front are the Lavan sands, which shift and are dangerous to cross. It is 3½ miles from Port Penrhyn. Pop. 560. *Aber* signifies, in British, the mouth of a river, and the place so named, or with *Aber* as a prefix, is generally connected with a haven thus situated.

ABERAERON, or ABERAYRON, in Cardigan bay, South Wales, E ¼ S., 3¾ miles from New Key head, a haven formed by two piers. From this

B

haven the coast turns N.E. by E. ¼ E. to Aberystwith: half a mile W. of the bar is Carregrock, round which there are 2 fathoms. Four miles E. is Sarn Cadwgan, a dangerous shoal, running out half a mile. Two tide lights on the shore must be brought in one to cross the bar, bearing S.E. ¾ S. It is high water at 7ʰ 30ᵐ. Rise 15 feet. Pop. 300.

ABERARTH, or LLANDDEWY, at the mouth of the river Arth, in Cardiganshire, with a small bar harbour in Cardigan bay, close adjoining Abereeron to the N.E.

ABERAVON, a town near the entrance of a bar harbour in Glamorganshire, South Wales, now called Port Talbot. Pop. 573.

ABERBROTHWICK, or ARBROATH, Scotland, 7 miles from the bar of the Tay, in Angus county, a small dry harbour at the mouth of the Brothwick river. Its chief trade consists in importing and spinning flax. Pop. 6,660. Lat. 56° 32′ N., long. 2° 34′ 15″ W.

ABERCASTLE, an inlet round St. David's head, Pembrokeshire, to the W. of which is Danllyn island; close to the shore, and half a mile N.N.E. from the isle, is Abercastle shoal. A mile from the shore all the way to Strumble head are 19 and 20 fathoms. Keep St. David's head well open off Peneleggr point; bearing W. ¾ S. will avoid the shoal.

ABERCONWAY, or CONWAY, Caernarvonshire, North Wales, a town, with a bar harbour fit only for small vessels, in lat. 53° 23′ N., long. 4° 30′ W. The channel is difficult, and the river not to be entered until at least three hours' flood. With S. or E. winds a vessel may stop off the mouth of the Conway, in about 1½ mile of the land, in 3½ or 4 fathoms water. The sands N. of the channel to the harbour stretch a long way out, and are 2½ miles in extent. There are two ways through them; to go clear to the W. of them, bring a round hill open W. of Penmaen Mawr, bearing S.W. by W. On sailing into the river, at the entrance of which are three black buoys, and above a red buoy, and a mile above a perch erected on rock marking the turn of the river to the S., steer for the outer or N.W. black buoy, leaving it and the second on the larboard; the third black buoy lies S.E. by E. ¼ E.; this with the red buoy and perch leave all to starboard. At the perch haul to the S. and continue for a mile, and anchor off Treenconway point on the larboard below Conway castle. The bar is 4 miles S.E. ¾ S. from the E. point of Puffin island, and S.W. ¼ W. nearly 3 miles from the W. end of Great Orme's head. The castle here was built by Edward I., and is a noble ruin. The trade of the place is small, consisting of bark, timber, corn, and slate. The town joins Caernarvon, Criccieth, Nevin, and Pwllheli, in returning members to parliament. It has a market weekly and several annual fairs. Pop. 1,245. High water 10ʰ 30ᵐ.

ABERDARON BAY, within the S.W. point of Caernarvon, South Wales, N.E. of Bardsey island. The best anchorage is in the middle of the bay in 4 fathoms, Aberdaron church bearing N. A heavy sea sets in with winds W. or S.W. It is high water at half-past 8.

ABERDAW, or ABERTHAW, a small harbour on the E. side of Breaksea point, in Glamorganshire, South Wales, 4 miles W. of Burry island.

ABERDEEN, NEW, a town and port of the county of that name, on an eminence at the mouth of the Dee, 108 miles N. of Edinburgh. The lat.

of Girdle Ness lighthouse, at the S. point of the bay, is 57° 8′ N., and the long. 2° 3′ W., distant from the N. pier of Aberdeen S. by W. 1,220 yards. There are two fixed lights here, 115 and 185 feet above the sea level, one over the other, visible 13 to 16 miles, but, at a distance, appearing as one light. Off the Ness lies a rocky shoal called the Girdle. There are two leading lights to enter the port, not exhibited when it is unsafe to enter the port. These lights are red, visible, when the weather is clear, 5 or 6 miles off. When first seen, coming from the N., they bear W.S.W.; from the S. due W., and when in a line W. ¾ S. nearly. The harbour is close in with the N. side of Girdle Ness. There is a beacon fixed on a rocky reef from the S. shore, and half a cable's length E. from the beacon is another reef, called Shock Ness. Hence a bar runs across the harbour's mouth about half a cable's length outside the pier head, with only 2 feet at low water, at high 16, spring tides. The marks are the N. side of the S. pier in sight, and old Aberdeen church, with two pointed steeples, on with the E. side of Broad hill. A pilot should always be taken. Aberdeen road is just round N. of Girdle Ness, the Ness S. by W., or S.S.W., the two Aberdeen steeples in one, in 7, 8, or 9 fathoms. Go no nearer than 7 or 9 fathoms between Aberdeen and Newburgh. The flood runs to the S. at Girdle Ness when strongest 2½ miles an hour, until 2ʰ 30ᵐ. Aberdeen is the principal city of Scotland N. of the Forth. It has a college called the Marischal College, and carries on a considerable trade with Greenland, the Baltic, Levant, and West Indies. The manufactures are woollen, linen, cotton, thread, and printed goods; there is also considerable ship-mending. The harbour is protected by two batteries. Old Aberdeen is a mile N. of New Aberdeen. It was once a bishop's see. It returns one member to parliament. The pop. of the borough and parish is 58,019.

ABERDINAS ROAD, on the E. side of Dinas head, in Newport bay, Pembrokeshire; an anchorage in W. winds in 3 or 4 fathoms, 2 cables' length from high-water mark.

ABERDOUR, or ABERDOWER, a village of Scotland, in Fife, on the Frith of Forth, 10 miles N.N.W. of Edinburgh. Pop. 1,751.

ABERDOUR, Scotland, a village on the N. coast of Aberdeenshire, 8 miles from Fraserburg; a fishing place only, the harbour having been ruined through neglect. Pop. 1,548.

ABERDOVY, or ABERDYFI, a haven at the bottom of Cardigan bay, the mouth of the Dyfi, or Dovy, a river which runs up to the town of Machynilleth. It is 8 miles N. of Aberystwith, safe for vessels not drawing more than 15 feet. Pilots attend, as the sands shift. It is the first land opening S. of Cader Idris mountain. Off the bar, 1¾ mile W. of the village of Aberdifi, is a black buoy, and ¼ mile further a red one. It is high water at 8ʰ, rise 15 feet.

ABERDSYNWY, an opening of the river Dsynwy in Merionethshire, North Wales, near Towyn: it is not navigable for vessels.

ABERFIORD, a small seaport on the coast of Norway, S.W. of Christiania 48 miles.

ABERFRAW, Isle of Anglesey, 13 miles E.S.E. of Holyhead. A fishing place where small vessels may lie aground. In moderate weather a ship may anchor in the bay in 4 or 5 fathoms, clean sand, about two cables' length from the N. side.

ABERGELE, Denbigh, North Wales. About 7

miles W. of Little Orme's head, and three-quarters of a mile from the shore, are Abergelé roads, opposite the town, which is half a mile inland. Anchorage here is only fit for small vessels during S. winds, the water being shallow. The shore, from Abergelé, runs E. ¼ N. 3½ miles to the river Clwyd, the sand extending dry half a mile from the land. In thick weather never shoal the water below 4 fathoms, between Orme's head and the land at the E. of the Dee entrance. Abergelé is a market-town, with a pop. of 2,506.

ABERILDUC, France, a small town within 4 leagues of Ushant, E., where there is a small harbour. It lies S. of the Boureau shoals and rocks, in the most westerly corner of France.

ABERISTWITH, or ABERYSTWITH. This harbour lies on the coast of South Wales, in Cardigan bay, 5¼ leagues E.N.E. from New Key head, and 9 E. by N. from Cardigan island. The harbour has 16 feet water at spring tides, protected by a pier 260 yards long, in a direction from S.S.E. to N.N.W. The river Rhydiol, which forms the harbour at high water, covers about 20 acres. On the pier there is a powerful capstan, and off the end of the pier are large transporting buoys, in 4½ fathoms spring tides; signals and lights also are exhibited. The harbour may be known by Pen-y-dinas hill at the S. end rising abruptly, and a ruined castle at the S.W. The time of high water is 7ʰ 30ᵐ, rise 15 feet. The town has become a fashionable watering place; the situation is pleasant, but the streets are uneven and steep. The rivers Ystwith and Rhydiol join here; there is a bridge of five arches uniting the shores of the Rhydiol. The new buildings are numerous, and the pop. above 4,000. There are lead mines in the vicinity. The exports consist of lead ore, corn, bark, slates, and timber, in vessels, some of which are of 200 tons burden. Herrings and cod are taken here, but not in such quantities as formerly. The castle was erected by Gilbert de Strongbow, rebuilt by Edward I. in 1277, and destroyed by Cromwell. There is a market twice a-week, in which some articles are sold by the pound of 18 oz. The flood tide runs with the coast from W. to E., but is weakened by the indraught, and the coast is dry at low water a considerable distance out. The lat. of the town is 52° 25' N., and the long. 4° 3' 19" W.

ABERKIBOR BAY, on the coast of Pembrokeshire, between Pendrwy head and the mouth of the Tyvi, in South Wales.

ABERLADY BAY, W. of Gulleness, in the Frith of Forth, Scotland, with soundings from 6 to 10 fathoms, clean ground and good anchorage everywhere. Gulleness point should be brought E. by S., and Port Seaton S.S.W., or S.W. by S. Ships often stand into this bay going to Leith roads, till North Berwick Law comes on the high land within Gulleness, keeping the Law open N. of the notch in the S. end of the high land. High water at 2ʰ.

ABERMAW, see *Barmouth*.

ABERMELIN BAY, between Fishguard bay and Strumble head, on the coast of Pembrokeshire.

ABERMENAI, the mouth of the Menai straits, between Anglesey and Caernarvon.

ABERPORTH ROAD, South Wales, lies E. of Cribach head, which last is 4½ miles E.S.E. ¼ E. from Cardigan island. The best anchorage is about 1½ cable's length from Ogavraih head, Cribach head bearing N.W. by W.

ABERTAY SAND, on the S. side of the entrance of the Tay on the E. coast of Scotland, stretching 4 miles E.S.E. from Tentsmuir Ness. Not to be approached nearer than 6 fathoms.

ABERVRACH, on the coast of Britany, in France, W. of Correjou, a harbour difficult of access from shoals and rocks. It is 8 leagues W. of the Isle of Bass, and nearly due S. of Plymouth. There are three channels to the port among the rocks, that of Malons to the E., Pedante in the middle, and one among the rocks E.S.E. There is 8 fathoms within, then 5, 4, 3, and 2. It is high water here at half-past 4. Turning to windward off the coast, come no nearer than 40 or 45 fathoms water, for it is deep even close to the rocks. Abervrach is in lat. 48° 32' N., and long. 4° 20ʹ W. A flashing light stands 2 miles E.N.E. from the outer anchorage here at the E. of Vierge island, 108 feet high, visible 15 miles, and varied by a red flash every 4 minutes. There are also two small lights, one red, fixed on Vrach island, visible 4 miles; the other bright, and fixed on Plougerneau church, 4 miles S.E. by E. by compass from the former, 226 feet above the sea, visible 10 miles.

ABIL ROCK, E. coast of Ireland, 7 miles N. from Howth head. It is 4 miles from the land. E. by S. of Skerries. Ships may sail round it.

ABLARIAK BAY, in Ungava bay, Hudson's straits, on the S.E. side, S.W. from Chudleigh island, Labrador.

ABO, Swedish Finland, in lat. 60° 26' 58" N., and long. 22° 17' 31" E., at the angle or S.W. point of Finland and Gulf of Bothnia, on the banks of the Aura Jocki. It had a bishopric and university, but these, with nearly the whole city, were destroyed by fire in 1827. The city had a pop. of 13,000. There is a castle at the entrance of the river, and vessels drawing 9 or 10 feet water go up to the place; larger vessels anchor 3 miles to the S.W. of the river, and their cargoes are conveyed in boats. Abo exports corn, iron, wood, fish, tar, pitch, salt provisions, furs, hides, coarse linen, &c., and has a considerable trade.

ABOUKIR BAY, Egypt, W.S.W. ¼ S. 18 miles from the entrance to Rosetta; celebrated for Nelson's victory over the French fleet in 1798. In the bay is the island of Aboukir, 2½ miles from which is an old castle. Three rocky islets lie off it, and a reef on the side of the island, but in the channel between the island and the main are 6½ and 6 fathoms. A reef extends also from the N.E. islet, called the Culloden's reef. From the point of Rosetta the bay forms a semicircle. On the W. side there is a dangerous bank, S.S.W. from the larger island, between which and its S.W. end there is a shoal of 2 fathoms. There is depth in the bay for the largest fleets, generally 6, 7, and 8 fathoms. Rosetta, its E. limit, is in lat. 31° 25' N., long. 30° 28' 20" E.

ABROLLA, West Indies, a shoal E. of Turk's island, in lat. 21° 33' N., long. 69° 50' W.

ABSECAN BEACH, North America, on the coast of New Jersey, 16 miles from Little Egg harbour.

ABYDOS, a town and castle of Asia Minor, now the S. castle of the Dardanelles; in lat. 40° 16' N., long. 27° 36' E.

ACADIA, North America, the name given by the French to Nova Scotia, and first applied to a tract from 40° to 46° of N. lat., granted to M. de Mons, of France, November 8, 1603, by Henry IV.

ACCAS, an island off the Ancobar river, on the Gulf of Guinea, in Africa.

ACCRA, Africa, West coast, a fort 10 leagues from Ningo river. Lat. 5° 32' N., long. 11' 30" W.

ACEYTERA, a dangerous reef off Cape Trafalgar, N. and S., a mile long, and two cables' length in breadth, 2 miles from the cape. On the N. end Trafalgar tower is in one with the Boqueron, and on the shoalest part of the danger this tower is in a line with the highest part of the high land of Meca.

ACHAMISH LOCH, Scotland, at 2¼ miles W. of Achane; passing Caliach head opens this loch, an anchorage only fit for fine weather; near the head, two or three cables' length only from the sand, a reef stretches two cables' length from Inivay point.

ACHASTIL LOCH, in Knapdale, Scotland, a fine harbour for vessels of any size. There is a rock midway, only covered at spring tides. On the W. side is the island of Donna, near which W. is an inlet called Loch Nakili, where vessels may stay a tide, except with a N.W. wind. There are several islets on the N.E., the outermost of which is Ruiskel island; about a mile from which, E.N.E. is a sunken rock, about a cable's length in circumference. In sailing towards Loch Achastil there is a small rock on the N. side of the two M'Cormick islands to be avoided; it is dry at spring tides only.

ACHEPI HARBOUR, North America, on the N.W. side of the island of Cape Breton, 4 leagues from its N. cape, which is in lat. 47° 7' N., long. 60° 3' W.

ACHILL HEAD, or ACHILL ISLAND, on the W. coast of Ireland, the W. point of the island, in lat. 53° 58' 20" N., and long. 10° 16' W., lying due N. from Sline head. It is high land 11 leagues S.W. from Killala. Between Achill head and Sline it is high water at 58 minutes past 4; springs rise 12, and neaps 6 and 7 feet. Achillbeg island and Achill island are separated by a channel called Achill hole, where a vessel drawing 10 or 12 feet may ride in all weathers. The stream of the tide here runs 3 miles an hour.

ACHINON PASS, Greek Archipelago, between the island of Negropont and the main, off its N.W. end, S. of the Gulf of Volo; in lat. 39° 40' N., long. 23° 55' E.

ACHORI LOCH, Scotland, on the W. shore of Linneheloch, W.S.W. of Ruchalochan point; vessels may ride here safely in moderate weather, and opposite to Shuna island is a good and safe harbour.

ACHUNE LOCH, Scotland, three-quarters of a mile to the W. of Mingary loch. It has no shelter against N. and N.W. winds. In the midst of it is a rock.

ACHUS, or AHUS, the port of Christianstadt, in Sweden, 10 miles from it inland. Vessels bound to Achus anchor behind the sand-banks at the entrance of the river, which boats only navigate, lying in 7 and 8 fathoms.

ACI REALE, Sicily, a mole, called the Marina of Aci, formed out of a mass of lava called Point Tocco, is the port of Aci Reale, to the W., built upon torrents of ancient lava. It is healthy, regular, and clean, with a pop. of 14,000. To the S. is Cape Molino, 2½ miles from Point Tocco, a promontory of hard lava, and W. is the cove of Trezzia, on the N. of which is the village of St. Anna. On the S. of the bay are rocks called the Cyclops. This cove or bay has tolerable anchorage in 12 and 7 fathoms, except near the rocks of Trezzia.

ACKLIN'S CAYS, or KEYS, in lat. 22° 26' N., long. 74° W. See Bahamas.

ACKMETCHID GULF, on the W. side of the Crimea and N.W. part of the Black sea. · Its W. cape is in lat. 45° 35' N., long. 32° 35' E.

ACLADI, on the N.W. side of the Achinou pass, in the Greek isle of Negropont, having a good coast.

ACOLA, CALA BINI, the most S. cape of the isle of Minorca, S.W. of the entrance of the harbour of Mahon 7 miles.

ACOMACK, a county of Virginia, North America, bounded W. by the Chesapeake bay.

ACQUIDAH, coast of Africa, a Dutch fort, 4 leagues from Dixcove.

ACRA, the S.W. point of Hadshilar port, on the E. side of the Crimea, situated in lat. 44° 42' N., long. 37° 16' E. Ada gulf lies on the E. side of the straits.

ACRE, or ST. JEAN D'ACRE, a bay and city of Palestine, small and well fortified; in lat. 32° 54' 35" N., long. 35° 6' 20" E. The port is neglected, unfit for large vessels, the anchorage rocky and bad.

ACUL, BAY OF, on the N. coast of St. Domingo, or Hayti, near Port François.

ACUNHA, see Tristan d'Acunha.

ADAIA, or D'ADAIA, a port on the N.E. coast of Minorca, sheltered from the N.W. winds, a little E. of Mount Toro.

ADALIA, at the bottom of the gulf of that name, in Karamania, the outer road of which presents excellent anchorage in 20 and 15 fathoms. The port is too small for ships of very moderate tonnage.

ADALIA ESKE, the ancient Sidé, is about 7 miles E.S.E. of a river, in 31° 18' E. long. No inhabitants are here, no wood nor water to be obtained; but there are the remains of harbours filled up, many ruins, and the theatre of Sidé, the most important remains of antiquity in Asia Minor. The theatre ruins are in lat. 36° 45' 37" N., long. 31° 25' 47" E. It may be known by a cleft in the mountains, through which passes the river Manaogat.

ADALRUK, Iceland, near Cape North, on the Arctic line.

ADAN COAST, windward coast of Africa, between the ivory and gold coast, best known as the Quaqua coast, in lat. about 5° N., and long. 4° W.

ADANA, a town on the river Choquen, in Karamania, in lat. 37° 26' N., long. 36° 12' E.

ADDA, or ADONA, Africa, a village on the Gold Coast, close to Cape St. Paul.

ADELAIDE ISLET, Fernando Po, African coast. Lat. 3° 34' 48" N., long. 8° 47' 17" E.

ADELPHI, a small island in the Greek Archipelago, 1 league from Scopolo, E.S.E.

ADESSA, or ALDEA ROAD, Teneriffe, on the W. side of the island, 6 leagues E. of Gomera island, over against it, open to the S.W.

ADMIRAL'S COVE, Newfoundland, within the harbour of Formosa, 7 or 8 leagues from Cape Race, N. by E. Vice-Admiral's cove is within the same harbour.

ADOUR, see Bayonne.

ADRA, Spain, in the Mediterranean, having a beach affording shelter from E. to N.W. winds, in 7 or 8 fathoms, the castle bearing N. ¾ E., about 2¼ cables' length from the shore. Two miles from the river of Adra is a tower called Aljamilla, near which are rocks. Lat. 36° 42' N., long. 2° 37' W.

ADRAMYTTI, GULF OF, in the Ægean sea, on the coast of Asia. The isle of Mytilene stands W. of it, and there are two passages in; one on the E., the other on the N. of that island; the E. passage is 10 miles wide, the N. about 5: there is deep water in both. From the N.E. point of Mytilene to the head of the gulf is 26 miles. Four miles inland stands the ancient town of Adramytti, now called Ydramit, a miserable place, in lat. 39° 34′ 50″ N., long. 26° 57′ E.

ADRIA, an ancient name for the Méditerranean sea.

ADRIAN, CAPE ST., Spain, W.N.W. ¼ W., 7 leagues from Corunna lighthouse. Off this cape lie the Cisargas islands, W. ½ S. from Cape Ortegal 50 miles.

ADRIATIC SEA, see *Gulf of Venice.*

ADVANCE BAY, North America, on the E. side of Hudson's bay.

ÆGADEAN ISLANDS, Mediterranean, off Sicily, namely, Maritimo, Favignana, Levanso, Fomiches, and Porcelli, with the banks of St. Catherine. The W. isle, Maritimo, has its castle in lat. 38° 1′ 10″ N., long. 12° 3′ 55″ E., and is nigh 7 miles in circumference. It has a few scattered houses called the village of San Sinone. The castle has a garrison of 40 Sicilian soldiers. Vessels bound to Palermo from the W. should keep St. Vito on the larboard bow; it is of a high, conical form. If bound to Trapani, having made Maritimo, they should pass close by the N. point of Levanso, E.S.E. ¾ E., for the lighthouse of Columbara at the entrance of Trapani. The island of Favignana is 14 miles in circumference, pop. 3,000. Its N. point lies E. by S., 11 miles from Maritimo castle. 1¼ mile to the S. is the port, a fine cove, with San Leonardo, a town, on one side, and a small fortress on the other. The castle is in lat. 37° 56′ 36″ N., long. 12° 17′ 45″ E. The banks of St. Catherine lie S. of Favignana 2 or 3 miles. There are two, divided by a channel a mile broad. Levanso is 3 miles N.E. ½ N. from Favignana, nearly inaccessible. The Fomiches are two rocky islets, a league E. ½ S. from the S.E. point of Levanso, frequented by the tunny fishermen. Porcelli, or the Pigs, are a dangerous reef 4½ miles E. by N. from Point Grosso, the N. point of Levanso, and N.W. from the lighthouse of Trapani.

ÆGEAN SEA, THE, now best known as the Greek Archipelago, extending from the Levant or E. end of the Mediterranean to the Dardanelles.

ÆGELSTAWIK, a harbour half a mile from the town of Sodertledge, in Sudermanland, Sweden.

ÆGINA, or ENGINA, an island of Greece, in the middle of the gulf of the same name, triangular in form, and about 6 leagues in circumference, having its port on the W. side. Near the N.E. point are the remains of the Temple of Jupiter, consisting of 23 columns. The principal town is on the site of the ancient Egina. Pop. 10,000. At the S.W. point is the island of Mogni. Lat. 37° 39′ N., long. 23° 33′ E.

ÆOLIAN ISLANDS, see *Lipari.*

AFFACUS, or SFACUS, on the Barbary coast, a small port, with a mole opposite the Karkenna islands, S. of Cape Africa.

AFRICA, CAPE, on the Barbary shore, nearly opposite the island of Lampedoso, in lat. 35° 36′ N., long. 11° 10′ E.

AFRICA, CONTINENT OF, one of the great divisions of the globe, extending from its most N. point in the Mediterranean, at Cape Blanco, in lat. 37° 25′ N., to lat. 34° 29′ S., its most southerly, at the Cape of Good Hope. In breadth it extends from Cape de Verde to Cape Guardafui, between 17° 32′ long. W., to 50° 12′ 0″ E. It is consequently about 4,250 miles long, by 3,900 broad. At once the most ancient in the cultivation of the arts, and the most barbarous; the least cultivated, and in certain spots the most fertile.

It is peopled with a race of a peculiar temperament, inured to the excessive heat and the pestilential nature of a territory fatal to the health of the inhabitants of other countries, even of those who inhabit soils equally torrid. This physical difference is the more remarkable, inasmuch as that while the inhabitant of the hottest parts of Asia sinks under the baneful yellow fever of the West Indies and Africa, the negro is proof against the disorder. In Africa, the animal kingdom is in a manner its own. In soil, part fertile and part deserts of burning sand, rivers peculiar in their characteristics, in some parts with shores rank from a vegetation that in decomposition spreads death far and wide; these things make it more singular on every account than any of the other great divisions of the globe. Here the arts of civilization have made no progress; the ages that have passed, and the generations that are no more, however for a moment they marked some insignificant spot with their temporary influence, time soon saw relapse again into primeval non-existence. The vast shores of this continent, so prolonged from temperate to torrid and from torrid to temperate, are washed by the more celebrated seas of the globe; the Mediterranean on the N., the Red sea on the E., the Indian ocean and the North and South Atlantic, break alike upon its burning shores. To the seaman the coasts are everywhere objects of some kind of peril; flat sands stretching far out from the shores—rivers, the anchorages of which teem with pestilence—a heavy surf raging upon a large portion of its coasts, and a paucity of good ports, render it particularly uninviting, save to that commercial enterprise which defies all hazard in the pursuit of lucre. The country of ivory and gold, it scatters over the world that insatiable craving which braves all dangers, while it remains to the seaman rather the trial of his better qualities of endurance under suffering, than of proud hopes or satisfactory recollections.

The inhabitants of this great continent are of various races. In the N., the Egyptians, supposed to be the present Copts, the Abyssinians, and the Moors; in the central regions, the negro races; and in the S. the Caffre and Hottentot, are well known and recognised. The general creed in religion is the Mahommedan in the N., and the pagan southward, while among some races no religious belief is traceable. The rivers, in proportion to the land surface, are few, and marked by peculiarities of their own, as the Nile by its rise and fall, and the Niger by its singular course. The mountains known to Europeans are many of them lofty, but not so high as those in Asia or America. But the most remarkable feature in this continent is its vast deserts of arid land, of which the chief is the Sahara or Zaara, the name for 'desert,' stretching from the confines of Egypt to the Atlantic, a length of 2,400 miles, and at least 720 broad—a region of burning sand. There are numerous other and smaller deserts scattered over its surface. There is little to interest the seaman on the shore of Africa on the side of the North Atlantic or the Mediterranean; the headlands of the last, and the long, sandy, shallow shores of the first, offer few ports worthy of regard, and tender a reception almost always inhospitable, and where, in some degree, the rites of hospitality are barbarously exhibited, disease and death attend their acceptance. Morocco, extending from the Straits of Gibraltar to Agadeer, offers few attractions. At El

Araich provisions are procurable for the seaman, but the coast is often low, and in winter rough S.W. and S.S.W. winds prevail. On all the Morocco coast, Mogador alone maintains a commercial intercourse with Europe.

Passing Cape Aguluh, from Capes Juby and Noon to Bojador, it is well to keep off shore, the weather being often hazy, and the currents, as far as Cape Bojador, running in the direction of the coast, from 4-10ths to a mile an hour, increasing in rapidity at Cape Juby. Here every effort is made by the natives to plunder wrecks and get possession of the crews. From Cape Bojador to Cape Blanco the space is divided by the tropic, about half-way being in the torrid regions of the globe. On land it is included in the vast desert of Sahara, composed of loose fine sand, driven about by the winds, some of which has been carried 300 miles over the ocean by the N.E. gales. The water off this part of the coast is deep in some places. S. of Cape Bojador, about 3°, the Rio Ouro opens into the sea, in 23° 36′ N. lat., near which the shore offers places of anchorage. St. Cyprian's bay, Cape Barbas, and Cape Corvoeiro passed, Cape Blanco rises, covered with sand. Then succeed the bank of Argouin, the currents about which set N. and S., Cape Mirik, Angel Hillock, and Angel Bank, are all places of the African character, arid and harbourless. These passed, Portendick appears, marked by two palm trees, or rather its supposed site. The trade here was negligently abandoned to the French in 1814. A vessel may run along the shore here 2 miles from the beach, in from 9 to 14 fathoms. The bar of the Senegal is not stationary. The winds on this part of the coast are not dangerous. It is 31 leagues from Senegal to Cape Verde, the most W. point of Africa. The same character of country prevails to the Gambia river, except that the land is covered with vegetation, but, if possible, increases in unhealthiness as the Gambia is approached. Rivers having muddy shores, from which putrid vegetation diffuses pestilence, a paucity of harbours and many anchorages, mark the coast of the continent as far as the river Congo. The coast is generally divided into the gum, the grain, the ivory, and the gold districts, as far as Benin, in the bight of which the once mysterious Niger is now found to disembogue its waters. The atmospherical temperature in the country N. of Sierra Leone, on the coast of Africa, and in the Rio Nunez, about lat. 10 N., is in March and April, 6 a.m., from 75° to 84°, at noon from 84° to 94°, and at 9 p.m., from 81° to 83°. Unwholesome fogs are often found upon this part of the coast. The winds from Cape Blanco to Sierra Leone blow in common rather from the N. to N.W. than from N. to E. From Sierra Leone to Cape Palmas the winds are mostly W.N.W., and beyond that Cape from W.S.W. to S.W. and S.S.W. In the Gulf of Guinea the winds generally blow from the S. and S.S.W. towards the coast, but in a S. lat. they take a more W. direction, close to the land, to the S.W. and W.S.W., between Cape Lopez and Benguela, becoming more S. as the distance from the coast increases. The Windward Coast is the term given to the whole of Africa, extending from Cape Mount to the river Assinee, where the Gold coast begins, and includes the Grain, Ivory, and Adan or Quaqua coast. From Assinee to the river Volta, on the Gold coast, the wind in January blows from the S.W., becomes stronger in February, sometimes to a hurricane; at the end of March,

tornados, rain, thunder, lightning, and even earthquakes are experienced, and continue to the end of May. In the rainy season, or May and July, there are no land winds, but it blows from S.W. and W.S.W., raising a great sea. In the months of December, January, and February the harmattan blows, coming on at any hour, and blowing sometimes only for a day at a time, at others for twelve or fourteen; it is not as strong as the sea breeze, which last sets in during the fair season from the W., the W.S.W., and S.W. It prevails on the coast between Cape Verde and Cape Lopez, or between lat. 14° 44′ 30″ N. and long. 17° 32′ W., and lat. 36′ 12″ S. and long. 8° 45′ 17″ W. The harmattan blows from the N.E. on the Gold coast; at the Isles de Los, N. of Sierra Leone, from the S.S.E.; and at Cape Lopez from the N.N.E. This wind is always accompanied by a gloomy haze, the sun is of a red colour, seen only an hour or two at noon; the fog is less thick on the sea, and is lost at 4 or 5 leagues from the shore, though the wind is felt for 10 or 12. The peculiar property of this wind is its extreme dryness; evaporation is increased by it from 64 to 133 inches: it is conducive to health. A wind up the coast blows from S.S.W. to S.S.E. for three weeks, from February to March. The natives distinguish the general winds in the W. of Africa, as the Harmattan or Aherramantic, blowing about 10 weeks, from December 1 to the middle of February; the Inakera, a coast wind from S.S.W. to S.S.E., from March 1 to the equinox; the Pempina, or tornado season, from March until May, lasting 12 weeks; the Abrenama, in June and July, enduring 8 weeks, which is the rainy season; the Atukogau, or strong winds and squalls, to the middle of August, three weeks; Worrobokorou, or three weeks' cessation of rain; Mawarrah, or foggy weather, in September, three weeks; Boutch, or no land breeze, the wind blowing six weeks down the coast; and lastly, the Antiophi, or frequent S. tornados and rain, called the latter rains, four weeks, to the beginning of December, when the Harmattan commences. From Cape Bojador to the Isles de Los no rain falls for eight months, and regular winds blow, but a peculiar haze at all times hangs on the African coast. The coast from Cape Lopez to the river Congo or Zaire, which last is in lat. 6° 6′ S., long. 12° 10′ E., may be run past in 16 fathoms, at 3 leagues from the land. In fine weather the winds are generally from S.W. afternoon, to N.E. early in the morning. The current sets with great rapidity near the land. From the Congo to the S. anchorages will be found. That river itself is navigable 80 leagues E. by N. The seasons are later than more to the N. The winds from September to March are commonly between S. by W. From Cape Padron to Loanda San Paulo, there is much the same character of coast as that more to the N. Loanda is a Portuguese settlement, the town containing 7,000 inhabitants, a third of whom are white. From hence there is nothing differing from the usual character of the African coast down to lat. 34° 27′, where the territory included in the Cape of Good Hope colony commences, the description of which is familiar. Off the extreme S. of the African continent Cape Aguillas is found, and the extensive bank of that name, E.S.E. from the Cape, from which it is necessary to steer to a considerable distance to avoid. On the E. side of the continent of Africa, along the Mozambique channel and as far N. as the Red sea, the same singularities mark this continent as on a greater degree are found on the W. side. From Delagoa bay, as far as the Straits of Babelmandel, the same verdant shores and unhealthy coasts or sands and burning deserts prevail, continued N. as far as Suez. The currents proceed from the same cause, the action of the wind, as on the other side, but have not yet been so well observed. From the discoveries made in this continent, its want of inland seas is not less singular than the paucity of its coasts in ports of tolerable excellence, and the marked and indeed distinct character of its population, the number of which there is no mode of calculating with any probable approximation to truth. The inhabitants of this least blessed of the grand divisions of the earth's surface have been condemned to be the slaves of the inhabitants of other countries; the physical condition of man and his locality here seeming to be in a species of unison alike singular and lamentable.

AGALIMAN, on the coast of Karamania, a good harbour for coasting vessels, and large ships might find shelter in the bays on each side of the little peninsula. Above it is a ruined fortress. It is 9 miles from Selefkeh, now a poor village. Lissan el Kabeth, a low point, rises to the S.E., which must have a good berth. The only point between this and Syria is Karadish Boornoo, a white cliff 129 feet high, in lat. 36° 32′ 40″ N., long. 35° 21′ 26″ E.

AGAMISCO ISLAND, North America, James bay; lat. 53° N., long. 75° 6′ 30″ W.

AGANLY RIVER, Sea of Azof, on the S. side of the N. branch of the Kuban river, the town of Ashuyer being on its W. side, in lat. about 45° 30′ N., long. 38° 3′ E.

AGATHONISI, a small island of the Greek Archipelago, 1 league S. of Samos.

AGATTAN, Africa, Guinea coast, 80 miles S. of Benin, in lat. 7° 20′ N., long. 7° 6′ E.

AGAY, France, a small port of the department of the Var, 2½ leagues E. of Frejas, at the entrance of a river of the same name.

AGDE, France, a town and port in the department of the Herault, 5½ leagues E.S.E. of Beziers, with a pop. of 7,500. It stands on the left bank of the Herault, a league from the sea. The port will receive vessels under 200 tons, but is principally used by fishing vessels. The tower of the cathedral is in lat. 43° 18′ 40″ N., long. 3° 33′ 22″ E. On the E. side of the mole vessels may anchor for a tide near the Cape or Conqua rocks in 3 or 4 fathoms, but they must not come too near Cape Brescou; there are dangerous rocks between the mole head and Fort Brescou, also to the W.; therefore, coming from the E., range along the mole head in from 17 to 20 feet of water. There is a fixed light on the end of the E. jetty, elevated 20 feet, and visible in clear weather 3 leagues. Above 120 vessels, from 100 to 200 tons, belong to this port.

AGDENAS BAY, Norway, in the district of Drontheim.

AGEN, BEC DE, France, the larboard point at the entrance of the river Soulle, which forms the port of Coutances, E.S.E. of St. Heliers, in Jersey.

AGERROE, Denmark, a small island in the Great Belt, 2 leagues S. from Corsoe.

AGHENISH, Ireland, an island in the Shannon, 16 miles below Limerick.

AGHRIS POINT, Ireland, W. coast, in Sligo, lat. 54° 17′ N., long. 9° 22′ W.

AGNES, ST., Cornwall, an exposed opening among perpendicular cliffs veined with metallic ores, near which rises St. Agnes' beacon, a hill serving as a sea mark, in lat. 50° 18′ 27″ W., long. 5° 11′ 56″ W.

AGNES, ST. See *Scilly Isles.*

AGNUNI, a village on the E. coast of Sicily, 10 miles from Catania, an unhealthy place, whence corn, oil, and rice are exported.

AGON ISLAND, Sweden, Gulf of Bothnia, in lat. 61° 20′ N., long. 18° 10′ E.

AGOSTA or LAGOSTA ISLAND, Gulf of Venice, nearly S. from the W. end of Meleda island, and W. of Augustina reefs and shoals: it has a good roadstead. Lat. 42° 45′ 1″ N., long. 16° 51′ 45″ E., the most S.W. of all the Dalmatian islands.

AGRICO, Africa, on the Gold coast, 37 miles W. of the river Volta; the ground is foul near the shore.

AGROCA ROAD, W. of Bastimantos, near Porto Bello, in South America, where a few ships may lie landlocked by several islands. To come out there is a narrow channel, but safe. The wind blows off shore at night, and E. during the day.

AGROPOLI, a small town of Italy, in Naples, at the entrance of a river of the same name, near Cape Licosa, which last is in lat. 40° 13′ 50″ N., long. 14° 53′ E.

AGUADILLA BAY, in the W. part of the island of Porto Rico, belonging to Spain. The town, of the same name, is situated in lat. 18° 25′ 10″ N., and long. 67° 7′ 17″ W. It lies N.E. of Cape Francisco 7¼ miles, at the end of a sandy bay. The anchorage is good. A reef running out from Cape Palmas, a mile S. of Point Brujuere, the N.W. point of the island, is the only danger to be avoided.

AGUILLON, POINT, a long narrow point to the N.E. of the Isle of Rhe, near Rochelle, on the coast of France, 3 leagues N. of that city, a little W. Between this and the land, 3 or 4 miles to the S., is 3 fathoms water, being a part of Pertuis Breton, a channel N. of Ré island.

AGUILLONES, Spain, a cluster of large rocks W.N.W. ½ W., 7½ miles from the river Barquero, in which is clean, deep, good anchorage, defended from S.E., S.W., and N.W. winds. A little to the W. of these rocks is Cape Ortegal. Lat. 43° 46′ 30″ N., long. 7° 28′ 15″ W.

AGULIA ISLES, Mediterranean, off Cape Serrat, near Galita island, mere islets. The S. end of Galita is in lat. 37° 30′ 50″ N., and long. 8° 54′ E.

AGULUH, CAPE, Africa, a slight projection of the coast, in lat. 29° 49′ N., long. 9° 48′ W.

AGUSA, PORT, in the Isle of Candia, on the S. side, lat. 35° 3′ N.

AHONI, a small port of Africa on the coast of Benin.

AIGUES MORTES, France, department of the Gard, a town 2 leagues from the sea, which once flowed up to it, and with which it now communicates by canal. The land round the town is now a marsh, but the town is neat though small; the pop. 2,600. Lat. 43° 34′ 47″ N., long. 4° 14′ 2″ E. Vessels do not anchor off the bight of Aigues Mortes without great caution, the current causing frequent wrecks. There is a mole off the shore having a fixed light, which is varied by flashes every five minutes, visible 5 leagues.

AILSA, an insulated rock on the coast of Scotland, near the Isle of Bute, 2 miles in compass and 900 feet high, accessible only on the N.E.

AILY, CAPE D', France, 5 miles W. by N. ¼ N. of Dieppe, with a bottom foul 1½ mile from.

the shore. Keep off half a mile. There is a light-house here 5 miles from Dieppe, 80 fathoms from the cliff, revolving and darkening every 80 seconds, seen 7 leagues off, 305 feet above the sea. As the lights are totally darkened it cannot be mistaken for Cayeux light.

AIR, or AYRE, POINT, the N. point of the Isle of Man, in lat. 54° 25′ N., long. 4° 26′ W. It is high water at 10ʰ 30ᵐ. King William's sand begins 7 miles E. of this point, and extends E.S.E. and S.E. by E. 6 leagues.

AIR, or AYR, a port of Scotland, in the Frith of Clyde. Lat. 55° 28′ N., long. 4° 3′ 7″ W. It has a pop. of 7,606, and a considerable trade in coal, with soap and leather manufactories. Ayr harbour has one light of a red colour, and a tide light 8 feet.

AIR, POINT OF, Flintshire, North Wales, the W. entrance of the river Dee, S.E. by E. ¾ E., 14 miles from Great Orme's head. On the point is a round tower striped red and white horizontally, having two fixed lights; the upper, of a natural colour, to be seen 11 miles off, 49 feet above high water, between the points of W. by N. seaward to N.W., and E. to S.S.E. ½ E. The lower light, of a red colour, is visible from N. by W. to N.N.E., 12 feet above high water, seen at 6 miles distance, intended for the navigation of the Welsh channel, or Mostyn, S. of the W. Hoyle bank. Lat. 53° 21′ 26″ N., long. 3° 19′ 14″ W.

AIMONTE, or AYAMONTE, on the E. side of the river Guadiana near its mouth, the boundary of Portugal on the W., in lat. 37° 5′ N., and long. 7° 15′ W. It has a deep bar, and is one of the best havens on the coast, with 3 fathoms at half-flood. Some shoals may be avoided by bringing a tree on the E. side of the river's mouth right on with the town, which direction is E. of the shoals. The coast to Port Maria when within is N.W. with 6 or 7 fathoms water. Tavila is 5 leagues W.S.W., and St. Michael 7 leagues E. by N. It is 34 miles W.S.W. of Seville, in the Gulf of Cadiz.

AIOMANO, GULF OF, in Romania, towards the N.W. corner of the Greek Archipelago. Cape Drapano is its N.E. extremity. Its S. point is in lat. 39° 50′ N., and long. 24° 40′ E.

AITKIN'S ROCK, W. of the N.W. of Ireland. This is one of those dangers in the ocean called Vigias, the existence of which is yet very uncertain. The first notice of this rock dates as far back as 1644. Its position was then said to be in lat. 55° 18′ N., and long. 11° 14′ W. from Tory island. Other accounts make its position 55° 15′ N., and long. 10° 40′ W. It is said to have been seen in 1826. In 1830, the Onyx and Leveret gunbrigs were sent in search of it, but nothing was discovered.

AIX, ISLAND OF, France, 4 leagues N.W. from Rochfort, about 6 square leagues in surface, and having 450 inhabitants, principally fishermen. Vessels from Rochfort generally descend to its road to complete their equipment for sea. The island is fortified and has a fixed light on the fort or S. point of the island, elevated 56 feet, and visible 3 leagues. Aix is in lat. 46° 0′ 36″ N., long. 1° 10′ 33″ W. The Longee sand is S.W. 2 miles from the S. end of Aix, and 1¼ mile E. of Oleron, and extends N. 2 leagues along the land of Oleron, uncovered at each end at low water. It has a bay on the W. with 5 fathoms, but there are rocks off its S. end, as well as from the N.W. round the N. point to the N.E. At springs it is high water soon after 3 o'clock.

AIXO, or AIXOS, South America, the shallows that lie within the entrance of the harbour of Carthagena stretching out towards the mainland S.S.W. Aixos is a general name for *shallows* or *flats* along the Spanish American coasts.

AJACCIO, the chief town of Corsica, walled and defended by a citadel, with a pop. of 8,000, standing at the bottom of a gulf of the same name. The N. point of the gulf is Cape Sanguinario, the S. Cape Muro, distant from each other S. by E. ½ E. about 10 miles. On coming from the N. the islands of Sanguinario are seen, the largest having a round tower on its highest part, and at the S.W. end an old square one. To the E. are two rocks called the Brothers and Little Sanguinario. There are dangers about all these, and great care is required in the navigation by strangers in entering the bay near either point. There is deep water, 10 fathoms, close to some of the rocks. At Cape Sanguinario there is an intermitting light established. The citadel of Ajaccio stands in lat. 41° 55′ 1″ N., and long. 8° 44′ 4″ E. In entering the harbour, particularly at night, a ship goes no nearer the shore than 25 fathoms. This town is celebrated as the birthplace of the Emperor Napoleon.

AJAZZO, a port of Natolia, near the angle of the coast at the N.E. end of the Mediterranean, 10 leagues N. of Antioch, N.E. from Cyprus, in lat. 37° N., long. 33° 10′ E.

AKERMAN, or BIELGOROD, a town of Bessabaria on the coast of the Black sea, and mouth of the Dniester, 36 leagues S.W. of Otchakoc, and 31 W. of Cherson. Lat. 46° 8′ N., long. 31° 14′ E.

AKERSUND BAY, Norway, 10 leagues W.N.W. from Frederickstadt.

AKITAKI, or LITTLE COMMENDA, Africa, Gold coast, 26 leagues E.N.E. from Cape Three Points, and 6 nearly E. from Secunda. High water ¼ past 4. Separated from Commenda only by a river; the lat. of the last is 4° 54′ N.

AKKIA, an island near the W. coast of E. Greenland, in lat. 60° 38′ N., long. 46° W.

AKLIMAN, a small harbour of the Black sea, 8 miles W. of Sinope. It is good, but difficult to enter. It is close to Cape Indjeh on the E. side. The cape is in lat. 42° 8′ N., long. 34° 56′ E.

AKTEBOLI, a small haven in the Black sea, in lat. 42° 4′ N., long. 27° 59′ E., 6 miles from Rezveh, and 69 from the Bosphorus.

ALACRAN, or THE ALACRANES, a shoal in the Bay of Campeche, having on it three islands, Perez, Chica, and Pajaros. There is a harbour S. and E. from Perez, formed by the reefs. Lat. 22° 23′ N., long. 89° 42′ W.

ALADRA, Spain, in the Bay of Almeria, before which there is anchorage in 10 or 12 fathoms, not far from the shore, towards the E. end of the bay, 6 leagues E. of the town so called.

ALAN BAY, Corsica, on the W. side, where good anchorage is found.

ALAN RIVER, the same as the Camel river, Cornwall, on the N.W., at the entrance to which is the harbour of Padstow. See *Padstow*.

ALAND, Russia, Gulf of Finland, an archipelago of 80 islands and rocks; the principal is 20 leagues in circumference. On it is the old fortress of Castle Holmen, one of the prisons of Eric XIV.

ALAND BAY, Ireland, Waterford, close by the W. head of that river.

ALAND ISLAND, Norway, about W. of Bergen; when in the lat. of Bergen, standing in for Jelliford, this island will be larboard. On the S. point is an iron ring, to which vessels becalmed attach their cable to prevent driving N.; the rock is clear all round.

ALANDSHAGE, Amack island, Denmark, a cape at its S. end, 8 miles from Copenhagen.

ALANPO, Africa, Gold coast, E. of Agrico and W. of the river Volta, a good anchoring place, with Sedghill bearing N. by W. a little westerly, and the vessel off a little to the W.

ALATAMAHA, or ALTAMAHA, America, United States, a river of Georgia, having an outlet in Alatamaha sound, 60 miles S.W. of Savanna, among beautiful islands. The Alatamaha is navigable 300 miles from the sea for boats of 30 tons. There is a bar at the mouth with 14 feet at low water. It is principally navigated by steam-vessels.

ALAYA, the next town on the coast E. to Eske Adalia, on the shore of Karamania; it has neither pier nor harbour. It is E. of the ancient Ptolemais, from whence the coast is bold and straight. There is good anchorage in the road from four to two cables' length E. by S. from the octagonal tower; small vessels may haul close to the beach, which is stony. The mariner must not make too free with this road except in summer. The peak of a mountain nearly 5,000 feet high, 3 leagues to the E., easily distinguishes the place when coming from the S.W.

ALBACK, on the W. coast of Africa, 35 leagues nearly S.S.W. from the river Oroodus, and E.S.E. from the island of Fortaventura, one of the Canaries, in lat. 27° 15′ N.

ALBAN'S, ST., HEAD, Dorsetshire, England, in lat. 50° 4′ N., long. 2° 10′ W.; a ledge of rocks runs off here full three-quarters of a mile. It is high water $8^h 30^m$, full and changes.

ALBAN'S, ST., Jersey, lat. 49° 10′ N., long. 2° 25′ W., within a deep bay E. of St. Alban's island, and W. of St. Helier. High water ½ past 12 or ½ before 1.

ALBATEL, CAPE, Barbary, 12 leagues E. from Cape de Tenes, within which are several good roadsteads, but no harbour.

ALBEMARLE SOUND, North America, coast of North Carolina, 60 miles in length, and from 8 to 12 in breadth N. of Pamlico sound, with which it communicates; the passage from it into the sea is called the Roanoke inlet, in lat. 35° 50′ N., long. 76° 10′ W.

ALBENGA, Italy, a seaport of Genoa, 37 miles S.W. of that city. Lat. 44° 4′ N., long. 8° 3′ E.

ALBERNAS PUNTA, the N.W. point of Flores, one of the Azores or W. islands.

ALBIANA, CAPE, Cyprus, the N.W. point. Lat. 35° 20′ N., long. 32° 18′ W. It is 91 leagues S. ½ S. from the E. point of Candia.

ALBO, Africa, see *Lagoa*.

ALBORAN ISLAND AND SHOALS, in the Mediterranean; nearly in the meridian of Adra, in lat. 35° 56′ N., long. 3° 0′ 40″ W., lies the island, about ½ a mile in length and ¼ in breadth. It is surrounded by breakers, and may be seen at the distance of 3 or 4 leagues. Half a cable's length from its E. side is an islet, where anchorage may be found in 25 or 30 fathoms. The current sets strong to the E. near this island.

ALBOUZEME, Barbary coast, lat. 35° 10′ N., long. 2° 54′ W. There is good anchorage at the bottom of a bay here, from 3 to 6 fathoms near the shore, and at N.E. from the mouth of a river which falls into it there is 10 or 12. Cape Mourou forms the W. point of the bay, and Cape Quilate the E., it is also called the Bay of Buzema or Buzemar; there is a fort on a small island before the town.

ALBRIDA, or ALBRADAR, a fort belonging to the French on the river Gambia in Africa.

ALBUFERA, the term used in Spain for a lagoon or sea-water lake.

ALCAZAR LEGUER, a town of Fez, coast of Africa, S. side of Gibraltar straits, lat. 35° N., long. 5° 30′ W. Ships may here anchor within a shallow bay in 10 fathoms, on the W. of which are two long narrow isles parallel to the coast. From Alcazar to Ape's hill the ebb, which begins to run W. at 4 o'clock, is narrow along shore.

ALCOL, Barbary coast, E. side of Cape Tenes, under which there is a small bay open to the N. and N.E., from whence ships should move round to the W. side if a gale is expected.

ALCUDIA, BAY OF, in Majorca, a bay on the E. side, in which large ships may anchor, running midway between Alcana and Torre Major, 3 or 4 cables' length from the land, on fine gravel.

ALDAN, BAY OF, W. coast of Spain, a good summer anchorage in parts. The harbour of Aldan in this bay is deep and clean, but it is advisable to moor E.N.E. and W.S.W., as the winds from the N.N.W. bring in a dangerous sea.

ALDBOROUGH BAY, Suffolk, between Ridge and Sizewell, with anchorage in offshore winds in 7, 8, and 9 fathoms. Orfordness lower light S.W. ½ W. Aldborough church N.W. by W. ¾ W.

ALDBOROUGH KNAPES, a shoal extending 2½ miles N.E., and S.W. 2½ miles. The N.E. end lies with Aldborough church W.N.W. ½ W. 4 miles distant, and Orford low light W.S.W. ½ W. 6½.

ALDE, or OLDE, an island on the W. coast of Norway, in lat. 61° 25′ N., long. 5° 9′ E.

ALDERHOLM, an island at the entrance of the river Geffle, 80 miles N. of Stockholm.

ALDERNEY, one of the Channel islands, on the coast of Normandy, about 8½ miles W.N.W. from Cape la Hogue, 3½ miles long by 1 broad, surrounded by rocks. The road of Braye communicates with St. Anne's, the principal town, having a tide harbour with a pier on the W. side 408 feet long, but affording poor shelter. The anchorage before the port is not safe in winter, though having a fine sandy bottom with 5 and 10 fathoms. In the middle of the roads is the half-tide rock, very dangerous to the navigation. Longy is a port on the S. side near the E. end, having a fort and flagstaff. The island is high, and may be seen in clear weather 5 leagues off. It is high water at 12 o'clock, and the island is in lat. 49° 48′ N., and long. 2° 15′ W. This island is 7½ leagues N.E. from Guernsey, and 3½ W.N.W. of La Hogue, between which is the Race of Alderney, deep and clear of danger, except when wind and tide meet, in gales of wind, when it should not be ventured. There are sunken rocks off the N.E. end; the furthest out is the Blanchard, with from 4¼ to 12 fathoms on it, E.S.E. from the E. point of the island. S. by W. of the Blanchard, within 1½ mile, are other shallows, called the Basse d'Alderney and Basse du Raz, with 3 fathoms at lowest. S.E. from the S.W. end lie within ¼ of a mile of the shore, the Coquelihou rocks, and W. ½ W. about ½ a mile, the Noire Pute rocks. In short, they are so numerous that much attention to the chart is required. There is the Coquelihou bank, the Basse de Milieu S. by W. ½ W. 4½ miles from the E. end of Alderney, the day mark for which is the rock Ortach in one with La Fourque, one of the Noires Putes. Then, a mile to the N.W. of the W. of

Alderney, are the Burhou rocks, with channels for small vessels between them. At their W. end is the Ortach, high above water, steep, and clear to the W. 1½ of a mile from this is the Burhou, the largest of the whole, while the N. of all is called the Verte Tête. To clear the whole of these to the N. the mark is the Caskets open N. of the Verte Tête bearing W. ¼ N. The passage between these and the island is called the Passage du Singe, and is only two-thirds of a mile broad in its narrowest part, having from 9 to 16 and 18 fathoms. About W.¾ S. 13½ miles from Cape la Hogue, and 6 miles from the W. end of Alderney, is the N. end of the Schôle bank, extending S. by W. ¼ W. about 3 miles, with only 15 feet in its shallowest part at low water spring tides. There are 9 or 10 fathoms at each end. It lies almost in a line from the middle of the Race of Alderney to Great Russel channel. The mark for the shoal parts is Les Nannets, seen between Burhou and Les Etats. The overfalls and ripplings about the island are numerous, and some dangerous. The Caskets lie 3½ miles W.N.W. of the Ortach, and 5½ N.W. by W. from the W. point of Alderney. On these rocks there are three lighthouses triangularly placed, revolving, and alternately presenting a bright light every way, 120 feet above the sea level. On a bearing S.E. ¼ E. the three lights will appear as two and may be seen 5 or 6 leagues off. The N.E. and S.E. lights come in one when bearing S.W. ¾ W. These rocks have 25 and 30 fathoms all round, and vessels may pass safely amongst them. There are two boat harbours on these rocks, one on the S.W., the other on the N.E. side. Steps are cut in the rock, and means provided for hauling up a boat. A flag directs which harbour to approach, a red flag the N.E., a blue the S.W. harbour. Loss will attend inattention to these signals. These, the only triple lights in the channel, cannot be mistaken. Coming up channel, when they bear E., steer N. If they bear S., all the dangers of the Channel islands are passed. If N. then a vessel is among the islands. Guernsey will be from S.S.W. to S.W., and Alderney N.E. If unable to get W. of the Caskets, the best course is to steer for Cape la Hogue, and so through the Race of Alderney with the flood tide. There is a lighthouse on a rock near Cape la Hogue 175 feet above the sea, having a bright fixed light. The Passage du Singe is safer than the Passage d'Ortach.

The French call the island the Isle d'Aurigny. It is 7½ leagues from Guernsey. The pop. is about 1,000, and the island is noted for its fine breed of cows, its pasturage and corn. The inhabitants are principally farmers, who all live in the town.

ALDERNEY RACE, or THE RACE OF ALDERNEY. This is a violent and dangerous ripple, when violent winds meet the flood tide between Cape la Hogue and the island of Alderney. For gales from the N.W. and W.N.W., if no alternative offers but passing through or lying to, the last course is best, to prevent getting into the race in an ebb tide, and thus wait the beginning of flood. When the wind and tide meet, the sea, though deep, is tremendous. At half-past 10 on full and change it is high water in the stream. Ships should be in mid-channel, and keep at least a mile both from the lighthouse at Cape la Hogue and from the island of Alderney. The tide on great springs runs 8 miles an hour. Vessels deeply laden should be on their guard passing. It must be recollected

too, by vessels from the eastward, that the flood runs as strong as the ebb. On the N. side of the Caskets and Alderney the tide runs until half-past 9; W. of the Caskets only till 9 o'clock. In the Race of Alderney, and in some of the French ports opposite, as well as in Jersey, the sea is 5ʰ 30ᵐ rising, and 6ʰ 30ᵐ falling. At the Caskets and Alderney it is high water at 6ʰ 45ᵐ, at Guernsey 6ʰ 30ᵐ, at St. Aubin's, Jersey, 6ʰ 10ᵐ. Spring tides rise from 33 to 39 feet. The whole French coast from Granville to Omanville, and the Race of Alderney, exhibit many irregularities in the tidal currents, which should be particularly studied by mariners.

ALDERTON, POINT, North America, on the larboard of the entrance to Boston harbour. It has many rocks and shoals near it, so that ships passing into Nantasket road must stand over to Lighthouse island on the starboard; this point is 15 leagues S.W. by W. of Cape Ann. Lat. 42° 18′ N., long. 70° 50′ W. High water 10 o'clock.

ALEBURGER, 9 leagues from Zeabuy, in the Cattegat, S. by W. The Scaw is the W. entrance, from thence passing the Holme's islands to starboard 4 leagues from it, the course is S.W. until athwart Zeabuy. Pilots must then be taken to Aleburger, which is 4 miles up the river.

ALEGOA CIDADES. See *Azores.*

ALEGRANZA ISLAND. See *Canaries.*

ALESSIA, Turkey in Europe, in Albania, near the mouth of the Drino. Lat. 42° 8′ N., long. 20° 9′ E.

ALESSIO, POINT ST., E. coast of Sicily, a bold promontory S. of Cape Grosso, having a tower and telegraph upon it.

ALEXANDER, ST., Russia, N.W. shore of Ritzart's island, round its S.W. point, the channel to St. Petersburgh being near the S. side of the island. At low water the beach, on the starboard side, extends considerably from the main land. It lies nearly E. and W. till within this island, when varying a point to the S. will lead up to St. Petersburgh. A fort and tower guard the passage. There are 4, 3, and 2 fathoms of water near the shore of the island, and on the S.E. shore, within the island, 5 fathoms.

ALEXANDRETTA. See *Iskenderoon,* or *Scanderoon.*

ALEXANDRIA, North America, a city of Virginia, on the S. bank of the Potomac river, 290 miles from the sea, from which there is a navigation. Lat. 38° 45′ N., long. 77° 10′ W.

ALEXANDRIA, Egypt, founded by Alexander the Great, 125 miles N.W. of Cairo. The lighthouse is in lat. 31° 12′ 35″ N., long. 29° 53′ 27″ E. Standing on the isle of Pharos, an isthmus in the shape of the letter T forms two harbours, the N.E. called the New Port, and the S.W. the Old. The New is circular, with the city on its S.W. side, on the N. point of which is the Pharos. The Old Port is much larger, being 6 miles along shore, having the city at its N.E. end, and on the seaside numerous shoals of rock, sand, and mud. The extreme points of the old harbour are Point Eunestos and the Isle of Marabout, the former bearing from the latter N.E. by E. 4¼ miles. Point Eunestos is a rocky point with a fortress, and has a lighthouse close to it. The N.E., or Pharos point, bears from it N.E. ¼ E. 1¾ miles, the coast between bordered by rocks. Abreast of the Pharos is a sand-bank, on which are from 7 to 4½ fathoms, the shallowest part bears from the Pharos lighthouse N. by W. ⅞ of a mile. It is 2 miles long, but not more than ¼ broad. The New port has a square castle on the Pharos, having twelve 32-pounders and six mortars. On the opposite point is the Pharellon, or Little Lighthouse, lying E. ½ S., distant ⅞ of a mile; both points encumbered with rocks and shoals: shallow water runs all round the harbour, but there are spots between the shoals and channels with 5, 6, and 7 fathoms. Vessels bound for the city commonly make for the tower of the Arabs, in lat. 30° 58' 15" N., long. 29° 34' 20" E.; between two dark-looking hills on the W. of which is a castle, generally visible 6 leagues off, where are 90 fathoms water. The depth should not be decreased at night to less than 25. Proceeding along the coast a small round hill becomes visible, and then Pompey's Pillar. The soundings will distinguish the land further. The Old Port has three channels, the Djerme, Central, and Marabout. To enter by the Marabout channel bring Pompey's Pillar to E. by N., and the ruins upon Marabout island S. ¼ W., distant ¾ of a mile, when a S.E. by E. course will carry clear of the rocky reef which stretches off the N.E. end of the island. When within ¼ of a mile of the coast proceed, at that distance, along shore all the way to the town, or within the W. arm of the Pharos island; here anchor in 6 or 7 fathoms. The Central, or Boghaz channel, is 2½ miles E. of the tower of Marabout. The Djerme channel is about a mile S.W. of Point Eunestos, and runs narrow between two rocks, with 19 feet of water. The New Port, or E. harbour, is known by the Pharos, square castle, and light at the extremity of the mole. A cable's length off is the Diamond rock, above water, to be left on the starboard a cable's distance. Anchor off the W. shore in 2½, 3, and 3½ fathoms. The space for anchorage is limited, the best is with the Pharos tower N.W. by W. ¾ of a mile distant, in 5 fathoms, but exposed to N. winds. The ground is rocky. The old city of Alexandria is a heap of formless ruins, the modern not occupying one-fifth of the site. Remains of the old walls still exist, and include, besides ruins and various buildings, two strong forts. Outside these is Pompey's Pillar, the shaft of granite, the whole 99 feet high. The Needles of Cleopatra lie ⅛ of a mile E. of the parade. There are two obelisks; one has fallen, and one remains erect, 70 feet high, consisting of a single stone.

ALFAQUES, PORT or HAVEN OF, on the E. coast of Spain, running in to the N.E. 5½ miles. Within is the Fort St. John, beyond which is a shallow inner haven. For vessels of moderate size there is shelter here in 3 and 4 fathoms. The islands here are called the Alfaques of Tortosa, to approach which, at night, from the S., caution is necessary, the shore being low and shallow. The Alfaques comprehend all the flat shore from Rapita to Ampulla.

ALGAGIOLA, a seaport of Corsica, lat. 42° 30' N., long. 9° 1' E.

ALGARROBA, POINT, Porto Rico, West Indies, is the N. point of the Bay of Mayaguez, and lies with the S. point nearly N. by E. and S. by W., the distance between 4 miles. This point is in lat. 18° 14' N., long. 67° 7' 30" W. There is good anchorage from N. winds in this bay for small and middling-sized vessels.

ALGEZIRAS, a town agreeably situated in the Bay of Gibraltar. The best anchorage here is off the middle of Algeziras town, toward Palmones river, the ground, sand or mud, is good. The best positions are between the Point of Rocadillo and the river Palmones. It is high water at 2 o'clock; the tides commonly rise 5 or 6 feet. See *Gibraltar*.

ALGHERO, Sardinia, a port in the gulf of that name, 6½ miles E. of Porto Conte. It has a large bay and good anchorage. Lat. 40° 32' N., long. 8° 40' E.

ALGIERS, Africa, bay and port. The bay is formed by Cape Matafou on the E. and Cape Caxines W., distant nearly E. and W. 10 miles. It is about 6 miles deep, and on its W. side is the city of Algiers, built upon the declivity of a hill, about 1½ mile in circumference; the pop. 110,000. It is surrounded by a wall 12 feet thick and 40 high on the sea-side, but only 30 towards the land. It has five gates and two castles. The works on the land side are of small moment. The harbour was formed with enormous labour, by two moles, one running N. the other S.W., and meeting at a little island, on which is a lighthouse, with a revolving half-minute light, the flashes of which are seen 5 leagues off, but the light does not wholly disappear at less than 3. The citadel is octangular, on an eminence to the S.W. There are forts and batteries all round the bay. The mole is defended by the lighthouse battery, which mounts three tiers of guns. The bay is free from danger, except off Cape Matafou, where they show themselves. A bank of coral rocks, with 5 feet water, is said to exist 2 miles N. of Cape Matafou. Algiers lighthouse stands in lat. 36° 48' 30" N., long. 3° 1' 20" E. The best anchorage is with the mast N., land N.W. ¾ W., the lighthouse W.N.W. ¼ W., the entrance of Haratch S. by E. ¾ E., about 2¼ miles from the head of the mole.

ALICANTE, Spain, the castle is in lat. 38° 20' 41" N., long. 0° 29' 57" W. This city is situated in the province of Valentia, in the bay of the same name. The castle is on a rock 900 feet high. The chief exports are barilla, salt, almonds, wine, raisins, silk, alum, &c., but the trade has much declined. The pop. 18,000. There is a pier, but the water is too shallow for any but very small vessels. There is a light on its head 95 feet high, seen 15 miles at sea. Large ships anchor further out in 7 fathoms. The road is bounded by the Capes Santa Pola, and De la Huertas, and is open to all winds, from E.N.E. to S. by W., but the ground holds well, and vessels may ride with S.E. and S.W. winds, if the swell is only moderate. From the N.W. and N.E. they are sheltered by inland mountains; mooring N.E. with small bower S.W., 1 mile from the shore, in from 11 to 5½ fathoms, or further in with 3 or 2 fathoms, bringing the Castle of Santa Barbara N. by E., or N. by compass. The approach to Alicante may be known by four hills. The depth of water decreases towards either of the Capes. The Cape de la Huertas goes down low and flat to the sea, rising inland with a white rocky appearance, and near it stands the tower of Alcora. A reef runs out from the Cape nearly E. ¼ S. 2 cables' length; a little further out are 5¼ fathoms. No other anchorage than that mentioned can be recommended. On steering from Cape Polas to Alicante keep well up to the N.

ALICATA, a town on the S. coast of Sicily, in lat. 37° 4' 3" N., long. 13° 55' 54" E., distant from Point Tenda, the S.E. point of Palma bay, E.S.E. ¼ E. 9 miles. The ground around it is shallow and rocky: it is a neglected place, though the want of a port there is much felt.

ALICUDI, one of the Lipari islands. See *Lipari Islands*.

ALIN LOCH, Scotland, opposite Caranihenach, and S.E. is Attornish bay. Loch Alin is a well-sheltered harbour, the ground good, and depth moderate, but the entrance is narrow. Arras anchorage is not far distant, proceeding up the Sound of Mull, but near several small green islands, in the middle, there is a shoal in the way; the best passage is to the N. Arras anchorage is good beyond 2 cables' length from the shore.

ALLIGATOR POND, KAY, and REEF, Jamaica, a dry kay, 2 or 3 miles in length, with a reef around it, 4 miles from the shore, and 7 leagues W. ½ N. from Rocky point, the E. end of Portland. There is good anchorage between that and the main in 6 and 7 fathoms.

ALLIPATAK ISLAND, at the entrance of Ungava bay, Hudson's straits, in lat. 60° N., lying W. of Ablariak bay, near the N. point of Labrador.

ALLOA, or ALLOWAY, Scotland, Frith of Forth. The frith narrows there from 7 to 2 miles, and widens beyond with deep water. It has a good harbour, with water for vessels of the larger class, a dry dock, and considerable exports, particularly of coal. It is 30 miles N.W. of Edinburgh. Vessels should take pilots for the passage to Alloa, as the frith is winding, and the tide rapid. Lat. 56° 10′ N., long. 3° 46′ W.

ALLUM BAY, Isle of Wight, Hampshire, round the Needles' point, having good anchorage, and water of good depth, not far from the bottom of the bay, out of the strong run of the tide. The Needles' light is in lat. 50° 39′ 54″ N., long. 1° 33′ 55″ W. High water at half-past 9.

ALMADRONES, BAY OF THE, Africa, without the Straits, under the lee of Cape Spartel on the S. The road is safe and clean, with 12 and 13 fathoms water, sheltered from the N. and E. winds, but exposed to the S. and W., so that ships must be ever ready to put to sea anchoring here, for fear of the dangers of a lee-shore. Strict attention to the weather will enable them to clear the road, and shelter round the point in Tangier bay. There is good anchorage before a hummock on low ground near the strand, that appears like the ruins of a castle a league S. of the Cape. It is high water here, and round Cape Spartel, soon after 3 o'clock.

ALMAFI, or ALMALFI, Italy, Gulf of Salerno; it is high and sheltered from E. and N.E. winds, but has no harbour, simply a road near the point; S. and W. winds blow dead in. Here, it is said, the mariner's compass was first discovered. It lies round a point of land E. of the Gulf of Naples, in lat. 40° 28′ N., and long. 14° 45′ E.

ALMAZARRON, Spain, Mediterranean, on the W. part of Almazarron bay. It affords protection from W.S.W. or S. winds by running to the N. side of the tower, and making fast to the rocks with an anchor out E., in 7 fathoms, ready to sail with the wind coming E. The town is a league inland.

ALMENADRANA, Spain, E. coast, N.N.W. from Cape de Gat, at the distance of 8 leagues. There are several small bays and anchorages between, particularly the Gulf of Mohaca.

ALMERIA, BAY and TOWN; this port was once one of the most commercial in Spain, but is fallen to nothing: the exports are barilla and lead. It lies in lat. 36° 51′ 29″ N., long. 2° 32′ W. From the town to the river the shore runs E. by S. 2 miles. To find shelter from E. winds vessels should moor S.S.W. of the bastion of La Trinidad, at the E. angle of the city, in from 9 to 14 fathoms, ½ a mile from the shore.

ALMERIA, North America, Mexico, a small port, 20 leagues N. of Vera Cruz. Lat. 20° 20′ N., long. 97° 50′ W.

ALMNUECAR TOWN, Spain, to the E. of Herradura. There are anchorages on each side of the rocky point of St. Christoval; that on the E. side is adapted for W. winds, a cable and half from the rocks in 8 or 9 fathoms. Lat. 36° 30′ N.

ALNMOUTH, Northumberland, 4 miles N.N.W. from Coquet island, the entrance to the river Aln, leading to Alnwick, where small vessels take in corn. Rocks extend nearly a mile from the shore towards Sunderland point. Creswell hall opened S. of Hauxley point S.S.W. ¼ W., clears Alnmouth rocks, Bulmer bush, and Bulmer stile.

ALPHEUS RIVER, Greece, in the Morea, on the W. side of which it falls into the Ionian sea, nearly S. by E. of Zante. The town of Longinico, to which small vessels can go up, is in lat. 37° 3′ N., long. 22° E.

ALPRECK, CAPE or POINT D', France, 2 miles S.W. of Boulogne, having a lighthouse on it painted black, the light 154 feet above high water. This light intermits, being bright for 1½ minute, then dark, then flashing red, then dark, occupying 2 minutes, and seen, when clear, 4 or 5 leagues off.

ALSEN ISLAND, Denmark, S. of Funen island, near the main of Holstein. The channel is narrow and called Alsin sound, 6 leagues long and 2 broad. A pilot is necessary to reach and pass it. Lat. 55° N., long. 9° 55′ E.

ALTAR, THE, and the ALTAR FLAT, at the entrance of Harwich; the first a shoal with 7 feet water, W. by S. ¼ a mile from the cupola of Landguard fort. The Flat lies N. of the Altar ⅓ of a mile, having a black buoy marked "West Altar" on the W. edge of the shoal in 3 fathoms.

ALTAVELA, or THE LITTLE MOUNT, a high and rocky islet, 5 leagues S. ⅓ W. from Beata or Little Mongon point, the S. extremity of Hayti; in lat. 17° 28′ 20″ N., long. 71° 40′ W. It is peaked, and appears at a distance like a dome emerging from a fog. N.W. from it is a small rock, and 1½ mile further a shoal, with a depth of 16 fathoms. Two leagues N. by E. ¾ E. lies the S. end of Beata island, with a good channel between.

ALTEA, BAY OF, Spain, in Valencia, E. of Benidorme. On the S. point of the bay stands the tower of Bombarda. There is anchorage S.E. from the town, in 12 fathoms, ¾ of a mile from the shore, but the ground further in is rocky. A fort protects the town, which is 30 miles E. of Alicante. It has a small traffic in wines, honey, and silk. Lat. 38° 34′ N., long. 15° E.

ALTEN, otherwise the GULF OF ALTEN-BOTTEN, on the coast of Finmark, near Wardhuys, Norway.

ALTENA, or ALTONA, a city on the N.E. of the river Elbe, going up to Hamburg, from which it is distant 2 miles, in lat. 53° 36′ N., long. 9° 58′ E., and has high water at 6 o'clock.

ALTENSOLEN ISLAND, E. of the S. point of Maggeroe island, on the coast of Norway. It has a good road, and shelter for all winds. Vessels may run for it within Surroy island, through Suyer sound, leaving all the islands except two on the larboard, and then run out to sea again without going round the cape. It is about 71° N. lat., and 26° E. long. High water about 3 o'clock.

ALTOS, or HEIGHTS OF MECA, Spain, N.E. of Cape Trafalgar; high land in the interior, the summit level, but divided into two parts. The tower of Meca is 3½ miles E. ½ S. from the Cape.

ALVARADO, a bar harbour in the Gulf of Mexico, in lat. 18° 45′ N., long. 95° 42′ W., and 36 miles S.E. ¼ E. true from Vera Cruz, admitting vessels drawing 12 and 13 feet, and well sheltered. The shores on each side of the entrance for a considerable distance are very foul and rocky.

AMAK, an isle of Denmark, on the E. coast of Zealand, separated from Copenhagen by a narrow channel, across which it communicates by bridges. It is defended by strong batteries. Lat. 55° 35′ N., long. 12° 42′ E.

AMAL, Sweden, province of Elsborg, on Wener lake, having an excellent harbour and a good trade in timber and tar; goods pass out by way of Gottenburgh, in lat. 59° 6′ N., long. 12° 40′ E.

AMAN, Africa, a seaport between Cape Cantin and Cape Gerard.

AMANSO, CAPE, in the S.E. part of Corsica, the N. limit of Bonifacio bay. It is easily known by a large castle, and by two rocks lying just before the harbour. The town of St. Bonifacio lies S.W. from it 3 or 4 leagues, and Cape Sigli, distant 5 leagues, with a bay between.

AMANTEA, Italy, in Naples, a port 20 miles S.W. from Cosensa, in lat. 39° 12′ N., long. 16° 8′ 40″ E., near the bay of St. Eufamia.

AMAPALLA, a port of North America, in the province of Nicaragua, on the W. side of the gulf of that name, in lat. 13° 10′ N., long. 91° 10′ W. The entrance to the bay lies between two projecting points of land which closely approximate. It is roomy within, being about 30 leagues from N.W. to S.E., and from 3 to 10 in breadth. Both the town and port of Amapalla are sometimes called Fonseca. There is a high pointed hill on the W. side of the bay, with a port at the foot, called Martin Lopez. There are from 10 to 12 feet of water everywhere in the bay. Ships may sail freely in, when these islands will be seen. Two of them lie E. and W. from each other; that on the W. is called Canchagua, and is near the continent; let the lead go all the way in, and anchor near it anywhere. The tides run strong, and sound cables are required. In the Gulf of Amapalla there are also two islands; one, called Maugera, is high round land, surrounded by rocks, with a sandy creek on its N.E. side; the other, called Amapalla, is 2 miles distant. Beyond this island, though the gulf runs far up, it is not deep enough in water for ships even of moderate burden.

AMASTRA, or AMASSERAH, in the Black sea, seeming at a distance a group of islets. It is joined to the main by an isthmus 100 fathoms broad. The town is 48 miles to the N.E. by N. of Cape Baba, and about 150 from the Bosphorus. Here is a sheltered anchorage, but with exposure to a swell from seaward. To the W. of Amastra there is another harbour opens from the N.W. to the W., with from 10 to 2 fathoms. The commerce of the place is inconsiderable.

AMATIQUE BAY, in the Gulf of Honduras, North America, near the W. extremity of the S. coast, formed by the land of Cape Three Points, and the peninsula between that and the Gulf of Dolce, where it runs far up S. The river Guanacos falls into it, and the port of Amatique is on the W. coast of its extreme end. Give Cape Three Points a good berth, and keep clear of the rocks and shoals on the opposite coast by remaining in mid-channel; the point being passed, the entrance into this bay is easy. Lat. 15° 23′ N., long. 89° W.

AMAXICHI, isle of Santa Maura, the capital; pop. 6,000. It has a trade in salt; Port Delpene is the best harbour.

AMBELAKI, COVE OF, between Kamiche Boroun and Akboroun; there is here only 18 feet water ¼ a mile from the shore. In 1828, vessels going to the Sea of Azof performed quarantine here, afterwards done at Kertche, a town 2 miles W. of Akboroun. The anchorage of Ambelaki is very bad with the wind at N.E.

AMBERGRIS KAY, on the coast of Yucatan, 19 miles long and 3 broad, running N.N.E. and S.S.W. The centre is in lat. 18° N., and about 87° W. long.; the N. end admitting a passage for boats between that and the main.

AMBLETEUSE, a port in the department of the Pas de Calais, at the mouth of the little river Selaque; pop. 400. It is 12 miles S.W. from Calais, and defended by batteries. A reef, long and narrow, runs off from St. John's road, which lies to the N., called the Bassare; it runs S.W. and S.W. by S., with from 4 to 6 fathoms. High water at 11 o'clock. Lat. 50° 48′ N., long. 1° 36′ E.

AMBO, Africa, western coast, a creek about 2 leagues from Cape St. Bernard, and not far from a second creek N.E. of that cape, named Canne.

AMBOISE, CAPE and ISLAND, Africa, on the coast of Benin; in lat. 3° 58′ N., long. 9° 15′ E. It is high land, with steep cliffs both on the main and islands, among which ships can sail, but should not anchor in less than 7 fathoms. S.S.W. from the cliffs sail may be made for the island of Amboise, to make Point Bato, and when there are 8 or 9 fathoms above that run for Cape Cameroon, to the S.E. by E., by which a ship may get into the river. The most E. of the island is very lofty, and violent gusts of wind rush down so suddenly from the heights that it is not wise to anchor near them. The river Key, or King's river, is to the N.E. of Amboise. To pass from that river W. of the island, never keep in less than 12 fathoms.

AMBROOK ISLAND, a small island in Russia, in the province of Livonia, in the Baltic, under Oesel, 7 leagues N.E. from Domesness, which last is in lat. 57° 39′ N., long. 22° 16′ E. This island is 5 leagues E.N.E. northerly from the S. point of Oesel, where ships may best anchor, the ground near Ambrook itself being foul, sharp, stony, and uneven, changing from 3 to 4 and 5 fathoms at once.

AMEEREN ISLAND, N.E. from Heligoland, and N. from the entrance in the Hever, on the W. coast of Heligoland.

AMELAND ISLAND, on the N. coast of Friesland, kingdom of Holland, with a narrow channel between itself and the main. It is one of a succession of islands, and is 3 leagues long from E.N.E. to W.N.W., and hardly 1 broad. There is anchorage on the S. side of the W. end of the island, called the Ameland channel; to enter it, the shore of Schelling must be made, having 15 or 16 fathoms, with a bottom of white sand. The beacon on the E. end of Schelling must be brought right over the steeple of Hoorn, bearing S.S.W. westerly, till the beacon on Ameland and the steeple of Hoelm, also on the W. end, come one over the other. Thus, having done with the first, make sail in upon the second, until the ship falls in with the outermost buoy, which is within the outermost point of the Born reef. In coming from the W. sail along the shore of Schelling, in 5 fathoms to 5½ at low water, as far as the outermost buoy, when the church and steeple on Ameland will be one over the other, bearing E. southerly. The second buoy lies athwart the W. end of Camper sand, and from the first to the second buoy there will be from 5 to 6 fathoms; passing this, bring the steeple a little N., and go E.N.E. to the third buoy on Gerritshonden, to avoid a flat towards Born reef; the third buoy must here be left to larboard, and a white to starboard, and run S.E. in 7 fathoms to a fourth buoy; after passing this, sail on S.S.E., in 8 to 12 fathoms, until Hoelm steeple comes over the S. point of Ameland, where the boats lie, and a vessel will be in the bight. In the Ameland channel it is high water at 9 o'clock, but outside it at 7ʰ 30ᵐ. At sea the island may be seen in 12 fathoms. Lat. 53° 30′ N., long. 6° 15′ E.

AMELIA ISLAND, on the coast of East Florida, 13 miles long and 2 broad, 7 leagues N. of St. Augustin. Lat. 30° 26′ N., long. 81° 10′ W. A second island of the same name lies on the coast of Georgia, North America, between which and Cumberland inlet on the N. is that of the river St. Mary, called Amelia sound, commanded by a fort on the S. end of Cumberland island, which lies N. of it.

AMERICA, CONTINENT OF, the largest and latest known of the four quarters into which the superficies of the earth has been divided by geographers. It is separated into N. and S.; the whole extending from the N. Pole to Cape Horn, in lat. 55° 58′ S., or above 9,000 miles, varying in breadth from 5,000 miles to 50, where the two grand divisions are connected by the Isthmus of Darien. The coasts necessarily embrace every variety of climate. They abound in magnificent rivers and the finest harbours in the world. On the N., America is washed by the Polar seas, or inlets from the N. Atlantic, and Baffin's bay on the E.; and the N. Pacific, through Behring's straits, on the W.; the N. and S. Pacific equally with the N. and S. Atlantic break on its extended shores. Its islets and isles are numerous, including those of the West Indies, commonly so termed. Upon this continent nature seems to have produced the rivers, lakes, and mountains on a larger scale than in the old world; even plains in America are upon a level with the highest European mountains. From icy Greenland on the E., in lat. 75° N., down to Cape Farewell, thence ascending on the W. side to Ross's bay, at the N. point of Baffin's bay, thence down the E. coast of that bay as far as Melville island, and back again through Hudson's straits and bay, then down the coast of Labrador, the isles of Newfoundland, and Gulf of St. Lawrence, thence along the shores of the United States to the Gulfs of Florida and Mexico, as far as Trinidad, from this point of South America round Cape Horn up again to the pole—what an extent of coast and magnitude of space for nautical exploration!

The wilful destruction of the native people in the S. by the Spaniards, has been accomplished or is accomplishing in the N. not less effectively by the introduction among them of the vices of the whites. The aboriginal inhabitants of North America are fast hastening towards extinction. In the N. part of the continent of North America the mariner, amid considerable difficulties of navigation, has to contend near the Gulf of St. Lawrence, around Newfoundland, and to the N., with fogs and ice. The danger from the last is not so great as from the first of these causes. If beset in the ice there is seldom to be encountered anything worse than temporary delay, but it is otherwise when enveloped in the dense fogs of this part of the coast, particu-larly in the early part of the summer, with winds between S. and W. Though clear in the upper part of the gulf, the fog may exist in the lower. These fogs last for several days; the deep sea lead and chart are the only refuge for security. St. Paul's island, 10 miles from North Cape, was once noted for wrecks, but that is now lighted. The Magdalen islands, and the coast down to the United States, beyond Nova Scotia, demand as much care as the coast of Labrador to the N. Everywhere along the coast there are fine harbours, but particular attention must be paid to the currents, which set according to the direction of the winds at sea. The river St. Lawrence allows of the navigation of the largest ships as high as Quebec. It is necessary to attend particularly to the variations of the compass on this part of the coast of the American continent. During the navigable season, the winds are generally up or down the St. Lawrence, while for a portion of the year the climate of this part of America is so severe that navigation ceases; the ice remaining for months on the waters. The temperature of Quebec, in lat. 46° 48′ 39″ W., is often 29° below zero in the winter months, and in summer above 100°; the winter lasting five months. This extreme of temperature does not appear to exist on the Pacific upon the W. side of the same continent, making good what is observable on the E. and W. coast of Greenland in regard to the difference of temperature between shores situated with those aspects. Extremes prevail far S. in the United States; the degrees of heat and cold felt in New York, in lat. 40° 42′ N., being unexampled in the severest winters or warmest summers ever known in England or Scotland, 15 or 16° nearer the pole; hence arises the necessity of an attention to the seasons in navigation on the E. shores of America, that is felt nowhere else so far to the S. Particular attention must be paid also to the winds in the Gulf of St. Lawrence; in October and November the N.W. winds blow violently in heavy squalls, with sleet and snow showers. The strong winds here do not veer suddenly to an opposite point, but generally die away to a calm before such a change occurs. Along the coast of the United States, down to Florida, when the winds blow hard from the N.E. without rain, they generally continue so for some days; but if attended with rain, they shift to the E., E.S.E., and S.E.; N.W. winds bring in general clear weather. Thunder-storms, easily observed in their approach, must be carefully watched for, and sails be clewed up; there is seldom time to hand them before the gust comes. From November 1 to the end of February the hardest gales experienced blow upon the shores of the Carolinas. The N.E. storms of North America are supposed to begin in the S.W. parts, and move in succession even up to Newfoundland. The W. winds have been remarked as being in the proportion of 2 to 1 in New Jersey, and of 4 to 1 in Lower Canada. The hurricanes that ravage the West India islands sweep also along the mainland and coasts of the United States, of course with diminished energy. The peninsula of Florida, on the W. side, skirts the Gulf of Mexico. Here new peculiarities are observed as the Gulf stream flows out N.E. from the vast bay into which the Mississippi and other immense rivers pour their waters, described under their respective heads hereafter. A current flows into the Caribbean sea, a most important circumstance to be remembered on the American coasts, which there is reason to fear has not always been kept in mind; ships have been lost by not making due allowance for it, passing along the Columbian coast and among the islands. Vessels, too, have been discovered thrown far out of their way by a cause that had never been considered at all. In the Gulf of Mexico and Caribbean sea the currents are modified by the direction of the wind. The proper navigation among the American islands is a most important point in considering the American coast from Florida to Trinidad, and the mouth of that vast gulf, with the Bahama islands, at the W. ingress of which the navigation is so peculiarly intricate. With the Island of Trinidad commences the navigation S. of the S. continent of America, near the mouth of the Orinoco, which has a course of 1,300 miles, and the coast of Guiana; the last held by five different nations: Columbian Guiana to the N., British Guiana, next Dutch, fourthly French, and lastly Portuguese. Portuguese Guiana, very little of which is known, stretches from the river Oyapoc and Cape Orange to the Maranon or Amazon river at the equator, which has a course of 4,000 miles; so great is the torrent of water it pours forth that it makes the sea fresh 80 miles from its mouth; the different streams by which it enters the ocean are bounded by alluvial soil and low sand banks: these have been lately explored. The river is navigable to a great distance from the sea. On the coast here there are two seasons observed dry and two rainy, being their alternate annual quarters. The winds on the coast are from E.N.E. to N.E., or E.S.E. to S.E., and the current running always W.N.W., makes the lesser lat. to windward. This relates to the general tendency of the sea; that which is produced by the rivers prevails over the land, and reaches about 12 miles from the shore, or into 9 fathoms water. On full and change the tides flow at Cape North at 7 o'clock, at Cayenne at 5, Surinam at 6, at the mouth of the Essequibo at ½ past 6, at that of the Orinoco at 6, and in the Gulf of Paria at ½ after 4. The water is shoal near the coast in many places, but the only part where a vessel is likely to be lost is among the rocky islets off Cayenne. From the mouth of the Amazon the coast runs S.E. to Cape St. Roque, N.W. of Cape St. Augustin, in S. lat. 8° 30′, long. 35° 40′ W.; then continuing S. for some distance, it takes a S.W. direction to Bahia, nearly due S. to Cape Frio, and S.W. again to Rio Janeiro, continuing nearly in the same point past the Rio Grande to the river La Plata, and city of Buenos Ayres. Much of the coast from the La Plata to the Straits of Magellan is barren, but the navigation on the coast offers no difficulties of moment. To enter the Pacific, the passage round Cape Horn is deemed preferable to that through the straits, on account of baffling winds. The coast of South America in the Pacific is easy of navigation by attending to a few simple rules. The chain of the Andes, seen far off at sea, furnishes a continued succession of land marks. It is necessary, however, to attend to the course of the prevailing winds, and in some cases to keep away from the land a considerable distance, to avoid their effects. The entire navigation from Cape Horn to the N. may be pronounced one with fewer impediments to judicious navigation than can be found on the E. coast, or on the coasts of any of the old continents, though the extent is so great and the difference of climate so varied. Little interrupted by island groups or extensive indentations, by deep gulfs, or projections of mag-

nitude, there is a uniformity of coast less calculated to cause currents and engender storms than on countries differently formed. In winter, as in all countries, hard gales are met with, both in the S. and N.; about Valdivia there is much rain, and in Chili the weather becomes unsettled, hard N. gales blowing for several days together; S. winds are sometimes frequent during summer. More to the N. the customary weather encountered everywhere within the tropics, but less in excess, is met with off that part of the coast. From Valparaiso to Lima the lofty coast is overtopped by the Andes beyond, visible, when the atmosphere is clear, the whole way. Some portions of the coast next the sea are sandy and desert, others are bounded by lofty cliffs, on which a heavy surf continually breaks. At the distance of 80 or 100 miles from the shore the Andes are seen in all their magnificence, and are excellent beacons to the seaman. The road of Callao is one of the best anchorages on the coast within the tropics. The climate on the W. coast of America is accompanied with those inconveniences, and, though in a less degree, with those diseases which affect tropical precincts in the other parts of the globe. Thus, at Guayaquil the weather is intolerably sultry, and the climate on the coast unwholesome, continuing thus as far as the N. tropic, whence the country to the northward is little known, but generally reported to be mild and much more congenial to health than on the E. side of the continent, or even upon the shores of the United States.

The original inhabitants of the American continent probably came from Asia, according to the evidence collected by the more diligent observers. The Peruvians and inhabitants of the southern parts of America differ considerably in language and personal appearance from those Indian tribes that roam over their hunting-grounds in the N. The Indian of the S. is a short, strong made, quiet, and in his present state an ignorant and incapable slave. In the N. he is warlike, noble of gait, lofty in feeling, and independent in spirit. The productions, both animal and vegetable, of this vast continent, were many of them unknown in the old world; some are a great addition to its luxuries and comforts, as the product of the sugar-cane, the fruits and cotton it so lavishly produces.

While the Spanish colonists seem destined to possess the S. continent, or to hold it in conjunction with the Portuguese of Brazil, the northern already seems within the grasp of the British race and language, by whom it has been principally peopled, and by whom, before long, it will be wholly subjected; England and her American descendants dividing the whole of this vast territory, there to increase and multiply one people and one language, to an extent to which the world, in the history of past ages, never exhibited a parallel.

AMFAR BANK, France, at the mouth of the Seine, the W. of which lies ½ a league S.W.S. from the entrance of Havre de Grace, 9 miles long from E.S.E. to W.N.W., and 2½ miles broad near the E. end, is a shoal dry on the ebb.

AMFROQUE, GRAND and PETIT, rocks near the N. coast of France; to be left to the E. by ships from La Hogue point to the Little Russel, at least ½ a league, to avoid the bad ground which runs off N.E. from the Petit Amfroque. The sunk rock called Boufress is ½ a mile N.N.W. from the Grand Amfroque.

AMHERST ISLAND, see *Magdalen Island*.

AMIRANTE, or CARNABACO BAY, North

America, on the N. coast of the Isthmus of Darien, in lat. 9° 5′ N., long. 82° 30′ W.; on the N.W. of the Bocca del Toro and the Bay of Conception, from which it is only separated by a mass of rocks near the coast. To enter this bay, the situation of the channel requires to be well known.

AMITOK ISLAND, S.E. of Chudleigh islands, at the entrance of Hudson's straits, in Labrador.

AMLWCH, or AMLOCK BAY and TOWN, Anglesey, North Wales. The bay is 6½ miles to the E. of Camlin bay, and 1½ W. of Point Lynas. The harbour lies in the bay; the entrance is ⅓ a mile wide, with a warping buoy moored there. There is anchorage in offshore winds here in 5 or 6 fathoms; small vessels may run up to the head of the creek if forced. The harbour lies between two steep cliffs; it has a stone pier and a dock. There are 20 feet water at spring and 13 at neap tides. With a N. wind a heavy sea sets into the bay, but the mouth of the harbour is protected artificially. There is a white lighthouse at the end of the N. pier, with a fixed light, seen 4 miles off, in bearings between S.W. ½ S. and W. ½ S.; another light is shown on the opposite side. The lights are only shown when the harbour is clear. When a ball is hung there is no access. The port will admit 30 ships of 200 tons. The town, principally inhabited by miners, contains a pop. of 6,235. It has risen from its vicinity to the Parys copper mine. Lat. 53° 23′ N., long. 4° 19′ 17″ W.

AMONA RIVER, South America, W. coast of Guiana, W. of Arwacas bay. It is deep and navigable, but straight, and the ebb tide runs so violently, that in the midst of the current anchors will not hold. Lat. 6° N., long. 55° 30′ W.

AMORGO, Greece, an island E.N.E. ¼ E. from the S. end of Nio. It has two ports, St. Anne in the N.E. and Port Vathy to the N.W., but they are out of the usual track of vessels. The E. end of Amorga is in lat. 36° 53′ 30″ N., long. 26° 6′ 0″ E.

AMORGO POULO, or LITTLE AMORGO, Greece, is in lat. 36° 36′ 55″ N., and long. 25° 32′ 44″ E., lying in the centre between the islands Anaphi, Santorin, Nio, and Amorga.

AMPENIA, Africa, on the Gold coast, 1½ league E. from Commenda, from whence a reef runs out parallel to the shore, within which is good anchorage for small vessels.

AMPLIMONT ROCKS, a vigia said to exist in lat. 42° 51′ N., long. 24° 15′ W., first seen in 1735, having two points, separated 30 feet above water. These rocks were again seen in 1829, and in 1842, in lat. 32° 41′ N., long. 44° 3′ W. They were distinctly visible a good distance off.

AMPULLA, CAPE, Spain, in the Mediterranean, 7 leagues E.W.E. and N.N.E. from Alfraque or the Alfaques, which is the term for all the low land of the Ebro from Repita to Ampulla, in the midst of which the river empties itself into the sea. Sailors call the ports of La Sofa and Fangal the Alfaques erroneously. The N. part of the Bay of Ampulla is open and unsheltered. All the coast here should be avoided in bad weather, or not approached nearer than 2 leagues, and the lead be kept going, as the sands continually shift. In fine weather there is no danger, there being a bottom of mud and sand for half a mile out, with 3½ to 5½ fathoms beyond that distance. See *Alfaques.*

AMPURIAS, a seaport of Spain, at the mouth of the Fluvia river, 60 miles N.E. from Barcelona. Lat. 42° 5′ N., long. 3° 6′ E.

AMRON BEACON, N. off Heligoland, 60 feet high, on a sand bank, 4 miles S. of Amron, and N.W. of the Isle of Pillworm. It may be seen 12 miles, bearing N.E. by E. ¼ E. from the Heligoland light, distant 25 miles. This beacon is of great use in the navigation of the Small Deep, Hever, Eyder, and Lister.

AMRON ISLAND, Denmark, W. coast, in lat. 54° 40′ N., long. 8° 15′ 10″ E.

AMSTERDAM, Holland, the principal city of that new monarchy, situated at the conflux of the rivers Amster and Wey, where they form a port capable of receiving 1,000 vessels, about 2 leagues from the Zuyder Zee. The port is near ½ a league in length by 1 mile in breadth, always filled with shipping. The sides of the port are secured by large piles driven into the ground and joined by beams in a horizontal direction. It is high water here at 3 o'clock. Lat. 52° 22′ N., long. 5° 32′ E. The pop. 220,000. It is one of the most considerable commercial cities of the globe. The buildings are fine, although their foundation and that of the whole city is upon piles; the stadthouse alone stands upon 13,659. The river Amster falls into the Wey or Y, and the channel into it is called the Pampus, being a branch of the Zuyder Zee, communicating with the lake of Haerlem by a lock, the waters of the lake being 5 feet higher than those of the Wey or Y. The chief security of the city consists in the facility with which the country round may be inundated. Amsterdam at one period possessed three-fourths of the commerce of Holland, and had 500 sail of merchant vessels, its own property. At Saardem, on the other shore of the Wey, there is an extensive ship-building establishment, in which Peter the Great of Russia worked as a common shipwright.

ANAFE, or ANFE, POINT, Africa, on the W. coast, from Sallee S.W. by S. 10 or 12 leagues. Ships might when eligible anchor to the S. of this point, off Algassa bay, on the S.W., in 18 and 25 fathoms.

ANAGADA ISLAND, see *Anegada.*

ANAMOUR, CAPE, on the coast of Karamania, in lat. 36° 0′ 50″ N., long. 32° 50′ 58″ E.

ANANES ISLANDS, Greek Archipelago, small islets, or more correctly rocks, lying S.W. of Milo island, about lat. 39° 36′ N., long. 24° 51′ W.

ANAPA, or ANACOPIA, on the N.W. side of the Black sea, 27 miles W. of Soudjouk Kaleh. On passing the cape vessels should not come within a mile of it, on account of a ledge of rocks which ends in the bay as far as the landing wharfs. Running N.W., steer to the E. until the landing wharf and the gate of Anapa bear S. ¼ E., in order to avoid a rock that has only 6 feet of water over it; then make the course for the wharf, and cast anchor in 5½ or 6 fathoms; or anchor more to the N.W., but not so close in, on account of the danger above mentioned.

ANAPHI, or NANPHIO, Greece, an island lying E. 12 miles from St. Stephano, the S.E. end of Santorin, 2 leagues long. The summit of its highest land is in lat. 36° 22′ 21″ N., long. 27° 47′ 9″ E. This island has no port. Pop. 1,500.

ANARFIORD, Iceland, a deep inlet on the W. side S. of Skalaruk.

ANARSTAPEN, a bay on the W. side of Iceland. Lat. 65° N., long. 15° 1′ 10″ W.

ANASTASIA, North America, a small island close to the coast of East Florida, S. of Mastances inlet, where the river forms two islands of the same name at the mouth. This island is bounded on the N. by St. Augustine's bar, and is 18 miles long; on its N. side is a high white tower, where a look-out is kept. The bar of St. Augustin has only 10 feet water at spring tides and 5 at low. The mark for anchoring in the bay is when the new barracks are open of the N. point of the island in 10 fathoms good ground.

ANCHOR, POINT, Newfoundland, North America, in lat. 51° 15′ N., long. 56° 47′ W. It is high water here at 9 o'clock.

ANCIO, CAPE, Italy, S.E. by S. from the mouth of the Tiber 6 or 7 leagues. See *Anzo, Cape.*

ANCOBAR, or the RIVER OF GOLD, Africa, on the coast of Upper Guinea, falling into the Atlantic in the direction of N. by W. It forms a good harbour at its mouth, with from 7 to 9 fathoms; near the coast on each side are from 8 to 10. The Dutch factory of St. Anthony is on the E. side of the river. An island lies in the middle, called Accas, which renders the channels on each side very narrow. This river is also called the Axim, Cobra, and Snake river, from its winding course. It is nearly W.N.W. from Cape Three Points, distant 10 leagues. Lat. about 4° 48′ N., long. 40′ W.

ANCONA, a free port of Italy on the Adriatic, between two hills, with a convenient harbour. The town is fortified, and has a pop. of 20,000. A citadel stands on one of the hills over the town, and a cathedral on the other. The best arsenal in the Papal dominions is found here. The principal exports are corn, wool, silk, sail cloth, soap, alum, sulphur, fruit, and ship biscuit. The inhabitants are industrious, and all engaged in trade. On approaching Ancona, Mount Conero is seen, and the church of St. Syriac within the town, bearing a semaphore, by which it is easily recognised. On entering or leaving the port, owing to currents and rocks, it is proper to stand out a good way from the point of the mole. With a wind not favourable, it is difficult to enter, for want of room to tack. The port will accommodate ships of war as well as merchantmen, and the Lazaretto channel can admit ships of 200 tons. The port is formed by the mole erected by Pope Clement XII. on the ruins of that built by Trajan; it is 2,000 feet long and 100 broad, with a lighthouse at the termination. The same pope also erected the Lazaretto. Near the mole stands the triumphal arch of Trajan, the most perfect Roman remains existing. On the side of the Lazaretto, the water is shallow and the ground rocky. When the mole is rounded with a good offing it is not difficult to steer in, but vessels drawing 15 or 16 feet water should anchor instantly on passing the lighthouse, where frigates entering must moor head and stern, there not being room to swing. Ships ride well in Ancona roads with offshore winds, 1 or 2 miles N.W. of the lighthouse, in 5, 6, and 7 fathoms. The lighthouse is in lat. 43° 37′ 42″ N., long. 13° 20′ 26″ E.

ANCONES, South America, on the Spanish main, a lofty black point of land on the E. side of a white shore on the W., seen in making the coast of St. Marta, in a direction N. and S.; other points between take the same name. Between them, in lat. 11° 15′ N., and long. 74° 4′ W., lie the city and port of Santa Marta.

ANDALUSIA, Spain, a maritime province, the most southern in that kingdom, containing the ports of Cadiz, Gibraltar, and Malaga, and having the Mediterranean and the Atlantic for its S. and W. limits.

ANDALUSIA, South America, more correctly the Gulf of Paria, between Trinidad island and the main. See *Trinidad.*

ANDAYE, France, department of the Basses Pyrénées, a small port, at the distance of '9 leagues from Bayonne, at the bottom of the Bay of Biscay, on the French side of the river Bidassoa. It is in lat. 43° 25′ N., long. 1° 45′ W., and has a strong fort defending the entrance of the port, and is noted for its excellent brandy and the expertness of its seamen. The time of high water is half-past 3 o'clock.

ANDERNO, Greek Archipelago, Island of Scarpente, on the E. side, in the run from the E. end of Candia to Rhodes to the N.E.

ANDERO ISLAND, North America, S.E. by E. from Cape Gracios a Dios, the point of Honduras furthest to the E. S.W. from this island is the entrance that leads to the Lake of Nicaragua, about 40 leagues off. Lat. 12° 35′ N., long. 81° 43′ W. The same with St. Andrew.

ANDERO, ST., Spain. See *Santander.*

ANDONY, RIVER, Africa, Benin coast, E. from Cape Formosa, between the New and Old Calabar rivers. Vessels making the port of Barry to escape the shoals and sands at the mouth of the river, run E. for the river Andony, and afterwards keep W. along the shore until near Cape Rough Point, in 4 fathoms, near which anchorage will be found in 10 or 12 fathoms, secure from E. or S. winds. It is high water here at half-past 6 o'clock, full and change. Lat. 4° 35′ N., long. 7° 22′ E.

ANDRAKI RIVER, Turkey, N. of Andraki point, upon the coast of Karamania.

ANDREA, ANDRI, or ANDROS ISLAND, in the Greek Archipelago, at the S. end of Negropont, 8 miles long and 2 broad, with a pop. of 4,000. It was anciently called Contandros, Cauros, Epigris, Hydrusia, Lasia, and Nonagria. It is in a fair way channel for the N.E. coast or for Constantinople, passing between its N.W. end, S.E. of Negropont. The N. end is in lat. 37° 50′ N., long. 25° 25′ E.

ANDREA, or ANDREW'S CAPE, the N.E. point of the Island of Cyprus, in lat. 35° 41′ 40″ N., long. 34° 37′ 25″ E.

ANDREA, CAPE ST., Sicily, 5 miles S. of St. Alessio, forming the N. side of Taormina bay, on which is a telegraph; in lat. 37° 48′ 15″ N., long. 15° 17′ 40″ E.

ANDREA, ST., an island in the Gulf of Venice, 14 miles W. of Lissa, high and covered with trees, 2 miles from which is Brusnic rock, above water, and off the W. point a rock called Kammick, above, and also a sunken rock; hence there is no passage between.

ANDREW, ST., CAPE and FORT, South America, in Berbice river, opposite Crab island.

ANDREW'S BAY, Scotland, between the Forth and Tay; the shore is steep and rocky, with 7 fathoms near the land, a safe and clean anchorage. St. Andrew's harbour, in the bay, is dry, having a pier, good for the shelter of small vessels; it has from 12 to 14 feet at spring tides, 7 and 10 at neap. The entrance is on the S. side of the pier head. Lat. 56° 19′ N., long. 2° 42′ W.

ANDREW'S, or ST. ANDRE'S BAY, Gulf of Mexico, on the S. coast of Florida, between Santa Rosa bay on the N.W. and St. Joseph's on the S.; sheltered from all winds. Lat. 30° 15′ N., long. 85° 45′ W.

ANDREW'S, ST., RIVER and POINT, N.N.E. from Cape Palmas, on the coast of Africa, about 25 leagues in the Red cliffs. There is only 12 fathoms water at 2 leagues from the shore, but anchorage

C

is found close under the point in 3 fathoms. Lat. 4° 57′ 8″ N., long. 6° 3′ 47″ W.

ANDREW'S, ST., ISLAND, or ANDERO, on the Mosquito coast, in the Sea of Guatamala, lat. 12° 35′ N., long. 81° 43′ W. It bears N.W. by N. 215 miles from Chagres; two kays lie S.S.W. of it and E.S.E., the former 20, the latter 14 miles.

ANDREW'S, ST., the S.W. point of Navy island, in New Brunswick, North America. Lat. 45° 3′ 30″ N., long. 67° 7′ W.

ANDREWS, THE, a shoal near Harwich harbour, lying within the Platters, and extending about half a mile S.S.E. from the Stony beach. It forms the W. boundary of Harwich harbour; on the edge it has a black buoy in 4 fathoms.

ANDRICHOL CAPE, Majorca, woody and steep, E. of Cape del Llamp.

ANDROPOS, CAPE, in the Ægean sea, N. of Thaso isle, on the main.

ANDROS ISLAND, N. of the Cyclades, in the Levant.

ANDROS ISLANDS, see *Bahamas.*

ANEGADA ISLAND, N.W. point; lat. 18° 50′ N., long. 64° 27′ W. See *Virgin Islands.*

ANGEL, or ANGLE'S ROAD, Milford Haven, South Wales. See *Milford.*

ANGEL HILLOCKS, Africa, land elevations so called near Cape Blanco. Lat. of the southern, 18° 29′ 30″ N., long. 16° 2′ W.

ANGELO, CAPE ST., or MALEA, Greece, in lat. 36° 26′ N., long. 23° 12′ 30″ E. The land here runs N.N.W. ¾ W. from Cape St. George, distant from which 4 leagues is the town of Napoli di Malvasia; all the way is rocky, and 4 miles from Napoli there is a village called Agiolindi, but no harbour. The pop. of Napoli is about 2,000; it has no trade, but an open roadstead exposed to all winds from the S. and E.

ANGELO MONTE, ST., a small port within Cape Felice, on the W. side of the Gulf of Venice. Lat. 41° 42′ N., long. 16° 11′ E. There is good anchorage here in W. and N. winds, in the bay of Manfredonia, the only shelter with deep water. This port lies in Apulia, 8 miles N. of Manfredonia.

ANGER ISLAND, or ANZERSK, 13 miles W. of Rovestra island, in the White sea; with other islands it lies in the way of the navigation of the entrance of the Gulf of Onega. Reaching Rovestra island, the customary passage to Onega is between that island and Anger, a S.W. by S. course.

ANGIA, GULF, on the E. coast of the Morea, in Greece, N. from Cape St. Angelo. It runs within 6 miles of the Gulf of Corinth. Within this bay are Napoli di Romania, Malvasia, and Vechia.

ANGISTRI, Greece, a small island in the Gulf of Athens.

ANGOLIN POINT, France, the N. point of Chataillon bay, S. of Rochelle, and N. of the Isle of Aix.

ANGOSTURA, or SAN TOME, in Columbia, in the province of Cumana, on the river Orinoco, situated at the foot of a hill, about 60 leagues W. from Point Barima, at the mouth of the river. In front of Angostura the Orinoco is 3,850 fathoms wide, and at low water 65 feet deep. The mean rise of the river at Angostura is 25 feet; the common rise of tide at the bar of the Orinoco is 12 feet at common times, but in rainy seasons upwards of 40. Point Barima is in lat. 8° 44′ 30″ N., and long. 60° 3′ W. The rising of the waters

of the Orinoco commences in April and ends in August. The stream runs at the rate of 8 miles an hour. The town is built on a rock, ascending from the river towards a fort on the brow of the hill, commanding the anchorage. The custom-house is close to the river. When the waters subside, this city is very unhealthy, owing to a neighbouring swamp, and the heat is excessive. La Soledad, on the opposite bank, is, however, a healthy station.

ANGRA, a town on the river Angra, in Africa, with a bar harbour; it is best known as the river "Danger," from the difficulties surrounding it. Angra is near Cape St. John, which is in lat. 1° 9′ 40″ N., and 9° 21′ 35″ E. long. High water at 6 o'clock.

ANGRA DE CINTRA, or CINTRA BAY, African coast, at the distance of 3 leagues from the S. end of the Fisherman's cliffs, or those of the Rio Ouro. This bay is sheltered on the N. by a low, sandy point. The depth of water is only 4½ fathoms, with a sandy bottom; at 2 miles from the beach there are 9 and 16 fathoms. Vessels from the Canary islands visit this coast to fish. A reef projects from the S. extremity. N. point, lat. 23° 7′ N., long. 16° 9′ 15″ W.

ANGRA DE LOS CAVALLOS, African coast, between Angra dos Ruivos, so called by the Portuguese, and the N. entrance of the Rio do Ouro. Lat. 24° 8′ 12″ N., long. 15° 36′ 18″ W.

ANGRA DOS RUIVOS, so called by the Portuguese, is on the African coast, often called by the English Garnet bay. It is marked by seven small hills, and is N. of the entrance of the Rio do Ouro, in lat. 24° N.

ANGRA, see *Azores.*

ANGUADA, CAPE, the W. point of the island of Porto Rico, in the West Indies, distant from the nearest land at the N.E. end of Hispaniola, 22 leagues. Ships may anchor here in 23 fathoms, soft ground, or go within a mile of the shore on coming from the N. to the S. of the N. point, and then anchor in 10 or 12 fathoms.

ANGUILLA, one of the Caribbee islands, belonging to England. Lat. 18° 8′ N., long. 63° 11′ 45″ W. It lies N. of St. Martins, distant due N. 4 miles. From the S.W. point it runs N.E. 13 miles. The inhabitants breed cattle and procure salt. There are anchorages on the S. and W. sides, and off the N.E. lies an island called Anguilletta, from the E. end of which a reef runs far into the sea. There are other rocks and islets about. Off the S.W. are two, called the Great and Little Anguilla Kays. Vessels may sail S. of Anguilletta and between Anguilla and Sandy island, keeping mid-channel, or between Anguilla and St. Martin's, avoiding the rock off the N.E. point of St. Martin, where the water shoals to 10 fathoms each side, deepening to 22 in mid-channel. There is no passage between the kays. There is a reef off the N. end of Anguilla. This island was discovered and colonized by the English in 1650. The country is flat; in the centre is a salt lake, yielding annually 3,000,000 bushels, of which a great part is exported to America. The climate is healthy. The pop. about 360 whites, 320 coloured, and 2,451 African. A deputy is sent to St. Kitt's assembly, and the island is governed by a magistrate confirmed by the Government of Antigua.

ANGUILLA BANK, or SALT KAY BANK. See *Bahamas.*

ANGUILLE BAY and CAPE, on the W. end of Newfoundland, in the Gulf of St. Lawrence, 10 leagues N.N.W. from Cape Race. The N. entrance is by a cape of the same name. Lat. 47° 54′ N., long. 59° 17′ W.

ANHOLT, an island W. of the S. of the Little Middle Ground off Zealand, with a light on its E. end, 120 feet above the sea; the high light flashing; the lower fixed, 65 feet under the upper, only used when the light-vessel is not in her station. There is a floating light-vessel, schooner-rigged, red, with a white cross, E. ½ S. from the lighthouse on Anholt, a quarter of a mile from the reef at the end of the island. The vessel remains at her station until the 21st of December, and resumes it in March as early as possible. Near the end of the reef spoken of before there is a rock, always above water, called the Knoben, 2 leagues from the island, joined to it by a neck of sand; at a quarter of a mile from it are 23 and 25 fathoms. The soundings near Anholt are very irregular, and require great attention; 3 miles S. of Anholt is a sandy ridge, called the Little bank. A shoal of sand surrounds the whole island, in one place shallowing out from the shore 3 miles, then turning N. again, and ending in the "N.W. Reef," as it is styled, marked by a black buoy in 4 fathoms water, bearing N.W. by W. from the lighthouse 10 miles. Lat. 55° 41′ N.

ANIADA, see *Iniada.*

ANIMABO, or ANNAMABOE, Africa, on the Gold coast, 3 leagues N. by E. from Fort Maurice. The anchorage here is safe in 7 fathoms. When the W. of four hills seen above Cape Corse is brought to bear N.W., Cormantine is 2 leagues E.N.E. Lat. 5° 10′ 12″ N., long. 1° 7′ 12″ W. One league from Animabo, between that and Fort Maurice, there was once an English factory called Anisham. Lat. 7° N.

ANKHIALOU, or HIALOU, on the Gulf of Bourgas, in the Black sea, on a projecting point, 6 miles S.S.W. ¼ W. from Messemvria, between which is the bay of Raveda. S. and S.W. winds prevail here. A reef 4 cables' length broad runs out from the point of Ankhialou for 600 fathoms. Vessels anchor here in from 15 to 7 fathoms.

ANN, CAPE, North America, United States, a point which forms the N. side of the Bay of Massachusetts, opposite to Cape Cod, on which the town stands. Lat. 42° 45′ N., long. 70° 17′ W.

ANN, CAPE ST., Africa, S. by E. of Sierra Leone river, on the W. coast. This cape is the N.W. point of the island, a slip of land 80 miles from W. by N. to E. by S. The channel is variable. At the pass there is a bar with only 10 feet at low water; a pilot is therefore necessary. The river Sherbro or Cores is near the E. end of the island. It is high water here at 7ʰ. The inner passage is called the Furno of St. Ann. See *Sherbro.* This cape is in lat. 7° 34′ N., long. 12° 57′ W.

ANNA DE CHAVES, Africa, a town on the Island of St. Thomas. Lat. 25′ 30″ N., long. 6° 46′ E.

ANNA, SANTA, town and lagoon, in Tabasia, Gulf of Mexico, South America. Lat. 17° 54′ N.

ANNA, SANTA, DE CORO, a town in the Gulf of Coro, in Columbia.

ANNA SHOAL, a vigia, reported to have been seen at noon, in lat. 39° 30′ N., long. 50° 50′ W., but wants confirmation, and was not definitely described.

ANNAPOLIS, North America, a river of Nova Scotia falling into the Bay of Fundy through the basin of its own name, on the side of which is the town and port of Annapolis Royal. On Point Prim, at the entrance of the basin, is a lighthouse on the W. side, 76 feet high. Lat. 44° 41′ 30″ N., long. 65° 48′ 18″ W. It is navigable for large ships 10 miles, and for small vessels further. The harbour is one of the finest in the world, having nowhere less than 4 or 5 fathoms, and in some places 16 and 18. The bottom is good, and there is shelter from all winds. The tides and currents run strong here, and there are frequent fogs, particularly at the mouth of the Bay of Fundy. The entrance or gut is about three-quarters of a mile wide, and 1½ mile long. High water about 10 o'clock, with 28 feet rise.

ANNAPOLIS, United States, North America, the chief town of Maryland, on the W. bank of the Chesapeak. Lat. 39° 5′ N., long. 76° 43′ W. It stands on a peninsula formed by the river Severn and two little creeks. Pop. 2,792.

ANNE, CAPE QUEEN, W. coast of Greenland, in lat. 64° 15′ N., long. 50° 35′ W.

ANNE'S BAY, ST., Labrador coast, North America, between Davis's inlet and Cape Charles, the N. point of entrance into the Strait of Belleisle, on the S. Lat. 54° N., long. 57° W., or nearly.

ANNE'S HARBOUR, island of Martinique; a secure port with good anchorage.

ANNE'S HEAD, Isle of Man, the N. limit of Derby haven.

ANNE'S ISLAND, one of the Scilly group. See *Scilly.*

ANNE'S, QUEEN, FORELAND, North America, on the N. main from Hudson's straits, lat. 64° 8′ N., long. 74° 41′ W.

ANNE'S SHOAL, ST., Africa, W. coast; the N. end in lat. 8° 10′ N., about 12 leagues W.S.W. ½ S. from Cape Sierra Leone. The outward edge extends 12 leagues due S., and then S.E. 17 leagues, where it is separated from Turtle island by a swash half a league broad, with 3 or 4 fathoms of water.

ANNE'S, ST., BAY, Jamaica, with an anchorage of 6 and 9 fathoms, the entrance narrow between two reefs, with 10 or 11 fathoms in mid-channel. It is on the N. side, 7 miles W. of Ocho Rios.

ANNET ISLAND, Cornwall, one of the numerous isles of Scilly, W.N.W. from St. Agnes island and lighthouse. St. Agnes and this island are of moment in entering Broad sound, to go into St. Mary's road. See *Scilly Islands.*

ANNHAMNEREN, Drontheim's Leed, Norway, the channel where the W. and S. entrances to Drontheim unite, between Hitteren island and the main.

ANN'S, ST., a port on the E. side of Cape Breton island, North America; small, and principally used by fishing vessels. It lies on the W. side of Labrador lake. Lat. 47° N., long. 60° W.

ANN'S, ST., BAY, Africa, N.W. coast, 11 leagues S. by W. from Cape Barbas, and 5 leagues N. by E. from Cape Corvoeiro.

ANN'S, ST., HEAD, or POINT, in Pembrokeshire, at the entrance of Milford Haven. The high lighthouse is in lat. 51° 40′ 59″ N., long. 5° 10′ 2″ W. There are here two lighthouses; the high light 192 feet above the sea level, the low light, at the extremity of the head, 159 feet high. See *Milford Haven.*

ANNOT SAND, Scotland, running off from the entrance into the river Montrose, N.E. of the Frith of Tay, for nearly 5 leagues.

ANOTTA BAY, Jamaica, N. side of the island, exposed to the N. and N.W. winds. There is anchorage in 7 fathoms. There is a shoal in the bay, called the Schoolmaster, which should be avoided. The town of Anotta is in lat. 18° 19' N., long. 76° 33' W.

ANQUETTES, PETITE and GRANDE, Isle of Jersey ; two rocks, S.E. from St. Clement's, the S.E. point of the island. There is a beacon on the Grande Anquette. The tides and currents here are extremely dangerous.

ANSER, CAPE, France, N.N.E. from Havre, on the N. point, at the entrance of the Seine river.

ANSIDONIA, POINT, Italy, in the Roman states, N. of Clementino river.

ANSLO BAY and PORT, Norway, in a gulf on the S. coast and N. side of the Cattegat. It is on the E. side of Christiania, in lat. 59° 24' N., long. 10° 14' E. The harbour is good, and goes up far inland ; the mouth is in lat. 59° about, and it is N.N.W. from Frederickstadt 10 leagues.

ANSTRUTHER, EAST and WEST, Scotland, on the S.E. coast of Fife, 9 miles S.S.E. of St. Andrew's. It is a small tide haven, in lat. 56° 12' N., long. 2° 26' W. Pop. 1,437.

ANTARCTIC SEA, the S. expanse of ocean comprehended within the limit of lat. 66° 30' S., as far as the nether pole.

ANTEBANICO, or ANTIBONITE RIVER, at the W. end of St. Domingo, in the West Indies, at the bottom of a gulf between Cape Nicholas on the N.W. and Cape Dame Mary on the S.W. of the W. point of the river. It is a good road, but the river itself is nearly dry at low water. The road is to the W. of S. of Petit Guave harbour.

ANTEFER, CAPE, or DE CAUX, France, about 11 miles from Cape la Hève. There is above 5 fathoms water half a mile from the coast thence all the way to La Hève.

ANTEGOSO, Greece, a low, small island on the W. of Gozo island, 11 leagues S.E. by E. from Cape St. John, the S.W. point of Candia.

ANTHONY'S HILL, ST., or MOUNT SAN-TONA and PORT, Spain, about 17 miles W. a little N. from Bilboa, on the coast of Biscay, in the Bay of Laredo, and a league N.N.W. from that place. It is lofty, and vessels may sail near its E. point in 10 fathoms water, and go round it to the S. The other point of the bay is called Cape Rastrillar, high, dark, and rugged, with rocks and shoals around it. Between these two points lies the town of Laredo, whence a small circular beach of sand terminates at the entrance into the harbour. To sail in approach Point Frayle, part of the hill of St. Anthony, but keep clear of the Merano shoal, lying N.N.E. by E., about 2 cables' length from the point. Run along the shore at some distance, until the S. point, on which stands the battery, comes in one with the convent of St. Francis de Ano, all on the W. side. No sail must be made S. of this mark until the bar is passed, on account of the Doncel, a bank or shoal lying S.E. ¼ E. a mile distant from Rastrillar point. There is also another shoal to be avoided, called the Pittoro, distant from Passage point E. from a quarter to half a mile, from Point Carlos S.W. by W. a quarter of a mile. The mark already given must be kept near to Point Carlos, then steer W. by S. towards the steeple of Cicero, but not 1¼ cable's length to the S. ; then pass the battery and point of Martin, and stand in mid-channel for the town of Santona, or St. Anthony, where is anchorage in 6, 7, and 8 fathoms, sand and mud. Spring tides rise 12 feet, neaps 8. This harbour affords shelter to vessels that dare not attempt the bar of Bilboa in gales and high seas, and to those that cannot make Santander to the W. Lat. 43° 26' 10" N., long. 3° 1' 22" W.

ANTHONY'S POINT, ST., Cornwall, about 13 miles N.E. of the Lizard, on which there is a light-house to direct vessels into Falmouth road. It is high water there at 5ʰ 30ᵐ. The lighthouse stands in lat. 50° 8' 35" N., long. 4° 59' 31" W., at the entrance of the harbour.

ANTHONY'S, ST., BAY, Island of St. Lucia; it has from 15 to 36 fathoms, and a good harbour. See *St. Lucia.*

ANTHONY'S, ST., CAPE, Newfoundland, America, E. coast, 4 leagues S. of Cape Humble-don, near the N. end, in lat. 51° N.

ANTHONY'S, SAINT, FORT, Africa, lat. 4° 52' 18" N., long. 2° 14' 45" W.

ANTHONY'S, ST., ISLAND, see *Cape Verde Islands.*

ANTHONY'S, ST., ROCKS, France, W. by N. of St. Maloe's 3 or 4 miles. There is a channel for vessels within them.

ANTIBES, France, an ancient town in the department of the Var, the most E. of France in the Mediterranean, well fortified. The town comes down to the sea. Before it is a mole, with a half-moon battery and a small square fort on the N. point. There is little water except about the mole, which will only admit small vessels ; the entrance is 150 fathoms broad, and the depth of water within from 15 to 18 feet. About 100 fathoms N.N.E. from the point of the great bastion, on the middle of the mole, is a ledge of rocks, with only a foot of water over them, and near them from 16 to 20 feet. Vessels may pass between the mole and this ledge with from 20 to 25 feet of water, by keeping a little nearer to the half-moon battery than to the ledge ; but the safest way is to pass E. of the rock and run in towards the N. shore, on the top of which stands Fort Quarre, then steer for the port, and anchor at the bottom of the harbour alongside the mole, with the stern to the town ; the N.W. wind alone can disturb a vessel anchored thus. The trade of this town is trivial ; it consists in salt fish, wine, olive oil, oranges, figs, and dried fruit. The pop. is 5,150. There is a tower here of Roman construction and some fine remains of fortifications. Lat. 43° 34' 40" N., long. 7° 7' 50" E. There is a light at the extremity of the E. mole, fixed, but varied by two minute flashes, the fixed light seen in clear weather 3½ leagues distant.

ANTICOSTI ISLAND, America, Gulf of St. Lawrence. The E. point in lat. 49° 8' 45" N., long. 61° 42' 59" W.; the W. point lat. 49° 52' 20" N., and long. 64° 35' 80" W. The lighthouse on the S.W. point stands in lat. 49° 23' 53" N., long. 63° 38' 47" W., and the entrance of the river or Bear bay in lat. 49° 30' 30" N., long. 62° 27' 29" W. It is 100 feet above high water, and is lit from March to December. This island is rocky, wooded, and from its position has been too frequently the cause of vessels having been lost upon its inhospitable shores, where no sustenance is obtainable for the support of life, even of the very coarsest nature. It requires, therefore, to be carefully regarded on entering the St. Lawrence by either of its two channels. It has only one harbour that can afford shelter to vessels. Boards are placed in different situations to direct unfortunate mariners shipwrecked here where they shall find relief. The S. shore is very dangerous. Cape Henry, the S.E. extremity, is in lat. 49° 47' 50" N., long. 64° 25' 44" W.

ANTIGUA, a fertile island of the Caribbean group, in lat. 17° 9′ 40″ N., long. 61° 52′ 30″ W., about 20 miles long and 54 in circumference, and containing 69,277 acres. It was discovered by Columbus in 1493, and named by him Santa Maria de la Antigua. It is the third oldest colony of England in the West Indies. The white and coloured pop. 35,000. The annual medium temperature 79·68° Fahr. Annual fall of rain 35·58 inches. The revenue is about 16,000*l.*, the expenditure nearly as much. No island of the group possesses such excellent harbours, but except St. John, English harbour, and Falmouth, they are difficult of access. These last are Freeman's, Rendezvous, Five island, Lydesenfis bay, Parham, Nonsuch, Willoughby, and Indian creek. The coast is irregular, indented, and surrounded by islets, rocks, and shoals, which require great care to approach, except on the S.W. The exports are about 170,000*l.* in value annually, but fluctuate according to the seasons. The entries of shipping inwards and outwards about 28,000 tons per annum. English harbour, on the S. side of the island, will receive the largest vessels. It has a dockyard supplied with every necessary for the service of the royal navy, and no merchant vessels unload or load there. There is anchorage outside the harbour, with Fort Barclay, on the N.W. side of the entrance, bearing N.N.E., and the extreme point E. ½ S., in 13 fathoms. In entering give the Old Horse-shoe a good berth, and steer in mid-channel into Freeman's bay, the S. part of the harbour, where there is good anchorage as far up as the storehouses, in 3, 4, 5, and 6 fathoms; there are four moorings in this bay, where large ships ride just within the harbour's mouth. A pilot is advisable to go in. No wood or water is to be obtained there. Falmouth harbour lies 1½ mile W. of English harbour; it is larger and the entrance wider. To enter, go close to the W. or Proctor's point, a ledge of rocks on the starboard running off one-third of the passage, called the Bishops. When passed, the channel being to the W., there is good anchorage in from 3 to 6 fathoms. On the W. side there is a battery on an islet, behind which there is indifferent water. The best anchorage is with the end of Black point S. by W. ½ W. and the battery W. ½ S. St. John's road and harbour are to the N.E. of Five Islands' harbour, with water for merchant vessels and security in all weather. The town of St. John, the island capital, is at the bottom of the harbour. The road is about 2½ miles E., in lat. 17° 10′ N., long. 61° 52′ 30″ W.; it lies between the Ship's Stern S. and Corbison's fort N., their distance being 2½ miles. About 1 mile from the Ship's Stern is a dangerous ledge, called Warrington, with only 3 feet of water. There are other dangers here, which make a pilot needful. Five island harbour is so called from a number of islands off the S. point. The entrance is three-quarters of a mile wide, the water decreasing in depth. Carlisle bay is a small harbour 3¾ miles W. of Falmouth, at the entrance of which a vessel may ride in 2, 3, or 4 fathoms. Willoughby bay is on the E. part of the S. coast of the island; its entrance not wide, and encumbered with shoals, nor is it much visited by vessels from Europe. Two forts protect the anchorage. Nonsuch harbour is on the E. of the island, and has Green isle on its entrance; it is so clogged with shoals few ships enter it, and those only belonging to the inhabitants. Parham harbour lies to the N.W. of Nonsuch, extensive, shallow, and dangerous, so that only small vessels enter it, running into the S. part of the bay, where the town of Parham stands; there they ride securely in 2 or 3 fathoms, protected by Fort Byam on the W. and Old Fort on the E. From Falmouth harbour to Carlisle bay the coast of the island is foul, sending out numerous rocky reefs.

ANTILLES, THE, Greater and Smaller, another denomination for the West India islands. From Barbadoes to Porto Rico, better known as the Caribbees, or Caribbean islands, from their original inhabitants, and again subdivided into the Windward and Leeward islands, because both winds and currents set towards Porto Rico. The word Antilles is derived from Antilla, "forward," they having been originally called "Antillas" by the Spaniards.

ANTIM RIVER, Africa, on the S. coast, 16 leagues E. by N. from Cape Three Points, which cape is in lat. 4° 44′ 30″ N. Ayobo stands on the W. point, before which ships may enter in 14 fathoms in foul ground, but must go no nearer. It is 4 leagues W. of Secunda.

ANTI-MILO, a small island on the N. side of Milo island, in the Greek Archipelago, almost closing the harbour, due E. from Cape St. Angelo.

ANTIOCH BAY and TOWN, Syria, at the entrance of the river Orontes, the chief Syrian river. Antioch is on the S. bank, in lat. 36° 11′ N., long. 36° 9′ 30″ E. S.W. from the entrance of the Orontes is Cape Possidi, the S. point of the Bay of Antioch. From this cape the land runs S.S.W. ½ S. 23 miles to Cape Ziaret, a high promontory 30 leagues E. by N. ½ N. from Famagousta, in Cyprus.

ANTIOCHE PERTUIS, France, the channel of Antioche, at the N. of the Isle of Oleron, on the W. coast, sometimes called the Antioche passage, lying between that island and Ré. The channel is 5 miles wide, and leads straight into Basque roads, where there is good anchorage in 9, 8, 7, and 6 fathoms. Vessels entering this channel should never approach the Isle of Ré nearer than 10 fathoms, as the ground is rocky and broken, and should equally avoid sailing within 2 miles of the Isle of Oleron, but go in mid-channel, on account of some shoals or rocky grounds off the last island, called the Antioches, extending 2 miles E. of the lighthouse, for though there are anchorages within the rocks, strangers should not attempt to find them. Yet further on are the Long banks, extending from the Antioche rock to Boyart bank, the S. end of which is nearly midway between Oleron and Aix. On the N.W. end of Oleron is the tower called Chassiron, with a fixed light 164 feet above the sea. It may be seen 6 leagues off. This new lighthouse is 328 feet E. by S. from the old tower. The light is in lat. 46° 2′ 52″ N., long. 1° 24′ 30″ W.

ANTIOCHETTA, a port of Karamania, on the Mediterranean, 88 miles S. of Konieh, in lat. 36° 5′ N., long. 32° 15′ E.

ANTI-PAROS, Greece, an island W. of Paros, separated by a channel 1½ mile wide, encumbered with rocks. This island is 7 miles long by 3 broad, and is a solid mass of marble. It has one village with 300 inhabitants. It is celebrated for a remarkable grotto, a scene of extraordinary beauty. The island summit is in lat. 36° 59′ 40″ N., and long. 25° 3′ 27″ E.

ANTIPATER, Syria, on the coast, now named Caphar Saba, little more than ruins, at the mouth of a small river about 2 leagues S.S.W. from Cæsarea. Lat. 32° 30′ N.

ANTI-PAXO, an island on the coast of Greece, divided by a channel from Paxo, one of the Ionian islands. It is small and uncultivated. Its S. end is in lat. 39° 8′ 30″ N., long. 20° 15′ 45″ E.

ANTIPSARA ISLAND, Greek Archipelago, 2 miles from Ipsara island.

ANTIVARI, Dalmatia, a port on the Gulph of Venice, S.E. from Budea, and N.W. from the Gulf of Lodrin, in lat. nearly 42° 10′ N.

ANTON LIZARDO, North America, Mexico, a most capacious and sheltered anchorage, the best and safest on the Mexican coast. It is within 15 miles of Vera Cruz, and is formed by shoals and reefs. The best anchorage is N.E. and E.N.E. of Point Antonio Lizardo, in 10 fathoms. It is needful in going in from the N. not to approach the W. side nearer than 27 fathoms, until within 6 miles of land. Then pass the inner reefs in 16 and 18, and endeavour directly to get a sight of Isle Blanquilla, which is of sand, and very low, so as from the N. not to be easily distinguished from the main. The Point de Collol cannot be mistaken, as it is the S. point of the anchorage, and when it bears S. 40° E., Blanquilla will be seen open of the point; while N.W. approach no nearer than three-quarters of a mile, until it bears E. by N. ¼ N., when Medio island will be one with it, and when this island opens to the S., haul up E.S.E., passing Blenquilla at 2 cables' distance, which must not be brought to the W. of W. by N. until three-quarters of a mile to the E. of it, then haul more to the N., and anchor as convenient. There is unfortunately a want of supplies and of water here; fish are in plenty. The site of Blanquilla island is lat. 19° 5′ 15″ N., long. 95° 58′ 5″ W.

ANTONIO, CAPE, Spain, in lat. 38° 49′ 50″ N., long. 8′ 8″ E. It is high, steep, and rises almost perpendicularly from the sea. The Bay of Xavia within it is sheltered from S. and S.W. and N.W. winds. Ships generally anchor in the N. of the bay, near St. George's tower, in the Rincon, a place free from rocks.

ANTONIO, CAPE ST., the W. end of Cuba, low, and covered with wood; lat. 21° 51′ 40″ N. long. 84° 58′ W. The cape is bold, and while bearing to the W. may be approached. The coast is bold to the E. as far as Point Hollandas, off which are sunk rocks. The cape may be doubled within 2 miles safely. There are banks between 4 and 6 leagues from the cape; one at 6 leagues, called Sancho Pardo.

ANTONIO ISLAND, see Cape de Verde Islands.

ANTONIO, PORT, Jamaica, lat. 18° 14′ 40″ N., long. 76° 31′ W., 16½ miles E. of Anotta bay, and 6 miles W. of the N.E. end of the island. To enter the E. harbour, bring the E. part of the Blue mountains S.S.W., and so steer to Folly point, then bring the church to the W. of the E. wharf, bearing S.S.W. ¼ W. In approaching the fort on the W. side on the point of Titchfield, open the church to the E. of the wharf. The fort at N.W., anchor in 9, 10, or 11 fathoms. The E. harbour is not so secure as the W., but the channel of the W. is not more than 70 fathoms wide.

ANTONIO, PORT ST., on the S. side of the Isle of Lemnos, in Greece. It has a good road, sheltered from all but S.E. and S.S.E. winds. The riding is good close to the W. shore.

ANTONIO, ST., see Cape de Verde Islands.

ANT'S ISLAND, see Azores.

ANTWERP, Belgium, on the river Scheldt, 26 miles N. of Brussels. The river here is 400 yards wide, and large vessels come up to the quay and into a large basin. On the opposite side of the town are docks for ship-building, well defended. The citadel is strong. This city stands in lat. 51° 13′ 16″ N., long. 4′ 22″ E.

ANVALES BANKS, W. of Cape Peñas, on the N. coast of Spain, over which the sea continually breaks.

ANZARON, N. coast of Spain, a small island near Ribadeo.

ANZO, CAPE, Italy, 28 miles S.E. ¾ E. from the mouth of the Tiber. Here is a light, which is darkened at intervals, standing in lat. 41° 26′ 30″ N., long. 12° 42′ E. The lighthouse is upon a square tower. There are rocks about the cape. The tower of Caldarno, a few miles N., is round.

APALACHICOLA BAY, between St. George's island and the main of Florida. The N.W. entrance is between the island of St. Dionysio and the N.W. point of St. George's island, on which a lighthouse is erected, with a fixed light to indicate the entrance. The island thence runs N.E. ¾ E. 21 miles to the E. point, between which and Dog island is the entrance into the bay. There is a light on the W. point of Dog island that revolves once in 3 minutes, 55 feet above the level of the sea. A black streak is painted round the lighthouse, to distinguish it from St. George's and St. Mark's. It bears from the bar N. ½ E. 3½ miles. This light is in lat. 29° 45′ N., long. 85° 23′ W. St. Mark's light is a fixed light at Point Casinas, at the E. side of St. Mark's river.

APAM, a Dutch settlement now abandoned, E. of Accra.

APANOSMIA HARBOUR and TOWN, Greek Archipelago, in the island of Santorin, on the N.W. coast, 7 miles N.W. of Scauro, in lat. 36° N.

APENRADE BAY, South Jutland, Denmark, on the E. coast, in the Baltic, between that coast and the N.W. end of the Isle of Funen. The entrance is easy, and has a good depth of water for ships of burthen, and the harbour is very secure. The town, situated at the bottom of the bay, has a very good trade. Lat. 54° 52′ N., long. 10° 7′ E.

APE'S HILL, or MONTE SINGE, on the African side of the Straits of Gibraltar, between Tangier and Ceuta, a point nearly S. from the S.W. point of Gibraltar bay, on the European side. A small island lies off the W. part of this elevation, near which there is 15 fathoms water. Lat. 35° 48′ N., long. 5° 22′ W.

APOKERA, or BECUR, CAPE, Egypt, N.E. by N. from Alexandria, distant 4 leagues, more generally known to Europeans as Aboukir, being the extreme point of the bay in which Nelson defeated the French fleet in 1798. See Aboukir.

APPEE, African coast, a fort; lat. 6° 22′ N., long. 2° 25′ E.

APPLE ISLAND, North America, in the river St. Lawrence, on its S. side, between Basque and Green islands. It is uninhabited. High water in Chignecto bay near 11ʰ; rise 32 feet.

APPLECROSS BAY, Scotland, Inverness, on the S.W. part of the coast, in which there is tolerable anchorage.

APPLEDORE, Barnstaple bay, Devonshire, within the bar on the starboard side. See Barnstaple.

APPOLONIA, CAPE, Africa, Guinea coast, 15 leagues W. by N. from Cape Three Points. It is not safe to approach nearer than 15 fathoms water, the ground being foul. The coast is low land to Axim, E. by S. 5 leagues; no boats can land there, from the violence of the surf. Lat. 4° 58′ 45″ N., long. 2° 35′ 5″ W.

APROUAK, APROUAGUE, or APPROBAQUE RIVER, French Guiana, South America. Eight leagues N.W. of Cape Orange is Garimore point; it is the E. point at the entrance of that river, which is but little known. The entrance is 2 leagues wide, with 3 or 4 fathoms of water. Cape Orange is in lat. 4° 10′ N., long. 51° 15′ W.

AQUAMBOE, Africa, Guinea coast, E. side of the river Volta.

AQUIDAH, Africa, W. coast, lat. 4° 45′ 27″ N., long. 2° 8′ W.

AQUIN BAY, E.N.E. ½ E. from Orange kay, in St. Domingo, is Moustique kay, and between that and Pigeon kay is a deep passage, through which ships sail into the bay of Aquin, where the Diamond rock stands in lat. 18° 13′ 48″ N., long. 73° 20′ W.

ARAB'S TOWER, Egypt, a remarkable object in making the ports of Alexandria, being visible 6 or 7 leagues off; it stands in lat. 30° 58′ 15″ N., long. 29° 34′ 20″ E. A rock little known exists, with only 9 feet water over it, the Arab's tower bearing E.S.E. ½ E. distant 5 or 6 miles, the hill to the E. of the tower E. ¾ S., and the extremes of land in sight E. by N. ¼ N. and S.W. ¾ W.

ARABAT, coast of the Crimea, in the S.W. angle of the Sea of Azoph, on a narrow peninsula, on the S. part of which stands the ancient fortress of Arabat, in a little bay open to the N. and N.E.; near it are from 19 to 24 feet of water, with a muddy bottom. At this place the sandy part of the coast terminates, and 18 miles to the N.E. ¼ E. is Cape Kezandibi, forming a peninsula, curving to the N.W., high and steep, having, at the distance of a mile, 34 and 36 feet, with mud. Lat. about 45° 5′ or 45° 54′ N.

ARABIA, GULF OF, Mediterranean, the name formerly in use for the African coast, S. from the E. end, or nearly so, of the Island of Candia, about lat. 31° 20′ N.; it is now disused.

ARACHAT, see *Breton Island.*

ARACLEA, see *Heraclea.*

ARAICHE, or EL ARAICHE, coast of Africa, S.W. from Arzilla, a town with a pop. of 2,500, having some little trade with Gibraltar; lat. 35° 12′ 50″ N., long. 6° 9′ W. Supplies may be obtained here. The best anchorage is with the town between S. and S.S.E. At low tide the mouth of the river has only 5 or 6 feet water over it, but there is a rise and fall of from 9 to 12 feet.

ARBA, an island in the Adriatic, near Fiume, 5 miles S. of Veglia, 11½ miles long, with a pop. of 5,000, and some trade in fruit, cattle, and corn. It is subject to Austria, and has numerous other islets in its vicinity.

ARBA, ISLAND and TOWN, on the E. side of the Gulf of Venice, about 5 miles from the coast of Dalmatia.

ARBONE, CAPE DEL, Gulf of Ufabassi, W. coast of Anatolia.

ARBORA, CAPE, in Anatolia. Lat. 37° 20′ N.

ARCADIA, GULF OF, Greece, in lat. 37° 38′ 12″ N., long. 21° 20′ E.; a shallow open bay, the N. point formed by Cape Catakolo, the S. by Cape Konello, which lies W.S.W. from the port of Arcadia about 5 miles, is in lat. 37° 12′ N., long. 21° 36′ 20″ E.

ARCADIA, Greece, a seaport in the Morea, at the foot of a rocky hill, on which stands a castle.

It is 22 miles N. of Navarino. Lat. 37° 22′ N., long. 21° 42′ E.

ARCAHAIS, PORT and TOWN, Haiti, N.W. of Port au Prince.

ARCAS, three islets in the Bay of Campeche, visible 5 miles off, 27 leagues W. ¾ N. from Campeche; between them is a good harbour. The anchorage is superior in N. gales to that of Campeche.

ARCAS, N. coast of Spain, a rocky point W. between Cape Peñas and the river Aviles, which last is shallow, narrow, and rarely visited.

ARCASISLE, one of the Bijoogas, lat. 11° 41′ 15″ N., long. 15° 39′ W.

ARCHER'S GUT, West Indies, island of St. Kitt's, a passage near the N.W. extremity of the island, sometimes called Wheelwright's Gut, between William's bay to the E.N.E. and Cooper's bay to the S.W.

ARCHIPELAGO, clusters or groups of islands, generally applied to such as are found in straits or gulfs, but properly to be understood of a similar series in any situation. It is most commonly applied to the Greek islands which lie N. of Candia, in the Mediterranean, as far as the entrance to the Dardanelles, anciently called the Ægean sea. Seamen frequently clip the word, and call it the Arches.

ARCHIPELAGO DE LAS MULETAS, see *Muletas.*

ARCHACHON, BASIN OF, France, about 19 leagues from the entrance of the Gironde. The land is low and sandy, with a clump of trees here and there. A little inland are two lakes, called the N. Grayan lake and the S. Canau lake, the last communicating by a small stream with the basin of Archachon. The entrance to this basin, greatly impeded by sandbanks, has only two channels. One of these is near Cape Feret, and is only navigable for boats, having but 3 feet water on the bar at low tide. The other and only entrance for vessels has 14 feet at low water, being the S. entrance. To make this channel, the two beacons erected there are to be brought into one. Within the N. entrance, there is a semaphore and a small battery, and on the S. two beacons, a battery, semaphore, and some fishermen's huts, the distance between the two points being 4 miles, and forming the bar, which stretches out 3 miles from the shore. The basin is triangular in form, about 2 leagues wide, but so intricate from shoals that a pilot is necessary. In the basin is the little island of Teste, and several towns and villages. There is a fixed light nearly 2 miles N. of the N. entrance, in lat. 44° 38′ 43″ N., and long. 1° 15′ W., 167 feet above the sea level, visible 18 miles. The tides flow at 4 o'clock, a little before full and change, the time of high water being about the same as at the entrance into the Gironde, which is from half-past 3 to 4 o'clock.

ARCHANGEL, or ARKHANGEL, Russia, the capital of the government of that name, lying in lat. 64° 31′ 40″ N., and long. 40° 47′ 30″ E. It stands on the river Dwina, 4 miles from the White sea. It was once the only seaport Russia possessed, and British traders resorted to it as long ago as 1553. The trade has diminished since the erection of St. Petersburgh. Making Cross island, or Sosnovitz, the arctic circle is re-crossed, then the course is made for Cape Katness, or Blue Nose, 22 leagues, but owing to the current, the course is generally taken S.W. ½ S. for 6 or 7 leagues, and then S.S.W. for Katness. The land is then kept parallel to the middle of a point called St. Nicholas, where there is a white tower 84 feet high, and at the bottom the pilot's house, who takes the ship over the bar up to Archangel. If ships are driven off they must endeavour to get a sight of the tower, and bring it N.E. by N., keeping it until they bear S.E., where a large red cone buoy, with a flagstaff, marks good anchoring ground. It is high water at 6 o'clock. Between the pilot's house and Archangel, upon Mudoska island, there is a stone lighthouse, 140 feet above the sea level, in lat. 64° 55' N., and long. 40° 17' E., visible August, September, and October every year, from N.N.W., westward, and S.E. by S. for 16 miles. The articles of trade here are those of the N. ports of Russia in general.

ARCTIC SEA. The sea comprised within the arctic circle, lat. 66° 30' N., and from thence to the North pole. Within this limit are included the Northern, Icy, Frozen or Arctic seas, the White sea as a tributary, though not wholly within the arctic limit, the seas of Kara and Obi, the N. part of Baffin's bay, and the sea N. of Behring's straits.

ARDALSFIORD GULF, on the W. coast of Norway, N. of Stavanger, called sometimes Buckenfiord.

ARDGRUME HARBOUR, S. side of Kenmare river, W. coast of Ireland, opposite Sneem; it has from 12 to 13 feet of water. Keep on the W. side of Carrickavanheen rock to go in, which is always above water. Vessels drawing 10 feet should wait until half flood to enter. Anchor in the creek on the W. side in 4 or 5 fathoms.

ARDNAGLASS, Ireland, on the W. coast, a bay 2 leagues S.W. from Sligo.

ARDOIS, Nova Scotia, 13 miles N.W. from Halifax, the highest land in the province, overlooking the Basin of Minas, Bay of Fundy.

ARDRA, or ARDRES, Africa, on the S. coast, in lat. 5° 5' N., long. 4° 10' E. To anchor before it bring the two round-topped trees within the village N. by E. and N.E., when 7 and 8 fathoms of good sand will be found. Cape Lagoa lies about 8 leagues N.E. by E. from this place, which is now little visited by Europeans. It is at the bottom of the Gulf of St. Thomas.

AREAS, ISLAND OF, America, W. coast of the Gulf of Mexico, in lat. 20° 45' N., long. 92° 40' W. See *Arcas*.

AREBOR, Africa, Guinea coast, at the mouth of the Formosa river, where the Dutch had a factory, in lat. 6° N., long. 5° 5' E.

ARENAS, ISLA DE, or SANDY ISLAND, America, off the N. coast of Colombia, lying in the middle of the Bay of Galera de Zamba, the navigation about which is intricate on account of the sand banks. The best anchorage is with the island of Zamba S.S.E. 2 leagues from the shore, in from 15 to 25 fathoms. Some reckon four islands of this name; they are mere sand banks. Arenas is a Spanish word signifying sands.

ARENAS, ISLA DE, Gulf of Mexico, about 90 miles N.W. from Punta Piedras, a low island, 3 miles long by 2 broad, rocks all round. Lat. 22° 10' N., long. 91° 21' W.

ARENAS KAY, W. off Cape Isabella, on the N. coast of Hayti, which cape is in lat. 19° 58' 40" N., long. 71° 6' 30" W.

ARENAS KAY, N. America, Bay of Honduras, lat. 71° 12' N., long. 88° W.

ARENAS, PUNTA DE, or SANDY POINT, at the entrance of the Gulf of Darien, lat. 8° 33' N., long. 76° 56' 15" W.

ARENDAL HARBOUR, Norway, 14 miles N.E. of Homborg Oe. The water is deep enough for the largest vessels. It is a place of considerable trade. Mærd-Oe is an island off the S. entrance of the port; on the N. is an island called Skudholm, between which and Mærd-Oe is anchorage in 12 fathoms upon sand. Heavy ships go out by the W. passage only. Arendal fixed light is on Great Tonningen island, in lat. 58° 23' 15" N., and long. 8° 52' 30". On Little Tonningen island there is a fixed light, bearing from the light on Great Tonningen N.N.E. 1,237 yards, both 130 feet above the sea level. There is a fixed light on Sandrig point, at the entrance to Arendal, 42 feet above the sea level. The buildings of the three lights are white. Two miles from land bring Sandrig point light N. ¼ E., and keep along the land E. of Little Tonningen, for Sandrig point light; the distance is about a mile from Little Tonningen. Three cables' length leads to a good anchorage in 12 or 16 fathoms. Passing Great or Little Tonningen, keep a quarter of a cable's length from Great Tonningen. When Sandrig light bears N. by E. ¾ E. steer towards it, and when within a quarter of a cable's length, bring up. The light of Mackoe, on the S. of Norway, has been discontinued.

ARENTSBURG, see *Ambrook*.

ARES, BAY OF, N. coast of Spain. It is shallow, with a tower on its W. shore, between Ferrol and Corunna, from whence the land turns S.S.W. to Betanzos. This bay is extremely dangerous, as well as that of Betanzos, being open to winds and seas.

ARGAGNA, an extensive shoal, 5 miles below the St. George's branch of the Danube, stretching across the river, with only 9 feet at times over its deepest part, dangerous to vessels coming down deeply laden into the Black sea.

ARGENT, CAPE D', on the E. side of Newfoundland, the extreme N. point of White bay, in Machigonis river, in lat. nearly 50° N., long. 55° 15' W.

ARGENTARO, MOUNT, Italy, in Tuscany, a lofty promontory, visible 10 leagues off, appearing to be an island seen from N. to S. A small island lies off its N.W. point. It lies E. of the island of Monte Christo, with that of Giglio between, in lat. 42° 35' 10" N.

ARGENTIERI, Greece, an island 5 miles long by 4 broad, separated from Milo by a channel half a mile wide. It was the ancient Crinolus. It is hilly; the highest mountain is in lat. 36° 49' 20" N., long. 24° 33' 23" E. The channel between this island and Milo has 7, 6, and 5 fathoms.

ARGOS, THE GULF OF, often in modern times called the Gulf of Nauplia. It runs in to the N.N.W. from Port Botte 44 miles. Nauplia, or Napoli di Romania, has been called the Gibraltar of Greece, and lies at the head of the bay, having an excellent harbour, as well as anchorage, all over the gulf. The town is unhealthy. It stands at the foot of an abrupt rock, and the Palamedi castles rise impregnable over its summit. The pop. has been estimated at 15,000. The city of Argos is 4½ miles N.W. of Nauplia. The houses are generally of wood. The pop. about 8,000. The environs furnish corn, cotton, tobacco, figs, grapes, and rice in abundance, and are watered by the river Xera, the ancient Inachus. Nauplia is in lat. 37° 33' 50" N., long. 22° 47' 30" E.

ARGOSTOLI, or AROGOSTOLI, the harbour of Cefalonia. Cape Aji, forming the S.W. side of the bay of the same name, is in lat. 38° 8' 40" N., long. 20° 23' 30" E. From Cape Aji to Cape St. Nicolo is 4½ miles, and two-thirds from Cape Nicolo is the island of Guardiani, having a lighthouse upon it. Within there is the entrance to the port of Argostoli, running N. ¾ W. for 8 miles, having at the further end a sandy beach called Livadi. The town is on the E. side of the port, on the W. of a little cove, consisting of 2,500 houses, and about 8,000 inhabitants. There are several dockyards here. On entering the port it is always preferable to pass in S. of the island Guardiani, giving it a berth of a mile. The rise and fall of water here is between 1 and 3 feet, and the current sets in and out every 24 hours, but irregularly.

ARGUIDAH, Africa, W. coast, a Dutch fort, 4 leagues from the English fort of Dixcove, near Cape Three Points, at the E. point, abreast of which is an anchorage in 14 and 24 fathoms, between 3 and 4 miles off. Cape Three Points is in lat. 4° 44' 30" N.

ARGUIN, BANK OF, 4 leagues from Cape Blanco, coast of Africa, a great shelf 30 leagues long, reaching S. of Cape Mirik, the N. point in lat. 20° 33' 12" N., long. 10° 56' 30" W.; the coast between this bank and Cape Blanco being full of shoals. The most important is the Bayadere, about 1½ mile S. of the cape, having only 20 feet water. Another shoal lies W. ¾ N. 3 miles from the cape, and a third 8 miles S.S.E. ¼ E. from it, on which 20 feet water have been found. The channel that conducts to the E. anchorage off Cape Blanco, lies N. of these shoals. The bank at Arguin is a hard sandy flat. Its outer edge has been fixed at 8 fathoms, and this limit cannot be overrun without risk. Close to the breakers, on some parts of the bank, not more than 10 feet water are lying. The dreadful wreck of the French frigate La Meduse happened on this bank. Between the N. point and W. extremity, in lat. 20° 6' 20" N., and long. 17° 7' 30" W., on approaching from the sea, the soundings decrease. At 10 leagues W., from 40 fathoms they fall to 8 with a gentle ascent, and to the S. of this parallel the bottom becomes more uneven, while from the point where the Medusa perished, lat. 19° 53' 42" N., and long. 17° 0' 35" W., still greater irregularity prevails. A strict attention and incessant use of the lead is indispensable in this neighbourhood.

ARGUIN, or AGHADEEN, ISLAND, BAY and TOWN, Africa, 10 leagues S.E. from Cape Blanco. The bay is so called from the island. It is in lat. 19° 20' N., long. 17° 20' W. There are two other small islands to the W. of the larger, and several shoals on the N. side. There is a small island also near the point of the bay, called Terra Gorda.

ARGUNA, a town on the river Benin, or Formosa, Africa, 13 leagues from Benin. Since the abolition of the slave trade, the towns and establishments here have become comparatively deserted.

ARIMINIUS, the old name of the Italian river now called the Mareccia, upon which stands the city of Rimini, about 16 leagues from Ancona. Vessels of small burthen were once able to sail up to the city, and vessels even rode under its castle in 6, 7, and 8 fathoms. The sea has since retired above a mile, and the harbour is choked up with sand. A canal had been cut to communicate with vessels, but that too has become neglected, and boats alone use it for carrying fish to the city, whence it is distributed to the interior towns.

The pop. under the French was 16,000. A small lighthouse stands at the entrance of the river, in lat. 44° 4' 35" N., long. 12° 34' 10" E.

ARIPOKE ISLAND, South America, off the mouth of the river Arawari, on the coast of Portuguese Guiana, 1° 15' N. lat.

ARISCH, EL, BAY OF, Africa, on the coast of Egypt and Palestine, exactly at the line which separates the two countries. It stands on a small eminence, half a mile from the Mediterranean, 36 miles S.W. of Gaza, and 110 N.E. of Suez. It is enclosed by a high, thick wall, loopholed for defence. It is memorable for the treaty made by Sir Sidney Smith with the French for the evacuation of Egypt, which the ministry at home cancelled, and in consequence sent armies both from England and from India to subdue the French, who, after several sanguinary actions, capitulated at last upon nearly the same terms as Sir Sidney Smith had obtained without the expense and loss incurred by a sanguinary warfare. The town is about 37 leagues due E. from Damietta, in Egypt.

ARKADI ISLAND, Greece, between Ithaca and the main.

ARKLOW BANK, Ireland, off Arklow bay, 2 leagues from the land, 10 miles long from N.E. by N. to S.W. by S., and from half to three-quarters of a mile broad. The N. end is shallowest, with only 3 feet water, the S. from 8 feet to 3 fathoms. It is to be approached no nearer than 30 fathoms. The mark for the S. end is Arklow rock, N.W. ½ N. distant 7 miles, in one with the declivity of the copper ore mountains. The marks for the N. end, the top of Carrig Mc Rely just open to the N. of Maughey point, bearing N.W. ¾ N., and the Great Sugar Loaf open to the E. of Wicklow head N. ¼ E. Wicklow head, bearing N.N.W., clears it. A light-vessel is moored 1 mile S. of the S. end of the Arklow bank, showing a light from the mainmast only. Wicklow head and lights bearing from it N.N.E. ¼ E. 17 miles; Arklow rock N.W. ¾ 7 miles; the Tuskar rock light S.W. ¼ S. 32 miles. This light is steady and bright, 25 feet above the sea, visible 8 miles off. A flag is hoisted in the day, and in foggy, dark weather, day or night, a gong is kept sounding. Shoals have been reported as forming between the Arklow bank and the shore.

ARKLOW TOWN, on the coast of Wicklow, Ireland, has a haven for small craft, or rather boats, at high water. About a mile E. of the river's mouth, S.E., there is good anchorage in from 5 to 8 fathoms. High water 10ʰ 15ᵐ, rise 3 feet.

ARLINGTON BAY, Ireland. See *Carlingford*.

ARMIROS, BAY and PORT, Greece, on the S.W. side of the Gulf of Volo, near its W. end, and 30 miles N.W. from the island of Negropont; lat. 39° 42' N., long. 23° 30' E.

ARMUYDEN, Holland, a seaport town of Zealand, strongly fortified. It stands on the E. side of the island of Walcheren. There is an old and new port, and a good depth of water, but the harbour is choked up. It is 3 miles E. of Middleburg; lat. 51° 31' N., long. 3° 42' E.

ARNO, Italy, a river of Tuscany, falling into the Mediterranean, 4 miles below Pisa, to which city small vessels can go up. Florence stands further up on both sides of this river; lat. 43° 42' N., long. 10° 24' E.

AROBO and AROBY, Africa, small places upon

the S. coast, the first 4 leagues from the river Benin, the second between Suma and Little Commenda, on the Gold coast. Rocks lie off the shore here, which make it dangerous of approach.

AROSA BAY, W. coast of Spain, about 5 miles S.E. from Cape Corrobedo, a dangerous place for vessels of every kind. High water $3^h\ 45^m$, rise 12 feet.

ARQUES, France, a small river falling into the channel at Dieppe, a league above which town it passes the Chateau d'Arques, now a fine ruin, where Henry IV. defeated the army of the League.

ARRAN ISLAND, on the W. of Scotland, in the Frith of Clyde, between Cantyre and Cunningham, 23 miles long by 11 broad; in lat. 55° 40′ N., long. 5° 10′ W.

ARRAN, NORTH, ISLE OF; this island lies off the coast of Donegal, Ireland, in lat. 54° 48′ N., and long. 9° 4′ W. Arran, or Arranmore, is about 7 miles round; there are a number of islands and small rocks S. and E. of it. There was formerly a light on its N.W. coast, discontinued since one has been established on Tory island. There is anchorage on the E. side, under a little island called Calf, in Arran road. The S.W. coast is distant 3¼ miles from Croby head, S.W. There are many dangers around this island; a pilot is absolutely required.

ARRAN, SOUTH, ISLES OF, three islands off the mouth of Galway bay, Ireland, called Arranmore, Innismain, and Innisheer. The inhabitants are 3,000 in number. These islands stretch nearly 7 leagues N.W. by W. from the W. side of the Bay of Galway, and have channels on all sides; that between Innisheer, the smallest island, and the main, is S. The sound, 5¼ miles wide, with 6 and 27 fathoms water. N.W. Foul sound divides Innisheer from Innismain, the second island in size; a rocky ledge extends one-third over, with only 6 feet of water. On the N.W. the only danger is a rocky ledge, 2 cables' length long, between Arranmore, or as some call it Killeny, the largest island, and Innismain. This passage is called Gregory's sound. Arran harbour, or Killaney bay, at the N.E. end of Arranmore, is clean ground on the W. side of Straw island, in 4 or 5 fathoms, for large ships, but exposed to E. and N.E. winds; in mid bay are 2¼ fathoms. The Branack islets lie N. of Arranmore, and there are rocks between. The stream of tide runs about 2 miles an hour. About a third from the S.E. end of Arranmore there is a revolving bright light, attaining its magnitude once every 3 minutes, visible all round for 15ʹ, and obscured 2ᵐ 45ʹ, 498 feet above high water, and visible 8 or 9 leagues. Lat. 53° 2′ to 8′ N., and long. 9° 30′ to 42′ W. Arranmore is about 5 miles long by 2 broad.

ARRANMORE ISLANDS, see *North and South Arran Islands*.

ARRECIBO, Porto Rico, West Indies, 9 leagues to the W. of the harbour of San Juan; a small town on the N. coast, upon a good river, but exposed to the N. winds, and therefore little used. St. Juan is 8¾ leagues to the W. of Cabeza de San Juan, the N.E. point of the island.

ARROE ISLAND, Denmark, in the Baltic, Funen being a little N., and Dulcen island on the S. Lat. 55° 10′ N., long. 10° 20′ E.

ARROJOKI, Sweden, a river in the Gulf of Bothnia, nearly opposite to Aland island, where it forms a bay. The town of Abo stands at its mouth. See *Abo*.

ARROWSMITH BANK, North America,

E.S.E. 22 miles from Mugeres, or Women's island, in lat. 21° 12′ 20″ N., long. 85° 40′ and 87° W.

ARSAKENA, or ARSACHENA SOUND, an opening S. of Caprera island, in Sardinia, on its N.W. point.

ARSOUF, or APOLONIUS, Asia, Mediterranean, a small port of Palestine, 6 miles N.E. of Joppa.

ARTA, BAY and TOWN, Albania, on the E. side of the Gulf of Venice, 20 miles N.N.W. from Lepanto. Lat. 39° 28′ N., long. 21° 20′ E. Pop. 5,000. Its port is called Salona.

ARTA, BAY OF, island of Majorca, Mediterranean; between Cape Ratche and Point Amer, and within the bay, is the small town or port of Vey, shallow, of little use to shipping, and open to the S.W. and S.E. winds. About 4 miles N.N.E. is Cape de Pera, the E. point of Majorca, in lat. 39° 42′ 20″ N., long. 3° 30′ 18″ E.

ARTAKI, Turkey, Sea of Marmora, on the S. side, 40 miles E. of Gallipoli, a small port. Lat. 40° 18′ N., long. 27° 39′ E.

ARTHINIWINNIPEC, East Maine, Hudson's bay. Lat. 56° 30′ N.

ARTIBONITE RIVER and POINT, W. coast of Hayti, West Indies.

ARTIGONISHE BAY, North America, Nova Scotia, having Cape Lewis to the N.W. and Cape Fronsac or Canso to the E., in the S. part of the gulf, and 20 fathoms in some parts within the W. cape and bay. St John's island lies to the N.W. of it. The middle of the bay is in lat. 45° 36′ N., long. about 62° W.

ARUBA ISLAND, South America, a small dependency on Curaçoa. Lat. 12° 30′, long. 67° 39′. This island is N. from Cape Romano, the E. point of the Gulf of Venezuela. Under the N. side of Aruba island there is a good road; it is near a smaller island, that must be left on the N.W. going into the road, to come to an anchor in 5 fathoms. The island is near 5 leagues in circumference.

ARUN RIVER and ARUNDEL, Sussex, 10 leagues N.W. by N. from Beachy head, and 12¾ miles W.N.W. ¼ W. from the entrance to New Shoreham harbour. Arundel town is 4 miles S. by E. from the entrance of the river, where the town of Little Hampton is situated, and stands in lat. 50° 55′ N., long. 0° 29′ W. It principally exports timber and bark, has a pop. of 2,803, stands on the side of a hill, and returns one member of parliament. Here is a fine castle of the Dukes of Norfolk, as old as the reign of Alfred, who bequeathed it to a nephew; it gives the title of earl without further creation. The town communicates with the Thames by a canal.

The entrance to the river is between two wooden piers, over a bar, on which there are 11½ feet of water at spring tides, and with W. gales 13 or 14. In bad weather, when boarding is impracticable, a pilot waves a flag from the E. pier, to direct the vessel in over the breakers, while another boards. Three flags are also hoisted to denote the depth of water: one white, half-mast, 8 feet on the bar; white union, under 10; union and blue pennant, under 11; blue pennant, 12; blue pennant union, under 13; when the flags are at the top of the mast, half a foot more is signaled. The signals are hauled down at high water. At full and change it is high water here at 11 o'clock. A light is sometimes shown at night from the flagstaff. The flood runs with great velocity until an hour after high water, and sets strong to the W. outside the

E. pier head from half flood to half ebb. Open the piers to enter, and steer between the warping posts; note the remarks on the tides. To anchor in the road, open the piers and bring Savington mill on with the brow of Pepperscombe hill, or the chalk pit of Cisebury hill, which will place the vessel in 3¼ fathoms.

ARZEO, BAY OF, Africa, between Cape Ivi and Cape Ferrat, or Ras al Mishaf, the distance 10 leagues W.S.W. between the land forming the bay. It is so named from a village S.E. of Cape Ferrat 7 miles. It is 58 miles from Algiers.

ARZILLA, coast of Africa, 20 miles S.W. of Cape Spartel, a small fortified town close to the shore, between which and Cape Spartel there is good anchorage with an E. wind, in from 10 to 15 fathoms on sand, a mile or two from the shore.

ASCENSION BAY, North America, Gulf of Honduras, on the E. coast of Yucatan. Cozumel island lies off the N. part of it, which island is in lat. 20° 16′ 45″ N. This bay is spacious, in lat. 19° 36′ 15″ N., long. 87° 30′ W. The bay of Guanacos lies S.W. and W. from it.

ASCENSION RIVER, South America, about 8 leagues W. of the river Allabrolies, and 30 leagues from Porto Bello, on the Spanish main. It is navigable for small vessels 20 leagues, but not for those of burden. Samballa point is just 15 leagues from the W. point, at the entrance of this river.

ASFEE, see Saffi.

ASHDOTH, or ESDOD, Palestine, 9 leagues S. of Jaffa, upon a small river which will only admit boats.

ASHTON ROCK, a vigia, said to have been seen in 1824, between the Bermudas and Cape Hatteras. The sea broke over it. The lat. given was 33° 48′ 50″ N.; long., inferred from a lunar the day before at 19ʰ 31ᵐ, 71° 41′ 20″ W. This is supposed to be the same rock said to have been seen by the Orion in lat. 34° 51′ N., long. 72° 28′ W.

ASIA, one of the four quarters of the world, extending from the Arctic ocean to the S. cape of Malacca in one direction, and from the Dardanelles to Corea and Japan, being in English miles about 7,583 in length, and in breadth 5,250. It is limited everywhere but on the W. by the ocean. The Indian ocean, Sea of China, the Yellow sea, the Sea of Okhotsk, or Lama, Behring's straits, on the E. side or side of America, and the Arctic ocean, branches of the Indian or of the Pacific seas, wash its shores on the E. and S.; while the Red, Mediterranean, and Black seas bathe its western coasts. The indentations of the shores of this continent are great in comparison with the other three continents. On the E. coast the Gulf of Anadir, near Behring's straits, down to the promontory of Kamschatka, one of the most remarkable peninsulas of the globe, encloses the deep sea, or rather gulf of Lama, or Okhotsk, and terminates in 50° N. lat., including 20° of deeply indented coast on the E. and N.E., almost doubled on the W. side of the Kamschatkan peninsula, in the shores prolonged to the channel of Tartary. Scarcely through the Sea of Japan, the Yellow sea, or Whang Hai, describes another great extent of coast indentation, convex in form, down to the Gulf of Tungquin. Again projected and again retiring, the coast forms the deep Gulf of Siam. On the W., the Malayan peninsula, shooting S. 14°, and carried into the midst of the Indian Archipelago, a mere tongue of land, extends its western coast, with the single interruption of the Gulf of Martaban, to above 20°, or from the line to the mouth of the Ganges. An immense length of coast, full of indentations, prolonged from the line to 70° of N. lat. The coast from the delta of the Ganges, round the peninsula of Hindostan into the Persian gulf, and that of Arabia in the Red sea as far as Suez, is tolerably well explored. The coasts of Palestine, Syria, Karamania, and Anatolia, to the E., around the Euxine sea, are also known, less intricate, vast, and varied than those on the E. of the continent. Further from Cape Shalatskoi to the Sea of Kara, an immense extent of icy bays and desert headlands pointing to the pole, remains nearly unknown to the western European. The appendages to this enormous coast line are those of the dependent isles properly attaching to this quarter of the globe, and equally unexplored with the continental shores. Here is yet a vast field of exploration open for the adventurous in almost every known latitude. The phenomena presented by the winds and waves are various, and in many respects different from those in other parts of the world. The typhoons of the Chinese sea, and the storms that carry destruction to vessels S. of Ceylon, the prodigious surf that thunders on the shore of the Indian peninsula, the immense levels and morasses at the mouths of the large rivers, the flat, far-extending sandy levels and shallows so dangerous to the mariner; in fine, every variety of object known to the seaman, with others new and strange to the profession, call for his utmost skill, energy, and resources upon the shores of this most ancient and extended of the continents of the old world; peopled, too, by races as yet scarcely known to the European. All these general points ask the attention of the navigator in the numerous details which they involve amid the longest navigations to which the European seaman is subjected, and amidst proportionate perils.

ASINARA, called also SASSARA, Sardinia; 13 leagues W. from Cape della Testa, is the island of Asinara, and within it on the E. is the bay or gulf of Sassara; the town is some distance inland; it is the capital of the N. province, well built, walled, and has a pop. of 30,000. Wine, oranges, and silk are produced in the vicinity. There are anchorages along this shore sheltered from N.E. winds, in 8 and 10 fathoms water. Lat. 41° 6′ N., long. 8° 24′ E.

ASP ISLAND, coast of Norway, lat. 63° N., long. 8° E.

ASPEROSA, Greece, Romania, on the N. side of the Archipelago, and about N.N.E. from Cape Monte Santo, the S.W. point of the Gulf of Contessa, in lat. 40° 58′ N., long. 24° 50′ E.

ASPEROSA, CAPE, in the Ægean sea, N.E. of the isle of Thaso, on the main.

ASPO, Sweden, in the Baltic, a small island 2 miles S.W. from Carlscrona; passing which, between that and Tjurko, one entrance goes to the port. There is also a passage W. of Aspo; the distance of the last island from Carlscrona is 2 miles.

ASSENS, a town of Funen, Denmark, on the Little Belt, S.W. of Odensy, in lat. 55° 17′ N., long. 10° 2′ E., in the Little Belt passage. On the point W. of the town there is a light, forming a triangle with that of Aro to the W., placed on an island, having Bago light on the N.W.

ASSINEE, Africa, a river with two towns on the W. coast, named Great and Little Assinee, or Grand and Piqueno, 2 leagues distant from each other; lat. 5° 35′ N., long. 3° 12′ 7″ W. In 1814 one of these towns that stood on the shore was demolished by a neighbouring tribe. Here the current is rapid to the E. The anchorage is abreast of the river in 10 and 13 fathoms, but the river is only navigable for boats. The Gold coast is said to begin at the W. town, and continue to Cape St. Paul's. Lat. 5° 44' 30" N., long. 0° 52' 18" W.

ASTREA SHOAL, lying between Huen and the shore of Denmark, discovered by the British frigate Astrea, having 4, 4½, and 5 fathoms over it.

ASTURA, on the S.W. side of Italy, a point of land near the sea, on the marine border of the Pontine marshes, 12 or 14 leagues from the mouth of the Tiber. It has a tower upon it, known as the tower of Astura.

ATALAYA POINT, Spain, Bay of Muros; it is high; the harbour of Muros runs in N.N.W. and N. by W. more than a mile.

ATHENS, a city of Greece, of great ancient renown, again called into existence by the modern resuscitation of the Greek nation. Situated 5 miles from its port, it was the great naval power of ancient Greece, and by erecting double walls to the haven, the port was, under all circumstances, in ancient days, effectively connected with the bay. At N.N.E., just 11 miles from the N.E. point of Egina, is the entrance to Port Lion, more correctly the ancient Piræus or harbour of Athens. It is small, well-sheltered, circular in form, having a narrow entrance, but capable of containing 300 or 400 small vessels that draw but little water. Upon the shore is a custom-house and a small church. But few relics of ancient times are visible about the port. There are the remains of two beacons, one of which is visible, in a tolerable state. To a modern seaman no idea is conveyed, by what remains connected with his professional pursuits, of the great naval achievements of antiquity when contemplating the space which the navies of the most renowned people of ancient times required to achieve their glories. Yet the difference of mechanical means duly considered, actions which have never been surpassed in heroism, that the purest moral feeling—the most exalted patriotism —called into action, were performed here. At the entrance of the port there is a large figure of a lion, sculptured in marble. The N. winds, except in winter, very rarely blow here, but when they do they bring very severe weather. In summer S.W. winds prevail. From the Port of the Lion, which modern phraseology has substituted for the Piræus, the Gulf of Egina turns S. towards Cape Colonna. Many of the points of land here require attention, for near them there are snug anchorages to be found, while a want of care may lead to very disastrous consequences. Greek pilots are always to be found, the most dexterous in their own seas of any race of mariners. Athens is in lat. 38° 5' N., long. 23° 43' E. Under a small island, due W. from the mouth of the harbour not quite a league, good anchorage is found in an excellent road in 18 fathoms. In fact there are several good havens in the vicinity of Athens, so that none, with proper management, need ever be at a loss for a place of refuge when navigating near the great naval port of the most celebrated seamen of antiquity.

ATHENS, or EGINA, GULF OF; the S.W. extremity of this gulf is formed by Cape Skylli, in lat. 37° 27' N., long. 23° 32' 10" E. The opposite point is Cape Colonna, in lat. 37° 39' 15" N., long. 24° 1' 34" E.; distant 27 miles.

ATHERFIELD POINT and ROCKS, Isle of Wight, on the S. side towards the W., the westerly point of Chale bay, extending N.W. from the S. point of the island, in the direction of the coast. Off this point there is a sand bank in the form of a triangle, nearly dry at low water. It must not be approached within a mile in any direction. Brixton bay is to the W., and this forms its E. limit. A line of sand runs parallel with the coast here from the S. point of the island, at 2 or 3 miles distance, where there is anchorage in 7 or 8 fathoms, but not to be depended upon at a hazardous moment. High water 9½ʰ. Lat. 50° 34' N., long. 1° 24' W.

ATKINS' KEY, a small rock among the Bahamas. Lat. 22° 7' N., long. 74° 31' W. See *Bahamas.*

ATLANTIC OCEAN. This ocean washes the coasts of Europe, Africa, and through some of its tributary seas, a portion of Asia. In common it is divided into the North and South Atlantic. It is worthy of the greatest consideration, as being of the most important character among the oceanic divisions of the globe, since its shores comprise those of the more civilized nations. The distinction preserved here separates this great sea into two parts, including tributary and inland waters which have with it a navigable communication. The North Atlantic extends from the N. side of the equator to the arctic circle, or from the parallel of lat. 0° to 66° 30' N., and from the same parallel of lat. 0° to 66° 30' S., or to the antarctic circle for the South Atlantic. Within the limits of the North and South Atlantic are included generally several minor seas, as the German ocean or North sea, the Baltic and Gulf of Bothnia, the Mediterranean, Black sea, and Sea of Azof, the S. part of Baffin's bay, Hudson's bay, the Gulf of Mexico, which are small seas of themselves; in fact, all the navigable waters tributary to its vast superficial extent and unknown depths. This distinction is preserved throughout the present work. In the Atlantic, as in all seas, the heat of the land affects the prevalent winds, attracting them by the greater rarity of the air. The trade winds are the consequence of this attraction of the air nearer the poles. In the North Atlantic, W. and S.W. winds are observed to be most common, and the more violent storms are generally from those points. The trade wind on the American or western side of the North Atlantic, extends 4 degrees further N. than on the African side, or to 30° of N. lat. The trades are not so steady in the Atlantic generally as in the Pacific, where they are less interrupted by islands and salient points of land. Between the limit of the trade winds and those more N., which are continually varying, there is a space probably of 3°, limited about lat. 29° N., over which great and sudden changes of weather happen, with frequent electrical phenomena; sometimes called the Horse Latitudes. Beyond this limit, the W. winds are those most frequently experienced. The limits of the trade wind continually differ in extent on the side of the equator, sometimes ceasing in the parallel of 10°, sometimes in that of 3°. The wind differing upon the different coasts, and affecting the seasons upon the land, is common to all parts of the globe, and is a copious subject for examination. The storms on the American coasts, or rather in the West Indies, denominated hurricanes, seem on many points to possess a peculiar character, not only for their violence, but the marked track which they follow; they seem to enter among the West Indian Archipelago on the S., or the Gulf of

Mexico, and sweeping in a curve among the islands, it is probable of no great breadth, to pass out and bear to the N.E., along the American coast, following, in most respects, the course of the oceanic currents. Of these last there are many in the North Atlantic, but perhaps the most peculiar in itself, differing from any other known, is that denominated the Gulf stream. A current sets into the Caribbean sea from the E., but the Gulf stream sets outward from the Mexican gulf, passing between Cuba and Florida, a perfect river amid the ocean, N. from thence to the Bahamas, and then along the American coast in a direction N.N.E., N.E., and E., falling into the S. in a meridian W. of the Azores. Its course may be computed at 3,000 nautical miles; in some places it is reckoned to be 75 miles broad, in others not much more than half that distance. It runs with most force in August and September, and with varying velocities. Its temperature is also a peculiarity of this current, decreasing as it flows into higher latitudes; in its early course it reaches 87°, then 84° and 83° of Fahr.; and on every account demands a careful study by the navigator, as there is reason to believe its effects upon the atmosphere are as important as its course is to the direction of the vessel. The tides in the North Atlantic are not different from those in other seas, being affected in the same way by straits and headlands. Thus, while at the Land's End, in England, the tides rise 18 or 19 feet, at the Holms in the Bristol channel they rise to 38 feet, and about the islands of Guernsey and Jersey several feet above this limit. In some parts of the Mediterranean the rise is not more than a foot, and seldom exceeds 3. The drift of icebergs from the N. into a comparatively low latitude, between 40° and 50° N., is another incident in the North Atlantic worthy especial notice; but as these allusions must be exceedingly general, it will suffice to notice only one feature more of that vast expanse in the Sargasso sea, a space between the parallels of 37° and 18° N., and between the meridians of 33° and 43° W., or $2^h 12^m$ to $2^h 52^m$, covered thickly with the *Fucus natans*, or gulf weed. The space is 1,200 miles long by from 50 to 150 wide. Some think it grows on the surface of the ocean, on which it floats by means of its air bladders, just where the vast ocean currents form an eddy. It was seen and noticed by Columbus. The temperature of the Atlantic, in lat. 11' N., and long. 81° 55' W., is about $82\frac{1}{2}$°, and in lat. 48° 11' N., and long. 11° 58' W., it has been found $57\frac{1}{2}$° in June. Near Cuba, at the surface, 83°; at 1,000 fathoms, $45\frac{1}{2}$°; difference, $37\frac{1}{2}$°. The greatest heat is from 82° to 84°.

ATTAH, Africa, E. side of the river Quorra, lat. 7° 6' N., long. 8° 30' E.

ATWOOD'S KEYS, lat. 23° 22' N., long. 73° 33' W. See *Bahamas*.

AUBIN'S, ST., BAY, Jersey, English channel. Situated at the bottom of it stands the town of the same name, defended by a fort near the S.W. end, out to the E.; lat. 49° 7' N., long. 8° 14' W. It is here high water at half-past 12. See *Jersey*.

AUDIERNE, or HODIERNE BAY and HARBOUR, France, situated on the W. coast, in the department of Finisterre, extending from N. to S.W., between the points of Raz and Penmark, a space of about 10 leagues. The harbour is a tide port, only to be entered at high water. Before the port there lies a rocky bank, called La Gamella, a mile from the shore. High water at 4 o'clock. On the point of Penmark is a revolving light, 134 feet above the sea, darkened every half minute, visible 22 miles off.

AUGUSTA, a city of Sicily, in a bay of the same name, $1\frac{1}{4}$ mile W. of Point Grosso Longo. There is good anchorage abreast of the town, in from 11 to 7 fathoms. A martello tower, on Magnisi, the S. boundary of Augusta bay, bears from the lighthouse S. $\frac{1}{2}$ E. $3\frac{1}{4}$ miles. The lighthouse is in lat. 37° 12' 50" N., and long. 15° 13' 15" E. There are several shoals and rocks to be avoided in the bay, coming to an anchorage. The town is large, well fortified, with regular streets, but low houses; the pop. 8,000. The trade is in salt, honey, oil, and wine. The fortress is upon the isthmus that connects the town with the main land, commanding both.

AUGUSTA ISLAND, Gulf of Venice, on the Dalmatian coast, near Ragusa; lat. 42° 35' N., long. 17° 40' E.

AUGUSTA REEF, lat. 20° 37' 30" W., long. 70° 13' W.

AUGUSTINE, ST., FLORIDA, North America, E. coast, 30 leagues S. of the river Alatamaha, and 80 leagues from the Gulf of Florida. The harbour is good; lat. 29° 53' N., long. 81° 24' 30" W. High water $8^h 4^m$, rise $6\frac{1}{2}$ feet. Here is a square stone tower, painted white, with a fixed light at 75 feet, on the N. part of St. Anastasia island.

AUGUSTINE'S, ST., a river and port on the coast of Labrador, on the N. side of the Gulf of St. Lawrence, 8 leagues from the N.W. end of Meccatina, or Canat island, near the coast, and not far from the straits of Belleisle, opposite St. John's bay. In the harbour are two small islands, and about 2 miles S.W. there runs a chain of small islands, called St. Augustine's chain, the outermost of which is a remarkably smooth rock; it is situated about 25 miles from Meccatina island. Lat. 51° 10' N., long. 58° 50' W.

AUKPATUK, an inlet in Ungava bay, Hudson's straits, on the W. side, N. of Kôksoak river.

AULONA, or VALONA, a town in the Adriatic, in the bay of Aulona, the entrance to which is formed by the island of Saseno, which divides the entrance into two excellent channels, without danger. The environs are mountainous; the town contains 5,000 inhabitants. The custom-house is in lat. 40° 27' 15" N., long. 19° 26' 20" E.

AURAY, or AUVRAY, France, a port in the department of Morbihan; pop. 3,020. It has a port for small vessels, and exports coastwise principally corn, iron, cyder, salt, honey, and butter. It is 9 leagues E.S.E. from L'Orient. High water $3^h 45^m$. It lies in lat. 47° 40' N., long. 2° 25' W.

AUSTIN'S BAY, and TOWN, Barbadoes, West Indies, round the S.E. point of the island. There are many rocks on the W. side of the town, but the bay is of little importance to the seaman.

AUTHIE, France, a small river in the department of the Somme, falling into the channel. It is navigable for small vessels to Nampont, 3 leagues distant. There is a sandbank at the entrance, dry at low water. On the N. side of the entrance there is a provisional fixed light, 56 feet above high water, visible 6 miles, on Point de Berck.

AVALLON, North America, a peninsula of Newfoundland, near the S.E. part, having Placentia bay on the S. and Trinity bay on the N. The E. part is encompassed by the Great bank, and has Conception bay on the N. and St. Mary's and Trepassy on the S. It contains several good harbours.

AVEIRO, Portugal, a river with a shifting bar, rendering a pilot necessary. Two stone pyramids, one on each side the bar, 70 feet high, mark the entrance. The town contains 4,000 inhabitants, and many English reside there. It exports oil, salt, and fish. It is about 10 leagues S. of Oporto. There is a lighthouse on the S. side of the river-entrance; there are 14 feet of water over the bar at spring tides, but at neap only 11.

AVER ISLAND, on the coast of Norway. Lat. 63° 6′ N., long. 7° 44′ E.

AVERTAY LAND, Scotland, on the left of the entrance into the river Tay, dry at low water; it extends from the S. side of the river E.N.E. 2¼ miles. Another, called the Gaa, projects on the N. side, marked by three chequered buoys, while that of Avertay is marked by three red ones. See Tay.

AVERTO, a little island on the Friuli coast, in the Gulf of Venice. Lat. 45° 46′ N., long. 13° 32′ E.

AVES ISLANDS, or BIRD ISLANDS. The same name is applied to rocky islets frequented by sea-birds in different parts of the world. Thus, on the E. coast of Newfoundland, lat. 50° 5′ N., Aves island occurs, and Aves island, or the Island of Birds, West Indies, W. from Dominica, and S. from the Virgin islands. A shoal runs off this last to the islands of Saba, St. Eustatia, and St. Christopher, 2 leagues broad, and having soundings of 10 and 20 fathoms; lat. 15° 26′ N., long. 66° 20′ W. Aves Islands, in the Lesser Antilles, near the coast of the Spanish Terra Firma, S.E. from Bonair island; lat. 11° 57′ 30″ N., long. 67° 28′ 20″ W. On the N. side of these last Bird islands there is a tolerable harbour and wells of water. Three miles from this is a dangerous reef, extending from E. to N., and then standing to the W. See Avis.

AVILES, Spain, N. coast, 2 leagues W.S.W. from Cape de Peñas; broad, steep, and high, the ground looking white, encompassed with rocks and shoals, over which the sea continually beats, so that vessels ought not to approach it nearer than 4 or 5 miles. There are several rocky points between Aviles and this cape. Such are those of Arcas, Lamporo, La Homa, and Cape Negro, the last being the E. point of the river Aviles, which is seldom entered, being shallow and narrow. To the W. are the shoals of Robillo and Anvales banks, over which the sea breaks. N. by W. 8 miles from the river Pravia, and nearly W. by N. 18 miles from Cape Peñas, lies Cape Video. Between these are the little harbours of Cadillero and Artedo; the first only fit for small fishing-boats, the second having good holding ground and anchorage with 5 or six fathoms, with partial shelter from S. and W. winds, but sometimes from N.W. the swell of the sea will be found annoying.

AVIS, AVES, or BIRD ISLAND, nearly S. ½ W. from Saba, 40 leagues distant. It rises about 16 feet above the water, is 800 paces long, and 120 broad, with two rocky islets near, on the S.W., all white with birds' dung. It is low, and not seen far away. Lat. 15° 40′ 56″ N., long. 63° 40′ 6″ W.

AVIS ISLAND, West Indies, a small island among the Virgin islands. Lat. 18° 30′ N. See Virgin Islands.

AVOLA, a town of Sicily, about 9 miles from Syracuse bay. It stands on a well-wooded eminence, and has a pop. of 7,000. It carries on a considerable tunny fishery, and exports wine, corn, oil, honey, fruits, and sugar, from the only plantation of the cane in the island.

AVON, North America, a river of Nova Scotia, falling into the North Atlantic a little E. of Halifax and S., navigable as far as Port Edward for vessels of 400 tons, and higher for smaller craft. The St. Croix river falls into it 7 miles from its entrance.

AVOVA, CAPE, on the coast of Asia Minor, on the W. side of the Gulf of Adalia, bearing from Tekrova N.N.E. 5 miles distant, with a bold cliff of white rock.

AVRANCHES, France, in the department of La Manche, at the foot of a hill near the mouth of the river Sees, 22 miles S. of Coutances; pop. 6,960. It is half a league from the sea, and has a port for small craft only. Lat. 48° 40′ N., long. 1′ 20″ W.

AXIM, Africa, coast of Guinea, 6½ leagues E. by S. from Cape Appolonia, and 10 leagues W. by N. from Cape Three Points. It belongs to the Dutch. This town stands on the E. side of the Rio Cobre, and has a fort and factory belonging to Holland; there are rocks near it inshore, and a good landing-place inside. There is anchorage in 10 fathoms, with the fort N.N.E. Cape Appolonia is in lat. 5° N., long. 2° 35′ 5″ W.

AXMOUTH, Devonshire, between Lyme Regis and Sidmouth, 5 or 6 miles W. of the former place. Here the bay formerly offered a good shelter for ships, and there was a creek at Axmouth, but a singular convulsion of nature changed the character of the coast on Christmas-day, 1839. A piece of land, extending E. and W. about a mile, and some hundred feet broad, sunk down with a roaring noise, and left a chasm of 208 feet in depth. This chasm lies parallel with the shore, and detached the above portion of the main land many yards in a S. direction towards the sea, inclined somewhat from its former level, and the whole extent in front of this detached portion, a mile long of the bed of the sea, was lifted 40 feet above the oceanic surface, and to a very considerable distance from the original line of coast, forming reefs and islands where none existed before, within which are bays and coves where boats now enter and find good soundings. The reefs thrown up are covered with marine productions. The W. basin thus made resembles the Cobb at Lyme, but is larger. The E. basin is entered through a narrow channel, which widens into a lagoon. It must be borne in mind that Axmouth and Exmouth, though both in the same county, are different places, the last being at the mouth of the river Ex, on which the ancient city of Exeter is situated.

AXO, Greece, Republic of the Seven islands, Isle of Cefalonia, 2 leagues from Argostoli, on the summit of a hill, a fortress intended as a refuge for the inhabitants during the descents of the corsairs near that port, now useless from the island being under British protection.

AYAS, CAPE OF, on the N. side of the Gulf of Iskanderoon, in the Bay of Ayas, on which is the village of that name. In the Bay of Ayas there are 7 fathoms water; 8 leagues E. from the village of Ayas, at the further end of the Gulf of Iskanderoon, is Bayassa, or Payas, near which begins the province of Syria, separated by the mountains of Taurus from Asia Minor.

AYLOCK, LOCH, Scotland, on the west coast, a league beyond Terna. A vessel may ride on the S. side, E. of the island called Gower, in 4 or 5 fathoms.

AYMOUTH, or EYEMOUTH, between Berwick and St. Abb's head, a tide haven and fishing-place. Harbour lights are erected here; one 26 feet from the ground, seen 6 miles off, the other on the pier head, affording, when in a line, the best marks into Aymouth bay. High water 2ʰ 18ᵐ, 15 feet spring, 10 neap.

AYR, Scotland, Frith of Clyde. See *Air*.

AYRE, ISLAND OF, the S.E. end of the island of Minorca, between which and Minorca is a good channel, with 6¼ fathoms, keeping midway. The hill of Toro, seen far inland from the coast, is in lat. 39° 53′ 20″ N., long. 4° 11′ 15″ E.

AYRE POINT. See *Air*.

AYRE'S CREEK, Antigua, West Indies, at the bottom of Nonsuch harbour, N.W. from Fry head, which last is the most E. point of the island.

AZAMOR, Africa, a small town of Morocco, on the coast, lat. 33° 17′ N., 8° 15′ W., about 18 leagues W.S.W. from Anafee. A little beyond it is a river, and several towers on low land, and 5 miles W. the town of Masagan. Pop. 1,000.

AZORES, islands so called, lying from lat. 37° 16′ 50″ N. to long. 25° 33′ W. These islands were once called the Isles of the Hawks; they are subject to Portugal, being 257 leagues from Cape St. Vincent, in Portugal. They are nine in number, Santa Maria, San Miguel or St. Michael, Terceira, St. George, Graciosa, Pico, Fayal, Flores, and Corvo. They are high and rocky. They were discovered in the 15th century, by one Vanderberg, of Bruges, who found them uninhabited. Prince Henry, of Portugal, colonized them in 1449, although the Flemings had sent a colony to Fayal, whence they were sometimes called the Flemish islands. The city of Angra, in Terceira, is the seat of government, in lat. 38° 38′ 33″ N., long. 27° 12′ 33″ W. The inhabitants are a plain, simple people; their climate the most delightful in the world, clear and serene, the thermometer never being above 80° nor under 50°. The fruits of both Europe and the tropics attain perfection, and are in profusion. The islands generally are subject to earthquakes. They may be seen at sea a great distance off. The origin of all of them is decidedly volcanic. The winds when most violent approach them from the N.W., and disappear at S.E., after blowing over them. Vessels caught in storms or hurricanes near these islands, would soonest get out of their limits by steering, when it is possible, to the N. St. Michael's island is covered with verdure. The W. point is in lat. 37° 54′ 15″ N., long. 25° 55′ 15″ W. The E. point in lat. 37° 48′ 10″ N., long. 25° 10′ 5″ W. The Cidades hills are extinguished volcanoes; the highest is 5,000 feet above the sea, so that it may be seen at a great distance. In 1811, a volcano in the sea elevated an island which was named Sabrina island, between 200 and 300 feet above the surface of the water. By degrees it disappeared, until 15 fathoms of water were found upon the spot where it had stood. There are in the island, 1 city (Ponta del Gada), 5 towns, 54 parishes, and 80,000 inhabitants. There is a militia of armed peasantry, and about 350 regular troops. The castle of St. Bras, close to the sea, is the principal fortification, mounted with 24 guns. Trade is principally confined to the city of Ponta del Gada, where there is a mole for the protection of small vessels; large ones are compelled to anchor in the roadstead, which is quite open, and there is no better in the island. Lighthouses have been established on the top of the cathedral of Ponta del Gada, at the E. point of the bay, called Ponta de Galleca, and another S.E. by E. from the cathedral light about 9 miles, while a third is placed on a peak at the S.W. quarter of the island, near Punta de Ferreira. These seem to have been neglected, and to have dwindled to bad lights on the cathedral tower and the Punta da Galera, neither of which can be seen above 6 or 7 miles' distance. It is high water at 2ʰ 15ᵐ, rise 7 feet. The road of Villa Franca is sheltered by a remarkable volcanic rock, called the Porto de Ilheo; the entrance has 7 feet water. The distant view of St. Michael is considered sometimes to be deceptive, which must be carefully regarded by seamen. St. Mary has one town and port, on the S.W. side; the road is exposed. There are three villages, with a population of 4,500. The coasts are clear and bold, and may be approached with safety. There are two high points of land seen on making the island that appear to rise together out of the water. The town and port of Santa Maria, or St. Mary, is in lat. 36° 58′ N., long. 25° 12′ 18″ W. The Formigas or Ants is in lat. 37° 16′ 50″ N., long. 24° 54′ 3″ W. It is N.E. of St. Mary, and consists of seven or eight lofty rocks, running N.N.E. and S.S.W. about three-quarters of a mile, with sunken rocks near. The highest is 60 feet high, and has some resemblance to the sails of a ship; there are 7 fathoms almost close to them. The Formigas have a frightful appearance, the breakers flying higher than a ship's mast-head. The sea runs from the W.; no bottom has been found off the E. side with 50 fathoms of line, until within 30 of the rocks. All the dangers, however, are visible, and vessels may pass on the N. or S. of them, as convenient. About 10 leagues from them is the reported Tilloch reef, the existence of which is disputed. To the S.E. of the Formigas is the Dollabarat's shoal, distant S. 44° E. true, from the Formigas 3½ miles, in lat. 37° 13′ 20″ N. A most insidious danger, with but 11 feet at low water, then only showing a swell. Terceira island is fertile and pleasant, very productive in fruit, having 30,000 inhabitants. The coast is high, and almost impregnable with rocks. On the W. side is a mountain running E. and W., of which the W. end, called the Pico de la Serrata, is the most elevated. It may be known by a break in the E. side. The island has suffered much from earthquakes. A good look out should be kept for shoal water on approaching Terceira, at from 15 to 20 miles from it. The Bay of Angra is open to all winds, from the S.S.W. by the S. to the E. The swell here from the S.W. is enormous. It is high water 2ʰ 30ᵐ, rise 6 feet. Ships should always moor W. of Fort St. Antonio. At the commencement of winter, on the least threatening of bad weather, ships should instantly put off from the coast. Praya bay has been reported as the largest and safest bay in the Azores. It lies on the E. side; but the E. coast in general is said to be rocky and dangerous. Praya bay is sheltered from the S. by the W. round to the N., but is open to the E. The safest anchorage is with Point Malmeranda in a line with the N. islet, Cameiros, and the highest tower or steeple of the town, the most northern, open to the W.; there are here 25 fathoms and a good bottom. Care must be taken not to mistake Praya for Angra. The difference between Mount Brazil and the Goat's rocks is sufficiently distinguishable. A vessel at Angra should have good chain cables.

Pico island is a remarkably regular peak of volcanic origin, nearly 7,000 feet high, and seen 24 leagues off. It has a pop. of 22,000, and is in lat. 38° 26′ 15″ N., long. 28° 27′ 58″ W. The wine of this island is considered the best in the Azores. The principal places are Lagens, Pico, Santa Cruz, St. Sebastian, Pesquin, St. Rocca, La Playa, and Magdalena. The S.E. point is low and sloping, named Ponta della Isla, and a reef extends from it E. a cable's length. On the N.W. side, breakers extend for nearly a league during a gale. The N. coast is mostly inaccessible. It is from the little port and isle of La Magdalena, on the W. coast, that the produce of the island is shipped, in small row-boats, to Fayal, for exportation. FAYAL is a fine island, with a temperature remarkably mild. The S.E. point, De la Guia, is in lat. 38° 30′ 12″ N., long. 28° 41′ 37″ W. The chief town is Villa Orta, on the S.E. side. Those who run for this island should not depend on the peak of the next island for a guide, as it is often obscured. Fayal has a good bay opposite to Pico. The common anchorage is opposite the town of Orta, in the bay of that name. It is the best anchorage in the Azores, but open to winds from the N. to N.E., and from S.E. to S.W.; the S.E. wind is very fatal in winter. It is high water at 2^h 15^m, rise 4½ feet. The safest anchorage is 1¼ mile from the town, in 35 or 40 fathoms, upon sand, with the point of João Diaz a little open to the right of Point Espalamanca, and the Company's college in the town a little to the S. of the Carmelite convent. ST. GEORGE island is 3 leagues from Pico, lat. 38° 29′ 22″ N., long. 27° 50′ 27″ W. From Graciosa it is separated by a channel 8 leagues wide, and is 9 leagues long and 1 broad. On the S. coast is the town of Villa das Velas, where small vessels find shelter in all winds. GRACIOSA, said to have its name from its beauty and fertility, is the most productive of all the islands. It has a pop. of 8,000, living in two towns and two villages. It is only 8½ miles long. The chief town is Santa Cruz, on the N.E. side. The best anchorage is with the islet Abajo, near the S.E. point, in a line with the W. part of Praya isle, or a little open, off the extremity of a slope of land extending towards Point Josef Ferrer; there are here from 30 to 40 fathoms, on sand, and vessels load and unload, sheltered from S. by the W. nearly to N. All produce from Santa Cruz is shipped here. Close to the town, on the S.W. side, there are three small hills, and a church on the highest part of each; they are good marks for the N. side of the island. CORVO island has a small port, and about 750 inhabitants. Between this island and Flores there is a great depth of water, and no danger of any kind. In FLORES, at Santa Cruz fort, the lat. is 39° 27′ 3″ N., long. 31° 8′ 37″ W. The lat. of Corvo is 39° 40′ 7″ N., long. 31° 8′ 0″ W. Flores is more populous than Corvo, having about 7,000 inhabitants. It has two towns, Santa Cruz and Lagens, and five villages. There is a remarkable peak in this island. Much orchilla, in use for dying, is gathered here. The land is well cultivated. Ponta Ruiva, the N.E. point of Flores, is high, and at its feet is an islet, called the Pan de Azucar. In a bay to the W. of this islet, there is anchorage in 25 fathoms, on sand, sheltered from winds from the S.E. by the S. to W.S.W. It is resorted to for water. The point of Santa Cruz is 2½ miles S. ⅜ E. from Point Ruiva, low and rocky. Between is the islet of Alvaro Rodriguez, very near the coast, and to the S.E. of this is the anchorage, in 36 fathoms, sandy bottom, sheltered from the W. and S.W.; about 1¾ of a mile from this point is the castle of Santa Cruz, near the town, the principal port of the island. On the N. side of Lagens point, in this island, is the town of Lagens, the church of which being large, is a useful mark on the coast. A vessel may anchor in this bay with the wind between W. by W. to S.W. by W., in 25 fathoms, sandy ground. This anchorage is much used, since a vessel can much more easily get under weigh here than at Santa Cruz, as there is better space for working out. Punta Albernas is the N.W. point of Flores island, of a red colour, moderately high; between this and Point Fanaes is the islet of Maria Gadella, high and round; W. ¾ N. from this islet is anchorage in 30 or 40 fathoms, sandy ground.

AZOV, AZOPH, or AZOF, THE SEA OF, the entrance to which is on the N. of the Black sea, through the Straits of Kertche. It is situated between the parallels of 45° 20′ and 47° 18′ N., and the meridians of 35° and 38° E.; the entire distance from the Strait of Kertche to the mouths of the Don, on the two bearings, being 168 miles. The N. coast is generally 100 feet above the sea, of a reddish colour; a few small hills only are to be seen upon the surface. Its E. coast, inhabited by Cossacks, is low from the Don to Temrouk. Its W. is formed by the sand, called the Tongue of Arabat, which separates it from the Sivache, or Putrid sea. The greatest depth of the Sea of Azof on the N. side, between Jenikalek and Bielosaria, is 7¾ fathoms, and the depth diminishes towards the Gulf of the Don. The bottom is mud mixed with shells, generally black, though red on the E. coast. The water is muddy, almost fresh, and might sometimes be found fit for use as far as 20 miles below Taganrog. There is no current of moment in this sea; with a strong N. wind it does not acquire a motion of more than 1 mile an hour. The navigation ceases from November to March, earlier or later, according to the setting in of the frost. Cape Kamennoi, or Stony cape, forms the E. point, at the entrance of the Strait of Kertche; a reef extends N. of it. This cape is steep, of a reddish colour, moderately high. Further to the E. the coast winds E. by S. ¼ E. for 25 miles, within which distance is a great marshy plain, 5 miles from the shore; the water varying in depth from 25 to 40 feet. The Gulf of Temrouk extends S.E. ¾ E. about 9 miles. At the entrance, which is only a quarter of a mile wide, an islet divides the passage. Inside there are only 4 and 5 feet water. On the W. side is a village, with a church, fortress, and some hills. The E. side is overgrown with reeds. Beyond Temrouk the coast is low and sandy, forming at 23 miles N. by E. ¾ E. the point of Atchouief. Here the depth a mile from the land is from 17 to 25 feet only. E. of this point an arm of the Kouban runs into the sea, and 23 miles N. by E. ¾ E. from this branch, is the Gulf of Akhti, 8 miles in extent, in a S.E. direction. To the E. of this gulf, the land becomes more elevated, and two hills are seen to the N.W.; 8 miles from the land are only 12 and 15 feet water. In fact, the Sea of Azof offers a proof of the slow but certain changes operating on the earth's surface in so many places. It is gradually filling up, and in time it is probable the Don will carry its outlet between banks to the Strait of Kertche. The shallowness of this sea, and its reedy marshy inlets, render the air on its borders exceedingly unhealthy.

AZUA, or AZUCA, St. Domingo, or Hayti, in the West Indies, at the bottom of a deep bay, an inconsiderable town. Cape Beata is 20 leagues to the S.W. Point Salinas lies S.S.E., at 10 or 12 leagues distance, being in lat. 18° 12' N., long. 70° 49' 30" W.

BAAL'S RIVER and BAY, W. Greenland, between Bear Sound on the S.E. and Delft's point on the N.W. coast of Greenland, N. of Ameraglick river, in lat. 64° N., where the Danes established a colony in 1721.

BAALSTED, Sweden, lat. 56° 30' N., on the coast of Gothland.

BABABARA, BABA, or BAHABORA, CAPE, Greece, a headland N.E. of the Isle of Mytilene, in the Archipelago, about 3 leagues; it is upon the continent of Asia, in lat. 39° 30' N., long. 26° 4' 30" E. Some accounts make it 26° 25' E. It is a lofty point, frequently called Cape St. Mary, and is the first land seen sailing N. to the Dardanelles.

BACCALIEU, or BACALEO ISLAND, E. coast of Newfoundland, 9 leagues from Cape St. Francis. The S.W. end bears with Green bay E. by N. and W. by S. distant 1½ league. It is about a league from the main, with a fair channel between. The N. point is in lat. 48° 9' N., long. 52° 50' 34" W.

BACHAASH, Scotland, a small isle near the N.E. end of N. Uist.

BACHASSY GAP, S. coast of Africa, E. of the entrance of Old Calabar river. Lat. 4° 29' N., long. 8° 32' E.

BACK ISLAND, Scotland, one of the small isles 11 miles S. of Coll.

BADA HEAD, North America, Gulf of Mexico, on the S.W., 4 leagues N. from Macaton head. Lat. about 15° 30' N., long. 98° 20'. Six leagues from Port Salines. The coast near is all dangerous.

BADAGRY, ROAD OF, African coast. Lat. 6° 20' N., long. 2° 47' 48" E.

BADCAAL, LOCH, Scotland, W. coast, to the N. of Callaway, sheltered by small islands, with good anchorage in the middle of the bay, of 10 and 12 fathoms; 4 miles N.E. is the island of Handa. It is not much above 10 miles from Cape Wrath.

BADELONA, Spain, a town 10 miles E. of Barcelona. Lat. 41° 12' N., long 2° 20' E.

BADIS, Livonia, a fortress on the S. side of the Gulf of Finland, 7 leagues E. of Revel, in lat. 59° 15' N., long. 24° 36' E.

BAFFA, anciently PAPHOS, island of Cyprus, a town nearly in ruins. The bay is large, but the port unsheltered and not safe, the mole leaving the S. open and only protecting the E. and W. A ledge of rocks extends out a league near the entrance. The anchorage may be either before the town or the castle, in 8, 7, or 6 fathoms water. It is situated 23 miles S. of Cape Salizano, which is in lat. 35° 6' 20" N., long. 32° 16' 30" E.

BAFFIN'S BAY, a deep gulf running N.E. beyond the arctic circle, between the S. point of Greenland and Hudson's straits. It lies between the parallels of 60° and 80° N., though some give the name only to the N. part, and denominate the S. half Davis's straits. It communicates with the North Atlantic. It is surrounded by the inhospitable shores of a far northern climate, is deeply indented with harbours, and covers that portion which is S. of the arctic circle, the W. side of which portion, or S. of lat. 66½°, is occupied by the broad entrance into Hudson's bay, and the islands that border upon it, together with Frobisher's straits, Cumberland straits, and the Metainco islands.

BAFFIN'S STRAITS, North America, the passage between James's island and the most E. of the Cumberland islands.

BAFFOU POINT, African coast, near Cape Palmas. Lat. 5° 9' 10" N., long. 9° 17' 30" W.

BAG, or BAGGY POINT, Devonshire, the N.W. point of entrance in Barnstaple bay. Lat. 51° 10' N., long. 4° 52' W. It is high water at 5ʰ 20ᵐ, or 1¼ʰ before it is full tide at Bristol.

BAGENBON HEAD, Ireland, S.E. coast, N.W. by N. 8 miles from the S.W. of the Great Saltee island. It is the W. limit of Bannow harbour, and has Ingard point N. The Saltees light-vessel on Coningbeg rock is 3 miles S.W. ¼ S. from the Great Saltee, in lat. 52° 2' 18" N., long. 6° 38' 15" W. According to some the above head is in lat. 52° 9' N., long. 6° 48' S.

BAGNARA, Naples, a village and small port 8 miles S. of Palma. Lat. 38° 15' N.

BAGNOL, CAPE, France, the S. point of Port Vendres harbour. See *Port Vendres*.

BAGROS RIVER, Africa, on the main, at the back of Sherbro island, nearly facing Macaulay's isle, about lat. 7° 23' N.

BAHAMA BANK, LITTLE—GREAT and LITTLE ABACO, Abaco kay, and the Hole in the Rock, are on the Little Bahama bank. The Hole in the Wall is a perforated rock at the S.E. point of Abaco. Great Abaco is 23½ leagues in length N.N.W. and S.S.E., and about 4 leagues in breadth, and over its W. end is Little Abaco. The Hole in the Wall is in lat. 25° 51' 30" N., long. 77° 7' 45" W. The light tower here is 160 feet above high water, visible 15 miles, revolving once a minute. The N.E. point of Abaco island is in lat. 26° 30' N., long. 76° 57' W. Maternillo bank, N. end of, lat. 27° 50' N., long. 79° 10' W. The same, N.W. end, lat. 27° 34' N. The W. end of Grand Bahama island, lat. 26° 41' N., long. 79° 3' W. The PASSAGE ISLANDS.— Little St. Salvador, W. point, lat. 24° 36' 22" N., long. 75° 58' W. Columbus, or S.E. point of St. Salvador, lat. 24° 8' 30" N., long. 75° 16' 48" W. Samana or Attwood kays.—E. Low kay, lat. 23° 5' N., long. 73° 36' 43" W.; Planas or Flat kays, centre, lat. 22° 35' 10" N., long. 73° 33' W. CROOKED ISLANDS.—The N.E. breaker, lat. 22° 43' 30" N., long. 73° 47' W. Miraporvos bank, 11½ miles from S.S.E. to N.N.W., shoals very dangerous, especially on approaching from the S.E. The current sets over these shoals 1 mile an hour. N. rock, lat. 22° 7' 50" N., long. 74° 32' 40" W. E. end of E. reef, lat. 22° 18' N., long. 72° 33' 16" W. The CAYCOS.—Cape Comet, N.E. point, lat. 21° 42' 50" N., long. 71° 27' 38" W. Little Cayco, S. point, lat. 21° 37' 30" N., long. 72° 28' 33" W. GREAT INAGUA, N.W. point, lat. 21° 7' 30" N., long. 73° 39' 30" W. LITTLE INAGUA, E. point, lat. 21° 29' 15" N., long. 72° 55' 33" W. TURK'S ISLANDS.—Endymion reef, lat. 21° 7' 15" N., long. 71° 18' 18" W. Grand kay roadstead, lat. 21° 28' 10" N., long. 74° 7' 30" W. SILVER KAY or PLATO BANK, E. end, 10 fathoms, lat. 20° 35' 20" N., long. 69° 21' 53" W. SHIP BANK, or Bajo de Navidad, lat. 20° 14' N., long. 68° 51' 18" W., at N. extremity of the bank. There is a lighthouse on the N.E. point of Grand Turk island, and one is erected at the N.W. extremity of Anguila, or Salt kay bank, on the highest of the detached rocks, known as Double-headed Shot kays, in lat. 23° 56′ N., long. 80° 27′ 38″ W. It is a fixed light, 100 feet above the sea level. From the lighthouse, the S.W. of the Double-headed Shot kay bears S.S.W. ½ W., distant 3½ miles. The navigation over the Great bank of Bahama can only be made by vessels that draw under 11 feet of water, and it demands great care. The bottom of the bank is stony, with sand banks in places, having very little water over them, extending W. from Andros 5½ leagues. Large vessels not being able to cross the Great bank from the Berry islands to the Roquillas, run along the edge of it from the Berry isles, until they reach Isaac kays, when, having doubled these, they sail southward to the Roquillas. All about these banks, reefs, and islands, requires the most careful navigation. The Bahamas and islets, some mere rocks, but 12 or 13 large and fertile, are said to be 500 in number. High water about 7^h 30^m.

BAHAMA ISLANDS, BANKS, and STRAITS, North America, off the coasts of Cuba and Florida. They form the E. side of the Bahama strait, as once called, better known as the Gulf of Florida. The old Bahama strait is between the N.E. coast of Cuba and the Grand Bahama bank, sometimes called the "Old channel," the fair way of which is about 8 leagues broad, narrowed on the Cuba side by rocks and small islands, called kays. From the S. point of the shoal or bank it treads N.W., and is edged with rocks on the S.W. and N. side, and is lined with islands on the N.E. as far as lat. 23° 4′ N., whence it turns due N. and becomes the E. side of the entrance into the Straits of Bahama or Gulf of Florida, through the Santaren channel on the N., formed by the Salt kay bank on the W. and the Greatbank on the E. The Salt kay bank divides it, having on the N.W. the channel of St. Nicholas, through which a current sets into the old channel or strait. In the Santaren channel the currents are irregular. N.W. of the kay the tide crosses from the N. side of Cuba and of Salt kay bank toward the Florida reefs with great rapidity. The Florida reef, generally, from the Tortugas, or from the W. end of the reef, at first E., then ascending to the N. as high as lat. 28°, forms the N. and W. sides of the Gulf of Florida. The S. and E. sides are bounded by Cuba Salt kay bank, the Great bank, and Little bank, as high as to the Manterillo reef. At the head of the Great bank, a channel called Providence North West channel separates it from the Little Bahama bank on the N. of the Great bank. This last is 120 leagues long and 45 broad at the S. end, being in lat. 21° N. and terminating in lat. 27° N. Bahama island is 20 leagues from the coast of Florida, and 10 W. from the island of Lucayo, being 28 leagues long and 3 broad. The islands comprise a great number of islets, kays, banks, and rocks. The banks are generally sandy, low, intermingled with porous rock, the soil light, producing small trees. Cattle are reared and cotton cultivated in the islands. Water is scarce. The climate for the lat. is healthy. The pop. in 1831 was 16,788. The total revenue 22,399*l.*, the expenditure 40,333*l.* Nassau, in New Providence, is the capital. This singular group, termed the Lucayos as well as the Bahamas, extends in the form of a crescent from 27° 50′ N. lat. and 79° 10′ E. long. as far as Turk's islands. NEW PROVIDENCE is the chief island and seat of government, about 21 miles long and 7 broad, flat, and having extensive lagoons. It is upon the Great Bahama bank. On the N. side is the harbour of Nassau for vessels drawing 12 or 13 feet, sheltered on the N. by Hay island. It has a harbour light 70 feet above the sea. The town is in lat. 25° 5′ 10″ N., long. 77° 21′ 4″ W. It has a considerable trade. ANDROS ISLE lies W. of New Providence 7 leagues; it is long and irregular. Morgan's Bluff, the N.E. point, is in lat. 25° 10′ 24″ N., long. 78° 1′ 30″ W. A tongue of the sea runs in S.E. as far as lat. 23° 24′, called the Gulf of Providence. This island, or rather chain of islands, is almost uninhabited, from the difficulty of access through the reefs that surround them. The Grassy creek kays are off the W. end, with the Ghost isles between. The BERRY ISLES are an irregular group, containing small harbours, but are seldom visited except by the people of New Providence. The S.E. end is called the Frozen kay, and the N. the Stirrup kay. There is anchorage off the N. of the last, on the bank, in lat. 25° 49′ N. The S. Stirrup kay, the N.W. point, is in lat. 25° 25′ 5″ N., long. 77° 56′ 13″ W. Great Stirrup kay, E. point, lat. 25° 49′ 40″ N., long. 77° 53′ 45″ W. Centre Holm's kay, lat. 25° 37′ 40″ N., long. 77° 44′ 0″ W. The GREAT and LITTLE ISAACS are W. ¾ N., 40 miles from Little Stirrup kay, the centre in lat. 26° 2′ N., long. 79° 6′ 20″ W. The BEMINI ISLANDS are the N.W. on the Great bank, 6 leagues S.S.W. from the Great Isaac. The ground about them is shoal and rocky, but there is good anchorage on the S.W. in 6, 7, and 8 fathoms. These islands were surveyed in 1810 by the Moselle. From the N. point of the most N. to the S.W. of the S. island is about 6½ miles. The last is called Hog island, about 4½ miles long, N.E. by E. and S.W. by W. The Moselle bank, on which that man-of-war grounded, is upwards of a mile in length, N. by E. and S. by W., and about 60 feet wide, with not more than 10 feet of water at high tide in some places, with 3 or 4 fathoms close to both sides. The Moselle reef is in lat. 25° 29′ 10″ W., long. 79° 17′ 30″ W. On Gun kay, connected by a chain of kays and rocks from S. Bemini, are the Turtle rocks, 3 or 4 miles S.W., and the Kat kays, the N. of which is named the Dog kay, the second the Wolf kay, and the third S. the Cat kay. These are 2½ and 4 miles from the edge of the bank. S.W. of Cat kay 1 mile, are some islets named the S.W. rocks. On the Gun kay, about 250 yards from the S. extreme, a lighthouse is erected in lat. 25° 34′ 35″ N., long. 79° 18′ 24″ W., 80 feet from the sea level. The light revolves once a minute, and may be seen in all directions except S. by W. ¼ W. and S. ¾ E., where 8 miles off it is intercepted by the Bemini islands. Vessels should not bring the light to the S. of S.E. The flood tide sets strongly to the E. through the intervals of the kays, where it is high water at 7^h 30^m, rise 3 feet. Two miles from the Cat kays are the three rocky kays. The Riding rocks are a double range, quite bare, 4 or 5 miles within the edge of the bank. Lat. 25° 14′ W. The NARANJOS, or ORANGE KAYS, are seven in number. The middle is in lat. 24° 56′ 30″ N., long. 79° 9′ 24″ W. They lie 5 leagues S. of the Riding rocks. From lat. 24° 6′ N. the edge of the bank trends S.E. by S. to the S.W. corner, then turns E. by S. 8 leagues to Ginger kay, in lat. 22° 47′ 0″ N., under which, on the S.W. side, there is an anchorage. From Ginger kay to Wolf kay is 11¼ miles S.E. ¾ E. ELEUTHERA, one of the largest of the islands, is of irregular form. It bears N.E. ½ N.

12 leagues from the end of New Providence. Its E. and N.E. shores are on the Atlantic. On the W. side are the settlements of Rock sound. SAN SALVADOR, or CAT ISLAND, lies 6½ leagues E.N.E. ¾ E. from Powel's point in Eleuthera, extending S.E. 12 leagues, with a breadth of 3 to 7 miles. On the E. it is inaccessible, from a reef on the S.W. side. There is good anchorage here. This was the first land seen by Columbus, on the 12th of October, 1492, and named by him St. Salvador. At 11 miles off S.E. is Conception island, 3 miles long. A reef runs off from it 7 miles N.E., as far as lat. 23° 55′ 30″ N., on which the Southampton man-of-war was wrecked, in 1812. The S. end of this island is in lat. 23° 48′ 46″ N., long. 76° 6′ W. From W. Rum kay, the S.W. end of this island, bears S.E. 4 leagues. The lat. of the W. end of Rum kay is 23° 39′ N., long. 74° 56′ 35″ W. YUMA, or LONG ISLAND, is 19 leagues in length from S.E. to N.W. The N.W. end is 5 leagues from Cat island; S. by W. ½ W. from Long island, 17½ leagues from its S. point, is Green kay, or Cayo Verde, in lat. 22° 1′ N., long. 75° 10′ W., near the S. end of the Great bank. The EXUMA ISLES lie W. of Long island, called Little and Great Exuma. They are S. of Cat island, and their chief produce is salt. Harbour island is 5 miles in length, about 1½ in breadth, about 1¼ mile from the N. side of Eleuthera. Egg island is small and covered with brushwood, in lat. 25° 31′ N.; a reef runs off from it N.N.W. ½ W., above 3 miles. Douglas passage is a channel S.W. ½ S. two-thirds of a mile from the W. end of Booby island, and N.N.E. ½ E. a good mile from the end of Rose island, where lie the Douglas rocks, on which two beacons are erected, bearing from each other S.E. ¾ E., and contrary. These islands are nearly all on the GREAT BAHAMA BANK, they are all low, and supposed to be the work of the coral insect. The ocean close to the isles is of an unfathomable depth, bounded by walls of coral. The appearance is inviting and agreeable. The total area of all the islands is estimated at 2,842,000 acres, of which only 408,486 are granted. The thermometer ranges from 58° to 94° during the year. The lat. and long. of the great isles not given above are as follow:—The Brothers, E. rock, lat. 22° 1′ 30″ N., long. 75° 41′ W. The JUMENTOS, Little Ragged isle beacon, lat. 22° 9′ 30″ N., long. 75° 41′ 30″ W. Ragged island flagstaff, lat. 22° 11′ 40″ N., long. 75° 44′ 17″ W. Racoon kay beacon, lat. 22° 21′ 50″ N., long. 75° 49′ 39″ W. Channel kay, lat. 22° 32′ 15″ N., long. 75° 32′ 50″ W. Jamaica South kay, lat. 22° 42′ 56″ N., long. 75° 54′ 46″ W. Man-of-war kay, lat. 22° 47′ 20″ N., long. 75° 54′ 0″ W. Flamingo kay hill, lat. 22° 52′ N., long. 75° 53′ 6″ W. Water kay, S.W. point, lat. 22° 58′ N., long. 75° 45′ 3″ W. YUMA, or LONG ISLAND, S. point, lat. 22° 50′ N., long. 74° 52′ W. Great harbour entrance, lat. 23° 7′ N., long. 74° 52′ 30″ W. Michael bank, 12 fathoms, lat. 23° 9′ 15″ N., long. 74° 45′ 30″ W. North end of the island, lat. 23° 41′ 37″ N., long. 75° 19′ W. EXUMA, the beacon, lat. 23° 32′ 30″ N., long. 75° 46′ W. Galliot cut, on the bank, lat. 23° 55′ N., long. 76° 15′ W. ELEUTHERA, S.E. point, lat. 24° 37′ N., long. 76° 9′ 23″ W. Governor's harbour, lat. 25° 11′ 15″ N., long. 76° 14′ 53″ W. James's Cistern, lat. 25° 21′ N., long. 76° 23′ W. Harbour island, lat. 25° 30′ N., long. 76° 39′ W. Egg island, reef end, lat. 25° 34′ N., long. 76° 55′ 30″ W. Isles on the N.W. FLEEMING CHANNEL beacon, lat. 25° 16′ 45″ N., long. 76° 53′ 3″ W. Douglas channel entrance, lat. 25° 7′ 30″ W., long. 77° 2′ 45″ W. ANDROS, High kay, E. coast, lat. 24° 39′ 30″ N., long. 77° 42′ 50″ W. Golding kay, lat. 24° 13′ 40″ N., long. 77° 37′ 20″ W. Green kay, in the Gulf, lat. 24° 2′ 12″ N., long. 77° 10′ W.

BAHAMA OLD CHANNEL, S. of that called the Gulf of Florida, which last is the narrow sea between the coast of America and the Bahamas, 135 miles long and 46 wide. The currents here run strong, from 2 to 5 miles an hour, among islands and shoals.

BAHIA ESCOCESA, or Scotch bay, N. coast of Hayti, next to Samana bay, which last is in lat. 19° 15′ 40″ N., long. 69° 6′ 15″ W.

BAHIA HONDA, Florida reefs and kays, West Indies, 7 miles N.E. ½ N. from Love kay, in lat. 24° 38′ N. It has a large entrance and fine channel, of 4 or 5 fathoms, though it shoals within, and the bottom is hard ground. Three small islands mark the W. of the entrance.

BAHIA HONDA, or DEEP BAY, Cuba, 7 leagues E. from the banks of Isabella, less than 8 from the Colerado reefs, and 40 leagues N.E. of Cape San Antonio. Its entrance is in lat. 23° 45′ N., long. 83° 12′ 30″ W. It is a spacious, well sheltered harbour, but the points which form its entrance are bordered with a reef, as well as the interior coast, on the edge of which the water is shallow. The breadth between the reefs at the entrance is not more than 2½ cables' length, and entrance should not be attempted without a leading wind. BAHIA BANK is 6 leagues N.W. by W. from Bahia Honda, about a league in length, and half a league in breadth.

BAHIA HONDA, Great and Little, coast of Colombia. The mouth of the larger is 3 miles wide, with one shoal in the entrance, distant from W. point 1 mile. Some parts are very shallow, but there is good anchorage. Bahia Honda Chica is very shallow, without shelter. Lat. of entrance of the larger bay 12° 20′ N., long. 71° 48′ 35″ W.

BAHNACALICH, Scotland, E. side of Rona island, where are two small well sheltered bays, available for anchorage, if needful.

BAHR, a word adopted as a sea term in the E., from the Persian, signifying a great bay or gulf.

BAHUS, Sweden, a town on a rock, 10 miles N. of Gottenburg, lat. 57° 52′ N., long. 11° 42′ E.

BAIA, BAIÆ, or BAJA, a bay of Italy, Bay of Naples. It has a castle and an anchorage within it, towards the N.W. corner. Lat. 41° 6′ N., long. 14° 45′ E.

BAILLIES, Ireland, a long narrow sand, W. ½ N. from the Tuskar rock, between the rock and the main. The Tuskar is E.S.E. ½ E. 6 miles from Carnsore point, which last is in lat. 52° 11′ N., long. 6° 23′ W. The Black rock, and the Barrells, the former 10 miles W. by N. from the Tuskar, the latter a mile S.E. from the Black rock, are dangers to be avoided here. There is a revolving light on the Tuskar N. ¾ W. from the Smalls 11½ leagues. The light has three faces, one of which is refulgent every two minutes. Lat. 52° 12′ 7″ N., long. 6° 12′ 38″ W.

BAJA NUEVO, or NEW SHOAL, Campeche bay, Gulf of Mexico, a head of sand, showing at low water. The sea breaks upon it in ordinary breezes. It lies a little E. midway between the Triangles and Los Cercos.

BAJO DE NAVIDAD, West Indies, a bank or shoal, in lat. 20° N., and E. of the St. Domingo channel. See *Bahamas*.

BAJOLE, CAPE, Minorca, the most N. point of that island.

BAKER'S DOZEN ISLANDS, North America, in Hudson's bay. Lat. about 57° 30′ N., long. 81° W. They are N. of an opening which goes E. and N.E. as far as the S. end of Hudson's straits, S. of some small islands called N. and S. Sleepers.

BAKER'S ISLAND, North America, United States, in Massachusetts. It is off Salem harbour 5 miles E.N.E., and 4 miles from Marble head. It has two towers, on which are fixed lights at unequal heights, in a line N.W. ¼ W.; the S. or high light visible 6½ leagues. Salem is in lat. 42° 31′ 19″ N., long. 70° 53′ 56″ W.

BALACKLAVA, Black sea, separated from Sevastapol by a narrow peninsula. Only one ship can enter the harbour at a time, from the extreme narrowness of the ingress.

BALASIA, Coast of Africa, W. of Cape Bugaroni, a very small, mean town.

BALBRIGGAN HARBOUR, Ireland, E. coast, 3 miles from the Skerries. It is sheltered by a pier, at the head of which is 12 feet water at spring tides. On the pier is a white lighthouse with a fixed light, visible 11 miles W.S.W. to S.S.E. ½ E. Lat. 53° 36′ 46″ N., long. 6° 10′ 53″ W.

BALD, CAPE, Africa, W. coast, lat. 13° 22′ 30″ N., long. 16° 49′ 20″ W.

BALD, CAPE, Africa, 11 miles W.S.W. from Cape St. Mary.

BALD HEAD, North America, United States, part of Cape Fear, in North Carolina, which lies S.S.E. of Bald head. On the last is a lighthouse on a black tower, a mile from the sea, having a fixed light elevated 110 feet. Lat. 33° 51′ 45″ N., long. 78° 0′ 40″ W.

BALD MOUNTAIN, North America, Gulf of St. Lawrence, a noted mark on the main, 30 leagues W. from the nearest or N.W. point of Anticosti.

BALDOCK'S ISLAND, N. coast of Africa, called also Isle Ashâg, 23 miles E.N.E. ¼ E. from Cape Tenez, a good mile from the shore, with a passage inside it. Cape Tenez is in lat. 36° 33′ N., long. 1° 45′ E.

BALDOYLE, Ireland, a creek on the N. side of Howth, fit only for small craft.

BALDRIN'S, ST., CASTLE, or WHITBERRYNESS, Scotland, 2¼ miles N.W. from Dunbar, having Tyningham bay or flats between them.

BALEARIC ISLES, Mediterranean, off the coast of Spain, comprehending the islands of Ivica, Majorca, Minorca, Cabrera, Dragonera, Formentera, Conejera, and upwards of 50 islets. These islands were called Baleares by the Romans from the inhabitants having been skilful in the use of the sling. Majorca is nearly square, about 50 miles long by 40 broad, and has a superficies of 1,440 square miles. It is rocky and mountainous; the climate is temperate and generally healthy; the population 160,000. The most conspicuous points are those on the coast forming the angles of its square form, as Cape Pera, N.E.; Cape Salinas, S.E.; Cape Dragonera, S.W.; and Cape Formentor, N.W. The Puig de Galatzo, a high peak to the W., in lat. 39° 36′ N., is the most remarkable eminence. The produce is similar to that of Valencia on the main. Off Point Rebagada, in this island, is the island of Dragonera, in lat. 39° 34′ 25″ N., and long. 2° 22′ 13″ E.; about 1¾ mile long, having two watch-towers upon it, and distant from the Punta den Sierra, in Ivica, 50 miles. There is a channel between this island and Majorca, having from 12 to 20 fathoms water, and half a mile wide. Two islets lie within, and some rocks under water, the largest called St. Elmo, behind which vessels sometimes seek a shelter from all but S.W. winds. Mytyana is the name of the other island. Under the N.E. point of Dragonera there is a cluster of rocks, called Calafetes. Cape Andrichol is high and wooded. Malgrat island is lofty and steep, having between it and Cape Regrete, the small isle of Conejos. From Malgrat island S.S.E. ¾ E. 2 miles, is Toro island, and 1¼ mile S.E. by E. from Toro island is Cape Cala Figuera, on one side of the Bay of Palmas, the other being formed by Cape Blanco. There are signals made by day, and a light shown at night, from a tower near its extreme point. The town of Palma, the island capital, has 6,000 inhabitants. There is good anchorage in this bay, save with S. winds. Near Pi-Port here was the old arsenal and lazaretto. From Cape Blanco the next point is Negra point, and near is Port Compos, shallow, and only fit for small craft. Cape Salinas lies 3½ miles S.E. of Port Compos, the most S. point of Majorca, lat. 39° 15′ 45″ N., long. 3° 5′ 3″ E. There is a tower upon it, called Negosta; a reef runs out from the point beneath. The Cabrera islands lie off this point, of which Conejera is the highest; off its N.E. point are four islets, three called the Planas, the fourth named Furadade. Cape Ansiola is the S. point of the Isle of Cabrera, near which is a cove, fit only for barks. Returning to the large island, beyond Cape Salinas N.E. lies Port Pietro, having a square fort on its S.W. entrance; the channel is a cable's length wide, with 15 fathoms. In bad weather, the entrance is not recommended; the port is open to S. winds. At Cala Longa, 1¼ mile from Port Pietro N.E., avoiding a shoal, frigates might ride in 4 or 5 fathoms, moored head and stern. Four miles more N. is Port Colon, with some coves between; it has a narrow entrance; it is exposed to S.W. and S. and S.E. winds. Cape Manacor is 7 miles N.N.E. ¼ E. from Port Colon, and may be known by a tower over its W. point; and 4 miles N.E. ½ E. is Point Amer, on which there stands a fort, to the N. of which is good anchorage with from 9 to 12 fathoms water. Beyond Cape Amer is Cape Ratche, low, dark, and short. The Bay of Arta, with the little port of Vey, lies between. Cape Bermejo is the next point, and 1½ mile further is the Point Fon de la Cala, and 2½ miles beyond again Cape de Pera, the E. point of the island, in lat. 39° 42′ 20″ N., long. 3° 30′ 18″ E., having a tower on the top. The small bay of Ratyada is close to this point, and 2 miles from Cape Pera is Cape Freu, dark, steep, and its extremity looking like an island. On the S. side is a remarkable cave, and on the point is the tower of Son Yaumel; other small points, with watch-towers, intervene as far as Cape Farach, the S. point of the Bay of Alcudia. This point is of moderate height, and within it is another point, and further on is Cala Mata, whence the coast runs S.W. by W. 2½ miles towards Py de la Entrada, and thence rounding W. to the little isle of Porros, 4¾ miles distant. Three parts and more of this distance is the entrance to the lake or Albufera, of considerable extent. The whole way the land is low, clean, with anchorage off shore in 3 fathoms at a cable's length, and further out from 9 to 23. About 1¼ mile from the entrance to the Albufera is the quay of Alcudia. The town has but 800 inhabitants. To anchor large ships, run in midway between Alcana and Torre Major, 3 or

4 cables' length from the land, in 5, 6, and 7 fathoms, with a bottom of fine gravel. Smaller vessels may go opposite the tower, in 4½ or 5½ fathoms, mooring N.W. by S.E. At the distance of 1½ mile from the quay of Alcudia is the little isle of Alcana. The bay is open from the N.E. to the E., and those winds bring in a heavy sea, but the bay is preferable to that of Pollenza, which is round Cape Minorca; the town, with a pop. of 6,000, lying between Capes Pinar and Formenton. This bay, though deep enough to afford the largest ships a shelter, is not large enough to contain any number well sheltered, without exposure to the E.N.E. winds. The best place for a large ship is 2 cables' length W. of the point of Castle Pollenza, in from 5½ to 7½ fathoms, sandy bottom. Cape Formenton is in lat. 39° 57' 15" N., long. 3° 17' 53" E. Port Soller is on the N.W. coast, narrow, and difficult to discover; in fact, from the isle Dragonera as far as Cape Formenton, the N.W. side of the island has no haven of safety, though the land is steep, and generally free from hidden dangers. The tides, that rise and fall about a foot, are very irregular here. There is a doubtful bank said to exist 4 leagues to the N.N.W. of Port Soller, in lat. 39° 56' 30" N., about 31 miles W. from Cape Formenton, with only 11 or 12 feet water over some parts, but it has not been well ascertained. MINORCA is the second in size of the Balearic isles, about 28 miles long, and 8 or 9 broad. The surface is uneven. Its only remarkable eminence is Toro, nearly in the centre of the island. The principal harbours are Port Mahon, Fornello, and Ciudadella. The pop. is about 34,000. Cape Dartach, the S.W. point of the island, is in lat. 39° 56' N., long. 3° 51' E., and about 26 miles, bearing E. a little S., from Cape Formenton. It is low, yet with deep water close up to it. Cala de Santa Galdana is the best port on the S. of Minorca for vessels of moderate size, except in W. winds. Ayre island, off the S. of Minorca, is low, but the channel between that and the main island is good, with 6½ fathoms. Vessels may anchor along the S. coast, with N. winds, in from 15 to 18 fathoms. The mountain Toro, with a convent on its summit, should always have its point only kept in sight above the shore land; its lat. is 39° 58' 20" N., long. 4° 11' 15" E.; if a great part of the hill appear, the ground will be rocky. Port Mahon, the principal harbour, is upon the E. side. The entrance from Port St. Phillip to the opposite side is not more than a quarter of a mile wide, and further in is narrower. The town, at the head of the harbour, is in lat. 39° 52' 55" N., long. 4° 22' E. To enter, the three rocks lying off Cape Mola should be brought into one, until the mouth of the harbour open; or if at a distance, bring Mount Toro to bear N.W. ½ N., which will lead to the entrance. If off-shore winds prevail, to anchor before it, bring the Mountain Toro into the middle of the opening, and Cape Mola N.W. ½ W., when waiting; anchorage will thus be found in 17 fathoms, with a sandy bottom. To sail in or out, keep the large farm-house of San Antonio just open of the Lazaretto point, bearing N.W. ¾ N.; this will clear the rocks at the entrance, into 8, 9, and 10 fathoms. When within these points, the mid-channel kept past Quarantine island, and anchoring before reaching Hospital island, from 13 to 6 fathoms will be found, in good ground, safe from all winds. Cape Mola, the N.E. point of this port, is in lat. 39° 52' 45" N., and long. 4° 24' 15" E. Several capes and bays of little

moment intervening, with the isle of Colon and road of Sesllanes, Cape Musene Vives and the little havens of Dadaya and Moli, the port or haven of Fornello, on the N. side of the island, is reached. It has a narrow entrance, round Point Ponza from the E. There are shoals within, but the anchorage for large ships, near the old castle, where the town stands, has from 7 to 11 fathoms of water. Cape Minorca, called also Cape Bayoli, is on the W. of the island, in lat. 40° 2' 50" N., long. 3° 50' 43" E. On it is a tower, and it has deep water all round. S. ½ E. from this cape is Point Den Banicus, and the same distance E. by S. is the entrance to the harbour of Ciudadella, which runs in N.E., and then turns more E. At the entrance is 6 fathoms, and half way up only 2, all sand and gravel, but so narrow that a brig can hardly find room to tack in it. When the wind blows from W. or S. a heavy sea is sent in. With other winds ships may ride well enough at the entrance in 18 fathoms. Ciudadella is esteemed the capital of the island, and is fortified, as well as Port Mahon, and the less important havens of the island. It contains barracks for cavalry and infantry, a cathedral, two churches, and four convents. IVIZA, or IVICA, is the ancient Ebusus, and is 20 miles long by 10 broad, having a pop. of 13,000. Yvica, the only town, is on the S.E., and has about 400 houses, and a pop. of 2,700. It is fortified. The port is spacious, well sheltered, and possesses about 60 xebecs of its own. Port Portinache, on the N.E., is a cove with 5 fathoms of water near its head, which is sheltered from the E. to N.W. round by the S. Porto Magno, or St. Anthony's bay, is on the W. coast, and runs in 2½ miles S.E., sheltered by the Cunillas islands, except from N.W. winds; hence in summer it is a safe anchorage for large ships, but not in winter, when N.W. winds prevail. Capes Grossa, Juen, Falcon, and others, are prominent from the coast, as well as the island Beder, close to Bedernella island, and the island of Tayomago. Formentera island is 2½ miles S. of Ivica, the passage between is called the Channel. It is 8 miles long, and of irregular breadth, having a pop. of 1,200. It has no good port, but small vessels anchor in the cove of Saona, at the W. end, which is only open between N.W. and S.W. Salt is made and exported from this island. The lesser islands round are three, Conejeras, Bleda, and Esparta, off the W. coast of Ivica, uninhabited, but affording pasturage. The Bledas are five islets W.S.W. 2½ miles from the Great Conejera. The Columbrettes are islets in the Gulf of Valencia, E.N.E. of Morviedra, 10 leagues from the coast, so called from their having been reported uninhabitable from serpents in the olden time; hence the Greeks called them Ophiusa, and the Romans Colubraria. On the N.E. side of the most N. is a bay affording shelter for small vessels.

BALEINES ROCKS, France, off the Isle of Ré, and dangerous, extending half a league from the S.W. side, and diminishing thence to the S.E. side of the island. There is an excellent light here, showing unequal flashes, visible 6 leagues, on the N.W. point of the Isle of Ré.

BALINA HEAD and COVE, North America, Newfoundland, between Cape Broyle and the Bay of Bulls; a mere fishing place.

BALINCAILACH, Scotland, a cape on the W. end of Benbecula island.

BALIZE PASS, North America, Gulf of Florida, one of the mouths of the Mississippi river most used by shipping, and protected at the entrance by a fort. There is a bar of mud across the river here, obliging the pilots to sound continually.

BALLA PORT, Africa, Great Scarcies river, N. of Sierra Leone.

BALLAGHEN, POINT, Ireland, E. coast, the S.W. entrance to Carlingford bay. Lat. 53° 58' N., long. 6° 4' W.

BALLANTREE, Scotland, Co. Ayr, near the mouth of the Stinchar, 24 miles S. of Ayr; pop. 1,506; a small port off the W. coast.

BALLARD, CAPE, Newfoundland, E. coast of Newfoundland, 4 leagues N.N.E. from Cape Race. Lat. 46° 46' 46" N., long. 52° 59' 11" W.

BALLARD, POINT, coast of Ireland; lat. 52° 42' N., long. 9° 32' W.

BALLENDEN, POINT, in the bottom of Donegal bay, on the N.W. coast of Ireland, N.W. by W. from Sligo harbour.

BALLIELA BAY, Ireland, 12 miles S.E. from the South Arran isles, on the W. coast. Lat. 52° 35' N., long. 9° 20' W.

BALLINAKILL HARBOUR, W. coast of Ireland, S.S.E. about 5 miles from Davillion island. The entrance on the starboard shore is free from rocks, the harbour well sheltered, and the water deep.

BALLINSKELLIGS BAY, Ireland, W. coast, N. by E. ¼ E. between Bolus head and Hog's head, exposed to the S.W. winds, and little frequented. Lat. 51° 42' N., long. 10° 6' W. High water 3ʰ 46ᵐ.

BALLYCASTLE BAY, Ireland, N., with a pier on the E. part. Lat. 55° 12' N., long. 6° 6' W.

BALLYCOTTON BAY and ISLAND, Ireland, S. coast, 6 miles S.W. from Cable island, the W. point of Youghal bay. Anchorage may be had in this bay with the wind from S.W. to N.N.E. by the N., anchoring with the Government house S.S.W. to S.W., and the outer island S.E. to S.S.E., in 3 fathoms sand and clay.

BALLYGELLY HEAD, Ireland, E. coast; lat. 54° 54' N., long. 5° 44' W.

BALLYHALBERT, Ireland, E. coast, near Donaghadee. Lat. 54° 29' 30" N., long. 5° 28' 10" W.

BALLYHAVEN, Ireland, within the entrance of Strongford haven, on the E. coast.

BALLYLEIGH, a name of Kerry head, S.W. of Ireland.

BALLYQUINTON, POINT, E. coast of Ireland, at the entrance into Strongford lough. Lat. 54° 19' N., long. 5° 26' W.

BALLYSHANNON HARBOUR, Ireland, N.W. coast; only adapted for small vessels, having only 3½ feet on the bar at low water. The access is dangerous with W. winds. The bar cannot safely be taken by vessels drawing more than 8 or 9 feet, and then at high water; the springs only rise 10 feet. High water 5ʰ 30ᵐ on the bar, at the Town quay 5ʰ 10ᵐ. Lat. 54° 30' 11" N.

BALLYTEIGH BAY, Ireland, near the S.E. extremity of the Irish coast, W. of Carnsore point. It is round the E. point of the entrance into Bannon bay, where is a good anchorage for small vessels. At the S. end is a small island called Inch island.

BALLYWATER, Ireland, E. coast, the mouth of the entrance of Carrickfergus bay, and the opening of Belfast river. Some pilots give the name to the sea S. along the E. coast of the peninsula, of which Strangford lough is the W. side.

BALLYWATER, Ireland, E. coast, the entrance into the Bay of Carrickfergus. Lat. 54° 42' 35" N.

BALTA SOUND, see the *Shetland Isles.*

BALTHSKOI, or PORT BALTIC, Russia, on the S. coast of the passage into the Gulf of Finland, 12 miles W. of Revel. See *Port Baltic.*

BALTIC SEA, Europe, E. of that portion of the North Atlantic called the German Ocean or North sea, the entrance being between Norway on the N. and the N. point of Jutland in Denmark, commonly called the Scaw, on the S., through the Sleeve or Skager Rack, the Cattegat, and Sound. The Baltic extends from the Sound and S.W. point of Sweden, along that country on the N. and on the W., or from about lat. 55° N. as far as Tornea, in Lapland, in lat. 65° 50' N. On the S. from Mechlenburg bay and the Gulf of Lubec, in lat. 54° N., along the shores of Pomerania and East and West Prussia, Samogitia, Courland, Livonia, as far as St. Petersburgh, and thence on the E. side, along the shore of Finland to Tornea. The northern portion is generally denominated the Gulf of Bothnia, proceeding N. from the Island of Aland as far as Tornea. On the E. side the Gulf of Finland runs up to St. Petersburgh. Through the straits between Dago island and the main, and Oesel island and the main, the Gulf of Livonia or of Riga opens, and following the coast southward, the Gulf of Dantzick and the Bight of Pomerska are found on the S. shores of West Prussia and Pomerania. Hanno bay is the only considerable bay on the Swedish shore. The Baltic contains numerous islands, with innumerable creeks and harbours. Its waters are fresher than those of the larger seas, having only from 1-30th to 1-40th of salt, owing to the discharge of the rivers it receives into the Cattegat through the sound between Denmark and Sweden, and the channels called the Belts, between Jutland on the W. and the islands of Funen and Zealand. There are no tides in this sea, unless when the N.W. wind impels the waters of the Atlantic into it. In winter much of it is frozen, owing to its great freshness, and then it is saltest. It has been supposed that this sea has been growing shallower for a long succession of years, but there does not appear to be any just ground for such an opinion. The entrance to the Baltic is difficult from shoals and variable currents. The entrance of the Sound is between masses of rock on both sides; the narrowest part of this entrance is 2,840 yards broad; it then widens, and between Copenhagen and Carlscrona is 6 or 7 leagues. The greatest depth where narrowest is 19 fathoms. The greatest length of the Baltic is 240 leagues, from Tornea to the Isle of Wollin. Between Aland island and the main the space is called by the Swedes Aland Hof or Aland sea. The greatest usual depth of the Baltic is about 50 fathoms, though one or two spots are said to be from 100 to 110. The bottom is very irregular and rocky. Some deny the rise of the Baltic waters to be owing to currents driven in by N.W. winds, and ascribe it to other causes. The currents in the Baltic are strong, caused by the rivers and winds, and are observed to be most prevalent in the Gulfs of Bothnia and Finland, that receive the more rapid rivers. There are superior and inferior currents in this sea. The irregular winds of the Baltic cause its waves to be short and confused, and consequently more disagreeable than those of the ocean. The needle was once said to lose its polarity near Hango island, at the entrance of the Gulf of Finland, in this sea. The rigour of the climate has greatly diminished. The Cattegat was formerly frozen over, as well as the Belts, a thing unheard of for a century and a half. In Uleaborg, the thermometer in the mean through the year is 29°, or 3° below freezing; at Stockholm it is 10° above. Hence the Gulf of Bothnia is still for the best part annually frozen. The rivers which fall into the Baltic are large and numerous. There are the Trave, the Wornow, and Peene, the Oder, Vistula, Elbing, Passage, Pregel, Niemen, Duna, Peman, Narrowa, Aa, Wuoxen, Kymena Kumo, Aleo, Kemi, and numerous others. Its trade is great, and considerable fleets annually pass out of its waters, besides the vessels employed in its great internal commerce. The Danes exact dues from all vessels passing through the Sound, although incapable of enforcing them, a vexatious tax upon the navigation of this sea, levied upon all nations, and wholly unfounded in reason or justice.

BALTIMORE, America, province of Maryland, the third city in population and fifth in commerce of the United States, on the N. side of the Patopsco river, 14 miles from its entrance into the Chesapeake bay. Lat. 39° 17′ N., long. 76° 36′ W. Pop. 102,213, of which 3,199 are slaves. The harbour consists of three parts; the entrance is 600 yards wide, with 22 feet water; the width increases to Fell's point, where it contracts to a quarter of a mile, which is the entrance to the second harbour; that has 12 feet water, widening to half a mile broad for the space of a mile to the third harbour, which has 10 feet water, and goes into the heart of the city. It is defended by a strong fort. Vessels of 200 tons lie up at the town, and those of 500 and 600 at Fell's point. The amount of tonnage belonging to the port in 1840 was 76,022.

BALTIMORE, TOWN and HARBOUR, in the E. part of Baltimore bay, the E. point of which bay is 5 miles from Cape Clear, the extreme of the S. coast of Ireland. Lat. 51° 18′ N., long. 9° 14′ W. The town is celebrated for the ruins of its fine abbey. High water 4ʰ 10ᵐ.

BALTRUM ISLAND, coast of East Friezland, 4 miles long and 1 mile broad. Lat. 53° 47′ N., long. 6° 56′ E.

BAMBOROUGH, a small town and harbour on the coast of Northumberland, 14 miles N. of Alnwick. Pop. 3,949. It has a castle on a high rock, said to have been built by King Ida in 560. It was bought in 1715 by Lord Crew, bishop of Durham, who left it for benevolent purposes to shipwrecked seamen. Fern island lies to the N.; the island is 2 miles from the castle, and there is a good channel, with 7 and 10 fathoms between them. Signals are used from the castle to indicate danger to any vessel appearing in distress. A gun is fired if a ship is stranded or wrecked near; two guns when behind or N. of the castle; three when S., to cause assistance to be given by the proper persons. In tempests two men patrol the coast from sunrise to sunset, to give an alarm of accident, and a premium is given for the first notice; that reward from midnight to 3 in the morning is doubled. Signals are made on Fern island to vessels in distress, that relief will be sent as soon as possible. In bad weather a flag is kept up, and a morning and evening gun fired, and a rocket thrown up until relief is sent. By these signals the Holy island fishermen can put off from the other islands, when no boat can leave the main for the breakers. Premiums are given to the first boats that put off, which are always to contain food and liquors. In fogs, a bell is rung on the S. turret, as a signal to the fishing boats, and a gun on the E. turret is fired every quarter of an hour,

as a signal to the ships without the islands. A large weathercock is placed on the top of a flag-staff, and a speaking trumpet is used when ships are near the shore or aground. On the E. turret is an observatory, whence a look-out is kept from sunrise to sunset in winter, to watch if ships want assistance, and all captains are requested to make the usual signals of distress in such cases. Rooms and beds are provided for shipwrecked seamen. Cellars for depositing property saved from wrecks, or portions of the wrecks themselves, of which an account is kept in a book for the purpose. All sorts of instruments are provided that can assist stranded vessels. Chains and machinery are ready for raising sunken vessels, to be lent gratis to persons wanting them within 50 or 60 miles along the coast, on security being given to return them to the trustees. Mooring chains are provided, and dead bodies cast on shore are put into coffins and buried free of cost.

BAN RIVER, Ireland, E. of Lough Foyle, and between that and the Skerries. It leads to Coleraine, and is only fit for small vessels, as there is but 3 feet on the bar at low water, 10 at springs, and 8 at high water neap tides. The depth within the bar is the same. Lat. 55° 10′ N., long. 5° 7′ W.

BANANA ISLAND, Africa, near the mouth of Sherbro river, not far from the Bengal rocks, 3 leagues from Cape Chilling, round which are several sunken rocks. This island is 4 miles long and 1 broad, and has two islets at its W.S.W. end, surrounded by rocks, and two small harbours on the E. The tides rise here at the equinoxes 9 or 10 feet, at other times a foot less. These islets are in lat. 8° 5′ N., long. 13° 15′ 12″ W., due S. from Sierra Leone. The highest peak on the large island lies in lat. 8° 5′ 48″ N., long. 13° 16′ 12″ W.

BANCHES VERTES, France, Bay of Biscay, N.N.W. of Roche Bonne, a shoal with a rocky bottom, dangerous in stormy weather. It has the isle of Ré on its W. side, 10 or 12 leagues. In some places there are only 2 or 3 fathoms water, in others 30.

BAND HAVEN, Ireland, a little S.W. of Port Rusch, in Skerries island, on the N. coast, 3 leagues S.E. by E. of the entrance of Lough Foyle. The harbour is only fit for boats.

BANDAL, France, 6 miles E.S.E. ¼ E. from Verte island, 4 leagues W. of Toulon, a port much visited by the small craft of the coast. But a small distance from it to the S. is Brusc road, 2 miles across, with deep water.

BANDOLS, France, department of the Var, 4 leagues W. of Toulon, a very small port, but much used by little vessels. Pop. 1,660.

BANDON RIVER, Ireland, S. coast, forming the harbour of Kinsale. The S. light here is in lat. 51° 36′ 45″ N., long. 8° 32′ W.

BANDT, coast of East Friezland, a small island in the German ocean. Lat. 53° 36′ N., long. 6° 33′ E.

BANES, Cuba, 11 miles from the Bay of Nipe, in lat. 20° 53′ 30″ N., long. 75° 34′ W. There is good anchorage for all classes of vessels. The entrance is in the midst of a bay formed by the coast, and there is nothing to fear but what is seen, but it is exceedingly tortuous; and it is difficult to leave because its mouth is presented to the sea-breeze.

BANFF, or BAMFF, Scotland, on the S. of Murray·Frith, at the mouth of the Deveron. The harbour is small and very indifferent, changing in storms with shifting sands. Lat. 57° 40′ N., long. 2° 12′ W.

BANGOR, a city of the United States, North America, at the head of the navigation, on the W. side of the Penobscot river, 60 miles from the sea, province of Maine. Pop. 8,627. The harbour is at and below the Kenduskeag river, is spacious, and the tide rises in it 17 feet. Its principal trade is lumber, employing 1,200 vessels of above 100 tons. The river is open 8 months in the year.

BANGOR BAY, North Wales, more correctly, perhaps, Beaumaris bay, leading up to the Menai straits, and having a large extent of sand E., called the Lavan sands. The town of Bangor is situated on the E. side. See *Menai Straits*. The harbour here, where principally slate is shipped, has been artificially improved. Pop. 4,751. Lat. 53° 12′ N., long. 4° 12′ W.

BANGOR CASTLE, S. side of Carrickfergus bay, N.E. coast of Ireland, a bay and town with a little pier, fit only for small craft. Lat. 54° 39′ 20″ N., long. 5° 42′ W.

BANISTER'S BAY, West Indies, Jamaica, S.W. side. Lat. 18° 25′ N., long. 78° W. The coast to the N. is rocky.

BANISTER'S KAY, or SHOAL, West Indies, St. Domingo, in the Bay of Samana, at the N.E. end, where it bears W.N.W. a mile; a vessel may safely anchor in 14 fathoms. Samana cape is in lat. 19° 15′ 40″ N., long. 69° 6′ 15″ W.

BANJAARD, THE, an extensive flat, lying off the mouth of the East Scheldt, and W. of the W. end of the Island of Schonen. Some parts are dry at low water, but there are swash-ways through it, as well as a channel of 3 and 3½ fathoms, between the Banjaard and Schonen or Schowen. It is also separated nearly in the middle by a channel, called the Middle or West Gat, within which runs the passage to the S. of West Schonen, and to the N. of Neeltje Jans and Rug Plaat, called the Hammon channel. The S.W. of the Banjaard is a triangular sand, 5½ miles in length, the W. point of which stretches towards the Room Pot, and has a buoy on it, with 1¼ fathom water. The N. point extends to the Middle Gat, and has 2 fathoms over its extreme end. The N.E. of the Banjaard is a mile from the Noord Land, and extends 4½ miles E.N.E as far as the entrance of Brouwershaven, running out to seaward full 6 miles. The S. side of this flat has three black buoys along its edge. The outer one in 4½ fathoms, 1¼ mile from the shore, with East Kapelle steeple bearing about S. ½ W. The second and third lie in a direction E. by S. from it; the second 2¼ miles from the outer buoy, and the third about 4¾ miles from it. Midway between the two last mentioned buoys, on the opposite side of the channel, there is a red buoy on the W. extremity of the Onrust, a large bank, projecting 2½ miles from the W. end of N. Beveland, separating the Room Pot from the Veer Gat. This buoy is rather more than one-third of a mile from the Walcheren shore.

BANKS, HOW, Isle of Man, the N.E. point of Douglas bay, whence the coast runs N.N.E. 2 miles to Day head.

BANNE, France, department of Finisterre, a small island a league S.W. of Ushant.

BANNON BAY, Ireland, see *Ballyteigh Bay*.

BANNON, PORT, Ireland, S.E. coast, 11 miles E.S.E. from Waterford; lat. 52° 12′ N., long. 6° 50′ W.

BANOS, Spain, S. coast, W. of Almeria, near the Point of Moro, a small bay or haven, shallow, with sand banks, from the E. point of which a ledge of rocks runs out S. half a mile.

BANTRY BAY, N. of Dunmanus bay, on the S.W. coast of Ireland, running inland N.E. for more than 10 leagues, from 8 to 3 miles wide; lat. 51° 36′ N., long. 9° 25′ W. It is large, safe, and commodious for all classes of vessels. Hungry hill, on the N., 2,160 feet high, with a pyramid at the top, is a conspicuous land-mark. At the bottom are two anchorages; that on the S., within Whiddy island, commonly called Bantry harbour, and Glengariff harbour on the N., opposite Whiddy island, with a narrow entrance, having 6 fathoms. To go in, keep the E. shore, to avoid some rocks off the island, then anchor off the town in 3 and 5 fathoms. In summer the largest ships may ride without the island, at the harbour's mouth, in 7 or 8 fathoms, good ground. The pyramid on Hungry hill is in lat. 51° 41′ 5″ N., long. 9° 47′ 6″ W. Bear island signal tower is in lat. 51° 37′ 43″ N., long. 9° 53′ 40″ W. BEAR HAVEN harbour is good, safe, and has water for the largest ships, with two entrances, one E., the other W. of Bear island. The E. entrance is the safest anchorage, and from 5 to 11 fathoms are found on the N. side of the island; it is an excellent rendezvous for a fleet. The tides on this part of the coast flow until ¾ past 3, with some variations to 4; rise 12 feet.

BANTRY, African coast, a Dutch trading fort, between Dixcove and St. John's river.

BANY RIVER, S.W. coast of Africa; there are shoals and sands about the mouth, whence the practice has been to run E. nearly to Andony river, and then keep in 6 fathoms by the shore, running nearly into 4 fathoms, and passing through Point Cape; round the point there is anchorage in 10 and 12 fathoms, safe from E. and S. winds. The coast runs nearly E. and W. here from Cape Formosa.

BANYAN POINTS, *see Gambia*.

BAR HAVEN, coast of Nova Scotia, North America, 3 leagues S.W. from Cape Canso.

BARABALEMI RIVER, African coast, 6 leagues E. from the river St. Barbara, and E. of Cape Fermoso. At the distance of 2 or 3 leagues from the shore, vessels can pass along in 8 fathoms. It was once known by two stakes on its E. side, like masts of ships.

BARACOA, Cuba, S.E. coast, a harbour in the bay, formed between the points of Majano and of Baracoa, on the W. of which is the mouth of the harbour, in lat. 20° 21′ 36″ N., long. 74° 29′ 31″ W. It is secure and well sheltered.

BARADAIRE'S BAY, Hayti, close to Roitelet's point, at the W. end of the island; lat. 18° 42′ N., long. 73° 37′ W.

BARATARRIA, PORT, near the mouths of the Mississippi, Gulf of Mexico, the entrance of a large lake, communicating with that river by two creeks, in the rainy season of great depth; 22 miles from the Isle of Timbalie, N.E.

BARATTO, POINT, Italy, about 11 leagues S. by E. ½ E. from the lighthouse of Leghorn. It is a high point, on which is a small town.

BARBADOES, West Indies, 17 miles long and 12 broad; the lat. of Bridgetown, the capital, is 13° 5′ 30″ N., long. 59° 37′ 35″ W., and contains a pop. of 102,000. Its principal exports are rum, sugar, coffee, and cotton. It is the most E. of the Caribbee islands, and may be seen 12 leagues off at sea. In making the island from the E., it is always advisable to get into the lat. of 12° 56′ N., and make the S. end, the eastern side being rocky. From the E. to the S. point, a reef runs along nearly 2 miles from the shore, called the Cobbler, a remarkable rock forming one of the number. These rocks terminate in a flat spit, running S.S.W. 2½ miles; ample berth must be given them, especially at night. The E. end will first appear to be lower than any other part of the island, but the N. end being brought to bear W. by N., and the S. point nearly W.S.W., it will then appear the highest part. The S. point has been mistaken in the night for Needham's point and fort, which, coming from the E., do not appear until the spit is rounded, when the extreme E. point of the island bears N.N.E., and the S. point distant about 3 miles, Needham's point and fort will then bear N.W. by W. distant 5 miles. Bridge-town, in Carlisle bay, is the best harbour on the S.W. part of the island, open to the W., and secure from the N.E. wind, except from the latter end of June to August, the hurricane season. The highest part of the island is not more than 1,100 feet above the sea. The island is divided into 11 parishes, which have an area of 162 square miles. The whites number about 14,000, and the coloured 7,000; the remainder are blacks. The value of the exports is about 460,000l. per annum. The revenue 16,300l., and expenditure 18,565l. The temperature ranges from 72° to 89°, the monthly mean being about 80·25°. This is the most healthy island of the Caribbees, and was first settled in 1605.

BARBANO, Gulf of Venice, a small island in the N. part, on the coast of Friuli. Lat. 45° 45′ N., long. 31° 14′ E.

BARBANOLA, CAPE, Levant, the S. point of the Gulf of Smyrna, 9 leagues S. by W. from Porto Gero.

BARBARY, POINT, Africa, the W. point of the river Senegal, lat. 15° 38′ N., long. 16° 32′ 10″ W.

BARBARY, Africa, a large region bordering upon the ocean on the W., along the whole extent of the Mediterranean on the N., as far as Egypt, being nearly 2,200 miles long by 500 broad. It contains the countries known in modern times by the names of Barca, Tripoli, Tunis, Algiers, Fez, Morocco, Tafilet, and Biledulgerid; but anciently as Mauritania, Numidia, Africa Proper, and Lybia.

BARBAS, CAPE, Africa, 26 leagues N. from Cape Blanco, lat. 22° 15′ N., long. 16° 39′ 12″ W.

BARBEL and COCKLE SANDS, on the W. side of the Cockle Gat; the Scroby sand, on which are two white and one red beacon buoy, forms the E. side; the channel being three-quarters of a mile wide. It forms a passage often used out of Yar-mouth roads.

BARBUDA, West Indies, one of the Caribbee isles, about 15 leagues N. from Antigua. The island is low and flat, with a sandy shore; it is in-habited by about 1,600 persons, and is very fertile. Its greatest extent is from N.W. to S.E., about 15 miles. Its highest land cannot be seen further than 6 leagues. Its shores are rocky, foul, and dangerous, and the currents round it very deceiving. A ship may have, near the coast, 4 or 5 fathoms astern, and 50 or 60 under her bow. On the S.E. a reef extends full 12 miles into the sea, and rocky soundings half way to Antigua. On the N.W. side, reefs stretch out for 7 miles, on which have been several wrecks. On the W. side there is a road with good anchorage in 9, 12, and 14 fathoms, or within the reef 3½, 4 miles from the S.W., or Palmeta point. The town and castle are on the W. side of the island, on the edge of a beautiful lagoon;

the castle is in lat. 17° 45′ 24″ N. The best an-chorage is off the fort, at the S.W. of the island, called the River, the fort N., in 7 and 8 fathoms, sand and mud, 1¼ mile from the shore. The island is 18 leagues from St. Kitt's, in 17° 50′ N. lat., and 61° 50′ W. long.

BARCELONA, Spain, the capital of Catalonia, with a pop. of 150,000, including Barcelonetta. It has a large export trade in a great variety of articles of home production, and imports from all the European countries. It is strongly fortified. It may be known from a great distance by the hill Montjuic, and its fortress S.W. of the city, which is in lat. 41° 22′ 51″ N., long. 2° 9′ E. The lighthouse of Barcelona bears from it N.E. ¼ E., the river Llobregat S. ¾ W., and Fell's castle W.S.W. ¾ S. It has a mole running out from its E. side in a S. direction; the mole has a light-house and batteries erected upon it. Merchant vessels may lie secure within it from the S.E. winds. From Barcelona the coast is low, and runs N.N.E. ½ E. to the little river Besos, beyond which is the town of Badalona, with the fortress of Mongat built upon it, 7½ miles from Barcelona. Further on is Blanes, near which vessels may shelter in 7 or 8 fathoms, S. of Blanes, but near the S. part of St. Anne's point, on which stands a convent; on the S. part of this point is a dangerous sunken rock.

BARCELONA, in the department of the Ori-noco, Colombia, S. America, situate near the sea, in a plain; it has a considerable trade, and a population of about 13,000. The Morro is in lat. 10° 13′ 15″ N., long. 64° 43′ 45″ W. The Morro is high land, lying N. and S., about a mile in extent. The W. coast is foul, and should not be passed nearer than two cables' length.

BARDSEY ISLAND, Caernarvonshire, North Wales, N.E. ¾ E. 70 miles from the Smalls' light-house, about E. ½ N. 20 leagues from the Tuskar light, and W. ¾ S. from Pen y Kill, the S.W. point of Aberderon bay, separated from the main by a channel 1¾ mile wide, having from 15 to 25 fa-thoms. On the S.W. point of this island is a bright fixed light, visible 20 miles off, building white. The island is the N. point of Cardigan bay, lat. 52° 58′ N., long. 5° 5′ W.

BARFLEUR, France, department of La Manche, a mere fishing port, close to Cape Barfleur, which has a high lighthouse, in lat. 49° 41′ 52″ N., long. 2° 15′ 48″ W. It has an intermittent light, 236 feet above high water; a flash appears every half mi-nute, visible 8 leagues off in clear weather, but not wholly disappearing when within 4 leagues. The harbour, dry at low water, will only admit a draught of 9 or 10 feet. The flood tide sets directly here for the mouth of the Seine. There are two small fixed harbour lights, 33 feet high, visible about 5 miles off. Cape Barfleur should not be approached nearer than from 25 to 22 fathoms. High water 8ʰ 45ᵐ. The N. jetty head light is in lat. 49° 40′ 14″ N., long. 1° 15′ 26″ W.

BARGAZAR, POINT, coast of Iceland, lat. 66° 20′ N., long. 16° 37′ W.

BARI, Italy, 10¾ miles from Mola, on a low shore, having a castle and two churches, by which the city may be known. In sailing along, the castle is shut in with the churches. Pop. 30,000. The road is good; large ships may ride in 16 and 18 fathoms water, good ground; but nearer the shore foul and rocky. There is a haven before the city for small vessels, at the mouth of which are rocks.

BARIEL, or BARNACLE, POINT, the S.E.

limit of Winthorp's bay, on the N.E. coast of Antigua island, West Indies, on the W. side of the entrance into Parham harbour.

BARIZO, Spain, a harbour for coasting vessels, near Cape Villano, and behind Point Nerija; the W. shore should be kept here, the harbour getting shallow upwards.

BARKING CREEK, Essex, river Thames, by which small vessels pass up to the town of Barking. It is in Woolwich reach.

BARLETTA, Italy, Adriatic, once a large city, now in a ruinous state. Here is a mole for small vessels, with a breakwater and a lighthouse, low and neglected. The anchorage in the road is good with off-shore winds, at from 1 to 3 miles N.N.W. ¼ W. from the lighthouse, in from 7 to 13 fathoms.

BARLEY COVE, S. coast of Ireland, a small haven on the E. side of Mizen head, in lat 51° 24′ N., long. 9° 40′ W.

BARMOUTH, North Wales, Merionethshire, 11 miles N. of Aberdovy, with the shoal called Sarn y Bwch between them. It is only fit for vessels drawing 9 or 10 feet water; these must sail over the bar at spring tides. There are two channels, the S. is the narrowest and best. Off the bar, at 1¼ mile from the town, is a black fair-way buoy, and inside, near the beacon porch, a red one. The last is to be kept on the left running up for the harbour; 10 feet is all the water to be relied upon for spring tides, as there is always a great swell. High water 8^h, rise 14 feet. Pop. of town, 1,450. Lat. 52° 38′ N., long. 4° 10′ W.

BARNABY ISLE, in the St. Lawrence, North America. The N.E. point is in lat. 48° 29′ 43″ N., long. 68° 35′ 2″ W.

BARNACK, Ireland, a small island in Black Sod bay, near the W. coast.

BARNARD SAND, between Orfordness and Yarmouth; it commences off Covehithe-Ness and extends N.E. ½ N. 2½ miles. It is joined to the Newcome by a ridge. There is a narrow buoyed passage between Covehithe point and the S. end of Barnard sand.

BARNET'S REEF, Musquito shore, Gulf of Mexico, 17½ miles N.N.E. ½ E. from Cape Gracios a Dios.

BARN POOL, in the port of Plymouth, W. of Drake's island, and under Mount Edgecombe.

BARNIRE, Guernsey, a dangerous rock, E. by S. from the ledge called Rock Dove, 7 leagues W.S.W. from Guernsey island, dry at low water.

BARNSTAPLE BAY, Devonshire; it is common to Barnstaple and Bideford rivers, the N. point of which, called Morte Point, bears from Hartland point E. by N. 15 miles. No vessel should enter this bay unless bound to Clovelly or Bideford bar, since there is great danger of being embayed, or driven on shore by sudden gales from the N.W., and there is no escape with the wind between W.N.W. and N.E. Barnstaple bar, continually shifting, is E. ¾ S. 11 miles from Hartland point, and vessels drawing more than 12 feet water should not attempt the passage. Two beacons have been erected to guide vessels, consisting of two fixed lights, near high water, made upon going in on the larboard side, their distance apart being 1,000 feet. The upper or inner lighthouse is a white tower 74 feet high, and the light being 88 feet above high water, may be seen 5 leagues off. The lower lighthouse is a square wooden building, white, and moveable, to keep the line of the two lights right as the sands shift. Signals are made indicating the depth of water. A passage must not be attempted earlier than a quarter flood, as in fine weather there is a cross breaking sea. The lights must be kept in a line, and in desperate cases, run on to with 200 fathoms of the outer light, then opening the high light to the W. of the low light, haul over to the S., and act as circumstances show to be best, going in as far as possible. None ought to proceed further without a pilot, except vessels well accustomed to the navigation. At Barnstaple it is high water at $6^h\ 30^m$, but in the bay at $5^h\ 30^m$, rise 25 feet. Barnstaple has a pop. of 7,000, Bideford of about 5,000. Barnstaple is in lat. 51° 15′ N., 4° 5′ W. The buoys here were last altered in 1844, by authority of the Trinity House, from changes in the sands.

BARNSTAPLE BAY, North America, S. part of Massachusetts bay, across an isthmus N.E. from Buzzard bay. The mouth of the bay has a bar, with 6 or 7 feet water. The town of the same name in the bay, capital of Barnstaple county, has a pop. of 4,301. Lat. 41° 36′ N., long. 52° 30′ W. Barnstaple town court-house is in lat. 41° 42′ 7″ N., long. 52° 30′ W.

BAROCHES, British channel, two reefs close by the W. end of Alderney, in the direction of the Caskets.

BARQUERO, Spain, N. coast, a river between Point Ventose and Cape Vares, where there is clean anchorage, safe from S.E., S.W., and N.W. winds. The sea only becomes heavy with N.E. gales. Cape Vares rounds to Point De la Estaca, the most N. land of Spain, W.N.W. ½ W. 7½ miles from which lie the large rocks called Aguillones, and W. of these is Cape Ortegal. Point De la Estaca is in lat. 43° 47′ 50″ N., long. 7° 38′ 50″ W.

BARRA, one of the Hebrides, S. of South Uist, of an irregular form, about 4 miles across. There are several small islands round it, as the Isle of Fiarray, joined to Fuddia by a ridge of sand, dry at low water, with a passage for small craft. Vatursay lies S. of Barra, and near it are the islets of Sanderay, Pabbay, Mingalay, and Bernera. A lighthouse is erected on Barra head, or Bernera, or Long island, in lat. 56° 48′ 0″ N., long. 7° 53′ W.; it is intermittent, appearing like a star of the first magnitude for 2½ minutes, and suddenly eclipsed for half a minute. It is 680 feet above the common sea level, and is open from N. by E. to E.N.E., and intermediate points W. and S., visible 32 miles.

BARRA, POINT, see *Gambia.*

BARRACAPE, W. coast of Africa, 7 leagues E.S.E. from Cape St. Mary. Vessels may ride off except when a S. wind blows.

BARRACOE, sometimes called BERKER, W. coast of Africa, 6 or 7 leagues W.S.W. from Accra, easily known off at sea by two very high mountains behind it, one of which is double at the top. Rocks lie off at sea before it, and form a species of haven. The point is in lat. 5° 29′ N., long. 24° W.

BARRACONE'S BAY, Trinidad, nearly in the centre of the Gulf of Paria.

BARRAS NOSE, Cornwall, a point E. of Tintagel head.

BARREE BAY, or BAXA TERRA, Africa, W. coast, 4 leagues S. of the river Ouro, in lat. 23° 30′ N. It is large, and has good anchorage on the N. side.

BARRELS, or BARRELS AND HATS, South Wales, off the coast of Pembrokeshire, rocks situated between the Smalls and Grassholm. Three of these rocks on the Barrels' side are a little dry after half tide. They lie W. ¾ N. from the centre of

Grassholm, distant exactly 3 miles. Though shown by the breaking sea, the Barrels must be approached with caution. Near high water they can scarcely be seen. The tide sets very strongly over them, especially at spring tides. They bear from the Smalls S.E. by E. 4½ miles. High water at springs half-past 7.

BARRET BANK, France, at the S. end of the isle of Oleron, forming the N.W. side of the Manmuson passage, as Point de Gardour on the main land forms the S.E. side.

BARRETINI ISLANDS or ISLETS, Sardinia, on the N. of the island, near those called the Barolinos, which partake of the same character, and are close to the Strait of Bonifacio.

BARROW, CAPE, Africa, the S. point of the isle of Fernando Po.

BARRY ISLAND, South Wales, the most W. of two islands off Cardiff point, on the coast of Glamorganshire.

BARSOUND, in the Baltic, 15 leagues N. by W. from the N. end of Oeland island, and 9 leagues from the Westerwyk channel; beset with rocks, so that pilots must be taken for the passage at Oeland.

BARSSALLACH, POINT, Luce bay, Scotland, near the S.W. end, 8 miles N. of Burrowhead.

BARTHOLOMEW BAY, Kent, just without the North Foreland, between that and Ramsgate.

BARTHOLOMEW LEDGE, see *Scilly Isles.*

BARTHOLOMEW, SAINT, West Indies, one of the Caribbees, 8 leagues N. of St. Kitt's; the E. point in lat. 17° 53' N., long. 62° 52' W. It has numerous rocks and rocky islets around its shores. From E. to W. it is about 6 miles long by 2 broad; the interior high, the soil dry, and the shores level; the pop. about 8,000. The N. shores are particularly dangerous, and must be navigated with a pilot. On the W. side there is an excellent harbour, called Le Carenage, where ships of any size may lie in good holding ground not far from a town called Gustavia. The course from the W. point of St. Kitt's should be nearly N. The white inhabitants are principally Swedes. Vessels bound to Gustavia may lie on and off the harbour's mouth from sunrise to sunset, but a permit must be obtained to do so from the commandant of the port. The customary fee is two dollars for a ship, descending intermediately to six reals for a small vessel. The flood tide at full and change runs S.E., where it is high water at half-past 10. The islets around are Cocoa kay, Sugar Loaf, Grenadier's, the Frigate, Bon Homme, Boulanger, La Fourchu, Table, Gnooper rocks, and Le Bœuf. There are passages between them all; bring steep to, with from 15 to 20 fathoms close. In sailing from Gustavia there are two passages among the islets, called by the Dutch the Long and Short passages.

BARTON, Lincolnshire, a little creek of the river Humber, from which the packets pass and repass to Hull. Lat. 53° 42' N., long. 0° 22' W.

BARTRACK ISLAND, Ireland, in Killala bay, 2 miles N.E. of Killala, about 2 miles long and half a mile broad.

BAS DE LIS, on the coast of France, a dangerous rock between the Bec du Raz and St. Matthew's point, 5½ leagues to the S. ¼ E., nearly in a fair way but not in a right line with it.

BAS, ISLE DE, France, on the N. side of the department of Cape Finisterre, about 1½ leagues long by three-quarters wide, surrounded by formidable rocks, which render approach very difficult, and defended by two forts and five batteries.

The soil is sandy. Its inhabitants, 850 in number, are nearly all mariners. It has but one fresh water spring, which is covered every tide. Lat. 48° 44' 45" N., long. 4° 1' 29" W. The rugged rock of Morlaix bears from this island E. by S. and E.S.E. between 3 and 4 miles. It has a town called Roscrow. It is 4 miles S.W. by S. from the W. end of the island to the river Ploughgoulin. The Morlion's rocks are on the E. side. There is a lighthouse on this island, 223 feet above high water, having an intermittent bright minute light.

BAS, France, department of the Seine Inférieure, a small town with 3,200 inhabitants, on the sea shore, close to the salt marshes, and possessing a small port, very favourable for fishing. The French spell both this place and the Isle de Bas as Batz.

BASILUZZO, coast of Italy, one of the Lipari islands. See *Lipari.*

BASQUE ROAD, W. coast of France, S.E. from the Isle of Ré, N.E. from Oleron, N. from the Isle of Aix, S. from the W. point of entrance into Rochelle, and directly W. with the bay of Chatetailon. High water 3ʰ 45ᵐ; rise 20 feet. See also *Pertuis d'Antioche, Bec du Raz, &c.*

BASS, or BIAS BANK, South Wales, 3 miles from St. David's head, extending E.N.E. and W.S.W. about 5 miles, and a quarter of a mile broad. Over one part of it the sea sets very heavily.

BASS ISLAND, a rock off the coast of Scotland, insulated near the entrance of the Frith of Forth, shaped like a sugar-loaf on the S. side, overhanging on the opposite, about a mile in circuit. It is perforated by a cavern. It is N.W. 3 leagues from St. Abb's head. Lat. 56° 3' N., long. 2° 35' W.

BASSA, African coast; lat. 5° 54' 50" N., long. 10° 4' 5" W., and E. from Cape Monserrado; two native towns. There is good anchorage in 12 fathoms, muddy bottom, off Grand Bassa, about 2 miles down the coast from the river of St. John's.

BASSAM, GRAND, Africa, Gold coast; a town distant from the river Assinee about 6½ leagues.

BASSAMUER ROCK, France, N. coast, on a shoal, a league N. by W. from La Clarte church, near the point so called, S. from the Seven islands. 6 leagues W. from Brehat island.

BASSE DE CHATS, a bank near the Isle de Groa. See *Isle de Groa.*

BASSETERRE, Guadaloupe, West Indies, applied to the S.W. part of the island, separated from the other by Salt river. It is a general name for low land, and is applied equally to the N.W. part of Martinique.

BASSETERRE ROAD, St. Kitt's, West Indies, on the S.W. of the island, towards its E. end. The best anchorage is in from 9 to 7 fathoms, the bottom coarse sand; the Half-moon battery N.W. by W., Basseterre town N.N.W., and Fashion fort N.E., on with Londonderry fort, at the E. end of the town, distant half a mile, and the W. point of the bay W. by N. There is good anchorage in 10 fathoms, sandy bottom, with the Nag's head in one with the S. point of the mountain of Nevis. There are lat. and long. as St. Kitt's.

BASSIN'S RIVER, Labrador coast, North America, nearly opposite the N. point of Newfoundland.

BASSURE SAND, France, Ambleteuse, a little S. of St. John's road, stretching out to the S.W., with 7, 6, and 4 fathoms upon it, though near it are places with 12, 15, and 25 fathoms; the deepest water is on the N. side. See *Bassurelle.*

BASSURELLE BANK, France, S. of the Quemer sand, or outermost of the two sands, 4 leagues N.W. from the river Somme. The Bassure sand is the most N. of the two. There are but 2 fathoms water on the middle of the Bassurelle bank, which is near 2 leagues W. from the S. point of the Somme.

BAST ISLAND, coast of Norway, 5 leagues N.W. by W. from Sister's island, and 5 leagues S.S.W. from Point Roge.

BASTAVOE BAY, Yell, on the E. side. See *Shetland Isles.*

BASTIA, Corsica, the capital, and a small port, adapted only for vessels of a shallow draught of water. The entrance is narrow and difficult; it is also exposed to the land winds, but with proper precautions it may be safely used, fastening a cable to some of the stone pillars erected on shore for that purpose.

BASTIMENTOS ISLANDS, North America, Gulf of Mexico, near the Isthmus of Darien, and W. of the Samballos islands, at the mouth of the Bay of Nombre de Dios. Lat. 9° 30′ N., long. 79° 45′ W.

BATA, a river on the African coast, falling into the Bight of Biafra. Lat. 1° 50′ N.

BATH, United States of America, on the W. bank of the Kennebec, 12 miles from the sea. It is one of the principal places of commerce in the state of Maine, in Lincoln co., and much ship-building is carried on there. Pop. 5,141. It is 165 miles from Boston. N. lat. 43° 49′.

BATHURST TOWN, Africa, river Gambia, in lat. 13° 28′ N., long. 16° 35′ 18″ W.

BATORIM, in the Euxine, a large white cliff at the base of high mountains, 63 miles E.N.E. ¾ N. from Rizeh. The anchorage is spacious and good, with 15 and 18 fathoms, deep enough for a line-of-battle ship to lie near the shore. The anchors should be ashore on account of the land breeze.

BATSTA BAY, Sweden, having a very small town of that name upon its shore, in the province of Warberg.

BATTEN MOUNT, the S. point of the harbour of Catwater, within the port of Plymouth, on the E. side of the sound, the citadel hill forming the N. point.

BATTENSTEIN, African coast, a Dutch fort about a league E. of Dixcove, near Antem river.

BATTERBY BAY, W. coast of Ireland, 2 miles N.E. from Convict islands, within Slyne head, on the S. side, and nearly E. from it, within some small islands on the N. shore of the coast, trending towards Galway bay. Lat. 53° 19′ N., long. 10° 22′ W.

BATTLE ISLES, Labrador; lat. 52° 15′ 44″ N., long. 55° 35′ 19″ W., the S.E. isle.

BATTOA, or BATTOW, Africa, W. coast, S.E. from Cape Cavallos, near the W. point of a river, opposite to Zino or Suino on the E. The river entrance has two rocks under water, half a league E. of the E. point, on which the breakers may be seen approaching a league off. Lat. about 5° 5′ N., long. 8° 30′ W.

BAUDSEY HAVEN, England, Suffolk coast, at the mouth of the river Deben, a league E. of Languard fort, on the E. side of the Orwell and Stour.

BAUDSEY SAND, Suffolk, the middlemost of three sands, nearly parallel to each other in the direction of N.E. Its S.W. end is nearly E.S.E. from the point of land at Baudsey, to the N.E. without the haven. When this cliff or point is brought to bear about N.W., that part of the sea is called the Stedway, and has the S.W. ends of Baudsey sand and Sheepwash sand N.E.; a ship must come no nearer than 6 fathoms. The S.W. end of Baudsey sand is 4 or 5 miles from the nearest shore, within which is Horsley bay, and the N.E. end about 7 or 8 miles from Orford haven at E.S.E.

BAULD, CAPE, Newfoundland, lat. 51° 38′ 10″ N., long. 55° 28′ 18″ W.

BAUME BAY, Hayti, N. coast, 6 leagues W. of Old Cape François, a good shelter for vessels, in lat. 19° 42′ N., long. 69° 35′ W. It is sometimes called Balsam bay.

BAXO NEGRO, off the W. coast of Porto Rico, West Indies, or the Black shoal. It is 3¼ miles from the nearest coast, S. ¾ E. from Punta de Figuero. Baxo Gallado is 6½ miles distant from the Punta de Malones, on the same coast; it is 3 cables' length in extent, and has 3 fathoms water, with a rocky bottom. It lies S. by E. ¾ E. from Desecho island, which last is in lat. 18° 23′ 48″ N., long. 67° 27′ 40″ W.

BAXO NUEVO, or the NEW SHOAL, off the Mosquito shore, said to be 7 miles from N. to S., and 14 from E. to W. All along the E. side is a steep reef. On one part is a rock with only 7 feet water, and another with only 4. Lat. 15° 53′ N., long. 78° 38′ 30″ W.

BAXOS, BAXIOS, sometimes corrupted to Bassas, low shoals, reefs, or banks, from the Spanish tongue.

BAXOS CAPE, or LOW CAPE, Africa, W. of Cape Formosa and Palmas island. Also E. from the river Volta about 12 leagues, by the turning of the coast, and 2 leagues from the Quatre Montes or Four hills.

BAY OF BULLS, Spain, N. of Cadiz, within Rota point, a good shelter from the N.W. to N.E., about N.N.W. from Cadiz bay.

BAY OF GOOD FORTUNE, North America, on the N. coast of Chaleur bay, in the Gulf of St. Lawrence, and on the N.E. coast of Nova Scotia.

BAY OF ISLANDS, Gulf of Newfoundland, on the W. coast, about 8 or 9 leagues to the S.W. from Belle bay, and 14 or 15 N.N.E. from Porta port, S. of Cape Gregory. Lat. 49° 6′ 12″ N., long. 58° 15′ W.

BAY OF ISLANDS, North America, Nova Scotia, 6 leagues S.W. from Cape St. Mary, all the way to Chibonectu bay, for 15 or 16 leagues to the S.W., being lined with islands. Cape St. Mary is in lat. 44° 5′ N., long. 66° 15′ 48″ W.

BAY OF SEVEN ISLANDS, North America, W.N.W. from Moisie river, and 18 leagues N.N.E. from Trinity point, on the N. shore of the Gulf of St. Lawrence. The entrance of the Moisie river is in lat. 50° 11′ 24″ N., long. 66° 7′ 41″ W.

BAY OF ST. LOUIS, North America, Labrador, having Cape St. Louis N. and Cape Charles S. The centre of the bay is in lat. 52° 23′ N., long. 55° 23′ W.

BAYA, or BAHIA, joined to Honda, deep, in many places, as Bahia Honda, a deep bay, from the Spanish and Portuguese.

BAYA, African coast, W.S.W. from the river Volta. It is low, marshy land, scarcely inhabited.

BAYAHA, or PORT DAUPHIN, HARBOUR OF, Hayti, N. side; a fine port, with clay bottom and 12 fathoms; the entrance is narrow, too narrow to be of extensive service. The tide flows at 7 A.M. Springs rise 5½ feet, ordinary tides 3½.

BAYAMO, or PORT ST. SALVADOR, Cuba,

West Indies, at the mouth of a river on the S. coast. It is the name of a channel between the main and the islands called the Queen's Gardens. Lat. 16° 5′ N., long. 76° 46′ W.

BAYAMOS, the name of certain gusts of wind that blow from the land on the S. side of Cuba, West Indies, so named from being more severely felt off the Bight of Bayamo or Buena Esperanza, than off any other part of the coast. Their approach is preceded by heavy and dense clouds seen over the mountains; thunder, too, generally precedes these gusts, on hearing the most distant sound of which, all sail should be taken in as fast as possible. Sheet and forked lightnings follow, of a blue indescribable colour. The sea is covered with foam, and the rain falls in torrents, unequalled on any other occasion.

BAYASSA, once Payas, Asia, a town of Syria, at the bottom of the bay of Iskanderoon, fallen into decay, and little more than a heap of ruins, in the N.E. corner of the Mediterranean.

BAYENETTE, CAPE, West Indies, Hayti, W. ¼ S. 5 leagues from Cape Jacquemal. The anchorage is on the N. side, but not sheltered.

BAYONA, Spain, S. of Vigo bay, a fortified town, with a castle at the foot of a mountain; the harbour is small and full of shoals. The islets of Bayona are in the same bay before Vigo town.

BAYONNE, France, department of the Basses Pyrénées, a city and port, in lat. 43° 29′ 26″ N., long. 1° 27′ 44″ W. High water 3ʰ 30ᵐ. It is situated at the conflux of the rivers Adour and Nive, about a league from the entrance or bar of the former river. Pop. 13,500. It is divided into Great Bayonne, Little Bayonne, and the Faubourg St. Esprit; the first situated on the left bank of the Nive, the second between that river and the Adour, and the third on the right bank of the last river, communicating with the rest of the city by a long wooden bridge. It has some trade, but its port is difficult to enter, although very secure. It is fortified, and has a strong citadel. The Adour is navigable for 50 miles, and the river 18, and the timber and tar of the Pyrenees is brought down by these rivers to the city. The bar of the Adour shifts continually, and has not more than 5 or 6 feet at low water. These changes are watched, and a moveable flagstaff adjusted by the chief pilot accordingly. Boats are always ready to aid vessels over the bar, but if the sea breaks much they cannot venture out. If a vessel draws no more than 14 feet, the tide rising 16 at spring tides, a square tower with a flagstaff upon it will be seen, and near it, on the shore, the moveable flagstaff. Bring the staff on the sand to bear S.S.E. ¼ E., then steer E.S.E. ⅛ S. between the two piers or entrances of the river; having passed them, all danger is over. Three miles S.W. from the bar is Biaritz, or Cape St. Martin, on which a lighthouse is erected, with a revolving light, the flashes succeeding each other every half minute. It is 114 feet above the sea, and is visible 7 leagues off, but the eclipses only 10 miles. Great care should be taken to keep N. of the bar whenever the wind has previously been from N.N.W. to E., and to keep to the S. when the wind has prevailed for five or six previous days from S. to W.N.W. Signals of approach are made by different flags, which it is necessary to understand. The bar should not be attempted when there are freshets in the Adour. In cases of difficulty run for Port Passage, in Spain. The signal

tower of the Adour is in lat. 43° 31′ 36″ N., long. 1° 29′ 53″ W.

BAZANE BANK, on the Bourean shoal, W. coast of France, upwards of a mile from Le Four bank, and N.W. from St. Matthew's point, in Quiberon bay.

BEACHY HEAD, Sussex, a remarkable headland of high bluff chalk, well known by the seven white cliffs W. of it, 9½ leagues W. ¼ N. from Dungeness. Lat. 50° 44′ 21″ N., long. 15° 15′ E. Upon the summit of the second cliff, to the W. of Beachy head, is a first class lighthouse, with a revolving light, visible 6 and 7 leagues. Vessels from the E. open the light N.W. ¾ W. a little westerly, and going up or down when E. of this head, within 3 leagues of it, by keeping the light open, will pass S. of all the shoals that are near. High water 11ʰ 19ᵐ, rise 18 feet.

BEAR BAY, or LITTLE PORT, North America, at the E. end of Anticosti river, near the mouth of the St. Lawrence. It has good anchorage in 4 fathoms, but the best for large ships is to the N.W., in 12 fathoms. Lat. 49° 30′ 30″ N.

BEAR ISLAND, GREAT, North America, Hudson's bay, in lat. 54° 34′ N., long. 81° 20′ W. It is high water at springs here 12 at noon.

BEAR ROCKS, NEW, West Indies, Jamaica, S. by W. from the extreme point of the island. Lat. 16° 20′ N., long. 78° 55′ W.

BEAR SOUND, Greenland, W. coast, in lat. 63° 20′ N., long. 49° W.

BEARD SHOAL, or RED BEARD SHOAL, S.W. from the W. end of the Cutler sand about a mile, having only 6 feet at low water, off Baudsey cliff, on the coast of Suffolk.

BEARHAVEN, see *Bantry Bay*.

BEARN, CAPE, France, about 875 yards from Port Vendres; it has a lighthouse elevated 722 feet above the sea, having a fixed light, visible at the distance of 7 leagues. It is in lat. 42° 32′ 5″ N., long. 3° 5′ 56″ E.

BEAR'S BAY, or WHITE BEAR BAY, S. coast of Newfoundland, towards the W. end, being the fourth bay from Cape la Hune, the S.W. point of Chaleur bay to Cape Ray, or the S.W. point of Newfoundland.

BEAR'S CAPE, North America, the S.E. point of St. John's island, near Nova Scotia. Lat. 45° 59′ N., long. 62° 28′ 16″ W.

BEAR'S COVE, E. coast of Newfoundland, a small fishing place between the harbours of Fermowes and Renowes.

BEAR'S POINT, North America, Nova Scotia, between Port and Cape de la Heve, to the N.E. of Cape Sable, the S.W. point of Nova Scotia, distant 28 leagues.

BEAR'S POINT, WHITE, North America, the E. point of St. Peter's river, on the coast of Labrador. Lat. 51° 55′ N., long. 55° 30′ W.

BEATA, CAPE, West Indies, the S. point of the island of St. Domingo, in lat. 17° 42′ N., long. 72° 2′ W. Beata island is 4 leagues from the cape S.W. by S., and has rocks on the N. side.

BEAUFORT, United States of America, a port and capital of Cartaret county, 11 miles N.W. of Cape Look-out. Pop. 1,100. The harbour admits vessels drawing 14 feet water. Lat. 34° 47′ N. On the E. side of Gore sound, also, a port of South Carolina, 50 miles N.E. of Savanna, on the W. bank of Port Royal river, a narrow branch of the sea; the harbour is distant 16 miles from the sea, but a bar prevents ships entering that draw more than 11 feet water. The site is unhealthy during autumn. The trade of Beaufort is very limited. The town stands in lat. 32° 26′ N., long. 80° 55′ W.

BEAUFORT BANK, lat. 42° 37′ N., long. 41° 45′ W. A bank, discovered by Lieut. Saint-hill, R.N., in the Beaufort, in 1832. Ground rocky, depth 100 fathoms.

BEAULY FRITH, Scotland, W. coast, the entrance to the river of the same name leading to Inverness. There is a bar at the entrance, with only 15 feet at low water. Strangers take a pilot at Cromartie.

BEAUMARIS BAY, Anglesea, North Wales, between Point Trywn Du and Great Orme's head, distant 6½ miles from each other. Near the extremity of the reef running off Trywn Du, or Black point, the N.E. extremity of Anglesea, a light-house has been erected for the guidance of vessels into the Menai strait, the entrance of which is in this bay. It is a fixed red light, visible from N.W. ¾ W. seaward to S.W. ¼ W., except where eclipsed by the intervention of Puffin island. Beaumaris town is in lat. 53° 15′ N., long. 4° 15′ W. It is high water at 10ʰ 26ᵐ.

BEAUVOIR SUR MER, France, department of La Vendée, a small town, with 2,160 inhabitants. It stands a league from the sea, with which it communicates by the channel of Calhouette. It is 13½ N.N.W. from the sands of Olonne. Lat. 47° 2′ N., long. 2° 5′ W.

BEAVER HARBOUR, New Brunswick, North America. The S.W. point is in lat. 45° 3′ 30″ N., long. 66° 47′ W. High water 8ʰ 45ᵐ, rise 7 feet.

BEBEK, on the Bosphorus, on the European side, a little N. of Constantinople.

BEC DU RAZ, France, 2 miles from the Saintes rocks, separated by a channel. On the highest part is a lighthouse, showing a fixed light 259 feet above the sea, seen 6 leagues off. Lat. 48° 2′ 22″ N., long. 4° 43′ 50″ W. The Sein light bears from it N.W. ¼ W. 5¼ miles, which is the general direction of the whole chain of rocks called the Chaussée de Sein. High water 3ʰ 35ᵐ, rise 21 feet.

BEDEL, or BEDEO BAY, North America, Gulf of St. Lawrence, on the S.W. coast of the island of St. John's.

BEDFORD'S BAY, or TORRINGTON BAY, North America, coast of Nova Scotia. It is entered at America point, 3 miles N. of the town of Halifax. It has at its mouth from 10 to 13 fathoms. Winter cove is at its S.E. angle, and there is a small island off the W. point of the creek. Summer cove is in the N.E. angle, and has three small islands off the S.E. point of its entrance. Halifax stands on the S. side, sheltered from N.E. winds, in lat. 44° 39′ W., long. 63° 37′ 26″ W.

BEE, GREAT and LITTLE, France, rocks, a little W. of N. from the point of St. Maloes. The Little Bee is W. of the Great, and both lie about gunshot from the town, N.W.

BEEF ISLAND, see *Virgin Isles.*

BEEF ISLAND, North America, Bay of Campeche, in the S.E. angle, close to Trieste island, E. It is 7 leagues long, and from 3 to 5 broad. It has a fine sandy bay, where ships may ride in 7 or 8 fathoms. Lat. about 18° 30′ N., long. 95° 15′ W.

BEENY SISTERS, Cornwall, N.W. coast, rocks almost close to Fire Beacon point, N.W. of Boscastle.

BEE'S HEAD, ST., Cumberland, near Whitehaven. There is a lighthouse here in lat. 54° 30′ 50″ N., long. 3° 38′ 7″ W.

BEG ISLAND, N.W. point of Ireland, the most

N. of the four islets called Balleness. Lat. about 55° 9′ N., and long. 8° 45′ W.

BEG LOCH, or LITTLE LOCH BRIM, Scotland, W. coast, 1½ mile E. by N. from Groinard island, well sheltered. There is one small rock at the W. side of the entrance, dry only at springs. The best anchorage, of moderate depth, is at the head of the loch.

BEGIA, or BEGGIA, Africa, 22 leagues W. from Tunis, a port with a castle, in lat. 36° 42′ N., long. 11° 30′ E.

BEGIN BAY, or HARBOUR, a small anchoring place near the E. end of Valentia island, on the N. coast of the W. peninsula of Ireland. It lies N.E. from the rocks called the Skelligs, and S. by E. from Blasques sound.

BEGU, CAPE, Spain, 3½ miles N.N.E. from Cape Sebastian. Lat. 41° 56′ 38″ N., long 3° 9′ 30″ E.

BEIROUT, Palestine. The cape bears from Cape Madonna S.S.W. 32 miles, and from Cape Gatte, in Cyprus, E.S.E. ¼ E. 43 leagues. The land E. of the cape is the Bay of Beirout. The town is the best of all the towns in Palestine, surrounded by a wall with five gates. The port and harbour are choked up; there are now only the remains of an ancient pier or mole, giving shelter to boats; even merchant ships must lie off at a distance. The anchorage off the town is good in summer. In winter the best is off the mouth of the river to the E., 2 miles from the town. It is 14 miles from Seyde, or the ancient Sidon, which lies from it S.S.W. Beirout, Bairut, or Beerut, was the ancient Berytus, a port of Syria, in the pachalic of Acre. The port so much neglected is clogged with sand. Raw silk, wine, and oil are the principal exports. It stands on the N. side of a broad promontory, 60 miles N.N.E. of Acre. Pop. 8,000. Lat. 33° 45′ N., long. 35° 36′ E.

BEKIA, or BECONYA, West Indies, a small island among the Grenadillas, between St. Vincent N. and Grenada, 18 leagues S. or S.W. of it. The French call it Little Martinique.

BELANCE ISLAND, France, one of the islets between Ushant and St. Matthew's point, at the entrance of Brest roads.

BELCHERS, THE, North America, a cluster of islands in Hudson's bay. Lat. 56° 10′ N., long. 80° 33′ W.

BELCRAN ROCK, Balearic isles, off the W. point of the island of Ivica, in the Mediterranean.

BELEM, CAPE, Spain, N.N.E. from Cape Finisterre, having between them the projection of Cape de Toriane, which bears S.W. 4 leagues from Belem.

BELEM TOWER, in the Tagus, a league from Lisbon. On the W. side ships must anchor before the village of St. Joseph, in 12 and 13 fathoms, under the N. shore, in lat. 38° 40′ N., long. 9° 40′ W.

BELFAST, Ireland, E. coast, a town situated in lat. 54° 36′ N., long. 5° 56′ W., in the bottom of Carrickfergus bay, 5 leagues S.W. from the bar, and 3 from Carrickfergus. The bay is only three-quarters of a mile broad. High water at 10 o'clock. This is one of the principal commercial towns of Ireland, belonging to the county of Antrim, at the mouth of the Lagan river. It returns two members to parliament, and has extensive manufactures of linen, canvas, glass, cotton, and earthenware. The river Lagan is connected by a canal with Lough Neagh. It is 80 miles N. of Dublin, and contains a pop. of 50,147.

BELFAST, North America, a bay and town of the United States, in the district of Maine, at the mouth of the Penobscot river, on the W. side. The bay, on the N.W. side of which the town stands, runs up by three short creeks into the land. Illsborough island lies in the middle, and forms two channels leading to the mouth of the Penobscot river. The harbour will take the largest vessels. In 1840 its tonnage was 38,218. The pop. 4,186. The island of Illsborough is in lat. 44° 14′ N., long. 68° 48′ W.

BELGOROD, Russia, a town at the mouth of the Dniester, on the Black sea, 80 miles S.E. of Bender. Lat. 46° 30′ N., long. 31° 10′ E. It is 80 miles S. of Bender.

BELGRAVE BAY, Guernsey, N. of St. Peter's port, on the E. side of the island.

BELINGSGATE, North America, on the W. side of the peninsula, of which Cape Cod is the N. point, within Cape Cod bay, 10 miles S. of Cape Cod harbour.

BELIZE RIVER and TOWN, in the S. part of Yucatan, Gulf of Mexico. The fort islet is in lat. 17° 29′ 20″ N., long. 88° 11′ 30″ W. Large vessels cannot go up to the town in consequence of the intricacy of the Grennel channel, unless by the aid of steamers. The road for shipping is a safe and well-sheltered anchorage in all points; the anchor holds well, and vessels can only be driven ashore upon mud. Fort George here is in lat. 17° 29′ 20″ N., long. 88° 8′ 20″ W.

BELL ROCK, Scotland, a rock opposite the mouth of the Tay, on which is a lighthouse, the base of which is covered at high water. It is 115 feet high, and stands in lat. 56° 26′ 3″ N., long. 2° 23′ 6″ W. This light revolves horizontally, bright and red alternately, both showing in a revolution of two minutes. In thick weather a bell is tolled.

BELLAVISTA, CAPE, island of Sardinia, the S. point of the roadstead of Ogliastra, Cape Monte Santo, its S. point, being 8½ miles distant.

BELLE ISLE, France, on the W. coast. Lat. 47° 18′ 43″ N., long. 3° 13′ 30″ E. Pop. 7,630. This island is nearly 10 miles long by 3 or 4 broad, high, and seen a great way off. The N. end lies S. ¼ E. from the isle of Groa, distant 16 miles, and its S. E. part, Point Lesquel, or Du Canon, bears from Hedic isle W. ¼ N. 9 miles. The N.W. point is surrounded by rocks, some above and some under water. There are several roads here for small vessels. The town and harbour of Le Palais are most used, but Sanzon is a better harbour. There is a powerful bright light on the S.W. point of the island, varied by minute flashes, elevated 276 feet above the sea level, visible 9 leagues. It is more brilliant than that of the Du Four rock. The elevation exceeds that of the Du Four 197 feet. It presents between the flashes a fixed light, seen at 10 miles distance, at which distance the eclipses of the Du Four light are total. A harbour light is also shown at the head of the mole on the S. entrance of Le Palais, 29 feet above the sea, visible 9 miles.

BELLE ISLE, North America, N. of the N. point of Newfoundland, near the midst of the strait to which it gives its name, between the Atlantic and the Gulf of St. Lawrence. The N. point is in lat. 52° 1′ 16″ N., and the S. in lat. 52° 21′ 24″ N., and the first in long. 55° 19′ 4″ W.

BELLE ISLE, North America, E. coast, near the entrance into the Bay of Conception. Lat. 47° 43′ N., long. 52° 52′ W. Also an island on the E. coast in lat. 50° 48′ N., long. 55° 30′ 18″ W.

BELLO POULO, Greece, off Cape St. Angelo, an island about 2 miles in diameter. Its N. point is in lat. 36° 57′ 15″ N., long. 23° 26′ 35″ E.

BELLOWS ROCKS, W. coast of Ireland. Lat. 53° 19′ N., long. 10° 4′ W.

BELTS, Denmark, the straits among the Danish islands called the Great and the Little Belt, between the main land and Denmark, and the islands of Funen and Zealand. The Great Belt, where deepest, has 22 fathoms of water, and is 6 or 7 leagues across. It leads out of the Baltic into the Cattegat between Zealand and Funen, but is not as commodious for navigation as the Sound. The Little Belt is to the W. and is very crooked, 9 miles wide and 27 fathoms deep where deepest. It leads to the same point. To proceed through the Little Belt into the Baltic from Samso island requires a pilot. There is a light on the Island of Kyholm, near the N.E. of Thuno, a revolving light, which gives a flash every half minute, elevated 56 feet. A channel runs in S.W. to the town of Frederica, where a toll is paid, and to Middlefort, Kolding, Christiansfield, Maderslaben, Apurado, Assens, and other places in the Little Belt. Its navigation is intricate. It has several lights, but there are shoals and great coast indentations. There is a considerable depth of water for every class of vessel, having from 9 to 25 fathoms all the way. The navigation to the entrance of the Great Belt accomplished as far as Reefness point, the E. point of the entrance is attained. Upon this point there is a light 70 feet above the sea level. The navigation to this point, both from the coast of Sweden and from the N., is intricate. A shoal runs out from Reefness W.N.W. nearly a mile. This point must have a good berth, and be approached no nearer than 8 fathoms. When Callundborg town comes open of the land, a ship will be S. of it, and may then steer S.S.W. for Romsoe; or being abreast of Reefness, in 10 fathoms water, a vessel may proceed until a square-steepled church, in the island of Fyen, is brought in one with the red and white cliff, which will deepen the water from 16 to 22 fathoms, muddy ground. There are several shoals in this passage; the first on the W. side, almost opposite Reefness point, named the Bolsakken shoal, nearly E.N.E. and W.S.W.; it is dangerous, composed of stones, and is 3 miles long and 1½ broad, with only 4 feet water in the shallowest part, and 9, 17, and 18 close to it. It may be cleared to the E. by attending to the above direction, or by bringing Fyen mill open of Bogebierg church. There is a brown beacon on the E. side of the Bolsakken. The Dictator's shoal has 3 or 4 fathoms water, and lies off Luus harbour, at the S. end of Samsoe, and is out of the way of vessels passing through the Great Belt. Round Reefness, and between that and Asnæs point, is a deep bay, called Callundborg, at the further end of which is the town of that name. There is a fixed harbour light there on the pilot house, visible 1 or 1½ league. This bay has rocky shores. Asnæs point is surrounded by shallow water. The Lyse ground is a long, narrow bank, S.W. from Asnæs point 4 or 5 miles long, with from 1 to 4 fathoms of water upon it. This shoal is cleared by keeping Ulstrop church within Reefness well open with Asnæs point, and the lead going. S. of the Lyse ground S. point there is a small knoll of 3 fathoms, the position doubtful. The Elephant shoal is directly between Reisoe and

Romsoe, a mile in length and three-quarters broad; on its centre are 12 feet, and round 4 fathoms. A mill on Fyen, touching the N. part of Romsoe, leads directly to the N. part. Kierteminde steeple touching the S. point of Starvres head, leads on its S. part. A long white house on Zealand, S. 40° E., being open of the S. point of Musholm, will clear it to the S. Roms island is to the starboard, between which and the Lyse bank there is good anchorage 1¼ mile from the N.E. of the island, in 12 fathoms, but no nearer. The passage W. of Roms island is narrowed on the W. by a reef; vessels using it must borrow near the Fyen shore to avoid it. There is also a sandy reef running out on the S.S.E. three-quarters of a mile. Kierteminde church just touching the high land most E. on Fyen island, will clear this reef. Then steer S. by W. towards the Isle of Sproe. When between that and Nyborg, a light will be seen to the S.W. directing to Nyborg. Sproe is midway between Nyborg and Corsoer, about a mile long, and narrow, with a channel on each side. On Knudes head and Halskow head, on the E. coast, near Corsoer, there are fixed lights, and two harbour lights at Corsoer harbour. There is a flashing light on Sproe island, revolving four times in a minute. In the W. channel round Sproe there are 12 and 15 fathoms water with an uneven bottom. The mid-channel must be kept, as a reef stretches out from Knudes head and a flat from Sproe island. When the opening to the W. of Langeland bears S.S.W., and the N. end of that island S. by W., the course is S. until Vresen island opens N. of the N. end of Langeland. The outermost shoal from Sproe bears S.W. by S. from it, and has 19 feet over it. The turning marks for the W. passage are Nyborg telegraph on with the N. white house, bearing N.W. ½ W. Nyborg telegraph and church-steeple in one are the turning mark for the S. side of the passage. The E. channel of Sproe is not so much used as the W. This passage is narrowed greatly by the shoals off Sproe. Sproe being passed through the W. passage, Vresen island is seen ahead, low, with a sort of hummock at its S. end, extensive reefs around. It lies midway between Knudes head and Langeland. Having passed the Sproe channel, the course should be S. by E., or more E., until the E. hummock on Sproe bear N., then run on due S., which will clear Langeland flat, that stretches out 3 miles N.E. from Langeland N. point. The Vengeance shoal is to the E. of this passage, a narrow ridge with 4 or 5 fathoms water, connected with what is called the Bridge all the way to Langeland flat, with 5, 6, and 7 fathoms, and two spots of 4½ and 4¾ fathoms upon it. The mark to clear the Vengeance shoal to the E. is Corsoer church on with the telegraph near it. The Langeland shore is all clear, except the N. flat abovementioned; the currents off it generally set to leeward. A vessel should run along the Langeland coast in 8 or 9 fathoms, lest the current set her between Zealand and Laaland. Run on until Langeland light bear W.N.W., when the church on Femeren island will bear S., distant 10 miles. Steer about S.E. along the S. coast of Laaland, in 7, 8, and 9 fathoms, or deeper, to 15 and 16 in mid-channel. When passing between the S. end of Langeland and Laaland, a vessel must bring Langeland lighthouse W.N.W., to avoid a flat off Westerness. The light here is a fixed light, 125 feet high, visible 16 miles. From off Westerness the course is about S.E. for 7 leagues, then S.E. by E. for 7 or 8 leagues, to go to the S. of the

Trindelen shoals, which are dangerous, and then an E.N.E. course will round them. There is a light on the N.E. side of Femeren island, near the dangerous reef of Ruttguards on the Oldenburgh Huk; it revolves, while all the other lights in the vicinity are stationary. From the light on the Trindelen point, called also Giedsbye head, or, in Danish, Giedser Odde, there is a lighthouse with a fixed light elevated 44 feet, visible 11 miles; a quarter of a mile from the extreme point of the land, on the N. of which are two churches, Giedesbye and Skielbye. Sailing S.E. they will come in one with the point. The most N. is red, with a short, pointed steeple; the S. white, with a square steeple. A good wide berth must be given rounding the head, as reefs and knolls lie off to a good distance, the most dangerous part S.E. ¼ E. from the light. A beacon, with a black barrel, is erected, to guide vessels there during daylight; when in a line with the lighthouse a ship is in the direction of the reef, and must keep out to avoid it. When going to the E., if the barrel is open to the W. of the lighthouse, the reef is cleared, and when to the E. the reef is cleared on the opposite side. Three triangular-shaped sands, with 4 or 5 fathoms water 8 or 9 miles from the land, are near the extremity of the reef, called the Superb's shoals. The S. of them bears from the lighthouse S.E. by E. distant 9 miles, and has 4 and 4½ fathoms: the middle shoal is N.E. of this, and bears S.E. from the lighthouse with 4¾ fathoms; the third is N., and is E.S.E. from the lighthouse distant 8 miles, and all surrounded by 7, 8, 9, and 10 fathoms. Darser Ort, or Dars head, S.E. by E. with Trindelen shoal, are the two points on each side the outlet of the Belts into the Baltic.

BEMBRIDGE POINT, Isle of Wight, E. extremity, off which is a ledge that makes it necessary to give the island a good berth, as it runs out E.N.E. There is now a light-vessel here, in lat. 50° 41′ 40″ N., long. 1° 1′ 40″ W.

BEMICARY POINT and BAY, Jamaica, N.W. by W. from Portland, at the S. extremity of the island. The E. limit of the bay is in lat. 17° 55′ N., long. 77° 17′ W.

BENBECULA, one of the Hebrides, between N. and S. Uist, about 7½ miles in diameter. The channels which separate it from the above islands are narrow and sometimes dry.

BENDER, an Eastern term, signifying a harbour.

BENET CAPE and BAY, Hayti, on the S. side of the W. peninsula, forming, with the line to Petite Goave, on the N., the narrowest part of the isthmus. Lat. about 18° 20′ N., long. 72° 47′ W.

BENGASI, a town and port of Tripoli, Africa, N. coast, having a castle in lat. 32° 6′ 50″ N., long. 20° 2′ 56″ E. To enter the port, the N. islands must be passed pretty close, then run out from the castle point, and the larboard shore must be kept. The S. part of the port has shallow water before it, and the passage is narrow, but there are from 2 to 2½ fathoms water within.

BENGO BAY, Africa, S.W. coast, 8 leagues from the river Danda; the land is low near by, and the shore steep. Lat. 8° 30′ S., long. 12° 30′ E. There is anchorage here in 10 fathoms.

BENGORE HEAD, N. of Ireland, in lat. 55° 15′ N., long. 6° 29′ 35″ W., on the N. side of which is the celebrated Giant's Causeway. Sunken rocks extend full a quarter of a mile from the head; N.N.E., or rather more, lies a rock with only 2 feet on it at low water, and 10 fathoms close by it. This head is composed of several capes or bays bounded by columnar basalt. That called Pleskin is a colonnade of 60 feet high, with a lower gallery of 50. The lower range contains the finest columns. The columns extend into the sea, and are lost in deep water. From Bengore head to Fairhead the distance is 11½ miles S.E. ¾ E., with Sheep island between. Beyond Sheep island, between that and Fairhead, are Ballycastle bay and harbour, the last 3 miles from Fairhead, formed at the mouth of a rivulet, but neglected and half buried in sand. Ships may ride in summer on the W. side of the bay in 9 or 10 fathoms, the Glass-house bearing S.S.W., Sheep island shut in with Kinbane head, and Clare house bearing on the top of the cliff near it. There are small patches of foul ground near the anchorage. It is not prudent to lie long, and should the wind blow from the N. or N.W., the anchor should be weighed at once and Fairhead rounded. The basalt columns here are coarser than at Bengore head. The columns in this bay are asserted to be magnetic, and to affect the compass, which must be carefully regarded. High water at Bengore head at 6ʰ. The following is a description of this basaltic head, considered as a natural curiosity. It is composed of pillars having from three sides to eight. The eastern point terminates in a perpendicular cliff formed by the pillars, some of which are 33 ft. 4 in. high; each consists of several joints one upon the other, from 6 inches to a foot in thickness. Some of these joints are so convex, that their prominences are nearly three-quarter spheres, round each of which is a ledge, which holds them together with great firmness, every stone being concave on the other side, and fitting the convexity of the upper part of that beneath it. The pillars are from 1 to 2 feet in diameter, and generally consist of 40 joints; one may walk along on the tops of them as far as the edge of the water. But this is not the most singular thing, for the cliffs themselves are more surprising. From the bottom of a black stone, to the height of about 60 feet, they are divided, at equal distances, by stripes of a reddish stone, resembling cement, about 4 inches in thickness; upon this is another stratum of the black stone, with a stratum of red 5 inches thick. Over this is another stratum 10 feet thick, divided in the same manner; then a stratum of the red stone 20 feet deep, and above that a stratum of upright pillars; above these pillars lies another stratum of black stone 20 feet high; and above this again, another stratum of pillars, rising, in places, to the top of the cliff, in others not so high, and in others above it. The face of these cliffs extends 3 miles.

BENICARLO, or VENICARLO, Spain, N. from Peniscola about 3½ miles, in the Mediterranean; it is noted for its red wine. Pop. 3,000. It stands on a bay N.E. of the Bay of Valencia. This last is in lat. 10° 11', long. 27' W.

BENIDORME ISLAND and TOWN, E. coast of Spain, E.N.E. ¾ E. from Villa Joyosa; the town is built on a rock, and the island is distant 2 miles S. ½ E. from the town.

BENIEN BAY, sometimes called St. Michael's, France, on the N. side of the Isle of Ouessant, or Ushant; it is a clean bay with deep water.

BENIGUET ISLE, France, one of a cluster of islets running N.N.W. ¼ W. and S.S.E. ¼ E. for 2 miles between Ushant and St. Matthew's point, Brest roads. There is a flashing light upon

the Teignouze rock. At their N. extremity is the Ecrevière sunken rock, forming the boundary of the Teignouze channel, and from thence to the Teignouze rock is about 1¼ mile, between which, midway, the soundings vary from 6 to 24 fathoms. There are also two shoals near the entrance of this channel which must be avoided.

BENIN RIVER, Bight of Benin, Africa, the lat. of the N.W. head is 5° 43' N., long. 5° 4' E. It was once called Rio Formosa, and is the outlet of the Niger, the river so long sought, and discovered by the brothers Lander, of Truro. The channel is close by the S.E. point. It is 40 leagues from Whydah N.E. The tide here has a rise of 5 or 6 feet, and flows at 4¼ʰ, full and change. The Bight of Benin is a bay in the Gulf of Guinea into which that river flows, extending, on the S.E., to Cape Formosa, and on the S.W. to Cape St. Paul. The river Benin is nearest to the S.E. cape. The best season to enter the river, as well as the most healthy, is between September and the middle of February.

BENITO, ST., a river and town upon the African coast, in the Bight of Biafra, in lat. 1° 35' N.

BENNET'S, ST., France, between Cape Frehel and the Island of Briack, or Brehaut, 8 leagues N.W. by W. of St. Maloes.

BENODET BAY, France, close to the Bay of Andierne, and near the Penmarks, not to be approached without a pilot. The river Quimper is in this bay, 2½ miles from the entrance to that of Pont L'Abbé. In this bay, at Quimper, Pont L'Abbé river, and at the Penmarks, it is high water at full and change from half-past 3 to 4.

BEOTIA, CAPE, Greece, Isle of Negropont, about E.N.E. from the bottom of the Gulf of Leponto.

BERBICE, Guiana, a river which falls into the Atlantic, in lat. 6° 24' 30" N., long. 57° 22' W. It is a league wide, and vessels can go up it to the British settlements, 45 miles from the entrance, at Crab island, where there is a light. Spring tides rise 11 feet, at the equinox 14 and 15. A light-vessel lies 10 miles N.N.E. of the river, showing a ball at the mast-head by day and a fixed light at night.

BEREBY, S. coast of Africa, N.E. of Cape Palmas, a town on the Ivory coast, in lat. 4° 39' 3" N., long. 6° 54' 30" W. There is anchorage off the shore in 13 and 15 fathoms, but the bottom is foul.

BEREZON, Black sea, an island opposite the mouth of the Dnieper, about 500 fathoms long by 126 broad, nearly inaccessible.

BERGEN, Norway, the capital, in lat. 60° 24' N., long. 5° 38' E., having a pop. of 18,010. It is situated at the bottom of a deep bay. It is large, fortified on the land side. The public buildings are of stone, but most of the dwellings of wood. It carries on a considerable traffic in eider down, skins, firwood, and dried fish. The harbour is sheltered and secure, but access to it is difficult and even dangerous without a pilot. It is the depôt of the produce of the fisheries of Finmark and the Nordlands, 126 Nordland yachts, each manned by 10 or 12 hands, having been seen there at one time. The value of the fish exported from Bergen alone is 958,000 dollars. Vessels proceed to the harbour through Carm sound, passing between the Skudesness and the Huiddings light-houses, and then steer N. through various intricate channels and fiords, which are accurately laid

E

down in the Danish surveys, without which this navigation should not be attempted. The tides here are much influenced by the weather in the North sea, rising and falling in general 4, 5, and 6 feet. There is a light, called the Hoyvarden, seen on steering N. between Carm Oe and Luden, on the Isle of Carm. Some of the fiords have from 200 to 300 fathoms within them, and the pilots make the vessels fast to rings placed in the rocks for that purpose. The currents in this intricate navigation require much attention.

BERGEN, the capital of Rugen, in the Baltic, on the coast of Pomerania, belonging to Prussia, lat. 54° 15′ N., long. 14° 2′ E.

BERGEN OP ZOOM, Holland, on the river Zoom, at its influx into the Scheldt, on the E. shore of that river, very strongly fortified. It is the capital of South Brabant. Lat. 51° 30′ N., long. 4° 18′ E.

BERGUES, THE, one of the Flemish banks. A mile N.N.E. from the E. end of the inner Ruytengen, begins the S. edge of the Bergues, and between them is a channel with 10 fathoms. They lie 15 miles from Dunkirk steeple N.N.E., and are the more dangerous for their distance from the land.

BERKER'S ARIAN CREEK, Holland, a sand to the seaward, and to the S. of Blenk, or the S. sandhill, on the coast of Holland, S. of the Land Deep channel into the Texel, the coast here trending N. and S. from the Maes to the Texel.

BERLENGAS, or THE BERLINGS, islands N.W. of Cape Carvoeiro, Portugal; the S. point of the Great Berling being N.W. by N. 5 miles from the cape. The island may be seen 20 miles off. On the N. are the high rocks called the Farilhoens. This island is level on the top. There is a lighthouse on the S.E. side of the Great Berling, 365 feet above the sea level, with a light revolving every 3 minutes, when it shows its brightest. Lat. 39° 24′ 40″ N., long. 9° 31′ 11″ W. The Estella rocks, and some sunken ones, lie on its N. side, and 5 miles N. is the Great Farilhoen. The light is on the S.E. of the island. Peniche light bears S. 32¼° E.; Cape Roca S. 29° W.; the top of the Great Estella rock N. 42¼° W.

BERMEJA ISLAND, Campeche bank, West Indies, an island of doubtful existence near the edge of the soundings, said to be in lat. 22° 33′ N., long. 91° 22′ W.

BERMUDA, or SOMERS' ISLANDS, a group or cluster of isles and islets fenced round with innumerable visible and sunken rocks, in all about 400. They lie in lat. from 32° 15′ to 32° 22′ N. Cape Hatteras, on the coast of the United States of America, is the nearest land to them. The tides flow here at St. George's from 7 to 8 o'clock, and rise from 5 to 6 feet. Bird island, Cooper, St. David's, St. George's, Ireland, Long island, Nonsuch, and Somerset, are the most considerable islands. The currents among them are very strong, and the sea, at times, tempestuous, being subject to sudden storms, from being near the limit of variable winds. They extend N.E. by E. and S.W. by W. about 9 leagues; rocky reefs lie round at 8 leagues distance, and render them very difficult of access. The climate is healthy. The winter lasts from November till March. The mean temperature is 60°. The islands are divided into nine parishes. The principal towns are St. George and Hamilton. There are four signal stations on the islands. One at St. George's, the head-quarters, on the E.; one centrical, at Mount Langton, near Hamilton; another on the W. coast,

at Gibb's hill. Here a lighthouse of cast-iron is erected, made in London, 120 feet high. Vessels must be very cautious of approaching these islands in hazy weather, or in the night, lest the winds or the current run them into some inextricable channel; they must be particularly cautious in approaching from the S.W., as upon the rocks off this end of the islands, from S.W. to W.N.W., many ships have been lost. Vessels make these islands from the West Indies and America, by running in their lat. from the W., for which purpose lat. 32° 8′ N. is the best, always having regard to a possible current in the way the wind blows. In running for them from the E., the best parallel is between lat. 32° 10′ and 32° 20′ N. Wreck hill forms the W. point of these islands, and St. David's head the E. The former is in lat. 32° 15′ 20″ N., long. 64° 50′ W. The latter, the eastern extremity, lat. 32° 21′ 25″ N., long. 64° 35′ 40″ W. The town of St. George is in lat. 32° 22′ 23″ N., long. 64° 37′ 40″ W. St. George's island contains 500 stone houses. The soil, not very fertile, is covered with cedar trees, and cotton is grown to some extent. Murray's anchorage, though exposed from N.E. to N.W., is nearly the only port which will admit line-of-battle ships, though a dangerous and narrow channel in the reef, but the ground holds well. The rain-water, received in a cistern in Tobacco bay, is used for watering the vessels of the navy. Murray's anchorage lies on the S.W. side of Catherine point. A pilot is generally got off Castle harbour, or ships may run as far as St. David's head; when to the N.E. of the head, stand no further to the N. than to bring the head N.E., when a white sandy bay will be seen S. of the head, between it and Castle harbour. In standing N. shut no part of this bay in behind St. David's head. The W. land of Bermuda will be shut in behind the land over this bay before this mark comes on. In the night, when waiting for a pilot, keep the lead going continually, for, if not going too fast, the ground will be struck in time to avoid danger. The entrance to Murray's anchorage is regularly buoyed. The banks to the S.W. of the Bermudas were surveyed carefully, in 1829, by H.M.S. Columbine. The N. extremity of the inner bank lies in lat. 32° 6′ N., and long. 64° 53′ W.; the S.W. bank in lat. 32° N., and long. 65° W. The least water found was 29 fathoms, bottom corally and rocky; on the edges 40 fathoms. In the S.W. is another called the Outre bank. The least water found on this was from 33 to 47 fathoms, rock and coral.

BERNAL, North America, a hill on the W. coast of New Mexico, 4 leagues W.S.W. of the volcano of St. Salvador, about lat. 13° N.

BERNARD, THE, the S. end of a small sand between Southwold and Lowestoff, on the coast of Suffolk, of which the N. end is called the Newcome, a short mile from the shore. Small vessels may pass within in fine weather. Ships should not approach this sand nearer than 5 fathoms. It has not above 3 or 4 feet at low water.

BERNARDO, BAHIA DE, sometimes called St. Louis bay and river, in the N.W. part of the Gulf of Mexico, North America. A long, narrow island lies before the mouth of this bay, with a channel at each end. The bay is in lat. 28° 45′ N., long. 96° 20′ W. The mouth of the river is in lat. 28° 51′ N., long. 95° 49′ W.

BERNARDO, ISLES OF SAN, coast of Columbia, province of Carthagena, 11 in number, W. of the point of San Bernardo. Tintipon, or N. isle, is in lat. 9° 48′ N., long. 75° 51′ 30″ W.; Santiago de Tolu, 9° 30′ 45″ N., long. 75° 36′ W.; Fuerte, lat. 9° 23′ 30″ N., long. 76° 11′ 15″ W.; Puerto de Cispata entrance, lat. 9° 25′ N., long. 75° 48′ 5″ W.

BERNENT BAY, France, at Quiberon point, the N. limit of the great bay thus denominated. The coast from hence turns round an E. point to the N. and N.E. into the larger bay.

BERNERA ISLAND, Scotland, near the W. coast of Lewis, separated by a channel, called Loch Bernera, about a mile wide. Lat. 58° 25′ N., long. 7° 3′ W.

BERNERA, LOCH, Scotch isles, Isle of Lewis, 22 miles from the Butt of Lewis, formed by the main of Lewis and the isles to the W. There are several bays or lochs here, besides that of Bernera, where vessels may ride in safety; that of Bernera is perfectly secure.

BERRY HAVEN, Ireland, W. coast; a port for small vessels, 5 miles N.E. of Ballyshannon.

BERRY HEAD, Devonshire, the S. point of Torbay. It is a mark S., S. by E., or S.S.E., and Brixham church on with the pier head, of 6, 7, 8, or 9 fathoms. Lat. 50° 24′ 2″ N., long. 3° 28′ 14″ W.

BERRY HEAD, Nova Scotia, W. side of Torbay. Lat. 45° 10′ 57″ N., long. 61° 22′ W.

BERRY'S ISLAND. See *Bahamas*.

BERSIAMITES RIVER, St. Lawrence, North America; entrance lat. 48° 55′ 31″ N., long. 68° 40′ 30″ W. High water 11ʰ; rise 12 feet.

BERTERBUI BAY, W. coast of Ireland, N. of the Skirds, having anchorage for the largest ships.

BERTHEAUME POINT, France, Brest harbour, 1¼ league E. from St. Mathew's point, without the entrance into the inner road. On the W. side of Bertheaume bay are the rock and castle of the same name. High water 3ʰ; rise 21 feet.

BERTON ROAD, Ireland, within Dalkey island, at the S. point of the entrance into Dublin bay.

BERVIE, or INVERBERVIE, a town of Scotland, in Kincardineshire, at the mouth of the Bervie, 12 miles N.E. of Montrose; its harbour is only adapted for fishing-boats. It is 7 leagues from Aberdeen. Lat. 56° 44′ N., long. 2° 4′ W.

BERWICK, a town on the N.E. bank of the river Tweed, having a spire upon the town-hall, and being fortified. It stands in lat. 55° 46′ 21″ N., long. 1° 59′ 41″ W. To the N. of the town is a building called the Old Bell-tower, with the Magdalen Field House forming conspicuous seamarks. The land should never be made here in thick weather, nor in less water than 35 fathoms. When clear weather, ships should not go into less than 18 fathoms until the proper time of the tide, shown by a red tide-light, when there are 10 feet water and upwards. A pilot, however, is indispensable. High water 2ʰ 18ᵐ, rise 15 feet. The flood sets S. from St. Abb's head, past Eyemouth and Berwick.

BERYORT SAND, Holland; it has a buoy on its E. point, and in sailing for the point of Schonen island it must be left on the larboard; on the starboard is another buoy, with a tail on the point of Poolvert, between Schonen and Oresand.

BESSACK ROCK, St. Ives' bay, E. side, S. of Godrevy island and the Stones, W. of the Sands of Guythian, coast of Cornwall.

BESSERNA, Denmark, a small island, 2 miles S.E. from Veyeroe.

BESSIN, PORT-EN-, France, department of Calvados, at the entrance of the river Drome into the Channel, 2¼ leagues N.N.W. from Bayeux, a fishing village, having about 600 inhabitants. Grandchamp, near on the river Vie, is a fishing village with 260 inhabitants. Both have temporary lights; that of Grandchamp flashing.

BESTOE ISLAND, Christiania sound, Norway, 5 leagues N.W. by N. from the Sister's island, on the starboard entrance into the sound. See *Christiania*.

BETANZOS BAY, Spain, adjoining the Bay of Ares. With the appearance of harbours and places of safety, these are not to be trusted, being dangerous, and lying open to wind and sea. None but small coasters venture within them, anchoring E. of Fonta castle, in 3, 4, or 5 fathoms, or to the E. of Ares, in 4½; S. winds are particularly dangerous to this anchorage. There is a shoal in this bay, with only 2 fathoms over it, near the Miranda rock. When Point Deixo is one with the Tower of Hercules, at Corunna, a ship is upon it. The channel between that and the shore generally breaks, though there are 7 or 8 fathoms within it. If a large ship be driven in, she must bring to in the best situation circumstances will admit.

BETSY'S ROCK, a vigia said to have been seen, in lat. 18° 7′ N., long. 50° W., by the brig Betsy, bound from Greenock to Jamaica, in 1808, described as a flat rock.

BEVER HEAD, North America, Nova Scotia, S.E. coast. Lat. 44° 42′ N., long. 62° 20′ W.

BEVERLAND, or BEVELAND ISLES, Holland, denominated N. and S. Beverland, the last is 28 miles long and 8 broad. It divides the E. from the W. Scheldt, the two points of its W. end approach the channels opposite the ports of Armuyde and Flushing. N. Beverland, S. of Schonen island, is 6 miles long and 4 broad, and is on the S. side of the channel of the E. Scheldt, that here runs W. into the sea.

BEVERLY, North America, United States, a town of New England, in Massachusetts, N. of Salem, to which it is connected by a bridge 1,500 feet long. The town is nearly N.N.W. from Marble head. It has some manufactures, but the inhabitants are principally employed in commerce and in the fisheries. Pop. 4,689; about lat. 42° 31′ N., long. 70° 50′ W.

BEVERS, LITTLE, W. of Point de la Hune, on the S. coast of Newfoundland, between Cape de la Hune on the E. and Cape Raye on the W., about 20 leagues asunder.

BEXHILL, Sussex, in Pevensey bay. The church is in lat. 50° 50′ 45″ N., long. 28° 48′ E.

BHURDA, Gulf of Bomba, N. coast of Africa, an island, N. by W. ¼ from which 11 miles is Cape Razatin.

BIAFRA, BIGHT OF, Africa, a gulf which may be said to comprise the angle where the African continent bends S. Cape Formosa, which lies in lat. 4° 28′ N., long. 5° 41′ 30″ E., is the N.W. point, and Cape St. John, lat. 1° 9′ N., long. 9° 21′ 35″ E. The Island of Fernando Po is in its central point. Both this bight and that of Benin are within the Gulf of Guinea. Some extend the bight to Cape Lopez, and include in it the islands of Prince, St. Thomas, and Anna Bom. Two currents meet in the vicinity of Fernando Po; one from the W., along the coast of Guinea, the other from the S., along the coast of Angola: the last is the most powerful. It is therefore advisable to bear in a direct line between the islands.

BIANCA STRAIT, Asia, 3 leagues wide, be-

E 2

tween the Island of Scio and the coast of Anatolia, in the Greek Archipelago, Gulf of Scala Nuevo.

BIARRITZ, France, Basses Pyrénées, about a league from Bayonne, a village N., 43° W. of the church on which stands the lighthouse. Lat. 43° 29′ 38″ N., long. 1° 33′ 36″ W. See *Bayonne.*

BIBAN ROCK, or ALBIBAN, N. coast of Africa, in Tripoli, 48 miles W. from the little port of Tripoli Vecchia, and 18 miles from the Port of Bucseala W.N.W. ¼ N. It is at the entrance of a large lagoon, in which are from 3½ to 2 fathoms water.

BICQUETTA, ISLE OF, near the Isle of Bic; on its W. end is a revolving light of the first class, 130 feet above the sea. The light revolves every 2 minutes, and is shown from April 15 to Dec. 15.

BIDDASOA, or VIDASOA RIVER, Spain, near Fontarabia, the boundary between that country and France. High water at the entrance half-past 3. The entrance, 15 miles W.S.W. from Bayonne, is between Point de las Areta and Cape Higuera. The fortress of Fontarabia is on its W. bank. The bar forbids access to all but small craft. Outside the bar there is anchorage in 6 and 8 fathoms, but care must be taken not to approach too near either of the capes. It is distant 6 miles E. from the entrance of Port Passages.

BIDDEFORD, York co., Maine, United States of America, on the S.W. side of the Saco river, connected by a bridge. It extends to the sea, and there is a revolving light off the mouth of Saco river, in lat. 43° 50′ N.

BIDDEFORD, Devonshire. See *Barnstaple.*

BIDSTONE, Cheshire, a parish in which is a lighthouse belonging to the corporation of Liverpool, supported by dues from shipping that enter there. Lat. 53° 24′ 6″ N., long. 3° 3′ 46″ W.

BIE, ISLE OF, in the river St. Lawrence, North America.

BIERVLIET, Holland, a small island in the Scheldt, E. of Cadsand, and 4 leagues E.N.E. of Sluys.

BIGBURY BAY, Devonshire, S. coast, midway between the Bolt Tail and Stoke point, into which the Erme and Avon rivers empty themselves. Off the entrance of the last is Burr or Borough island, which is dry at low water.

BIGG'S BAY, Jamaica, on the S. side, E. of N. of Poland point, the most S. point of the island.

BIGHT, a term used to express the shallow course of any coast of a depth too inconsiderable to be denominated a gulf, and yet of considerable extent.

BILBAO, Spain, the capital of Biscay, on the river Ibaicabel, 6 miles from the sea. The town has a considerable trade in wool, iron, oil, saffron, and wine. Point Galea is the E. point of the entrance to the river, and Point Lucerno is its W., bearing from each other W. by N. and E. by S. nearly 3 miles. Half a mile from Point Galea, on the E. shore, is the castle of Galea, and a mile farther are the battery and Point of St. Ignacio. The point is red, and has sunk rocks off its shore, one of which is called the Pilot rock. The river entrance is formed by two piers. On the W. is the town of Portugalete, where large ships should anchor. The bar of Bilbao is dangerous, a heavy swell sometimes setting in; when the bar cannot be passed, a ship may come to between Lucerno and Galea, in 14 or 15 fathoms. Santurce is the residence of the pilots on the W. side of the bay. The largest vessels that can go over the bar must not draw more than 15 feet water, and even then the sea must be smooth. High water at Bilbao 2ʰ 53ᵐ,

rise 13 feet. Lat. 43° 13′ N., long. 2° 53′ W., that of Portugalete 43° 20′ 10″ N., long. 2° 58′ 20″ W.

BILL HEAD, Cornwall, 2 miles N.W. of the Lizard point.

BILLE, CAPE, Greenland, lat. 62° 1′ N., long. 41° 57′ W.

BILLS, a rock near the W. coast of Ireland, 6 miles N.W. from Clare island, and 6 S.S.E. of Achill head.

BILOXI BAY, between the Bay of St. Louis and that of Mobile, in the Gulf of Florida; it has very little depth of water.

BIMBIA, and BIMBIA RIVER, Africa, S., under the Camaroon mountains, falling into the same bay as the river of Camaroons. The land here is high and bluff. Lat. 3° 57′ N., long. 8° 32′ E.

BIMINI ISLANDS. See *Bahamas.*

BINARD ISLAND, France, N. coast, E. of Roteneuf point, with channels at the E and W. end, and a bay within. It is to the E. of St. Maloes, towards Cancale.

BINAROSA, or VINAROSA, Spain, E. coast, 5 miles N. of Pensacola, and 20 miles S. from Tortosa. It has a small harbour and an anchorage about a cannon shot from the town.

BINGUT, CAPE, E. by N. from Algiers, on the African coast. There are some rocks or islets off the cape, which also gives its name to a bay. The town is at the bottom. The cape is about 12 leagues from Algiers.

BINNIQUET, France, a small island on the coast, a league S.W. of Le Conquêt, and three-quarters of a league W. from St. Matthew's point. See *Beniguet.*

BINWY HEAD, N.W. coast of Ireland, Mayo co. Lat. 54° 20′ N., long. 9° 36′ W.

BIOMSEE, Denmark, a small island near the S. coast of Funen.

BIORKA ISLAND and SOUND, Gulf of Finland; on the S. end of the island is a beacon tower, 40 feet high, to mark the entrance to Biorka sound, leading to Wyburg. Lat. 60° 17′ N., long. 30° 21′ E.

BIORKO ISLAND, Sweden, 3 miles from Stockholm, on which once stood the royal palace of Birka.

BIORNEBURG, Finland, on the E. of the Gulf of Bothnia, 27 leagues N. of Abo. Lat. 61° 27′ N., long. 22° 10′ E.

BIRD ISLAND, Plymouth co., Massachusetts, North America, near Sippican or Rochester harbour. It has a stone tower, elevated about 30 feet. The islet is low; the light may be seen 5 leagues. It is a revolving light. At first the time of darkness is double that of light, but the interval decreases as 40 to 1.

BIRD ISLAND—the islands so called are many. A small island in Strangford lough, 8 miles S. of Newton, Ireland. A 2nd, one of the Bermudas. A 3rd, on the N. side of Antigua, West Indies, towards the N.E. corner. 4th. Several on the coast of Newfoundland, a league W. by N. from Flower's point, on the E. side of the island, between 2 and 3 miles N. from Cape Largent. 5thly. Bird island, N.W. from Cape St. Mary, in Fundy bay, Nova Scotia. 6th. In the West Indies, among the Virgin isles.

BIRD ISLAND, Africa, river Gambia; lat. 13° 39′ 12″ N., long. 16° 40′ 30″ W.

BIRD ROCKS, Tripoli, Mediterranean, near Gharra island.

BIRD, THE, on the Norfolk coast, a dangerous bank on the flat that stretches along from Wells harbour and off the Scald heads; Burnham flats are to the W. It is partly dry at low water.

BIRDS' ROCKS, North America, Gulf of St. Lawrence, near the W. coast of Newfoundland, in lat. 47° 51′ N., long. 61° 12′ 11″ W.

BIRON, or BRION ISLAND, Gulf of St. Lawrence, North America, 26 leagues W. of Cape Anguilla. Lat. 47° 47′ 58″ N., long. 61° 27′ 33″ W.

BIRTENBUY BAY, Ireland, W. coast. Lat. 53° 20′ N., long. 9° 50′ W.

BIRVIDAUX ROCK, France, 2½ leagues S.S.E. from Groa island, off Quiberon point. It is very dangerous, and is sometimes left dry; there are other rocks near it which never appear. At the E. end of Groa island is Point de Croix, and when Point Ley, which is to the W. of Port Louis, is right over that point and the most N. windmill on Quiberon E. by N., a vessel is athwart this rock.

BISCAY, or THE GULF OF GASCONY, France and Spain, a vast bay, extending from Ushant S. as far as Cape Ortegal. It is principally noted for the heavy seas which are produced in it by N.W. winds, for its depth, and its S.E. current. The coast from Ushant to the Gironde has numerous islands, bays, and river entrances. The principal bays are those of Douarnenez, affording good anchorage in 10 fathoms for the largest fleets in N. and in N.E. winds: Audierne, or Hodierne, which is open: Quiberon, formed by a peninsula on the W.: Morbihan, with several islands, formed E. by the peninsula of Rhuys, and the Bay of Noirmoutier. From the mouth of the Gironde to the Adour, a distance of 60 leagues, the coast is wholly composed of *dunes*, or, as metamorphosed in English, sandy downs. Some of these sands roll inland, and overwhelm whole villages. There are lakes or lagoons on this part of the coast, that communicate with the sea through channels called *Bocautes:* the principal is the Basin of Archachon, 15 leagues in circuit. A portion is sheltered by a pine forest from S. and S.W. winds, and 100 sail of the line might anchor in the basin for depth or space; but the channels, owing to winds and waves without, are continually shifted. From about the boundary of France and Spain the shores of this bay extend W. to Cape Ortegal, and as along the coast of France, the depth increases from 20 to 40 and 60 fathoms gradually from the shore, and beyond to greater depths. Off the N. coast of Spain it becomes deep water from 150 to 200 fathoms almost immediately. The high seas in this bay have also become matter of remark among seamen. A current also is found to set in strongly from the Atlantic W. along the N. coast of Spain, sweeping round to the northwards, or N.W. or N. The soundings in this bay, S. of the Gironde river, are observed to show no mud, which is found everywhere to the northward, as if swept round there by the current's action. Vessels, too, making for Port Passage, in Spain, standing off shore, with a gale right on, and their sails blown away at night, have unexpectedly found themselves in the morning far off shore to the N.W. The fact is admitted, that there is such a current, though not yet in its details sufficiently examined to render the fact indisputable in all its bearings. The action of a current out of the N. part of the bay athwart the British channel is not controverted.

BISCAY, BAY OF, Newfoundland, between Capes Race and Pine, on the E. side of a deep and large bay, of which these capes are the limits. Lat. 46° 48′ N., long. 53° 5′ W.

BISCAYNO KAY, in the Florida reefs, West Indies; the S. end, in lat. 25° 41′ N., extends N. about 5 miles, and is nearly 2 wide. It lies N.W. of the entrance to Hawke channel, 1½ mile from the main. On this kay is a tower, which once exhibited a fixed light, 70 feet above the sea level. It was burned by the Indians, 1839. There is a bank off its E. side. The coast to the N. of this kay has foul ground for 4 or 5 leagues, and the sea breaking on it has a fearful appearance. By keeping 5 or 6 miles from the shore there is mostly 5 or 6 fathoms water, with a fine sandy bottom. On approaching the end of the reef, haul in towards the kay, giving the reef a good berth without, on account of sands and bars on its inner edge. There is not less than 3 fathoms till abreast of the S. end of the kay, where there is a small bank of 11 feet only: give the kay a good berth now, as a large flat stretches from it. Then steer E. of S., and pass to the E. of the shallow surrounding Oswald's kays, the course will then be more W. by the edge of the bank. It is a good rule to have a careful man at the mast-head, who, in a clear day, will see all the heads and shoals a mile off. Go no nearer the Soldier's kays than 12 feet, nor further off than 18. A mile E.S.E. from *Sander's* cut there is a small round bank, with only 9 feet, and from thence to Black Cæsar's creek there are sunken rocks, and the bar of that creek runs a great way out. Abreast and N. of the bar is a very fine anchorage in 22 feet, close to the back of the reef. The care to be taken about here cannot be too great. It is high water at the N. entrance of the Hawke channel, opposite Soldier's kay, at 5½ʰ; spring tides only rise 2 feet 6 inches.

BISCIE ROCK, Sardinia, a reef of 20 feet water, about a mile from E. to W. Its most dangerous part bears from Biscie island N. by E. 1½ mile distant. This reef should not be approached nearer than 15 fathoms water.

BISHOP AND CLERKS, South Wales, N.N.W. of Ramsay island, off the S.W. of Pembrokeshire, exceedingly dangerous. On the S. Bishop there is a revolving light, easily distinguished by its brilliancy at short intervals, from the other lights near this perilous coast. The N. Bishop is 3 miles N.E. ¼ from the S., its lat. 51° 54′ 10″ W., long. 5° 23′ 6″ W.

BISHOP AND HIS CLERKS, Norway, a cluster of rocks, 5 miles S. of the Naze of Norway, exceedingly dangerous to mariners.

BISHOP'S ISLAND, Ireland, a small island on the W. coast, in lat. 52° 38′ N., long. 9° 35′ W.

BISHOP'S ROCK, promontory of St. John, Newfoundland. Lat. 49° 55′ 30″ W., long. 55° 29′ 18″ W.

BISSAGOS and BIJOOGA ISLANDS, Africa, an archipelago or group, between the parallels of 10° 2′ and 11° 42′ N. lat., and between 15° and 17° W. long. These consist of islands and shoals, of which the interior is little known. Jatt's island, Bassis, Bissao, Sorcerer's isle, Arcas, Boolam, Gallina or Hen's island, Parrot isle, Cazegoot or Point island, Corbele, Carasche, Warang or Formosa, Cayo, the Three Sisters, Soga, Rooben, Hog's island, Bawak, Kanabec, Orange, and a number more inconsiderable are enumerated among or near them. The inhabitants being hostile, nothing of the interior is known. The channels that bound the group are on the N. the Jeba, or great channel of Bissao, and on the E. that of the Rio Grande. These channels have been partially surveyed. At Bissao the Portuguese have an establishment, and a fort 100 fathoms from the shore, a square redoubt flanked. Water is to be had here, wood, and excellent bullocks, as well as other provisions. The island is about 5 miles long by 2 broad, and has been a great hold of the slave trade. The Bolola channel is another name for that of the Rio Grande. Boolam island may be approached within a mile. It is well wooded, and has good anchorages, where the tide rises from 12 to 15 feet. Water may be had on the S.W. side. It is 30 miles from Bissao, and claimed both by the Portuguese and English. It is a most unwholesome spot, and on that account was abandoned, until a Portuguese slave dealer settled there again in 1829, but this nest of men dealers was rooted out in 1838, by Lieut. Kellett, of the British navy. High water among these islands is 10ʰ 17ᵐ A.M., rise from 16½ to 18 feet. Captains Belcher and Owen, R.N., have surveyed and given the best extant account of the islands. There are many volcanic marks about these islands and shoals. The navigation is intricate and often dangerous. The N. end of the shoals of Rio Grande is in lat. 11° 40′ to 11° 43′ N. Bonn and King's isle are near Bissao. The river Nunez on the main, or river of Nuna Tristao, is a place of trade for ivory; its entrance is in 10° 36′ N. lat., and 14° 42′ W. long. Alcatraz islet, with its reefs, is in lat. 10° 37′ 12″ N., long. 15° 26′ 30″ W. Sandy island is in lat. 10° 36′ 37″ N., long. 14° 42′ W. Bissao fort is in lat. 11° 51′ N., long. 15° 37′ 6″ W.

BIZERTA BAY, N. coast of Barbary, Africa. Cape Zibeet is the N.E. point; its W., called Ras al Abead, or Cape Bianca, bears from Cape Zibeet W. by N. ½ N. distant 10¾ miles. The bay is 4 miles deep, and has on its W. side the town, port, and bay of Bizerta. There is good anchorage at a distance from the shore in 5, 6, and 7 fathoms. This bay is 10 leagues W. from the ruins of Carthage, in lat. 37° 5′ N., long. 9° 47′ E., and by some is said to have been the port of Carthage.

BJORKA ISLANDS, Russia, three small islands in the Gulf of Wyborg.

BLACK BAY, Labrador, S.E. coast. Lat 51° 30′ N., long. 56° 20′ W.

BLACK CAPE, E. coast of Newfoundland, S.E. from Cape St. Francis, between that and Red head, with which it forms a bay called Torbay. Cape St. Francis is in lat. 52° 20′ N.

BLACK COVE HEAD, Ireland, N.E. coast, in lat. 54° 53′ N.

BLACK GROUND, Cyprus, Levant, a bank so called 12 leagues S. of Salinas bay, with only 6 feet water.

BLACK-HALL HEAD, Ireland, S.W. coast, at the N. of the entrance into Bantry bay. Lat. 51° 32′ N., long. 9° 55′ W. Also on the S. coast within the Old head of Kinsale. Also on the S. side of the entrance to Galway harbour.

BLACK HARBOUR, Scotland, W. coast, near Loch Melfort, a small creek on the W. side of Loing, E. of Blada island and Duskar rock. It is sheltered by rocks, to which small vessels make fast, and ride safe in 3 fathoms water.

BLACK HEAD, Newfoundland, on the S. coast, to the W. of Cape Race, the S.E. angle of the island, and half a league further W. from Cape Pine, which forms the E. limit of the deep Bay of St. Mary. Cape Pine is in lat. 46° 37′ 14″ N., long. 53° 35′ 50″ W.

BLACK HEAD, E. coast of Ireland, entrance into Belfast Lough, 4 miles N.E. from Carrickfergus. Lat. 54° 56′ N., long. 5° 42′ W.

BLACKHEAD, Cornwall, between Falmouth and the Lizard, W.S.W. from the Manacles. Lat. 5° 0′ 27″ N., long. 5° 6′ 35″ W.

BLACKHEAD, Cornwall, the N. of Mevagissy, and S. head of St. Austle bay.

BLACK ISLANDS, Labrador, near the E. coast. Lat. 51° 8′ N., long. 56° 30′ W.

BLACKNESS, France, N. coast, 4 leagues W. of Calais, a foul shore all the way. Hence the coast trends to Boulogne, distant S. 3 leagues. At a league from the Ness the coast forms a kind of sandy bay, which is a tolerable secure road in E. winds, with 5, 6, 10, and 15 fathoms water. The steeple, or village, over the midst of the houses, is the best anchorage. Off this Ness the flood-tide sets the first quarter to the land, on the N.N.E., but between it and Boulogne the flood sets N. by E. by the land, and N.N.E. in the offing.

BLACK POINT, Caernarvonshire, North Wales, N.E. entrance to the Menai strait. It has a stone lighthouse, showing a fixed light, of a red colour, for guiding vessels into the strait.

BLACK POINT, North America, S.W. of Cape Elizabeth, the bay of that name being N.E. of it. It is also the limit of Saco bay. There are rocks off this point.

BLACK POINT, Africa, W. coast, between Capes Cavallos and Palmas, 3 leagues W. from Little Setry, and 8 leagues from Great Setry. There is anchorage here in 18 and 20 fathoms.

BLACK POINT, Labrador, S.E. from Cape Chudleigh, the N. point of the Labrador coast. Lat. 59° 20′ N.

BLACK POOL, a town in Lancashire, 25 miles from Lancaster, on a straight coast, with a broad sandy beach; much used as a watering-place, but without any shelter for the smallest vessels.

BLACK RIVER, Jamaica, in Elizabeth parish; it flows through a flat country, and is navigable for flat canoes and boats for 30 miles. It has a harbour at its mouth. Lat. 18° 1′ 10″ N., long. 77° 53′ 15″ W.

BLACK RIVER, Mosquito coast, Gulf of Mexico, known by the La Cruz mountains, after passing Nicaragua. The anchorage is an open road, and demands considerable attention to the weather from October to February. The bar of the river is very dangerous.

BLACK RIVER, at the mouth of the Tonto, 20 leagues E. of Cape Honduras, North America, Gulf of Mexico. It is the only harbour from Rattan isle to Cape Gracias a Dios.

BLACK RIVER, Lorain co., United States of America, a small town on the river of that name, adjoining Lake Erie, with a good harbour and rising trade. Pop. 668.

BLACK ROCK, Scotland, in the navigation of Loch Eil from Curran ferry to Fort William. This rock, and a sunken rock with 18 feet at low water, within a quarter of a mile W. by S. of it, are the only dangers to be feared. Both are on the S. shore, 500 feet from high water mark, with 6 and 7 fathoms inshore of them. The Black rock is always visible, bearing N.E. by E. ¼ E. from Curran point, distant 2⅔ miles.

BLACK ROCK, Ireland, 2 leagues W.N.W. from Achill head, on the W. coast, and S.S.W. 6 leagues from an out point S.W. by S. from the Stags. Ships may sail between this rock and Achill head, and also between it and Broad Haven island. Lat. 53° 55′ N., long. 10° 52′ W.

BLACK ROCK, Erie co., New York, United

States of America, a town near the bottom of Lake Erie, 2¼ miles below the city of Buffalo. The harbour is formed by an immense stone pier projecting into Niagara river, for sheltering boats and vessels from the lake.

BLACK ROCK, Antigua island, West Indies, between Humphrey's bay and Boar point. Lat. 17° 5′ N., long. 61° 58′ W.

BLACK ROCK LIGHTHOUSE, N.W. side of the entrance to the Mersey river, England, having a revolving light 88 feet above the sea, alternately two bright lights and one red, at minute intervals. See *Liverpool*.

BLACK ROCK POINT. See *Cork*.

BLACK SEA, the ancient Pontus Euxinus, it lies between 41° and 46° of N. latitude. It is about 260 leagues long by 80 broad. It is entered from the Mediterranean through the Channel of the Dardanelles, the Sea of Marmora, and the Bosphorus, or Channel of Constantinople. It receives the waters of the Sea of Azoph, or Azof, by a strait on the N. It receives also a great number of the European rivers, as well as some of the Asiatic. By the Danube it collects the waters of Germany, Hungary, Bosnia, Servia, and Transylvania; by the Dniester and Dnieper, those of a part of Russia and Poland; by the Phasis, those of Mingrelia; by the Sangaris and others, those from the coast of Anatolia; and of the Kuban and Don, from the Sea of Azof. This sea is so fresh that it freezes with a small degree of cold, and its waters are turbid with the mud it receives from the rivers, which it is calculated discharge into it nine times more than passes out into the Mediterranean through the Bosphorus. The river-streams cause strong currents in this sea, in summer particularly, it being increased in that season by the melting of the snows: and when strong winds act against these currents, a high chopping sea is produced, sometimes with fogs, which renders the navigation at times somewhat hazardous. The climate is in winter humid and very cold. The variations of temperature are exceedingly great, even at Constantinople. The Sea of Azof (see *Azof*) pours its muddy waters into the Euxine, through the Strait of Kertche and Jenikaleh; but though there are numerous islets in this sea at the mouth of the Danube, the rocks and shoals are few. One which stands off there, some distance from the main, is called the Serpent's island, and contains a lighthouse. The shores are varied, often rocky, frequently they consist of sandy downs. The outlet to this large body of water is the narrow strait called the Bosphorus, though a constant current sets to the southward, which is but natural when it is considered that it is the outlet of a superficies of so great an extent as 260 leagues by 100. It is everywhere deep. In several places the shores become mountainous inland almost immediately. The currents vary according to the violence of the wind. The N.W. and W. winds are observed to bring humidity and a thick, foggy atmosphere, and a N.E. to bring clear weather in summer and in winter cold. N. winds are prevalent in summer, and in autumn and spring alone are S. winds experienced. The harbours on the N. side of the Euxine are often frozen, and sometimes the sea itself for ten miles distance from the land. The harbour and bay of Odessa are almost annually frozen, and some portions of the Danube. The whole difficulty in the navigation of this sea seems to be that of making the opening of the Bosphorus. This arises from the practice of getting to Iniada,

and there waiting for a fair wind, then a gale has sprung up unexpectedly, placed them on a lee shore, and they have been wrecked. Whereas, had they kept out to sea, and run for it when it bore S.W. by S., or S.W. by W., there would have been no hazard, as it may be seen from 25 to 30 miles off in clear weather. This sea is far from being as stormy or dangerous as it has been represented. The Black sea has no tide nor rise of water but what is caused by wind or current, or both. The approach to the Bosphorus is always with a current, which may be accounted for from its being the great outlet of all the waters that flow into the large expanse of the sea itself. The outlet is marked, on the point of Europe, at the very entrance, by a lighthouse and fort, and the same is repeated on the Asiatic side, in a double light, one over the other. The lat. of the European light is 41° 14′ N., the long. 29° 7′ E. There is a false entrance 15 miles E. of the real, which is deceptive. See *Bosphorus*. The W. point of the Black sea, from the foregoing outlet, curves as far as the Gulf of Ainada, or Iniada, in lat. 41° 53′ N., about 62 miles. Then succeeds the Gulf of Bourgas, next the Gulf of Varna, where the coast has turned N. Passing the Bay of Kavarna and Cape Kallagrich, the coast runs somewhat N.E. to the mouths of the Danube and Dniester, thence by Cape Fontan, on which there is a lighthouse, as far as Odessa; thence nearly E. to Otchakoff, off which there is a light, at the entrance of the estuaries of the rivers Bourg and Dnieper, leading up to the town of Nicolaef. The coast then for a short distance from Fort Kilbourn passes the gulf of that name, as far as the point on the Tendra isle, where there is a light which revolves, continues yet more S.E., and rounding the Gulf of Kerkinet, at the upper end of which is the Isthmus of Precop, it turns S.W. past Cape Karamnoun to Cape Tarkan, where there is a fixed light. Turning S.E. again from this point, passing the Cape and town of Kosloff, Cape Katcha, the town of Sevastapol is reached, seated on a deep bay. One point of this bay is Cape Chersonesus, on which is a revolving light, 170 feet high, of a red colour, in lat. 44° 34′ 24″ N., long 33° 21′ E. From hence the coast rounds to Cape Aitodor, and then goes N.E., with several indentations, as far as the entrance to the Sea of Azof, or Strait of Jenikaleh. From this strait the coast runs S.E. to Gouria, and soon after, curving westward by Rizeh, Treboli, Sinope, Cape Kerempels, Amasserah, Cape Baba, and Cape Kirpel, stretches to the Asiatic entrance of the Bosphorus, the S. shores presenting far less indentations than the N., and having deep soundings. The headlands E. of the Bosphorus are of a whitish hue, those on the W. are rugged, high, and irregular.

BLACK SOD BAY, near Achill head, W. coast of Ireland, a place with tolerable shelter and clear ground, with water for any vessel, but the bottom does not hold well in gales from the W. or S.W. The best anchorage is on the W. side of the bay, a quarter of a mile S. of Barnack; further off large ships may lie in 4½ or 5 fathoms, hard sand bottom. Lat. 54° 7′ N., long. 9° 48′ W.

BLACK STONE, Devonshire, S. coast, a rock near the entrance of Dartmouth harbour, 200 fathoms within a rock called the Home stone, which has only 4 feet of water over it. The Black stone is bold to, and sailing in from the W. with a leading wind, give the Coomb rocks a good berth, and steer E. until Kingsware castle be open

E. of the Black stone, which will clear the Home stone, and when Stoke church is shut in, run for the Black stone pass, as it is steep, and steer in with the castle on the larboard bow.

BLACK TAIL, a beacon 3 leagues from the Nore, in the entrance of the Thames, to the E.N.E., being the point of a long sand which runs down on the N. coast E., between which and the Nore buoy it is all clean sailing. See *Thames* and *Nore.*

BLACKWATER, Essex, below Malden, on the S. of Mersey island; it forms a tolerably convenient haven.

BLAESDALE'S REEF, about lat. 0° 57′ N., long 41° 6′ W., a coral reef, on which the brig Richard of Ulverston struck in 1819. The ship remained fast 10 minutes. No breakers were seen, and close by no bottom was found at 150 fathoms. Three holes were afterwards found in the ship's bottom, nearly through.

BLAKELEY, North America, United States, co. of Alabama, a town on the E. side of the Tensaw, the E. outlet of Mobile. The harbour admits vessels drawing 11 feet of water.

BLAKENEY HARBOUR, Norfolk, is the best harbour on the coast for refuge during gales upon the shore. It is a bar-harbour and buoyed, but the sands frequently change their position, and it is therefore hazardous to strangers. The tide flows until 6ʰ in the harbour, but outside runs S. 3ʰ longer. Signals are established to guide vessels, to which it is necessary to pay exact attention on making the entrance.

BLAKENEY SAND, or KNOCK, Norfolk coast, a small bank 7 or 8 miles N. from Blakeney, on the W. of the larger sand called Cromer Knowl. It has but 3½ fathoms at low water, but 4 within it, and on the outside, or N. of it, 5 or 6 fathoms.

BLANCA, CAPE, Greece, the N.W. point of the Isle of Negropont.

BLANCA ISLAND, North America, Mexican gulf, a small isle 6 leagues S.S.E. from Vera Cruz.

BLANCH, CAPE, France, a headland, N.W. of Calais, in the English channel.

BLANCH ISLAND, France, often called the White island, between the Seven islands and Marion island, near the shore N.E. of Morlaix, in Britany.

BLANCO, CAPE, or BIANCO, and SHOAL, Corfu island, S. end, lat. 37° 15′ N., long. 20° 12′ 8″ E. On the W. end of the shoal is a black buoy.

BLANCO, CAPE, Africa, W. coast, the southernmost, in lat. 20° 47′ N., and 17° 4′ 36″ W.

BLANCO, CAPE, Palestine, N. of Acre, 13 miles S. of Tsour or Tyre.

BLANCO, CAPE, Tunis, called also MARABUT. Lat. about 37° 15′ N.

BLANCO, CAPE, Island of Corsica, one of the two points forming Cape Corso, on the extreme N. of the island, 3 or 4 leagues from Cape Segri, between which two points there is a bay, where ships may be secured from a W. or S.W. wind.

BLANCO, CAPE, Majorca, on the S.W. side of the island, answering to Cape Cuba Figuera, the Bay of Palmas between. See *Balearic Islands.*

BLANCO, CAPE. W. coast of Africa, the northernmost, lat. 33° 8′ N., long. 8° 38′ W.

BLANCO, CAPE, South America, Caraccas province, St. Jago de Leon lying S.W. of it. It is in lat. 11° 36′ N., long. 67° 6′ W.

BLANCO, CAPE, on the S.W. of the island of Cyprus.

BLANCO ISLAND, South America, off Margaritta island from the main.

BLANCO, POINT, North America, S.W. coast of Mexico. Lat. 10° 1′ 40″ N., long. 83° 5′ 42″ W.

BLANES, or BLANDA, E. coast of Spain, in Catalonia, 7 leagues S. of Girona. Lat. 41° 40′ N., long. 2° 50′ E.

BLANK MOINS ROCKS, France, off St. Matthew's point, at the N.W. point of the entrance towards Brest road; there is a good road between these and the Finisterre rocks under St. Matthew's abbey.

BLANK, THE, or THE WOOLSACK, Holland, a white sandhill on the island of Schonen, upon its W. side, standing higher and looking whiter from seaward than the others.

BLANKENBURG, coast of Zealand, between Ostend on the S.W. and Cadsand island to the N.E. Off this coast there is a peculiar set of the tide, for when bearing S.E. by E., the tide to the E. turns about against the sea, though it does not do so from thence to the W. It is 4 leagues N.E. from Ostend. Lat. 51° 18′ N., long. 3° 24′ E.

BLANQUILLA ISLAND, coast of Cumana, South America; 40 miles from the N.W. point of Margarita are the Hermanos isles, and 7 miles from the S. of these is the island of Blanquilla, 6 miles from N. to S., and 3 from E. to W.; flat and wholly desert.

BLAS, CAPE ST., or BLAISE, E. from the Mississippi; in lat. about 29° 47′ N., long. 86° W. The W. limit of Apalachicola bay.

BLAS, ST., BAY, North America, 4 miles S. from Sombello point, W.N.W. from the Gulf of Darien, a point or two more to the N. About lat. 9° 34′ 36″ N., long. 78° 57′ 40″ W. It is 18 or 20 leagues E. of Porto Bello.

BLASIUS ISLET, Spain, near the entrance to the harbour of Ferrol.

BLASKETS, W. coast of Ireland, a cluster of rocks lying W. ½ S. 3½ miles from Dunmore head, and extending in several directions to others differently named, as Ferriter's islands, the Foze and Tiraught rocks, and others. The Foze rock is in lat. 52° 1′ N., long. 10° 39′ 40″ W. Blaskets sound lies within these islands nearly N. and S., and has 10 fathoms water. There is here a strong ebb and flood tide. High water at half-past 4.

BLASQUES ISLAND, North America, W. coast of Newfoundland. ·Lat. 47° 30′ N., whence the coast trends N. by E. and N.N.E.

BLASQUES, France, see *Basque Road.*

BLAVET, France, sometimes called Port Louis, and properly the name only of the river where it enters the sea. It is in the department Côtes du Nord, passes the road of L'Orient, and joins the sea at Port Louis. It is rendered navigable by artificial means as far down as Hennebon, and to Hennebon the tide flows up. The part artificially made navigable is called the canal of Blavet.

BLAYE, France, department of the Gironde, on the river of that name, near Bourdeaux. It has a citadel upon a lofty rock, and vessels bound up the river leave their guns and powder there until they return. There is an island in the river, defended by batteries. It stands in lat. 45° 7′ N., long. 45′ W. It is high water at 3 o'clock. Blaye is on the E. bank of the river, 7 leagues from Bourdeaux. Pop. 2,880.

BLAZEY, ST., BAY, on the S. coast of Cornwall, more generally denominated Tywardreath bay. Its E. point is Gribbin head, and it has the Black head to the S.W., which last is also the limit of St. Austle bay. There is no navigation here, except for small vessels. The extensive sands at the head of the bay, called Par sands, are dry at low water.

BLEKINGEN, see *Carlscrona*.

BLEWFIELD'S LAGOON, and BLEWFIELD'S BLUFF, Mosquito country, Gulf of Mexico, North America. The point is in lat. 11° 19′ 20″ N., long. 83° 40′ 18″ W.

BLIGH BANK, one of the banks 2¾ miles E. of the East Hinder, from half to three-quarters of a mile broad, running N.E. 10 miles.

BLIND HARBOUR, Ireland, a bay on the S. coast, in lat. 51° 31′ N., long. 9° 40′ W.

BLIZARD'S BAY, West Indies, Antigua island, just where the coast begins to fall off to the N. It is S. by E. from Beggar's point, with a sand running up to the N.N.E. from a point on its S.E. side. This sand and bay are on the W. side of the channel at S.S.E. and S.E. by S., between Maiden island and Long island, towards the N. Sound point.

BLOCK ISLAND, North America, United States, near Rhode island. Lat. 41° 9′ N., long. 71° 32′ W.

BLOCK ISLAND, United States, North America, Newport co., on the Atlantic, 14 miles S.S.W. of Point Judith. It is 8 miles long and 3 broad. The inhabitants are principally employed in fishing. N. lat. 41° 8′. On the N.W. point are two lights, 25 feet from each other, fixed lights, to mark the direction of the reef; they are in a line bearing S.

BLOCKS, THE; these are two rocks in the way to Bommel from Wybrant's Eye, on the coast of Norway, the course N.W. by N. and N.N.W. 5 leagues, all broken land, islands, and rocks on the W. side, except that in the midway there is a channel clear to the N. or N. by W. into the sea. The Blocks are level with the water, and must be left on the larboard; in sailing to the N. and the E. shore, run along pretty close, having Longholm island in a bay near the shore on the starboard.

BLOO BARRA, Barbarra factory, Sinou, coast of Africa; lat. 4° 59′ 15″ N., long. 9° 2′ 5″ W.

BLOCKZYL, Holland, at the mouth of the old Aa, having a good port, in the Zuyder Zee, in the province of Overyssel. The harbour will contain 200 sail of vessels. This port is in lat. 52° 44′ N., long. 5.° 39′ E.

BLOODY FARLAND'S CAPE, N.W. coast of Ireland, 12 leagues W.N.W. of Londonderry. Lat. 55° 10′ N., long. 8° 11′ W.

BLOODY ISLAND, Minorca, the largest island within the harbour of Port Mahon, almost in mid-channel, the small island of Golgotha W. of it, in a line nearly towards the mouth of English cove, to the W., on the larboard shore going up towards the town.

BLOODY POINT, St. Kitt's, West Indies, on the S.W. side of the island, and S.E. from Old Road bay, and the N. point of Palmetta bay.

BLOOM ROCK, off Falconera, Greek Archipelago, N.E. by E. from Cape Angelo, a reported shoal. It was sought by Commander Brock, of the Royal Navy, without success.

BLOWING POINT, Jamaica, W. of Anotta bay.

BLUEFIELD'S BAY and RIVER, West Indies, Jamaica, on the S.W. side of the island, a roadstead, contracted at the entrance. A shoal runs off S. by E. from Cabarita point. Lat. 18° 10′ 30″ N., long. 78° W. The watering place is on the lee side of the bay, N. of Bluff point. Large ships anchor outside the reef or shoal in from 8 to 5 fathoms, with Crab Pond point about S.E. by E., and the tavern E.N.E. or E. by N.; the lead must be kept going, the water shoaling of a sudden.

BLUE MULL SOUND, see *Shetland Isles*.

BLUFF ISLAND, Africa, S. side of the entrance of Cameroon river, the N. entrance lying in lat. 3° 7′ N., long. 9° 20′ E. A chain of heavy breakers lies 6 miles N.W of it, called the Dogs' heads.

BLYDONES ISLAND, Spain, off the mouth of Porto Verde, S.E. by E. 4½ leagues from the river Roxo. Ships may sail round it, though on the N. side there is only 5 or 6 fathoms water.

BLYTHE SAND, river Thames, close to the S. shore, in the passage from Tilbury fort to the Nore. The steeple of Tilbury church about a sail's breadth without Swansheet point will escape it. At night keep the lead going, and come no nearer the shore than 7 fathoms at high and 5 at low water; this will keep all clear to the Nore head. A permanent standing beacon marked by a staff and ball, has been placed on the N. westernmost spit, 1½ cable's length from the old beacon, N.W. by W. ¼ W.

BLYTHE, coast of Northumberland, at the mouth of the river Coquet, 3 miles N. from Seaton sluice, or Hartley haven. It is high water at full and change here at half-past 2 o'clock. The flood tide sets S.E. by S., and at springs rises nearly 18 feet, and usually 16. The island at the mouth of the river is in lat. 55° 13′ N., long. 1° 36′ W.

BOAT, coast of Durham, a sunken rock with only 2 feet water upon it at ebb tide; the Dogger Rock near it appears above water; both are 2 leagues N. of Hartlepool, about a mile from the land.

BOAT ISLAND, Gulf of St. Lawrence, North America, near the coast of Labrador. Lat. 50° 2′ N., long. 60° 55′ W.

BOB'S, or HOB'S NOSE, Devonshire, the E. point of Torbay, or a little N. of the N.E. limit. 50° 27′ 50″ lat. N., 3° 26′ 43″ long. W.

BOB'S ISLAND, Africa, at the entrance of Sherbro river.

BOCA CHICA, or THE LITTLE CHANNEL, the narrow entrance into the harbour of Carthagena, South America. It is about 8 miles S. of the Boca Cerada, and has an island just within which is fortified. The entrance is strongly defended. Lat. 10° 19′ 30″ N., long. 75° 36′ 17″ W.

BOCA DE DRACO, South America, commonly called the Dragon's mouth, one of the entrances into the Gulf of Paria, between Trinidad island and the main. See *Bocas of Paria*.

BOCA ESCONDIDA, Yucatan, North America, in the bay of Campeche. Lat. 18° 50′ N., long. 91° 46′ W.

BOCA GRANDE, North America, on the S.E. coast of Costa Rica, in Guatemala. Lat. 10° 50′ N., long. 83° 26′ W.

BOCA GRANDE, Florida reefs, West Indies, E. of Cayo Marques, or Marques kay, the W. of the Florida reefs, 14 leagues E. ½ N. from the S.W. kays of Tortugas, the last are in lat. 24° 31′ 30″ N. It is a large opening 2 miles wide going N., with 9 feet water at the lowest, but it is not to be recommended to strangers.

BOCA TORA, BAY OF, Veragua, Mexico, North America. Lat. 8° 58' N., long. 82° W.

BOCANAO RIVER, Cuba, West Indies, E. of the Havanna, the Cojimar river being between.

BOCARETS, France, the name of the channels which communicate between the sea and the lagoons of the Landes, in the department of that name.

BOCAS, mouths of rivers, passages or channels between islands, originally applied by the Spaniards to such localities. The word is most commonly used in the Americas and West India islands.

BOCAS OF PARIA. The Boca de Meno, or Ape's channel; Boca de Huevos, or Egg channel; Boca de Navios, or Ship channel; and Boca Grande, or Great channel. These are denominated generally the Bocas del Drago, or Dragon's mouths, between Trinidad and the main of South America. The Boca de Huevos is sometimes called the Umbrella channel. There are different entrances by the E. end of Trinidad in succession by the islands denominated Apes, Huevos, and lastly Chacachacarreo, with the Point or Cape de la Pana, near Cape Salinas, on the opposite coast. The Ape's channel, which is nearest to Trinidad, though narrow and dangerous, has 25 fathoms water, but with a strong and irregular current, subject to calms, and least estimable of all the Bocas. There are rocks near Meno island that have 8 fathoms close to them. High water, full and change, in this Boca at 3 o'clock. The Huevos, or Egg channel, named the "Umbrella" from a curious tree growing near, is about a mile wide between the islands Meno and Huevos, and above 100 fathoms deep. It is considered the shortest and best passage by which to enter the gulf. There is a rock at the S.W. point of Meno island, easily avoided by keeping closer to Huevos, besides preventing being becalmed by Meno island, which is high. This passage should not be attempted with scant wind. The Boca de Navios is best for vessels bound outwards. It is from three-quarters to 1½ mile wide. Keep the E. land on board. The water almost always runs out of this channel at flood at the rate of 1½ mile an hour, and with ebb to 4 miles, and therefore it is not to be used for entering. The Rio Grande passage is 5 miles broad; it is clear of dangers, except one rock, about 40 yards in circumference, which has only 9 feet of water, with very deep water close around it. There is in some places no bottom to be found at 100 fathoms. The best time for entering the Bocas is two hours after the tide begins to flow, and to go out one hour after it commences. The Boca de Navios, or Ship channel, is in lat. 10° 41' 45" N., long. 61° 45' 30" W.

BODKIL, Holland, one of the channels between Belgium and Walcheren island in Zealand, deep enough for large vessels. These channels are often generally described by the term of the Weikings, from the name of the largest. To sail in or out of the Bodkil, or the E. Gat or Land deep, to the N. of it, the passage is N. or S. of the Stone bank, and when Middleton comes over Domburgh, it is athwart of this bank. When West Chapel is E.S.E. in a black valley, is the fair way into the Bodkil, before which the tide commences running an hour before low water. Spring tides have high water at 12 o'clock.

BOGGY POINT, Barnstaple bay, Devonshire. See *Bag Point.*

BOGHAZ CHANNEL, Egypt, the central entrance into the port of Alexandria. See *Alexandria.*

This word thus used is a superfluity, "Boghaz" being Turkish for "Channel," and it is found used ¹n that sense in many parts of the East expressive of such localities.

BOGNOR ROCKS, coast of Sussex, between the mouth of the Arun and Pagham. Their direction from Bognor is S.E. by S. Dunnose open of Selsea bill clears them.

BOGOE, or BOOG ISLAND, Denmark, between the islands of Moen and Falster.

BOHEMA, or RAMSEY SAND, Isle of Man, off Ramsay bay at the N.E. end, breaking off the sea; it forms a tolerable harbour, where ships may anchor in 20 or 25 fathoms. At the bottom of the bay is the little river of Selby, where small vessels may run in at high water, and lie dry on ebb.

BOIS, CHEDE, France, the N. point of the entrance into Rochelle, 2½ leagues S. by E. from the Isle of Aix.

BOIS LE DUC, Belgium, nearly surrounded by the rivers Aa on the S.E., and the Dommel on the S.; in winter only approachable by boats. Lat. 51° 40' N., long. 5° 16' E.

BOJADOR, BOIADOR, or BOGADOR, W. coast of Africa; lat. 26° 7' 2" N., long. 14° 30' 24" W. False Cape Bojador, lat. 26° 25' 15" N., long. 14° 12' 30" W. High water 12ᵇ.

BOLOLA TOWN, Bijooga islands, in the Rio Grande. Lat. 11° 35' N., long. 15° 2' 18" W.

BOLSWAERT, Friesland, Holland, a town near the Zuyder Zee, 10 miles N. of Sloten, and a league from the Fly or Vlie Stroom, or channel, along the E. coast, within the islands going into the Zuyder Zee. Lat. 53° 3' N., long. 5° 25' E. The port is greatly obstructed by sand.

BOLT HEAD, Devonshire, signal stations, in lat. 50° 13' 15" N., long. 3° 48' W. There are two points, called the Bolt Head and Tail, 5 miles distant from each other. The Head is a little S.E. of the Tail and the W. point of Salcombe harbour entrance.

BOLTSACK, or BOLTSACKEN ROCKS, Denmark, at the N. entrance of the Great Belt, 5 miles S.E. from Samsoe island. Lat. 55° 48' N., long. 10° 40' E.

BOLUS HEAD, Ireland, S.W. coast, 38 miles S.E. of Killarney. Lat. 51° 44' N., long. 10° 12' W.

BOM FELIX SHOAL, BONETTA SHOAL, BOUVET'S SANDY ISLAND, dangers reported to have had an existence, the first near the Maria rock, which last is said to be in lat. 19° 45' N., long. 20° 50' W., but neither can be found. The Bonetta is supposed to have been taken for Hartwell's reef, in sight of Bonavista. Bouvet's Sandy island in like manner cannot be found, after careful search. The same may be said of Baxo das Garcias, lat. 13° N., long. 29° 50' W.

BOMBA, Tripoli, a gulf on the coast, of which Seal island is the S. point. It has a second island in middle of the gulf called Bhurda, and a third named Zouzra Mezrata, or Menelaus, where there are 11 and 12 fathoms water. The centre island, or Bhurda, is distant from Cape Razatin 11 miles; that cape being in lat. 32° 34' N., long. 23° 13' 10" E.

BOMBA, S.W. coast of Africa, a river which falls into the Sherbro, within Cape St. Anne, S. by E. of Sierra Leone. There is good anchorage on both shores at its mouth.

BOMBORN ISLAND, Africa, an islet on the N. side of Princes island, Bight of Biafra.

BOMMEL ISLAND and SOUND, Norway, on the coast N. of Scute Ness, its S. point is named Bommel head, and S. of that is Bommel sound, into which two or three small rivers fall. The N. point of the island is in lat. 59° 25′ N., long. 5° 57′ E.

BOMMEL-WAART ISLAND, Holland, province of Gelderland, formed by the junction of the Waal and Maas. It is 14 miles long, and 4 broad in its widest part.

BOMMENE, Zealand, on the N. shore of Schonen island, opposite Goree, and a league N.E. of Browershaven. Lat. 51° 50′ N., long. 4° 2′ E.

BON, CAPE, N. African coast, N.N.E. of Tunis, in lat. 37° 6′ N., long. 11° 2′ 35″ E. It is high land, stretching towards the islands off the W. end of Sicily, to which a bank of soundings extends. On the N. part of this bank are Keith's reefs and the Skerki rocks. The middle of the former being in lat. 37° 50′ N., and long. 11° 9′ E., dangerous for vessels bound to Malta. The Skerkis are two reefs, 3½ miles N.N.E. and S.S.W., in lat. 37° 46′ N., long. 10° 47′ E. Cape Bon may be seen 15 or 16 leagues off. From Cape Bon the coast runs W. more than 2 leagues.

BONA, Algiers, coast of Africa, called Blaid el Aned by the natives. Lat. 36° 53′ 30″ N., long. 7° 48′ 20″ E. It is 5 miles from Cape Mavera. The harbour was once commodious, but it is now choked up. Vessels may anchor between Bona and Cape Mavera, in the little cove of Genovese, where there is from 15 to 7 fathoms of water, but exposed to N.E. and E. winds.

BONACCA, or GUANAJA, an island, 8 leagues N. ½ W. from Point Castilla, Mosquito shore, Gulf of Mexico. Lat. 16° 30′ N., long. 85° 47′ 37″ W. The harbours in this island are excellent.

BONAIR ISLAND, South America, on the N. coast, 10 leagues W.N.W. from the islands of Aves, and about 14 E. from Curaçoa, about 18 leagues in circumference. There is a good bay and harbour on the S.W. side, near the middle of the island. Lat. 12° 26′ N., long. 68° 18′ W.

BONANDRIA, CAPE, N. coast of Africa, on the shore of Barbary, with a small town E. of it, in lat. 32° 20′ N., long. 23° 14′ E. It is E. by S. from Cape Razatin.

BONAVENTURE, CAPE and PORT, E. coast of Newfoundland, S.W. of Bonavista cape, forming the N. entrance into Smith's sound, whence the coast runs S. by W. into Trinity bay. The port is 2 miles or more from the head of the cape N.W. and S.E. The cape or head is in lat. 48° 16′ 30″ N., long. 53° 23′ 58″ W.

BONAVENTURA ISLAND, South America, on the starboard side of the entrance into Porto Bello harbour, Columbia. Lat. 79° 28′ W.

BONAVENTURE ISLAND, New Brunswick, North America, N.E. of the Bay of Chaleurs, a league from Gaspe bay. The N.W. point is in lat. 48° 30′ N., long. 64° 11′ 12″ W.

BONAVISTA, see *Cape de Verde*.

BONAVISTA, CAPE, Cuba, West Indies, the extreme N.W. of Cuba island, covered with rocks and shoals; about lat. 22° 23′ N., long. 84° 2′ W.

BONAVISTA, CAPE, E. side of Newfoundland, lat. 48° 41′ N., long. 53° 5′ W. It forms the S.E. limit of Cork bay. It is a bluff, bold head. Gull island is three-quarters of a mile from it N. by W. Lat. 48° 42′ 40″ N., long. 53° 5′ 8″ W. On Cape Bonavista there is a lighthouse, with a revolving light at two minute intervals, alternately red and white, 150 feet above the sea.

BONCASSIN, Hayti, N. of Port au Prince, about lat. 18° 9′ N.

BONE AND GRISTLE, shoals within the bar of Harwich harbour, in the middle of the channel; they are both buoyed.

BONETTA SHOAL. See *Cape de Verde*.

BONIFACIO, Corsica, a port, known by its white cliffs, houses, and fortifications. It is in lat. 41° 23′ 11″ N., long. 9° 9′ 16″ E. The harbour is fine, but the entrance narrow, and it can only be entered in fair wind and weather. Pop. 4,000. The STRAIT OF BONIFACIO is bounded by Corsica and some small islands on its N., and on its Sardinian side by other islands. It is safe for all kinds of vessels, except in one small spot, where the sea runs high, S. from Lavessi island about a mile. There is a lighthouse, with a revolving light, upon Cape Pertusato, at the W. entrance of the strait, in lat. 41° 22′ 10″ N., long. 9° 11′ 20″ E.

BONNE BAY, Newfoundland, W. coast, lat. 49° 35′ N., long. 53° W.

BONNE NUIT, a headland on the N. side of Jersey island.

BONNY RIVER, S.W. coast of Africa, of which Rough corner is the S.E. point; the town is 4 leagues from Rough corner up the river, on the starboard side. It is about 21 leagues E. from Rough corner to Old Calabar river, which last is in lat. 4° 23′ 40″ N., long. 7° 7′ E.

BOOBY ISLAND, West Indies, island of St. Kitt's, off the S.E. end, opposite Mosquito bay, about half a mile from the shore. It is also off the N. end of Nevis, within a bank interspersed with rocks on its N.E., for almost 3 leagues parallel to the coast of both islands, from N.W. to S.E.

BOOM KITTAM RIVER, Africa, near Sherbro river, in lat. 7° 14′ 24″ N., long. 12° 8′ 36″ W.

BOOMLAND, or LALAND BANK, the second from the shore, ending nearly opposite Nieupert, in Flanders, between that and Ostend. The four banks beginning off Ostend are called the Flemish banks.

BOON ISLAND, United States, coast of Maine, between York river and Cape Neddock. On its N. part is a stone tower, with a fixed light, 70 feet high, visible 6 leagues. This light is in lat. 43° 7′ 30″ N., long. 70° 27′ 30″ W.

BOON'S POINT, West Indies, Antigua, the most N. point of that island, from which the coast trends a little S. of E. on that side, and S.W. or more S. on the other. Lat. 17° 7′ N., long. 62° W.

BOOSAIDA, African coast, Gulf of Sydra, lat. 31° N., just where the bank of Kudia commences, off which a danger, as yet undetermined, is reported to exist.

BOOTH BAY, United States, North America, coast of Maine, 2 miles W. of Pemaquid point, lat. 43° 42′ N. At Burnt island, near this bay, is a tower, with a fixed light, elevated 56 feet.

BOOTHBAY, Lincoln co., Maine, United States of North America, a town between the Sheepscot and Damariscotta rivers, each of which may be called arms of the ocean, and having the sea on the S. It has an excellent harbour. Pop. 2,641. Lat. 43° 42′ N.

BORA, Gulf of Venice, or Adriatic sea, the name of a wind which blows from E.N.E. to N.N.E., or off the E. coast nearly direct, and more prevalent there than in any other part of that sea. It is chiefly formidable for its impetuosity and the suddenness of its gusts, which come rushing through the defiles of the mountains, and from between the numerous islands. Vessels are thus all at once placed in danger of being blown upon the Italian coast or of losing their masts, and this almost without previous warning. A sheltered anchorage on the E. coast affords the only security from being driven on the W. shore. The bora blows sometimes for 3, often for 12 or 20 days. It is most furious about Sebonico, Kattaro, Narenta, and Macarska, and most violent of all near Quarnero, and in winter all vessels must be cautious near the entrance of that channel. A few dark clouds over the mountains rising up rapidly are the only indications of the approach of this wind. The S.E. wind sometimes alternates with the bora, and is very dangerous to ships between Aulona and Ragusa.

BORANO, CAPE and ISLAND, Gulf of Venice, near the N.W. extremity, and the N.E. point of the cluster of isles in the fair way to that city, as well as the limit to the road before the city, from which the coast trends away E. towards Trieste. Lat. 45° 16′ N., long. 12° 30′ E.

BORCUM, an island of the North sea, on the coast of Groningen, opposite to which is the channel into the West Ems, or Dollart river. A reef, called Borcum reef, runs off 6½ miles to the N.W. of the island, which is 2½ miles long, lying E. ½ S. and W. ½ N., having two beacons and a lighthouse upon it, the last 150 feet above the sea, visible 6 leagues, and elevated 150 feet. Lat. 53° 36′ N., long. 6° 18′ E.

BORDIGHERA, Piedmont, 3½ miles E. of Venti Miglia, a small town, near which is a point of the same name, and close to which sail may be made, avoiding some rocks near the land above water.

BORDOE ISLAND, see *Faroe Islands.*

BOREA RIVER, Africa, Bight of Pannavia, behind the island of Fernando Po, on the main, lat. 3° 35′ N.

BORGO, Russia, a town in the Gulf of Finland, in Wyland, to the W. of Lovisa, at the entrance of the river Borgo. The navigation to the town is difficult, and the trade trifling. 24 miles E.N.E. of Helsingfors, which last is in lat. 60° 10′ N., long. 25° 6′ E.

BORMES ROAD, France, 2 miles N. of Cape Benat, department of the Var; the town is a league from the sea. It is 9½ leagues E.N.E. of Toulon. The ground is good, with from 6 to 9 fathoms, except in S. or S.E. winds.

BORNHOLM, Denmark, an island, 7 leagues long and 4 wide, surrounded by dangerous reefs, 6 leagues from the extreme point of Zealand, and 2 from Ystad in Schonen. It has no harbours. There is a lighthouse at Hammeren, the N. point, 280 feet above the sea, with a fixed light, visible 23 miles, an excellent guide for passing between Bornholm and Sweden by night. The N. end of the island is so bold that first-rate men-of-war may go within a cable's length of it without danger. A beacon stands on a saddle hill at this end, and above it are the ruins of a castle. There is a white church near the centre of the island, visible 8 leagues off. Vessels ride under the N. end of the island, half a mile out, in 14 fathoms. There are dangers at the S. end. On the N. only one shoal is dangerous, having 9 feet water, E. by N., three-quarters of a mile in length, due N. from the lighthouse 9 miles. The lighthouse should therefore be passed at a distance of only 2 or 3 miles. There is a variable current on the E. of the island, 6 or 8 miles from the land. The island is in lat. 55° 15′ N., long. 14° 55′ E.

BORRACHA, or DRUNKEN WOMAN ISLAND, Orinoco, South America, the N.E. point, in lat. 10° 19′ 40″ N., long. 64° 44′ 40″ W. About 2 miles long and 1½ broad. It has two islets near, called Los Borrachitos.

BORROWSTON NESS, BO NESS, or BOW NESS, Scotland, on the S. side of the Frith of Forth, with a safe harbour; a town of considerable trade in coal and ship-building, 5 miles N. of Linlithgow. Pop. 2,809.

BORSOE, Denmark, a small island 2 leagues N.E. from Apenrade, on the E. coast of Sleswick.

BOSA, an island, town, and river, S. by E. of Alghero gulf, 11 leagues N. of Oristano, in lat. 40° 15′ N., long. 8° 30′ E.; on the W. coast of Sardinia.

BOSIN ISLAND, Ireland, the third of the Balleness islands, near the N.W. point of Ireland.

BOSPHORUS, or the CHANNEL OF CONSTANTINOPLE, Turkey; this celebrated strait, about 15 miles in length, and varying in breadth from 1¼ mile to half, extends from the city of Constantinople to the Black sea, in a crooked channel, having throughout a great depth of water and a rapid current. At its S. entrance there are two lighthouses; one on the right, upon a hilly point of land, a mile S.W. of the ruins of Karac Serai; the other, on the left hand, a little S. of the celebrated mosque of St. Sophia. At the entrance, where the strait is about a mile wide, the city of Constantinople is on the left, and on the opposite side of the strait are the suburbs of Galata, Pera, and Tophana. The mosque of St. Sophia is in lat. 41° 0′ 12″ N., and in long. 28° 59′ 2″ E. The strait here is a little more than a mile wide. The Bosphorus being opened, anchorage is found S.W. of the Fanar lighthouse. On the E. is a rocky islet, upon which stands what is called the tower of Leander. From hence, if going up the strait, a ship may beat across, but in all cases great attention must be paid to the set of the current. The stream coming down sets upon the Seraglio point, and there divides, a part going strongly up the harbour on the side of Constantinople, and a part along the Seraglio walls into the Sea of Marmora. The distance it runs up the harbour or towards Tophana scarcely extends beyond the custom-house there, and then runs down the other side, along the Tophana shore and up the Bosphorus as far as Arnoudkeni. It sets on the point of Efendi Bornou, and then into the stream, in what is called the Devil's current. Thus one sets up, another down. The two currents where the eddy forms, will not allow a clear anchor for more than a few minutes. Ships should therefore go up higher and make fast to the shore. The harbour has 20 fathoms water throughout. The current leading as far as Arnoudkeni, thence a vessel may work her way, and be towed during calms. The shoals are of little moment, as they are marked; but a good breeze is essential to stem the current, and some care and activity in reaching different anchorages for fresh starts. The current is so strong in one part that scarcely anything can beat up against it. In what is called the Devil's current it runs sometimes 5 and nearly 6 knots an hour. The European side is the best to keep near on the entrance into the Black sea. Along the whole length of the shore of the Bosphorus there are villages and country residences. The castles are well armed, and fully protect the channel against being forced from the north.

BOSSELS, or BOSCH ISLAND, North sea, on

the W. side of Borcum island, off Rottum island, on the coast of Groningen. There is a channel on the W., and also on the E. side, but shallow and full of shifting sands. The whole island is nearly overflowed at high water spring tides.

BOSSERNE, or BASSER, entrance of the Great Belt reefs, a stony island, three-quarters of a mile long, rising 6 or 7 feet above the sea. Near it on the W. is Vaero or Vegeroe, another small island, having a bank attached to it with 2½ and 3 fathoms over it. Lat. 55° 54′ 10″ N.

BOSSESSAME, or TOMBELLY, Bijooga islands, N. point lat. 11° 29′ N., long. 15° 30′; S.W. point 11° 19′ 24″ N., long. 15° 32′ 12″ W.

BOSSU ISLAND, N. coast of the Gulf of St. Lawrence, N.E. from Trinity island, and near Egg island, which last is in lat. 49° 38′ 21″ N., long. 67° 3′ 10″ W.

BOSTON, and BOSTON DEEPS, Lincolnshire, a town and port for vessels of 100 tons, the entrances to which are divided from those to Lynn by several sands and shoals. There are two passages to the entrance of the Witham, called one the South channel and the other the Boston Deeps, and these are joined by the Maccaroni channel. The navigation is buoyed. The lofty tower of Boston church serves for a mark at sea. It returns two members to parliament, and has a pop. of 12,000. Lat. 53° 1′ N., long. 25′ E. The entrance to Boston Deeps is divided from that of Lynn by the Dog's head, Roger, and Gatt sands. The immediate entrance is formed by the Dog's head, and Outer and Inner knocks. Farther in is the Long sand, joined on the W. to a large sand called the Roger. These form the N. side of Lynn wall, and the S. of Boston Deeps. The Wainfleet sand here is the N.E. of extensive flats, continuing along shore all the way to the entrance of the Witham river. The Scurlig is another sand, narrow, and parallel with the Wainfleet and other flats; beyond it is the Long sand, having a channel between, called the Outer channel, one-third of a mile wide, with 6 or 7 fathoms water. The bar at the entrance of Boston Deeps often shifts, and a direct course over it sometimes varies 4 or 5 points. Pilots must be obtained for this navigation. The pilot sloops always have a broad vane, and when they desire a ship to sail towards them, they place a blue or red jacket under the vane. Pilots may be obtained in Lynn W. channel, Wisbeach Eye, and Clay Hole, for Lynn, Wisbeach, and Boston.

BOSTON, North America, United States, in Suffolk co., the capital of Massachusetts province, situated on a peninsula at the W. extremity of Massachusett's bay, is in lat. 42° 21′ 22·7″ N., long. 71° 4′ 9″ W. The tide rises 14½ feet; time at full and change 11ʰ 31ᵐ. The distance from New York is 207 miles, and the pop. in 1840 was 93,383. The harbour, N.W. of Cape Cod, is formed by Point Aldertus on the S. and Nahant point on the N. It is capacious enough to contain 500 sail of vessels at anchor in a good depth of water, and yet the entrance is so narrow as scarcely to admit of two ships sailing in abreast. It contains more than 40 islands or islets, for not more than 15 can be styled anything more, the others being mere rocks or patches of sand scantily covered with verdure. Some of these islands are used for pasturage. Castle island, 3 miles from the town, once called Castle William, defends the entrance of the harbour; and a garrison is kept there. Only seven of the islands are in the town jurisdiction: Noddle, Hog, Long, Deer, Spectacle, Governor's and Apple islands.

The view of the city is highly picturesque, lying in a circular form around the harbour. The town was first colonized in 1631, and here commenced the revolution which terminated in American independence, while it boasts of having given birth to the celebrated Franklin. The communication of the peninsula on which the city stands with the main land was called the Rock, and is a little more than a mile in length. The command of this neck enabled the English to cut off the communication of Boston with the main land in the revolutionary war. There are a number of bridges in the town, one of which is 3,432 feet long. Boston consists of Old Boston on the peninsula; South Boston, formerly a part of Dorchester; and East Boston, formerly called Noddle's island. South Boston extends 2 miles along the S. side of the harbour. The commercial wharves are very extensive, one 1,000 feet long, is devoted to Liverpool and its steam ships. Long wharf is 1,650 feet long, and Central wharf 1,240. Boston is the second commercial city of the United States. It possesses large steam-vessels, and has sailing-packets in constant communication with Europe. The public and commercial establishments are imposing. The tonnage of this city was in 1840, 220,243 tons. The imports were valued at 16,000,000 dollars, and the exports at 10,000,000. It possessed 25 banks and 28 insurance companies. It publishes 30 newspapers, 7 of which are daily, and is governed by a mayor, 8 aldermen, and 48 common council. The harbour is formed by Point Elderton on the S. and by Nahant point on the N. Castle island, 3 miles from the town, defends the entrance of the harbour. The outer light stands on a tower 82 feet high, light revolving, bright for 40 seconds and obscured 20, alternately; at 7 leagues distance the interval of darkness is equal to twice the duration of the light. Thence it decreases on approaching, until the greatest power becomes as 24 to 1. There is also a tower on Cedar point, with two fixed lights, 15 feet apart, 86 feet above the sea, and seen at 5 leagues distance; the lower of the two of a blood-red colour. N.W. winds prevail here from October until February, and they blow excessively strong and cold. Great regard must be had to the currents on this coast, and to the setting of the tides. To strangers pilots are necessary. The entrance from the N.E., called Broad sound, runs direct for Nick's Mate island. There is a shoal from Deer's island, on the starboard, to Lovel's island, which makes it impossible to be used, except by small vessels.

BOSWICK, Sweden, on the E. coast, in the Baltic, near the Karels; a pilot is requisite in this vicinity.

BOTALLACK HEAD, Cornwall, 2 miles N.W. of Cape Cornwall.

BOTHNIA, GULF OF, called the Bothnic, because the land on each side of it is called East and West Bothnia. It is the N. branch of the Baltic sea, commencing at the Isle of Aland, in about lat. 60° 15′ N., and extending to lat. 65° 30′ N. See *Baltic.*

BOTTE ROCKS, coast of Naples, between Ventodena island and Ponza, about 7 miles from the latter island, and from the W. point of the former 14 miles. They rise some feet above the level of the sea. These rocks are in lat. 40° 50′ 20″ N., long. 13° 6′ 10″ E. The sea breaks heavily over these rocks when the wind blows.

BOTTOMLESS PIT, Africa, on the W. coast, 13 miles E. from Cape la Hoa, near the W. end of the Quaqua coast, not far from the mouth of Mayo river; it has its name from its great depth. There is one place, however, with 35 fathoms, 2 miles from the shore, where vessels may ride.

BOUC, PORT OF, France, near the town of Martiques, at the mouth of the Rhone, 2 leagues E. It has a narrow entrance, and on its starboard side a fortress, with a square tower of white stone in the middle, standing on the point of an island. Here is a lighthouse, 98 feet 5 inches high, visible 5½ leagues. There is also one on the mole head, visible 3 leagues, 26 feet 3 inches above the sea, distant 300 yards from the other. It is the depôt of salt from the salt-marshes of Berra.

BOUDROUN, GULF and TOWN OF, in Anatolia, over against the island of Stanchio, on the N. side of the gulf, 45 miles W. by S. of Malassa.

BOUG, THE RIVER, in the Gulf of Dnieper, in the Euxine, 16 miles E. of Otchakov. At its mouth it is 3 miles in width, and has a depth of 3½ fathoms. The city of Nicolaef stands on the E. bank, 20 miles from the entrance, at its confluence with the Ingool. The depths vary as far as Nicolaef from 20 to 60 feet. Men-of-war are launched here into the Ingool, and pass thence into the Boug, by a channel from 20 to 27 feet deep. In an opening on the N.W. of the town is the observatory, the lat. of which is 46° 58′ 55″ N., long. 32° 0′ 21″ E.

BOULDER BANK, Sussex coast, 2½ miles to the S. of the old entrance to Rye harbour, and a mile to the E. of Cliff's End point. It is a bank with 2 and 2½ fathoms, three-quarters of a mile long, parallel with the shore. Its S.W. end lies with the S. point of Fairleigh cliff W. by N. One mile S.W. by S. from the S.W. end of the Boulder is a spot with only 10 feet upon it. Morris's cliff W. by N., and Rye Tower on with the public-house near the beach, bearing N.E. ½ N., are the marks.

BOULEY BAY, Jersey island, on the N. side, W. of the point called the Tour de Rosel.

BOULOGNE HARBOUR, France, department Pas de Calais, a seaport town, with a pop. of 20,000; lat. 50° 43′ 56″ N., long. 1° 35′ 18″ W. West Jetty head. Two tide-lights are exhibited at the head of the S.W. jetty, one 40, the other 31 feet high, visible 2½ leagues off, as long as there is sufficient water. As soon as the tide begins to flow into the harbour, a signal is made with lighted straw. When there are 8½ feet on the bar, the first lantern is lighted, and at high water a second under the first. At the end of the N.E. jetty a red light is kept burning as long as the upper light on the S.W. jetty; it is 20 feet above high water, seen only at 1½ mile distance. A new entrance has been opened to the W. of the old, the direction of which is S.E. by S., then S.E. into the harbour. The one to the S. westward extends out 1,970 feet; the other to the N.W., parallel with, but not so far out as the S.W. one. Off the end of the pier a buoy marks where it is rocky. High water at $11^h 1^m$, rise 19, lowest neap 11 feet. The object of late improvements has been to make Boulogne a harbour of refuge, and the signals for day or night are perfect, and the entrance easy, so that seamen who have once been in take no pilots. Various shoals and banks lie off the harbour, which must be carefully noted. At Point d'Alpreck, 2 miles from Boulogne to the S.W., there is a lighthouse painted black, 154 feet above high water mark, displaying a flashing or intermittent light for 1½ minute, then an eclipse, then a *red* flash, and then another eclipse, the whole occupying 2 minutes, and seen in clear weather 4 or 5 leagues off. Four leagues from Boulogne, on the E. side, to the entrance of Etaples, is a fixed harbour light, seen 2 leagues off, and another on the Point de Berck, or Haut Banc, at the mouth of the river Authie, on the N. side, 66 feet above high water, visible 6 miles off. Lat. 50° 23′ 50″ N.

BOUNDEN NESS, TOD HEAD, or DOUNEFIT, for it is known by all these names, is in Scotland, about 4½ leagues N.E. by E. from the bar of Montrose, and 6 leagues from Red head, in the same direction. It is low, and not soon perceived, the high hill of Craig Daire being often mistaken for it. It lies S.W. ½ W. 38 miles from Buchanness.

BOUND'S CLIFF, Cornwall, in Portissick, or Port Isaac bay, on the N. coast.

BOURBON, Africa, an islet on the W. coast of Africa, near the shore of Senegal. Lat. 11° N., long. 15° W.

BOURDEAUX, France, department of the Gironde, an opulent mercantile city, with a pop. of 94,000. It is situated on the left bank of the Garonne river, which is capacious enough to moor 1,000 vessels. It has an extensive commerce with all parts of the world, particularly in the export of its choice wines. It has some fine public buildings, and a bridge over the Garonne of 17 arches. Lat. 44° 50′ 16″ N., long. 33′ 42″ W. The entrance of the Gironde is recognised easily by its celebrated lighthouse, called the Tower of Cordouan, standing on a rock, nearly in the midway entrance of the river. It exhibits a revolving light, 207 feet above the level of the sea, seen 8 leagues off. The eclipses succeed each other every minute, each bright flash being succeeded by one less brilliant, but the eclipses do not appear total until at a distance beyond 3 leagues. Just within the Point de la Coubre there is a battery, semaphore, and two beacons. A harbour light is shown on the point 2½ leagues N. by W. from Cordouan lighthouse; a fixed light, elevated 65 feet, and visible 10 miles. At 6½ miles distance, on the same shore, is the Terre Negre light, 118 feet high, and visible 10 miles off, designed to lead through the Passe du Nord, and clear the shoal called the Barre à l'Anglaise at night. It is only visible S. of the towers of the Terre Negre and St. Palais when in one. There is also a N. channel, but no attempt should be made to enter the Gironde at night. There are beacons and semaphores, with other marks, all the way up the river beyond these mountains. A vessel carrying a fixed light 33 feet above the river level, is moored on the E. edge of the Tallais bank, in 4 fathoms low water, with the fixed light on Point de Grave N. by W. ¾ W. 5,047 fathoms; Talmont steeple E. ½ N. 3,527 fathoms; and the red harbour lights at Richard S. ¼ E. 4,484 fathoms. Tallais bank forms the W. side of the channel, and Talmont bank the E., and may be seen 9 miles from the deck of a pilot-boat. The light is so marked, it cannot lead into the channel W. of Tallais bank. It is readily distinguished by having a skeleton ball at the mast-head. 46 feet above the water. A bell is rung on board during a fog. High water at the entrance of the Gironde from 3½ to 4 o'clock. Steam-boats may be had to tow ships up to Bourdeaux, at prices rated according to tonnage. A pilot is absolutely necessary. Vessels may often easily work up to Bourdeaux without losing time or even waiting spring tides. Ordinary spring tides rise 15 feet, at the equinoxes 17 and 18.

BOURDEAUX HARBOUR, Newfoundland, St. John's. Lat. 47° 45′ 28″ N., long. 52° 58′ 30″ W.

BOURDINOT ROCK, France, a sunken rock a mile off St. Cas point, on the N. coast of France, 3 leagues W. from St. Maloes; never dry.

BOUREL, BRILO, or BURELA, POINT or CAPE, Spain, N. coast, about 6 leagues W. from Ribadio. Off this point are several rocks and headlands, the most remarkable of which are the Farallones of St. Cyprian, behind which, under pressing circumstances, a vessel might find shelter in winds from the N.W. round by S. to E., over a bottom of sand, with good holding at 4 fathoms.

BOURG, or CAYENNE, French Guiana, South America, the capital of the colony. Lat. 4° 56′ 15″ N., long. 52° 20′ 30″ W.

BOURGAS, GULF OF, in the Euxine. The N. point is Cape Emoneh, and Sizopoli, a town of that name on the S., distant 17 miles, encloses the bay, which is open to the E. On the N. side is the little town of Massemvria, on a small peninsula. The little bay to the W. of the town has good anchorage in from 7 to 10 fathoms; that on the E. cannot be recommended.

BOURGNEUF, or NOIRMOUTIERS BAY and ISLAND, France, department of La Vendée, near Nantes, about 1 league from the shore and 11 in circumference. Pop. 7,000. The most productive part of the island is 12 inches under the level of the sea, which is kept out by great labour. The inhabitants are sailors and fishermen, for the larger part engaged at a distance on the W. side of the island. It has a port with 12 feet water, and a road called La Chaise, which will receive vessels of any size. It is 13 leagues E.S.E. from Belleisle to this island, which is S. of the S. point of the Loire.

BOURGNEUF EN RETZ TOWN, France, Loire Inférieure; pop. 2,330, at the bottom of the bay of its own name, S.E. of Nantes. High water in the bay at 4ʰ. The bay is very dangerous.

BOURLOS, CAPE, Egypt. The cape is at the entrance of the lake from the sea, E. of Alexandria. It is 35 miles from the entrance to Damietta, on the Eastern Nile.

BOURLOS LAKE, Egypt, near Alexandria.

BOUROUN, the Turkish appellation for a cape or headland, as Karabouroun, the Cape of Kara; Samoursac Bouroun, Cape Samoursac.

BOUROUN, Roumelia, in the Bay of Lagos, E. of the Gulf of Contessa.

BOVENA ISLAND, France, one of the Hieres islands, in the Mediterranean, E. of Toulon.

BOVENBERGEN, or BOVBIERG, on the W. coast of Jutland, Denmark, N. of Numet island, 10 leagues from Kincopper Deep, or Namet, which last is in lat. about 55° 50′ N. The whole coast is a clear strand, with small white sand hills; the blue double hills inland being seen a great way off, called the Holms. The banks off this coast have all 7 and 9 fathoms of water over them.

BOVI ISLES, Greek Archipelago, nearly midway between the islands of Nixia and Nicaria, N.E.; they are mere rocky islets.

BOVIN ISLAND, France, an islet S. of the river Loire, at the bottom of Bourgneuf bay.

BOVISAND BAY, Plymouth sound, opposite the E. end of the breakwater.

BOWNESS, Cumberland, W. coast, forming a promontory into Solway frith.

BOW ROCKS, Scotland, E. coast, a rocky ledge three-quarters of a mile from the harbour of Elle, and a quarter of a mile from the shore. Elle is a small town and harbour near Fifeness. The rocks are dry at half ebb. They lie W.S.W. from the entrance, which has only 12 or 13 feet water at spring tides.

BOWLING ROCK, W. coast of Ireland, a mile S. of Kilkerrin point, 2 cables' length from the E. shore, having an iron perch upon it. The rock is uncovered at the last quarter of spring tides. A shoal surrounds the rock a full cable's length on all sides, leaving a narrow channel of 2½ fathoms between the shoal and the main.

BOXY POINT, Newfoundland, in Fortune bay. Lat. 47° 36′ N., long. 55° 40′ W.

BOYER'S BAY, West Indies, at the N.W. end of the island of St. Kitt's. The Larionitz river falls into it.

BOZA TOWN, RIVER, BAY, and ISLAND, Sardinia, E. of Alghero. There is anchorage before the town in 8 and 7 fathoms. Another islet lies S.

BRAAK BANK, bearing from Dunkirk N.W. ¾ W., distant 4 miles, and from Mordyck N.E. ¾ N., the same distance. It is only divided from the Snow by a narrow channel of 9 feet water.

BRAAKE, THE, North sea, S. of the mouth of the Elbe, the indentation of the coast that receives the rivers Weser and Jahde, within Heligoland Deep, on the S.E.

BRAAWIKING BAY, W. side of the Baltic sea, in Sweden, 30 leagues S.W. from Stockholm.

BRACADALE LOCH, Scotland, W. coast, a large bay, in which are several smaller ones forming good harbours. Loch Harpart is the most S., a perfectly safe anchorage. The mouth of Loch Harpart is between Oronsa isle and Bracadale.

BRADA HEAD, the N. point of Port Iron, in the Isle of Man.

BRADING BAY, Isle of Wight, almost opposite to Portsmouth harbour, at the E. end of the island, not far from St. Helen's road; it is dry for a large part at low water, and has only a narrow channel for small vessels when the tide is in, having a bar at the entrance.

BRADORE HILLS, Labrador. N.W. hill, lat. 51° 35′ 11″ N., long. 57° 17′ 6″ W.; South hill, lat. 51° 34′ 2″ N., long. 57° 4′ 32″ W.; Middle W.N.E. hill, lat. 51° 34′ 57″ N., long. 57° 13′ 50″ W.

BRAGMAN'S BLUFF, Gulf of Mexico, North America, upon the main; it affords shelter from N. and W. breezes, and is formed of steep red cliffs, 30 or 40 feet high, extending 3 miles, and peculiar in aspect. The N. part is in lat. 14° 3′ N., long. 83° 31′ 40″ W.

BRAH, or BRAY HEAD, Ireland, on the E. coast, 4 leagues N. of Wicklow, and 2¼ leagues from Dalkey, the S. point of entrance into the Bay of Dublin. The fishing town of North Castle, having a flat point, is between Brah and Wicklow, from which point to Brah there is 13 or 14 fathoms, but at North Castle only from 4 to 6. S. of Brah head, and opposite Sugar-loaf hill, there is a foul bank, with only 4 fathoms on it. Lat. 54° 22′ N., long. 6° 1′ W. High water 3ʰ 30ᵐ.

BRAHE'S TOWER, Copenhagen, named from the astronomer Tycho Brahe, a mark for entering that city from Elsineur.

BRAINTREE, Norfolk co., Massachusetts,

North America, a town lying on a bay, 14 miles S. of Boston; pop. 2,168. It has several manufactories, and is much occupied in the fisheries of that coast.

BRAKE, THE, a sand before Dunkirk, N. and N.W. from it, and nearly parallel with the coast. See *Braak.*

BRAKE, off the coast of Kent, an extensive sand, lying N.E. by N. and S.W. by S., about 4½ miles long, having a buoy at each end and one in the middle. The N. buoy, red, lies W. ¼ N. 1½ mile from the white buoy of the Gull, with St. Lawrence's church on with the N. Cliff point at Ramsgate N.W. by N., and the North Foreland lighthouse N. by E. northerly; the Middle Brake buoy now lies in 5¼ fathoms water, with Upper Deal church cupola in line with the windmill next S. of Sandown castle, S.W. by W.; St. Lawrence windmill, its apparent width E. of Ramsgate pier lighthouse, N.; N. Brake buoy, N.E. ½ N.; Fork buoy, S. ¼ E.; Gull light-vessel, S.E. ¾ E. The S. Brake buoy now lies in 4¾ fathoms water, with Waldershare monument, in line with a barn, midway between the two windmills next S. of Sandown castle, W. ¾ S.; North Foreland lighthouse in line with the Preventive station-house on Broadstairs E. cliff, N.N.E.; Middle Brake buoy, N.N.E. ½ E.; Fork buoy, S.E. ½ E.; S. Deal bank buoy, S.W. ¼ S.; Gull light-vessel, E. by N. ½ N.; S. Sand head light-vessel, S. by W. ½ W.

BRAMAO POINT, South America, Surinam, on the N.E. coast, 12 leagues from Cupanama, E. by N. See *Bram's Point.*

BRAMBLE, THE, a shoal at the entrance of Southampton water, with from 1 foot to 2 fathoms over it. Four buoys lie on its edges; the E. white, N.E. chequered black and white, N.W. red, with a staff and ball, and W. red. The E. lies half a mile from the shoal; between this and the W. buoy there are 3 fathoms, a ship's length off 6, and in mid-channel 11 at low water.

BRAM'S POINT, Surinam, N.E. coast of South America, 12 leagues from Cupanama, E. by N. Lat. 5° 56′ 20″ N., long. 55° 12′ 48″ W. There are two iron vessels with masts and two copper buoys laid down here as a guide to the river of Surinam.

BRANCO ISLAND, Gulf of Mexico, bay of Campeachy, or Campeche, starboard in the fair way from Cape Condecedo, across the bay, S.W. by S. and S.W., to Vera Cruz.

BRANDARIES, THE, coast of Holland, on the W. end of Schelling or Scheveling, remarkable by its tower and beacon. To go through the western Boom Gat or channel, bring these both in one at E. and E. by S., keeping them so, and running past the first buoy in 6 fathoms at low water as far as the second; then bring the tower a little southward of the beacon, and then run between a white and a black buoy, where is 17 feet at the shoalest; and from thence, in 6 or 7 fathoms, in the fair-way of the Boom channel, the N. side of which is deepest.

BRANDON BAY and HEAD, W. coast of Ireland, 7½ miles from Ballydavid head; it is the E. point of Smerwick bay, and Kerry Head is the extreme W., distant 11½ miles E. ½ N. Lat. of the head 52° 22′ N., long. 10° 1′ W.

BRANDON HARBOUR, North America, United States, on the N. side of Long island, New York, 3 leagues W. of Smith's town.

BRANDY POTS, river St. Lawrence, North America; the S.E. point is in lat. 47° 52′ 22″ N., long. 69° 43′ 47″ W.

BRANDYWINE CREEK, United States, North America, enters the Delaware river, and forms the harbour of Wilmington. See *Wilmington.*

BRANFORD, New Haven co., United States, North America, in Long Island sound. It is seated in Branford river, navigable only for small vessels, 10 miles from New Haven. Pop. 1,322.

BRANSOE, Denmark, an island situated in the Little Belt, 5 miles N.N.W. from Assens.

BRASSA and BRASSA SOUND. See *Shetland Isles.*

BRASS ISLAND. See *Virgin Islands.*

BRAUNSBERG, Prussia, in Oster Prussen, on the river Passarge, near its entrance into the Frisch Haff, 22 miles N.E. of Elbing. Lat. 54° 22′ N., long. 20° 6′ E.

BRAYE HARBOUR, Alderney island, N. coast, capable of admitting vessels of 200 tons, but unsafe from exposure in heavy gales.

BRAYE ROCKS, 2 miles N.E. from the N.E. point of Guernsey island.

BRAZIL BANK. See *Liverpool.*

BRAZIL ROCK, on the coast of Nova Scotia, in lat. 43° 24′ 15″ N., long. 65° 23′ 48″ W.

BRAZIL ROCK, N. lat. 51° 10′ N., long. 16° W., a vigia, doubted by many to exist at all. It is marked by the French in charts of 1742, and is said to have been seen in 1791 by the master and crew of an English merchant ship, as a high rock, bold to.

BRAZZA ISLAND, Gulf of Venice, on the coast of Dalmatia, 10 leagues long and 3 broad, between the island of Lesina and the main, in lat. 43° 19′ 29″ N., long. 16° 26′ 27″ E.

BREAK, CAPE, or BREAK HEART POINT, Newfoundland, the E. point of entrance into Trinity bay, which last is in lat. 48° 21′ 30″ N., long. 53° 22′ 38″ W.

BREAM BAY, N. of Rosemullin head, Cornwall, in Falmouth bay.

BREE, or BROAD BANK, a sand, one of those that lie across Brewer'shaven channel, on one side of the Isle of Schonen, in Holland. It is at a good distance from the land, and has 7 fathoms upon it, and 19 within it. It is the outermost of the three.

BREEDT, THE, or BROAD BANK, the largest of the Dunkirk banks, divided into E. and W. by only 3 fathoms. The whole are distant from Dunkirk from 5 to 8 miles N.N.E. and W. ½ N.

BREHAT, ISLE OF, N.E. point in lat. 48° 51′ 54″ N., and long. 2° 59′ 8″ W. On the N. coast of France, about 8 leagues N.W. by N. across a bay from Cape Frehel, Brieu haven, in the bottom of the bay, between them. It is almost surrounded by rocks, ledges, and small islets. The Heaux shoal is N.W. from Brehat, and 2 or 3 from the main. On the Heaux de Brehat is a lighthouse, in lat. 48° 54′ 33″ N., long. 3° 5′ 4″ W. The town is small, containing 1,570 inhabitants. It is almost surrounded with rocks and islets. Menmare is 4 miles E. from the N. end; the Horaine shoal about a mile E. from Garvo; Lobrasses shoal is about 2 miles E. from the N.E. end of Brehat, and the Eschaudees are a gun-shot N.E. from that, with many shoals and rocks between them and the island. On the N.W. side are the islands of Tuscées and St. Maude, and the N.E. channel between them and the island, where ships may ride in 8, 10, or 12 fathoms. Doran and Zearn islands are almost 2 miles N. of these, but without, and to the N.E. into the sea, all is foul and rocky.

The Heanx shoal is also about 4 miles N.W. from Brehat, and 2 or 3 miles from the main; and the Noires rocks are about a mile W.S.W. from that; and on both are several above water. Within a mile or two N. of these is all foul ground, and full of rocks.

BREHAT, or BREHA SHOAL, Newfoundland, having only 6 feet of water. Lat. 51° 25' 40" N., long. 55° 27' 48" W.

BREHAR ISLAND, or BRYER. See *Scilly Isles.*

BRELADE'S BAY, ST., Jersey island, S. side of the island, W. of St. Aubin's bay.

BREMEN, a strong town, situated on both sides of the river Weser, in lat. 53° 5' N., long. 8° 40' E. It has a pop. of 48,000, and is one of the four free cities of Germany, bounded in its territory N.E. and S. by Hanover, and W. by Oldenberg, divided by the Weser into the old and new towns. The harbour is 9 miles below the town. It has an extensive trade. It is high water at 6 o'clock. The Weser is regularly buoyed at the entrance. There is a lighthouse on the Wenger Oog, at the W. entrance of the river Weser, intermitting, visible and invisible every minute. A light-vessel is moored off the mouth of this river, having seven lantern lights at night round the mast, 28 feet higher than the deck, visible 3 miles, bearing from Heligoland N. ¼ W.; from Wenger Oog light-tower W. ⅜ N.; Minsen church S.W. by S.; and the Bremer beacon S. by E. A vessel bringing the light to bear S. by W. should steer directly towards it, when intending to enter the N. Weser. There is a second light-vessel in the Wurster channel, when the navigation is clear of ice; it lies S. by E. ½ E. 7¼ miles from the first light-vessel. Passing W. of the light-ship No. 1, she must immediately be brought to bear N. ¼ W., steering S. for 3 miles, at which distance the light-ship No. 2 will bear S.S.E.; then steer direct for it, pass it on the E. side; steer S.E., keeping the light-ship N.W. about a mile; then come to an anchor at low water. There is no proceeding further without a pilot. On the W. side of the river is Blexum, and nearly opposite on the E. Bremerlehe, where a new harbour has been constructed, called Bremer haven.

BRESS, or BRASS, Africa, an old trading fort near the River Nun. Lat. about 4° 30' N.

BRESSONS, Cornwall, two small rocks, three-quarters of a mile W.S.W. from Cape Cornwall, and N.E. ¼ N. from the Long Ships 3½ miles. A sunken rock, called the Venus, lies two-thirds of a mile from the Bressons.

BREST, France, department of Finisterre, after Toulon the principal naval arsenal of France, strongly fortified. St. Louis is in lat. 48° 23' 20" N., long. 4° 28' 27" W. The time of high water is 3ʰ 25ᵐ; the rise 21 feet. The pop. of Brest town is 28,900. It is situated on the side of a fine roadstead, capable of accommodating 500 sail, with anchorage in 8, 10, or 15 fathoms at low water. The entrance from the W.S.W. is narrow and dangerous from being rocky. The Fillettes and Mingan rock are always concealed, and are a little on the starboard of the middle, between the main on either side, except at the lowest ebb of spring tides. A ship must therefore keep one-third more to the N. shore, in a parallel nearly with the coast, which will be a good general direction for passing through into the inner road; but high winds often cause a very rough sea, and occasion ships at anchor to labour greatly. The entrance and coast on both sides are strongly fortified. To make the mouth of the entrance from the N. or English channel, run S. round the islands to the southward of St. Matthew's point, upon which is a revolving light, the eclipses succeeding each other every half minute, visible 6 leagues. The distance to St. Matthew's point from the Passage du Four, the channel that lies between Ushant and the main, is 12 miles S.S.W., and the tide runs very strong. There are several rocks here to be noticed and avoided. From the islands round St. Matthew's point run E. a little S. near 2 leagues, by which the Coq and other small rocks will be left on the larboard, being about half a league from the shore. When abreast of the windmill near the S.E. point of Bertheaume road, observe the general direction given to sail through the passage. Ships coming from the W. or S.W., if they do not make St. Matthew's point, should secure the N. point, near a mile to the E. of this mill, and attend to the same caution. The town of Brest is 2 miles from the N. point of the entrance into the road, and on the N. shore. About 1¼ mile E.S.E. from the Coq, and S. by W. from Bertheaume point, is Busec rock, very dangerous at low water; a ship may go on either side of it, but the best way with large ships is to keep the N. end of Beniguet island open of St. Matthew's point, until Point Bertheaume bears N.N.E.; then steer E. northerly, for Point du Minou, and keep along the N. shore to avoid the Fillettes, Gourdon, and Mingan shoals, which lie in mid-channel, on the latter of which is a beacon. As soon as Brest opens of Point du Portzic, steer for it, keeping 2 cables' length off the shore, and anchor in 8 or 9 fathoms. In Bertheaume bay, about 4 miles to the eastward of St. Matthew's point, is good anchoring, with northerly and N.E. winds, in 10 or 11 fathoms. Opposite Bertheaume bay, and about 4 leagues S.S.E. from St. Matthew's point, is the great bay called Dovarnenez, or Poldavid bay. Between these bays lie many rocks, which must be carefully avoided. Two lighthouses have been erected, one on the point called Petit Menou, the other on Portzic, 6,400 metres N. by E. of the former, or 20,000 English feet. The dockyard and magazines of Brest, the artillery, stores and materials for ship building, have ever been of the first order in naval service, and the accommodations are in every respect complete. In 1841, the number of vessels of war here in ordinary or commission was 53. The stores were valued at 5,641,029 f.; the marine hospital at 1,439,571 f.; the provisions in store at 1,538,875 f.; the artillery works at 20,029,356 f.; the naval works at 92,286,153 f.; altogether the value of the public property here was estimated at 144,530,820 f., being 12,000,000 f. more than any of the other four French arsenals contained. A great deal of the labour is performed by convicts, the value of whose labour was estimated for Brest at 140,644 f. for 1840. The town and neighbourhood of Brest, except from their connexion with the naval arsenal, possess nothing worthy of observation. The climate is bad, and heavy storms in winter prevail, with a deluge of rain from the S.W.

BRETON, CAPE, or CABRETON, France, at the bottom of the Bay of Biscay, at the N. end of the lake N. of Bayonne harbour. It is on the main land, and is known by a flat tower. Here, too, are beacons, in lat. 43° 39' 26" N. See *Fosse of Cape Breton.*

BRETON ISLAND, often called CAPE BRETON, North America, in the Gulf of St. Lawrence,

II.   F

about 100 miles long and 50 broad. It is separated from Nova Scotia by the Strait of Canso, about a league broad, and it is 7 leagues across to Newfoundland. All the harbours are open to the E. going round to the S. for 50 leagues from Port Dauphin to Port Toulouse, near the entrance of Fronsac strait. The sea here is often covered with thick fogs till near the middle of April, and the coast choked up with ice. Sidney is the capital town, and a free port. The harbour is large and secure. The island pop. is 20,000. The climate is similar to that of Nova Scotia. It is needful to be cautious in approaching this island, on account of the currents which set alternately about Cape North, according to the winds at sea. Arachat, in the Isle Madame, has a fine harbour, where the Jersey fishing establishments are situated, a free and warehousing port. Cape North is in lat. 47° 2' N., long. 60° 26' 20" W. ; Sydney harbour lighthouse, on flat point, is in lat. 46° 18' 15" N., long, 60° 8' 30" W., a fixed light ; Scutari lighthouse, with a revolving light, on the N.E. point, lat. 46° 1' 30" N., long. 59° 40' W.; Louisburg lighthouse, lat. 45° 54' 30" N., long. 59° 55' 30" W.; Albion cliff, S. side of Isle Madame, lat. 45° 28' 12" N., long. 61° 2' W.; Ship harbour, Gut of Canso, lat. 45° 36' 24" N., long. 61° 21' 25" W. On the W. side of the Gut is a fixed light, 115 feet above the sea. There is good anchorage under the lighthouse, with offshore winds. Off Cape North, on St. Paul's island, there are two lighthouses of wood, octagonal, one near the N. and the other near the S. extremity, of which one is always open, unless ships are near the central rocks. The N. light is 140 feet above the sea. The S. lighthouse is a revolving light, at the same elevation as the N. one.

BRETON PASSAGE, France. See *Pertuis Breton*.

BREYER ISLAND, coast of North America, lat. 44° 19' N., long. 66° 25' W. High water at springs 9½ hours.

BRIDE'S BAY, ST., South Wales, on the S.W. of Pembrokeshire. It is about 8 miles deep, with good anchorage from 17 to 10 fathoms, but winds S. or W. set in a heavy sea, and care should be taken not to be caught there. High water 6ʰ. On the N. of the bay is the little city of St. David's, the cathedral of which is in lat. 51° 52' 56" N., long. 5° 14' 53" W.

BRIDGPORT, United States, North America, Fairfield co., on the W. side of an arm of Long Island sound, a town with a pop. of 4,570. The harbour is small, and has a bar, with only 13 feet at high water. It is principally engaged in the coasting trade and in fisheries.

BRIDGTOWN, West Indies, the capital of the Island of Barbadoes; it stands on the N. part of Carlisle bay, in lat. 13° 5' 30" N., long. 59° 37' 45" W. The harbour will contain 500 sail of shipping. The town extends nearly 2 miles along the bay, and is defended by the fort of St. Anne. This town has suffered much from hurricanes, when the shipping has sometimes been driven high and dry up on the land, at others received great injury. N.N.W. from the N. part of Carlisle bay there are banks and reefs, called the Pelican and Half-acre shoals. On hauling in towards the bay, bringing Needham's point S.E. ¼ S., the N.W. point of the bay N.N.W., and the church N.N.E., there will be found 25 fathoms, sandy ground. Anchorage will be found in 12 fathoms or less, with Willoughby fort N. by W. ¼ W. and Charles fort S. by E. ¼ E.

BRIDGEWATER, Somersetshire, on the river Parret, 12 miles from the sea, in lat. 51° 7' 41" N., long. 2° 59' 30" W. To enter the river and navigate it, a pilot is necessary. If driven to the bar after two hours' flood, after an early sight of Brent knoll brought to bear E.S.E. ¼ E., and having led in until Burnham high light be three times the breadth of its tower, open to the S. of the low light, and so continue until Flat Holm light opens to the E. of Steep Holm island, when the two lights may be brought in one, and the high light be gradually opened N. of the low light, to clear the Stert and Lark sands. These lighthouses stand on low ground on the larboard side of the entrance, and are 1,500 feet apart ; the upper light is intermittent. The tides here rise sometimes 30 feet, and so rapidly in some cases, as to come in with a head of 2 fathoms at once, called a "Bore." Vessels of 200 tons can go up to the town, the pop. of which is 8,000. It has a considerable coasting trade, and sends two members to parliament. The celebrated Admiral Blake was a native of this town.

BRIDLINGTON, or BURLINGTON BAY, Yorkshire, on the E. coast; it is encumbered with a sand, called the Smithic, the N.E. end of which is 1 mile S.W. by S. from Flamborough head, the extremity of the bay, on which stands a conspicuous lighthouse, in lat. 54° 7' N., long. 5' W., 400 yards distant from its extreme point. The light is a revolving one, having three faces, each of seven reflectors. To distinguish it from Cromer and Tynemouth revolving lights, which show a face every minute, one face appears illuminated every two minutes, and the colour is red. The light from that face being diminished cannot in hazy weather be visible so far as the others; therefore when only two faces are seen, the interval of exhibition is two and four minutes alternately, which is a sufficient distinction. Cromer light bears from the Flamborough head light S. ¾ E., distant between 29 and 30 leagues. There is good riding in Bridlington bay generally, but the best is with Quay street open, bearing W.N.W., in 4 fathoms. There is a good tide haven here, secured by two projecting piers, and a harbour light at night at the town. The shores are bold and rocky.

BRIDPORT, Dorset co., England, a town with a harbour much contracted by the sand thrown up from the sea. Pop. 4,242. The manufactures are principally cordage, canvas, and nets. The harbour is 2 miles S. of the town, and will contain about 40 sail of small vessels. It is a borough, 134 miles W. by S. from London. Lat. 50° 40' N., long. 2° 55' W. High water, springs, 6ʰ 45ᵐ. It lies 5½ leagues from the Bill of Portland; the harbour has two wooden piers; the bar dries at low water.

BRIEUC, ST., *Anglice* ST. BRIEUX, France, department of the Côtes du Nord, a town and port, with a pop. of 9,960. It stands at the bottom of a bay, at the opening of the little river Gouet. The port will receive vessels of 400 tons; it is three-quarters of a league distant, at the village of Leque St. Brieuc. It stands in lat. 48° 31' 21" N., long. 2° 42' 52" W.

BRIGAS, CAPE and BAY, Newfoundland, in the Bay of Consumption, 1 league from Cape Broyle. The cape is high and rugged, and the bay deep. See *Cape Broyle*.

BRIGOLIA, Corsica, on the N.E. part, a deep, narrow, shallow bay, facing the N., a little S. of Bastia, only fit for small craft.

BRIGHTHELMSTON, or BRIGHTON, Sussex co., England, a large town, having the ports of

Newhaven on the E., and Shoreham on the W. It is on an open beach, between Beachy head and Worthing point, the shore steep and shingly. A chain pier has been erected nearly at the centre of the town, for the convenience of passengers embarking and disembarking from the steam-packets. This town is a borough, and has a pop. of 50,000. Packets continually pass from the town to Dieppe, running regularly from the pier, which is carried out 1,113 feet S. from high water mark upon piles. It has 23 feet at high water at its outer end at spring tides, and 5 at low. It is dangerous to approach the shore in foggy weather nearer than 5 fathoms. The outer end of the pier has a green coloured light in the night, and during fogs a bell is kept ringing. Lat. 50° 49′ 32″ N., long. 7′ 40″ W.

BRILL, or BRIEL, Holland, a seaport, and mark for the S. channel of the Maas, or Maes. It stands on the N. side of Voornland island, is fortified, and has a pop. of 4,500, chiefly pilots and fishermen. The large square tower of the church is to be brought E.S.E., or a little more E. or S., as the beacons there are disclosed, until they are brought into one, and so the run is made right in with them. Lat. 51° 54′ N., long. 4° 8′ E.

BRINDISI, Italy, W.N.W. ¾ W., and 3½ miles from Cape Cavallo. The harbour is blocked up, but the largest vessels may enter the road by two channels, but in both care must be taken not to approach the land. The anchorage is abreast of the beacons on the W. coast. What is called the port can only be entered by vessels drawing 8 feet water. There is a castle and batteries here, in lat. 40° 39′ 27″ N. Upon the mole of the fort, facing the sea, at the height of 123 palms, there is a lighthouse, illuminated in the old manner. Lat. 40° 39′ 17″ N., long. 15° 37′ 57″ E., meridian of Paris, or in Greenwich time, 17° 58′ 12″ E., and 3° 42′ 50″ E. of Naples. Reported as established in 1844.

BRIONI, ISLANDS OF, in the Gulf of Venice, near the coast of Istria. Lat. 45° 10′ N., long. 13° 51′ E. Two in number, called the Great and Little Icoglio, having some small rocky islets about them. On the N.E. side is the town of Brioni, before which is anchorage in 4, 5, 6, and 8 fathoms. The islands are remarkable for producing fine marble.

BRION ISLAND. See *Magdalen Isles.*

BRISAS, the name given to the trade winds that blow in the Gulf of Mexico from October to August, off the coast of Panama and Veragua.

BRISEIS' SHOAL, in the Great Belt, Denmark, N.E. of the Hasteen's ground, 5 miles N.W. from Hasel Oe, and S.E. by E. 13 miles from Foreness. It has only 3 and 4 fathoms water, and was discovered by the British frigate Briseis. Near its S.W. side are 5 fathoms, and 10, 11, and 12 at every other part.

BRISTOL, a city of England, in the counties of Somerset and Gloucester, situated at the confluence of the Avon and Frome rivers, which, uniting there, fall into the estuary of the Severn, 8 miles below, at a part called King's Road. This ancient city, for it was celebrated as a place frequented by shipping as long ago as A.D. 1139, is situated, by an observation taken at the cathedral, in lat. 51° 27′ 6″ N., long. 2° 35′ 29″ W. The configuration of the sea-coast on the side of Wales, and the noble entrance of the Severn, a river navigable without a lock for 160 miles by loaded barges, causes the tidal wave to rise to a great height in what is called

the Severn sea, or estuary, and hence the tributary rivers, of which the Avon is one, have depth at high water to float the largest vessels, the tide rising at springs in the Severn, near its mouth, 46 feet; on extraordinary occasions it has been known to rise to 50. The time of high water in King's road is 6ʰ 45ᵐ. Hence, although the channel of the Avon at Bristol has but a few inches of water at low ebb, at full tide it is capable of floating at the dock gates the largest vessel required for mercantile purposes. The docks in which vessels after entering are kept afloat, were completed in 1808, at an expense of 600,000*l.*, and cover between 70 and 80 acres. The number of vessels in 1829 returned as belonging to this port was 316, of the average burthen of 148. This gives, 316 × 148 = 46,768 tons. This was not equal to the return of 1787, which was 53,491 tons and 328 ships. In this year 416 vessels entered inwards and 382 cleared out, while in 1826 only 400 entered inwards and 312 cleared outwards; the commerce of the city extending to all parts of the world. But the prodigious increase of the coasting trade here far surpasses in consequence this diminution. In 1787 there were only 30 coasting vessels of 100 tons each, belonging to the port, and 7 fishing vessels. In 1826 no less than 5,196 coasting vessels, including 700 from Ireland, entered inwards, displaying the increase of its domestic activity in an extraordinary degree. The ancient history of Bristol is obscure, and exhibits nothing worthy of record; it was made a bishopric by Henry VIII., and has the usual municipal government of a place with a like designation, sending two members to the imperial parliament. The pop., including the adjacent suburbs, is 120,000. There are 16 parishes in Bristol, besides an extra-parochial district. Sebastian Cabot and Hugh Elliot, both renowned in the history of naval discovery or adventure, were natives of this city. It had the peculiar privilege for a commercial city possessing so many vessels, of being exempt by land and water from the jurisdiction of the Lord High Admiral of England. It was from hence that King Harold set sail to invade Wales. The city was a mart for English slaves until the sale was suppressed by William the Conqueror, and at a later period, before the African slave trade was abolished, it partook largely in that inhuman traffic, and was greatly enriched by it. The jurisdiction of the city by water extends to the King's Road, along the S. side of the Bristol channel, to the Holmes; eastward to Denny island, and thence back to King's Road. The most important manufactures of the city are those of brass, copper, glass, zinc, pins and earthenware. The Exchange is a handsome building of the Corinthian order. The Bristol channel to King's Road and the passage up to the city are generally made under the care of experienced pilots. Ships approaching the mouth of the Severn, after passing either to the N. or S. of Lundy island, must keep towards the S. shore as much as possible, to avoid the shoals scattered along the Welsh coast, which, however, are now all buoyed. Sailing N. of the island an offing of 3 leagues, not more, may be kept, when, if by night, the upper light of the lighthouse on the island will be seen. From the N. end of Lundy island to the Flatholm lighthouse is E.S.E. ½ E. about 20 leagues. If a ship is within 6 miles of the island, the ship's course will be S.E. by E. ½ E., and 10 or 12 fathoms water will be found, lessening E. Sailing

S. of Lundy island, Barnstaple bay must be avoided with a N. wind. Baggy and Morte points must have a good berth. A ship may then proceed along the S. shore safely until abreast of the Foreland, keeping 2 miles distant, in 14 or 15 fathoms. The course then is E. by S. ⅛ S. from the Flatholm lighthouse, keeping N. of the Culver sand, which is buoyed on the W. with a striped white and red buoy, and to the S. of the One Fathom bank, on the W. end of which is a black buoy. The leading mark between is St. Thomas's head, or Sand point, near Wood Spring, open to the N. of Steepholm, bearing about E. ¼ S. In the night it is in this part of the navigation that the Flatholm light becomes so useful a guide to ships bound to the E.; no one would be justified in proceeding without seeing it; the soundings are so uncertain and the tides so strong. In approaching Flatholm from the westward by day, the lighthouse should be brought to bear to the N. of E., and the island rounded on the S. side, at the distance of about a third of a mile or less, and then one care must be taken not to shoot so far as to bring the lighthouse to the westward of W.N.W., unless certain of being full half a mile to the E. of the island, in order to avoid the New Patch, a knoll of sand with little water upon it, and on which a white buoy is placed. But if it is near low water, and the buoy not seen, round Flatholm nearer to the eastward, then have Hayes windmill (which is white) appearing half way between Sully island and the main, bearing N.W. ¼ W. Vessels may anchor thereabout for a tide with the lighthouse bearing W. by S. The Flatholm is low, and has a lighthouse with a fixed light, seen 4 or 5 leagues, in lat. 51° 22' 35" N., long. 3° 7' 4" W. It is 77 feet high and the light is 156 above high water. From the Steepholm the Flatholm bears N. by E. ¼ E. about 2 miles. Shelving rocks extend 2 cables' length from the N.W. point of the Flatholm. The soundings about it vary much. The tides run violently between the holms; at Steepholm the tide flows, full and change, at 6ʰ 37ᵐ, and ordinary springs rise 38, equinoctial 41 feet. Steepholm rises 220 feet above high water. Few vessels pass S. of Steepholm. There is the shoal, called the New Patch, E. of the lighthouse on Flatholm, with only 9 feet water; it is buoyed in the shoalest part. Three rocky heads, called the Wolves, lie between Sully island and Flatholm, which show at low water springs; they are very dangerous. From Lavernock point they bear S. by E. 1⅔ of a mile, and a red and white check buoy is laid down in 5 fathoms, half a cable's length from them. The English grounds are the next danger, between Sand point and the Walton roads, extending from the shore more than 3 miles. A light-vessel with a bright revolving light, is moored in 6 fathoms on the border of this flat, about 10 miles W. of King's Road. The channel towards King's Road is bounded by the English grounds, and by the Welsh on the N. side of the channel; the last are most extensive and dangerous, the tide setting furiously over them. To sail from the Holms to the King's Road demands the skill of the pilot, nor should the passage be tried without one. Of the customary passage between the Flat and Steepholms, in which passage it has been already stated are from 5 to 8 fathoms water, the direct courses and distances are, from the Flatholm to N.W. elbow of the English grounds, E. by N. 5¼ miles; from Flatholm to the tail of the S.W. patch of the Welsh grounds, E.N.E. ¼ E. 6 miles; N.W. elbow of English grounds to N. elbow, E. ¾ N.

2 miles; N.W. elbow of English grounds to light-vessel, E. ¼ N. 1½ mile; N. elbow of English grounds to the Pigeon-house, E.S.E. ¾ E. 5 miles; Pigeon-house to Blacknose point, E.N.E. 1¾ mile; Blacknose to Portishead, E. 1¼ mile; Portishead to King's road, E. 2° S., 2 miles. The phenomena of the tides here are extraordinary. One of them is styled "the Bore," when the tidal wave advances in a ridge of 5 or 6 feet high with a roaring noise, appearing certain to swamp boats and small undecked vessels. From King's Road to the Bristol docks, up the river channel, vessels are taken at flood tide in perfect security, by those whose business it is to conduct them.

BRISTOL, North America, United States, Lincoln co., E. of the Damariscotta river. It has a number of fishing vessels, and a considerable trade. Pop. 2,945. Here the first settlement in Maine was commenced in 1625.

BRISTOL, Bristol co., United States, a seaport, in lat. 40° 40' N. It is 63 miles from Boston on the E. side of Narraganset bay, with a good harbour. Mount Hope rises in the town, a fine elevation. The coasting trade and fishery employ numerous vessels. Pop. 3,490.

BRITAIN, or GREAT BRITAIN, the largest of the European islands, extending from the Lizard point on the S., in lat. 49° 57' 40" N., to Duncansby head, in 58° 40' 22" N., and in width from 2° to 7° of W. and E. long., or from the North Foreland, long. 1° 26' 47" E., to the Land's End 5° 41' 31" W., together 7° 8' 18" of long. It is bounded S. by the English channel, N.E. by the North sea, N. by Pentland Frith and the North Atlantic, and W. by the Island channels, as the Irish sea, St. George's channel, and the open Atlantic S. of Ireland. It has many dependent islands, and coast indentations of great variety and extent, but irregularly dispersed. The climate is varied. In the S. and S.W. free from extremes, and having a mean which will suffer plants of more S. climes to flourish in the open air, the thermometer indicating in the extreme western counties great evenness of temperature and mildness, the extremes increasing as the advance is made to the N. The prevailing winds are from the W. and S.W. on nearly all the coasts, with a remarkable prevalence of the N.E. and E. in spring, which greatly retards the approach of summer, and detains vessels from entering the channel too often for a considerable time. The most violent storms occur from the W. and S.W., and rage with great fury upon the coasts exposed to their influence. The variations of the compass differ on the respective shores. Thus, off the North Foreland, in 1843, it was 23° 10' W.; in Cardigan bay, in South Wales, in 1838, the variation was 26° 10' W.; at Greenwich observatory, in 1844, the mean was 23° 18' 59", the magnetic dip 69° 0'. More to the N., in the Frith of Forth, the variation was in 1815, 28° 15' W. In 1833, on the W. of the Scotch islands, it was 28° 30' W., while off Dublin bay, in the Irish sea, it was found in 1845 to be about 27°. See *Variation of the Compass*.

BRITAIN, NEW, Labrador coast, bounded on the N. and W. by Hudson's bay, on the S. by Canada and the St. Lawrence, and on the E. by the North Atlantic.

BRITTON'S BAY, Jamaica, in the most N. part, lat. 19° 3' N., long. 77° 46' W., and W.N.W. from Platform bay. There are rocks along the greater part of the coast.

BRIXHAM, Devonshire, a town on the W. part of Torbay, on the side of Berry head. It has a tolerable harbour, and employs a good many coasting as well as fishing vessels. Spring tides rise 24 feet at the pier, on which there is a fixed light of a deep red colour. Pop. 5,015. It was here that William III. landed to aid in the great Revolution of 1688.

BRIXTON BAY, Isle of Wight, opens to the N.W. from the S. point of the island.

BROAD BANK, coast of Flanders, the S. end 4 or 5 leagues N. E. by N. from Calais cliff, and on the same end a shoal of only 1½ fathom water, while on the rest of the bank there are 3 or 4, and on the N. end 6, 7, and 8 fathoms. It extends till Nieuport bears E. by S., or the cloister Ten Duyn E.S.E. 5 or 6 leagues. On the W. side, between the Broad and Rattle bank, there is a channel of 18 or 19 fathoms. Along the Flemish banks the first of the flood tide sets over the banks towards the main for one-third of the tide, afterwards along shore, and the after flood along the sea to the N. When the flood begins to thwart off at sea, it is half-flood by the shore, and at this state or soon after is the time to enter any of the harbours on the Flemish coast.

BROAD FOURTEEN BANK, North sea, a sand 5 or 6 leagues N.E. from the Texel, between this place and the Brown bank. It is of an irregular form, and has a projection running N. from the Texel and N.W. from the Vlieland S. It runs along the Dutch shore opposite Catwyck, and the W. side stretches 11 or 12 leagues into the sea. There are generally 14 and 15 fathoms on this sand, except where it is nearest the land, when it shoals to 9 or 10. On this sand the tide to the North sea runs round with the sun, and falls but a quarter tide on the coast. The S.E. border of this sand is called the Maas bank, a narrow strip with 11 and 13 fathoms water, opposite the mouth of the river Maas, on the Dutch coast.

BROAD GROUND, THE, Denmark, in the Sound, the N.W. end lying 4 miles S. by E. from Dragoe buoy; on this part a double broom is placed on a large top buoy, in 4 or 5 fathoms. The highest steeple of Copenhagen on with Dragoe town, takes upon the body of this sand. On the W. edge are 3 fathoms, but further in not more than 6 feet water.

BROADHAVEN HARBOUR, Ireland, a small anchorage for a very few vessels of any size, on the N.W. coast. It is 5 miles S.E. of Enis head. Not more than half-a-dozen vessels of any size can be sheltered here with N. winds, and they must ride with not more than half a cable out E. and W.

BROAD HOCK, South America, a shelf or shoal which runs off N. by W. seawards from Cape Nassau, the E. point of the river Poumaron, Cape Nassau running up from the river Essequibo, in British Guyana.

BROAD SAND, Holland, on the W. side of the buoys of the channel in the Zuyder Zee, within the Texel.

BROAD SOUND, coast of Norway, near the N. point of Stadland, having the Flowack islands between. To enter, go round the most N. of those islands, and enter E. by S. 2 or 3 leagues from it, and Godon island will be seen on the larboard bow, where a pilot must be taken to go into the sound as far as Sudmer. To go to Romsdale river it is necessary to sail by the island of Luycke, the first of five large islands to the N. of this sound, where there is from 10 to 18 fathoms in midway. A reef lies across the channel after passing the island, where ships must edge off from the island on the

larboard side, and run close by the high head of Horrel, which will have 3 fathoms over the reef, and soon after plenty of water.

BROAD SOUND, France, between Point de Saut and the Isle of Ushant, extending seawards about 5 leagues, with 45 fathoms water, from St. Matthew's point, near Brest.

BROAD SOUND. See *Scilly Isles.*

BROADSTAIRS PIER, Kent, within the bight of the point between the North Foreland and Ramsgate, and nearly west from the Foreland. The tide rises and falls here 13 feet. It is 2 miles from Ramsgate. There are several shoals off this place, one W. of the pier, with not 20 feet at low water, another in the channel to the S. has not more than 14. This place is much visited for the benefit of the sea air.

BROADSTAIRS KNOLL, W. of the North Foreland, with only about 12½ feet of water, with 4 fathoms around. It lies directly in a line between the buoy of the Elbow and the North Foreland lighthouse. The marks for it are, Broadstairs mill on with the chimney of the middlemost house upon Crow hill, bearing W. a little N.; a small post near the public-house at Kingsgate just open with Ice-house point, bearing about N.N.W. ¾ W.; and the North Foreland lighthouse N.W. by W. distant 1 mile. It is a small round knoll, dangerous for vessels that draw more than 15 feet water. Within this are two others, the Fox and Goose, a quarter of a mile to the W. Many other spots of shallow water lie to the S., as the Thorn, Thistle, Spit, Crab, Boiler, Caldron, Burr, Bill, Splinter, Spur, Coburn, Cobb, &c., mostly situated on the flat which extends from the North Foreland to the Brake, met with in the passage to Ramsgate, and never to be attempted but by those well acquainted with the navigation. There is a black buoy upon the Thistle, which is also called Broadstairs knoll, in 5 feet water. Vessels drawing 9 feet water and above, should go E. of this buoy. Broadstairs is visited as a watering-place.

BROER'S BANK, coast of Holland, opposite the town thus named, and the cloister of Ten Duyn, running W.N.W. a league into the sea. When the cloister bears S. the innermost part of the bank is there, between which and the land there is a channel at low water. Vessels usually run W. of it.

BROEMESBRO, Sweden, a small town N. of Carlscrona, near the sea shore.

BROGUELES POINT, Cispata, one of the isles of San Bernardo, whence to N. Darien is 56 miles S.W. ¼ S.

BROKEN ISLANDS. See *Gambia.*

BROLUM LOCH, Scotland, Western Isles, a mile W. of Loch Valumis, reckoned unsafe.

BROOKLYN, a city and capital of King's co., New York, North America, on the W. end of Long island, opposite the lower part of New York. Pop. 36,233. Separated from New York by an arm of the sea, three-quarters of a mile wide, called the E. river; four steam ferries communicate with New York. The shipping in New York harbour is supplied with water from this place. Lat. and long., see *New York.*

BROOM LOCH, Scotland, Ross co., an arm of the sea, principally noted for its herring fishery, in the N.E. part of the Island of Martin, having a harbour and custom-house.

BRORA, Scotland, a town on the S.E. coast, in Sutherland, with a small harbour, at the mouth of the Brora, 14 miles N.E. of Dornoch.

BROSTER ISLAND. See *Shetland.*

BROTHERS, THE, West Indies, Bahama islands, a little N.W. of Acklin's kays; lat. of E. rock 22° 1' 30" N., long. 75° 41' W.

BROUAGE, France, department of the Charente Inférieure, a small port with a pop. of 800. It is situated opposite the Isle of Oleron, on a tide channel, in the midst of unhealthy salt marshes, which render it difficult of access by land. It exports much salt of the best kind. The bay or road is large, and deep enough for vessels of burden. To sail for it make for the S. end of the Isle of Aix, about S. by W. till past Isle Madame, which must be left 2 miles to the E., then sail S.S.E. direct for the port. It is high water before 4ʰ at springs. Lat. 45° 52' N.

BROUGHTON ISLAND, North America, United States, in Georgia, at the mouth of the river Alatamah.

BROW HEAD, Ireland, on the S.W. coast, lat. 51° 23' N., long. 9° 39' W.

BROWERSHAVEN, a port and town of Zealand, on the N. side of the Isle of Schonen, opposite to Goree, and only 2 leagues S. of that island, and 3 leagues S.W. from Helveotsluys. Lat. 51° 40' N., long. 4° 15' E. BROWERSHAVEN CHANNEL is the S. of the two channels between the Islands Schonen and Goree. A dry sand, called the Springer, separates it from the N. channel. This S. channel is the only one used by large vessels. Before this channel the flood turns about with the sun, so that the after-flood runs into the channel, and afterwards turns about to the southward, and continues to the running of the ebb; but the after ebb runs right out of the channel thwart into the sea, and then turns about to the northward with a fore-flood.

BROWICK, or BRAVIKEN GULF, Sweden, on the coast of East Gothland.

BROWN BANK, German ocean or North sea, a considerable sand, having on it from 16 to 20 fathoms water, and lying E. between Yarmouth and the coast of Holland, but W. of the Broad Fourteen bank.

BROWN'S BAY, West Indies, on the S. side of Nonsuch harbour, in Antigua, the middlemost of three bays on that side.

BROWN'S POINT, West Indies, S. extremity of the Island of Tobago, lat. 11° 10' N., long. 60° 4' W.

BROWNSTOWN HEAD, S. coast of Ireland, lat. 52° 7' N., long. 7° 7' W.

BRUGES, Belgium, 4 leagues E. of Ostend, from whence there is a ship canal. It is a place of considerable trade, and has manufactures of lace, linen and woollen stuffs. It has canal communications with Ghent, Sluys, Nieuport, Furnes, Ypres, and Dunkirk. Pop. 40,000. Lat. 51° 12' N., long. 3° 13' E.

BRUKSAL ISLAND, Africa, W. coast, near the mouth of the Senegal river.

BRUN, CAPE, France, the N.N.E. point of entrance into the harbour of Toulon, in the Mediterranean, having St. Margaret's castle and Garonne road to the E. of it.

BRUNET ISLAND, Fortune's bay, Newfoundland, lat. 47° 15' N., long. 55° 4' W.

BRUNET, CAPE, France, the S. point of the entrance to the Bay of Arcachon, on the W. coast, whence a point stretches seaward at W.S.W. The cape must have a moderate berth to go into the bay, and when in the entrance the land must be kept close on the S. or starboard side to the next point, in 7 or 10 fathoms.

BRUNSWICK, a town on the W. side of Cape Fear river, about 2 leagues from its mouth. Lat. 33° 58' N.

BRUNSWICK, North America, province of Georgia, lat. 31° 10' N., long. 82° W. It has a large harbour, with 13 feet on the bar at the lowest tides.

BRUSCA, Sicily, a small port, E.S.E. ¼ E. from Agnuni, which last is 10 miles from Catania; it has a small town seated on a rock, having a square castle with round towers at its extremity. The little port or cove seems more the work of art than nature; it is visited by small craft for wheat, tunny fish, and stone for building. The oysters are said to be very fine. There are 2½ fathoms of water under the castle. Lat. 37° 25' N., long. 15° 30' E.

BRUSTERVORT POINT, Baltic sea, 3 leagues W. from Kœnigsberg Deep, with Locksteed castle, a well-known sea mark, between. Memel Deep lies N.E. by N. from it, distant 16 leagues.

BRUZZANO, CAPE, Italy, on the extreme S., 8½ miles N.E. from Cape Spartivento. There is a watch-tower upon it, and a town of the same name inland, about 3 miles. There are several rocks off this point.

BRYER'S ISLAND, Nova Scotia, near Cape St. Mary, lat. 44° 14' 30" N., long. 66° 22' 48" W. There is a tower, with a fixed light, upon this island, 90 feet high.

BUA, a Dalmatian island in the Gulf of Venice, S. of Trau, and 20 miles W.N.W. of Spalatro.

BUARCOS, Portugal, province of Beira, at the mouth of the Mondego. Lat. 40° 12' N., long. 8° 54' W. The signal here is 720 feet elevation.

BUCCARI, or BUCHARI, a seaport of Illyria, on the N.E. of the Gulf of Venice, 12 miles S.E. of Fiume. The harbour is safe, though a little exposed to the S.E. wind.

BUCKHAVEN, Scotland, a small fishing town on the N. side of the Frith of Forth.

BUCK ISLAND. See *Virgin Isles.*

BUCK HARBOUR, on the S. coast of North America, United States, province of Maine. Lat. 44° 42' N., long. 63° 30' W.

BUCKIE, Scotland, a small fishing town on the shore of Moray frith, S.W. side.

BUCHANNESS, Scotland, a headland, somewhat of a peninsula, in lat. 57° 29' 15" N., long. 1° 47' W. There is a lighthouse here, 130 feet above the sea, seen in clear weather 6 leagues. This light twinkles like a star of the first magnitude, emerging every 5 seconds from a state of partial darkness to a momentary light, by which it is distinguished from the slow motion and red colour of the Bell rock light on the S., or the stationary light of Kinnaird's head on the N.

BUCKS, THE, Cornwall, sunk rocks near the land, S.W. of Lamorna cove.

BUCKHAIFA, ISLAND OF, Mediterranean, off the coast of Tripoli, near the bottom of the Gulf of Sydra. Lat. 30° 18' N., long. 19° 12' E.

BUCKSPORT, Hancock county, Maine, North America, a town on the E. side of the Penobscot river, 15 miles S. of Bangor. The harbour has sufficient water for the largest vessels, and is rarely much obstructed with ice. Pop. 3,015.

BUCSEALA, African coast, a small port on the coast of Tripoli, behind the headland called Ras al Mahabts. Its bar has only 9 feet water.

BUDE HAVEN, Cornwall, S.S.W. and S. ¼ W. from Hartland point. There is little or no tide in the small harbour here, which is seldom visited but by craft belonging to the place. At the flood tide by day a flag is hoisted, and at night a light is displayed on what is called the Chapel rock, and upon a cliff to the W. if vessels are to keep away.

BUDUA, in Illyria, Gulf of Venice, nearly E.S.E. from Ragusa, and 8 miles N.W. from Antivari. Lat. 42° 16′ 33″ N., long. 18° 50′ 27″ E. It is fortified and subject to Austria.

BUEN AYRE, coast of Colombia, South America, an island 27 miles E. of Curaçao, and 33 W. of Isles des Aves, or Birds' islands. There is a lighthouse on Point Rosa del Lacue, on the S. point, with a fixed light 70 feet above the sea, painted white, with red vertical stripes. Lat. 12° 2′ 30″ N., long. 68° 22′ 30″ W. Vessels are thus warned off that dangerous point, and ships bound to Curaçao may steer clear by it from Little Curaçao, which is level with the sea. The islet of Little Buen Ayre is on the E. side of the larger, in a bay, about a mile from which islet is the anchorage. The best passage in or out is S.W., west of it being the most free from reefs. There is a low one at the S. end of Buen Ayre, which has been fatal to many vessels.

BUENO RIO and DRY HARBOUR, Jamaica, 3 leagues E. of Falmouth, and 1¼ from Mavabona bay. It is open, exposed to all winds from N. to W.N.W., with indifferent anchorage. Dry Harbour, 3 miles E., is good for small vessels, having but 16 feet water.

BUENO, CAPE, Cuba, West Indies, E. of the Occa point or bay of the English charts, E. of Punta de la Puerto, reported as having good anchorage with from 37 to 5 fathoms water.

BUFF RIVER and BAY, West Indies, Jamaica, on the N.E. side; within it is Crawford's town. Lat. 18° 27′ N., long. 76° 32′ W.

BUFF ISLAND, Africa, coast of Benin; around it the depth of water is very irregular, but most gradual and even on the N. side.

BUFFALO, a town of the United States of America, capital of the county of Erie, upon Lake Erie, at the head of Niagara river, in New York. Its harbour is formed of Buffalo creek, and is spacious, having 12 and 14 feet of water a mile from its entrance into the lake. Upwards of 60 steamboats and 300 schooners and other craft are engaged in commerce from this lake port. The pop. is 18,300.

BUGG'S HOLE, West Indies, St. Kitt's, on the W. side of the island, round the point called the Nag's head. Inshore it is lined with rocks, but off it there are 10, 7, and 5 fathoms, more or less, approaching the land.

BUGIARONI, N. coast of Africa, called also Seven capes, E. off the Gulf of Bujeya. It is the E. point of the Gulf of Storah, and W. of Cape Ferro, with the Bay of Storah between.

BUGLES, THE, West Indies. Rocky islets, said to exist in lat. 15° 40′ N., long. 78° 20′ W.

BUJEYA, Algiers, African coast, in the Bay of Bujeya, built on the declivity of a mountain. The port is large, and there is anchorage off the town in 10 and 8 fathoms, but N. and N.E. winds send in a heavy swell. It is about 34 leagues E. of Algiers. There is a light off the port.

BULAMA, or BOOLAMA ISLAND, off the Bijooga islands, at the mouth of the Rio Grande. Lat 11° 34′ 42″ N., long. 15° 30′ 24″ W.

BULKHOFT LIGHT, Holstein, Denmark, on the entrance to Kiel, which last is in lat. 54° 9′ N.

BULL BAY RIVER, West Indies, Jamaica, E. of Port Royal harbour, between Cane river and Four Mile wood.

BULL ISLAND and HARBOUR, North America, S.W. from Cape Carteret. The entrance is at the N.E. end, having a small island within, and the passage between. Lat. 32° 56′ N., long. 78° 24′ W.

BULL AND COW, Newfoundland, rocks a mile S.E. from Cape St. Mary, St. Mary's bay, on the S. coast of Newfoundland, E. of the deep bay of Placentia. Cape St. Mary is in lat. 46° 49′ 25″ N., long. 54° 14′ 33″ W.

BULL ROCK, W. coast of Ireland, near Dursey island. Lat. 51° 36′ N., long. 10° 18′ W. It is 3 miles N.N.W. ⅓ W. from the Calf rock, the Cow rock between.

BULL SAND, a shoal off the river Humber, 1½ mile long and three-quarters broad, lying N.N.W. and S.S.E., with 4½ and 3½ fathoms over it. On its E. side is a red buoy. A light-vessel is moored off the S.E. end, having a red light from sunset to sunrise.

BULL'S BAY, sometimes called BABOUL'S BAY, E. coast of Newfoundland, 6 leagues S. of St. John's harbour. It has 14 fathoms water, and is land-locked. There is a rock 20 yards from Bread and Cheese point, and one with 9 feet of water off Magotty cove. It is 5 leagues from Cape Broyle to the N. by E. ½ E. The Bull head is half a league N. from the S. head of the bay. The head is in lat. 47° 18′ 1″ N., long. 52° 47′ 7″ W. Cape Broyle, N. point, is in lat. 47° 3′ 52″ N., long. 52° 53′ 15″ W.

BULLEN, CAPE, Africa, the N. cape of the island of Fernando Po.

BULLEN BAY, Ireland, W. coast, between Achill head and the coast of Mayo. Lat 54° N., long. 9° 59′ W.

BULLERS OF BUCHAN, Scotland, E. coast, a natural hollow in the rocks, 6 leagues N. of Aberdeen, in which small vessels have sometimes found a hazardous shelter.

BULLMAN ROCK, Ireland, E. of Kinsale harbour. This rock lies more than a cable's length S.S.W. ⅓ W. from Hangman's point, on the E. side of the entrance.

BUMBLE POINT, Cornwall, a headland beneath and a little E. of the Lizard lighthouses.

BUNASTRATHER BAY, Ireland, W. coast, 7 miles W. of Killala bay. Lat. 54° 19′ N., long. 9° 14′ W.

BUNCE ISLAND, Africa, a small island in the river of Sierra Leone.

BUNDAT ROAD, Ireland, N.W. coast, about 5 miles W. of Ballyshannon, where vessels may find a shelter in land winds.

BUNOW BAY, W. coast of Ireland, a place where small vessels might sometimes ride in fine weather in 2 or 2½ fathoms.

BUNT HEAD, Kent, near the middle part of the Goodwin sands, on the inner side S. from the N. Sand head, without the Brakes and other sands round the N. Foreland, the channel between the S. end of the Brakes and the Bunt head being called the Gull stream, and having 8 or 9 fathoms water, but deepest on the side of the Bunt head, which has 11 close to it. When between the Gull light-vessel and the Fork, vessels should not go nearer to the Bunt Head than 8 or 9 fathoms.

BUOY ISLAND, E. coast of Newfoundland, nearly in the fair way between Ferryland head and harbour and Cape Broyle.

BUR SALUM, or SALUM RIVER, Africa, W. coast, N. of the river Gambia. It has several large islands at its mouth. It lies S.E. from Cape

Verd, in lat. about 13° 52' N., long. 16° 30' W.

BURA. See *Orkney Isles*.

BURDJ KADIJA, CAPE, Africa, coast of Tunis. Lat. about 35° 8', long. 11° 10' E.

BURELA POINT. See *Bourel*.

BUREMAY BAY, N. coast of Africa, 7 leagues E. of Buttery bay, from whence Cape Three Points bears E.N.E. 18 leagues.

BURGEO ISLAND, White Bear bay, Newfoundland. Lat. 47° 30' N., long. 57° 30' W. Great Burgeo, or Eclipse island, is in lat. 47° 36' 6" N., long. 57° 36' 15" W.

BURGH SAND BUOY, Holland, W. by S. of the beacons on Voorgel sand; when Eyerland island is open with the E. point of the Texel, then a vessel is the length of this buoy, which lies on the S. shore, on a tail of sand, within which is a channel of 5 or 6 fathoms. There are four churches on the Texel island, of which Burgh is the most N. that has a steeple attached.

BURLINGTON, NEW, North America, United States, a town on an island in the Delaware river. Pop. 3,434. It is 17 miles from Philadelphia, in lat. 40° 5' 10" N., long. 72° 52' 37" W.

BURNHAM, Essex co., a town on the N. bank of the river Crouch, with a commodious quay. It is noted for the oyster-beds which lie off in the river estuary. Pop. 1,393.

BURNHAM LIGHTS. See *Bridgewater*.

BURNHAM FLATS, on the coast of Norfolk. These are marked by a red buoy on the N. end with a staff and ball in 4 fathoms, with Lynn light-vessel W.S.W. ¼ W., distant 10½ miles; Hunstanton lighthouse S.W. ¼ S., 12 miles; Brancaster high mill S.S.W. ½ S., 11¼ miles; and Holkham church S. ¼ E., 12½ miles. It is high water here at 6 o'clock, rise 16 feet. Burnham town has a good harbour, and is a league W. from Wells, within the middle part of the sand called Burnham flats, which extends W. from before Wells, parallel to the shore, about a league, dry in many places at low water.

BURNTISLAND, Scotland, Frith of Forth, on the N. side, Fifeshire coast, situated under a great rock, with an excellent harbour. Lat. 56° 8' N., long. 3° 5' W.

BURO, EAST, ISLAND. See *Shetland Isles*.

BURR ISLAND. See *Bigbury Bay*.

BURRA ISLAND. See *Orkneys*.

BURRAFORD HOLMS. See *Shetland*.

BURRELL'S GUT, West Indies, St. Kitt's, E., between Deep and Sandy bay, between which, at a little distance from the coast, is a rocky bank, parallel with the shore, inside which is a channel for small vessels from the N.W., with 5 and 3 fathoms; outside the bank are 16 and 20.

BURROWS, WEST, a sand in the Thames, E. of Black Tail beacon. See *Thames*.

BURRY HARBOUR, South Wales, Glamorgan. The entrance is N. of Burry Holms, which forms the S.W. point. It is not safe to enter without a pilot. Burry port, now called the Old harbour, is half a mile from Pembrey, or the New harbour. It has a wet dock and a lighthouse on the extremity of the pier, having a fixed light 30 feet above high water, kept lighted as long as there are 10 feet water in the harbour, and 18 feet over the bar of the river, which has a depth of 6 feet at low spring tides. High water at 6 o'clock, rise 30 feet. The harbour of Pembrey will contain 80 large coasters sheltered from all winds. Lat. 51° 41' 18" N., long. 4° 15' 23" W.

BURTON BAY and BURTON HIVE, coast of Dorset, extending from the village of Burton to the W. of Abbotsbury. It was formerly recommended for vessels unable to clear the bill and race of Portland, to run ashore on the beach at the ebb tide here, and wait a dozen seas breaking, when landing with safety was certain, but not to leave the vessel until such was the case, for instant departure would ensure destruction. This is dangerous advice, and not prudent to be taken. If anchoring from the state of the sea be impracticable, a vessel should run for Bridport pier, to the N.N.W., at which the ground is good, or a ship may run safely aground there on the E. side.

BURY ISLE. See *Shetland*.

BUSH KAY, or REEF, Jamaica, S. of the entrance into the harbour of Port Royal, between Salt Pond reef on the N. and Three Fathom bank on the S., lying N. and S., with from 5 to 6 fathoms water near its N. end, and from 7 to 8 on its S.

BUSO, CAPE, Isle of Candia, the N.W. point. Lat. 35° 36' 38" N., long. 23° 35' 30" E., nearly S. from Cerigitto about 18 miles, or S.E. ¼ E. The island of Grabrusa Agria lies off its point.

BUSTARD RIVER, river St. Lawrence, North America, having the Ozier's islands at its mouth. Lat. 49° 28' N., long. 68° 5' W.

BUSTO, CAPE, Spain, N. coast, W. by N. ¼ N. from Cape Video, 9½ miles distant, having several rocks off its point, and a rocky shore prevailing between the capes. There are several small harbours N.W. of Cape Busto, but of no service to shipping, until that of Ribadeo is made, where vessels drawing no more than 15 or 16 feet may find anchorage abreast of the castle of Damien in 5 fathoms, moored head and stern with four anchors, for S. gales blow here with prodigious fury.

BUTE, ISLE OF, Scotland, W. coast, between the peninsula of Cantyre and the main land of Argyle, 14 miles long and 7 broad. The S. end is level, but the N. mountainous.

BUTRINTO, a small seaport of Albania, opposite to the Island of Corfu, 11 leagues from Chimera. Lat. 39° 45' N., long. 20° 10' E.

BUTT, THE, or N. part of Lewis island, in the Hebrides, W. of the coast of Scotland, bearing from Cape Wrath nearly W.N.W. ¼ W. by compass, distant 40 miles. It has rocky islets near it.

BUTTER HOLE, Cornwall, N. coast, S.W. of Stepper point and the opening of the Camel river.

BUTTON'S, ISLES, coast of Labrador, S. entrance of Hudson's strait. Middle isle, lat. 60° 35' N., long. 65° 20' W.

BUTTON'S BAY and STRAIT, North America, in the N. part of Hudson's bay, between 60° and 66° N. lat.

BUTTONESS, Scotland, near the mouth of the Tay river; on the N. shore are two lighthouses, with bright fixed lights on separate towers, one higher than the other, visible like stars of the first magnitude 3 or 4 leagues, 85 and 65 feet above the water. The land at Buttoness is remarkable as being a red sandy down. The high light is in lat. 56° 28' 7" N., long. 2° 44' 53" W.

BUXEY BUOY, coast of Norfolk, the N.E. of Buxey sand, and nearly in the same direction from the Whittaker beacon, which sand there forms the N.W. side of the King's channel; the sand extends nearly to the Whittaker, between which and the S.W. end of the sand, is a channel directly W.

BUXEY FIORD, West Greenland, upon the W. coast. Lat. 64° 15' N., long. 49° 40' W.

BUYEN, coast of Holland, the E. point of the flat called the Hinder, lying between the two channels of the Quack's Deep to the N. and the Gorees Gat to the S.; on this flat there is not more than from 5 to 7 feet water. The Buyen passed, vessels go in to Goree island, or to the E. of Helveotsluys.

BUYUKDERE, a village on the W. side of the Bosphorus, Turkey in Europe.

BUZEE BANK, coast of France, a bank situated about half a league E. from the sunken rock called the Coq, and E. of St. Matthew's point, in the course towards Brest, E.S.E. from Point Craomar. Vessels may proceed into the gullet on either side of the Buzee.

BUZEMA, or AL BUZEMA, N. coast of Africa, a bay of a circular form, having a Moorish tower inland, and two rivulets running into the bay. On the W. side is an islet called the Garrison rock. The anchoring ground here is good, but the bay is exposed to N. winds. Both points of entrance are rocky, and must have a good berth passing. There is also a rock above water to the W. of Morro Nuevo, called the Frayles. Cape Tres Forcas, lat. 35° 27′ 55″ N., and long. 2° 57′ 40″ W., lies E. of this bay about 13 leagues.

BUZZARD'S BAY, Massachusetts, North America, on one side of the peninsula forming Cape Cod, near Roundhill point; on the W. side is a dwelling-house. On a cluster of rocks called the Dumplings, elevated 43 feet, is a fixed light, about N. lat. 41° 25′ and 41° 42′, and between 70° 5′ and 70° 10′ W. long.

CABALLO, PORTO, St. Domingo, West Indies, the first anchorage W. of Isabella point, or Isabelica, which last is in lat. 19° 58′ 40″ N., long. 71° 6′ 30″ W.

CABARRA, Turkey, Dardanelles, a village on the N.W. side, nearly opposite Bourgas, and due W. from it.

CABEIRO POINT, coast of Spain. See *Muros Bay.*

CABENAS, PUERTO DE, or PORT CABENAS, N. coast of Cuba, 4½ leagues E. of Bahia Honda, a fine deep bay, the entrance between two extensive reefs, with room inside for 100 sail of ships; 4½ and 5 fathoms water at the entrance, and 8 or 10 within, sheltered from all winds, and the anchorage good. It may be known by a hill called Pan de Cabenas, or, by the English, the Dolphin's Head. It is necessary to take care of the reefs from both sides of the channel at the entrance. It is 4½ leagues E. of Bahia Honda.

CABENDA, N. coast of Africa, a port of Tunis, in lat. 33° 40′ N., long. 10° 44′ E.

CABES, or GABES, N. coast of Africa, on a river near a gulf of the same name. Lat. 33° 40′ N., long. 10° 55′ E.

CABEZO DE COPE, the HEAD OF THE CAPE, Spain, S.E. coast. Lat. 37° 25′ 5″ N., long. 1° 33′ 27″ W. It has deep water near it.

CABEZOS, rocky shoals, 5 miles N.W. by N. from Tarifa point, the S. point of Spain, on which there is a tower, with a revolving light, in lat. 36° 0′ 50″ N., long. 5° 36′ 15″ W. The Cabezos shoal is 2½ miles S. by W. ¼ W. from Point Paloma, and 3 miles from the tower of Peña, with only 2 or 3 feet at low water, a little way from them 2½ or 3 fathoms, and a little farther out still 7 and 11. The greatest care must be taken of them and their locality. The dangers near here are the rock of Peña, with only 2¼ fathoms at high water, W.S.W. from the tower of Peña, 1½ miles, and N.W. by N. from the Isle of Tarifa 4 miles. The West Bank, or Arroya del Puerco, N.W. from the Cabezos, a mile long, narrow, having 3 fathoms over it, and between that and the shore 10, 14, and 18, and that and on the Cabezos from 8 to 10. W. winds make a whirlpool between this shoal and the shore. New Bank, W.S.W. of the Cabezos three-quarters of a mile, is a new bank of 7 or 8 fathoms, running W.N.W. ¼ N. and E.S.E. ¼ S. There is a channel of 10, 11, and 7 fathoms between that and the Cabezos, while outside it runs very rapidly into deep water. Thisbe's rock, so named from having been discovered by the Thisbe ship-of-war, lying S. by W. ½ W. nearly 5 miles from Point Paloma tower, W.N.W. ½ W. 5 miles from the island of Tarifa, and 1½ miles to the S.W. of the Cabezos, having only 14 feet water. Tarifa lighthouse, E. by S. or E. ½ E. will clear the Cabezos dangers. The existence of the Thisbe rock is doubted, and it is supposed to be the Cabezos on which the ship touched.

CABLE GROUND, Gulf of Finland, an extensive rocky shoal, in 5 fathoms water, 13 leagues E. by compass from abreast of Revel stone, and 9 leagues W. by S. from the N. end of Hoogland Island. It is buoyed and marked with a flag on the S. end.

CABLE ISLAND, Ireland, S. coast, at the N.W. entrance of Youghall Bay, off Ring point, and little more than a league from Youghall, which is in lat. 51° 52′ N., long. 8° 22′ W. A reef runs off both from the island and Ring point, and there is a channel midway between the extremity of the reef, with 3 fathoms water. It is high water here at 4½h. Inside Cable island there are some patches of rock, called the Black rocks, to the N.W. of the bay, but out of the track of vessels making for Youghall, as well as of the anchorage immediately within Cable island, but they have proved fatal to vessels that have been obliged to attempt running on shore there. The tides rise in the bay 12 feet spring and 7 on neaps.

CABO DE CRUZ, West Indies, Cuba, a headland on the S. side. Lat. 19° 57′ N.

CABO DE ST. JUAN, West Indies, Porto Rico, the N.E. point of that island, in N. lat. 18° 30′.

CABOLLAS, or PORT CEBOLLAS, Cuba, E. side, in lat. 20° 42′ 20″ N., long. 75° 2′ 35″ W.

CABONICO, PORT, Cuba; lat. of entrance, 20° 42′ 20″ N., long. 75° 21′ W.

CABRAS ISLAND, West Indies, St. Domingo, N. side of the island, W. of Isabella Point.

CABRERA ISLANDS. See *Balearic Islands.*

CABRITA ISLAND, a dependent of the Island of St. Thomas, off the African coast, in lat. 27° N., long. 6° 45′ E.; sometimes called Goat island.

CABRON, CAPE, West Indies, St. Domingo, peninsula of Samana, 22 leagues S. by E. of Old Cape Françoise. Lat. 19° 21′ 30″ W., long. 69° 14′ 13″ W. It appears vertical from the sea.

CACCIA, CAPE, Island of Sardinia, close to Porto Conte.

CACELLA, Portugal, a port 8 miles W. from the mouth of the river Guadiana. Lat. 37° 9′ N., long. 6° 40′ W.

CACHARIA, Isle of Nicaria, Greek Archipelago, a village on the N.W. shore of that island.

CACHEO RIVER, African coast, 6½ leagues S.E. of Cape Roxo; the entrance between 2 reefs.

The Portuguese have here an establishment of the same name; the country round is well peopled. The establishment is on the S. bank of the river.

CADAQUIE, BAY OF, Spain, E. coast, lat. 42° 17′ 15″ E. It is often used by vessels bound E., that making Cape Creux, 3 miles to the N.E., with a W. wind, find it has shifted E. On the N. side is a small creek called Port Lliegat, where small merchant-vessels anchor over a sandy bottom in 3 and 3½ fathoms, secure foom all winds but those to seaward.

CADE BAY, on the S.W. point of the Island of Antigua, in the West Indies, a little to the N. of W. from Carlisle Bay, also on the S. side of the island.

CADEDNA. See *Scilly Isles.*

CADER IDRIS, North Wales, a mountain of Merionethshire, 2,914 feet above the level of the sea at Barmouth, and serving as a conspicuous mark for vessels approaching Cardigan bay.

CADES POINT, Island of Nevis, West Indies, at the N.W. end, and the nearest cape to Major's Bay, at the S.E. end of St. Kitt's; Moreton Bay is a little to the W. of S. from it. To the N. and N.E. from it is a bank, at the end of which, bounding the channel between the two islands, are the rocks called the Cows. Off the point, at N. by W. and N.N.W. are 4 and 5 fathoms water, the former depth close to the W. side of the bank, but near to the S. is only 3 fathoms, but deepens so fast that at 2 miles W. of the S. end of Moreton Bay there is 47 fathoms, and at a mile S. of the Nag's Head in the Island of St. Kitt's there are 43 fathoms, and 79 a little more to the S.

CADIZ BAY, West Indies, on the N. coast of Cuba, in lat. 23° 2′ N., long. 79° 55′ W. It is 164 miles E. of Havanna.

CADIZ, TOWN, BAY, and HARBOUR OF, Spain, the Gades of the ancient Romans, by the Tyrians named Gadir; the principal commercial city of Spain, in the province of Andalusia, 420 miles from Madrid, and 40 from Gibraltar. It occupies the whole of a peninsula connected with the Isle of Leon by a narrow isthmus, which carries the only land entrance into the place. It contains 3,740 houses, 208 streets, and 33 public buildings, besides convents, having a pop. of 70,000. It is strongly fortified, indeed the whole is enclosed by a sea wall 35 feet high and 7,000 yards in extent, and except on the isthmus is everywhere washed by the sea. The configuration of the wall is not regular, but is ruled by the profile of the ground. The flanks are few, but the sea shoals defend its S. and W. fronts. Towards the harbour there are regular bastioned works, supplied with artillery and casemates. A lighthouse stands near the extremity of a reef of rocks, as well as the enclosed works of San Sebastian and Santa Catalina. The works which defend the isthmus are strong, and pierced with embrasures for 101 cannon, with most extensive bomb-proof accommodation. The sides of the land front are terminated by two walls of hewn stone, 90 and 95 feet high, descending into the sea, and having but one gate. There are but four gates towards the water. Towards the road on the N. the water is not deep enough for heavy shipping to attack the town. The naval arsenal is called the Carraca, and is situated on the S. shore of the inner road, 6 miles from the city. It has 3 docks and 12 building slips, and once employed 5,000 workmen. Before the close of the last war, after 1790, 1,000 vessels entered Cadiz, and in 1801 only 20. There was

recently a corps of pilots here, with a particular commandant, numbering in all 464, and ranking as officers, but divided into 4 grades. No part of the city is more than 200 feet above the sea level. The houses are high, built round square courts, each having a fountain in the centre, for coolness in the summer. The streets are narrow, but clean and well lighted. The bay is a very extensive basin, open only on the W., and a heavy swell rolls in from that point, straining ships at anchor; it is admirably protected over all the other points, and defended by 4 forts, San Sebastian, Sta. Catalina, St. Louis, and Puntales. The town of San Fernando stands on the E. side of the Isle of Leon, distant from Cadiz 6 miles; near it is the town of San Carlos, the residence of the captain-general and the public officials and buildings. A bridge leads from the main land to Cadiz. The town of Puerto de Santa Maria or Port St. Mary lies in the bay, N.E. of Cadiz 5 miles, on the shore of the river Guadalate, navigable, but having a bar at low water. Puerto Real stands on this, the inner harbour. The channel of Trocadero, which runs from Puerto Real towards forts St. Louis and Matagorda, has a dock for cleaning and repairing vessels of war and merchant ships; the channel is only 34 yards in breadth; vessels going in or out wait for the full tide, for at the mouth the depth at low water is only 7 feet, increasing inwards to 9, 10, and 12 feet. The tower or lighthouse of St. Sebastian, in lat. 36° 31′ 10″ N., long. 6° 18′ 42″ W., has on it an excellent light, 172 feet above the base, which revolves once in a minute, and is seen in fine weather more than 6 leagues: the mariner must not mistake the tower of Hercules, or Torre Gorda, for the tower of St. Sebastian; in hazy weather the error might be fatal. The former stands on the top of a little sand hill, 5½ miles S. by E., ¼ E. from the latter, a round building with a battery at its base. There is another tower called Casa de Beva, which lies N. ¾ E. from the lighthouse of St. Sebastian, distant about 4 leagues; this stands on a lofty ridge of hills, and may be seen at a great distance; square, situated between two large houses. From Rota to this tower the bearing is N. by E. ¼ E., and distant 2 leagues. There are several dangers to be shunned in this bay, as the Diamond, Coral shoals, La Galera, and others. The entrance to the harbour of Cadiz is between Rota town and the city itself. Vessels coming from the W. will, as they approach Cadiz, observe inland a ridge of the Sierra Ronda, having one hill among them higher and rounder than the rest; this is called the Moor's Head, and should be brought to bear E. ¾ S. direct for Cadiz. Pilots are always in readiness to conduct ships into the harbour. Cadiz observatory lies in 36° 32′ N. lat., and 6° 17′ 30″ W. long. from Greenwich; it is distant from Cape St. Vincent about 44, and Cape St. Maria 26 leagues. The steeples of the churches of Cadiz are seen 8 or 10 miles off in clear weather, like masts, and the tower of St. Sebastian, upon which the light is exhibited, is also seen, and with the soundings, will readily inform the mariner of his approach. To enter the harbour, coming from sea, with an E. or S.E. wind, give the point of St. Sebastian a good berth, to avoid the Olla rock; then bring the castle of Sta. Catalina del Puerto in a line with the road leading to Xeres, or on with a break in the N.W. part of Xeres hill, proceed in that direction until the 2 steeples of the Carmelite convent come open of one another; the ship will then be beyond the Cochinos. Haul up, and stand toward the shore, provided the castle of Sta. Catalina be open of the hill of Xeres; but with a large ship stand not so far as to bring Puerto Real open of Medina, but put about to the S. to avoid the Diamond. A small vessel may keep on toward the N. shore, if assured of making good way on the larboard tack, so that the Puercos may not come in one with the tower of St. Sebastian before Puerto Real is well on with Medina, for this latter will lead on the Diamond. Standing on the larboard tack, before clear of the Cochinos, the two Carmelite towers must not be got in one when the the tower of St. Sebastian bears S.S.W.; but so soon as the opening between the 2 towers is shut, put about, bringing the break before mentioned and Puerto Real church in a line; then the next board will go between the Diamond and the Puercos, with care not to bring Medina open of Puerto Real, until the steeple of St. Francis, in Port Sta. Maria, appears on the E. side of the castle of Sta. Catalina; then continue on to the N., the Diamond being passed; but when standing to the southward again, do not separate Puerto Real from Medina until St. Domingo steeple comes open of Point St. Philip; the Frayles, or Friar shoal will have then been weathered, and a ship may stand on until the steeple of Puerto Real is a little to the N. of Medina, when the vessel is within all the shoals, and the anchorage is from 4 to 6 fathoms. It is high water in Cadiz Bay at 2 o'clock, full and change. The tides rise 10 feet, and on the following days 11; the neaps only 6 feet. At the Puntal of Cadiz it is high water at 2^{h} 15^{m}. The quarantine regulations here require attention.

CADSAND BANK, West Scheldt, Belgium, lying directly at the entrance, with a S. buoy on its W. point, lying E. by S., nearly $2\frac{1}{2}$ miles from the E. buoy of the Paarde, in about 4 fathoms. Its marks are, the Orange mill, open with Flushing, to the S. of the harbour lighthouse, E. a little S. The Cadsand is a narrow shoal, about 2 miles long, in the middle part of which are 18 feet, and near each end 4 fathoms, water.

CADSAND, or CADSANO ISLAND, Belgium, at the mouth of the Scheldt, commanding the navigation of that river, on the S. side of the entrance into the W. Scheldt, opposite Flushing, about 12 miles long by 5 broad, having on it several churches and mills. A spire and telegraph are seen near the centre, attached to the church of Groede. The town of Cadsand is on the W. part of the island, and its shores form the W. part of the entrance to the Scheldt. The Belgian pilots here for the Scheldt are all invested with a medal, indicating their grade, station, and number, and furnished with instructions in English, French, Dutch, Danish, German, Spanish, and Italian. Their boats' sails have "Antwerpen" in black letters on them upon both sides, surmounted by the letter P, also the number of the boat, sewed in white on a red flag at the mast head, and the number painted on the stern. The advantages of taking these pilots must be obvious. The island is defended by forts.

CAEN, France, a town of Lower Normandy, situated in a valley at the confluence of the Orne (see *Orne*) and Odon, about 7 miles from the English channel, and 10 leagues from Havre de Grace, in lat. 49° 11′ N., long. 25′ W. High water at 11^{h} 3^{m}, rise 19 feet. The flood tide here sets S.S.W. up the river Orne to this town, which has a considerable coasting trade.

CAERNARVON BAY, HARBOUR, and TOWN, North Wales. The bay lies between Braich y pwll Head, and from thence nearly W. by N. $\frac{1}{4}$ W. to Holyhead island, includes the intermediate space. It has in most parts a clean sandy bottom, and little tidal current. A vessel may stop a tide with the wind off land conveniently, but the ground is not strong enough to hold, more especially with W. winds, which always send in a great sea. Porth Dinlleyn harbour is 12 miles from Braich y pwll Head, only fit for vessels that draw under 10 feet water. To the E. of this harbour is Nevin harbour, a place only used by small craft and coasters. In the bottom of the bay is the harbour of Caernarvon, at the entrance of the Menai strait. The bar has only a foot upon it at low water, but it never dries, and vessels drawing 12 feet water may sail over it at high water, neap tides, and ride in safety in the river, about one-third of a mile from Abermenai point. The bar is $2\frac{3}{4}$ of a mile distant from the low points which form the entrance of the strait. On the S. point is Belan fort, and on the N. Abermenai ferry-house. The S. sands, or the dry shoals, which form the S. boundary of the channel, run off W. from Belan point for nearly 2 miles, incline a quarter of a mile to the N., and then join the sand bordering the shore to the S. Caernarvon bar is buoyed. It should be sailed over at half-flood to gain the anchorage within Abermenai point; or if on an ebb tide, then with a brisk wind. The old walls of the town still remain, and the castle, built by Edward I., occupying $2\frac{1}{2}$ acres, and one of the noblest ruins in Wales. There are no manufactories at Caernarvon, but it has a considerable coasting trade. It returns one M.P. in conjunction with three other boroughs. The pop. is 9,000. Lat. 53° 6′ N., long. 4° 23′ W.

CAGADA, West Indies, a small island near the N.E. coast of Porto Rico, lat. 18° 33′ N., long. 64° 10′ W.

CAGADOS ROCKS, Africa, between the island of Palmas and Cape Formosa. They are two in number, large, and a reef runs from them to seaward. Large ships should not come nearer than 30 fathoms. The ground within 16 fathoms is foul along the coast, and every class of vessels would do well to keep at 20 fathoms along this part of the coast.

CAGLIA CAPE, or CAPE MATAPAN, the S. point of the Morea, in lat. 36° 21¼′ N., long. 22° 26′ E.

CAGLIARI, GULF and TOWN OF, lat. 39° 12′ 10″ N., long. 9° 6′ 36″ E. Cape Pula is the W. point, and Cape Carbonera the E., bearing from each other E. by N., distant 24 miles. The town is the capital of the island, near the mouth of the river Mulargia; pop. 30,000; it is fortified. There is good anchoring ground in the gulf.

CAHIR, Ireland, W. coast, 5 miles S. of Clare island, lat. 53° 44′ N., long. 9° 53′ W.

CAILLACH POINT, Scotland, on the N.W. coast of Ross co., 7 miles E. of Udrigil head.

CAIRN RYAN POINT, Scotland, Loch Ryan, upon the E. shore. On this point there is a lighthouse, in lat. 54° 58′ 23″ N., long. 5° 1′ 47″ W. A fixed light of the natural appearance. This light is open from S. by W. $\frac{1}{4}$ W. round to N. $\frac{3}{4}$ E. in a W. direction, 30 feet above the level of the sea, and generally visible 10 miles.

CAIRSTON, Orkney Isles, at the S. end of Pomona. See *Orkney Islands.*

CAISTOR NESS, N. of Yarmouth, within the sands that form the road.

CAITHNESS ORD, or ORD OF CAITHNESS, Scotland, at the N.E. end of Sutherland co., on the E. coast of Scotland, lat. 58° 12′ N., long. 3° 13′ W.

CALABASH BAY, Jamaica, on the S.W. side of the island, N.W. from the Pitch of Portland, between Flint bay on the E. and Swift's river on the W. It contains good anchorage. It is 11 miles from Pedro Bluff. Lat. 17° 53′ N., long. 77° 25′ W.

CALABRIA, Italy, Naples, the nearest part of that kingdom to Sicily, on the E. side of the Straits of Messina. It is divided into Further and Lower Calabria. Lower Calabria stretches from the Gulf of Policastro on the W. to the Gulf of St. Eufemia to the S. on the W. side; on the E. side it is bounded by the Gulf of Tarento, having on that side a very irregular line of coast round Capes Nau and Rizzuto into the Gulf of Squillace, in which the limits of Further Calabria begin on the E. side, pass round Cape Spartivento, ascend on the W. through the straits of Messina, and terminate at their junction with Lower Calabria, in the Gulf of St. Eufemia.

CALAFETES, Rocks of the Island of Majorca, close under the E. point of Dragonera island, one of the Balearic isles.

CALAGHRIAH, CAPE, synonymous with Cape Kaliakri, on the W. side of the Black sea, the E. point of the bay of Kavarna.

CALAIS, France, department of the Pas de Calais, a fortified town and harbour, with a road-stead; a place of some note in history, having a pop. of 9,500. Here several canals unite from the interior, and a considerable inland trade is carried on in corn, as well as a coasting traffic of some importance. The town is well built, and the ramparts form a pleasant promenade. It has shared with Boulogne, since the use of steam-boats, the station of the Dover packet boats, and the most commodious port of access from England. It is distant from Dover S.E. ¼ S. 22½ miles; S.E. by S. 20½ miles from the S. Foreland, and nearly S. 23 miles from the Goodwin Sand light-vessel. Three steeples are seen on approaching the town from the sea, one larger than the others, and several windmills are also observed. A light is shown on the tower of Calais, bright and revolving, making a revolution in three minutes. First the light increasing, and then full for 30 seconds; next decreasing, then dark for one minute, and thus in constant succession from sunset to sunrise. There is also a tide-light shown from Fort Rouge, W. of the harbour entrance, a fixed light 32 feet 10 inches above the level of the sea during the time there is a depth of 8 feet between the jetties. In the day a flag is hoisted during the same period of the tide, and in foggy weather a bell is tolled. The W. jetty extends 296 yards, and at the end is a small fixed light, visible 3 miles. In entering Calais harbour with N. winds, there is always some hazard. The mill at the E. end of the town must be kept on with the E. jetty head, a vessel must run in close by it, keeping the citadel on the W. The reef that runs from the W. jetty must be avoided, and when once within, a vessel must keep to the E. for Paradise basin, which is dry at low water. The tides run strongly, and it is not wise to enter until near high water, which in springs, at the jetty head, rises 21 feet, and within the entrance from 15 to 18 feet; according to the winds, at neaps the rise is 8 feet; high water full and change 11ʰ 30ⁿ. The run from Dover to Calais is always shorter than the reverse, the tide being more favourable. The canals communicate with St. Omer, Gravelines, Dunkirk, Ostend, and the towns of Flanders. The harbour within the jetties is nearly E. and W., and will contain about 100 vessels. The longest sides of the town are parallel with the sea, towards which it has one gate, and one on the land side. A mile W. is Fort Nieulay; in all five forts defend the harbour and road. There are suburbs without the ramparts to the S. The road of Calais lies N.W. of the harbour, sheltered by a sand bank, with 3½ fathoms, beginning at 2½ miles from N.W. by W. ¼ W. of Fort Lapin, and extending E.N.E. ¼ E. 3 miles; this is the "Ridens of Calais," as commonly denominated, and ships may anchor near this bank in from 10 to 13 fathoms. The best marks are, the great steeple on which the westernmost fort, and Blanc-nez 2 sails' breadth open of Calais land. The ground, composed of gravel mixed with mud, is excellent for holding. At N.W. by W., 3 miles from the entrance of the harbour, are the westernmost of the shoals, called the Ridens of the Road, with 3½ fathoms on them. To pass to the eastward of the Ridens, bring Calais to bear S. by E. There is also a good channel within them, nearly a mile wide, with from 8 to 10 fathoms in it. The lat. of Calais at the light-tower is 50° 57′ 36″ N., long. 1° 51′ 8″ E. The French give it in lat. 50° 57′ 32″ N., long. from Paris, 28′ 59″ W.; other statements make the town 50° 58′ N., long. 1° 49′ 30″ E.

CALA LONGA, Mediterranean. See *Balearic Isles.*

CALAMO ISLAND, Greek Archipelago, best known in modern times as Calamin, near the coast of Asia, in the vicinity of Mytilene.

CALAPADA CAPE, Candia, N. side of the island. It is to the E. of Cape Melaka, the W. point of a bay, which here falls open to the sea, with an island before it, and is the S.E. point of another bay, in which are several islets, opening round the point E. of Cape Maleka.

CALAURIA, Greece, a mountain N.E. of Poros, in the Bay of Egina, lat. 31° 37′ 12″ N., long. 23° 29′ 40″ E.

CALAVA, BAY OF, Sicily, N. coast, 6½ miles from Point Tyndaris, in lat. 38° 9′ 40″ N., long. 14° 54′ E. It is rather steep-to, with a number of red spots over it. There is deep water between this cape and the Lipari islands.

CALDAR, Cumberland, a small river falling into the sea S.S.E. from St. Bee's head, but it has no harbour at its entrance. Near it the shingle and sand is remarkably loose. Drig rock, seldom seen, is about a mile from the mouth of this stream, and is very dangerous.

CALDERO CAPE, in Colombia, on the N. coast of South America, 12 leagues E. by S. from the fort of Caraccas. It is called the White Cape, from its white appearance, or more frequently Cape Blanco. It lies a little S.W. from the W. end of Margaritta island.

CALDRON, THE. See *Coburn.*

CALDERONI ISLAND. See *Candia.*

CALDY ROAD and ISLAND, Wales; the outer road lies off the N.E. point of the island; ships may anchor in 6 or 7 fathoms, with the spire near the middle of the island, S.W. by W., or more W., and the E. point of the island S.W. by S., about three-quarters of a mile distant. Here the spire will appear over the High Cliffs. From the High Cliff, which is the N.E. point of Caldy, a ledge runs N.E. about 300 fathoms, with 1¼ and 2 fathoms over it, to the W. of which is the Inner Road sheltered from all but E. winds. The marks are, the house before mentioned on with the body of an old windmill, standing near a cliffy point, next W. of High Cliff, bearing S. ¼ W., or the spire on with the old windmill, bearing S. by W. ¼ W., and the N.W. part of St. Margaret's isle a trifle open of Old Castle point, bearing W.N.W. ¼ W. There are 3½ fathoms at low springs, good holding ground.

CALE'S BAY, West Indies, Island of Antigua, another name for Old Road bay.

CALEDONIA, North America, a port in the Isthmus of Darien, Gulf of Mexico, 25 leagues N.W. from the river Attata. Lat. 9° 30' N., long. 77° 36' W.

CALEDONIAN CANAL, Scotland, about 5 miles from Muirtown, near Inverness, entering Loch Ness with its E. end. The water of the Loch is very deep, in some places 129 fathoms. Through this loch, and by means of the canal, vessels pass across Scotland from the North sea into the Atlantic. The canal has 15 feet of water in its shallowest parts. The W. entrance is near Loch Eil, on the starboard side of the river Lochy, where stands the castle of Inverlochy, behind which is the mountain of Ben Nevis, 4,370 feet high, capped with snow. It begins near Corpach, on the N. shores of Loch Eil, whence it runs in a parallel direction to the river Lochy, as far as Mucomer, 6 miles distant, where it joins Loch Lochy. It commences again at the N. end of Lochy, and communicates with Loch Oich; at the northern part of Loch Oich, the canal is again continued to Fort Augustus, which stands on the shore of Loch Ness; on the banks of Loch Ness are Invermorrison, Allisay, Fyers, and Castle Urquhart, and near its northern end is Ness castle; to the W. of this latter, the canal again commences, and ends at Muirtown, near Inverness. At Curran point, or "the Narrows," Loch Linnhe, the tide is extremely rapid, and a vessel with a good breeze cannot stem it. Here is 15 fathoms water. This great strength does not extend for more than three-quarters of a mile, and is increased when heavy rains have caused a freshet from the Loch. Inverscardale bay and Cambus na Gaul bay are places where a vessel may anchor in this Loch, in fine weather. The ground is clear from the last named anchorage to Fort William, where the navigable channel becomes much contracted. It is high water at the entrance of the Caledonian canal about 5ʰ 30ᵐ, being 15 minutes earlier than at Curran Ferry; rise, springs 14, neaps 9 feet.

CALEROUSSE, CAPE, France, Hieres' islands, an Isle of the S. coast.

CALF OF MAN, Isle of Man, off the S.W. point, in the Irish sea, lat. 54° 1' N., long. 4° 43' W. High water at half-past 10.

CALF, Orkney Isles, N. of Eda. See Orkneys.

CALF ROCKS, Ireland, at the entrance of Bantry bay, near the S. end of Dursey island, on the S.W. coast.

CALF SOUND, Sweden, 4 leagues from Maelstrand, and an equal distance N. of Wingoe island. It is full of dangerous rocks, where no vessel should venture without a pilot.

CALIACH HEAD, Scotland, a cape on the N.W. coast of Mull island, in lat. 56° 36' N., long. 6° 25' W.

CALK INCH, or CHALKY ISLAND, Scotland, E. coast, just without Peterhead, N. of Buchanness.

CALLABAR CREEK, Africa, on the E. side of the river Benin, about 2 leagues from the entrance of that river.

CALLABAR RIVER, OLD, Africa, S. coast, the W. point of the entrance, called Tom Shot's point, lies in lat. 4° 35' N., long. 8° 19' E. NEW CALLABAR RIVER and TOWN are to the W. of the Old, in lat. 4° 22' 42" N., long. 7° E.

CALLAWAY LOCH, Scotland, a small bay on the W. coast, nearly up to Cape Wrath. It lies N.E. of Sark. The ground is good, and the depth sufficient for any ship, but it is exposed to the N.W. winds, and not safe in winter. W. of the Loch are the Isles of More, Callaway, and Callaway Beg, and yet further off, N. of W., are the Melland isles; off the E. a little S. is a rock, dry at low water. Between these and the Callaway isles there is a good channel of 27 fathoms.

CALLIAGUA, a haven and town at the S. end of St. Vincent's harbour, in the West Indies, the best in the island, though the town itself only contains about 60 houses; it is 3 miles from Kingstown, the capital, which last is in lat. 13° 8' N., long. 61° 17' W., near the S.W. extremity of the island.

CALLIBOGIE RIVER and SOUND, North America, United States, on the coast of South Carolina, formed by the outlet of the May and New rivers. The entrance is at the S.W. end of Gaskin's bank, which is to the seaward of the whole of Hilton Head island. Both these extend to the mouth of Port Royal harbour, South Carolina, and the S.W. point of Hilton Head island is called Callibogie point. Behind this narrow island is the sound. Lat. 32° 4' N., long. 81° 12' W.

CALLUNDBORG, Denmark, a deep inlet of the sea, between Reefness and Asnæs, having at the further end of the inlet, the town of Callundborg, before which there is anchorage in 3, 4, and 5 fathoms. There is a fixed light on the pilot-house in the harbour, visible from 1 to 1½ league. The bay has rocky shores, and should not be neared too closely. Asnæs, too, is surrounded by shallows, and should always have a good berth, though ships find good anchorage inside Asnæs. See *The Belts*.

CALLUTA ISLAND, W. coast of Mexico, North America, whence water is obtained for shipping. It is also called the Isle of Sacrifices, or de Sacrificios.

CALMAR TOWN and SOUND, Sweden; the channel between the Island of Oeland or Oland and the main land, is usually called Calmar sound. It is about 24 leagues long, and from 4 to 9 miles wide. There is no harbour of moment, save that of Calmar, which is small and secure; the town had once a considerable trade in timber, alum, hemp, and tar, but this has declined. There are buoys at Nyeket, Oswell, Torno Cliff, and near the Castle ground. The first is red, E.N.E. 1,300 yards W. of E. Swinoland point; the second, at Oswell, white, 1,800 yards W. of the Ship or Wharf Holms; the third, near the Torno Cliff, colour white, 1,059 yards S.S.E. ¼ E. of the town rampart corner; and a fourth red, near the Castle ground (Skansground), 350 yards E. ¼ S. of the S.W. rampart, corner of Grinskar. There is good anchorage in from 7 to 10 fathoms, and good shelter from N.E. and E. gales, under the Oland shore, from 1 to 2 miles, the lighthouse bearing about S., and nearly abreast of Ottomby church and village.

CALMINA, Greek Archipelago, an island near the coast of Asia, 7 leagues N.W. of Stanchio, in lat. 36° 56' N., long. 26° 46' E.

CALOERI, KALOERI, or THE MONKS, Greek Archipelago, two dangerous rocks above water, lying E.N.E. from Cape Guardia, in Andros, 32 miles, they have some sunken rocks about them, and deep water all round. They are in lat. 38° 9′ 33″ N., and long. 25° 18′ E. There are soundings about and between of 7 to 50 fathoms coral. To the W., three-quarters of a mile, there are 112 fathoms, and about the same distance S. of them, 70, 82, and 105 fathoms.

CALOO BANK, coast of Flanders, beginning one-third of a mile from the N. end of the Rassen, another bank of sand, N.E. of the Raen, stretching along in the direction of the W. shore of Walcheren, and so by the Dutch coast, between the Bodkil on the S., and the E. Gat channel to the N. It has not above 3 or 4 feet water, though there are 4½ in the Bodkil channel.

CALOT ISLAND, France, N. coast, S.S.E. of the Cow rocks, near the mouth of the river of St. Pol de Leon.

CALPE, BAY OF, Spain, S.E. coast ; 2 miles E. from Cape Negrete, is the Point de Fox, the W. point of the Bay of Calpe. The E. end is formed by a rocky ridge called by some Mount Ifak, but by seamen the Penon, or Great Rock of Calpe, distant from the W. point, or Point Fox, 2¾ miles. There is shelter here in summer only, from E. winds, under the hill, about 2¼ cables' length from the shore, and S.E. of the town in 6½ fathoms. There is anchorage too E. of the hill.

CALSHOT CASTLE and FLAT, Hants, off Southampton water. The castle is in lat. 50° 49′ 7″ N., long. 1° 18′ 6″ W. On the top of this castle there is a bright fixed light of two colours ; the one, of the natural hue, visible to vessels coming in from the W., when abreast of Stone point, and about 2 cables' length from the shore, the other visible on arriving abreast of the W. buoy of the Bramble, when a red light is seen until to the E. of the line of the chequered buoy of the same, and the black jack off the castle, when the brighter light again becomes visible, and will lead clear of Cadland point, in running up, if not shut in to the W. A floating light-vessel was moored near the spit off Calshot castle, in 1842, showing a revolving light at the entrance of the water.

CALVADOS, ROCKS OF, coast of France, E. of the rocks of Langrune, on the coast of Normandy, near the mouth of the Orne river.

CALVI, GULF OF, Corsica, between Capes Del Spano and Rivellata. The town is situated picturesquely, and is strongly defended. It has a pop. of 1,500, and is in lat. 42° 34′ 7″ N., long. 8° 45′ 16″ E.

CALVOEIRO ROCKS, Greece, E. islands, rocks near the Isle of Tinos, on the side of the Isle of Andros.

CALYMNOS, W.N.W. from Cos, in the Greek Archipelago, is near 9 miles from the N.E. point of that island ; it is triangular in shape, and about 11 miles in length from S.E. to N.W. It is mountainous, with a fertile soil, and noted for the excellence of its honey, but is seldom visited. The summit of its highest mountain is in lat. 36° 59′ 30″ N., long. 26° 59′ E. There are some rocks off its S W. point, and a little island, on its E. side, having a creek between.

CAMARET ROAD, France, department of Finisterre, outside the Bay of Brest, with anchorage in 8 or 10 fathoms on mud, and shelter from E.S.E. and S.W. winds. There is also a small dry harbour here, well sheltered, for fishing craft.

CAMARGUE, France, mouth of the Old Rhone; there is a lighthouse here called that of La Camargue, with a fixed light 118 feet above the level of the sea, upon a tower erected 60 yards N. 37° W. of the old lighthouse.

CAMARINHAS, BAY and TOWN, Spain, the N. point of the bay being Cape Villano, in lat. 43° 22′ 17″ N., long. 8° 47′ 30″ W. The S. point is Cape Cruces, on which is a chapel. Entering the bay, vessels with E.N.E. or E.S.E. winds, keep near Point Villano, avoiding the Bufardo shoal. There is a passage on both sides, and the cape bearing N.N.W., vessels run for the Chapel point, called N.S. de Barca, clearing the breakers off Point Cuerno, then steering for the point of N. Sen. de la Monte ; when 2 cables' length off, haul up for the point of Castille Viejo ; when this point comes W.N.W. go in 2 cables' length from the New Castle point ; make for the E. shore, and anchor abreast of the town, half a mile off, in 6 or 7 fathoms, mud, moor E. and W. It is high water here at 2ʰ 22ᵐ.

CAMAROONS, CAPE and RIVER, Africa, S.W. coast, about S.E. from Old Callabar river, in lat. 3° 50′ N., long. 9° 32″ E. The peak of the Camaroon mountain is in lat. 4° 13′ 30″ N., long. 9° 9′ 30″ E. The principal trade in the river is for ivory and palm-oil, the former of which has been obtained in abundance, and the last has increased with the demand. There are three large towns to which the ivory is brought from the interior, called Mungo, Batimba, and Belimba. The trade is all by barter.

CAM BASS, Cornwall, a sunken rock in Mount's bay, on which there are only 6 feet water, due E. from Paul's church, in Mount's bay. The low lea is due S. of it a quarter of a mile, and there are 10 fathoms between them. Large vessels generally anchor without them, between St. Michael's mount and St. Clement's island, in 12 or 13 fathoms.

CAMBRAE, or CAMBRAY'S ISLANDS, Scotland, within the Frith of Clyde, on the E. side of the Isle of Bute. The islands are two in number, styled the Greater and Lesser : both small. The greater has excellent stone quarries, and the ruins of an ancient cathedral upon it. There is a lighthouse on the S. or lesser island, a fixed light, visible at 3 or 4 leagues' distance.

CAMBRIDGE, North America, United States, in lat. 42° 22′ 21″ N., long. 71° 7′ 38″ W. It is 3 miles from Boston, and connected with that city by a bridge.

CAMBRISHAMN, Sweden, a town on the E. shores of Scania, a place of some small trade, but without a harbour ; vessels anchor in the offing, opposite the town, in 8 and 10 fathoms About 5 miles from the shore are the Long Ground and Stone, having only 16 feet water on them, and within a mile of the land lies the Nedian, with only 9 feet water. The town is a little N. of this. Pop. 700.

CAMDEN, North America, United States, a town situated on the W. side of Penobscot bay, having a good harbour and considerable coasting trade, as well as fisheries ; pop. 3,005.

CAMDEN, North America, United States, co. of Camden, on the E. side of the Delaware river, opposite Philadelphia. The ship-channel is on the Philadelphia side, and ships of the largest class come up to the lower village, and those of 150 tons to the central part of the city, at high water. The Camden and Amboy railroad terminates here.

CAMEIROS ISLET, Azores islands, off the coast of Terceira.

CAMELEON CAPE, coast of Africa, S. point of Orange island, one of the Bijoogas, or Bissagos, sometimes called Yellow cape.

CAMERON CAPE, Mosquito coast, North America; the E. extremity is in lat. 16° 30′ N., long. 85° 2′ 40″ W. It is a small, low headland.

CAMERON'S PLANTATION, South America, Guyana, between the Marongue and Surinam rivers, the only cultivated land between them. S. of Port Orange, in lat. 5° 55′ 30″ N., long. 54° 59′ W. Here a good look-out must be kept S. of Port Orange for a break in the land, and the house called Cameron's castle, a white mansion. Port Orange is in lat. 6° 1′ N.

CAMERS' BAY, Colombia, coast of Caraccas, S. of Cape Codera, the N. of which is Cape Corsanio.

CAMILLE, MOUNT, North America, on the St. Lawrence river, a well known mark, 7 miles inland, height 2,063 feet. Lat. 48° 28′ 44″ N., long. 68° 15′ 55″ W.

CAMINHA, Portugal, at the entrance of the Minho river, in the province Entre Douro e Minho; it is defended by a fort, and stands 12 miles N. of Viana, in lat. 41° 52′ 42″ N., long. 8° 44′ 30″ W.

CAMISCARRICK, LOCH, Scotland, W. coast, about 7 or 8 miles from the entrance of Loch Yew, an exposed place to the N. The E. side of the Island Groinard is the best anchorage, 2 cables' length from the shore.

CAM ISLAND, one of the Virgin isles in the West Indies, a mere rock N. of St. John's, in the King's channel, lat. 18° 20′ N., long. 63° 25′ W.

CAMLIN, or KEMLIN, BAY, 2 miles from Carnel's point, Anglesea island, North Wales. It is only fit for small vessels; the entrance is narrow, but the ground within is clean and well sheltered. It is high water at 10h 30m; rise of spring tides 24 feet.

CAMORRO POINT, Spain, near Tarifa Island, N. upon the main land.

CAMPANELLA POINT, E. side of the channel of Capri, Bay of Naples, lat. 40° 33′ 15″ N., long. 14° 19′ 50″ W. There is a light on this point 77 feet above the sea level.

CAMPANICHE CAPE, Mediterranean, the most E. point of the Island of Iviza or Ivica.

CAMPEBARIS, CAPE, S. coast of France, S.E. from Toulon, between that port and the Hieres islands, and nearly E. from Cape Sicie.

CAMPECHE, TOWN OF, and BAY, Yucatan, Gulf of Mexico; in lat. 19° 51′ 10″ N., long. 90° 28′ 15″ W.

CAMPBELL, PORT, called also ROSEWAY, on the coast of North America; in lat. 43° 40′ N., long. 65° 17′ W. High water 8¼h.

CAMPBELTOWN, W. coast of Scotland, in Cantire. Lat. 55° 25′ N., long. 5° 36′ 30″ W. There is here one harbour light fixed.

CAMPEN, Overyssel, Holland, near the mouth of the Yssel and Zuyder Zee, 23 miles N. of Deventer, in lat. 52° 35′ N., long. 5° 46′ E. The port is filled up with sand.

CAMPERDOWN, Holland, about a league S. of Honbosch. The N. end is remarkable for a white steep sand hill, high and sloping, the S. part grey and rough. It was off this part of the Dutch coast that Admiral Duncan defeated the Dutch fleet in 1797.

CAMPO BELLO, North America, New Brunswick. On the N.E. extremity of the island there is a harbour light for Head harbour, in lat. 44° 48′ N., long. 66° 57′ W.

CAMPO BELLO, United States of North America, a narrow island on the E. coast of Washington co., Maine, the N.E. of all the islands of the district. It lies at the mouth of a large bay, into which the Cobscook river empties itself, and has a communication with Passomaquoddy bay, on the N. by 2 channels. The N. end of this island lies in lat. 44° 48′. N.

CANA ISLAND, W. coast of Scotland, 5 miles N.N.W. of Rum, a narrow island, lying nearly W. by N. and E. by S., about a league in length, and half a mile in breadth. Off the E. end is Sandy Island. Between Sandy Island and Cana Island is a small, well-sheltered harbour, well situated for vessels bound either N. or S., and consequently much visited. On the S. side of the entrance is a rock, lying off a cove in Sandy Island, the N.E. end of which only is covered at high water. Above a cable's length to the W. and S.S.W. from Coraghan point there is a small rock within half a cable's length of the shore of Sandy island, which dries only with spring tides. When the E. house of Coraghan is hid by Coraghan point, a vessel will be past this rock, and when the house is seen clear outside of the point a vessel will be abreast of it. The best place to anchor is on the W. side of the bay, in 3 and 4 fathoms. There is also a rock on the W. side of Coraghan point, within the harbour, about pistol-shot from the shore, which dries at half-ebb, and must have a good berth. There are basaltic columns about a quarter of a mile above the harbour.

CANADA, North America, two British provinces, styled Upper and Lower Canada, lying between 61° and 81° W. long. from London, and 42° to 52° N. lat. Canada is about 1,400 miles long, and 500 broad. It lies on both sides of the river St. Lawrence, which is navigable for large sea-going vessels as high as Montreal, 170 miles above Quebec, which is itself 320 miles from the sea. The climate varies, being less rigorous towards the mouth of the St. Lawrence than at Quebec. Thus, in the district of Gaspe, the thermometer ranges from 20° below zero to 80° in the summer shade, while at Quebec it ranges from 29° under zero to 103° in the shade. The winters of Lower Canada generally commence in November, and the snow is seldom off the ground until April commences. In Upper Canada the winters seldom exhibit any degree of severity for more than a few weeks, at least in the S. parts, nor are the summers so hot as in Lower Canada, but much more prolonged. The temperature in the upper province, too, is rarely in any winter more than a few degrees below zero, and that only in the month of January; while in Lower Canada, Fahr. ranges at Quebec for the first three months of the year, and the two last, from 13° to 29° below zero. The mean minimum in Upper Canada, for the year, is 25·72°, the mean maximum 73·8°; mean of the year 48·37°. In Lower Canada it is 68·25°, and 11·75°, and the annual mean is 42·1°. The mean of the summer months in Upper Canada is 77·37°; of Lower Canada, 77·54°; in Upper Canada, during the winter months, 22·49°, in Lower Canada, 11·25°. The navigation of this great river is in consequence closed for no inconsiderable space of time every year. The entire pop. of Upper and Lower Canada, with the other British colonies on its borders, is 1,200,000. In 1830, the shipping trade of these colonies was 430,000 tons, employing 22,000 seamen, a large portion being taken up in the timber trade, and that to the West Indies. For the navigation of the St. Lawrence and its coasts, see the heads of the respective localities.

CANADA BAY, Newfoundland. Lat. of the entrance 50° 42' 30" N., long. 56° 9' 48" W.

CANAL OF HOLSTEIN, Denmark. This canal communicates between the North Sea and the Baltic ; it is cut across the Duchy of Holstein, from the river Eyder, by Rendsburgh, to about 3 miles N. of Kiel, at the mouth of the river Lerwensawe. The Eyder is navigable more than 6 miles above Rendsburgh, and the distance, from the western sluice of the canal at Rendsburgh, to its commencement, near Kiel, is 20½ English miles. The perpendicular fall of the canal towards the Baltic is 25 feet 6 inches; that towards the North sea, 23 feet, and vessels passing through are raised or let down by means of six sluices. The breadth of the cut is 100 feet at the top, and 54 at the bottom ; the sluices are 27 feet broad and 100 feet long, and the lowest depth of water is 10 feet. Merchant-vessels of 120 tons may therefore sail through this canal, and the distance from Tonningen to where the canal joins the Baltic is 66 miles. This canal was intended to facilitate the commercial intercourse between the towns of Bremen, Hanover, and Westphalia, which heretofore had been carried on by the Weser and Gluckstadt to Hamburg and Lubeck, and also to transport the merchandise of Holland and the North sea to the Baltic, but the number of shoals between Rendsburgh and Tonningen makes most mariners prefer the circuitous and dangerous passage round the Scaw.

CANALES ISLANDS, Panama, America, S. of Honda, or Deep Bay about 2 leagues, and within the limit of Golfo Dulce, and nearly W. from the Gulf of Panama.

CANANOVA, PORT, Cuba, E. side ; the entrance is in lat. 20° 42' N., long. 75° 0' 25" W.

CANARIA, or GRAND CANARY. See *Canary Islands*.

CANARY ISLANDS. These isles, consisting of seven principal islands and two or three islets, lie between lat. 29° 25' 30" and 27° 42' N., and long. 13° 18' and 18° 12' W. The two are S. of the Salvages, and lie off Cape Juby, on the W. coast of Africa, Fuerta Ventura, the nearest island to that continent, being separated by a channel about 70 miles wide, occupied as a fishery for bream and cod. The coast of Africa opposite is in general level, but rendered inaccessible by the raging serf which continually breaks upon it. The land of the Canaries on the continent is generally high, and studded with volcanic mountains, among which is the celebrated Pic or Peak of Teneriffe. These elevations cause a difference of temperature in the islands. During eight months of the year the summits of the mountains, except in Lancerota and Fuerta Ventura, are covered with snow, while in the valleys below fires are scarcely ever deemed necessary. A large proportion of the surface of these isles is covered with volcanic remains. The number of inhabitants in all the islands is reckoned to be 200,000. The passages between the islands are free for vessels of all sizes, as there is no danger except what is visible, save in a sunken rock marked in some charts as in existence between Teneriffe and Grand Canary, nearest to the last, and reputed of doubtful existence. Its position is given as 8 leagues E.S.E. of the S. point of Teneriffe, and 4 leagues W. of the centre of the island

of Grand Canary. The Salvages, between these islands and the Madeira islands, are, however, to be carefully avoided, particularly as they are doubly dangerous at night, whence in sailing from Funchal to Teneriffe, it is proper to keep well to the W.; or if that is not possible from the prevalence of W. or S.W. winds, a vessel may bear up and run to leeward of the Great Salvage, avoiding these shoals to the N. and E. of that island, if the swell be heavy more particularly. The N. of these is three-quarters of a mile N. of the island, the inner one N.E. 250 fathoms from it, and the outer in the same direction one and one-tenth of a mile. Two others lie with only 3 and 3½ fathoms about half a mile from the S. shore. See *Salvages*. The island of LANCEROTA, at Port Naos, is in lat. 28° 58' 30" N., long. 13° 32' 30" W., and Allegranza island is off the N. end, in lat. 29° 25' 30" N., long. 13° 30' 30" W. It is very high, and its mountains may be seen at a great distance. On a near approach it appears dark, waste, and rocky. From its N. end, which is in lat. 29° 15' N., precipitous inaccessible cliffs rise to the elevation of 1,500 feet, extending in a S.W. direction 7 miles, and terminating in a sandy plain, where volcanic eruptions have been experienced at very recent periods. The N.W. coast as far as the S.W. extremity of the island is high and precipitous. A cove here, once a harbour for small vessels, was converted into a salt-water lake by an eruption in the year 1765. The inhabitants are engaged in agriculture and in fishing. The pop. is estimated at 17,500. On the E. side of the island, the coast is lower than on the W.; and about the middle of it is the Port of Naos, with a little secure harbour, formed by rocky islets. It has two entrances, the N. having 12 feet, and the S. 17½ feet at low water; the rise of the tide is 9 feet. Two strong forts, one with 11 and the other with 12 guns, defend the entrances of this port. To the S. of the Port of Naos is the town of Arecife, with some good houses. The inhabitants are about 2,500. The highest land in the island rises to about 2,000 feet. The Port of Naos is the only convenient place for the repair of large vessels, as those which draw 18 feet may enter at high water spring tides, and be in complete security. Water is scarce, and camels are used on that account as beasts of burthen. There is another port on the W. side of Arecife, called Puerto de Cavallos; it has an excellent harbour, but there is no more than 12 feet of water upon it at spring tides. A square castle of stone stands upon an island between this port and that of Naos, and acts in defence of both. On the N. of the island is an islet called Graciosa, uninhabited, and between is a channel called El Rio. In the midway channel there are 6 and 7 fathoms water. It is about a mile wide, and is the only safe harbour in the Canaries for large ships, but here it is almost cut off from communication with the main island, as basaltic cliffs rise perpendicularly 1,500 feet, and their summit can only be attained on this side by a narrow path along the ridge of the precipice. Half way up is the only spring of fresh water in the island, useless from its situation. There are two remarkable vitrified rocks at the N.E. extremity of Lancerota, resembling the Needles at the Isle of Wight. While the trade wind blows, this is a good place for a ship to lie. A vessel coming from the E. must run some way in, and double a shallow point on the starboard, giving it a very good berth, by going no nearer than

4 fathoms, and when past edging towards Graciosa and anchoring in any convenient depth. This is a commodious place for careening large ships, even men-of-war at hostilities with Spain, for there is no opposition to be apprehended. The LITTLE CANARIES lie N. of Lancerota. Allegranza is the principal, a volcanic island rising 950 feet above the sea. It has a well-defined crater; the W. cliffs rise to 700 feet. There are about 40 persons on the island who collect orchilla. The only landing place is on the S. side in a little sandy bay. Graciosa is about 5 miles long by 2 wide, and forms the N. side of the Rio, or channel between its own shores and of those of Lancerota. Clara is another islet to the N.W. of Graciosa and the E. rock. Roca de Este, and the W. rock, are mere lava crags, upon which many ships have been lost. FUERTAVENTURA or FORTAVENTURA, lies S. of Lancerota, and is nearest to the African continent. The channel or strait between the two islands is called the Canal de Bocayna, 6 miles in length. This island is of a singular form, less mountainous than the other islands, though its N. and S. ends rise to 2,500 feet. Off its N.E. point is the islet of Lobos, or Seal island. It possesses two ports, Cabras on the E., and Tarajalejo on the S.E. Cabras has no more than 1,000 inhabitants. The anchorage is poor, and the place of landing worse. The exports are generally barilla, orchilla, corn, honey, goat skins and camels. The island has a very barren appearance, intermingled with fertile sports. The valley of Oliva towards the N. is very fertile, and there, in a village of the same name, the governor resides. The valley is 15 miles long and 2½ wide. Only two streams of pure water are found in the island. The population is about 17,000. This island has on the N. extinct volcanoes; one, Monte Mudo, on the N.E. rises to 2,160 feet. Date, palm, and fig trees are the only trees on the island. The S. extremity is a peninsula called Monte Handia, of the height of 2,820 feet. The channel on the N. between the islands, is deep and safe in the middle, but shoals towards Lancerota, near which are 5 fathoms. Near Lobos island, which is in lat. 28° 45′30″ N., long. 13°48′30″ W., is a good road for shipping, bringing its E. point to bear nearly N.E. by N., and anchoring half-way between that and Fuertaventura nearest to the last island. It is proper not to approach too near to Lobos. When there is a W. swell, the sea breaks terrifically on the N. end of Lobos. Some of the seas there have been witnessed breaking 60 feet in height, their noise has been heard like thunder at 6 leagues distance. The S.W. extremity of the island of Fuertaventura is Point Handia, a low point and rocky, in lat. 28° 3′ N., long. 14° 31′ W. There is a rock half a mile S.W. from it. CANARIA, or GRAND CANARY ISLAND, has its N.W. point, 16 leagues N.W. by W. ¼ W. from Point Handia. Its centre is very high, and surrounded by mountains so elevated above the clouds, as to stop the current of the N.W. wind prevalent here, so that when it blows hard on the N. side of the mountains, it is either calm or there is a gentle breeze from the S.W. At the N.W. end of the island is a peninsula called the Isleta, 2 or 3 leagues in circumference, and on each side of its isthmus is a bay exposed to the N.W. swell. On its opposite side is a spacious sandy bay, where there is a good road for shipping of any size, with all except S.E. winds, which but seldom blow there, and then not hard enough to endanger vessels. Palmas is the capital of Grand Canary.

The road is not good, and ships that are to unload discharge at anchor for better dispatch. Gando, in the middle and E. part of the island, is the next port of note; a good place for water and refreshments. Palmas is said to contain 18,000 inhabitants, and it has a very handsome appearance from the sea, with its cathedral and churches. This island has more anchorages than any of the others. In summer there is a constant N.E. wind. El Cumbre, the highest peak of the island, rises 6,648 feet above the sea. The Isleta here is in lat. 28° 11′ N., the S.E. point, and long. 15° 25′ W. The S. point, called Arguinequin, lat. 27° 44′ 55″ N., long. 15° 40′ 10″ W. The W. point or Point Aldea, 28° 1′ N. lat., and 16° 0′ 30″ W. long. The island of TENERIFFE is in lat. at the mole of Santa Cruz, 28° 27′ 30″ N., long. 16° 15′ 54″ W. The celebrated Pic, lat. 28° 16′ 24″ N., long. 16° 39′ W. Point Naga, the N.E. end of this island, bears N.W. ¾ N., 15½ leagues from the N.E. point of Canaria; but from the W. part of Canaria to the nearest part of Teneriffe, it is only 10 leagues. The celebrated Pic is in the centre of the island, and is called by the old and modern inhabitants, the Pic of Teyde. This mountain may be seen in clear weather at a great distance, or as much as 50 leagues, first looking like thin blue vapour, and very little darker than the sky. At a greater distance even this disappears, and the whole blends with the azure around it. Before losing sight of the peak, it seems at a considerable height above the horizon, when the upper part of the island around, which is very high, has disappeared beneath it. In general on approaching it, it looks, when the trade wind blows, like a haze in the sky or a cloud, until within 5 or 6 leagues, and the headlands show first like land. This island has a volcanic appearance at the first glance from the sea, but is less impressive of the fact when seen close. The peak is 12,176 feet high, an extinct volcano, the depth of the crater at present is only about 170 feet. Two-thirds of the mountain are covered with vegetation, the remainder is barren. The cone is very small in proportion to the rest of the mountain, being only 537 feet in height. The lower part of the island is basaltic. At a little distance from Point Naga, the N.E. point of Teneriffe island, is the bay and road of Santa Cruz. The best road for shipping is between the middle of the town and a fort or castle a mile to the N. of it. In this space vessels anchor about a cable's length from the shore in 6, 7, and 8 fathoms, to half a mile in 25 or 30. No part of the town on going in must be brought to the N. of W., lest calms be occasioned by the high lands under the peak, otherwise there is danger of driving upon the shore, and when there, no ground will be found on the opposite side of the ship with 200 fathoms of line. Anchors and cables therefore would be of no use. When a ship lies any time in this road, it is necessary to buoy the cables, which if good will secure ships in all winds. When running for the anchorage, both leads should be kept going, and the vessel should be brought up to the N. of the mole head, or the clock front of the square church with a cupola be brought to bear W.N.W., and then anchor with this mark on or to the N. of it. Ships may anchor when in less than 30 fathoms. Give a large scope of chain cable. When the N. fort or Fort Paso Alto bears N.N.E., the depth of the water will be 25 fathoms on the line pointed out. Never bring the mole head anything N. of W.N.W. The

II.

G

next port in goodness to Santa Cruz is Oratava, in lat. the N.W. side, 28° 25' N., long. 16° 34' W. It is on the W. side of the island, 8½ leagues S.W. of Point Naga. This is the fertile part of the island, where the wine is made and shipped off. From the beginning of May to the close of October the harbour is good. In winter vessels are too frequently obliged to slip their cables, and go off, lest they should be exposed to the heavy sea sent in by the N.W. wind. Such winds are rare, and there is generally good warning. Ships roll much at this anchorage from the N. swell. It is in 50 fathoms, 1½ mile from the shore, with the peak bearing S.W. A pilot should be on board while lying here. Oratava was destroyed by a hurricane in 1826, which committed dreadful ravages, and numerous lives were lost. There are regulations respecting vessels visiting Teneriffe, which every one should possess to prevent delay and inconvenience. Santa Cruz is the preferable port to visit for water and refreshments only. To Oratava the course from the Grand Salvage is S.W. by compass, distance 38 leagues. The ISLAND of PALMA, the E. town is in lat. 28° 43' N., long. 17° 45' W. at Santa Cruz. The lat. of Tassacorta on the W. side is 28° 38' 12" N., long. 17° 56' 55" W. It is 15 leagues from the W. part of Teneriffe to the nearest part of Palma. The summit of this island is even higher than the general level of Teneriffe, the Peak excepted. The chief place is Santa Cruz, on the E. side. On approaching the island in that direction it appears like a saddle, and the best course to take is to steer for the lowest part or middle of the saddle, until within a mile of the land, then keeping along shore to the S., the town and shipping in the road are quickly perceived. The road is within a short distance of the land, in 15 and 20 fathoms, but exposed to E. winds. Yet, when well-bound, ships remain in security with all winds, for the ground is clean and good, and the wind is repelled from the land. Thus the road of Santa Cruz in Palma is preferable to any in Canaria or Teneriffe. Santa Cruz de Palma is a good sized town. Near the mole is a castle and battery for defence of the shipping. Tassacorta, on the S.W. side of the island, is exposed to W. winds, and little visited except by boats. There are no other towns, but numerous villages on this island. GOMERA ISLAND, in its central part, lies W.S.W., 5 leagues from Point Teno, Teneriffe. The principal town is St. Sebastian, close by the sea shore, in lat. 28° 5' 42" N., long. 17° 7' W. It is in the bottom of a bay, on the E. side, where vessels are secure from all winds except the S.E. Here they may anchor in from 15 to 7 fathoms, but it is necessary for a ship to moor with a good length of cable, to prevent the danger of her being blown out of the bay. The sea is in general so smooth, that boats may load without danger. Ships of any size may hawl down, clean, and repair close to the shore, in a cove on the N. side of the bay. There is plenty of good water in the town. On the N. side of the cove is a chapel, and a battery for the defence of the port. The best place for a ship to lie, is where a view may be had along the main street of the town, at about a cable's length from the beach. It is requisite to moor as soon as possible on account of the eddy winds that at times blow into the bay. FERRO is the most W. of the Canary islands, but has no road or harbour worthy of being described. The principal town, Valverde. Port Hierro is in lat. 27° 46' 30" N., long. 17° 54' 22"

W.; and the S.W. point, Point Orchilla, is in lat. 27° 42' 20" N., long. 18° 9' 45" W. The coast rises very abruptly from the sea, and is so craggy on all sides as to render the ascent difficult. It produces many trees and shrubs, and better herbage and flowers than any of the other islands; honey abounds, but the wine is not good, and there are only three springs of water in the whole island. One peculiarity in the vicinity of these islands, is the calms which are met with by vessels that approach them. These are supposed to be caused by the heights of the mountains which rise above the clouds. For 20 or 25 leagues to the S.W., these calms, and at times eddy winds, are encountered. There are calms to the leeward of others of the islands besides Canaria, where they have been most remarked. Those of Teneriffe extend 15 leagues, of Gomera 10 leagues, and of Palma 30. Upon the edge of these calms, there is a dangerous swell of the sea, which breaks. This turbulence, dangerous for small vessels, is limited, since within it towards the island the usual appearance of a pleasant calm is found to prevail, or else a breeze at S. or S.W. always blowing in an opposite direction to the outer wind, beyond the belt which defines the limit of the two, where it foams and tosses like a boiling pot. N. or N.E. winds are most prevalent among these islands.

CANCALE BAY, TOWN, and POINT, France. The bay is on the N. coast, N.E. from Granville, and S.W. from Cancale, 4 leagues distant by sea, which is the breadth of the bay; 100 men of war and 500 transports might anchor safely here. There are only three rocks in the whole bay. The shore is fortified. There are from 13 to 4 fathoms water, down to 1. At low water it dries a long way out. To the E. of the town are three rocks, under which is anchorage in 10 fathoms. To go to the town, a ship passes between these rocks. The town is 6 leagues E. from St. Malo, in lat. 48° 49' N., long. 1° 56' W. The bay of Avranches is within this bay, and also Mount St. Michael, in the S.E. angle. The great road of Cancale is one mile S. of the Isle of Landes, between that and the Isles of Rimains and Chatellier. It offers excellent anchorage upon a bottom of clayey mud, 2 or 3 feet thick, with from 5 to 6 fathoms water. Mount St. Michael, with its little fortified town and abbey, lies 4 leagues S. ¼ W. from Granville. Cancale point, or the Groin of Cancale, is W. of N. from Cancale town and the Isles of Rimains and Chatellier.

CANCON ISLAND, near Cape Cartoche, Yucatan, North America, 7 miles long and very narrow, running N. ¼ E. and S. ¼ W., between which and the main is a lagoon, with only 3 feet water. The island is bold-to.

CANDELARIA, CAPE, Spain, S.W. coast; also a BAY, in the Gulf of Darien.

CANDELE, Syria, a village near the sea shore, a little N.E. of Cape Ziaret.

CANDIA, a large island of the Mediterranean, subject to Turkey, situated S.E. of the Morea, and S. of the Greek Archipelago, anciently Crete, and called by the Turks, Criti or Kirit Adassi. It lies E. by S. and W. by N., 46 leagues in length, in breadth varying from 11 leagues to 2. It is divided into the three districts of Canea, Retymo, and Candia, agreeing with the names of the principal towns, the chief of which is Candia. The greatest part consists of rocky mountains, the most remarkable of which is that now called Psilorite, anciently Mount Ida, near the centre of the island. The N. parts of the island are the most cultivated. The soil is fertile, and the climate healthful. Pop. 280,000, Greeks and Turks, with a few Jews. The vegetable productions are excellent in quality. There are manufactures of soap and silk, but all articles for exportation must go from the port of Candia, which augments their expense by the price of carriage thither. The oil and wax are for the most part carried by French vessels to Marseilles, soap and cheese to Constantinople, fruits to Syria and Egypt, and flax seed to Italy. The N. shores of the island are deeply indented with bays. Cape Busa is the N.W. point. The bay of Candia is situated between Point Fraschia and Cape St. John, bearing E. ¾ S. and W. ¾ N., distant 12 leagues. The city of Candia is in lat. 35° 21′ N., and long. 25° 8′ E., and contains 13,000 inhabitants, principally Turks; it is fortified; the houses are mean, and the neglected harbour can only accommodate 8 or 10 vessels. Two leagues from the city is the island of Standia, 4 miles in length and 2 in breadth, barren, high, and uninhabited. There are three small harbours on its S. side, the centre one the best, called the Port de la Madonna, and there vessels ride which cannot enter the port of Candia, discharging their cargoes into boats, riding made fast to the shore. Off the N.W. of Standia, there is a rock above water, called Petalida, steep all round, and E. by S. a rocky islet, called Paximada, and between them are 10 and 12 fathoms. Cape St. John lies 11 leagues E. ½ S. from the bay of Candia, a sharp headland projecting N.E. From the isle of Standia, N.E. by E. 20 miles, and N.N.W. ¼ W. from Cape St. John, is the rock called Ovo or Egg island, in lat. 35° 37′ 30″ N., long. 25° 34′ 30″ E.; it has deep water all around it. From Cape St. John, in lat. 35° 16′ N., the coast bends in S.S.W. for 11 miles. Capo Sidero is the N.E. point of Candia, in lat. 35° 18′ N., long. 26° 19′ E., bearing from Cape St. John direct E. 25 miles. It is the termination of a hilly promontory, stretching out to the N.E. The islands of Caso and Scarpanto are to N.E. of Sidero 27 miles. The S. coast is high and steep. At the S.E. of the island is a cluster of islets, called the Christiana or Kouphonisi islands, bearing from Cape Xarco S.W. by S., nearly 10 miles, but only 3 distant from the nearest land; a safe channel between. Cape Gialo is distant from these isles 4 miles, and 16 miles from Cape Gialo are the Gaidronisi or Calderoni islands, in lat. 34° 52′ 35″ N., long. 25° 43′ 15″ E., about 10 miles from the shore, consisting of one large island, some others smaller around, the town of Girapetra on the main, bearing N.N.E. from Cape Gialo to Cape Matala is 65 miles, W. and a little S. Cape Matala is in lat. 34° 55′ N., long. 24° 45′ E., and from this cape 9 miles distant are the islets of Paxamedes, three in number, lying in the bay of Messara, 7 miles from the nearest land. From Cape Matala W. ½ S., distant 32 miles, are the islands of Gozzo and Antigozzo, the last 4 miles to the N.W. The W. point of Gozzo is in lat. 34° 52′ N., long. 24° 2′ E., 4½ miles in length, and not 2 broad, elevated with rocky shores. The passage between Gozzo and Antigozzo is dangerous, owing to a reef said to run off from one to the other. Between these islands and Candia, however, there is a safe and good passage, 5 leagues wide. From Cape Matala to Cape Crio, the S.W. end of Candia, the distance is 64 miles W.N.W. ½ W. There is no shelter between. Cape Crio is in lat. 35° 15′ 45″

N., long. 23° 32′ 35″ E.; a little N. of it is the little isle of Serui, close to the land. The shore extends from thence to Cape Buso 20 miles. The channel between this cape and Cerigotto is 18 miles wide, and is the best passage into the Archipelago.

CANDLER'S ROCK, a vigia, once reported to be in lat. 39° 47′ N., long. 34° 29′ W., to the W. of Flores, is expunged from the charts as having no existence.

CANDURA, CAPE, Mediterranean, on the W. side of the island of Rhodes.

CANEA, BAY OF, Island of Candia, between Cape Spada and Cape Maleka, after some accounts in lat. 35° 26′ N., long. 23° 55′ E. It is wide and deep, and upon its S.E. part stands the town, containing about 8,000 inhabitants. The streets are regular, and furnished with fountains. It is fortified, and has a mole (with a harbour light 95 feet high) which will admit vessels of 200 tons each, but larger ships must anchor in the roads. The town has little trade, that in oil being removed to Candia; 5 miles W.N.W. from the town is the island of St. Theodore, to the S.E. of which is the road. The anchorage is clean and good, close to the island.

CANEIRO, N. coast of Spain, a river to the S. of Cape Busto, frequented only by small craft of the country.

CANET TOWER, France, S.E. coast, N. of Perpignan, on the river Tet, near the town of Perpignan.

CANEY CREEK, North America, near Matagorda island, Gulf of Mexico; lat. of the mouth 28° 38′ N., long. 97° 57′ W.

CANI ISLANDS, N. coast of Africa, in Tunis, lat. 37° 45′ N., long. 10° 30′ E.

CANISBAY, Scotland, a small fishing town, 11 miles N. of Wick.

CANJE, POINT, South America, entrance to the river Berbice.

CANNAVEREL, CAPE, North America, the extreme point of the rocks on the E. side of the peninsula of East Florida. Musquito inlet lies N. by W. from it, and a large shoal S. by E. This point shows the bound of Carolina, as fixed by charter of Charles II. Lat. 28° 27′ N., long. 80° 33′ W. By other observations the S. point is in lat. 28° 16′ 30″ N., long. 80° 27′ W.

CANNAY, Scotland, one of the Western islands, and S.W. from Skye.

CANNOUAN, West Indies, one of the Grenadines, a mere islet, S.E. of Carriacou, and N W. of the rocky islet called Castle island.

CANOA, PUNTA DE, or POINT, and BAY, Colombia, near Carthagena, lat. 10° 34′ 15″ N., long. 75° 33′ W. There are 5 fathoms water in the bay, but a large offing should be kept in going round the point from Carthagena. The Nigerillo shoals lie off its opening.

CANOT CAPE, West Indies, on the E. side of the island of St. Lucia, towards the N. end; a bay of the same name is on the S. side.

CANSADO, PORT, Africa, about 7 leagues N.N.E. from the S. Cape Blanco. Lat. of entrance 23° 2′ N., long. 12° 14′ W.

CANSO ISLAND, now GEORGE'S ISLAND, North America, Nova Scotia, it has a good harbour, 3 leagues deep, in which are several islands. It forms two bays, with good anchorage. It lies in the strait of Canso. Lat. 45° 20′ N., long. 60° 55′ W.

CANSO, STRAIT or GUT OF, North America,

G 2

Nova Scotia; the N. end is in lat. 45° 43′ N., long. 61° 29′ W. It is about 6 leagues in length and 1 in breadth, but the navigation is good. It separates Cape Breton island from the continent, leading into the great Gulf of St. Lawrence. It has the harbour of Tatmagouche on its S.W. side.

CANSO, PORT, North America, Nova Scotia, the S. point is called Cape Canso. There are many islands off the port for 7 or 8 leagues; it is on the Gut of Canso; lat. 45° 20′ N., long. 61° 0′ 18″ W., the S. entrance.

CANSO, CAPE, North America, Nova Scotia, and the S. point of Port Canso, to the S.W. of which 3 leagues is White point, which shelters Bar Haven. Lat. 45° 18′ 10″ N., long. 61° 0′ 8″ W. High water, springs, 8½ʰ.

CANT EDGE, Kent co., England, the name of a bank on the N. side of Sheppy island, on the Kentish side of the Thames, and round its E. end. Its edge is steep, and runs in an E.S.E. direction, suddenly deepening from 2½ to 9 and 10 fathoms water. About 3 miles from the land there is a white buoy, placed on its N. edge, in 4 fathoms water, the marks for which are Shottenden mill, just open of the high land of Sheppy, bearing S. by W. ¾ W.; the Nore light-vessel N.W. by W. ¼ W., and the W. buoy of the Oaze E. ½ N., distant 1¾ miles; this buoy must be left to port.

CANTIN, CAPE, African W. coast, to the S. of the northern Cape Blanco, S.W. by S., 13 leagues along which distance there is a strong current to the S.W. This cape is 17 miles from the bay of Saffia or Saffee, and 42 leagues from Cape Ghir or Geer, in lat. 32° 32′ N., long. 9° 21′ W.

CANTYRE, Scotland, a peninsula of Argyle co., 50 miles from N. to S., and from 5 to 8 wide. It terminates in the S. in what is called the Mull of Cantyre. It is joined to the main by a mountainous country, scarcely a mile across. The Mull is a rocky promontory, high, running S., and approaching within 12 miles of the coast of Ireland. On its extreme summit is a lighthouse, 297 feet above the sea level at high water, having a fixed light, which looks like a star of the first magnitude, visible to the S.W. and N.W. on all points of the compass, between N.N.E. ¼ E. and S. by W. ¼ W. for 22 miles in clear weather. Four miles N. of this lighthouse is Duninmore head, off the N. point of which, at half a cable's length, lies a rock that dries at half ebb, and further N. a rocky shoal, off Point Usid, at the Saltpans, a quarter of a mile from the shore, on which, at low water, are 4 or 5 feet. The sea breaks on this shoal in the calmest weather.

CANVEY ISLAND, England, Essex, 5 miles long and 2 broad; 5 miles N.W. from the Nore at the entrance of the Thames.

CAPE BAY, Africa, on the starboard shore of the river of Sierra Leone.

CAPE COAST CASTLE, Africa, windward coast, the chief British settlement on that coast, 3½ leagues from the Dutch fort of St. George del Mina. It is defended by a fort very inferior to its Dutch neighbour both in beauty and strength. The flag-staff stands in lat. 5° 6′ N., long. 1° 13′ 42″ W. This is considered the most healthy situation on this very unwholesome coast, but it is ill supplied with water.

CAPELINOS ISLETS, off the W. end of Fayal, in a S. direction; there is a passage for fishing boats between them and the island. See *Azores*.

CAPEMAYOR, Spain, N. coast, the W. point of the entrance to Santander, with a watch-tower and lighthouse upon it. The last is 330 feet above the level of the sea; the lights visible 7 leagues. The upper and lower parts show fixed lights, and between them an intermitting light once a minute. This light is in lat. 43° 30′ 10″ N., long. 3° 39′ 52″ W.

CAPERON, POINT, France, N. coast, S. from Cape Barfleur, and on its E. side. This shore must have a good berth of at least 3 miles, as it is foul and rocky, particularly off Caperon and Draguet points. Such an offing will clear all the rocks, except a sunken rock, 5 miles from the shore.

CAPERVADER'S ROAD, Holland, within the Texel island, N.E. from the E. point of the Helder. A vessel to enter must either go by the buoys in the Spaniard's Gat, to the S.S.W., then S., and then S.S.E., till the N.E. of the Helder bear nearly E., or by the new Slenk or Land deep channels into the same bearing; run along this point of the Helder till all the S.E. side of the Texel be open, and then run N.E. for the road, keeping the buoys on the port side, and anchor any where a mile from the shore, in from 9 to 6 fathoms. It is high water here at about 7¼ʰ.

CAPES, THE SEVEN, or SEVEN POINTS, Africa, W. coast, between Capes Bojador and Ollaquedo, or Laguedo; the central cape is in lat. 24° 41′ 12″ N., long. 15° 0′ 30″ W.

CAPPATCHES RIVER, Africa, a trading river, S.E. of Rio Nunez, and between the mouth of that river and Cape Verga. Lat. about 10° 30′ N.

CAPREJA ISLAND, Italy, coast of Tuscany, 10 leagues S.W. from Leghorn. Lat. 43° 5′ N., long. 9° 56′ E. Teglia is the only town, near which there is a small port.

CAPRERA ISLAND, Sardinia, near Maddalena island, off the N. coast, and near the strait of Bonifacio.

CAPRI, CAPE, Cefalonia island, 11 miles from Cape Kelia, the E. point of the bay of Samos; a bay affording convenient shelter, 2½ miles wide, with 12, 14, and 19 fathoms, before a broad beach on its S.W. extremity, on which are the remains of the ancient city of Samos. On the W. side is the small cove of St. Eufemia.

CAPRI, ISLAND OF, Naples, in the gulf of that name. This island is 3¼ miles long by 1¾ broad, at its W. end. From Cape Companella to the N.E. point of Capri, the course is W.S.W. ¾ W., and the distance nearly a league; near the N.E. end is a bay and sandy beach, along which are some curiously built fishermen's houses. Above the beach, on rising ground, is the town of Capri, and near it is an old castle. Upon the E. point of the island is an ancient building, said to be the remnant of the sepulchre of Tiberius Cæsar. If a vessel is pressed, she may anchor with S. and S.S.W. winds before the above-mentioned bay, at the N.E. end of the island, within a short cannon-shot of the landing-place, in 4 or 5 fathoms mud and weeds; but at a greater distance from the shore, there are from 15 to 25 fathoms. Close to the island there are rocks, both above and under water, but a ship may go all round at a little distance without danger, except at the S.W. point, from a reef running S. a considerable way. The S. shores of the island are lofty, and Mount Solaro may be seen a good way off. There is a good channel between Cape Companella and the island, having 15 and 20 fathoms near the Cape; in the middle 60, 70, and 80, and near Capri 14 and 15 fathoms. This island has springs of good water, and the valley between the E. and W. high land is covered with fig, orange, lemon, almond, olive, and such like trees. The climate is a perpetual spring. Pop. 1,500. It is noted for the number of quails caught there and exported.

CAPUTA ROCK, Italy, Lower Calabria, a rock 4 miles S. of Panicello, on which there is a tower, and 5 miles from Cape Suvero, nearly opposite to a tower on the main, and far from the land.

CARABASTA, Greek Archipelago, an islet off the S.W. point of Nicaria.

CARACA ISLAND, coast of Colombia, in Cumana, an islet W. of Port Mochima, and S. of the W. end of the Island of Margaretta.

CARACCA, Spain, a small port, 2 leagues E. of Cadiz.

CARACCAS, South America, a province, in which are situated the Ports of La Guayra and Porto Bello, with others of inferior note. It is bounded on the N. by the Caribbean sea, E. by the Atlantic, S. by British Guyana and New Granada, and W. by New Granada. It is divided into the provinces of Cumana, Caraccas, Coro, Maracaibo, Varinas, and Guyana. They form the united provinces of Venezuela, and the E. part of the Republic of Colombia. The capital is Caraccas, 12 miles from the sea, the port of which is La Guayra. It has a pop. of 30,000. Lat. 10° 30′ N., long 67° W.

CARAMANIA, or KARAMANIA, Mediterranean, S. coast of Asia Minor, N.W. from Syria. It comprehends the ancient Cilicia, Pisidia, and Cappadocia. Konieh is the capital.

CARASCHE BANK, coast of Africa, 2 miles S. of Cayo bank, near the island of that name, one of the Bijoogas isles.

CARATASKA SHOALS and REEFS, off the Mosquito coast, in the Bay of Honduras, North America.

CARBONERA CAPE, Sardinia, the E. point of the Bay of Cagliari, in lat. 39° 4′ 24″ N., long. 9° 32′ E.

CARDAGUI, or TCHERDAKLI, Turkey, in the Dardanelles, opposite Gallipoli, a little N.E. and N. of Lamsaqui, on the E. side of the Strait. It affords a temporary anchorage.

CARDIFF, South Wales, Town and Roads; these last are sometimes called Penarth roads, and are the safest in the Severn sea, upon the Welsh coast. They are bounded W. by Lavernock point, Penarth head, and the flats, and on the E. by Cardiff grounds. They afford excellent anchorage in 4 and 5 fathoms, good holding-ground. The best position is with the W. extremes of the two Holms in one, S. by W. ¼ W., and Leckwith house, a white house just in sight, as it opens from Penarth head, bearing N.N.W. ¼ W. Particular attention must be paid before anchoring to the state of the tide, as springs rise here from 36 to 40 feet, or a fathom an hour. For lack of this, some ships have at low water grounded upon their anchors. There are good berths in 4½ and 5 fathoms, at low water. There are also some dangers, as the Grounds, the Moulestone, and Centre Ledge. The town of Cardiff stands on a plain 2 miles N. of Penarth head. It is the county town of Glamorganshire on the river Taaff. A canal for ships, and a basin capable of receiving vessels of 600 tons, has been constructed here by the Marquis of Bute. It contains 200 acres of water; has stone wharfs of a mile in extent, and is entered by a canal 1,250 yards in length, varying from 19 to 13 feet in depth. The sea-gate of these docks is in

lat. 51° 27′ 46″ N., long. 3° 10′ 20″ W. The pop. of Cardiff is between 4,000 and 5,000. There is a light on the pier-head. It possesses a venerable castle, built in 1110, in which Robert Duke of Normandy was confined 26 years, in a room 15 feet square, after being robbed of his inheritance by his brother, Henry I. The town carries on a considerable trade, as vessels of 300 or 400 tons' burthen can come up to the quay. It also possesses iron and tin works on the banks of the Taaff. There is a light-vessel between the English and Welsh grounds, with a revolving light, 10½ miles W. of King's road.

CARDIGAN HARBOUR, South Wales, a harbour only fit for small vessels drawing less than 10 feet. The town is in lat. 52° 4′ 59″ N., long. 4° 39′ 17″ W. The W. entrance is Kenmaes, or Cardigan head, and the E. is at Cardigan island, in lat. 52° 7′ 54″ N., long. 4° 41′ 26″ W.; between these two points are Cardigan roads. It is high water at Cardigan bar, at full and change, 6ʰ 45ᵐ. Rise 15 feet.

CARDINALS' HATS, or ST. GEORGE'S ISLAND, Greek Archipelago, towards Cape Colonna and the Gulf of Athens, and S.W. S. from the Island of Macronisi. It is high and steep and easy to be known by the tops of the hills as above named. Vessels from Cape St. Angelo to Zea pass close within this island, which has no harbour.

CARDINAL'S ISLAND, North America, on the N.E. coast of Labrador, in lat. N. about 59° 30′.

CARDINAL'S ROCKS, France, Bay of Quiberon; when brought to bear S.S.W., S. by W., or S., there is good anchoring, with soft, clay ground, and even soundings in from 10 to 12 fathoms.

CARDRONAC BAY, coast of Cumberland, a dangerous spot, full of shifting sands. How point, a mile W. of this, is high land, but it then falls and becomes low, winding by Scargevil head to this bay. On the S. side of it is Holm abbey.

CARDY, a rock, in the Irish sea, near the E. coast, in lat. 53° 39′ N., long. 6° 9′ W.

CARENAGE BAY, West Indies, St. Lucia island, deep in to the E. nearly half across the island, between Grand Cul de Sac on the S., and Choque bay on the N., in lat. 13° 35′ N., long. 60° 50′ W.

CARENERO, PUNTA DEL, N. coast of Cuba, near Bahia Honda.

CARENTAN, France, department of La Manche, situated on the Taute, that forms a small port, which the lesser coasting craft visit. It is N. from Coutances, in lat. 49° 20′ N., long. 1° 16′ W.

CAREY'S SWAN'S NEST, North America, E. by N. from Cape Southampton, Southampton island, and a harbour at the S.E. extremity of the coast on the E. side of the Welcome sea, in Hudson's bay. The names given by the earlier discoverers of these parts are not all preserved in the modern charts. It was so named by Button, who followed the unfortunate Hudson, in 1612, into the bay which bears his name, on a similar voyage of discovery.

CARIBBEE ISLANDS, West Indies, sometimes called the Antilles, and divided into the windward and leeward isles. Under the denomination of the windward isles, the whole range, from the Virgin islands to Trinidad, is now generally included; and the leeward islands are those existing between Trinidad and the Gulf of Maracaibo. The old designation was faulty, including, under the name of "Leeward Islands," those only from Porto Rico to Dominica; and under that of the "Windward Islands," those from Martinique to Tobago. The islands thus classed are as follow (a minuter account will be found under separate heads)—Sombrero, Dog Isle, Anguilla Kay, St. Martin, Saba, St. Bartholomew, St. Eustatius, St. Christopher, Nevis, Redonda, Montserrat, Antigua, Deseada, Guadaloupe, Marie Galante, the Saintes, Dominica, Martinique, St. Lucia, St. Vincent, Barbadoes, Granada, Tobago, Trinidad, Testigos, Margaritta, Blanquilla, Tortuga, Orchilla, Los Rocques, Ilos dos Aves, Buen Ayre, Curaçoa, Little Curaçoa, and Oruba.

CARINO BAY, Spain, N. coast, an anchorage near Cape Priorino, not much used.

CARIPACHO POINT, S. coast of Cuba, opposite Anna Maria kays, about lat. 21° 25′ N., long. 79° W.

CARISTO, or KARYSTO, Greece, a town of Negropont, in the E. part, lat. 38° 4′ N., long. 24° 27′ E.

CARLINGFORD LOCH, Ireland, 2 leagues E. of the entrance of Dundalk harbour, a well sheltered bay, with the ground good and water for the largest vessels, but having several rocks and dangers to claim attention. The entrance is between Ballaggan and Cranfield points, bearing from each other E. by N. and W. by S., distant 2 miles from the lighthouse on Haulbowling rock, and E. ¼ N. from Cooley point. Carlingford lighthouse stands in lat. 54° 2′ N., long. 6° 8″ W. The hill, 1,580 feet high, is in lat. 54° 2′ 39″ N., long. 6° 13′ 9″ W. It is high water at Carlingford bar 10^h 40^m. Rise of springs, 16 to 18, neaps 12 to 14 feet.

CARLISLE BAY, Jamaica, on the S. shore of that island, in lat. 17° 47′ N., long. 77° 6′ W.

CARLISLE BAY, West Indies, Island of Barbadoes, on the S. side of Bridge Town point. It is the harbour of Bridge Town, the capital of the island, and is capable of containing 500 sail. The bottom, however, is foul, and the nearer the shore the worse becomes the ground. The depth is from 25 to 30 fathoms. It is necessary to buoy up hempen cables, and have them well secured, the ground being bad all over.

CARLSCRONA, Sweden, the principal naval arsenal of Sweden, E. of Carlshamn. This is the third city of that country, and the chief station of the Swedish navy, built upon several rocky islets, in lat. 56° 9′ 30″ N., long. 15° 35′ 12″ E. Pop. 13,000. The opening of the harbour is defended by two strong forts on two islands, commanding the entrance by a cross-fire. The rocky islands within are connected by bridges, and the port will contain 100 sail of the line. The admiralty buildings are of brick, on the island of Trosoe. On that of Lindholm there is a dock for four line-of-battle ships excavated in the solid rock 80 feet deep, out of which the water is pumped. An immense basin was begun here by Gustavus III., but not finished by that monarch. It is a staple town, and has numerous vessels employed in its foreign trade. It has a fine anchor-forge. The trade is principally in timber, tar, tallow, potash, and marble. Most of the islands near Carlscrona have deep channels between, through any of which sailing to the port is perfectly safe. Eleven miles from the entrance of Carlscrona stands the Utklipporna, or Outer Barrier rocks, and lighthouse. See *Utklipporna.*

CARLSHAMN, Sweden, N. of Hano island, a town having a dockyard, fortress, and harbour. Pop. 8,000. It is in the province of Blekingen, and makes large exportations of potash from the birch forests near. A number of small vessels are built here.

CARLSO ISLAND, Baltic sea, 7½ leagues N. ¼ E. by compass, from the S.W. point of Gothland island, and S. by W. ¼ W. from the light of Lands-ort.

CARLSOAR, Baltic sea, a small island near the W. coast of Gothland island.

CARMARTHEN, South Wales, a town seated on the river Towy, navigable for small coasters. It has a commodious quay for loading and unloading vessels. It lies nearly N. from Barnstaple bar. The channel into the river leads in E.N.E. on the E. side. There is a bank on the S. which lies off a great distance from the shore, in Carmarthen bay, into which the Towy falls. The outer side of the bank goes off flat. Vessels going in sound it along by the E. or N. shore to the shoalest part of the bar, where there are about 14 feet at half flood, and 4 fathoms along the bar. They then run by the lead along the S. shore until the river is fairly entered. The town is in lat. 51° 52′ N., long. 4° 23′ W. It is high water at springs here about $5\frac{1}{2}^h$.

CARMARTHEN BAY, South Wales, between Worm's head, in lat. 51° 33′ 56″ N., long. 4° 18′ 56″ W., and the island of Caldy. Lat. of lighthouse 51° 37′ 56″ N., long. 4° 50′ 57″ W. The N.E. part of the bay is a large level, through which runs up the channel to Carmarthen, bearing distant from Caldy island about E. 3½ leagues, and from Worm's head N. ¼ E. somewhat more than 3 leagues. It is narrow, has 14 feet over the bar at half-flood, and is only to be attempted with a pilot.

CARMEL, CAPE on the coast of Palestine, in the Mediterranean, now called St. Margaret de Amené. It is 5 leagues S.W. from Cape Blanco, between which is a large bay, and at the bottom is Acre, the ancient Ptolemais. A bank runs out to sea from Cape Carmel with 2 or 3 fathoms, behind which is a tolerable road between the bank and the shore. Both the shoal and the road are known by a flat tower on the top of a point. It affords a tolerable protection in S. winds, and here large vessels choose their anchorage, for the ground of Acre is foul, and cuts hempen cables. Nazaria, the ancient Nazareth, bears from Mount Carmel S.E. by E., distant 19 miles, and from Nazaria to Tiberia, on the edge of Lake Tiberias, the direction is N.E. by E., distant 13 miles.

CARM-OE, ISLAND, W. coast of Norway; at Hoivarde, in this sound, there is a fixed harbour light 63 feet high. Carm-Oe lies in a N.N.E. and S.S.W. direction, 16 miles in length and about 4½ miles broad at the S. end, but narrowing as it advances to the N. At its S.E. end, called Skudesness, there is a lighthouse, with a fixed light, that points the way into the sound by the W. entrance. Those who wish to enter the sound may distinguish the light of Skudesness from that of the Isle of Huiddings. The last is suspended between two poles and gives a blaze light; but the S. light of Skudesness exhibits a clear and steady light; and besides, this last light cannot be seen by those coming from the W. unless so much to the S. that the cliff of Gicetongen does not interrupt its view, or that it be in bearing N.E. by E.; and on this point of the compass the navigation is clear from the Isle of Huiddings; those, therefore, who have gained sight of one light, and are in doubt which, should steer a little E.; and if it be Skudesness light that appears, the blaze of Huiddings isle will shortly be seen, unless in hazy weather or a snow-storm. If, by steering E., another light soon appears, it must be that on the Huiddings isle, and a course may be set, in order to take a view of the other; and if the light on Skudesness be hidden by the high land, and a light appear more to the eastward than due N.E. by E., it is the Huiddings isle light. Having ascertained this, a vessel may steer for Skudesness with safety; and seamen may know to a certainty, by the light, where the bight is, and accordingly run under the land, and so into Carm sound. This island is in lat. 59° 30′ N., long. 5° 29′ E.

CARMOYLE ROAD, the place where ships ride, bound to Belfast on the N.E. coast of Ireland. It is 5 miles S.W. from Carrickfergus; but as a long spit runs out from the N. shore, ships must run up into 4 fathoms at low water almost to the White house to come into Carmoyle, which will bring Castle rock, on the top of a hill on the E. shore a little above Carmoyle, on the village by the water-side. Then they sail two-thirds over, and bear in. But in foul and hazy weather, when the White house cannot be seen, they sail near it in 4 fathoms at low water, and then run S.S.W. two-thirds over and bear in at S.W. after the same manner, where they anchor in 3 fathoms at low water; higher up, the water is still shoaler.

CARN, or CAM, BANK, coast of Flanders, W. of Broer's bank. It has 3 fathoms water, within which a vessel may ride in 8 or 10 fathoms by the land. Between this and Broer's bank there are 5 fathoms, and between the Carn, or Cam, and the main, in coming from the E., a vessel may sail along the Quad bank, till she is before Dunkirk harbour.

CARNEIRO or CABRITA POINT, Spain, E.N.E. of Point de Frayle, and E. of Tarifa. Pigeon island lies between Carneiro point and Point del Frayle. This island is also named Palomas, and has rocks about it. One-third of a mile from the point is Point Secreta, where there is a shelter for small craft. Cabrita point is the W. boundary of the extensive bays of Algesiras and Gibraltar. In case of necessity, vessels lie at single anchor about half a mile from Cabrita point, in from 18 to 27 fathoms, sand and small gravel, and from thence as far out as 38 fathoms the ground holds well.

CARNEL POINT, North Wales, the most N. point of Anglesea island. The Skerries rocks, on which is a lighthouse, lie N. by W. ½ W. from it, and the lat. of the light is 53° 25′ 18″ N., long. 4° 36′ 30″ W. It marks the N. limit of Church bay; Cleopatra Point, forming the S., at 4½ miles distance, the 2 points lying N.N.E. ¾ E. and S.S.W. ¾ W. from each other. The bottom of the bay is foul, and it is not above three-quarters of a mile deep, Out of a line drawn from point to point, 1½ miles N.N.E. from Cleopatra point is the Fenwick rock, having but 9 feet at low water, and marked by a buoy. Off Carnel head, within three-quarters of a mile to the W. are two spots with 3½ and 4 fathoms.

CARNERO, GULF OF, Istria and Croatia, surrounded by high mountains.

CARNESS, a cape on the N. coast of Pomona island. See *Orkney Isles.*

CARNSORE POINT, Ireland, the S. point of that island, in lat. 52° 11′ 20″ N., long. 6° 23′ W. It bears from Cape Cornwall N. by E. ¼ E. 42 leagues; from the Smalls light N. by W. ½ W. 37 miles; and from St. David's Head, Wales, N.W. ½ N. 42 miles.

CAROCO, CAROCHA, or the DUTCHMAN'S CAP, W. coast of Africa, an islet S. of Prince's island, in the bight of Biafra.

CAROUSAL ISLAND, North America, coast of Labrador, in lat. 50° 5′ 29″ N., long. 66° 26′ 35″ W.

CAROUZE, CAPE, West Indies, Hayti, on the N. side, in lat. 19° 59′ N., long. 70° 53′ W.

CARPENTER ROCK, African W. coast, about half a mile W. by S. from Cape Sierra Leone. It is seen at half tide, and the sea generally breaks over it. On going into Sierra Leone river it is left on the starboard.

CARPORI ISLAND, South America, Guyana, at the mouth of the river Arawary. Cape North, in this island, is the most W. point of Guyana, in lat. 1° 10′ N., long. 50° W.

CARR ROCKS, Scotland, a ledge E.N.E. from Fifeness, whence they project a mile into the sea; the outermost is dry 2 hours before low water, but the other is not seen except at low ebbs. A buoy and beacon are placed at the N.E. end. The course and distance from which to Babert Ness are N.W. by N. 4 miles; to the fairway buoy at the entrance of the river Tay N. ¾ E. 8¼ miles; to Aberbrothick, N.N.E. 15 miles; to Red Head, N. E. by N. 19 miles; and to the Bell Rock N.E. by E. ¾ E. 10 miles. It is well to come no nearer these rocks by night than from 15 to 17 fathoms, the depth N. of the rocks, there being from 22 to 24 fathoms to the S., keeping the square or tower part of Cryll steeple below the spire or point above the head.

CARRETO, PORT, Colombia, N. of Cape Tiburon, Gulf of Darien. Lat. 8° 47′ 15″ N., long. 77° 34′ 45″ W.

CARRIACOU, West Indies, one of the Grenadine Islands, the second in size, 5 leagues from Granada. It has a large, fine, and safe harbour, but no fresh water. Lat. 12° 30′ N., long. 61° 22′ W.

CARRICK ISLAND, Ireland, river Shannon, used as a guide for avoiding the S. end of Rinauna shoal, which lies S.W. by S. from the S. end of Scattery island, by hugging it close, or having the summit of Ray hill a little open to the S. of Kilkadran cliff.

CARRICKAMEEL ROCKS, Ireland, near the W. coast, in lat. 53° 15′ N., long. 10° 4′ W.

CARRICKAREED, Ireland, N. coast, a rocky islet, in lat. 55° 14′ N., long. 10° 4′ W.

CARRICKFERGUS BAY and TOWN, Ireland, near the N.E. extremity, 6 leagues broad at the entrance, and narrowing towards Belfast. The town lies 2 leagues within the bar on the N. shore, with a good road before it, and a pier at the town, which is dry at low water. At a mile distance, S. by W. there is a bank of sand which has 9 feet upon it at low water. To the N. and S. of it there are 2 fathoms, and further to the S.W. 3 and 5 in the middle of the road. The town of Carrickfergus is in lat. 54° 42′ 45″ N., long. 5° 49′ 15″ W.

CARRICK ROAD, Cornwall, at the entrance of Falmouth harbour, which last runs in from it to the W., between Pendennis and Trefusis. This road lies nearly N., a little E., and commences at the Castle point, St. Mawes, on the E. side. It extends for nearly 4 miles to Turnaware point, where the channel narrows, running up to the town of Truro, about 5 miles further. The best anchorage is in St. Just pool, on this side. On the opposite or W. side there are 2 creeks, one going up to Mylor, the other to Restronget. On both sides there are flats near the shores. In the road there are from 7 to 17 fathoms water; the road itself being about 640 yards wide in the widest part. That portion of the road opposite to Falmouth is called the Cross road, and the whole might accommodate about 40 sail of the line. Above Missic point, S.W., 60 fathoms, there is a small bed of rocks with 8 fathoms upon them at low water, and opposite Restronget creek are the Stangate rocks, dry at low water. Mylor creek or pool and Restronget, between Trefusis point and the entrance to Truro river, are safe places for merchant-vessels, with sufficient depth of water at all times. To enter the road by the E. channel bring Mylor point on with Killaganoon house, bearing N. ¼ E.; this will clear the banks through Carrick road until Budoc church bear W. ¼ N. over the rising land of Trefusis point; St. Kevern church will then be seen on with Pendennis point, which leads through the pool until Feock church and house are in one; this mark kept on until Restronget western ferry-house comes on with Restronget point, will bring Mylor and Pendennis points on with each other; this will lead into the mid-channel, when Trefusis house will be got in one with a high point of land to the N.; proceed until near Turnaware point, off which a shelf of gravel stretches nearly half-way over the channel, to be carefully avoided: the direction of the river will then guide further, keeping nearly mid-channel. To enter by the W. channel bring Feock house in one with Mylor point, bearing N. by E. ¼ E.; this mark will carry over Falmouth flat at half tide in 5 fathoms. Frigates may pass through the Western channel safely, and men-of-war will have sufficient water at half-flood, but neither should attempt it at low water. Vessels will sail in free with the wind at N.W. by W. on the larboard tack; and should the high land of Pendennis cause the wind sometimes to baffle, there will be no danger, having passed the Black rock. In dark weather it is proper to give St. Anthony's point a berth of 2 cables' length, running in with the land of St. Mawes a point open on the starboard bow; or steer for Mylor point, passing St. Mawes, keeping the lead going, and taking care not to get into less than 9 or 8 fathoms. When St. Mawes' castle bears E.S.E. a ship will have entered Carrick road, and may anchor. If with S. or S.E. winds, bring up with the small bower in 12 or 13 fathoms, rather inclining to St. Mawes' bank, and moor with best bower towards Falmouth bank, in 12 fathoms. With the wind to the westward of S. bring up with the best bower in 12 or 13 fathoms, more inclined to the Falmouth bank, and moor with the small bower towards St. Mawes in 12 or 13 fathoms; the upper part of the road is considered the best; but the Cross road is safe and has more room. Here you place your anchors so that you moor open to the southward; bring up with your best or small bower in 10, 12, or 14 fathoms. See *Falmouth.*

CARRICKVILAN ROCK, Ireland, W. coast, E. of the anchorage in Tarbert bay, about a cable's length from the opposite shore.

CARRIGEEN ROCK, W. coast of Ireland, 2 miles E. of the Long rock, near Tarbert bay.

CARRON LOCH, Scotland, a large and tolerably sheltered anchorage and good ground, but the entrance is so rocky it should not be attempted by a stranger. It lies in what is called the inner sound of Skye along the mainland of Scotland.

CARTAGENA, or CARTHAGENA, Spain, S.E. coast, one of the best harbours in the country. It is a basin of a large size, close to the town, and secure from all winds. The town is built on a kind of peninsula, contains a fine arsenal, and is well supplied with good water. It possesses an observatory, standing in lat. 37° 35′ 50′ N., long. 1° 0′ 37″ W. The pop. 30,000. The principal exports are wood and barilla; it possesses a manufactory of sail cloth, and extensive alum works. The hot springs of Archena are situated a few miles from Carthagena. The town is well defended. At the N.W. part of the harbour is the arsenal, protected by a battery of 40 guns. The suburb of St. Lucia extends down the E. side, which forms a species of bay, on the S. point of which is the powder magazine. From Cape de Gat to the entrance of the harbour is N.E. 25 leagues. The narrowest part of the channel, which lies between the batteries of Navidadad and Florentina, is 680 yards, the extent of the harbour inwards being more than a mile. The W. point, moderately high, is protected by the batteries of Podadera and of Navidadad, and within is the great careening place, a bay turning to the W., with from 3 to 5 fathoms of water, the common place of anchorage for ships-of-war, sheltered and close to the arsenal. There are several shoals near the entrance, or within the entrance of the harbour, and some to be avoided without. The Laxa, or Port shoal, lies at the W. side of the harbour entrance, N.E. by N. from the battery, or point of Navidadad, distant 384 yards, with 6 fathoms water round it. The shoal of Escombrera, small, with 10 feet water over it, and round it 6 or 7 fathoms, nearly W. from the island. In anchoring in the bay, the first point made is the middle of the harbour's entrance, directly for the Student's Grotto, passing the Laxa shoal to the larboard; if the wind is from the sea, anchor in 6 fathoms, and moor N.W. and S.E.; if the wind be from the land, an anchor should be let go first to the N.W., and then one to the S.E. Ships-of-the-line moor in winter in the Espalmador, secure from seas and winds, but not more than three sail can swing their anchors there, though 10 may lie moored head and stern with 4 anchors, and others in the Cano, or Little Espalmador, where 12 may lie made fast two and two. The Cano is the name of the channel leading to the docks, latterly called the New Deep, having 4 or 5 fathoms of water within it, but too narrow to turn in, so that a large ship must enter stern foremost, unless going into the docks. A sand bank extends from this channel E., where other vessels may moor, but not conveniently in winter. In summer they anchor in the middle of the harbour, as there the vessels are more healthy, from the free circulation of the air, and the winds are seldom strong. When ships seek the harbour in winter, with offshore winds, they will find the anchorage best on the N. side of Escombrera island, S. of the castle, in 27 fathoms, good holding ground.

CARTAGENA, or CARTHAGENA, Colombia, South America, the name of a province and town on the main land. The province is bounded N. by the Caribbean sea. The town is in lat. 10° 25′ 36″ N., long. 75° 34′ W., and 69° 19′ 37″ W. from Cadiz; it stands on a swampy island, connected with the main land by two causeways. Pop. 20,000. The town is well fortified. At three-quarters of a mile distance from the sea stands the hill of La Popa, on the summit of which is a convent of Augustine monks. In clear weather this hill may be seen at 10 leagues distance from the deck of a line-of-battle ship. The town is strongly fortified. The harbour extends 2½ leagues from N. to S., and has safe anchorage. The climate is hot and unhealthy, very fatal to strangers, the yellow fever raging there with more than its customary virulence. The little tongue of low land on which the town stands extends nearly 2 miles S. by W. ½ W. from the city, and by turning to the E. makes a basin, which is the harbour. A mile to the S. of the exterior point of this tongue of land is an island called Tierra Bomba; and the passage between the two Boca Grande. This last can be closed artificially, so that only boats or vessels drawing very little water can enter by it. This island of Tierra Bomba is 4 miles from N. to S., and its S. point is the N. point of Boca Chica, the only entrance to the harbour of Carthagena. Lat. 10° 19′ 30″ N., long. 75° 36′ 17″ W. There are castles on both sides to defend the entrance, called St. Joseph and San Fernando.

CARTAGO BAY, North America, on the E. side of Honduras, between the coast and Cape Gracias Dios. It has a fine sandy strand and good anchorage, having several islands across its mouth, and one in particular just in the entrance, behind which is a good road. Cape Gracias Dios is in lat. 14° 59′ 30″ N., long. 83° 12′ W.

CARTERET, CAPE, France, N. coast, 9½ miles S. by W. from Cape Flamanville. There is a lighthouse upon it, showing an intermittent light, with half-minute flashes, 262 feet above high water level, visible 6 leagues. This cape is N.E. from Jersey, and nearly E. from Guernsey and Serk islands.

CARTERET, CAPE, called also CAPE ROMAN, on the coast of North America, a small island between two others, a little S.W. from the Santee river, where the coast begins to take a more N. direction, and appears to form a point, in lat. 33° 1′ N., long. 79° 16′ W. It is about 15 leagues S.W. from Cape Fear, the coast clean all the way.

CARTHAGE, CAPE, Tunis, coast of Africa, S.W. from the port of Tunis. The site of the ancient city of Carthage is within the promontory of the same name, on the N.E. side of the lake of Tunis, about 10 miles from that city; the ruins remaining are nearly 3 miles from the sea, though Carthage is said to have extended to the sea shore. The ancient harbour is closed up, but retains its ancient name in El Marsa, or the Port. A lighthouse has been erected on Cape Carthage, with a three-minute revolving light.

CARTHAGENA, POPA DE, Colombia, South America, a remarkable hill to the N.E. of the city, which supplies an excellent land-mark; it looks like the quoin of a gun, and from seaward has the appearance of an island. It in is lat. 10° 26′ N., long. 75° 33′ 15″ W. This hill has a monastery on its summit.

CARTMEL, a town and small harbour on the sands between the S. part of Lancashire and Furness, with a harbour for small vessels. It lies between two bays, one formed by the entrance of the River Ken from Westmoreland, and the other by the confluence of several rivers into the Irish sea. Lat. 54° 12′ N., long. 3° 6′ W.

CARTMEL WHARF SAND, Lancashire, on the E. side of the channel to Ulverstone, at the entrance of the River Ken.

CARVASONE, CAPE, Africa, W. coast, 10 leagues S. by W. from Cape Barbas.

CARVEL OF ST. THOMAS, West Indies, a rock between the Virgin islands E. and the island of Porto Rico W. At a distance it looks like a sail, white, and has two points. Between this rock and the island of St. Thomas is the channel of Sir Francis Drake, once so named. Lat. 14° 58′ N., long. 61° 7′ W.

CARVEL SHOAL, Azores islands, a reef running off a cannon shot distance to sea from the Whale rock, N. by E. 20 leagues from the Ribeira Grande, on the middle of the N. coast of St. Michael's island. Lat. about 39° N., long. 25° 15′ W.

CARVOEIRO, or PENICHE, CAPE, Portugal; the lighthouse here is in lat. 39° 21′ N., long. 9° 25′ W. The great Berling, or Berlinghas, is N.W. by N. 5 miles from this cape. The channel between, nearly 6 miles wide, is clear of danger, having from 14 to 24 fathoms water. The light here is 182 feet above the sea, visible 5 or 6 leagues, with half-minute flashes, which distinguish it from the Berling light. The towns of Peniche, upper and lower, lie the first on the E. and N.E. side, the second on the E. and S.E. It is high water 1ʰ 54ᵐ. Peniche has been called the Portuguese Gibraltar.

CARYSFORT REEF, West Indies, part of the Grand Florida reef, and the most considerable of the coral rock patches to the N. of Rodrigues shoal; at its S. end there is a vessel with two lights, just at the bend of the reef, 50 and 60 feet above the sea level, and visible 3 leagues.

CAS, ST., BAY and POINT, the former on the N. coast of France, W. of St. Malo, N.E. of the village of St. Cas. It has a fine clear beach, which extends along shore N.E. and S.W. to a rock which lies N.W. from the village. The point is about a mile from the Bourdinot rock, 8 miles W. of St. Malo.

CASAMANZA, or CASSAMAS, RIVER, 4 leagues to the W. of Cape Roxo, on the coast of Africa, about 16 leagues S. of Bald Cape. Lat. 12° 35′ 30″ N., long. 16° 43′ W.

CASAMBEC, POINT, North America, round the N. point of St. John's, Gulf of Nova Scotia; it is S. from the N. point of the island, on the E. side, and the N. limit of a chain of islands along the coast.

CASCABEL ROCK, South America, off the coast of Colombia, in lat. 10° 55′ 10″ N., long. 75° 5′ 10″ W., not far from Port Sabanilla.

CASCAES, ROAD, TOWN, and FORT, Portugal, a league E. from St. Roque, on the N. point of the entrance into the Tagus. There is good anchorage abreast of the town, in 8 and 18 fathoms. The road is sheltered from N.W., N., and N.E. winds. In running in bring the town open with the castle and point, where a pilot may be taken up. S.E. ¾ E., 4½ miles from Cascaes fort, is St. Julian's castle, erected on a projecting point, which has a small reef before it; in the fortress a lighthouse is erected, a fixed light, 120 feet above the level of the sea; and this part may be considered the N. point of the Tagus river.

CASCARELLI, POINT, Italy, bay of Naples, between Point della Scala and Cape Bruno.

CASHEEN BAY, Ireland, W. coast; the ground on the E. side of Dynish island is good, sheltered, and deep. A shoal lies S.W. from Dynish about half a mile, with 8 feet water over it. The anchorage is the E. side of Dynish, the White sand at the point bearing W. 2 cables' length from the shore, in 5 and 6 fathoms water.

CASHE'S LEDGE, a dangerous reef, existing in lat. 43° 1' N., long. 69° 9' W., extending about half a mile each way; the soundings very irregular from 10 to 4 fathoms in a boat's length; there are 17 fathoms within a cable's length of it deepening to 90 on the W. side. The currents on it are rapid and devious. On the shoalest part there are only 12 feet at low water. It lies off Massachusetts bay, United States of America.

CASILDA, PUERTO DE, Cuba, entrance in lat. 21° 38' 15" N., long. 80° 2' 30" W.

CASKETS. These are rocks so called, 3½ miles W.N.W. from Ortach, and 5½ miles N.W. by W. from the W. point of the Alderney island. This cluster of rocks is about a mile in circumference, some above water and visible, and some below it. On the largest, or W. of the Caskets, there are three lighthouses, placed in a triangular position, exhibiting lights revolving, alternately, a bright ray of light in every direction, at 120 feet above the sea. Upon a S.E. ¼ E. bearing these lights appear as two. The N.E. and S.E. lights are in one when bearing S.W. ¾ W. The N. of the rocks, off the N. end of Alderney and the Caskets, bear from each other nearly E. by S. and W. by N. They are clean and steep-to, with 25 and 30 fathoms all round. Between the rocks to the W. and the lights, is deep water, vessels may pass amongst them all, though the velocity of the tide occasions the sea to ripple. Ships unable to weather on either side, may pass within the distance of an oar's length in safety; the passage to the W. is half a mile wide. There are two boat harbours at the Caskets, one on the S.W. side, the other on the N.E.; that to the S.W. is so formed by nature, that a frigate might lie in it as in a dock. Steps are cut in the rock, and means provided for hauling up a boat; the N.E. harbour is not so compact. Boats approaching the Caskets have signals shown to them, declaratory of which harbour they are to land at; a red flag directs to the N.E. and a blue flag to the S.W. Any one so imprudent as to neglect, or act contrary to the directions, will most probably be lost. It is high water here at 6ʰ 45ᵐ.

CASLEH, HARBOUR and BAY, W. of Galway bay, on the W. coast of Ireland, a safe shelter for vessels drawing only 9 or 10 feet water. The best anchorage is at the head of the bay, on the W. side, in 2 fathoms. A rock, called Rock Caninock, lies midway at the entrance of this bay, and must be avoided; it dries only at spring ebbs. A rock, called the Carrickcleshin rock, lies at the head of the bay, covered only at high water.

CASO, Greek Archipelago, N.E. of Candia, 27 miles, 11 miles long and 5 or 6 broad, uneven and rocky. On its N. side is a little bay, sheltered by some islets, called Caso Poulo, and a village, inhabited by about 100 fishermen. There is but one practicable landing-place, where there is a basin for the reception of boats. The S. point of the island is in lat. 35° 18' 20" N., long. 26° 52' 35" E.

CASSAN, PORT OF, Africa, 70 leagues up the River Gambia, to which place it is navigable for vessels of burthen.

CASSANDRA, a gulf of Greece, E. of the Gulf of Salonica, from which it is separated by a peninsula of high land. Its S.W. point is Cape Pailliouri, and its S.E. Cape Drepano, in lat. 39° 56' 53" N., long. 23° 57' 15" E. These capes bear from each other E. by N. and W. by S., distant 8 miles.

CASSASEZ, Africa, a settlement on the River Nunez, near Walkeria, from 70 to 80 miles up that river. The entrance of the Rio Nunez is in lat. 10° 35' N., long. 14° 41' 30" W. It is sometimes called the river of Nunez Tristão, and is noted for its trade in ivory.

CASSIPOUR, CAPE, South America, Guyana, in lat. 3° 55' N. The land from Cape North to Cape Cassipour is low all the way and thickly wooded. This part of the coast is best recognized by the little hill of Mayez, somewhat flat in shape, visible 5 or 6 leagues off, in 3° 5' N. lat., and vessels from Europe are above all things recommended to get a sight of it.

CASSIS, France, department of the Bouches du Rhone, a little port, pop. 2,220, situated in lat. 43° 12' 20" N., long. 5° 33' E. It is only adapted for small vessels, and has some trade in dried fruits and wines; it also carries on a coral fishery.

CASTEL A MARE, GULF OF, Sicily, N. coast, a spacious bay between Point Rama and Cape St. Vito, with deep water and good anchorage near its shores, but is not safe with N. winds, which send in a heavy sea. The town of Castel a Mare is 9½ miles from Point Rama, a dirty town of 5,000 inhabitants. The exports are wine, fruits, manna, grain, and sumach. Cape St. Vito is the N.W. point of Sicily, in lat. 38° 12' 26" N., long. 12° 45' 53" E. It should have a berth of 3 miles in passing, in order to go on the N. side of a rocky shoal, which lies off the cape.

CASTEL A MARE, Italy, Naples, a town bearing from Naples lighthouse, S.E. ¼ E. 13¾ miles, built at the foot of a mountain near the sea. To the W. there are 7 and 8 fathoms of water, about a mile from the landing place. There is a small river before the town running out N.E. ¼ E., where a few small vessels may lie. From Castel a Mare to Cape Orlando, the course and distance are W.S.W. ¾ W. 1½ mile; a rocky islet lies near this cape, on which account it must have a good berth. S.S.W. 1½ mile from the cape is the town of Vico. The coast between is high and steep, called the Secca Calvarasa.

CASTEL DEL UVO, or OVO, Italy, bay of Naples; it is crowned by strong works, and commands a good portion of the inner bay, next the city of Naples, upon a projecting point of which it stands.

CASTELLON DE AMPURIAS, Spain, near the N.W. angle of the Bay of Rosas in the Mediterranean, on the W. side of a cove in that angle. There is an island that almost blocks up its entrance. It is in lat. 42° 18' N., long 2° 58' E. The town stands 1 mile inland behind Ampurias lake, as the lagoon is styled.

CASTEL ORIZO, ISLAND, coast of Caramania, about a league in length, with a good harbour on the S. side, 90 miles E. of Rhodes. Lat. 36° 7' N., long. 29° 31' E.

CASTINE, North America, United States, a seaport, Hancock co. Maine, on the E. side of Penobscot bay, opposite Belfast. The harbour is spacious and good, accessible at all seasons, and capable of receiving the largest vessels. Castine is also the name of a river falling into the same bay.

CASTLE BAY, Scotland, at the S. end of Barry island, on the W. coast.

CASTLEHAVEN, TOWN and HARBOUR, Ireland, S. coast, a harbour only fit for vessels drawing 14 feet and under, and these must lie a quarter of a mile above Reen head, with the Stags of Castlehaven in sight, between the Horse island

and the main. There is a sunken rock between Adam's isle and Shillenragga head, on which there are only 12 feet at low water. When the wind does not blow from the S. or S.E. vessels may ride in the bay of the town of Castlehaven in 4 and 5 fathoms. The water shoals to 2 proceeding up the harbour. The Carrikananin rock, between Ragged and Squince islands is covered, except at very low ebbs, and there is a small rock, dry at half ebb, a cable's length S.W. ¼ W. from the W. Black rock. The Stag rocks, or Stags of Castlehaven, lie 3¼ miles S. by W. from the entrance of the harbour, and 9 miles E.S.E. ½ E. from Cape Clear lighthouse. The lat. of the Stags is 51° 28′ 15″ N., long. 9° 13′ 46″ W.

CASTLEMAIN HARBOUR, W. coast of Ireland, near Dingle bay. It is difficult of access, has a bar across the entrance, with only 1½ fathoms at low water, and the entrance should not be attempted without a pilot.

CASTLE TOWN and BAY, isle of Man, on the S.W. side. It has a castle, which the distance from the shallow and rocky harbour renders of no utility. It is in lat. 53° 55′ N., and long. 4° 38′ W.

CASTLES' HEAD, THREE, Ireland, a point of land a league N. of Mizen head, so called from three old towers upon its summit, and having another, called Sheep's head, to the E.N.E., the true limits of which form the points of Dunmanus bay, a league S.E. from Beerhaven. Mizen head is in lat. 51° 27′ N., long. 9° 50′ W.

CASTLES OF EUROPE AND ASIA. See *Bosphorus.*

CASTOR HAVEN, Nova Scotia, North America, 4 leagues S.W. from Cape St. Mary, and 2 leagues N.E. from the Bay of Islands, which last is about 15 miles N.E. from Chedabucto bay.

CASTRI, VILLAGE OF, in the bay of Lepanto and gulf of Corinth, built upon a mountain, above which there is a rock 200 feet high. The celebrated Pythian cave is here, and about a league S.W. ¾ W. from Castri is the port of Cirrha, in a state of decay. Near Castri are several sepulchres hewn in the rocks.

CASTRIES, France, Mediterranean, near Montpelier, in lat. 43° 37′ N. Pop. 2,400. It has a fine aqueduct, which brings to it the waters of Cadoule.

CASTRO, Italy, a city of Naples, 7 miles S. of Otranto, having a harbour and a small trade. It is 5 miles from Chianca.

CASTROMARIN, Portugal, on the W. side of the river Guadiana, opposite Ayamonte, 13 leagues E.N.E. and N. from Faro or Cape St. Maria. Lat. 37° 5′ N., long. 7° 16′ W.

CAT, Ireland, a rock near the S.W. coast, hardly a mile from Crow island.

CATA, COVE OF, coast of South America, W.S.W. ½ W., from La Guayra 23 miles, clean, with from 25 to 4 fathoms, 1¼ cables' length from the beach.

CATABARIZA STRAIT, W. coast of Africa, with the islands of Los Illgelos, or Ilhas dos Idolos, or Isles of Los, S.E. from a bay in the midst of them. Ships may run with a better ebb round a flat lying between both, and so to the coast by this strait. Factory island has an anchorage on the E. side, half a mile from the shore safe and good, except during the tornado season in April and May. The tide rises here 13 feet, and flows with a N.E. moon. The N. end of the outer island Tamara, is in lat. 9° 31′ N., and long. 13° 40′ 30″ W. There is good fishing here with the seine, and great plenty of turtle.

CATALDO CASTRO, Italy, about 9 miles E. of the Torre dell Orso, the lat. of which is 40° 15′ 35″ N., long. 18° 25′ 37″ E.

CATALINA, and LITTLE CATALINA, North America, Newfoundland, E. coast. The bay is 5 leagues from the N. point of Trinity harbour, having several creeks or coves between them. Ragged harbour, which lies in to the S.W. is a part of this bay, and so is the harbour called Catalina harbour, 2 leagues N. from the S. point of the bay, and 2 miles from Ragged harbour. It is a good and safe harbour, and has gradual shoalings from 8 to 3 fathoms; on the S. of which is a small island, which a ship may lead through with a large wind, though the channel is not more than a cable's length broad. In this passage is from 4 to 5 fathoms. To go without this island eastward, give it a small berth, keep the fairway mid-channel and hold it so for a mile. A shoal lies off from the S. point of the harbour to E.N.E., on any side of which a ship may go. It may be known by the breaking of the sea upon it. LITTLE CATALINA is N.N.E. half a league from the other, having from 7 to 10 fathoms all the way; beyond which is the N. head of the whole bay at N.E., easterly from the harbour, and half a league from the other. N. head here is in lat. 48° 32′ 28″ N., long. 53° 1′ 54″ W.; S. head, lat. 48° 27′ 38″ N., long. 53° 6′ 40″ W.

CATALINA, or ST. PROVIDENCE ISLAND, bay of Guatimala, North America. Lat. 13° 21′ N., long. 81° 22′ 20″ W.

CATANIA, GULF AND CITY OF, Sicily, E. coast. The gulf commences at La Trezzia, and continues as far as Cape Santa Croce, bearing from each other distant 17 miles S. by E. ¼ E. The land from La Trezzia bends in S.W., 4½ miles to Point Armisi, and within this point is the small town of Aci Castello, built upon a block of lava, and having about 800 inhabitants. The shores are black and full of caverns. There is a cove here, in which stands the town of L'Ognoro, whence lava is exported squared for building. The depth of water in the cove is from 7 to 3 fathoms. The bottom foul. The city of Catania is about a mile from Point Armisi, and is the second city in the island, noted for the elegance of its buildings and the polished manners of the inhabitants. It is the chief residence of the Sicilian nobility. The pop. 80,000. The port is generally full of small vessels, and there is trade deserving a more convenient harbour. The exports are principally corn, olives, silk, wine, potatoes, macaroni, fruits, cheese, soda, manna, cantharides, amber, lava, and snow. Between Point Armisi and Point Schiarra, vessels bring up in from 12 to 7 fathoms, but they are obliged to attend carefully to the wind, for the ground is for the most part rocky. The inner port is formed by a stream of old lava on the S., and by two short piers to the N. Its greatest depth is 3½ fathoms, and this does not go far into the bay. The mole stands in lat. 37° 28′ 30″ N., long. 15° 4′ 42″ E. The outer anchorages are clear, and a good berth may be found by the lead, riding safe enough in the summer season. The shores of the bay are formed by a sandy beach, which runs nearly 10 miles to the S. of the city.

CATCHOP, NORTH and SOUTH, two shoals at the entrance of the Tagus, in Portugal, They form two channels into the river, the N. or little channel lying between the N. Catchop and St. Julian's Land, and the S. between the two Catchops, constituting what is called the Bar of Lisbon. The N. is rocky, with 3 or 4 fathoms water, it extends W. 2¼ miles, and with a W. wind the sea breaks furiously over it. Its N. port is a musket-shot from St. Julian's, and it is its shoalest. The S. bank is more shoaly than the N., and its N.E. part is always dry. There is a light on the middle of this bank. See *Lisbon.*

CATFORD SOUND. See *Shetland Isles.*

CATFRITH VOE. See *Shetland Isles.*

CATHARINE'S, ST., ISLAND, Mediterranean, off the S. end of the island of Rhodes.

CAT HEAD, Newfoundland; the extreme point is in lat. 50° 7' N., long. 56° 42' 18" W.

CATHERINE BAY, Isle of Jersey, E. end, S. of the point so named, and a good road in W. winds to those who know the dangerous coasts of that island.

CATHERINEDAL LIGHT. See *Revel.*

CATHERINE'S ISLAND, North America, on the coast of Florida. Lat. 31° 36' N., long. 81° 41' W.

CATHERINE'S, ST., or CATALINA CAPE, North America, W. coast of Mexico, lat. 10° 40' N., long. 86° 50' W. It is the S.E. point of Papagayo bay, 8 leagues W.N.W. from Port Velas, and 22 N.W. from Point Guiones.

CATHERINES, ST., Isle of Wight, lat. 50° 35' 34" N., long. 1° 17' 51" W. The lighthouse at St. Catherine's point is of the first order, 105 feet high, having a fixed light at 178 feet, seen 6 or 7 leagues at sea in all directions.

CATHERINE SOUND, United States, North America, in a small island of the same name on the coast of Georgia. Lat. about 31° 10' N.

CAT ISLAND. See *Bahamas.*

CAT ISLAND, North America, coast of Florida, near the Gulf of Mexico. Lat. 30° 6' N., long. 89° W.

CAT-NESS, or GREY POINT, Russia, White sea, 20 leagues from Cross island, S.W., and 13 leagues N. or N. by E. from Archangel bar: a projecting point of the larboard coast in sailing S. round the Wardhus, and by the mouth of the river Kola for the White sea. Cross island is near the coast on the starboard side, where the shore forms a semicircle. On the W. side is from 7 to 10 fathoms, and a good situation with E. winds. N.E. of the point as far as the S. end of Mezene gulf or Cape Pentecost, it is good anchorage close to the shore. Lat. 65° 27' N., long. 38° 25' E.

CATONI, Greek Archipelago, an island S.S.E. from Samos, and N.N.E. from Patmos.

CATRARO, Italy, coast of Lower Calabria, a small town 14 miles from Belvidere to the S.

CATTEGAT, THE, a space of sea between the Skager Rock, or Scagerac, or space between the S. coast of Norway and the N.W. coast of Jutland, extending N.E. and S.W., and the Baltic entrances, or to the Sound and Belts, being the channel between the E. coast of Jutland and the coast of Sweden, a direction nearly N. and S. The Scaw, the N. point of Jutland, low and sandy, is the W. entrance to the Cattegat. About 5,000 yards W. by N., within the extreme point of the land, stands a lighthouse, in lat. 57° 40' N. It shows a fixed light visible 4 leagues. There are some cottages near the lighthouse, and a little S. is a sand-hill. 2¼ miles W. is Skagen church with a square steeple, and between is a windmill. A reef runs off the Scaw point, N.E. by E. 3 miles. The lead must therefore be kept going coming from the W., and the shore with off land winds not be approached nearer than 8 or 9 fathoms; if from the S., outward

bound, not within less than 12 or 14; in cloudy weather or at night, not in less than 18 or 20 fathoms. When the light and the old church are in one bearing W. by S., a ship is N. of the Scaw reef. When she is abreast of the reef, the steeple and lighthouse are in one; when the old church is brought to the S.E. of the village, she will be S. of the reef; and when the lighthouse is brought W., she is within the reef. When the light bears S. 6½ or 7 miles, the depth is 49 fathoms, muddy bottom. When there is drift ice in the Cattegat, visible from the lighthouse, a white flag with a blue perpendicular stripe in the middle, is displayed there in the day time. The word Cattegat or Cattgut is from the Dutch, signifying the Cat-channel. The currents here are strong and variable.

CATWATER, Plymouth sound, a branch of the harbour of Plymouth, on the E. side of the sound, as Hamoaze is on the W. The entrance lies between Mount Batten and the Fishes Nose, beneath the citadel, at the S.E. angle of the defences. See *Plymouth.*

CAUCEA, POINT and RIVER, coast of the province of Coro, in Colombia, the S. point of the Gulf of Coro, which opens into that of Maracaybo; on the W. side is the point of Caucea. The river is small.

CAUCEDO, PUNTA, St. Domingo, S. point, lat. 18° 22' N., long. 69° 38' W.

CAUCHUMILLI, a small Greek island subject to Turkey, perhaps Staupalia, 20 miles S.W. of Stanchio, lat. 36° 30' N., long. 26° 24' E.

CAUDEBEC, France, N. coast, on the River Seine, on the N. side of the river, and half-way up the river to Rouen. Lat. 49° 31' N., long. 1° 25' E.

CAUSAMUL and HISKERE, W. isles of Scotland, 2 miles W.N.W. from Arderunner point, the W. point of the isle of Uist. Causamul is a mere islet, with dangerous sunk rocks around it; and N. by E. ½ E. 3½ miles distant from Causamul is Hiskere, larger than Causamul, with sunken rocks about it, some as far off as 2 miles, no ship should approach the W. side nearer than a league.

CAUX CAPE, France, another name for Cape Antifer.

CAVA ISLAND, one of lesser Orkneys. See *Orkneys.*

CAVAGLERO, or CAVALGERO, Africa, a cape in the territory of Morocco, 3 leagues W. of Cape Blanco, and 6 leagues W.S.W. of Cape Cantin.

CAVALERI, Greek Archipelago, an islet between the S. end of Negropont island and the continent of Greece, lat. 38° 7' N., long. 24° 17' E.

CAVALHO ISLAND, African coast, one of the Bijooga islands.

CAVALIERE BAY, France, on the S. coast, a small bay nearly N. from the E. extremity of the Hieres islands to the N.W. from Cape Taillar, having from 5 to 3 fathoms within, and anchorage under the land between, in from 7 to 15 fathoms.

CAVALIERE, CAPE, coast of Karamania, a broad headland, having the appearance of an island, with cliffs of white marble rising almost perpendicularly out of the sea to the height of 733 feet, connected to the main by a narrow isthmus, which forms a small bay on each side; it is between 30 and 40 miles E. of Cape Anamour, and this last is in lat. 36° 0' 50" N., long. 32° 50' 58" E.

CAVALLERO BAY, West Indies, Haiti, on the S. side of the island, E. from the town of St. Domingo, and not far from its S.E. end; W.S.W. from Saona island; lat. 18° 3′ N., long. 68° 35′ W.

CAVALLO, South America, on the Isthmus of Darien, a post town on the N. coast of that continent, 25 miles N.E. from St. Jago de Leon. Lat. 10° 15′ N., long. 68° 12′ W.

CAVALLO, CAPE, Italy, in the S. near Cape Otranto.

CAVALLOS BAY, W. coast of Africa, 70 leagues W. by S. from Cape Bojador; a rugged coast, with rocky headlands, called the Seven capes or points.

CAVALLOS DE FAM, Portugal, two small islands near the coast, half a league S.S.W. of Esposenda. Lat. 41° 30′ N., long. 8° 36′ W.

CAVALLY RIVER, coast of Africa, 5 leagues from Cape Palmas, which is in lat. 4° 22′ 9″ N. Much rice and grain is produced in its vicinity.

CAVESAS ISLANDS, South America, E. of Cape St. Blas, near the isthmus of Darien, lat. 9° 35′ 36″ N., long. 78° 40′ 40″ W.

CAVILLAN ISLANDS, Scotland, W. coast, a little N. of Gigo island.

CAWIE ISLES, Labrador, North America, lat. of Little island, 49° 49′ 29″ N., long. 67° 4′ 57″ W.

CAWSAND TOWN and BAY, on the W. side of Plymouth sound, a place of secure anchorage from N.W. and S.W. winds, in from 5 and 6 fathoms. The best anchorage is with Penlee point S.W. or S.W. by S.; Cawsand town N.W.; the Bolt head in one with the Little Mewstone, and Plymouth new church open of Redding point. Penlee point, the S.E. head of the bay, is in lat. 50° 19′ 25″ N., long. 4° 10′ 40″ W. See *Plymouth*.

CAXINES, CAPE, Algiers, the W. headland of the bay of Algiers, between which and the opposite cape of Matafou the water is deep. It is 58 leagues from Cape Ferrat. The coast rocky, precipitous, and bold-to, having 50 or 60 fathoms a short distance from the shore, and 2 or 3 leagues from land, from 190 to 300 fathoms, while further out it becomes out of the reach of soundings.

CAXONES, Mosquito coast, a cluster of dangerous kays and reefs 65 miles N. of Cape Gracias a Dios, and 20 within the edge of the Great Mosquito bank, extending 12 miles W.N.W. and E.S.E., outside which, to the N. and E., the bank is free from danger.

CAY, or KAY, or CAYO, names given to small islands, generally sand-banks or shoals, which appear above water, probably derived from the Spanish, very improperly called keys by seamen and those not acquainted with the derivation of the word. Caycos is a Spanish term for small islands.

CAYAGUANEQUE, Cuba, a port on the E. side, in lat. 20° 30′ 30″ N., long. 74° 31′ W.

CAYCOS, THE, among the Passage islands, in the West Indies: the S. end of West Cayco is in lat. 21° 37′ 40″ N., long. 71° 44′ 33″ W.; of Little Cayco, S. point, lat. 21° 37′ 30″ N., long. 72° 28′ 33″ W. See *Bahamas*.

CAYENNE, South America, French Guiana, or Guyana, a town built upon an island called Cayano, in lat. 4° 56′ 15″ N., long. 52° 20′ 30″ W. A river runs on each side of the island, and the town is situated on the most N. river, called the River Cayenne, that to the S. being denominated the Oya. The town is mean and miserable in the midst of a fertile soil; the island on which it stands is about 6 leagues from N. to S., and its breadth about 4. The sea bounds it on the N. The town and fort stand on the N.W. part of the island, where there are hills and high ground, but the S. is swampy. The River Oya, which separates the island from the main land, is about a league wide, and has 3 fathoms water at low tide, the banks high and wooded. Most of the rivers on this coast require a good berth in passing, being encumbered, at their entrance, with shoals, rocky islets, and clay banks. The climate is very unhealthy.

CAYENNE BAY, West Indies, island of St. Vincent, 2 miles N.W. of Kingston bay.

CAYES, LES, TOWN OF, West Indies, St. Domingo, S. side of the island, N.W. of Vache island: lat. 18° 11′ 10″ N., long. 73° 44′ W.

CAYEUX SUR MER, France, department of the Somme, on the S. side of the entrance. Here is a tower, with a lighthouse, in lat. 50° 11′ 42″ N., long. 1° 31′ E. It exhibits an intermitting light, 92 feet above high water. The flashes succeed each other every four minutes. The faint light in the interval being preceded and followed by a short eclipse. There is also a small light on the Point du Hourdel, the S. point of the entrance to the river, just round to the S. of which is a small port. See *St. Valery sur Somme*.

CAYMAN, or CARMAN, ISLANDS, West Indies, a cluster of islands between Jamaica and Cuba, called Little and Great Cayman, with a low islet called Cayman-brack, situated W. of Little Cayman. They lie in lat. 19° 20′ N., and are between 30 and 40 leagues N.N.W. from Point Negire on the W. of Jamaica, the Grand Cayman being the most remote. Little Cayman and Cayman-brack are within 5 miles of each other, and 35 N.E. of the Grand Cayman. The Grand Cayman S.W. of both. Fort George is in lat. 19° 17′ 48″ N., long. 81° 24′ W. The S.E. point is in lat. 19° 16′ N., long. 81° 6′ 40″ W. A village on the W., formerly called Hogsties, is now dignified with that of Georgetown; the most populous village is Boddentown, on the S. Grand Cayman appears very low land, most like a group of trees in the ocean, seen 12 or 14 miles at sea. It possesses no harbour, but has a tolerable anchorage on the S.W. coast. On the N.E. it is defended by a reef of rocks, between which and the shore the inhabitants have pens for folding in their turtle. The soil is fertile, corn and vegetables are in plenty. The inhabitants, 1,500 in number, are supposed to be descendants of the English bucaneers, and are excellent seamen and pilots. These islands are said to be very healthy. The people choose their own chief officer, frame their own regulations, and have their justices of peace appointed from Jamaica.

CAYMAN ROCK, North America, Mosquito coast; 2 miles from the land, and 6 N. ¼ E. from Blewfield's Bluff, a barren rock, with 6 and 7 fathoms around it. The Bluff is in lat. 11° 19′ 20″ N., long. 83° 40′ 18″ W.

CAYMITES, West Indies, St. Domingo, an island 2 leagues long and 1 broad, and a bay called Caymites bay, in lat. 18° 40′ N., long. 73° 44′ W., on the N. side of the island, in the vicinity of other islands and rocks.

CAYO BLANCO. See *Isabella Banks*.

CAYO BLANCO, Cuba, W. of the isle of Pines: also on the N. side of the island near Matanza bay.

CAYO BLANCO DE SAZA, or ZARZA, Cuba,

S.W. coast; W. point, lat. 21° 22′ 30″ N., long. 79° 42′ 30″ W. The E. point, lat. 21° 31′ 20″ N., long. 79° 43′ 35″ W.

CAYO BRETON, Cuba, S. side of the island; lat., E. point, 21° 2′ 25″ N., long. 79° 29′ 20″ W.

CAYO CONFITES, or SUGAR PLUM KAY, N. coast of Cuba; the N. point is in lat. 22° 12′ 25″ N., long. 77° 37′ 50″ W.

CAYO CRUZ DEL PADRE, Cuba, N. coast; lat. 23° 17′ 20″ N., long. 80° 53′ 15″ W.

CAYO HUESO, KAY WEST, or THOMPSON'S ISLAND, Cuba, E. of Cayo de Boca Grande about 6 leagues; Cayo Canalette lies 1 mile N. of Hueso. The S.W. point of the last lies in lat. 24° 32′ N., and long. 81° 54½′ W., 28 leagues from the Havanna, N. by E.

CAYO ISLAND, coast of Africa, one of the Bissagos islands. See *Bissagos* or *Bijooga Islands*.

CAYO LARGO, LONG KEY, or WOOD ISLAND, S. of Punta del Carenero, N. coast of Cuba.

CAYO RATONES, South America, coast of Colombia, in lat. 9° 23′ N., long. 78° 16′ 15″ W.

CAYONA, another name for the island of Cayenne.

CAYONNE RIVER, West Indies, island of St. Kitts, where it falls into the sea from the W.S.W., near Little bay, S. of Madan's point.

CAYO ROMANO, Cuba, N. coast; S.E. point, lat. 21° 50′ 20″ N., long. 77° 35′ W.

CAYO SAL, or SALT KAY, Cuba, lat. 23° 39′ 8″ N., long. 80° 16′ 38″ W.

CAYOS DE VACAS, or the LOW KAYS; the island Cayo Vacas is about 3½ leagues from Bahia Honda on the N. coast of Cuba; there is also a cluster of these islets extending 13 miles N.E. by E., of which that called Duck kay is the most E. There is a channel between Cayo Hueso, the Cayos Vacas, and the reef.

CAYOVARNO, or CAYVARNOS, S. coast of Cuba, near the Pine islands, a harbour fit only for small craft, with 7 and 8 feet only of water. It is 4½ miles N. from Looe kay.

CAYO VERDE, or GREEN KAY, N. coast of Cuba, lat. 22° 7′ N., long. 77° 36′ 30″ W.

CAZUELOS POINT, Cuba. See *Colerados Reefs*.

CAZZA, an island in the Gulf of Venice, S.W. by S. from Point Kancira, distant 10½ miles, and 24½ S.E. ¾ E. from the S.W. point of Lissa. A vessel during a N. gale might ride, under shelter of the land here, in from 4 to 9 fathoms. The signal on the highest part of Cazza is in lat. 42° 46′ 2″ N., long. 16° 30′ 54″ E.

CAZZIOLA, an island in the Gulf of Venice, less than the Isle of Cazza, and lying between that and Lagosta, 8 miles from the first, and 4 from the second. There is good anchorage all about it. A little to the N. there is a rock above water called the Kopit; 1½ miles to the W.N.W. is another called Bielaz. To the W.S.W. of Cazziola, distant half a mile, is a little knoll of 4 fathoms, with deep water all round, between which and Bielaz there are 33 fathoms, and between Bielaz and Cazza 61, 68, and 66.

CEDEIRA, Spain, S.W. from Cape Ortegal 9½ miles, a small harbour with shallow water, good holding ground, and an easy entrance. Point Pantin, the starboard land, should have a good berth, clearing the Mexiones rocks, and keeping 1½ cables' length from the shore, until the White rocks are cleared. Run alongside Serreidal point till opposite Solveiras. Stand towards the centre of the harbour till the castle flagstaff is in a line with Point Pantin, then moor N. and S. in sandy ground, in 2½ and 3 fathoms. The harbour is exposed to N.W. winds.

CEFALONIA, CEPHALONIA, or CEPHALENIA, one of the seven Ionian islands, off the coast of Albania and Greece, the largest and most populous of the whole. Cape Viscardo, the N. point, lies in lat. 38° 28′ 40″ N., long. 20° 33′ E.; and S.W. from this, distant 9½ miles, is Cape Aterra, the N.W. point, in lat. 38° 21′ 30″ N., long. 20° 24′ 15″ E. The length of the island from Cape Viscardo to Cape Skarla is 28 miles; the breadth variable, and appearance hilly, if not mountainous. There are 3 towns, 75 villages, and 65,000 inhabitants. The coasts are indented with deep gulfs. The valleys inland produce wine and timber, and the fruit trees bear fruit twice a year. The soil, though fertile, does not produce corn enough for the inhabitants, but currants from a small raisin of Corinth, oil, muscadel wine, brandy, cotton, linseed, oats, raisins, honey, melons, citrons, oranges, and pomegranates, are abundant. It is said that 3,000 tons of raisins are every year exported. The oil exported is 30,000 casks, and of wine 35,000 per annum. Rosalia or maraschino, of great repute, is also among the exports. The chief imports are corn, woollens, linen, sugar, and hardware. The number of vessels belonging to the island is stated to be 400, navigated by 6,000 of the islanders. The capital is Argostoli. A vessel having passed to the W. of Cape Aterra, will be carried by a S.S.W. course for 6 miles to Point Ortalita, near which are some rocks above water. From thence steering 3¾ miles S. by W. for Point Kobbo; then S. by E. about 2 miles, Katudakia will be weathered; and then sailing S.E. ½ S. 2½ miles, a ship will be abreast of Cape Aji. The land almost all the way is precipitous from Cape Aterra, with here and there a little sandy beach, but no shelter. Cape Aji forms the S.W. side of the Gulf of Argostoli. The S. coast, at the entrance from Cape Aji to Point St. George, a distance of 3 miles, is low and bordered by a shelf ½ a mile broad. There are some rocks about Cape Aji, and a spit runs out S. from Point St. George for a full mile, which renders the passage between that and the Island Guardiani narrow and frequently unsafe; while, though the point and coast round St. Nicolo are also rocky, the reef does not run out. To enter the port of Argostoli, it is always best to pass to the S. of the Island Guardiani, giving it a berth of about a mile, which will clear the low ground that surrounds it. For this end bring the church of St. Nicolo on with the centre of the village above it, bearing E.N.E. ¾ E., keep to this until the town of Luxuri opens within the E. point of the gulf; then it is necessary to steer N., keeping the two north mills at Luxuri on with or open of the E. steeple of the town, which will clear the reef stretching out from Point St. Giorgio. The thwart-mark is a bluff point on the E. side of the entrance, half way between St. Nicolo point and Point St. Todera, on with a large white house standing upon a hill above it, near a village. The reef passed, the course is N. ¼ W. until beyond Point St. Todera, at the entrance of the inner port, low land, with a rocky shoal running from it to the S. The convent or lighthouse on Guardiani, on with Port St. Giorgio, will clear this until the town of Argostoli opens, when a vessel may proceed on in mid-channel and anchor.

CEFALU, Sicily, N. coast, 17 miles W. from Carnica. It is situated in lat. 38° N., long. 14° 3′ 57″ E. The town contains about 9,000 inhabitants, agriculturists and fishermen. Near it E. is a small cove called Calura, where coasters seek shelter. On the other side to the W. is a small bay, used as a summer anchorage by coasters, having from 3 to 7 fathoms water. There is a small mole here, and to the N.W. are two shoals, with 8 or 9 fathoms of water over them.

CENCHREA or KENCHREA. See *Corinth*.

CENIZA, BOCA DE, South America, Colombia, the mouth of the Rio Magdalena. lat. 11° 5′ N., long. 74° 53′ 45″ W.

CENTER, ISLE, North America, Bay of Fundy. This island, just open of Moffat bay, gives the best of the channel.

CENTINELA ROCKS, Colombia, coast of Caraccas, N.E. from Chuspa, about 12 miles from the main. There is deep water near them.

CEPET, CAPE, France, S. coast, the S. point of the entrance to the bay of Toulon. It is high land, upon which there is a signal tower. Off the point of this cape the sea breaks upon a rock called Le Rasias, about a cable's length from the land. There are some batteries round the cape. To sail into the bay, Cape Cepet must have a fair berth. There is good anchorage all over the bay or outer road, in from 6 to 15 fathoms, the ground mud and clay. It is needful to moor with an open hawse to the E. The N.W. winds are the most prevalent, and blow strongest, but there can never be much sea from that quarter.

CERANO, TOWER OF, E. coast of Italy, a mile N. of the castle of Silvi, which stands on an eminence, in lat. 42° 33′ 14″ N., long. 14° 6′ 48″ E.

CERBOLI, Italy, Tuscany, an islet or rock on the coast a little N. of Elba island.

CERIGO, ISLAND OF, one of the Ionian Islands, the ancient Cytherea, separated from the Island of Servi by a deep channel, 4½ miles broad. It is 20 leagues in circuit, and appears only an assemblage of confused rocky precipices. The interior is said to abound in fine porphyry. The island lies in a direction from Cape Spati to Cape Capella S. by .E. ½ E., and N. by W. ¼ W., in length 17 miles, and in breadth from Elaphonesa to Dragonera 11 miles. The pop. 9,000. It grows corn enough for its consumption, and abounds in game. Its produce is wine, oil, fruits, cotton, and silk. The climate temperate and healthy. The oil is deemed of excellent quality, and the fisheries are very productive. Cape Spati is in lat. 36° 22′ 20″ N., long. 22° 56′ 30″ E., having a chapel upon its E. extremity. Near it is Point Nicolo, on the W. side of which is anchorage in from 28 to 14 fathoms, and near the shore in from 9 to 8, but there is a sunken rock there. On the W. side there is no commendable anchorage.

CERIGOTTO, Greece, one of the Ionian islands, 4½ miles long, lying S.S.E. ½ E., and N.N.W. ½ W., bearing S.E. ¾ S. from Cape Capella, distant 16¾ miles. It is almost uninhabited. Keeping within 1½ ‡mile of Cerigotto, or 5 miles off in mid-channel with Cerigo, will avoid the sunken rocks called the Potamos, and others, on one of which a sloop-of-war was lost in 1809. Cape Apolitara, the S. point of Cerigotto, is in lat. 35° 50′ N., long 23° 17′ E.

CERINES, Cyprus island, a port in lat. 35° 22′ N., long. 33° 24′ E.

CERNOUAILLE or CORNOUALLES POINT,

France, the inner point of the gullet, or entrance to the road or harbour of Brest.

CERRADAS BOQUILLAS, or the CLOSED OUTLETS, Mexican sea, North America, lat. 25° N., long. 97° 45′ W.

CERVI, or SERVI, Greece, a small island off the coast of the Morea, 6 miles N. of Cerigo.

CERVIA, E. coast of Italy, N. off the entrance of the River Marecchio. It is a modern town, with good streets, and has a semaphore. There are also salt works in its neighbourhood. It is about 4 leagues S.E. from Ravenna, in lat. about 44° 30′ N., long. 12° 17′ E.

CERVICALES, Corsica, Mediterranean, four small islands so named, S. of Cape Della Chiapa. The nearest to the cape is called Cibiciani, and is the largest of the four. Cape della Chiapa and Cape St. Cyprian are the S. and N. boundaries of Porto Vecchio, and are 1¾ of a mile asunder. Porto Vecchio is in lat. 41° 35′ 29″ N., long. 9° 16′ 36″ E.

CESAR BREAKERS, a vigia, reported in 1730 to be in lat. 2° N. long., 22° 18′ W., but considered to have no existence.

CESAREA, Palestine, 5 miles S. of Tantura, once the principal port of Samaria, now only distinguished by the ruins in its vicinity. It is about 20 miles N. of Jaffa.

CESENATICO, Italy, Gulf of Venice, 16 miles from Ravenna S.E. It is a small fishing town, with a good harbour and a canal, at the entrance of which is a semaphore. The coast has shoal water and much sand in their vicinity.

CETTE, France, department of the Herault, a port and town, pop. 10,000, situated at the entrance of the canal of the South, or Languedoc, and the only place in the Gulf of Lyons that offers at all times a shelter to vessels. Fine quays traverse the town in its whole length, and not far off are salt marshes, from whence salt of a superior quality is collected. Its exports are principally wine, brandy, liqueurs, and some of the southern fruits. The lighthouse is in lat. 43° 23′ 42″ N., long. 3° 42′ 12″ E. This is the port too of Montpelier. It lies 11 miles N.E. ½ E. from Fort Brescou, and may be known by a mountain like Mount Agde, which, coming from the E., appears an island. Between Agde and Cette the shore is low and sandy. Port Cette lies E. of the above mountain, and has a long mole projecting on the larboard side, on the head of which is a battery. There is a heap of stones to the N.E., placed to stay the rolling of the sand, and another in the midst of the entrance. The mole extends to the entrance of the Great Canal at the bottom of the port, which canal runs into some lakes, on the banks of which are the baths of Balme and mineral springs of Frontignan and Montpelier. The town has a large lagoon between it and the land, standing upon a peninsula. To sail in, steer abreast of the entrance and go within half a galley's length of the mole head, then run in and anchor alongside it, with a fast upon the mole. There is 12 and 16 feet water, ooze, and sand to the further end of the mole; the starboard side is shoal. At the entrance, near the lantern, there are 3 and 4 fathoms, though a little way out there is a shelf with only 3 upon it, and a heavy sea with the wind from E.S.E. to S. W. The light here is fixed, and in case pilots cannot get off to vessels, a sea-mark has been erected on the foot of a hill a little S. of Fort Richlieu, a quadrangular pyramid 197 feet above the sea level, E. front white, a black lozenge in the middle for a day signal, and two lights, one above another, for the night. The light may be seen 4 leagues off, and the two lights distinguished 1½ mile from the port. To make the N.E. passage in daylight, bring the lighthouse on Fort Louis in a line with the pyramid, and keeping them on will lead into the right channel. The same to be done at night with the Fort Louis light and those on the two pyramids.

CETTINA, Dalmatia, Gulf of Venice, a river with a fortress, in which is a little port called Stobrez. There is a rock to be avoided on entering the river; it is on the starboard side, lat. 43° 25′ 53″ N., long. 16° 43′ 58″ E.

CEUTA, Africa, a fortified seaport town opposite the rock of Gibraltar, standing upon the same spot as the ancient Abyla, styled one of the pillars of Hercules, supposed to be Carthaginian in origin. It belongs to Spain, to which it was yielded by treaty with the Moors in 1685. It is strong from position, and by its fortifications almost impregnable. It is seated upon a peninsula called Almina, extending 1½ mile E. and W., and nearly surrounded by rocks running out a cable's length from the shore. The castle of Ceuta stands on the E. and most elevated part of the peninsula, locally termed the Acho, lat. 35° 54′ N., long. 5° 17′ 42″ W. The suburbs are on the W. skirt of the Almina; next comes the city, which is fortified, between the suburbs and the main land. The N. point of the Almina is called Punta Santa Catalina. It is low, and has some rocky islets lying off it, and between this point and some islets called El Campo, is the road in the smaller bay of Ceuta, with but indifferent anchorage in from 13 to 8 fathoms, with ground of sand or loose stones. To anchor, the Calle del Obispo, or Bishop's-street, must be brought open. This street commences on the N. side of the Governer's house, Moor N.E. and S.W., letting the longest cable be to the N.E. The point of Castillejo low, round, and covered with trees, lies S.W. by W. ¾ W. from the castle of the Almina, called the Acho. This point, low, round, covered with trees, bears about N.N.E. ½ E., distant a mile from a river having the same name. S.W. about 8 miles from the Acho or castle is Mount Negron, lofty, dark, and having a square tower upon it. S. by W. ½ W. from the Acho, distant 4 leagues, is Cape Negro or Porcas, of middling height, having numerous islets about it, and crowned with a round tower. This tower stands in lat. 35° 41′ 30″ N., long. 5° 17′ W. The great bay of Ceuta is between the Almina on the N. and Cape Negro on the S., 11 miles long and 3 deep, where ships may anchor in any depth beyond 1½ or 2 miles from the shore, open to all winds but those from N.N.W. to S.W. The time of high water at Ceuta is $1^h\ 55^m$. Ceuta is a better shelter than Tetuan bay. When the E. wind approaches, indicated by a swell or current, and Ape's hill and Gibraltar are covered with clouds, the E. wind is approaching, and ships must instantly get under way.

CEVERA, CAPE, Spain, S.E. coast, in lat. 37° 58′ 55″ N., long. 41′ 27″ W.; it has a round tower upon its summit.

CEZEMBRE, ISLE DE, France, N. coast, off St. Malo port, an islet about 4 cables' length in extent, and very high, E. of the rock called the Grand Jardin.

CHAASH, or BA CHAASH, Scotland, W. islands, a harbour at the N.E. point of Uist, sheltered on the E. and N. sides. It has room for the largest ships. There are several dangers in the vicinity.

CHACACHACARREO, ISLAND OF. See *Bocas de Paria.*

CHACHEO BANK, and CHACHEO RIVER, African W. coast, the E. point of the N. side of the river is 9 leagues E. by S. from Cape Roxo. The Chacheo forms one side of the Bissagos channel.

CHAFALAYA RIVER, North America, one of the branches by which the Mississippi discharges its waters into the sea. It is a considerable stream, and the land on each side of it is high and fertile. 10 miles S. by W. from this river's mouth is the Punta de Fierra, and the intermediate space is a large bay. On this Punta de Fierra or Point of Iron, there is a lighthouse with a fixed light, 70 feet above the level of the sea, serving as a guide for vessels entering Chafalaya bay. It bears from Belle isle S.S.E. 12 miles, and from the entrance of the river S. by W. 10 miles.

CHAGRES, PROVINCE OF, Panama, South America, in Colombia, a sea-port at the mouth of the river Chagres, defended by the castle of St. Lorenzo, on the E. of the city; in lat. 9° 18′ 38″ N., long. 80° W., or according to Lieut. Raper, in lat. 9° 18′ 36″ N., and long. 79° 59′ 12″ W. The long. of this site is important, as being the place of connection with Panama of the measurements between the E. and W. sides of the American continent. The lat. of the port is given in some charts 9° 21′ N., long. 80° 4′ 5″ W. from Greenwich, and 73° 30′ 45″ W. from Cadiz. The following are different observations upon this place:—Observation of Captain Foster, 3° 48′ 41″ W. of Morant point, or 80° W. of Greenwich. From Barracoa, 50° 30′ 6″ or 80° 0′ 37″. Captain Barnett, from Port Royal, 3° 9′ 22″, or 80° 0′ 7″ from Greenwich. The River Chagres is formed by the junction of several rivers which flow into it, and bounded by the Penon or great rock on the N., upon which the castle of San Lorenzo is situated, and the Punta Arenas, or Sandy point on the S. The width is little more than 2½ cables at the broadest part, and 1½ at the narrowest. There are about 2½ or 3 fathoms water at the meridian of the penon, a depth varying little for half a mile up the river. About 92 fathoms W. from the castle of San Lorenzo is a shoal named Laja, extending 35 miles from N.E. to S.W., rocky, with little water over it. It is very dangerous to enter or put to sea from this river, unless in vessels easily managed, and not drawing more than 12 feet water, and then with a fair wind. To anchor in Chagres road the best place is in 7 fathoms, with the castle about S.E., and Point Brujas about N.E. ½ N., about 1½ miles from the shore. The bottom is good for holding, but a heavy sea rolls in with a wind N.E., N., N.W., or W., and vessels then only ride at great risk. Vessels bound to Porto Bello commonly make Chagres in the first instance. Between the rivers Chagres and Coclet there are four remarkable mountains seen from the sea, two inland and two on the coast. The Caladeros altos de Chagres are two mountains on the river, well inland, appearing separate on the run from Porto Bello, and seeming only one, when they bear S.E. ½ E. The Piton de Miguel de la Borda is a sugar-loaf looking hill, 10 leagues S.W. by S. from Chagre. The Sierra de Miguel de la Borda is on the same coast, of moderate height, stretching N. and S. 17 leagues W. of Porto Bello, and the Sierra de Coclet, rather lower than the Miguel, bears from the Rio de Coclet S. by W. ½ W.

CHAGRES RIVER, Colombia, Mexican sea, to the N.W. of the Gulf of Darien. It is large and deep, and falls into the sea below the town of Chagres. See *Chagres.*

CHAGUARAMAS BAY, or PORT ROYAL, West Indies, Isle of Trinidad, N.W. peninsula of the island. In this bay vessels may anchor in any depth from 5 to 20 fathoms. The port is called Puerto d'Espagne, or Port of Spain. The town is in lat. 10° 38′ 42″ N., long. 61° 31′ 45″ W. Chaguaramas harbour is within an island called Caspar Grande, 1½ miles S. after going through the Monos passage. Between this island and the peninsula lies an islet called Gasparillo, forming two passages into Chaguaramas harbour, which is spacious, and has good anchorage, the most convenient in 12 fathoms, the E. end of Gaspar Grande bearing S. by E., and Delgada in a line with St. Joseph's point. High water by the shore at 3ʰ, rise 5 feet. At the anchorage the flood runs till 3¼ʰ. It runs E. only 5½ʰ; the ebb runs W.

CHALE BAY, England, Hants co., on the S. side of the Isle of Wight, round the extreme S. point of the island, which forms the S.E. limit of the bay. There is a sunken rock near the extremity on that side, and to the N.W. Atherfield point and rocks become the limit. The whole bay is so full of smooth rocks that there is no anchorage.

CHALEUR BAY, North America, Nova Scotia. It goes in from the N.E. point of Port David, 22 leagues W. and N.W., and varies in breadth from 5 or 10 to 3 or 4 leagues. The bay of Chaleur is to the W. from the S. end of Long island, coast of Newfoundland. Point Macquereau and Miscou island form the entrance, about 4½ leagues asunder. There is a good depth of water near the shores, and in the middle from 25 to 40 fathoms. Lat. 48° 1′ N.

CHALEUR BAY, S. coast of Newfoundland, W. of Hare's Ears point, running in about 2 leagues N.N.W. It is narrow, with deep water in most parts. There is anchorage inside a rock within a small island on the N. side of the bay, in 28 fathoms.

CHALK GROUND, Gulf of Finland, due E. from the Ekholm light, and dangerous, having but 5 feet water over it in one part. To go N. bring the light W. by S.; to go S. bring it W. by N. There are marks on its S. and N. end. The Ekholm light is a fixed light, 75 feet high, visible 15 miles. Lat. 59° 41′ 8″ N., long. 25° 47′ 58″ E.

CHALKI, or COPPER ISLAND. See *Prince's Island.*

CHALODA ISLAND, Gulf of Finland, a small rock above water, 4 miles S.E. ½ S. from the S.E. point of Lavenskar island, and 4 miles at S.W. by W. from it, a small shoal, with only 3 feet water upon it. There is another small shoal 3 miles S. by W. from it, having 5 feet water. Lavenskar island is 20 miles from St. Petersburg, in lat. 59° 54′ N., long. 27° 50′ E.

CHAMA, Africa, S. coast, a trading fort between Dixcove, in lat. 4° 47′ 45″ N., and St. John's river, to the last of which it is very near.

CHAMPLAIN, North America, United States, a lake between New York and Vermont States, 120 miles long and from half a mile to ten miles broad. It contains several islands, as North Hero, South Hero, Le Motte, and others. The Champlain canal connects this lake with the Hudson river, and the sea at New York. It is generally frozen from January to April. It is navigated by numerous steamboats and vessels from 80 to 100 tons burthen.

CHANGE ISLANDS, Newfoundland; lat. of N.E. islet 49° 41′ 35″ N., long. 54° 25′ 48″ W.

CHANONRY POINT, LIGHT, Scotland, at the entrance of the Frith leading to Inverness and the Caledonian canal, a fixed light of the natural colour, open from W. ½ N. to N. by E. in a S. direction; in lat. 57° 34′ 32″ N., long. 4° 5′ 28″ W. It is 40 feet above the level of the sea, visible 11 miles.

CHANNEL, the general term for a strait or narrow sea; sometimes for the deepest part of a broad current where the water has excavated the most navigable part in regard to depth. Sometimes it is put for the entire bed of a river, or the snaky course of fresh water among sands or mudbanks, where the tide has left dry all but the space subject to the land waters. In a general sense, it applies to a narrow sea between a considerable extent of land on either side, as the British channel between the land and Ushant on the W., as far as the straits of Dover eastward, or the Bristol channel within Hartland point and St. Bride's bay.

CHANNEL, THE, or THE ENGLISH CHANNEL, the space eastwards between the Land's End or the Scilly isles, and Ushant. To this space the term channel is more particularly applied. At the entrance the channel is spacious, and it scarcely contracts, sailing E., until the Start point is attained, when the Channel islands and the Caskets reduce it to half its previous breadth, and show how wise is the old caution "to keep the English shore on board proceeding up Channel." The Caskets, nearly S. of Portland, with the Shambles lying nearly N. and S., alter the configuration, and diminish the sea room almost in the centre of the navigation, which, however, again widens considerably, to be again contracted, N.E. of the entrance to the Seine, on the side of France, on which side the dangers principally occur, from the much greater irregularity of the coast. Beyond Dieppe, on the French side, half way as far as to the mouth of the Somme, the coast makes a turn to the N.E., and then inclining almost due N. as far as Cape Grisnez, to the N. of Boulogne, displays numerous shallow inlets and petty harbours, the line of coast being less indented as the straits of Dover are neared; but taking such a form as to show that the contraction of the water way is produced wholly by the French shore. A steamer might clear Scilly, pass the Lizard in deep water, run from thence on a line to Beachy head, and still without variation enter the North Sea through the straits, on the same point from the Lizard. The French coast, on the contrary, is so irregular that a line drawn from Ushant, clearing the Caskets, would run a vessel on shore W. of Beachy head, and the prolongation of such a line would cut the Thames after passing overland about the E. end of Sheppy. The entrance of the channel and the exact position of a vessel there when homeward bound, are most important things to ascertain, so important that they take precedence of every other consideration in the mind of the seaman who has a just feeling of his duties. Before seamanship had reached its present perfection this was a matter of much more difficulty than it is at present. Yet there was one means always within reach of the older mariner, had he benefitted by its adoption, which was the lead, and a close observation and notation of its results. This resource always remained when no observation could be taken, nor headlands seen amid fogs and wintry

II.

H

darkness. A want of due attention to its notices has occasioned more losses subsequently in the channel than any other cause. The attention that should paid to the lead—for it is impossible to suppose it will not be casually had resort to for the purpose of discovering whether a ship is in soundings or not—the attention paid to the lead must be second only in assiduity to that of the helm. It must be progressive and unremitting from the time the ground is first met until some well known point is sighted. It is not enough to notice the depth of soundings after they have been at the first possible moment attained. The lead must be hove at night as well as day, and the variations in the bottom be continually noted, the substances observed, more particularly their colour and changes. The greater depth of water from 7 to 10 fathoms at the entrance of the channel bespeaks the French side. The stones there are larger and coarser than on the English, and paler in colour than on the N. Thus suppose early soundings obtained between the lat. of 49° 15′ and 49° 25′ N., it is the position which above every other should if possible be obtained, the depth, quality of ground, and colour being better ascertained in that parallel than any other. The edge of soundings will be first struck about 65 leagues from Scilly, in a depth varying from 270 to 335 fathoms; the bottom will be a dark ooze, intermingled with sand. When the depth lessens to between 80 and 90 fathoms, muddy bottom, the vessel will be 5 or 6 leagues within the edge of the bank, and about 57 leagues W. of Scilly. Still progressing E. in the parallel of 49° 25′ about 6 or 7 leagues, 69 fathoms will show sand with specks. This will be on the outer edge of what is called the Great Sole bank, and 3 leagues further there are 70 fathoms on its inside edge, which suddenly increase to 90, with a muddy bottom. More E., about 10 leagues on the same parallel of lat., 80 fathoms will be found, and about a league further 76 fathoms, on white sand of a fine grain. St. Agnes' lighthouse, in Scilly, will be distant 40 leagues E. by N. Sailing E., about 3 leagues, 80 fathoms will be found with a sandy bottom, and about the same distance, yet further, 83 fathoms with mud. This muddy ground will be found to continue for 9 leagues further E. and about 1 league more 78 fathoms and sand will be discovered. This will be found to be in long. 8° 2′ W., being about 24 leagues W. ¼ S. from the Scilly light. At the distance of 5 leagues further E. soundings will be on the E. end of the Haddock bank, with 60 fathoms, sand, across which the water deepens to 75 and 70 fathoms, mud. At the distance of 6 leagues further 69 fathoms will be found, with sand. On the same parallel, in long. 7° there will be found 66 fathoms, sand. In the same direction for 9 leagues E. there will be discovered 61, 64, 63, 61, and 63 fathoms, brown and speckled sand. This last is in the meridian of Scilly, and continues 15 leagues; further E., 63, 62, 64, 61, 59, 58, and 55 fathoms will be found, all sand, the last depth being in the meridian of the Lizard. To the N., which some call the stream of Scilly, and between 50° and 50° 10′ N. there are 101 fathoms over a muddy bottom, in long. 10° 53′, to the W. of Scilly. Further E., about 4 leagues, there are 84 fathoms, and yet 4 leagues further 72 and 73, and 3 leagues further still 78, all sand. In long. 9° 50′, which is 3 leagues further E. there are 77 fathoms, mud, from whence the bottom is muddy invariably to within 4½ leagues of Scilly. The depths are 72, 75, 75, 74,

and 61 fathoms; the latter is on the S.W. edge of the West Bank, in long. 8° 39′. Four leagues yet further, 69 or 70 fathoms are found close to its E. end; afterwards 64, 75, 65, and 60 fathoms. The last depth, on the S. end of the N.W. bank, in lat. 50° 5′, long. 7° 43′ just 18 leagues distance to the N.W. of Scilly. From this bank, 62, 65, and 64 fathoms will be found further E.; the depth then decreases from 61, 58, so far as the muddy ground extends, and from thence to 53, 57, 50, and 46 fathoms, on coarse sand and gravel, which is within 6 miles of the Scilly rocks. Around Scilly, and 4½ leagues and 10 leagues S. and N. coarse gravel, shingle, decomposing rocky substances, and sand, are met with, but no mud whatever. Ten miles S.W. from the Scilly lighthouse there are 68 fathoms; at half that distance in the same direction 55, and within 5 miles of the rocks 54 fathoms. Ten leagues S. of the lighthouse there are 58 or 60, on a spot which has the name of the "Admiralty Patch," round which in every direction there are 63, 64, 65 fathoms, mixed soundings and coarse sand; at 5 leagues, 55 and 60 fathoms; at 4 leagues, 51 and 52; and at 4 miles distance in the same direction, 45, fine sand. There are 50 fathoms at 5 leagues distance S.E. from the lighthouse, and 45 fathoms, at 4 miles distance. Six leagues N.W. from the lighthouse, there are 57 fathoms, and at 3 miles from the rocks in the same direction, 44 fathoms. Five leagues N. from the lighthouse there are 47 fathoms, and at 10 leagues 52. N.E. from the lighthouse, distant 5 leagues, there are 48 fathoms; and 2 miles from St. Martin's head, 42 fathoms, and there are 42 within 2 miles of the rocks to the E. The E. end of Jones' bank lies 17 leagues from Scilly lighthouse, between lat. 49° 49′ and 50° 3′, and long. 7° 37′ and 8° 10′. It has from 41 to 64 fathoms upon it, sand, and from 68 to 72, mud close to it both S. and N. To the N. of the parallel of 49°, a little distance, in long. 11°, there are 208 fathoms, muddy bottom; in 10° 53′ about 162 fathoms, mud; 5 miles more E., there are 93 and 85 fathoms, both mud; and in long. 10° 42′ W., the S. tract of sandy ground begins uninterrupted by mud or ooze, as far as the coast of France. In lat. 49°, and long. 10° 12′ W., there are 90 fathoms, and at 10 leagues farther E. 85 fathoms. Six leagues beyond, in long. 9°, there are 90 fathoms; next, 86, 90, 87, 74, 75, 78, 75, 73, 72, and 70. This last, or 70, is in 6° 56′ W. long. Still going E., there will be found 73, 70, 75, 73, 71, 66, 69, 67, 59, and 60 fathoms, in a space of 22 leagues, or to the meridian of Ushant, the ground whitish and pale, like hard marl, having a mealy surface. In lat. 48° 24′, long. 9° 30′, and 59 leagues to the W. of Ushant, lie the W. part of the Little Sole bank, which, in detached patches, extends N.E. and E. 13 leagues. On the W. part are 84 fathoms, on sand; at 3 leagues farther W., no bottom at 200 fathoms. These banks consist of knowls, having 65 to 89 fathoms over, and from 90 to 100 fathoms between. Near the S. part, lat. 48° 15′, long. 9° 7′, there are 160 fathoms, sand. In lat. 48° 23′, long. 8° 34′, there are 100 fathoms; 10 or 11 leagues farther E., 92 fathoms, sand: then 87, 87, 85, 86, and 97 fathoms, all sand; the latter depth in a pit, about 2 miles across, in long. 7° 34′. Thence, for 24 leagues E., the depths are 87, 85, 86, 78, 72, 71, 68, and 66 fathoms, all sand; the latter depth being 8½ leagues to the W. of Ushant. In the parallel of Ushant, at 23 leagues off, there are 85 fathoms, white sand; at 18 leagues, 75 fathoms; at 16 leagues, 70, 71, and 72, with a coarse pale yellow bottom, having a mealy surface, with broken pieces of shells, and a substance like chaff. At the distance of 9 leagues, on the same parallel, there are from 63 to 66 fathoms, similar, and 65 fathoms within 3 leagues of the rocks; approaching Ushant, the sand is mixed with shells whole and broken. The following points should be particularly borne in mind. "Go no nearer Ushant nor any part of the French coast than 70 fathoms; small shells and hake's teeth are found in the soundings here. Do not keep over any where to the French shore, it is full near if it can be seen from the mast head. At 10 or 12 leagues off the shore, the flood tide sets S.E., and the ebb does not set N.W., but W. along shore. There is no harbour easy of access for ships in distress on the whole coast of Britany or Normandy, save the isles of Jersey and Guernsey, until Cherbourg is made. Keep on the English coast from 5 to 7 leagues distance, as high as Portland. Having passed Scilly run E. 10 or 11 leagues, coming no nearer the English shore than 53 or 54 fathoms, nor further out than 60. The Lizard will then bear N.E. by N. 8 leagues, and a vessel will be in from 53 to 55 and 56, sandy bottom. Make the Lizard before all things, as when the land there is sighted, the passage up the channel is to be made in the darkest, thickest weather. Unless this is done, no vessel can be secure of her position, and may get on the S. side of the channel, while supposed on the N. The indraught between the Channel islands and the French coast being very strong, remember, coming from the S. towards the channel in hazy or dark weather, not to get too far to the N. of account and fall into the Bristol channel N.W. of Scilly, as from the tides running nine hours to the N., and only three to the S. is very likely to happen, the stream of tide commencing 14 leagues W. of Scilly in this unequal manner. Come no nearer Scilly than 60 fathoms, nor the Lizard than 46, if it be night or the weather thick." Off the Lizard in the mid-channel the stream of tide runs E., on full and change days of the moon, until 55^m past 7^h., or until half ebb by the shore. Two leagues without the Lizard, the flood runs E., and the ebb W. Within that distance the flood runs S. of E., and the ebb N. of W. Vessels coming from the S.W., may shape a course for the Lizard direct, keeping a little more W. as events may demand. Time may be saved this way, and all fear escaped of being driven N.W. of Scilly, or forced perhaps into an Irish port, a vessel, too, will have nothing to fear from a N.W. current. In strong W. gales an E. current is frequently forced up the channel, but with steady E. winds the current has often been found setting out to the W., especially when the Bay of Biscay is open. There is a current, which setting constantly in by Capes Finisterre and Ortegal, enters the Bay of Biscay, and sweeping round the French coast, sets outward in a N.W. direction across the entrance of the British channel. When W. and S. winds do not prevail for a considerable period of time, this current is scarcely observable in its effects. On the contrary, when those winds have set in strongly for any time, this current has been calculated to have 60 miles of westing and 12 of northing in its course per day, in its most rapid part. Its entire breadth in lat. 50°, is sometimes increased to 30 leagues. This W. current, appearing to extend from 24 leagues W.S.W. of Scilly to more than 30 from Cape Clear, is sup-

posed to go off to the N.W. in the parallel of 51°, between 14° and 15° of long., to the S.W. of Ireland; but this is not determined. The middle of the current seems to preserve its original course more correctly than its borders, and to set to the N.W. by W., and the E. border more N., and the W. more W., so that the N. current is much stronger close to the W. of Scilly than further out. This current is called "Rennell's Current," from Major Rennell, who first explained it, and noted more particularly the fact of the high level of the channel waters, during strong winds from the W. and S.W., and the mode in which it must affect vessels as high up even as the straits of Dover. The force of the flood tide is thus augmented, and the ebb retarded, so that the water rises 10 feet above its ordinary level. The consequence is, that vessels having entered the channel with a strong S.W. wind, are driven ahead of their reckoning, and taking the first of the flood tide, will have it for 10 or 11 hours, and at 8 or 9 knots will be thus carried from the Start to Beachy head, or even Dungeness. Vessels entering the channel must in all cases remember to keep sufficiently S.; while those bound to the W., with a wind near S.W., should keep the larboard tack, to have the benefit of Rennell's current. Vessels bound up the channel and abreast of Scilly, but S. 5 or 6 leagues, will find the course to the same distance abreast of the Lizard is E.S.E. 15 leagues; they will haul in there and make the land. It may be added, that from the Lizard to the Start, the course is E. by S. 20½ leagues. A vessel should not, in running up, go into less water than 40 fathoms, as 35 fathoms is in the stream of the Eddystone; nor should 46 or 50 fathoms off the English shore be exceeded. In this depth there will be black sand, with small stones and shells, even from abreast of Scilly. The Caskets bear from the Start point S.E. by S., 19 leagues distant, and the Bill of Portland S.S.W. ¼ W., westerly 16 leagues. From the Start to Portland, a vessel may run up between 30 fathoms in shore, and 36 or 38 off South, mostly sand with shells, though inwards between 26 and 25 fathoms, there will be found ooze and sand. Not standing out beyond 36 or at most 38 fathoms avoids the strong indraught between the Channel islands to the S. Off the Bill of Portland the flood runs to the E., on full and change times, until $10^h 15^m$. The importance of making the Lizard on entering the channel must be here more insisted upon. If in coming up, the land has not been seen until as high as abreast of the Start, the coast of England should be made whenever possible, and the sighting it not be dispensed with on any account when it can be safely done. Being too far to the S., and the Casket lights being mistaken at the first view for those of Portland, has caused many shipwrecks upon the surrounding dangers; make the Start therefore if possible, if the land has not before been seen. If it be not possible and safe to do this, and if with scant wind the tide runs into every creek on the coast, the lead must be kept going, and upon finding the water deepening to 50 or 60 fathoms, a vessel must run at once to the N. into 40 or 35 fathoms, sand and shells, for the greater depth marks the stream of the Caskets; a vessel will thus get in a fair way. This far may suffice, as the channel is opened in the best mode. It may be added that from the Start to Dun Nose the course is E. by S. 32 leagues, with 30 and 32 fathoms to Portland. From Dun Nose to Beachy head, keep within 18 fathoms, without to 28. The course up is E.S.E. 7 leagues, clear of the Owers, and thence E.S.E. ½ E. 11 leagues to Beachy head. From Beachy head to Dungeness, is E. ¼ S. 9½ leagues: Beachy head must not be brought to the W. of N.W. ¾ W., until a vessel is 2½ leagues E. of it, to clear the Horse or outward shoal. From Dungeness to the S. Foreland are E.N.E. ½ E. 6⅔ leagues course and distance; inshore not less must be kept than from 14 to 12 fathoms, nor less off-shore than 16. From the Lizard to the Isle of Wight, the flood tide sets E. by S., and the ebb W. by N. From the Isle of Wight to Beachy head, the flood sets S.E. by E., the ebb N.W. by W. From Beachy head to Dungeness, the flood sets E., and the ebb W. High water at the Lizard, 4ʰ 55ᵐ, rise 18 feet; Isle of Wight, 10ʰ 45ᵐ, rise 15 feet; Beachy head, 11ʰ 19ᵐ, rise 18 feet; Dungeness, 10ʰ 54ᵐ, rise 24 feet; Forelands, 11ʰ 15ᵐ, rise 17 feet. The winds in the channel, at the entrance particularly, are at all times uncertain; yet in a general way, in the commencement of the year, N. and N.E. winds are common, but always interrupted more or less by those from the S. and S.W., though the last seldom continue long. In March N.W. winds frequently occur, but N. and N.E. winds are still prevalent. From the end of May to December the prevalent winds are W. and S.W. It is presumed that no vessel relies upon a single compass in making the channel. Error here is so likely to occur from various causes, that this danger should always be borne in mind. Local attraction should be considered, and no opportunity of an observation by day or night be passed by. Such are the improvements and the resources of modern navigation, that error of position is almost inexcusable, except in very rare cases of accident.

CHANTEREAU'S SHOAL, a vigia, said to have been first noticed in 1721, by Captain Chantereau, coming from Martinique. It was seen again in April, 1828, by Lieutenant Scott, in the Princess Elizabeth packet, at 3 P.M. The sea was green, and had an appearance of soundings, and on looking before the starboard beam, he saw under water, at 2 cables' length distance, a white sand bank or rock, on which the water did not break, but it was so plainly seen that there could not be much water upon it. The lead and line were ordered by Lieutenant Scott, but before they were ready the water had changed to a deep blue again, and it became useless to sound. There was a good breeze and little swell. The lat. of the shoal was 38° 16′ N., long. 39° 48′ 49″ W., observed with care.

CHAPEAU ROUGE, or RED HAT, North America, the W. point of the bay of Placentia, on the S. coast, 15 leagues W. by N. from Cape St. Mary, the E. point. Lat. 47° 49′ 46″ N., long. 53° 57′ 14″ W.

CHAPEL, or CHAPPLE, ISLAND, Ireland, a league S. by E. from the mouth of Youghall harbour, on the S. coast, and E. of Cork harbour. There is a road under it on the E. side, with anchorage at low water.

CHAPEL NESS, Scotland, a headland in the Frith of Forth, half a league W. of Elness.

CHAPELLE ROCK, THE, a vigia, stated to be in lat. 47° 43′ N., long. 8° 4′ 30″ W. In the French chart of the Atlantic, in 1786, it was noted that a rock called La Chapelle had been laid down on the chart of 1766, in lat. 47° 24′ N., and long. 7° 12′ W. It was said to have been first seen in 1764. It was reported as observed in a dark night, by M. Houel, and as showing above water. On the 27th of September, 1822, the sloop Favorite, returning from Malaga to Liverpool, noticed that the water appeared green in colour, and that the vessel was in soundings, and supposed the Chapelle bank must be in the vicinity. Bottom was got at 65 fathoms, about 47° 49′ 38″ by account; or as observed lat. 47° 49′ 49″ N., long. by account 9° 15′ 59″ W., at noon, 28th September. It was certain that the vessel passed over a bank. The French surveyors declared their search for a rock ineffectual, having only found in the place assigned to it an insulated bottom of small extent, having 80 fathoms water, and this the shallowest, obtained in lat. 47° 33′ 47″ N., long. 7° 20′ 12″ W., exactly determined. Now as the spot assigned to this rock was only just beyond the limit of the line of soundings, it was probable there might be an irregular spit thrown off thereabouts that might carry the mysterious danger so long recorded. At length, on the 9th of August, 1842, the Grace Darling, of Liverpool, fell in with the mysterious rock. At 1ʰ 30ᵐ P.M., breakers were observed close to the vessel, and a sunken rock was distinctly observed and repeatedly seen above water in the hollow of the sea, which broke much. It was seen by all the crew, and the vessel passed only her own length to windward of it, going at 7½ knots. It was 40 feet in circumference, looked like freestone, and no weed appeared upon it. All on board were much alarmed. The lat., from a good meridian observation, was 47° 43′ N., and long., from chronometric observations 9ʰ 30ᵐ A.M., and 3ʰ P.M., was 8° 4′ 30″ W. The chronometer had been obtained of Mr. Frodsham, and its rate ascertained only nine days before. Its accuracy was confirmed by solar and lunar observations, and by making Deseada on the 5th of September.

CHAPELLIO, ISLAND, South America, Bay of Panama, from which town it is distant 6 leagues, and which it supplies with fruit. Lat. 9° N., long. 80° 15′ W.

CHARBON CAPE, North America, on the E. side of Cape Breton, to the S. of Spaniard's bay, being the N. point of a bay between itself and the large bay of Morienne.

CHARDONIERE, West Indies, W. end of the island of St. Domingo, S. side, an anchorage in front of the La Hotte mountains.

CHARENTE RIVER, France, a river which rises at Cheronnac, in the department of the Haute Vienne, and forms the port of Rochefort, at its entrance into the sea, opposite the roadstead of the Isle of Aix. It has a total course of 100 leagues, and is navigable for barges 48 leagues. The tides are perceptible at Saintes, about 24 leagues.

CHARGE ROCK, Newfoundland coast; it has 6 feet of water over it, and lies in lat. 49° 18′ N., long. 53° 22′ 58″ W.

CHARLES TOWN, isle of Nevis, West Indies, near the S. end of the island. It is the principal town, with a good roadstead or bay. Lat. 17° 10′ N., long. 62° 45′ W.

CHARLTON ISLAND, Hudson's bay, North America, lat. 52° 8′ N., long. 79° 55′ W.

CHARLES, CAPE, North America, United States, the S. point of the state of Maryland and N. point of the entrance into the bay of the Chesapeak, in lat. 37° 3′ N., long. 76° 2′ W. Off this cape there is a small island, called Smith's island, and the coast N. as far as 38° has many small islands lying along its borders. Strangers without a pilot on board should go without the reef, that runs off from the cape nearly to the middle ground, and also keep without that, going for Cape Henry, the S. point of Chesapeak bay.

CHARLES, CAPE, North America, Labrador, is nearly S. of the S. point of the great bay of the Esquimaux, and N.E. from the straits of Belleisle, at the N. end of Newfoundland. Lat. 52° 25′ N., long. 55° 20′ W. The harbour is N.W. ¾ N., distant 4 miles.

CHARLES TOWN, North America, United States, Massachusets, N. of the town of Boston, between the mouths of Charles river on its S.W., and of Mystic river on its N.W. It is joined to Boston by bridges, and deemed the port of entry. It stands in lat. 42° 26′ N., long. 71° 5′ W. It holds one of the navy yards of the United States. This is in the N.E. part of the haven. It covers 60 acres, and has a dry dock of hewn granite. There are here a marine hospital, large warehouses, an arsenal, a powder magazine, and 2 large roofs for covering the construction of vessels on the stocks, adapted for those of first-rate dimensions. Pop. 11,484.

CHARLES'S GUT, West Indies, island of St. Kitts, at the mouth of a river, E. of Red Flag bay, not quite a league N.W. from Cayou river.

CHARLES, POINT ST., Labrador, North America; the extremity is in lat. 50° 15′ 25″ N., long. 65° 51′ 50″ W.

CHARLES ISLAND, Africa, River Gambia, an island about 2½ leagues W. by N. from James fort, called also the Isle of Dogs. A reef of rocks S. nearly 2 miles W. by S. of this island must be carefully avoided; it is 5 leagues up the river on the N. shore. Lat. of James fort is 13° 19′ 42″ N.

CHARLOTTE PORT, North America, Labrador, N.W. by N. from Point Spear about 3 miles, and fit for any ships.

CHARLOTTE RIVER and HARBOUR, North America, on the W. coast of Florida, in the Bay of Mexico, in lat. about 26° 40′ N., long. 82° 40′ W.

CHARLOTTE TOWN, West Indies, Island of Dominica, situated on the S. side of a deep bay.

CHARLOTTE TOWN, North America, Prince Edward's island; the church is in lat. 46° 14′ N., long. 63° 10′ 16″ W.

CHARLTON SHOAL. See *Thames.*

CHARLESTON, North America, United States, a city and seaport of South Carolina, the largest American city S. of the Potomac, and the 9th in pop., having 30,000 inhabitants, nearly 15,000 of whom are slaves. The lighthouse stands in lat. 32° 40′ N., long. 79° 49′ W. The bay formed at the junction of the Ashley and Cooper rivers, is 2 miles wide, and extends S. of E. 7 miles to the Atlantic, below Sullivan's island. The Ashley is 1,050 fathoms wide, and the Cooper 700, deep and navigable for the largest vessels. A sand bar, extending across the mouth of the harbour, has four entrances, the deepest of which, near Sullivan's island, has 17 feet at high water, which is here at full and change, 7ʰ 21ᵐ, rise 8 feet. It is defended by Fort Moultrie and by Fort Pinkney, 2 miles below the city, and Fort Johnson, 4 miles further down. The harbour is open to E. winds, and storms from that point are troublesome to the shipping when at the wharfs. The city stands only 7 feet above high water, so that at times it has been overflowed, when high winds and tides have raised the waters above the customary level. The trade of Charleston is considerable. In 1840, the tonnage was about 30,000 tons. Three lines of packets continually communicate with

New York. One line has six ships, sailing every five days. A second consists of 8 brigs, sailing every fourth day, and a third, of 6 brigs. It has also a canal of connection with Santee river, 22 miles long, and is connected by rail with Hamburg on the Savannah by a road of 136 miles. This city was first settled in 1680. Lieutenant Raper gives the lat. of the lighthouse 32° 41·2′, long. 79° 52·7′.

CHARROY BANK, France, between the castles of Oleron and Brouage, on the W. coast. It extends from the reef that runs off to the N.W. from Brouage, almost as far as to the sand-mark S. of the castle of Oleron.

CHARTIER'S ROCKS, France, N. coast; these rocks are only visible at very low ebb tides. They lie half a mile W. of the Barbotes, and half a league S. of the Curtois bank.

CHARYBDIS, coast of Sicily, near Messina, the celebrated whirlpool of the ancients. It lies at the back of the Braccio di St. Rainière, a spot of troubled water from 70 to 90 fathoms deep, which circles in eddies rapidly, occasioned by the meeting of the currents there. It is about 2 cables' length from the shore, in the midst of a kind of bay, between Tangara point and the shoals, which are two sandy shelves, stretching from the land near the great and little lights, having over them from 2 to 15 fathoms. The dangers of Charybdis consist in sometimes endangering small craft by the eddies, and even vessels-of-war have been turned round by them, but with care no inconvenience of moment need be apprehended, but it is necessary to be cautious lest the rapidity of the current here should carry a vessel on the shoals above mentioned; and to prevent such accidents, a small lighthouse has been erected, to be brought in one with the old. The dangers are reported to have ceased in a great degree since an earthquake that occurred in 1783.

CHASSIRON, France, at the N. end of the Isle of Oleron, where there is a fixed light of the first class, at an elevation of 164 feet, visible 6 leagues. Lat. 46° 3′ N., long. 1° 24′ W.

CHAT, or CAT CAPE, North America, at the mouth of the river St. Lawrence, the extremity is in lat. 49° 6′ N., long. 66° 48′ 19″ W.

CHATS, POINT DES, the end of a reef stretching out from the Isle de Groa, on the coast of France, department of the Morbihan, at the mouth of the Blavet; 1¾ miles from the point S. by E. is a rock called Basse de Chats, with only 13 feet water, called by the English the Venerable rock.

CHATEAU D'IF, France, S. coast, a strong fortress, and three islets, W.S.W. of Marseilles, and nearly opposite to that port, in the department of the Bouches du Rhone.

CHATEAU, POINT, Island of Guadaloupe, West Indies, 8 miles S.W. by W. from Desenda, easily known by a parcel of grey rocks, which stretch from the shore 5 or 6 miles.

CHATEAUBELAIR, BAY, West Indies, W. coast of St. Vincent's island. Lat. 13° 14′ N., long. 61° 17′ W.

CHATEAUDIN, ROAD OF, Hayti, the entrance to which is 4 leagues nearly N. from Point Abacou, which last is in lat. 18° 4′ N., long. 74° W.

CHATELAILLON, France, on the W. coast, on a point N.E. of the Isle of Aix. Its bay, which is E. of Basque road, and N. of the point, varies in depth from 6 to 2 fathoms, and even less, a part of the shore being dry at low water. There is anchorage and sufficient depth N. by W. from Aix.

A good berth must be given to the rocks of Cha-telaillon, some of which are sunk, and extend W. half a league. The whole coast is dangerous here, from rocks which extend from Point de Che to Point Courcil, the S. limit of Rochelle port. Cha-telaillon is 5 miles S. of Rochelle.

CHATHAM, England, Kent co., a suburb to the city of Rochester, on the river Medway, upon the S.E. bank, and on the N. side of Chatham, or Cetteham-hill. It is principally noted as a naval station and military depôt. The dockyard was begun in the reign of Queen Elizabeth, occupying nearly a mile in length, and was removed to its present situation in the reign of James I. and his son Charles I. Capacious storehouses were then erected, and docks constructed so as to enable ves-sels to enter afloat at high tide. It was further improved in the reign of Charles II. The various houses and offices are substantially and conve-niently built. There are numerous storehouses, one 600 feet long, containing all which is necessary for the purpose of naval equipment, arranged in the most complete order, and ready for use upon the shortest notice of an emergency. Masthouses, basins, rope-houses, sail-lofts, forges, saw-mills, docks, slips for building, in fact, all the conve-niences needful for a great naval arsenal. Chat-ham is defended by lines and forts against any sudden surprise from an enemy. It holds exten-sive barracks, hospitals, for the navy, and also for decayed shipwrights and artificers, with their wi-dows. One was founded by Sir John Hawkins, of Plymouth, in 1592. A fund, called the Chest of Chatham, was established at Chatham, by Sir Francis Drake and Sir John Hawkins, for the relief of disabled seamen. To this, after the defeat of the Spanish armada, the seamen of the navy agreed to contribute a portion of their pay. This has since been absorbed in the establishment at Greenwich, in consequence of some fault found with the management. A hospital also exists here, founded by Gundulph, bishop of Rochester, in the time of William II. The pop. is 18,000. Chatham returns one member to parliament, and with Rochester and Strood forms a continuous street of 2 miles in length. This town is 30½ miles from London. Here a portion of the royal navy is always laid up in ordinary in time of peace. Lat. 51° 23' 48" N., long. 35' E., at the dockyard.

CHATHAM BAY, North America, on the W. coast of Florida, within the Gulf of Mexico. It is spacious, having Cape Roman for its W. point, and being N.W. from Cape Sable. Lat. 25° 43' N., long. 82° 20' W.

CHATHAM ISLAND, North America, United States, Massachusets, on the S.E. coast. Lat. 41° 41' N., long. 69° 57' W.

CHAUME, France, W. coast, on the W. side of the Sables d'Olonne harbour. Here is a light-house on the quay 118 feet high, seen at 3½ leagues, and a harbour-light, at the head of the jetty, leading into the port. The light on the lighthouse is a fixed light. The harbour-light, which is a fixed light also, is on the E. side of the entrance, 33 feet high, visible 2 leagues. The light on the jetty, and that of La Chaume, lead into the Great channel.

CHAUSSEE DES BOEUFS, N. coast of France, a bank dividing the Passage du Senequet from that of Anquette, with many dangers near. The E. end is called the East Basse, having but 2 feet at low water. It lies on with a remarkable tuft of trees on Mount Huchon, 6 miles inland, bearing

S.E. by E., and Cape Carteret N. by E. ¾ E., while S.E. three-quarters of a mile from the same spot, or the East Basse of the Chaussée, is the Jourdan, a small shoal with about the same water, and having a channel between with 5½ fathoms.

CHAUSSEE DE SEIN, or THE SAINTES, France, W. coast, a group of islands, rocks, and dangers, lying in a W.N.W. ¼ W. direction from the Bec de Raz, its S.E. extremity, separated by a channel 2 miles wide. The W. part is distant from the Bec 14 miles. The largest island is near the E. end, and is called the Isle de Sein, or Saintes. It is low, flat, and inhabited by fisher-men. From hence several rocks extend, called the Pont de Chats, the most E. of which is gene-rally uncovered, bearing from La Veille 2 miles, nearly W. The space between is the Passage du Raz, with 13 and 20 fathoms of water, the tides strong; the flood setting N. and the ebb S.; time of high water 3ʰ 45ᵐ. There is a lighthouse on the N. point of the Isle de Sein, showing an inter-mittent light, 148 feet above the level of the sea, visible 20 miles. The flashes, seen every 4 minutes, are preceded and followed by short eclipses, which do not appear total within 10 miles. The weaker light of 3 minutes between the flashes, is visible, in fine weather, 10 miles. On the highest point of the Bec de Raz is a fixed light, 259 feet above the sea, seen in fine weather 6 leagues off. The two lights in one, on a E.S.E. ¼ S. bearing, point out the general directions of the Chaussée de Sein. The first light seen, in approaching the rocks from the W., is the flashing light on the Isle de Sein, which will show whether a vessel is to the N. or S. of the line of direction of the two lights. Even in clear weather the light on the Bec will not be seen until a vessel is within 4 or 5 miles of the W. extremity of the chain of rocks. When the course is to the S. of the rocks, the vessel should be steered so as to open the light on the Bec to the right or S. of that on the Isle de Sein. If it be intended to pass to the N., or to enter the Iroise, no time should be lost in opening the Bec light to the left or N. of that on the Isle de Sein. The intermittent light on the Isle de Sein presents the same appearance as the Penfret light (on one of the Glenan islands); but this cannot occasion any mistake, as the light of Penfret is within the horizon of the great light of Penmark, the flashes of which are at half-minute intervals, and which, in fine weather, are seen as far as the Bec du Raz. To the N. of the Raz de Sein is a bank, containing several rocks and islets, which divide the entrance to the Raz, from the northward, into two channels. The high-est of these is called the Tevenec, or Stevenet rock; it is large, and very conspicuous, bearing from the lighthouse on the Isle de Sein about E., distant nearly 3 miles; and from the extreme point of the Bec du Raz N.N.W. ¼ W., a similar distance, being surrounded with rocks under water. Great caution is required to avoid them, in consequence of the tide setting right over them; all the vicinity of these rocks is so dangerous that without a pilot even the chance of security is a piece of good fortune not to be tempted. The lighthouse is in lat. 48° 3' N., long. 4° 52' W.

CHAUSEY, ISLES OF, France; the principal island is 3 leagues N.W. ¼ W. from Granville, in the S. part of Chausey sound to the E. of the Grande isle. On the S. side of the group, coast-ing vessels find good anchorage, and two or three, drawing 13 to 15 feet of water, may lie afloat. The south-western of the Chausey group is the rock called the Corbière.

CHAUTAUQUE, LAKE, North America, United States, 18 miles long and from 1 to 3 broad. It is said to be 1,291 feet above the level of the ocean, and is the highest water on which the steam-boats of the United States navigate. It is in New York, in the co. of its own name, and is navigated by steamers from Mayville at the N., to the foot of the lake, thence by boats, through its outlet to Connewango creek, a tributary of the Allegany river; rafts proceed from it to the Gulf of Mexico.

CHAUVEAU, POINT DE, France, the S.W. point of the Isle of Ré. Lat. 46° 12' 26" N., long. 1° 21' 44" W. Upon this point there is a fixed light, visible 14 miles, 72 feet above the sea.

CHAVES, RIVER and BAY of, Colombia, Province of Caraccas, a little E. of Puerto Cabello.

CHE DE BOIS, France, the N.W. limit of the entrance into Rochelle, on the W. coast, 2¼ leagues N. by W. from the Isle of Aix. There is anchorage under it in 5 and 6 fathoms.

CHEBUCTO BAY, North America, Nova Scotia, a fine harbour on the E.S.E. coast, in lat. 44° 40' N., long. 63° 31' W. Its entrance is from the S. with a large island on the N.E. side, called Cornwallis's island; between which and the S.W. shore is a channel wide and deep enough for the largest vessels. George's island is 6 or 7 furlongs N. from the N.W. point of Cornwallis's island, on the W. side of which the channel is about three-quarters of a mile broad, with 12 and 14 fathoms. Sandwich or Hawke's river is on the larboard side going up, nearly W. from the N.W. point of Cornwallis's island. From the E. point a sand projects E., so as to have only 4 and 5 fathoms water; and to the N. of it is Spaniard's shoal, about 1½ miles S. from the town of Halifax. Byng's beach is N. of the town, and America point about 3 miles N. near the head of the bay stands the city of Halifax, the capital of the province, upon the W. side, in lat. 44° 39' N., long. 63° 35' 28" W., or, according to Lieutenant Raper, in long. 63° 37' 30" W.

CHEDABUCTO BAY, Nova Scotia, with Crow harbour and Rook island, lat. 45° 20' 45" N., long. 61° 18' 8" W. It is a large and deep bay on the E. part of the colony, at the mouth of the Gut of Canso; opposite to its mouth stands Isle Madame.

CHEPIDO, Greece, a small town N.E. of Marmora, on the bay of that name, in the Isle of Paros.

CHELIDONIA, or KHELIDRONIA, CAPE, Mediterranean, on the S. coast of Anatolia; the islet, lat. 36° 13' N., long. 30° 26' E.

CHELMER RIVER, England, Essex co. Upon this river is situated the town of Malden, 10 miles E. of Chelmsford; small vessels go up and unload at the town, but those of burden unload in Blackwater bay. Pop. 3,831.

CHEPSTOW, Monmouth co., England, situated at the confluence of the Wye and Severn rivers. Vessels of burthen come up to the town as the tides rise there 36 feet, and the flood flows with great rapidity. There is a bridge here over the Wye, 70 feet above the surface of the water at low tide. It is 2 miles from the passage over the Severn at Aust ferry, and 16 miles from Bristol. Lat. 51° 33' N., long. 2° 44' W.

CHERBOURG, France, 7 miles W. ½ S. from Cape Levi, a town, port, and naval arsenal. In lat. 49° 40' 16" N., long. 1° 34' 53" W., at the Pelée lighthouse; the lat. by the French observations, is 49° 38' 31" N., long. 3° 57' 18" from Paris, or 1° 36' 56" W. of Greenwich. The harbour here is formed artificially out of the road, at the bottom of a large bay, in the form of a crescent, between the Capes La Hogue and Barfleur, 19 miles distant from the former, and 16 from the latter place. The sea is upon the N.; on the E. is a large plain above 3 miles long; on the S. cultivated ground, and an eminence called the Rouls, on the top of which stood the forests of Brix and Tourleville. It is said to have had its name from William I. of England, who built a strong castle there, exclaiming, "Ly castell est un cher bourg per mi!" In 1696 it was contemplated by the French Government to make an important naval arsenal at Cherbourg, near La Hogue, on account of its position with respect to England, and Cardinal d'Ossat speaks of it in his existing letters to M. de Villeroy. It was said that the celebrated Vanban laid down the plan. A sort of dyke was formed by means of cones of timber, made hollow and sunk in the sea, being towed to the destined spot by vessels, and afterwards sunk and filled up with stone at an immense outlay of money. The construction of these was begun about the year 1736, when two jetties were completed, of workmanship amazingly strong, and sluices were dug to communicate with an inner basin, in which 400 merchantmen might have been received. In 1758 the works begun were attacked, carried, the piers and basin destroyed, with the batteries, forts, magazines, and stores, by an English expedition. The works, however, were resumed, and, in addition, the engineers blew up the neighbouring rocks and threw them into the sea, in one place, until low water-mark appeared, in a line at an angle of 45°. It was not, however, until the Empire, and under the energetic hand of Napoleon, that Cherbourg reached its existing eminence, although not yet complete in its entire design. The breakwater is 2 miles long, stretching across the entrance of a bay which in depth, to the entrance of Cherbourg harbour, is about 1,650 fathoms from the central angle of the breakwater. At the N.E. entrance stands Fort Pelée, on which there is a lighthouse, with two lights, 46 feet distant from each other. At Fort Querqueville, 4 miles distant on the other side of the bay, there is another lighthouse having a single light; at the N.W. passage into the road, and on Fort Central at the middle of the breakwater, there is a revolving light, seen at seaward between the two others, which are fixed. Fort Pelée stands on a rocky isle of the same name 2 miles E.N.E. ¼ E. from the town of Cherbourg, at the S.W. angle. Some style Fort Pelée, Fort Royal. Keeping the light on Fort Querqueville S.W. by W. ¼ W. until the light on Fort Pelée appears a little N. of the light on Fort Central, the W. entrance will lie open, and a vessel may steer S.S.E. for 1 mile, and then bear up for the road E.S.E. On entering, a ship will be 2 cables' length from the W. of the W. end of the breakwater, and E. of Chavagnac bank of 3 fathoms, and a spit about the same depth that runs out W. of Fort Querqueville. Buoys have been laid down to mark the dangerous points on the E. side of the E. entrance. The first, painted black, has a small beacon attached to it, situated N.W. of the Isle of Pelée in 25 feet at low water. Large ships, when the tide is low, pass this buoy to the W., at a quarter cable's length distance, in order to avoid touching the ground, irregular at this spot. The second has also a small beacon attached, is painted red, and placed at the W.

extremity of the rock called La Truite, in 23 feet at low water; but the bottom being irregular, from 15 to 22 feet, large ships must, when the water is low, pass at a small distance W. The E. extremity of this rock is half a cable's length from the other end, on which only 3 feet 6 inches are found at low water; large ships ought to avoid this extremity, lying in the direction of Octeville with the N. angle of the large rock of Cherbourg; both extremities of the rock are united by a rocky and irregular ground of 10 to 15 feet in depth. The best mark in daylight, to pass through the E. passage, is the beacon on the quay of the Hotel de Ville of Cherbourg in one with the Church of Octeville; this will lead between the two buoys to the E., and the buoy off the E. extremity of the breakwater, to the W., up to the end of the jetties. The mark for the S. limit of the Outer road is the Church of Nacqueville, in one with the high tower of the Church of Querqueville. This mark clears the Bank of Flamands on the E. side of the harbour, to the N., and passes just to the S. of La Tenarde, a bank of 7½ feet, situated one-sixth of a mile from the Fort du Homet. The mark for the Chavagnac bank is the Church of Nacqueville, one with the N. chimney of the house at the S.E. part of the Fort of Querqueville. High water at Cherbourg, full and change, at 8ʰ 2ᵐ; spring-tides rise 22 feet, neaps 15 feet (the establishment of the port gives it 7ʰ 40ᵐ). The town contains 16,000 inhabitants. The inner harbour is entered to the right of Fort d'Artois, erected upon what is called the Homet rock. Cherbourg is one of the five "ports militaires." The military and mercantile marine have separate establishments, the whole defended by eight redoubts. The dockyard and arsenal, opened in 1813, in their existing state, are excavated out of the solid rock. There are accommodations in the basin here for 30 sail-of-the-line, there being 25 feet at low water. The Cherbourg imports greatly exceed the exports. The average of several years gives about 3,000 tons of exports to nearly 30,000 of imports. The coasting trade employs only 14 vessels of less, on the average, than 100 tons, with about 142 boats and small craft. About 1,600 entries of vessels of all kinds take place within the year, most of them to seek shelter. Eggs, to the value of 1,000,000f. a-year, are exported from Cherbourg alone to London. The sum expended on the breakwater, between 1819 and 1837 only, amounted to 17,000,000f., and 800,000f. were then required to complete it. Cherbourg has an extensive victualling office, in which the value of the stores kept ready is 270,922f. The value of the public property is, in hospitals about 173,769f.; in provisions, 386,553f.; naval materials, 25,279,683f.; marine artillery, 3,771,961f.; *travaux hydrauliques* and buildings for persons employed, 83,545,492f.; plans, maps, and instruments, 18,838f.;—total, 113,176,296f. Lieutenant Raper gives the lat. of Fort Royal as 49° 40' 18" N., long. 1° 35' W.

CHERO ISLAND, Greek Archipelago, a mere islet, in lat. 36° 53' N., long. 25° 40' E.

CHERSO, ISLAND and TOWN, Gulf of Venice, in Illyria. The island is about 50 miles in circumference, between the coasts of Istria and Croatia. On the S. it is separated from Osero by a narrow channel, over which is a bridge. The harbour is good. The island abounds in oil, wine, honey, and cattle. Lat. 45° 10' N., long. 14° 45' E.

CHERSONESUS, CAPE, the Crimea, W. side;

a lighthouse 170 feet high has been erected upon this cape; a reef runs out from it a full cable's length. The light at night may be seen at the distance of 17 miles; it is of a red colour and revolves, being eclipsed one minute in four, and visible in the direction from *true* E.N.E. by W. to S.E. by E. Lat. 44° 34' 24" N., long. 33° 21' E. The distance to the harbour of Sevastopol is about 7 miles to the E.

CHESAPEAK BAY, North America, United States, one of the largest and safest bays in that country, 200 miles long, and from 7 to 20 broad, with a general depth of 9 fathoms. Its entrance is in the State of Virginia, between Cape Charles on the N., in lat. 37° 12' N., and Cape Henry, the lighthouse of which stands in lat. 36° 58' N., and long. 75° 56' W. It is from 12 to 14 miles broad, and extends 270 miles N., dividing Virginia from Maryland, and the last state into two parts, called the E. and W. shores. It affords many secure and convenient harbours. Its navigation is safe and easy. It has many fertile islands, generally along the E. side of the bay, though a few exist on the W. shore. Numerous navigable rivers find an outlet in the Chesapeak, as the Susquehannah on the N., the Potomac on the W., and the James, near its mouth, which are the largest streams. Besides these there are the Rapahannock, Patuxent, Patapsco, Chester, Elk, Choptack, Nanticoke, and numerous others. The entire surface drained by the rivers flowing into this bay is estimated at 70,000 square miles.

CHESIL, Isle of Portland, English channel, a bank of gravel and pebbles, thrown up by the sea, about 9 miles long, extending N.W. to Abbotsbury. There is a narrow inlet between the bank and the mainland, in some places half a mile broad. This bank is the isthmus which connects Portland with the main land.

CHESME, Anatolia, W. of Smyrna, and opposite the Island of Scio. The harbour lies near a promontory called Red Head, in the strait of Scio. It is a place of small moment.

CHESTER, CITY, England, Cheshire co., situated on the River Dee, pop. 22,000. It is said to have been once a place of considerable commerce, which is now principally confined to the coasting trade and Ireland. Its port became choked up with sand about the middle of the 15th century. In the beginning of the 18th a new channel was cut for the river, and the access was so improved that ships of 600 tons burthen could come up to the port, but the commerce of the city has not been restored. The Dee, below the city, expands into an estuary of considerable extent, to approach which there are two entrances, one called Chester bar, the other Helbre Swash. The first is the W. entrance, the second the E. The leading mark over Chester bar is the Point of Air lighthouse bearing S.E., which will lead N. of the middle Patch, and S.W. of the Hoyle Sands. The navigation thence is intricate, but it is buoyed. The passage or channel to Park Gate from the Chester bar is called Salisbury Gut, which divides the great Salisbury bank into two parts. The entrance lies opposite the S.E. Air buoy. The course through the Gut is S.S.E. ¾ E. The banks here change and shift, so that no description can be relied upon as permanently correct, and local knowledge can alone supply the want of space, depth, and the happier dispositions of other navigations. The Helbre Swash, or E. entrance of the Dee, lies between the E. and W. Hoyle banks, the entrance bearing S. ½ E. from the N.W. light-vessel in Liverpool bay, between the first buoy of the E. Hoyle bank, on the larboard, and the first Swash buoy on the starboard side; the distance between these buoys being three-quarters of a mile, and the depth on the bar from 11 to 13 feet. The channel extends from the bar about 4 miles to the three rocky islets called the Helbre, Little Helbre, and the Eye. On the first of these is a telegraph, and a quarter of a mile to the E. is a beacon; on the Eye also there is a beacon; these two beacons in a line, bearing S. ½ E., are the leading marks up the Swash way. In the channel the depths, from 10 and 13 feet on the bar, increase further up to 10 fathoms, and opposite the Helbre islet are 8 fathoms. Over the E. bar of the River Dee, called the Welshman's Gut, where the Helbre bears E.N.E., there are only 3 or 4 feet at low water. A series of buoys is laid down in the Helbre Swash, which must be carefully attended to. At Park Gate, a good way up the river, vessels must lie aground on the beach, the stream being so strong and the ground so bad, that a vessel would drag her anchor. Pilots are to be obtained at Dawlpool, as far as which Liverpool pilots are expected to take charge of vessels. At the Point of Air it is high water, full and change, at 10ʰ 44ᵐ; springs rise 28 feet, neaps 22. At Park Gate, below the city, vessels receive and discharge for Ireland.

CHESTER, North America, United States, a town on the River Delaware, S. of Philadelphia.

CHESTER, Nova Scotia, on the N. part of Mahone bay, a town, with good anchorages before it.

CHESTERFIELD INLET, North America, in the N.W. part of what was formerly called the Welcome sea, S. of Wager bay, and on the W. side of that sea. It lies N.W. from Cape South-ampton, on the E. point of the entrance into that sea, round the W. point of the peninsula between the N. part of Hudson's straits and the Welcome. The N. point of entrance into this inlet is in lat. about 64° 10′ N., and long. 88° 40′ W.

CHETECAN HEAD, North America, a cape on the W. coast of Cape Breton island. Lat. 46° 40′ N., long. 60° 45′ W.

CHENAL DE LA GRANDE ET PETITE CONCHEE, France, off the town of St. Malo.

CHEVRIER ROCKS, France, department of the Gironde, in the S. channel, at the entrance of the river of that name, rocks of the Plateau of Cordouan so denominated. The Passe de Grave runs between these and the rocks called Banc des Olives.

CHIARENZA, or KLARENZA, Greece, a port of the Morea, opposite the Island of Zante, in lat. 37° 57′ N., long. 21° 10′ E.

CHICAGO, RIVER and CITY, Illinois, United States of North America, a river, the outlet of which forms the harbour of Chicago, near the S. end of Lake Michigan. The S. branch has sufficient water to afford a secure harbour for vessels of any burden, 6 miles long. The city is the capital of Cook co., Illinois state, situated on level ground, but above ordinary floods, on both sides of the Chicago river, between the junction of its N. and S. branches, and three-quarters of a mile from its entrance into Lake Michigan, where the river is 70 yards wide and from 15 to 25 feet deep. There is a bar next the lake with only three feet water, but artificial means have been adopted to deepen it. Pop. 4,470.

CHICHESTER HARBOUR, Sussex, 6 miles N.N.W. ½ W. from Selsea Bill. The spire of Chichester cathedral is in lat. 50° 50′ 12″ N., long. 46′ 42″ W. A pilot must be taken to enter this harbour, as well as to go out, as there is a bar with only 4½ feet in some places, in others 5 or 6. The channel is extremely narrow, and liable to alter. Within the bar there are two channels, one N.E. to Itchenor, the other N. to Emsworth. Chichester is situated on the Lavant, a little river which reaches the sea-water about 2 miles below the city. There is a small trade carried on by the harbour, principally with London, through the coasters. The pop. of the town is 8,270. Here is a fine cathedral, the seat of a bishopric. The exports of the place are principally malt, flour, grain, and timber. Once celebrated for the manufacture of needles; that manufacture is now wholly abandoned. The Portsmouth and Arundel canal passes S. of the city. Chichester returns two members to parliament, and has a canal communication with London. At new and full moon it is high water in this harbour about 11¾ʰ, and low about 4¾ʰ, flowing in general more than 7 hours, and seldom ebbing more than 5. Ordinary springs rise 14 feet, and extraordinary sometimes 16.

CHICHIBACOA CAPE, South America, province of La Guajira, on the N.E. side of the peninsula, nearly at the entrance into the Gulf of Maracaybo. It is S.W. by W. from the Island of Amba; the rocks called the Monks lying between. Lat. 12° 16′ N., long. 71° 18′ W.

CHIENS PERRINS, France, rocks lying off the Isle Dieu or Dyeu, department of La Vendee.

CHIERI, or KIERI, CAPE, the S. point of the Isle of Zante, in lat. 37° 38′ 35″ N., long. 20° 49′ 30″ E.

CHIGNECTO, CAPE, Nova Scotia, North America, in lat. 45° 22′ N., long. 64° 51′ 18″ W. There is a bay of the same name within.

CHIKAPEQUE RIVER, North America, 7 leagues W. from Tabasco, in the Gulf of Mexico, in the bottom of the Bay of Campeachy. The shore between is a clean strand, but the land is low. There are several good roads here, and a clear coast where vessels may ride in any depth to 10 or 12 fathoms. The river is not more than 20 yards broad at the mouth, and has 8 or 9 feet water on the bar, but inside there are 2 fathoms more. The river Palmas is 5 leagues distant from its mouth.

CHILLING or SHILLING CAPE, coast of Africa, lat. 8° 9′ 30″ N., long. 13° 10′ 12″ W., about 30 miles S.S.E. a little E. from Cape Sierra Leone.

CHILMARK TOWN, North America, United States, Duke's co., Massachusets; it is on the S.W. part of the island called Martha's Vineyard. At the W. end of the town is a peninsula owned by the Indians, the N.W. point of which, called Gay Head, projecting into the sea, has a light-house on it, 150 feet beyond the sea level. It has a brilliant revolving light. It revolves once in four minutes, and is in full lustre twice every revolution. At 12 miles it is eclipsed three-fourths of the time; at 3 miles it may be seen dimly in its alternations. Gay Head light is in lat. 41° 21′ N., long. 70° 50′ 42″ W.

CHILTEPAQUE, BARRA DE, Campeche, Yucatan, Gulf of Mexico, lat. 18° 26′ 30″ N., long. 92° 59′ W.

CHIMANA, South America, between Cape Manare and the Morro of Barcelona, with several other islands called the Picudas, Chimanas, and Borrachas. The Great Picuda lies to the W. of the W. Caraca, with which it forms a channel above a mile wide, and clean, except a sunken rock, 2 cables' length E. from the E. point of Picuda. S.W. by W. 3¾ miles from the Great Picuda, is the second Picuda, an islet about 3 cables' length in extent, and clean. One mile S.S.E. from the last is the E. Chimana, a less islet, equally clean. Two miles further W. is the E. point of the second Chimana, about 1½ mile in extent, also clean. Off its end are islets, the outer about 5 cables' length distant. Two cables' length S.W. from the W. point of the second Chimana is the E. point of the Great Chimana, an island of irregular figure, 3½ miles from E. to W., and to the W. of it, one-third of a mile, lies the W. Chimana, united to the great one by a shoal of rocks and sand. There are other islets, but all are clean and navigable near, except the danger from the rock close to the Great Picuda, and the shoal in the channel between the great W. Chimana. Lat. about 10° 19′ N., long. 64° 51′ W.

CHIMLICO, Mosquito coast, North America, 6 miles S. by W. from the River Lua, and 10 miles from Porto Caballo.

CHIMNEYS, THE THREE, a vigia, said to have been seen first in 1729, July 10, by Captain de Clas Fernel, who approached it within 2 leagues, and remained in sight of it for two hours. It was reported to have been seen again in 1824, by the master of a merchant-vessel and a whole watch. July 29, 1842, Captain Roallons, of the brig Eagle, passed a rock above water, at the distance of 4 miles, at 8ʰ 21ᵐ 14ˢ A.M., the long. 28° 32′ W., lat. 47° 41′ 22″ N., the vessel then making a true W. course. When abreast of it, bearing S.S.W. by compass, leaving it in lat. 47° 37′ 22″ N., long. 23° 51′ W. It formed three distinct points. The highest to the W. appeared to be 80 feet high, the sea breaking violently over its lower part, but there being no appearance of shoal water around it.

CHINCOTEAGUE ISLAND and SHOALS, North America, 16 leagues S. ¼ W. from the entrance of the Delaware bay. Several shoals lie off this part of the coast. The N. shoal lies 7½ leagues N.E. by E. from the island, which is small, and has 10 feet upon it, while between itself and the shore there are 13 fathoms. Another small rock lies 3 leagues S. by E. from this, with 13 feet water upon it, 7 leagues E. by N. from the island. Between these shores are 7 and 6 fathoms, and round the latter from 10 to 13. Another lies 3 leagues E. ¼ N. from the island, with 9 fathoms between, and 4 miles S.E. ¼ S. from the island is the N. end of the flat that extends from the islands lying between this and Cape Charles. Near this end there are 5 fathoms. Chincoteague shoal is 12 leagues S. by E. from Assateague island, and 6 leagues from land, lying N.N.E. and S.S.W. From Assateague a S.S.E. course will clear it. After it is cleared, a course of S.S.W. ¼ W. 18 leagues will make Cape Charles, but the lead should be kept going. A vessel should come no nearer the shoals than 16 fathoms. When Cape Charles bears W.N.W. a ship is clear of these banks. A lighthouse has been erected, having a fixed light, upon the S.E. point of Assateague island, to show out the vicinity of these shoals.

CHINGUI, North America, on the W. coast of Mexico, N.W. from the Bay of Panama, between Point Barica and Puebla Nueva, within it, at the bottom of the bay. The mouth of this river is full of islands; a pilot is required to go up to the town.

CHIOGGIA, Italy. See *Venice*.

CHIPIONA, CAPE, Spain, S. point of the Guadalquiver, S.W. by W. from Cape Jacinto 4 miles. Lat. 36° 44′ 18″ N., long. 6° 24′ 15″ W. entrance to San Lucar. The tides rise here 6 feet; high water at 2ʰ.

CHIRAS ISLES, Cyprus island, on the N.E. shore, nearly S. of Cape Lissan el Kabeh, on the coast of Caramania.

CHIRIQUI LAGOON, America, Darien, in lat. 9° 16′ N., long. 81° 58′ W. It is 26 miles long and about 12 wide; it affords good anchorage.

CHISAMIL BAY, or BA CHISAMIL, Western Isles of Scotland, a safe and good anchorage, if the rocks at the entrance are avoided. It lies in the channel between Vatersay island and the S. shore of Barra, both narrow and dangerous.

CHITI, CAPE, Island of Cyprus, a low flat point on the S. side of the island; there is a tower on its summit. This point should always have a good berth. There is a rock reported to lie S. of it, and E. by S. of Cape Gatte 11 leagues. The sea breaks over it.

CHITRIANI ISLAND, Greek archipelago, a rocky islet off the S. end of Siphanto.

CHITRO, Greece, a town in the Bay of Salonici, in lat. 40° 30′ N., long. 23° 10′ E.

CHIVEYA, coast of Africa, one of the four Hog islands, near the entrance of the Rio Grande.

CHOCOLOTEROS, Mexican sea, a term applied in the Spanish West Indies to winds from the N., or as they are termed "Norths," which do not frequently occur in May, June, July, or August; but when they do happen, blow very furiously, and are in consequence called Nortes del Hueso Colorado, while the name of Chocoloteros is given to moderate breezes from the same quarter.

CHOCO POINT, America, W. of Candelaria bay, Darien; it affords good anchorage.

CHOLE, or SHOAL, BANK, Channel islands, the N.E. of which is 2½ leagues from Alderney, towards Serk, and a little E. of both islands. It extends 6 miles W.S.W., and lies in a direct line from the middle of the Race. It has only 12 feet. Its middle bears S.S.E. 4 leagues from the Caskets; 3 miles S.S.W. from Alderney windmill, and 3 leagues N.E. by E. from Serk.

CHOQUE BAY, West Indies, St. Lucia island, on the W. side between Gros Islet bay on the N., and Carenage bay to the S. The point of land on its S. extends W. considerably, but not so as to shelter it from S.W. to N.W.

CHOUCHOU BAY, Hayti, N. side near the Bay of Acul.

CHRISTIAN, CAPE, Greenland, lat. 59° 49′ 30″ N., long. 44° 5′ W.

CHRISTIANA, Greece, the Cyclades, two islets bearing from Cape Acrotisi, the S.W. point of Santorin, S.S.W., distant 8 miles. The summit of the largest is in lat. 36° 14′ 41″ N., long. 25° 18′ 45″ E. These islets are sometimes called Ascania.

CHRISTIANIA FIORD and CHRISTIANIA, the first extending from the Sleeve N. as far as the city, which is the capital of Norway. This fiord, or channel, may be said to begin within side the island of Faerdu, being the W. entrance, on which there stands a lighthouse, with a revolving light to guide vessels to Dram and Christiania. Between Faerdu light and that on Bastoe, there is one upon an islet E. of Notter island, then the Bastoe light N.E. of Aasguard, the next at Filved, 12 miles N.

of Bastoe, 24 feet high, left to the port on going in; the next light is 14 miles N. of Filfed at Steilness, 22 feet high, to be on the starboard going up, and the last is the harbour light, 23 feet high. All above Bastoe island are fixed lights, kept burning from the 15th of July to the 31st of May. The fiord, or gulf, is enclosed on each side by mountains, and is studded with a vast number of rocky islets. The city of Christiania is well built, and stands 20 leagues from the sea; the pop. about 12,000. The Quartal, or part nearest the sea, is inhabited principally by merchants and public officers. It is divided into three parts, the city, the fortress of Aggerhuus, and the old city of Apsloe, or Ansloe. The streets for the most part run at right angles. The castle of Aggerhuus stands on an eminence, and is well fortified. The port will accommodate vessels of the largest size, having from 30 to 40 feet water, close to the quays, and though the channels up the gulf require a pilot, they are not dangerous. The exports are considerable in plank, pitch, tar, soap, iron, copper, and alum. There are 140 saw-mills in and near the city, and glass, coarse woollens and linens, are manufactured there, and numerous vessels built. There is no perceptible rise of the tide here. The lat. of Christiania is 59° 54′ 6″ N., long. 10° 45′ E.

CHRISTIANBURGH, Africa, S. coast, 5 leagues E. of Axim, and 12 miles W. of Acquidah, lat. 4° 10′ N., long. 1° 55′ E.

CHRISTIANOPLE, PORT OF, Sweden, on the Baltic sea, 13 miles N.E. of Carlscrona; it is 4 leagues from the S. end of the island of Oland, to the point of this port, which is the breadth of the S. entrance of Calmar sound.

CHRISTIANSTADT, West Indies, Island of Santa Cruz, on the N. side, on a fine harbour. It is the residence of the Danish governor. Lat. about 17° 49′ N., long. 63° 33′ W. The Flagstaff fort Louisa Augusta is in lat. 17° 44′ 30″ N., long. 64° 41′ 32″ W.

CHRISTIANSUND, a town of Norway, built not with much regularity upon three rocky islands, in the province of Nedences, in lat. 63° 7′ 26″ N., long. 7° 42′ 10″ E. It is 36 miles N.W. of Drontheim. The principal trade is in timber and fish. The common entrance to the Sund is to the N. of the Ravnene shoal. There are lights to direct the navigator towards this port. The first at Quitholmen, a revolving light of 10 or 12 minutes duration, followed by an eclipse, but not a total one. It may be seen 18 miles off, in the direction of S.S.W. ¾ W., through W.N. and E. up to S.S.E. ¾ E. It stands in lat. 63° 1′15″ N., long. 7° 12′ 15″ E., 130 feet above the level of the sea. From hence passing along the coast to the entrance of the Sund, the next light is that of Staveness, a fixed light seen at 3 leagues' distance in all directions of the compass, from N.W. by W. ¾ W., through N., and E. to S.E. It stands in lat. 63° 7′ N., long. 7° 39′ 6″ E., and is 63 feet above the sea. The building is painted of a bright colour. As soon as Staveness light appears, ships alter their course from E. to S., and steer the same until the light bear S.S.E. ¼ E., and till arrived at the side where the light is visible, when the course is altered to S.E.; bound for Christian Sund the course is as above as soon as a ship has cleared the light within three-quarters of a league. The lights burn from August 15th to April 13th.

CHRISTIANSAND, Norway, a seaport, and capital of the government of the same name, celebrated for its iron mines. It stands at the mouth of the Torrisdals, opposite the Island of Flekker Oe, which forms a roadstead several miles in extent, with good anchorage in 8 and 9 fathoms. The town, 150 miles from Christiania, is built upon a sandy level, close to the sea. It is considered the fourth town of Norway, having about 5,000 inhabitants. Its exports are principally timber, iron, and salt fish. Ship-building is much followed here, and repairs are made with facility to the fullest demand of the mariner. A lighthouse stands upon Ox Oe islet, E. of the Isle of Flekker Oe, at the E. entrance to Christiansand, 135 feet above the level of the sea, visible 16 miles. It shows a steady light for 2ᵐ 55ˢ, then grows fainter, succeeded by a bright flash, then faint again, and then a steady light for 2ᵐ 55ˢ. It is lighted all the year. The lighthouse is white, as in the day it is a seamark. There is also a harbour light on Oder Oe island, 4 miles N. ¼ W. from Ox Oe. There are two harbours at Christiansand; the W. is most visited. The havens from Christiansand W. never freeze, and ships always find safety in them. Pilots are readily obtained, when they are required for the navigation, even off Flekker Oe. Allowance must be made for currents on this coast, there are no tides apparent. Beacons are everywhere erected along this part of Norway, as found upon the charts. Christiansand is in lat. 58° 8′ N., long. 8° 3′ E.

CHRISTCHURCH, HARBOUR and TOWN, Hants; the head lies in lat. 50° 42′ 38″ N., long. 1° 44′ 31″ W. The harbour entrance is three-quarters of a mile to the N.E. of the headland. There is good anchorage in the bay off the harbour's mouth, in 5½ and 6 fathoms. The head is N.W. by W. ¼ W. from Hurst castle, distant 7½ miles. Vessels can only enter at high water, and they must then be of very small burden.

CHRISTOPHER'S, ST., West Indies, an island, the property of Great Britain, often called St. Kitts; lat. of Basse Terre, the capital, 17° 19′ 30″ N., long. 62° 52′ 30″ W. By the Indians called Ayay, and by the Caribs Liamniga, the fertile island. It was discovered by Columbus, in 1493, and he was so pleased at its aspect that he gave it his own name. It was never colonized by the Spaniards. It is the oldest of the British settlements. Sir T. Warner was the first settler in the island, with his son and 14 persons from London. The shape of the island is irregular, and somewhat oblong. A chain of mountains runs through the centre from N. to S., in midst of which stands Mount Misery, 3,711 feet high, of volcanic origin. On the W. side is Brimstone hill, 750 feet high, having two peaks, the N. crowned with and called Fort George; the S. called Fort Charlotte, or Monkey hill. The works of defence are very strong. The climate is reckoned among the more healthy of the West Indian isles. The more prevailing winds are from the N.E. and S.E., and the rains are more frequent than heavy. The superficies of the island is 68 square miles, and it contains about 1,700 whites, 2,000 persons of colour, and 19,000 blacks. The exports and trade are similar to those of the other islands.

CHURCH BAY, Ireland, N. coast, on the S.W. part of Rathlin, or Rachlin, island, the S. point of which is about 2½ leagues N. from Fairhead. Church bay is a deep bight, directly opposite to Ballycastle bay.

CHURCH POINT, West Indies, a cape on the W. coast of Barbadoes island, three-quarters of a mile N. from Hole town.

CHURCH POOL, W. coast of Ireland, on the S.E. side of Inishkiel island, known by having a chapel upon its E. point, a mile to the E. of Dunmore head, where a vessel may run for shelter, if driven in between Teelin head and Arranmore. The island is a mere rock, half a mile in length, with some ruins on the S. and S.W. sides. The best anchorage is in the middle of the bay, to the E. of the chapel, in 2¾ to 3 fathoms, with the S. point of island bearing W. by S. The ground is good, and the anchorage sheltered by a reef that runs off from the E. end of the island, to which a good berth should be given on going in, and the mooring should be with one anchor towards the E. point of the island, and the other towards the shore. In seeking this place of shelter, some shoals near Roaninnish are to be avoided, and also Bulloconnel shoal, three-quarters of a mile N.N.E. from the N.E. end of Inishkiel, with only 10 feet upon it at low water.

CHURCHILL, CAPE and RIVER, on the W. coast of Hudson's bay. The cape directly E. from Churchill river, forming the S.E. limit of the smaller, called Button's bay. Lat. 58° 48′ N., long. 93° 12′ W. Churchill river is on the W. side of Hudson's bay, in lat. 58° 48′ N., and long. 94° 7′ W. Prince of Wales fort stands at the mouth; it falls into the sea to the W. of the cape of the same name. It is a noted place in the history of the fur trade, but surrounded with snow and ice for eight months in the year.

CHURCH-HILL FORT, North America, New North Wales, at the mouth of the Seal river, on the W. side of Hudson's bay, 120 miles N.N.E. of York fort.

CHURCH GUT, West Indies, Island of St. Kitt's, on the S.W. side, to the W. of Old Road bay, and nearly off St. Thomas' church, upon the shore; a small narrow sand lies along the coast here, with from 4 to 6 fathoms close to it, but the water deepens fast off land, so as to have 40 fathoms at the distance of 4 miles from it.

CHUSPA, BAY OF, South America, province of Caraccas, an excellent anchorage. The shore runs in about S.W. for 1½ miles from Chuspa point, the N.E point of the bay, as far as a river of the same name, on the E. bank of which stands the town of Chuspa. The lat. of Chuspa point is 10° 19′ 30″ N., long. 66° 20′ W. From this point to Point Cururau the shore is all clean, and the lead is the only guide necessary for anchoring; but beyond Point Cururau the shore is very foul, with a reef running out two cables' length, and so continuing as far as Punta del Frayle, which has a rock of the same name about a cable's length from it. From Punta del Frayle the coast trends to the W. inclining S. 29 miles to the anchorage of La Guayra.

CHUSPARA POINT, West Indies, Isle of Trinidad.

CILIDRONI or CHILIDRONIA, Greece, a narrow island lying N.E. by N. and S.W. by S., 10 miles long, 3 broad, and distant from the S. part of Scopolo about 3 miles. Midway there is a rocky islet called St. Elias, which rises a considerable height from the sea; it is lofty, and its shores precipitous. There are only 150 houses on this island, which pirates have continually plundered.

CIMBRISHAM, Sweden, a small sea-port of Schonen, in E. Gothland, in the Baltic, with a harbour, 8 leagues S. of Christianstadt. Lat. 56° 40′ N., long. 15° 25′ E.

CINQUE ISLES BAY, Newfoundland, S. of North bay.

CINQUE PORTS, England, in the cos. of Kent and Sussex, originally five in number, having in former times certain privileges granted to them on account of their position and activity in fitting out ships against France. They were subjected to the jurisdiction of the constable of Dover castle, and though their services are become valueless by time, and some of their ports are choked up, they yet retain their privileges and immunities. These ports are Dover, Sandwich, Rye, Hastings, Winchelsea, Romney, Hythe, and Seaford. They are presided over by a warden, called the warden of the Cinque ports, an office become little more than a sinecure.

CINTRA BAY, Africa, West coast, the N. point in lat. 23° 7′ N., long. 16° 9′ 12″ W.; the S. point in lat. 22° 56′ 36″ N., long. 16° 14′ 6″ W.

CINTRA, CAPE OF, Portugal, more commonly known as the "rock of Lisbon." It is on the N. side of the Tagus, and is celebrated for the peculiar natural beauties with which it abounds. Lat. 38° 46′ N., long. 9° 30′ W.

CIOPPETO, POINT, Bay of Naples, channel of Procida. Lat. 40° 46′ 30″ N., long. 13° 40′ 48″ E. Upon this point there is a fixed light of the fourth class.

CIOTAT, LA, France, Mediterranean, N. of Cape l'Aigle, three-quarters of a mile. The town is at the bottom of the port. Pop. 5,240. The traffic is confined to small vessels that come for the wine, oil, and dried fruits, for which the neighbourhood is noted. The port has two moles, between which vessels anchor in 10, 12, and 15 feet of water. There is a fixed harbour light on the port, on the starboard side, elevated 82 feet, and visible 3½ leagues, and one on the new mole head, on the E. side of the entrance, 120 yards S.E. of Fort Berouard light. This, though less elevated, is seen 3 leagues, and prevents any confusion with another fixed light at Port du Cassis.

CIRCELLO, MONTE, Italy, 21 miles S.E. ¼ E. from Cape Anzo, a high mountain running into the sea, having between them a deep bight and a low marshy shore. There are a number of towers along this coast, and among them the tower of Astura, 5 miles S.E. by E. from Port Neptune. Monte Circello is in lat. 41° 13′ N., long. 13° 3′ E.

CIRELLA, TOWN and ISLAND, Lower Calabria, Italy, a town on the main land, 8 miles S. of Casaletto, and opposite to the town is the island, about half a mile from the shore; it is small and rocky. Lat. 39° 37′ N., long. 15° 50′ E.

CISARGAS ISLAND, Spain, N. coast, W.N.W. ¼ W. from Corunna lighthouse, off Cape St. Adrian, distant from the light 6¼ leagues. This island, or islands, for there are rocks and shoals extending 2 miles N., one of which is generally thus named, is in lat. 43° 22′ 15″ N. The ground is rocky and bad here, yet there are several channels between the rocks with from 6 to 10 fathoms water. The current with the flood is strong, and sets E. Necessity alone should tempt a passage with W. or N.W. winds; or if from the E. with E. and S.E. winds.

CISPATA, PORT OF, Colombia, S. of the entrance of the Gulf of Darien. Lat. 9° 25′ N., long. 75° 48′ 5″ W.

CISTRY, or SETRY, or CISTROS, Africa, S. coast, LITTLE and GREAT. The former is W. of Great Cistros. The S. entrance of Cistros river is in lat. 5° 26′ 25″ N., long. 9° 34′ 45″ W. In the distance of 15 leagues S.E. by E. from Dron to Cape Palmas, on this coast, there are the towns of

Little Cistros, Baddou, Grand Cistros, Garraway, and Rock Town, all carrying on trade in the same mode that prevails from Grand Bassa, in lat. 5° 54' 50" N., and long. 10° 4' 5" W. to Cape Palmas in lat. 4° 22' 9" N., long. 7° 44' 16" W. Little Cistros is called Piquino and Picquinini Cistros, and is known by a cape that runs out with three black points, and by 2 rocks, one much larger than the other, standing far out in the sea, but clean and steep. A reef of rocks runs off the W. point of the river, and the town stands a little to the E. of the reef, near the point. The land within is high, and appears in many round hummocks. To anchor upon the coast bring the rocks of Cistros at N.W., the point called Wappo or Wappen at E.S.E., and Cistros town at N.E. This will secure good ground in 15 or 16 fathoms. There is foul and rocky ground further out, in 20 and 30 fathoms, in which ships will be apt to lose their anchors. Great Cistros is 5 leagues E.S.E. from Little or Piquino Cistros. Bring the river at N. or N. by E., and the land near the point, on which there are some trees like bare poles, to the N.E. by N., and anchor in 25 or 26 fathoms. This is 22 leagues N.W. by W. and W.N.W. from Cape Palmas. There are some creeks between, but no port for vessels, which generally ride off in the open sea, that is free of all hidden danger, a league or two from the shore.

CITRON BAY, West Indies, St. Lucia island, at the S. end, between Golden river and Old Fort bay.

CITTA NOVA, Italy, a town on the E. coast, E. of Ancona, upon a small river or creek. It is in lat. 43° 16' N., long. 13° 46' E.

CITTA NOVO, Gulf of Venice, Istria, a town upon the N. side of the port, in a state of decay from the unwholesomeness of its situation, inhabited only by a few fishermen. It has a cathedral and three churches still remaining. The entrance to the harbour is nearly a mile wide, and goes in about 1½, so that large vessels might if needful sail in, as there are 14 fathoms at the entrance, lessening to 2 in its remoter part within. Lat. 45° 36' N., long. 14° 2' E.

CIUDADELLA, PORT. See *Balearic Islands*.

CIVITA VECCHIA, Italy, papal territory, and its principal sea-port, having a pop. of 10,000; the houses well built and spacious. It contains an arsenal, and has a considerable trade, but the site is unhealthy. This town is in lat. 42° 4' 36" N., long. 11° 45' E. It may be known by the white tower of Corneto, and Cape Linaro to the S., which stretches further than any other into the sea. The tower of the lighthouse is soon seen, and the town itself. The light is placed on the mole, which extends 200 fathoms seaward, and was built by Trajan; opposite its ends are two forts. The light is seen 18 seconds, and eclipsed 12; it is visible 16 Italian or 23 English miles. There is a fine castle here. At the N. entrance are 3 and 3½ fathoms of water. The S. entrance is deepest, having 8, 6, and 4 fathoms. Ships anchor in the middle of the port, in 16 and 18 feet water. Cape Linaro is 5 miles from the lighthouse of Civita Vecchia, having a square tower upon its summit.

CIVITA VICTORIOSA, or IL BORGO. See *Malta*.

CLACKLAND, Scotland, an islet near the E. coast of Arran island.

CLAGGAN, POINT and BAY, Galway co., Ireland, lat. 53° 34' N., long. 10° 4' W. The

ground here is good, and the shelter tolerable. There is a sunk rock and a shoal in this bay.

CLAGGENCARROCK BAY, Scotland, Isle of Islay, off the W. coast. It is directly N. from the N. point of Ireland or Rachlin island. It is a rocky coast not to be adventured without a pilot.

CLAIR, ST., BAY OF, S. coast of Labrador, between Wolf's creek E., and Philipeau bay on the W., lat. 51° 25' N., long. 53° 10' W.

CLARA, CAPE, S.W. coast of Africa, lat. 30' 30" N., long. 9° 20' 30" E.

CLARE ISLAND, Ireland, W. coast; a lighthouse in lat. 53° 49' 59" N., long. 9° 59' W., stands upon its N.E. point. It lies in the middle of the entrance to Newport bay, and is 4 miles in length and 2 in breadth. The lanterns of the light are elevated 487 feet above the level of the sea, showing from the E. seaward to the S.E., visible 10 leagues.

CLARE ST., ISLAND OF, N. coast of Spain, at the entrance of St. Sebastian's port.

CLARENCE PEAK, Island of Fernando Po, African coast, a hill in lat. 3° 35' N., long. 8° 46' 30" E.

CLARENDON, or CAPE FEAR, RIVER, North America, United States, its mouth is on the W. side of Cape Fear, the S.E. point of which is in lat. 33° 48' N., long. 78° 9' W.

CLARK'S ROCK, a vigia, lat. 45° 40' N., long. 19° 17' W. A rock called the Mayda was stated to exist in 1705 in this lat.; and in the French chart of 1766, its lat. is given as in 46° 48' N., long. 19° 50' W., both being supposed deficient in accuracy. The original report was made in October 1705, by Pierre Nau, and the rock stated to be a little white island. In 1738, Captain Biden returning from Martinique, discovered the Mayda, which he found by observation in lat. 46° 10' N. He saw five heads of rock, and a breaker on them of 6 or 7 feet. The bark Hartly, Bradford master, from Sierra Leone to Plymouth, on the 26th of August, 1842, passed at half-past 5 P.M., in lat. 45° 40' N., long. 19° 17' W., at about three-quarters of a mile distance, a double-headed rock, which on the fall of the sea was uncovered 6 or 8 feet. One of the crew saw it first, and supposed it a whale's carcase, but the master thought it a rock. Its position was ascertained by the chronometer. It coincides, however, with the preceding accounts, and with a third, stating that on a passage to the West Indies in 1840, in lat. 45° 36' N., long. 19° 30' W., a rock was seen within 100 yards, of a conical form, which must have been the Mayda or Clark's rock.

CLATTISE HARBOUR, Newfoundland, S.W. by W. from the Burgoe islands 5 miles.

CLAUDE, ST., ISLAND, North America, near the main land of Nova Scotia, and very near Cape Herring on the N.W., upon the W. side of the sound, between the main and St. John's island.

CLAUDE, ST., CAPE, North America, W. of Nova Scotia, at the S.W. end of the Bay of Annapolis Royal, in the Bay of Fundy, 20 leagues E. from Cape Minas.

CLAY, England, coast of Norfolk, a village with considerable salt-works and some trade. It is near Blakeney, the harbour of which is the best on that coast. They are on opposite sides of the sea arm forming the harbour. Its lat. is about 53° 2' N., long. 1° E.

CLAY HEAD, Isle of Man, on the S. point of Laxey bay, near which on the N. point there is good anchorage.

CLAY HUTS, THE, England, off the coast of Lincolnshire, about 1¼ miles to the S. of Trust-thorpe church, running out three-quarters of a mile from the shore, with 4 and 8 feet water upon them, and 2 on the E. side.

CLEA NESS SAND, River Humber, part of the Bull sand, extending from Sandhaile flats, and running out E. from the Ness 2½ miles, then turning towards Grimsby, the inner part drying at low water. A black buoy lies on the point of the sand. On Clea common there is a beacon 60 feet high, octagonal, black of colour, clearing, kept in a line with Grimsby church, bearing N.W. by W. ½ W., the W. edge of the Inner Binks.

CLEAR, CAPE, Ireland, S. coast, lat. of the lighthouse 51° 26′ N., long. 9° 29′ W. The cape itself is in lat. 51° 24′ 56″ N., and long. 9° 25′ W. The lighthouse stands on the S.E. side of Cape Clear island. It is a brilliant revolving light, attaining its greatest brilliancy once every two minutes. The tower is white, 42 feet high, and elevated so that the light stands 455 feet above the level of high water: it is visible 8 leagues. Vessels may anchor on the N. side of Cape Clear island, in moderate weather, about a mile from the shore. Fastnet rock, 5 miles W.S.W. from Cape Clear lighthouse is 98 feet above the sea level. It rises in lat. 51° 24′ N., long. 9° 36′ 15″ W. It should not be approached nearer than a mile. The bottom W., S., and N.E. is shoal and rocky. To the N.E. there is a flat rock, a quarter of a mile from the Fastnet, with only 9 feet at low water. The Mizen Peak on with Brow Signal Tower, leads through midway nearly between the Fastnet and the Cape. The Peak, if kept open to the westward of Brow head, will lead half a mile to the S.W. of the Fastnet. These marks may be very useful in a partial fog, to which this coast is much subject. Between the rock and Cape Clear are from 12 to 35 fathoms. It is high water at the island about 4ʰ 30ᵐ, spring tides. The tide at flood sets E.N.E. along this coast, and the land may be seen by vessels in 54 fathoms, approaching from sea, and from the cape, W. to the Durseys, in 58 and 60 fathoms. In the night no ship should come nearer than 30 fathoms. Cape Clear is 17 leagues S.W. of Kinsale, and 54 from the Long Ships off the Land's End, Cornwall.

CLEAR COVE, Newfoundland, E. coast, one of the smaller coves of that large haven, on the N. side of the entrance into the road.

CLEENISH ISLAND, Lake Erne, Ireland, on the coast, 3 miles from Enniskillen.

CLEMENT'S, ST., ISLAND, Cornwall, on the W. side of Mount's bay, near the S.W. point. Here there is clean ground, and inside the island is a large sandy bay, where ships may anchor in 7 and 8 fathoms. It is open to the S.E. and S.S.E. winds, but land-locked against all others. It lies off Mousehole, and S. of Gwavas lake.

CLETA SKERRY. See *Orkney Islands*.

CLEU BASSEVEN, France, a rock S.S.E. of Ushant, having only 2 feet over it, upon which the British line-of-battle ship Magnificent was wrecked in 1804. It lies N. by W. 1½ miles from the Buffalo rock, and 16 from the S.W. point of Ushant.

CLEVELAND, North America, United States, the capital of Cuyahoga co., at the mouth of the river of that name, and the N. termination of the Ohio canal upon Lake Erie, in lat. 41° 31′ N., long. 81° 46′ W. Pop. 6,071. The harbour is one of the best on Lake Erie, and improved by piers extending on each side 425 yards into the lake, at the distance of 200 feet from each other, constructed of substantial stonework. This port is the great grain-mart of the principal corn state of the Union.

CLEVELAND REEF, said to have existed in lat. 35° 45′ N., off Cape Ghir or Geer, coast of Africa, 9 leagues from the coast. It has been sought for without success, and is believed to have no existence.

CLEW, called also Newport Bay, W. coast of Ireland, within the Island of Achil. Here the tide commonly runs about a mile an hour. There are several places in this bay where vessels may ride in good ground, and well sheltered, with sufficient water. In sailing along the S. shore, the Carrickadillish rocks must be avoided. These lie W. of Dourinsho island, about a mile from the shore; the largest dries at half ebb, and another only at spring tides. There are several shoals near them, but the breakers are a sufficient warning.

CLIBBERSTACH ISLE. See *Shetland Isles*.

CLIDI ISLANDS, Isle of Cyprus, off Cape St. Andrea, E. end, but there is no passage between these islets and the main. They lie in lat. 35° 41′ 39″ N.

CLIFF END BANK, Kent co., above Birchington and to the W. of Margate. It consists of three small spots, lying mid-channel between Margate sand and the shore, having upon the E. spot 9 and 10 feet water. The W. shoal lies with Birchington church and Seed mill, bearing S. by E.

CLIFF FOOT ROCK, coast of Suffolk, W. of the Altar. Between them is the entrance to the channel of Harwich harbour. Their E. edge is marked with a chequered black and white buoy.

CLIFF OF CALAIS BANK or THE EASTERN DYCK, one of the Flemish Banks. It is separated from the Middle Dyck by a channel a mile broad, with a depth of 4½ fathoms. The mark is the church at Dunkirk, midway between Bergues and Cassel, bearing about S. ¼ W.; but the mariner must be careful of running on in this direction, since the same mark will lead upon the shoal part of the Inner Ruytingen. The S.W. end of the Eastern Dyck bears N. ¼ E. from Dunkirk, distant 8 miles; and N.E. by E. from Gravelines, distant 12 miles, Cassel appearing 1° open to the westward of Dunkirk. The direction of the bank is N.E. by E., 13 or 14 miles. It continues exceedingly dangerous, full 12 miles from its southwestern part, or until Nieuport steeple bears S. by E. Its shallowest part is near the S.W. extremity, where there are only from 5 to 9 feet water, and this continues until Dunkirk bears S.S.W. The breadth is about three-quarters of a mile. The lead should be kept going here.

CLIFF SOUND. See *Shetland Isles*.

CLIFF'S END CHANNEL, in the Downs, between the Quern and Cracker, S. of the buoy of the Brake.

CLIFFS, THE SWAN, Sussex co., W. of Beachy head; athwart their W. end, there is anchorage in from 7 to 9 fathoms water.

CLOGHER HEAD, Ireland, E. coast, lat. 53° 47′ 40″ N., long. 6° 14′ W.

CLONEKILTY HARBOUR, Ireland, S. coast, 7 miles W. of Courtmaskerry bay, only fit for small vessels, and in a S. wind entering or coming out very hazardous. It has only 2 feet water on the bar at spring tides. The harbour in blowing weather ought not to be attempted, except under urgent necessity, and then only at three-quarters flood.

CLONMESS, Ireland, a village on the W. side of Sheep-haven, at the bottom of a bay, where there is anchorage in 4 fathoms, for such vessels as can go over the bar.

CLONTARF ISLAND, Dublin bay, midway from the Point of Ring's End on the S., and the village of Clontarf, a pleasant bathing place on the N., to the W. of the point of sand called the North Bull.

CLOUGH or CLOCH, Scotland, E. side of the River Clyde. Here is a fixed light N.E. of Tow-art Point light, on the W. side.

CLOUGHY BAY, Ireland, coast of Down co., on the N.E. of the island, opposite to which are two large rocks called the N. and S. rocks. The first is a mile from the shore, opposite the N. side of the bay; the other is a mile S. and two miles from the shore, nearly opposite to the S. side of the bay, and between them is a channel nearly a mile wide. The tide seldom covers the first rock, but the second is always covered at high tides, and dangerous shoals extend from it a mile further out. An offing of a league or more is necessary, till the channel in the bay is fairly open. There is a light on the S. rock. Lat. 54° 24′ N., long. 5° 25′ W.

CLOVELLY, Devonshire, N. coast, 5½ miles E. of Hartland point. It has a small pier to the W., and a roadstead, where there is anchorage three-quarters of a mile from the shore, in 6 fathoms, blue mud, with the E. end of the village on with the pier-head, bearing W.S.W., and the W. land near Hartland, closing with Gallantry Bower bearing N.W. three-quarters of a mile from the shore. The pier of Clovelly is capable of affording room only to half a dozen coasting vessels. Lat. about 51° 6′ N.

CLOWES' REEF, lat. 19° 17′ N., long. 65° 50′ ½ W. to the N. of Porto Rico, on which, in 1817, an American vessel struck. It was seen by Capt. Clowes, of the Caledonia, in April 1825, who assigned the above latitude to it. It was about a quarter of a mile in extent E. to W., with very little water upon it, and a quantity of sea-weed at each end, which appeared to be drifting to the S.W.

CLWYD RIVER, North Wales, the mouth of which falls into the sea between Orme's head and the Point of Air; the port of Rhudlan is near its entrance, accessible only to small vessels.

CLYTHNESS, Scotland, E. coast, 15 miles E.N.E. from Ordhead, a rugged shore extending all the way, without anchorage or harbour. Inland, the Paps of Caithness, two remarkable sugar-loaf hills, are observed, visible a great way off. The course is E. by N. A little S. of Clythness there is a great rock above water, with several lesser around. There are from 9 to 12 fathoms close to this rock, yet still it is prudent to give it a good berth. Off this ness, the stream diminishing in strength S. of Duncansby, the tide runs only 3 knots on spring, and 1½ on neap tides. It is high water about 11¼ᵇ. Lat 58° 20′ N., long. 3° 8′ W.

COAL SHOAL or ROCK, N. side of Anglesea island, having on it 2 fathoms water. It is 2 miles N. of the West Mouse, 2½ miles E. ½ S. from the Skerries light, and 1½ miles from the nearest shore. It dries at very low water spring tides, but with common tides has 1 or 2 feet over it. A black buoy, with a staff and ball, are placed a cable's length N. of the rock. To sail on its N. side the Skerries lighthouse must be kept W. ½ S.; to sail on its E. side, the Telegraph, near the N. Stack at Holyhead, must touch Carnel head, S.W. by W. ½ W. In sailing to the W. of it, stand no nearer than with Holyhead West Windmill just touching the extreme of Carnel point, S.W. ½ W. Point Lynas open of Llanillan head, S.E. ¾ E., also leads to the N. of the rock, one-third of a mile distant; and Llanbadrig church, one-third of a point open of Wylfa head, bearing S.E. by E., leads to the S. of the Coal rock, and to the N. of a gravel bank of 3½ fathoms, which lies nearly three-quarters of a mile N. by W. ¾ W. from Kemlyn point. The two beacons on the shore, and the West Mouse in one, bearing S.W., strike the rock.

COBECA RIVER and HARBOUR, W. Indies, Island of Porto Rico; it is W. from the principal port of the island.

COBEQUIT, or COLCHESTER RIVER, Nova Scotia, North America. This river has its source within 20 miles of Tatamogouche, on the N.E. coast; thence runs S., then S.W. and W. into the E. end of the Basin of Minas. There is a small bank at its mouth, and a fair channel on each side, which vessels of 60 tons may pass and go 40 miles up the river.

COBHAM ISLAND, Hudson's Bay, N. America. Its two extremities bear N. by E., and E. by N., in lat. 63° N.; long. E. from Churchill 3° 50′. Captain Middleton took this island to be the Brook Cobham of Fox.

COBRA or the RIO COBRE, or SNAKE RIVER, S. coast of Africa, 6 leagues E. from Cape Apollonia. Here are several towns, among others Axim; great supplies of gold are sometimes had here. Cape Apollonia is in lat 4° 58′ 45″ W.

COBURN, THE, a knoll of shallow water in the inner channel in the Downs; of the same character as the Crab and the Caldron.

COCAGNA ISLAND, Nova Scotia, off the mouth of a small river, 6 or 7 leagues N.W. by N. from St. Claude's island.

COCHE ISLAND, S. America, coast of Colombia. It stands in the channel formed by the Island of Margaritta and the main. Here are two islands, of which Coche island is the E., named El Coche by the natives; the other island on the W. is called Cuagua. Coche island is low, surrounded by a rocky shoal, and a reef extending from N.W. to S.E., about 1½ miles, forming two channels, that on the N. with the Island Margaritta, in its narrowest part 2 miles in width, and that on the S. with the main land of nearly the same breadth in its narrowest part. Both channels are clear, and the bottom good, while a ship may ride in them as securely as in a harbour. Cubagua or Cuagua island is less than the other, lying nearly E. and W. From the E. point a shoal runs out, as well as a reef, for nearly a mile. The N. and S. sides are clean. On the W. a rocky shoal extends one-third of a mile from the shore; like its brother island this forms two channels, one N. of Margaritta, the other S. with the main land; both are free. In the narrowest part of the N. channel, which is between the shoal and the reef, stretching off from the E. point of Cubagua and the shoal stretching off from Punta de Mangles of Margaritta, the breadth is 3 miles. It was here that the sloop-of-war Sapphire, on the 29th of April, 1821, struck on a shoal of 15 feet water, the Morro of Chacopata then bearing E.S.E.

¼ E., by compass, distance 2⅔ miles, and the W. end of Caribes island, about S. ¼ E. 2 miles. Lat. 10° 57′ N., long. 63° 10′ W.

COCHINOS BAY, Island of Cuba, S. side.

COCHINOS ISLANDS, off the Mosquito coast, N. America, two large islands, clean on the N. side, but foul on the S., 23 miles from the Island of Utilla.

COCK SCAR, or more correctly KOKSKAR, in the Bay of Narva, Gulf of Finland, a large bare rock, on which there is a lighthouse in lat 59° 42″ N., long. 25° 1′ 42″ E. It is 99 feet above the level of the sea, and visible 15 miles. This rock is 23½ miles from Ekholm, which lies from it E. by S.

COCKLE SANDS, and BARBER SANDS, Yarmouth Roads, on the W. side of the Cockle Gat, and the Scroby sand, on which are placed two white and one red buoys. The channel was three-quarters of a mile wide in 1846, but changes. The Cockle is N. of the Barber or Barbel, running out in that direction 2 miles, being in fact a continuation of it. On its S. part it has 9 feet of water, but to the N. narrows and deepens to 3 or 4 fathoms. There are generally three black buoys on the E. edge of the Cockle sand. A light-vessel is moored in the Cockle Gat way, in 8 fathoms at low water, ¼ a mile S. W. by W. ¼ W. Its position was changed to the present in 1843, and is liable to fresh alteration; at present it has not occasioned any alteration in the courses of approaching her from the N. or S, but great caution is necessary in all the navigation in that vicinity. A white buoy is on the W. spit of the Cockle sand, in 3 fathoms.

COCKLE SHELL BAY, W. Indies, Island of St. Kitts, at the S.E. extremity, between Majors and Musquito bays, in lat. 17° 22′ N., long. 62° 22′ W., opposite to the Isle of Nevis. The ground is rocky, but there are 2, 3, and 4 fathoms in the bay.

COCKERMOUTH, England, Cumberland co., situated between the rivers Derwent and Cocker, N. of Whitehaven. It has a harbour for vessels of some burthen, above Workington. Lat 54° 32′ N., long. 3° 20′ W.

COCKBURNPATH, Scotland, near Berwick, a fishing town, 8 miles S.E. of Dunbar.

COCKSPUR ISLAND, N. America, Savannah river just within; there is good anchorage on the S. side, in 4 and 4½ fathoms good ground.

COCOA NUT BAY, near Cape Samana, Hayti, West Indies, in the Sound within Point Graplin on the starboard, and Banister bay on the port side. The second anchorage within Point Graplin under the second point is preferable, and this is in Cocoa Nut Bay.

COCOROCUMAS KAYS, off the Mosquito shore, North America, 17 miles E. of the Seal Keys.

COCOS KAY, West Indies, on the S. side of the old Bahama channel, between that and Cuba.

COD BANK, coast of Suffolk, a quarter of a mile N.N.E. from the buoy of the W. Altar, and half a mile from Landguard fort, having 9 and 12 feet water at low water.

COD ROY ISLAND and ROAD, Newfoundland. Lat. of S. side, 47° 52′ 36″ N., long. 59° 26′ 42″ W.

COD, CAPE, North America, United States, Massachusets. Truro or Cape Cod lighthouse stands in lat. 42° 44′ N., long. 70° 4′ W. This cape curves inward, is 65 miles long and from 1 to 2 broad, containing the co. of Barnstaple.

CODA CAVALLO, CAPE, Sardinia island, lat. 40° 19′ 10″ N.

CODDLE'S HARBOUR, Nova Scotia, 3½ leagues from Berry head, only fit for small craft.

CODERA, CAPE, a round morro or hill on the coast of Caraccas, South America, on the N. of which a tongue of low ground extends to the distance of 1 mile, but so clean that half a cable's length off there are 9 fathoms water. This tongue of land forms a very fine anchorage to the W. called Puerto Corsarios, or the Privateer harbour. The high mountains of Caraccas are visible from Cape Codera, extending W. many leagues. This cape is in lat 10° 36′ N., long. 66° 4′ W.

COD'S HEAD POINT, Ireland, the second point of Kilmore bay on the E. side, within 2 leagues of the bay of Killarish; there is from 40 to 45 fathoms near this head.

COETZA ISLAND, South America, off the main, near Margaritta island, in lat. 11° N.

COFFIN ISLAND, near Liverpool harbour, Nova Scotia, North America. Here, in lat. 44° 2′ N., long. 64° 41′ W., stands a lighthouse, the tower painted red and white horizontally, 90 feet high, having a revolving light.

COFFIN ISLAND, River St. Lawrence, North America, one of the Magdalen islands; N.E. point is in lat. 47° 37′ 30″ N., long. 61° 26′ W.

COFRE DE PEROTE, North America, Mexico, a mountain seen on making the coast near Vera Cruz. It is distant from the nearest part of the coast 13 leagues. It is 13,992 feet high, and in lat. 19° 29′ N., long. 97° 7′ W.

COHANZY, North America, United States, a small river of New Jersey, sometimes called Cesaria. It rises in Salem co., New Jersey, and running through Cumberland co. falls into the Delaware, opposite the end of Bombay hook. It is navigable for vessels of 100 tons to Bridgetown, 20 miles from its mouth.

COHASSET ROCKS, North America, United States, 1½ mile from the land S. of Boston harbour. There is a passage within used by coasters.

COIMBRA SHOAL, Africa, W. coast, between Cape Roxo and Cacheo, from whence there is a reef to the N. bank, a sand running E. by N., and only discoverable at falling tide. The breakings begin upon it before the water is one-third fallen.

COJUMAR RIVER, Island of Cuba, West Indies, on the N. side, E. from the Havanna.

COKE'S BAY, Lancaster co., at the mouth of the River Coker, which falls into the Irish sea, on the S. side of the village so named. Much of it is dry at low water.

COL ISLAND, Scotland, W. isles, 13 miles long and 3 broad, and about 11 miles from Mull. Lat. 56° 42′ N., long. 6° 20′ W. It stands nearly in a line N. of the Skerries, in Ireland.

COLBERG, Baltic sea, 17 leagues E. from the N. deep, at the entrance of Cammin sound: a pilot is always taken to enter. It is 20 leagues N.E. from Stettin; lat. 54° 12′ N., long. 15° 50′ E.

COLCHESTER, Essex co., England, a town seated on the River Coln, which is navigable for small vessels to Hithe, a mile from the town, to which vessels of burthen come up. There are remains of a castle here, and it has a custom-house at Hithe. It is remarkable for its oysters. Pop. 16,167. Lat. 51° 50′ N., long. 58′ E.

COLD HARBOUR POINT. See *Thames*.

COLDING, Denmark, North Jutland, once the residence of the Danish kings. It is situated on the Thueths, at its entrance into a bay in the Little belt. The harbour is much encumbered, and the trade much diminished. Lat. 55° 33′ N., long. 9° 25′ E.

COLERAINE, Ireland, a town on the River Bann, 5 miles from its entrance into the sea; it has a good salmon fishery. Pop. 5,764.

COLLINS' BAY, North America, E. coast of Newfoundland, at the bottom of Consumption bay. It is too shoaly for anchorage.

COLINSA and ORONSA, ISLANDS OF, Scotland, W. coast. Oronsa lies E.N.E. 14 miles from Tonvore head, in Isla, and 7 miles N.W. from the entrance of Loch Tarbert, in Jura island. Off its S. side there are several rocks, one of which extends W.S.W. from Gudimel, a small island on the S.E. of Oronsa, called the Coryderna, dry at half ebb. Colinsa isle is to be avoided; there is no anchorage on the W., and very indifferent on the E. side, but there are numerous rocks and shoals around it. It is high water in the Sound of Isla and on the shore of Colinsa island at 3ʰ 30ᵐ.

COLLIOURE, France, department of the Pyrenees Orientales, a port N.W. of Port Vendre 2½ miles, with a pop. of 3,240. It is strongly defended, but is an old ill-built place, situated in a species of bay at the foot of rising ground, crowned with a fort. There is a castle to the W. standing on a rocky point. Off the starboard entrance to Collioure there is an island with a chapel upon it, and a passage for boats between the island and point. The port is safest with W. winds, riding in the middle before the town, in 3, 4, and 5 fathoms, but it is not a good anchorage, as the E. wind sends in a heavy sea, and therefore only small craft visit it. To the N. is the town of Perpignan, and not far distant is Cape Leucate, in lat. 42° 55′ 30″ N., long. 3° 5′ 10″ E.

COLLISTON, Scotland, a small fishing town, 2 miles E.N.E. from Newburgh, in the Bay of Aberdeen. A little to the W. of Colliston the shore is rocky, and so continues to what are called Cruden scars, 4 miles from it E.N.E.

COLNICK, Scotland, E. coast, a small fishing place without Peterhead, from which it is 2 leagues distance along shore N. to Ratterhead. A ledge of rocks runs up here E.N.E. a mile into the sea. To keep clear of it, the land must not be approached nearer than 12 or 13 fathoms.

COLOMBIA, South America, a republic divided into provinces, bounded on the N. by the Caribbean sea, E. by Guyana, S. by Brazil and Peru, and W. by the Pacific ocean. The maritime provinces of this republic on the E. are Cumana opposite Trinidad; Barcelona upon the W. of Cumana, S. of the Gulf of Mexico, on the Caribbean sea, opposite the Island of Tortuga; next Caraccas; W. of that, the provinces of Coro, Maracaybo, Guayra, Santa Marta, Cartagena, Darien, and Porto Bello. In these provinces there are numerous ports, or havens, the principal of which are, going W. from Cumana, on the Gulf of Paria, Cariaco, Cumana, Barcelona, Tocayo, Chuspa, Guayra, Porto Cabello, Port St. Juan, Guadalupe, Maracaybo, Bahia Honda, El Portete, Santa Marta, Pampetai bay in Margarita island, Port Mochima in Cumana, Corsarios road in Caraccas, Port Turiamo, Port Sabanilla, Cartagena, Port Ecosses, Porto Bello, and Port Chagres. This united Republic was formed by Simon Bolivar.

COLON, ISLAND. See *Balearic Islands*.

COLONEL ROCK, Scotland, E. coast, a mile

E. by S. ½ E. from Kinnaird's head, having 5½ fathoms of water upon it, and 14 fathoms close by.

COLONNA, CAPE, Greece, Gulf of Egina. This celebrated headland lies in lat. 37° 39′ 15″ N., long. 24° 1′ 34″ E., the N.E. boundary of the Gulf. On its summit are the ruins of an ancient temple, thought to have been dedicated to Minerva, of which 16 massy columns yet remain, elevated 200 feet above the sea, and visible at a great distance. There are caverns under the cape resorted to by pirates. The coast turns N.E. from Colonna, and on the E. is fronted by the Isle of Macronisi, or Long island, the S. end of which bears from Cape Colonna E. ½ S., distant 4 miles. It extends N. by E. ½ E. about 7 miles, and is 4 in breadth, a low island, with the ruins of an ancient city inhabited by a few shepherds only. Sunken rocks lie off its N. end.

COLONNA, PORT, or MEGALA CHORA, a haven of Samos, one of the Greek islands, sheltered on the N. by high mountains, on the E. by a headland, S. by the ruins of a mole, and having on the W. an extensive plain, on which stands amid ruins one column of a temple of Juno. The Kerti mountain in this island is in lat. 37° 43′ 44″ N., long. 26° 31′ 21″ E.

COLORADOS REEFS, Cuba, to the N. of the bank that extends N. from Cape St. Antonio, near to which bank lie their S.W. extremity, in a S.E. direction, and having between them and the shoals next the main a channel called the Canal of Guaniguanico, continued yet further to the N.E. by the banks of Isabella. The S.W. termination of the reefs, or Drowned kays, lies about lat. 22° 18′ N., and long. 84° 45′ W., and extends from thence about 35 miles. The channel within the Colorados reefs is dangerous, only admitting vessels of 11 or 12 feet draught, and requiring great knowledge and experience in the navigation. The most W., called Black key, alone shows itself above water, and looks like the hull of a ship, seen 4 or 5 miles off. The other Colorados, or Red kays, are only seen when the weather is calm, have only 2 feet of water over them; in less than a mile to the W. 100 fathoms, and soon after no bottom. The Colorados are, in fact, an enormous range of reefs and breakers, extending nearly to Bahia Honda, steep-to, with currents setting strong over them, commencing at Cazuelos point, Cape San Antonio. The dangers near them are great, and giving them a very wide berth gets a vessel too far to leeward. Cazuelos point may be rounded within 2 miles. Then steer N.W. by N. 10 miles; N. 15 miles; N. by E. 20; N.N.E. 10; and N.N.E. until a parallel is gained, secure from danger. The current on this course sets S. and W. from 12 to 20 miles in 24 hours, without altering until E. of Bahia Honda. A shoal, said to have been seen in 1797, and to lie a little N.W. of Cape San Antonio, about long. 85° W., and lat. 22° 9′ N., and that of Sancho Pardo, both of disputed existence, are the only dangers reported to exist in the course recommended by seamen of experience.

COLOURIS, or KOULOURI, Greece, an island 9 miles long, and 2½ broad. It is divided from the territory of Athens by a narrow and crooked channel, but its W. side forms a fine harbour, extending E. and W. for 5 miles. There are 18 fathoms water and excellent ground, with the town of Colouris bearing W. ½ N., the heights over Athens E., and the bulk of Colouris island to the

II.

I

S. On the N. side lies the Bay of Eleusis, in the strait leading to which the water shoals from 20 to 2½ fathoms, and deepens again inward. On the W. side of the entrance of this strait, 2 miles from the Piræus, is the Island Psyttalia, which has shelter in all winds, in 18 and 20 fathoms, hard ground. Colouris town was the ancient Salamis. It stands on the N. shore at the bottom of the gulf. Pop. 5,000. N.W. of Colouris is Megara, whence the coast runs W.S.W. for above 5 leagues to the Bay of Kenchreas.

COLSA ISLAND. See *Shetland Isles.*

COLSON'S POINT, coast of Colombia, North America, near the Rio Dolce, N.E. end, lat. 17° 4′ 15″ N., long. 88° 15′ W.

COLUMBARA, W. coast of Sicily, an islet, having upon its inner end a lighthouse, in lat. 38° 1′ 53″ N., long. 12° 30′ 18″ E., off the N.W. point of Trapani, called Sigia. There is also a battery upon the island protecting that harbour.

COLUMBIA, DISTRICT, North America, United States, a province of considerable importance, of which Washington is the capital, and within 6 miles of which vessels of the largest class can ascend the Potomac. Alexandria is the principal seat of its commerce. The Navy Yard of the United States is formed at the junction of the E. branch with the Potomac at Washington. It covers 27 acres, and is in every respect complete for all purposes of naval construction. It is in lat. 38° 32′ 54″ N., long. 77° 1′ 48″ W.

COLUMBINE SHOALS, Gulf of St. Lawrence, 2½ miles from Old Harry head, S.S.W. ¼ W., and S.E. by S. 2 miles from the E. end of the cliffs; the outermost has only 3 fathoms at low water. These shoals are very dangerous, and much in the way of vessels hauling round the E. point of the Magdalen islands with N. winds. To clear the E. side, the whole of the high N.E. cape must be kept well open to the E. of Old Harry. There are no good marks for clearing the W. side, or for leading clear outside. In this case the only guide is not to bring the E. point to bear to the E. of N.E., and for the former not to bring the W. end of Coffin island to bear to the W. of N.W. ¼ N.

COLUMBRETES, Spain, E. coast, a cluster of rocky islets, 14 in number, of which the largest lies 29 miles E.S.E. ¼ E. from Cape Oropesa, and has a kind of bay on the N.E. side, where small vessels sometimes anchor. S.S.E. of the largest islet the points of rocks are seen for 2 miles above water. These rocks are in lat. 39° 56′ N., long. 42′ 53″ E.

COLUMBUS, North America, United States, the capital of Lowndes co., Mississippi, on the E. bank of the Tombigbeer iver, 120 feet above the water at the head of the steam-boat navigation. Pop. 4,000.

COLUMBUS KAY, off the shore of Yucatan, North America, about 8 miles from Tobacco Kay.

COMABINDO, coast of Africa, on Gallinas river, lat. 7° 0′ N., a slave factory station, now destroyed.

COMACCIO, Italy, Gulf of Venice, at the most S. of the mouths of the Po, 5 leagues N.W. by N. from Ravenna. Port Magnavacca runs up to this place, situated in a marshy valley, and having a pop. of 6,000, principally employed in the adjacent salt-works. The current of the Po is so violent here that it is seldom prudent to approach it without a pilot.

COMAIGUA, otherwise VALLADOLID, North America, in Mexico, a town 280 miles E. of Guatimala, on the Gulf of Mexico. Lat. 14° 35′ N., long. 88° 20′ W.

COMANAGOTTA, South America, coast of Colombia, 10 miles W. of Cumana. Lat 10° 6′ N., long. 64° 21′ W. The shore here is flat and low.

COMANOE ISLAND, West Indies, lat. 18° 25′ N., long. 63° 4′ 30″ W.

COMASSOON, African coast, on the Gallinas, or Gallinhas river, lat. 7° 0′ 1″ N., long. 11° 38′ 5″ W. It was a noted slave establishment, which is now destroyed.

COMB MARTIN, Devonshire, a small town on the N. coast of that co., 7 miles from Ilfracombe, and 10 from Barnstaple, on the Severn sea. Lat. 51° 13′ N., long. 4° 2′ W. It has a mine of lead and silver.

COMEGRA, Africa, N. coast, 5 leagues S.S.E. from Cape Sasa.

COMET, CAPE, West Indies, Caycos isles, the N.E. point being in lat. 21° 42′ 50″ N., long. 71° 27′ 38″ W.

COMFORT, CAPE, North America, W. of Cape King Charles, the point of a large island facing Davis's straits, in lat. about 65° N.

COMFORT POINT, North America, United States, the S.E. part of Elizabeth city, co. of Virginia, formed by James's river, at its mouth in Chesapeak bay, 19 miles W. by N. of Cape Henry.

COMFORT POINT, in James's bay, the S. part of Hudson's bay: it lies S.E. from Charlton island, in lat. 64° 55′ N., long. 82° 30′ W.

COMINO, CAPE, Sardinia, E. coast, lat. 40° 31′ 10″ N.

COMMENDA, GREAT and LITTLE. S. coast of Africa, one English, the other a Dutch fort, and both between Cape Three Points, and Cape Coast Castle, E. of the Boosempra or St. John river, navigable for 20 miles. At the angle of the river with the sea, and between that and the Commendas, is a small round mount called Gold Hill, with a creek on its E. side. The coast is somewhat foul near the land, and should not be approached beyond 9 fathoms, where the anchorage is good, with a sandy bottom. The British fort bore N.W.; the Dutch settlement 8 miles E.N.E. They are not now used. Lat. 4° 54′ N., long. 34′ W.

COMMENTAWANA BAY, West Indies, on the N. coast of St. Vincent's island.

COMMEWINE RIVER, Dutch Guyana, South America, a branch of the Surinam, running E. for 16 miles, with a depth of 3 or 4 fathoms at high water, the tide rising 12 feet. The banks of this river are well cultivated.

COMPASS VARIATION. See *Variation of the Compass.*

COMPECHE, or CAMPEACHY BANK, N. of the peninsula of Yucatan. This bank is a vast shoal, extending from the N. coast of Yucatan, almost to the 24th degree of lat., from the coast of Compeche, as far as the meridian of the River Chiltepec. The depth and nature of the soundings are very irregular, so as to render it not possible to ascertain the situation by the lead. Yet the soundings towards the shore, from the depth of 18 fathoms, are regular enough for a safe navigation, since that depth being ascertained at 10 or 12 leagues from the coast, it is clear that it follows the direction of the land, as far as N.W. of Point Piedras, where it suddenly decreases 2 fathoms. The same regularity is observed in all the soundings, from the depth of 18 fathoms to 4, and the same decrease will be met at the N.W. of Point Piedras, caused perhaps by the spit or ledge of rocks that stretches off from that point, since the soundings are always upon flat stones. From 4 fathoms to the shore, on all the coast between the Cuyo at the River Lagartas and the Vigia or Lookout of Chuburna, there are various rocks and shoals dangerous to navigation, with a bottom from 18 fathoms to the shore, marked by no degree of regularity. It is a remarkable fact that the E. edge of the Compeche soundings is an excellent corrector of a vessel's log. It runs nearly N. and S., and all who strike its edge may consider themselves in 86° 17′ long. W. from Greenwich.

COMPONEE RIVER, Africa, situated to the N. of the N. mouth of the River Nunez about 10 miles. The mouth of the Componée is among a cluster of islets, and was discovered and ascended 12 miles by Captain Belcher, R.N., when it was found deeper, swifter, and promising larger branches than the Nunez. It flowed from the E. and showed extensive arms leading N. and W.

COMPOON'S BAY, West Indies, at the N.W. end of the Island of St. Kitts.

COMPOS, CAPE, Majorca, 3½ miles from Cape Salinas, the S. point of Majorca island.

COMPTESSES, LES, France, near Cape Frehel, rocks so called in lat. 48° 38′ 58″ N., long. 2° 34′ 23″ W., about 3 miles from Point Plenent.

COMUS SHOAL, West Indies, a shoal with from 3 to 10 fathoms water, 3 miles distant from land. It was found by the Comus ship-of-war in 1840, on her passage from St. Martha to Cartagena, with from 3 to 10 fathoms water, Cape Ajuja bearing N.E. ¼ N., extreme of land, W.S.W.

CONCARNEAU, France, department of Finisterre, a little port containing 1,510 inhabitants, 5 leagues S.E. of Quimper, situated on a small island at the bottom of a bay. It employs about 400 fishing vessels. It is high water here at 3ʰ. From Quimper here there are many dangerous rocks, as well as at the entrance of the port. The marks to enter are, Beuzec church and Concarneau in one, anchoring before Concarneau in 4 and 5 fathoms. Fort Cabellon defends the starboard of the entrance, and there are a signal post and battery on the W. point, called Begmeil point. The dangers around are numerous.

CONCEDEDO, or POINT DELGADA, North America, the E. side of the Bay of Compeche, W. from Cape Catoche. Lat. 19° 52′ N., long. 96° 24′ W.

CONCEPTION BAY, North America, Newfoundland, on the E. coast. It contains several coves, and on the E. side is the little Island of Belleisle. It is the next deep bay S.E. of Trinity bay, and the port N. from Cape Race, and is sometimes called Consumption bay. Lat. 47° 40′ N., long. 52° 40′ W.

CONCEPTION, PORT, North America, on the main 90 miles W. of Panama, with a harbour on the River Veragua, in lat. 8° 52′ N., long. 81° 51′ W.

CONCHEES, LES, ISLETS, Petite and Grande, France, N. coast, at the entrance to St. Malo. One of the passages into that port lies between these islet rocks, called le Chanal de la Petite Conchée, to the W. of which there are several rocks and islets, some of which are from 20 to 30 feet above low water.

CONCORD ISLAND, North America, 17 miles W.N.W. of Boston.

CONDA, or VILLA DO CONDE, Portugal, the bar-haven of the little River Ave. It has rocks scattered about its entrance, but vessels may sail among them on all sides. The narrowest channel is on the N. side, having from 5 to 6 fathoms. A bank further in crosses the haven, with only 2 fathoms at high water, but within this the haven is 5 or 6 fathoms deep; and round the mouth of the river there are 9 or 10 fathoms. A pilot is always necessary. A bank lies off this place, discovered by Captain Glascock, in the Orestes, Villa do Conde, bearing E. ½ N.; S. part E. ½ S. Offshore 3½ miles. Lat. 41° 21′ N., long. 8° 36′ 42″ W.

CONDOULIERE ROAD, France, S. coast, on the S. side of the peninsula, forming the sides of the two roads of Toulon.

CONEJERA ISLAND. See *Balearic Isles*.

CONEY ARM HEAD, Newfoundland, lat. 49° 57′ 30″ N., long. 56° 47′ 48″ W.

CONFLICT REEF, African coast, between the Alcatraz and Rio Nunez. Lat. 10° 30′ N., long. 15° 11′ W. Its W. edge is 14 miles S.E. from Alcatraz, and it is from 3 to 4 miles broad. Two other rocky banks to the S. come within the distance of 8 miles. The S. point of the last is in 10° 20′ N., and it has a depth of 11 to 16 fathoms. The Rio Nunez bears from this spot E.N.E. ¼ E. 10 leagues.

CONGER ROCK, West Indies, Island of Barbadoes, on the N.E. side, W. of Consets bay.

CONGIANUS, GULF OF, Isle of Sardinia, N.W. by N. from Cape Figari, 9 miles from which is Cape Libano, the two headlands of the bay. The ports of Marinella and of Congianus are in this bay, places of shelter for small vessels. At the N. entrance are several islands, among them Mortorio and Soffa, the former in lat. 41° 4′ 42″ N., long. 9° 36′ 5″ E. Here too are the Libano islands and the Isle of Biscie, in lat. 41° 9′ 45″ N., with rocks off its E. end.

CONGIE CHANNEL, France, N. of St. Malo, and S. and S.E. between the Mewstone rock, and other rocks to the E.

CONGREHOY, PEAK OF, Mosquito coast, North America, seen from the sea. Lat. 15° 38′ N., long. 86° 55′ W. This peak is 7,555 feet high.

CONGSBACK HARBOUR, Sweden, S. of Wingo sound, about 3 leagues; the Island of Mael-sound lies close to it.

CONIES, Cornwall, a ledge of rocks near the entrance of Fowey harbour. They are cleared by keeping Deadman's point open without the Windhead rock.

CONIL TOWER and SHOAL, Spain, W. coast, off Cape Trafalgar. There is a small river here admitting coasting vessels at high water. The shoal of Conil lies 2 miles from Castilobo tower, which is square, and 5 miles from Cape Trafalgar. The shoal is S.W. by W. ½ W. from the tower. Lat. about 36° 16′ N., long. 6° 18′ W.

CONINGBEG ROCK. See *Saltees*.

CONINGMORE ROCK, Ireland, S.W. coast; it is always above water, and lies 1½ miles S.S.W. ¾ W. from the S. point of Great Saltee island; in lat. 52° 4′ 45″ N., long. 6° 37′ 49″ W.

CONNAIGRE SHOAL, Newfoundland, in lat. 47° 23′ 57″ N., long. 55° 57′ 19″ W.

CONNANICUT ISLAND, North America, United States, Newport co., Rhode island, 3 miles W. of Newport.

I 2

CONNECTICUT RIVER, North America, United States, the most considerable stream in the E. part of the Union. It rises in the high lands, separating Vermont and New Hampshire from Lower Canada. There is a bar of sand at its mouth with only 10 feet water, and it has shoals which have only six feet water, where the tide does not rise more than 8 inches. This river is navigable to Hertford city, 50 miles from its mouth, and 200 above that place for boats. The vessels employed in trade, foreign to the river, are brigs of from 60 to 190 tons.

CONNECTICUT RIVER, Long island, United States of America, falling into a bay on the S. side of the island, 2 miles S. of Rockonhama pond.

CONQUET, France, a small port on the most W. point of France, about 3 leagues W. of Brest, in lat. 48° 22′ N., long. 4° 42′ W. To sail into Conquet haven go round within a cable's length of Vinotierre point until the haven be open. Avoid the sunk rocks called the Finisterres by running along close inshore until the haven be entered. The harbour is dry, and protected by small forts on each side of the entrance. Conquet point is a mark for sailing round St. Matthew's point towards Brest harbour.

CONSEIL ROCKS, France, round the W. side of the W. point of the bay or road of Camaret; on the E. side of the entrance into a bay which runs in S.E. towards the bottom of that bay.

CONSETS BAY, West Indies, Barbadoes, on the N.E. side, the coast of which is very rocky. This bay is 10 miles from Bridgtown, N.E., and forms as deep an inlet as any on the coast.

Conset point is the E. point of the bay of the same name, about 5 miles from the E. point of the island.

CONSTABLE BANK, on the N. coast of North Wales and Cheshire, off the point of Air, and stretching to within 3½ miles of Little Orme's head. The shoalest part has but 1 fathom at low water. It is a long narrow bank, extending N.W. by W. for 7 miles, though the shoal parts do not lie further to the W. of the buoy of the Tail than 6¼ miles. The E. end is about ¾ of a mile wide, whence it narrows going W. To clear this W. shoal ground, bring Penmanmawr Peak over the S.E. Toe of Great Orme, bearing W.S.W. ¼ W.

CONSTABLES, or GUNNERS, South America, N.N.W. of Cape Orange, 15 or 16 leagues, on the coast of French Guyana, and E. of Cayenne, two bare rocks, the outermost shaped like an orange, called the Great Constable, the other nearly level with the water, a mile distant from the larger, towards the main. Cape Orange is in lat. 4° 10′ N.

CONSTANTE REEF, Lat. 37° 56′ 20″ N., long. 34° 4′ 8″ W., not very distant from some sunken rocks, marked in the charts as doubtful, said to have been seen from the Brazilian brig Constante by the master for six hours, the surf nearly disappearing at high water. The brig was towed away from the danger, it being calm at the time. A second sunken rock was seen by Manoel Ferreira, the above master, which had been before reported as seen by Captain Robson, to the N. of Fayal. The sea broke upon it as he passed it to leeward, at the distance of 1 or 2 miles. This second danger was found by chronometer to be in long. 30° 25′ 10″ W. of Greenwich, lat. 38° 26′ 44″ N.

CONSTANTINOPLE, HARBOUR OF, Turkey in Europe. After passing through the sea of Marmora, 60 miles from the Lighthouse island, on which there are the remains of an old Pharos, the harbour of Constantinople, or rather the channel of Constantinople appears, and is continued through the Bosphorus into the Black sea, being about 15 miles long, and running N.N.E. At its entrance there are two lighthouses, one on the starboard side on a hilly point of land, a mile to the S.W. of the ruins of Kavac Serai ; the other on the larboard side, a little S. of the celebrated mosque of St. Sophia, in Constantinople. The strait being entered, here about a mile wide, the city appears on the left, triangular, having a wall on its S. side from 14 to 20 feet high, flanked at intervals with towers, and having six gates. On the W. side is the ancient wall of Theodosius, 3 miles in length, with seven gates. The remainder of the triangle runs along the interior harbour, and forms what is called "the Porte;" this has 13 gates. This port or harbour is not on the side of the sea, but is formed of an inlet running along the N.E. walls of the city, deep enough for the largest vessels, and capable of receiving 1,200 sail. The N.E. point here, at the turn of the harbour to the W. is called Seraglio point, and is the limit of the residence of the Ottoman porte. On the opposite of this harbour, between Seraglio point and Tophana, beyond where the port runs inland as far as to where it is entered by the little rivers of Quiaquidhana and Alibequien, are the suburbs denominated Galata, Pera, and Tophana. These are the principal seats of traffic and bustle. Tophana and Galata are along the harbour side, inhabited by mariners. On the same side is the arsenal, and other public establishments. The depth of water in the port is from 27 to 9 fathoms. The mosque of St. Sophia stands in lat. 41° 0′ 12″ N. long. 28° 59′ 12″ E. See *Bosphorus*.

CONTE, PORT, or PORTO CONTE, Sardinia island, on the W. side, the S.W. point of which is Cape Caccia. It is narrow, turning in at the entrance from S. to N.E., and shoaling from 15 to 5 and 4 fathoms water, with a safe anchorage. The opening of the bay to the port of Galera is round its E. point.

CONTOY ISLAND, N. of Cancun island, on the coast of Yucatan, North America ; it is N. by W. ¼ W. 12 miles from the N. end of Mugere island, and extends N. by W. ½ W. 5 miles. It is very narrow, and a reef of rocks above water runs off from the S. end of the island, S. by E. 6 or 7 miles, and a spit from the N. end, N. by W. 3 miles, with from 2½ to 3½ fathoms upon it. The N. point of this island, in lat. 21° 32′ N., and long. 86° 49′ W., is often mistaken for Cape Catoche, which lies 16 miles W. ¼ N. from it. Eight miles W. of Contoy, on the main, is a large building of stone, in ruins, which has stood for ages, and will probably serve for ages longer as a landmark for this celebrated cape.

CONTRARY HEAD, Isle of Man, a little S. of Peel Castle point, and 4 miles N. of Dalby point.

CONTRERAS ISLANDS, North America, in Bahia Honda, W. of the Gulf of Panama, on the W. coast of Mexico, 4 leagues S.W. from Puebla Nueva. Three miles W. of the Contreras are the Secas, or Dry islands.

CONVENT BAY and GUT, West Indies, at the N.W. extremity of the Island of St. Kitts.

COOG CHURCH, Dutch coast, a well-known mark for sailing out of the Texel. To find the outermost buoy of Keyser's flat, this church bearing E.N.E. easterly, the two beacons on Huysdown being come into one, a vessel will be athwart the outermost buoy.

COOK'S LOAF, Africa, a small round hill on the sea shore, 3 leagues W. from Accra, on the W. coast.

COOLLY POINT, Ireland, E. coast, the S. point of entrance into Carlingford bay, and E. limit of the Bay of Dundalk. Lat. 53° 57′ N., long. 6° 7′ W.

COOMBE POINT, England, Devonshire, S.W. from the mouth of Dartmouth harbour, between which the land is high and steep; to the W., round the point, the land trends away to Stoke Fleming, and circuitously to the Start point.

COOPER'S HILL, E. coast of Florida; 6 miles N. of the New Inlet is the Dry Inlet, or Rio Seco, and 8 leagues to the N., in lat. 26° 42′, is Grenville inlet, admitting no vessel drawing more than 5 feet water; it is distinguished by a high mount of sand and rocks thus named, close to the beach, which the Gulf stream approaches very near.

COOPER'S ISLAND, West Indies. See *Virgin Islands*.

COOPER'S RIVER, North America, United States, in South Carolina, close to the point of land on which Charleston stands.

COOPER'S ROCK, African coast, in the entrance from the S. to the road of the Isles de Los.

COPELAND ISLAND, E. coast of Ireland, E. by N. from Carrickfergus, open with the bay, and 2 miles from the land E. There is a sunken rock on the W. side of it, a quarter of a league N.E. from the Flat point, on which there is not more than 3 feet of water. Between the rock and the island there is a good passage, and the space between the island and the main is very good, having from 7 to 8 fathoms, but the ground is foul near the main. There is a sunken rock a quarter of a mile off from the next point. There are two islets, called the Cross and Mew, off the N. of Copeland, and between these and Copeland, good anchorage in 7 and 8 fathoms. The distance from Carrickfergus is 3 leagues W. by N. ½ N. The Plough rock is 8 miles S.S.E. from it. The North rock is N. by W. or N.N.W. from it 5 or 6 leagues, and the South rock half a league from that. Near the island it is high water at 10½ʰ, spring tides. There is a lighthouse here, called Copeland, or Cross Isle light, a fixed light on a white tower, 131 feet above the sea, seen 5 leagues off. It stands in lat. 54° 42′ N., long. 5° 32′ W.

COPENAME, or CUPENAMA, South America, Surinam, Dutch Guyana, 8 leagues W. of the Surinam; a shallow river, incumbered with rocks.

COPENHAGEN, Denmark, the capital, situated on the Island of Zealand, S. by W. a little W. from Elsineur, the shore all the way lined with a sandy flat, projecting from 1½ to 2½ miles in one or two places, with 4 fathoms on the edge. This city is in lat. 55° 40′ 6″ N., long. 12° 34′ 42″ E. It derives its name from "Kœbing-haven," the "merchant's port." It is the best built city of the North, constructed of brick and stone, and some of the edifices are noble and in the best taste. The citadel is a regular work of five bastions, having a double ditch filled with water, and advanced works. It is 5 miles in circumference, and has a pop. of 100,000. It possesses numerous manufactures of porcelain, woollen, cloths, canvas, and leather. The harbour is crowded with vessels of all nations, and the streets are intersected with canals. There are two royal dockyards, the principal being a complete naval arsenal. The port is one of the best in the Baltic, lying between the Island of Amak and the main land. The entrance, admitting only one vessel at a time, will receive 500 sail. Each ship-of-war has her particular storehouse. Copenhagen has been a free port, and 11,000 vessels and upwards have entered and cleared out in the year. It is 18½ miles S. by W. to the first black buoy of the ground from Elsineur, with generally 8 and 9 fathoms. After discovering the steeples of Copenhagen, and navigating towards the battery of the Three Crowns, where a light is established, 30 feet above the parapet to the N. of the flagstaff, while the navigation is unimpeded by ice, the city may be said to be reached. Before Copenhagen city there lie three sand banks, from E. to W. 7 miles in length, and from N. to S. 10 miles. These are called the Grounds; the first surrounds Amak island, and the most E. surrounds the Saltholm, while between stands the middle bank. The passage W. of the middle is the King's channel; that E. of the middle is the outer or E. gat; and the one where both unite to the S. of the middle is called the Gaspar, or Drogden channel, between Amak and Saltholm. These passages are buoyed from March to November. To sail through the E. gat, a vessel must keep nearer the middle ground than the Saltholm, the latter being so steep, that a vessel may be aground, and have 7 fathoms under the stern, between the first and second buoys. The last buoy is 1¼ mile S. of the first. The channel is very narrow here, and opposite to the buoy three-quarters of a mile, are small shoals, having only 12 feet water. The third buoy, sometimes called the Gaspar, lies on the S. extremity of the middle ground, more than 1½ miles S. by W. from the second buoy, the mark being the two steeples of Copenhagen in one. The two beacons on the cliff near the S. end of Amak, brought to bear S.S.W. ¼ S. will clear every danger through the E. gat, regarding carefully the steepness of the bank. A ship may stand either way in turning, until one of the beacons just reaches the houses at Dragoe, but they must not be brought on with the houses until the second buoy is passed. Dragoe town is on the S.E. point of the Isle of Amak, which bounds the W. side of the Gaspar, or Drogden channel. There is a light-vessel off Dragoe, placed every year, unless when prevented by the ice. In all cases a pilot is needful in the navigation about Copenhagen. The numerous buoys are altered, and marks are continually changed, owing to the differences wrought by the currents out of the Baltic, and the shifting of the sands; nor in any navigation is caution and experience more necessary than in that from the Scaw through the Cattgat, Sound, and Belts, into the Baltic.

COPINSHA. See *Orkney Isles*.

COPPAY ISLAND, Western Isles of Scotland, a mere islet, W. of Toe head, Harris.

COPPERAS BANK, coast of Suffolk, a flat extending along the shore of the main, 1 and 2 miles offshore from the W. rocks to the Buxey.

COPPERWICK, or SOEN WATER, sometimes called DRAM, in the N.W. branch of Christiana sound, Norway.

COQ, France, a sunken rock, E. of St. Matthew's point, at the entrance of Brest harbour, and half a gunshot from the shore. To avoid it, steer E.S.E. from the point, keeping the N. end of Beniquet island so long without the point that the mill on the N. land appears N. of the trees. The

S. end of the Isle of Beniquet upon the point is directly upon the Coq. To sail within it to the N., a vessel must keep along by the N. land from the point so as to shut in the Island of Beniquet behind the great rock which lies off the point, right with the W. point of Bertheaume's bay, till the mill upon the N. land comes N.W. by N., and the trees N.N.W., when the Coq will be passed.

COQUAR PLUMP BAY, Jamaica, S. coast, E. of Devil's point, and W. of Calabash bay; the soundings are from 5 to 10 and 17 fathoms, within 2 leagues of the shore.

COQUAR TREE RIVER, Jamaica, at the E. end of the island, S.E. from Plantain river.

COQUET ISLAND, off Northumberland, in lat. 55° 20′ N., long. 1° 32′ W., a small rocky island, two-thirds of a mile from the main, 8½ miles N. by E. ½ E. from Newbiggen point, and nearly 20 miles N. by E. from Tynemouth lights. There is a lighthouse here, having a bright fixed light, visible from N. by E. ½ E. to S. by W. ½ W. A light, inferior in power, is shown landward in all directions. The lantern is 80 feet above high-water mark. Six buoys of direction for the anchorage have been placed within the island. The safest approach to these is N. of the island, between the N.E. Coquet and Pan Bush buoys, there being only 8 feet in the S. entrance, or between S.W. Coquet and Sand Spit buoys. High water 2ʰ 45ᵐ, rise 12 feet.

CORAL ISLAND, African coast, an islet S.S.E. from the S. end of Tamara isle, one of the Isles de Los, which last lies in lat. 9° 31′ N., long. 13° 40′ 30″ W.

CORBELLE, African coast; this island forms with that of Carasche, a large bay, on the opposite side of the channel from Jatt's island, or to the S. and W. of Bissao.

CORCAN, PORT, Isle of Man, one of the creeks subordinate to Douglas.

CORCUBION, RIVER and BAY OF, Spain, a league W. of Cape Finisterre, once called Corcovia and Seche; the bay is formed by the points of Galera and Cé, distant from each other about a mile. The bay runs in N. about 2 miles, the shores high and free from danger, when past the points, off which are rocks. The town of Corcubion is on the larboard side, and on the starboard is the Fernelo river, and further on, the town of Cé. The place of anchorage is between Prince's and Cardinal's forts, in 10 or 11 fathoms. There is anchorage in 7 or 8 off the beach of Fernelo, but this is not to be recommended, on account of S. winds, which render many places on this shore dangerous. It is high water here, as at most of the river entrances S. of Cape Finisterre, about 3ʰ, the rise 13 feet.

CORCHUNA, or CAPE SACRATIF, Spain, S. coast, in lat. 36° 41′ N., long. 3° 27′ W., high rugged land, with a round tower on its summit, E. of Almunacar town about 16 miles. On the E. of Cape Sacratif is a small cove, called Cala del Chuco, and 2 miles distant is the castle of Corchuna, near the Tower de los Llanas.

CORDOUAN, France, department of the Gironde, and mouth of that river, the finest lighthouse in the world, built by Louis de Foix, in the reign of Henry IV., anno 1611. It was originally 169 feet high, but is now higher, being 206 feet, having a revolving light of the first class; flashes alternately, weak and strong, but the eclipses not total within 3 leagues; range 9 leagues. Lat. 45° 35′ N., long. 1° 10′ W.

CORE BANK, North America, United States, coast of North Carolina. It is above 30 miles long, and scarcely 2 broad. Lat. 34° 22′ to 55′ N., long. 76° 26′ to 50′ W.

CORENTYN RIVER, South America, Guyana, in Surinam. The Nickerie river battery on the E. is in lat. 5° 57′ 30″ N., long. 56° 52′ W.; Mary Hope on the W., lat. 6° 14′ 30″ N., long. 57° 2′ W. At the entrance of this river there are three islands, one above another, and within these it is of considerable breadth.

CORETT, coast of Africa, one of the four Hog islands, off the Rio Grande, and S.W. from Boolam.

CORFU, one of the seven Ionian islands, and the most important; it lies with its dependants Merlera, Fano, and Samotraki, opposite the shores of Albania, and is the most N. of all the islands. Corfu is the seat of the government. It is of irregular form, 17 miles wide at the N., and at the S. part only 2. Its length from Cape Drasti to Cape Bianco is above 35 miles. The surface is varied, generally hilly, and from the sea the aspect is mountainous, the hills rising so high in the air. On its S. part, which appears white, and is therefore called Cape Bianco, there is a lighthouse, which at a distance looks like a sail. The island is subject to earthquakes, and some parts are esteemed unhealthy. The soil is very fertile, producing corn, cotton, wine, honey, wax, olives, figs, oranges, lemons, and currants. It exports besides much fish and salt. The pop. is estimated at 62,000. The island is everywhere well cultivated. Many of the inhabitants are seamen. The island makes on an average 800,000 jars of oil annually. Horses and cattle are imported from the continent; sheep and goats are numerous, bred on the island. The different oils are divided into eating oil, kernel oil, and a species called morgu. The produce of the vines is about 14,000 barrels in the year, but the wine is not carefully made, and is apt to grow acid. Upwards of 15,000 measures of salt are annually exported. Under the Venetians the revenue was only 17,000l. The islands of Fano and Merlera lie off the N. of Corfu island, the first of which is said to be the Isle of Calypso. The chief town, or rather the city, of Corfu, stands on the E. of the island, defended by three fortresses, and having a mole, which shelters small craft, in 2 and 2½ fathoms. Two forts flank the town, and the port is defended by the little Island of Vido. The town is situated within 5 miles of the Albanian coast. The citadel is to the E. of the town, upon a ragged rock, separated from the main by a deep ditch, which admits the sea. The streets are regular, and the suburbs adorned with gardens and vineyards. There is a good road a little more than 3 miles to the E., which leads to a battery, overlooking a strait, that admits the sea into a lake, called the Limne, a great source of malaria. There is a lighthouse upon the citadel of Corfu, situated in lat. 39° 37′ 6″ N., long. 20° 6′ 12″ E. The pop. of the city is about 15,000, partly Greek and partly Italian. On the W. of Corfu island are the two low uninhabited islands of Samondrachi. In making the N. channel of Corfu, vessels coming from the S.W. with the wind to the N. of W. endeavour to make the Island of Fano, passing to the N. of both Fano and Merlera, and then steering E.S.E. or S.E. by E. for the entrance of the channel. They might pass to the S. of Fano as well, upon giving the Island of Samondrachi, or Samotraki, a wide berth, and going either between Fano and Merlera, or bearing in a line with the N.

point of Samotraki, and the N.E. point of Fano, about midway, steering E. by N. to the S. of Merlera, between this isle and the N. point of Corfu; the channel being clear, except of the rocks and shoals about Samondrachi. Cape Drasti, the N.W. point of Corfu, is in lat. 39° 48′ 15″ N., long. 19° 38′ E. There are official instructions for navigating the channel of Corfu issued by the government, which are efficient, and necessary to be in the hands of all who navigate near the shores of the island.

CORINTH, CITY and ISTHMUS OF, on the S. of the Gulf of Lepanto, about 1½ miles from the shore, W. of the isthmus. The town is still of some extent. The castle is strong, and surrounded by a wall 2 miles in circumference. It is situated upon the Acro-Corinthus, a high mountain to the S. The town consists of about 1,300 houses and 4,000 inhabitants. There are some vestiges of antiquity still existing. In ancient time there were two harbours, one of which forms the existing port, where the remains of an ancient pier still remain. The other harbour was in the Gulf of Egina, and is now little visited; it is at present called Kenchreaa. The Gulf of Lepanto is much sheltered. Lat. of Corinth 37° 54′ N., long. 22° 54′ E.

CORISCO ISLAND, S.W. coast of Africa, having 5 miles E. from it a smaller island called Little Corisco. The N.W. point lies in lat. 55′ 54″ N., long. 9° 19′ 15″ E. Six leagues and a half S. from the S. point of the larger island is Cape Esteiras, on the S.W. side of the River Danger, which last is in lat. 1° N.

CORK CITY and HARBOUR, S. coast of Ireland; the city stands upon an island on the Lee river, over which it has a connection with its suburbs, by means of five stone bridges. It is the principal port of Ireland for trade, with the exception of Dublin. Vessels of 120 tons come up to its quays, but larger vessels unload 6 miles below, the city standing 14 miles from the sea, and 150 S.W. from Dublin. Its exports consist principally of provisions of all kinds produced in the country, and its importations of much foreign as well as British produce. The harbour is commodious, called the Cove of Cork. It is 6 miles below the city and well defended, capable of receiving the largest vessels, and the only port in the S. of Ireland, except Bantry bay, admitting ships of the line, which may enter at all times of the day, with a leading wind, by attending to the different marks that lead clear of the Harbour rock and Turbot bank. Cork is a city, a bishoprick, and returns two members to parliament. It has a pop. of 103,000, and stands in lat. 51° 54′ N., long. 8° 28′ W. Roche's lighthouse, at the entrance of Cork harbour, stands in lat. 51° 47′ 20″ N., long. 8° 16′ 32″ W. The light here is fixed, of a clear red colour towards the sea, but towards the harbour it appears bright. The tower on which it is fixed is white, 26 feet in height, and the light, 92 feet above high water, is visible 14 miles seaward, from all points, between S.E. by E. and N. by E. Cork head bears from it S.W. by W. ¼ W., distant 4 miles. Roche point may be approached within a cable's length, after passing the Stag, or Cow and Calf rocks. Dog's Nose, the next point further in, should have a berth of 1½ cables' length given it in passing. The space between Dog's Nose and Roche's point is White bay, where a flat, 2 cables' length, runs off the shore. The Western shore of the entrance may be approached within a cable's length until a vessel reaches Turbot rock,

to the N. of which a spit runs off, having only 4 fathoms upon it, full 3 cables' length from the land. On this the first or outermost white buoy is placed. The Harbour rock with 2½ fathoms, surrounded by a shoal of 4 fathoms to nearly a cable's length, lies 3½ cables' length N.W. ¼ N. from Roche point. This rock lying near the middle of the entrance is the first danger to be avoided upon going in. There are two buoys upon this shoal, lying E. by S. ½ S., and W. by N. ½ N. of each other, distant 1½ cables' length; the E. buoy white with a red rim, the W. red with a black. The Turbot rock has 19 feet, the least water over it, and lies one-third of a mile N.N.E. from the Harbour rock; it extends a cable's length from E. to W., having 4 fathoms on its outer edges. There are two buoys on this shoal a cable's length apart, bearing W.N.W. ¼ N., and E.S.E. ½ E. of each other. There is a good channel with 6 fathoms on either side of the Turbot rock, and between that and the Harbour rock 8 or 9 fathoms. Between the Turbot rock and the town of Cove there are 9 white buoys, marking the W. bounds of the harbour, to be left on the port-hand going in. Between Fort Carlisle and the Bar rock, at the entrance of the passage to Cove, there are six black buoys, all to be left on the starboard going in. Between the third and fourth of these, and nearly E. from Spike island, is the Man-of-War's road, where there are 8 or 10 fathoms in the channel. The Bar rock, at the entrance of the passage to Cove, has 16 feet water, and on the N. side of this rock the seventh white buoy is placed. Large vessels, at low water, should pass to the E. of the white buoy. No. 7, and after rounding it, at the distance of a quarter of a cable, haul suddenly round to the westward. The best place for large ships to anchor in is the outer road between Fort Camden and the buoy of the spit, to proceed till Cove church is just shut in with the eastern angle of the new citadel on Spike island, and take a position, in from 12 to 7 fathoms. Merchant-ships may ride off Cove in smoother water and less tide : they may go also further up the harbour of Passage, and ride anywhere between the first houses and Ronan's point. The tide runs 2 and 3 knots off the anchorage at Cove. It is high water at full and change at 5ʰ 3ᵐ; the rise is from 10 to 12 feet. There is a little creek within the entrance of Cork harbour on the W. opposite the Dog's Nose, where small vessels may ride. Haulbowline island is in lat. 51° 50′ 24″ N., long. 8° 19′ W.

CORK KNOT, a rocky shoal, about 1½ miles from the Cork ledge, apparently joined to it. It bears N.E. from the buoy of the W. rocks, distant 4 miles, with 19 or 20 feet over it.

CORK SAND, in the passage to Harwich towards Orfordness, lies the E. end, N. by W. from the E. part of the W. rocks, distant about 3½ miles; thence extending S.W. by W. about 1½ miles : it is narrow and a large part dry at low water. On the N. side of the Cork ledge a light-vessel is stationed, carrying a bright revolving light.

CORMACHITI, CAPE, Cyprus, N. side of the island, S.E. of Cape Anamour, in Caramania.

CORMANTINE, Africa, windward coast, once the most considerable establishment of the English in the Gold country. Tantumquerry point is 5 leagues to the E., and the coast is bold. A chain of the Cormantine mountains approaches the fort. Lat. of flagstaff 5° 10′ 30″ N., long. 1° 5′ 36″ W.

CORNACCIA, CAPE, Italy, Naples, the N. point of the island of Procida.

CORNE, or HORN, ISLAND, America, 70 leagues E. of the mouth of the Mississippi, on the Mobile. It is within a league of Isle Dauphin, and the Isle Ronde, or Round, is N. of it. It is 10 miles long, and 1 broad. Lat. 30° 11′ N., long. 88° 32′ W.

CORNER BUOY, Dutch coast; there are three here belonging to the channel of the Fly, Vlie, or Stortelmeck. These are called the Outmost, Middlemost, and Inmost buoys. A tail of sand runs from the shore between the outmost and middlemost, which reaches almost to the buoy, and may be sailed over from without; there the stream breaks through the sand, and scours the tail over to the N. shore. From the innermost buoy a vessel runs by the sand of the Fly, or southward of the buoy upon the plat.

CORN ISLANDS, Mosquito shore, North America; these are two in number, bearing from each other N.N.E. ½ E., and S.S.W. ½ W., apart 7 miles. The S. is the largest, being 2 miles from N. to S. The N. island is 1½ miles from N.W. to S.E. The larger has three hills upon it, the central one is highest, and its coasts are foul. Little Corn island is clean on its W. side. There is an anchorage called the Brigantines near the N. island, but in making it the coast must not be approached nearer than 2 miles, or a vessel must not get out of 10 fathoms in approaching it until the middle hill bear E., then make for the island on that course, and anchor at any convenient depth where there are 4½ fathoms, at 2 cables' length from the beach. In the S. part of the S.W. anchorage there are three holes of good water. Lat. 12° 9′ 12″ N., long. 83° 3′ 42″ W., Great island.

CORNWALL, CAPE, at the extremity of the county of Cornwall, 2 leagues N. by E. of the point of the Land's end, from which the coast runs N.E. by E., and E.N.E. 5 leagues to St. Ives. When it bears E.N.E. 3 leagues off, it is a fair point of land; but when due N. and the Longships at N. by W. 3 leagues off, it shows broken, disjointed, and rugged. The Bressons' island is a mile W. from the cape; and the Seven Stones are at W. ½ S. 6 or 7 leagues, which are not above water till half ebb. It is in lat. 50° 4′ 8″ N., and long. 5° 41′ 31″ W.; high water on full and change days at $4\frac{1}{2}^{h}$.

CORNWALLIS, ISLAND, North America, Newfoundland, on the N.E. side of Chebucto harbour, Nova Scotia, near the entrance, a large irregular island with a small one near it, of the S. point, called Red island. The channel between the two islands is wide and deep.

CORO, GULF OF, and TOWN OF SANTA ANNA DA CORO. The town is situated in lat. 11° 24′ N., long. 69° 47′ 50″ W. It stands N. of the Isthmus of Medanos, connecting the Peninsula of Paragua with the main, and N. of the peninsula, but S. of Coro is the gulf, both in the province of the same name. Vela de Coro lies E. of the city with the River Coro between, and the S. end of the Isthmus of Medanos. Vela de Coro lies in lat. 11° 26′ 30″ N., long. 69° 40′ 5″ W.

CORON, in the Morea. See *Koron*.

CORRALES OF REGLA, Spain, near Cape Chipiona and passing it, on which account the shore must have a good berth, a bottom covered with rocks and stones. It is 6 miles from Point Candor, from which a reef runs out all the way to Rota, 1½ miles distant.

CORREBEDO, CAPE, Spain, 24 miles S. of Cape Finisterre. Lat. 42° 35′ N.

CORREJOU, France, a small port of little moment; from Abervrach distant W. about 4 miles.

CORRIENTES, CAPE, island of Cuba, on the S.W. part of the island, about 10 leagues from Cape St. Antonio, which it much resembles and has been often mistaken for by strangers coming from the E. It stands in lat. 21° 45′ 20″ N., and long. 84° 31′ 3″ W. It is remarkable from the E. as a low point, with clumps of trees upon it like buildings, and one much resembling a tower, when it bears W. by N., appearing to be on its extreme point. On approaching the land it is much more dissimilar than when thus seen at a distance, and round Cape Corrientes there is a remarkable bluff headland, of a black rugged appearance, seen as soon as the cape is doubled, which will decide whether a ship has run the length of Cape Antonio, to which the course will then be about W. ½ N. 31 miles.

CORROBANO POINT, South America, British Guyano, at the entrance of the rivers Demarary and Essequibo, on the S. point of the River Demarary, in lat. 6° 49′ 20″ N., long. 58° 11′ 30″ W. It consists of a fixed light, 100 feet above the sea, on an octagonal tower, seen on a clear day about 5 leagues off. There is a signal-staff on the tower. The lighthouse is striped white and red. A light-vessel is also moored off the shore S.W. by S from the lighthouse, 12 miles distant, off the river-bar, in 3½ fathoms, near which light-vessel ships heave to for a pilot. It is high water here at $4^{h}\ 45^{m}$, P.M., full and change; springs rise 9 feet, neaps 3.

CORSEWELL POINT, Scotland, entrance of Loch Ryan, N.W. angle of Wigtonshire. There is a revolving light at this point, 112 feet high; time 2 minutes; clear bright and red alternately.

CORSICA, ISLAND OF, France, in the Mediterranean. It is situated between the 41st and 43rd degree of N. latitude, and extends from N. to S. about 34 leagues, and from 9 to 10 in breadth, forming a great number of gulfs, bays, and capes. The principal gulfs are those of Aliso, St. Florent, Ajaccio, Porto Vecchio, Santa Manza, Sagone, Porto, Valinco, and Vintilegne. The ports are those of Ajaccio, Bastia, Bonifacio, Calvi, Centuri, Isola Rosa, San Fiorenza, Macciano, and Porto Vecchio, the most considerable of all. There are many small islands dependent upon Corsica, among others Capusa, Finochiarola, Giraglia, Gargalo, Sanguinaria, Monachi or Fourmis, Cavallo, Lavessi, and Cervicales. The island forms a department of its own name, covering 484 square leagues, and having 185,079 inhabitants. The island is in many places unhealthy. The interior is covered with mountains, the highest rising to 9,900 feet. Cape Corso is the N. point of the island. Off its extremity is the little island of Giraglia, lat. 43° 2′ N., long. 9° 24′ E., having a watch-tower on its summit, and a passage of 7, 8, and 9 fathoms between it and the cape. Giraglia bears from the lighthouse of Genoa S. by E. ½ E. 85 miles, from Palmaria island, in the Gulf of Spezzia, 65 miles S.S.W. ¾ S., and from Leghorn S.W. ½ W. 51 miles. Cape Bianco, the N.W. point of Corsica, is 2¼ miles from Giraglia, and 11 miles from San Fiorenzo, to which the coast runs S. ½ W., and before reaching which is seen the tower of Fanisale. Capes Martello, del Spano, Riveleta, Gargani, and Omigno, are the principal headlands.

CORSINI, PORT, Italy, E. coast, the port of Ravenna, up to which the river ascends. There are a semaphore and battery at Port Corsini.

CORSO, CAPE, Isle of Corsica, the N. point.

Lat. 43° 0′ 35″ N., long. 9° 22′ 55″ E. It has two points, denominated Cape Blanco and Cape Segri.

CORSOER, Denmark, a town 52 miles from Copenhagen, S.W. on the W. side, off the Island of Zealand, on a small peninsula in the Great Belt channel. It has a good harbour for small vessels. Lat. 55° 20′ N., long. 11° 15′ E.

CORTEEMO ISLAND, W. coast of Africa, 4 miles S.E. of Yellaboa, in the mouth of the River Scarcies. Yellaboa island is in lat. 8° 55′ 42″ N., long. 13° 17′ 45″ W.

CORTEZ, BAY, Cuba, E. of Cape Corrientes, on the S. side of the island, within and E. of Point Piedras.

CORTON SAND, England, coast of Suffolk, joined to the Holm, a large sandy flat, both lying to the E. and N. of the Newcome and Stanford passage. These sands stretch in a direction parallel with the shore full 6 miles, and form the E. boundary of Lowestoff and Corton roads. The whole are carefully buoyed with nine buoys of various designating colours. The chief passage for large ships into Yarmouth roads having always been between the Corton sand on the W. side, and St. Nicholas bank or Kettle Bottom on the E., commonly called St. Nicholas gat.

CORUNNA, Spain, N. coast, a seaport surrounded by a wall and defended by a citadel. It is divided into the Upper and Lower towns. It is an arsenal, and has an antique tower, now made a lighthouse, remarkable for its height and strength. The town stands S.W. from Ferrol, on the opposite side of the bay. The harbour is secure and spacious, built in the form of a crescent, and has a handsome quay. It is the capital of the province, and the Governor-General of Galicia resides there. The castles of Santa Cruz and St. Diego, and the forts of Domideras and St. Antonio defend the port. The lighthouse stands in lat. 43° 23′ 36″ N., long. 8° 19′ 35″ W. It is high water here about 2ʰ 30ᵐ. The light on the Tower of Hercules, as it is called, is a revolving light; the intervals of light and shade are of equal duration, visible 8 leagues. In good weather and a fair wind from the N.E. or N.W. the course in is towards point Seixo Blanco and Mount Mera, until the Castle of St. Diego comes open E. of that of St. Antonio, which will clear the Basurel reef and bank. Past Antonio fort, steering for St. Diego, and between the castles the anchorage may be chosen. With a large ship, St. Antonio should be brought N.E. by E., anchoring in 7 fathoms, ooze and mud. Small vessels may stand in further. With violent winds from N.N.E. or N.N.W., the best entrance is under the lighthouse, between the Basurel bank and the land, keeping the N.E. side of the tower, near enough to see its foundation; past Point Rebaleyra and the Ox rocks; the point is not dangerous at two cables' length; then going S.E., St. Diego castle must be kept on with St. Antonio as above. Between the Point of Seixo Blanco and the above bank the passage is not so good as through the W. channel. There is a battery on Pedrido point, and there runs off it a small reef to the E. The Castle of St. Antonio stands on a rock separated from the land, and has some shoals extending 60 fathoms from its S. side: between the point and castle are the rocks and shoals of Pedrido, partly seen at low water. To the eastward of this reef lies a most dangerous shoal, with only 3 fathoms over it; it is not safe nearer the Pedrido rocks than 2½ or 3 cables' length; but as they bear W.N.W. and N.N.W., it is safe within a stone's throw of them. A cable's length N. of St. Diego

castle is a small shoal, with only one fathom over it, and N. by W. from the same castle, and S.W. from St. Antonio's castle, is another rock with only 3 fathoms. From the former castle towards the interior of the harbour, the whole shore is encumbered with shoals: the best anchorages, therefore, are from the middle of the harbour to the S.W. shore. Between Santa Cruz and Mount Mera is a large bay, with good anchorage, but a bad situation, with N. and N.W. winds. Care must also be taken of the shoal of La Tonin, which, although there are 10 and 11 fathoms over it, always breaks with heavy gales. The marks for this shoal are, the N.W. point of Canaval island in one with Great Priorino cape, and the Porteli rock on with the Mosori chapel. N.W. ¾ W. from Seixo Blanco point, and E.N.E. ¼ E. from the Tower of Hercules, lies the centre of the Basurel and Jacentes banks, extending E.N.E. and W.S.W. about a mile, with from 6 to 17 fathoms. The sea breaks on them with a heavy swell from the N. and N.W. Seixo Blanco point on with the N.E. end of some old walls on the top of a hill adjacent, bearing S.E. by E.; and the Cota rock, on the top of a mountain in one with the Nota rock, on the N. side of the Tower of Hercules, bearing W. by S., are the marks for the Jacentes, in 8 fathoms, rocky ground. For the Cabanes bank, the Dormideras fort must be open to the N. of Hercules tower, or the steeple of St. Francisco on Carboeiro hill, a ship will then be on the S.W. end of the shoal. The current off the coast of Spain here must be carefully attended to.

CORVO ISLAND. See *Azores*.

CORVOEIRO, CAPE, Africa, W. coast. Lat. 21° 46′ 44″ N., long. 16° 56′ 40″ W.

CORYDERVA ROCK, Scotland, Oversay isles, among the Western Isles, at the W. extremity of the Rhinns or Runs of Isla. There is no passage but for boats between this rock and Oversay, in which last island there is a light, on the W. side Isla, bearing N.N.W. ¼ W., 10 miles distant from the Mull of Kinho. This light at a distance appears like a star of the first magnitude, producing a bright flash every 5 seconds, without the intervals of darkness characterising the other revolving lights upon the coast. It is 150 feet above the sea level.

CORYVRECKAN, GULF OF, Scotland, formed by the N. end of Jura, and the Island Scarba, distinguished by a violent breaking of the sea, with whirlpools, sometimes inundating a ship's deck, occasioned by a rapid stream running over a steep rock, near the W. point of Scarba. The waves break, and the current is not less in velocity than 12 or 15 miles an hour, forming eddies. On the Jura side, at ebb and flood, the sea rages and forms whirlpools, but they do not rise so high as on the other side of the gulf, but at slack water all is smooth. Should a vessel be becalmed near the E. entry of the sound, with flood and spring-tide, and should there not be a brisk breeze, it is useless to attempt to pass the Gulf of Coryvreckan, by sailing or towing: the best way is to secure the hatches, and endeavour, by the sails and helm, to steer the vessel through the middle, so that the tide may take her between the most violent breakers, which lie on each side. If the tide should carry her near to the Jura side, it is not advisable to attempt to clear it, but keep so near that the tide may take her between the E. small island and Jura, by which, if the wind be favourable, she may get into a small bay in Jura, opposite the island, where the ground is clean, and the shelter good. To avoid being carried through Coryvreckan, coming from the S. with the flood, keep near Risantru island, and a moderate breeze will be sufficient to pass the Sound of Scarba; or, unless it is about an hour before high water, the tide will do it. In the gulf it is high water at half-after 4.

COS, Greek Archipelago, called also Stanco island. It is of an irregular form, long and narrow, lying E.N.E. and W.S.W., 7 leagues long and 2 broad, at the entrance of the Gulf of Boudroun. The town lies in a bay on the N.W. side, in lat. 36° 53' 11" N., long. 27° 16' 37" E., having a pop. of 9,000. The port is formed of heaped up stones, enclosing a space 300 feet long and 50 broad, into which boats alone venture when the wind is not N. This island is badly cultivated, though very fertile, and the climate is temperate, but fevers and agues are very prevalent in May, June, and July, arising from causes which might be easily removed. The N. extremity is a low sandy point, on which there are several windmills. A flat bank, with 3 fathoms upon it, runs out to the W. A large vessel should not approach the island nearer than 10 or 12 fathoms water, for there are shifting banks in its vicinity. On the E. of the bank, ships working into the Bay of Cos, should keep the black minaret in the town open with the E. angle of the fort, abreast of which is the best anchorage, in 10 or 12 fathoms. The whole of the bay is clean sand. The town is of small extent, but defended by some heavy guns in the fort, three faces of which are to the water. This port is much visited by merchant-vessels, and supplies are secured with facility.

COS, GULF OF, known as that of Boudroun, in Anatolia. This is a deep bay, at the N. part of which is a town of the same appellation, giving the modern name of Boudroun to the gulf. On entering the bay, there are two sunken rocks, small, lying in the midst of the channel. When a boat is anchored on the outer rock, the remarkable round hill upon the Isle of Cos is twice its diameter, open to the left of the extreme point of the W. island off Boudroun bay; and a whitish rocky hummock is just open to the S. of the tongue or low black cliff. This rock has 2½ fathoms on its summit, and 20 fathoms close round it. The inner rock is about a cable and a quarter from the outer, N. ¼ E. It has 2½ fathoms on its shallowest part, and deep water all round. When over this rock, the foot of the round hill on Cos just touches the S.W. point of the bay, and the hummock is shut in to the N. of the point. N. of these rocks, there are others above water, with foul ground from the adjacent point, but no ship has any need to approach the shore so near. Vessels, those dangers past, may anchor outside the harbour, in from 20 to 10 fathoms good bottom. The harbour is small, but well sheltered, and has the remains of piers. There were 3 fathoms water on the W. side. The fortress of Boudroun is well situated on a projecting rock, the walls high and solid, mounting 50 guns. It is supposed to occupy the site of the ancient Halicarnassus. The harbour is visited by Turkish cruizers. The palace stands on the margin of the harbour. There are some ruins in the neighbourhood. The tower stands in lat. 37° 1' 21" N., long. 27° 25' 18" E. There are about 2,000 houses of the Turks, and 110 Greek in the town. A ship-of-war, a tribute to the Turkish navy, is constantly building here, at the expense of the Pachas of Anatolia.

COSCIA DI DONNA, Sardinia island, W. coast, is a rocky islét, surrounded by a reef, 6 leagues W.S.W. of the Gulf of Bosa.

COSMA, CAPE, Anatolia, on the N. side of the Gulf of Boudroun.

COSTA RICA, North America, a province, bounded on the S.E. by Veragua, on the N.E. by the sea, on the N. by Nicaragua, and on the W. and S.W. by the Pacific ocean. It has a few good ports, but is thinly peopled. Its principal port is at the mouth of the River Matina, in lat. 10° 1' N., long. 83° 7' W.

COSTA RIVER, Africa, S. coast, W. of Grand Bassan, which is near its E. bank. This river is easily known, coming from the W., by a cliff or rock on the beach of its W. point. It is 7 leagues to Assinee, from hence E.S.E. and S.E. by E.

COTTENTIN ISLAND, N. coast of France, on the N. side of which the port of Cherbourg is situated, near Cape La Hogue.

COUBRE, POINT DE LA, France, at the entrance of the Gironde, a low point, with sandhills inland, carrying a small battery, two beacons, and a semaphore. There is a harbour light on the point 2½ leagues N. by W. from Cordouan lighthouse; a fixed light, 65 feet high, visible 10 miles. Lat. 45° 41' 30" N., long. 1° 15' 12" W.

COUDRES, ISLE AUX, N. America, St. Lawrence river, in the midst of the channel below Quebec, about 15 leagues N.E. from that city, and the same distance from Tadousac port. The best channel up is the S. side of this island, called the Pass of Iberville, though it is more usual to pass by the W. channel. The N.W. end is in lat. 47° 24' 48" N., long. 70° 28' 30" W.

COULIHAUT, W. Indies, Dominica island, 16 miles S. of Portsmouth, lat. 15° 30' N., long. 61° 29' W.

COURANTIN RIVER, South America, the boundary between British and Dutch Guyana. See Corentyn.

COURCIL POINT, France, the S. point of Rochelle harbour, on the W. coast of France.

COURLIN ISLANDS, Scotland, two islets near the W. coast, 4 miles E. from the Island of Scalpa.

COURONNE, CAPE, France, 4¾ miles S.S.E. ¾ E. from Bouc, in lat. 43° 18' 54" N., long. 5° 3' E., the W. extremity of the bay of Marseilles, opposite Cape Croizette on the E. There are sunken rocks off Cape Couronne, called the Regats and Muets, over which the sea frequently breaks.

COURSEULLES. See River Orne.

COURTMACSKERRY BAY and HARBOUR, S. coast of Ireland, between the Old Head of Kinsale and the Seven Heads, a distance of 6 miles. In the N.W. part of the bay is the harbour, in which vessels may anchor near the quay in two fathoms; but vessels drawing 8 or 9 feet water must go in at half-flood, the water being shoal off the point next the quay. Two rocks, called "The Barrels," lie near the middle of the bay. The S. is small and dries, the other is larger, half a mile to the N. of the first, and seldom seen above water. At the southern Barrel Rock the end of the Old Head of Kinsale bears S.E. by E., and the Horse rock always above water W. by N. Cork harbour is distant from the Longships lighthouse at the Land's end N. by W. ¼ W. 46 leagues, and from St. Ann's point, Milford haven, N.W. by W. ¼ W., nearly 39 leagues. Coming from the S., Knockmedown hill should be kept about N.E. by E. until the Old Head of Kinsale is seen, a bluff point with a lighthouse upon it. From this point the entrance to Cork Harbour lies E. ¾ N., distant about 5 leagues. When off the harbour, Roche's tower and lighthouse, on the E. point of the entrance, will be recognised. A little outside of that point, E. of the entrance, lie the Stag rocks, or Cow and Calf.

COURTOWN KAYS, or E.S.E. KAYS, North America, Bay of Guatimala, in lat. 12° 24′ 15″ N., long. 81° 28′ W.

COUTANCES, a port of France, on the N. coast, in lat. 49° 3′ N., and long. 1° 27′ W. It is 13 leagues S. of Cherbourg, about E.S.E. from the Island of Jersey.

COVE, Ireland. See *Cork*.

COVE POINT, Ireland, the W. point into the harbour of Kinsale, about a league N. of the Old Head. It is foul ground round this point.

COW and BULL, Bahama islands, 4 leagues S.E. by E. from Harbour island. A mark on the Island of Eleuthera, so called, being the form assumed by broken land.

COW and BULL ROCKS, Newfoundland, on the S. coast, about 1 mile S.E. from Cape St. Mary, in lat. 46° 49′ 25″ N., long. 54° 14′ 33″ W. They stand fair above water, though there are some sunken ones, near which caution is required.

COW and CALF ROCK, Ireland, S.W. coast, lat. about 51° 22′ N., long. 10° 35′ W.

COW and CALF ROCKS, Cornwall, on the N.W. coast of that county. They are four in number, and are sometimes called the South Rocks, situated half a mile due W. from Trevose head, on an iron-bound coast, where, however, there are no dangers but what are visible. Lat. 50° 33′ N.

COW HEAD, North America, coast of Newfoundland, lat. 49° 55′ 12″ N., long. 57° 51′ 16″ W.

COW, or VACHE, ISLAND, West Indies, Hayti, E. of Point Abacou, and near the W. end of the island, on the S. side. Once it was a noted rendezvous for privateers during war.

COW ROCK, Ireland, in Dingle bay, on the W. coast, off the W. point. Vessels may sail round it without danger; at high water, spring tides, it is under water, but may then be readily discovered by a look-out.

COWES, Isle of Wight, Hampshire, near the N. extremity of the island, a port for merchant-vessels and the passage-boats from Portsmouth to the island. It is in lat. 50° 46′ N., long. 1° 17′ W. Time of high water, $10^h 45^m$; rise, 15 feet.

COZUCAY, Isle of Candia, Mediterranean, one of the Yanis or Janissary islands, the most N. of the four, in the Gulf of Settia, about 18 miles from Cape St. John.

COZUMEL, North America, off the coast of Yucatan, an island in lat. 20° 16′ N., long. 86° 59′ 39″ W. for the S. point. The N.E. is in lat. 20° 35′ 30″, long. 86° 44′ 34″ W. It is 8 leagues in length, low, flat, and covered with trees, much resembling other parts of the coast. There are rocks lie off the S. point, about a mile from it, and extend to the point N.N.W. of it, half a mile offshore. Along the W. side it is clear of danger, and may be approached within a quarter of a mile. This side of the island is 9 or 10 miles from the coast. There is good anchorage off the N. side. To the N.N.E. of the island lies a bank, near the parallels of 19° 50′ and 21° 30′ N., and long. 87° 5′ and 85° 40′ W.

CRAB, THE. See *Coburn*.

CRAB HOLE, Island of St. Kitt's, West Indies, round the extreme S. point of the island, to the N.E., and in the strait between that island and Nevis. It is the W. part of what is called Major's bay, which has from 5 to 2 fathoms, shoaling gradually to the bottom of the bay. Towards the points it is somewhat foul, but in the middle it is tolerable ground.

CRAB ISLAND, West Indies, on the S.E. side of the Island of Porto Rico. It is fruitful in citrons and oranges, and the valleys are well planted. The E. point is in lat. 18° 14′ N., long. 67° 8′ W.

CRAB ISLAND, South America, Guyana, at the entrance of the River Berbice. See *Berbice*.

CRAB VALLEY BAY, Antigua island, West Indies, 2 miles S. from Reed point.

CRADOO LAKE, S. coast of Africa, Bight of Benin, a long narrow water, with many native towns on its shores, running parallel with the sea for 15 leagues, from which it is separated by a narrow slip of land, called Curamo island, at the W. end of which the River Lagos discharges itself into the sea, in lat. 6° 24′ N., long. 3° 22′ E., the E. end terminating at the outlet of the River Benin.

CRAGUISH LOCH, a little N. of Loch Crinan, with better shelter and good ground. In summer any part of it is safe; but under apprehension of bad weather, the best anchorage is at the E. end of Ilan-ree, between the point of that island and Ormeg.

CRAIG DORIE, Scotland, a cape on the E. coast of Kincardineshire.

CRAIG ENDIVE, Scotland, a small island on the W. coast, 4 miles E. from Jura island.

CRAIG GAG POINT, Scotland, on the E. coast of Sutherland, 16 miles N.E. of Dornoch.

CRAIGHEAD POINT, Scotland, co. of Elgin, upon which stands a lighthouse, called the Cover-sea Skerries lighthouse, in lat. 57° 43′ 21″ N., long. 3° 20′ 14″ W., bearing from Tarbet Ness light S.E. by S. ¼ S., distant 16½ miles. It exhibits a revolving light, attaining its brightest state once a minute, then declining till, to a distant observer, it disappears. It is 160 feet above the sea, visible 6 leagues, the colour natural from W. by N. ¼ N. to S.E. by E. ¼ E., and red from S.E. by E. ¼ E. to S.E. ¼ S.

CRAIG LEITH, Scotland, E. coast, 1 mile from the shore, a rocky islet, steep all round, W.N.W. 2⅜ of a mile from the Bass Rock. The passage is good between this and Berwick rocks.

CRAIG LEITH ROCKS, Scotland, 3 small islands between the Island of Bass and Gulleness, on the E. coast.

CRAIG LOGAN, Scotland, a cape on the N.W. extremity of Wigtown co., 9 miles N.N.W. from Stranraer.

CRAISTER HELEN ISLAND. See *Shetland Isles*.

CRAKE SAND, coast of Zealand, in Holland, N. from Sluys. To avoid it, when a vessel is near Sluys harbour, bound for Flushing, she must steer N.E. somewhat N. along the shore, and when the Castle of Sluys and the steeple of St. Lambert's come in one, sail E. right in with Flushing.

CRAKENISH POINT, Scotland, W. coast of the Isle of Skye, 6 miles N.N.W. of Dunan point.

CRAMOND, Scotland, a small green island in the Frith of Forth, lying nearly S.W. by W. ¼ W., a mile from Mickry island; S.W. by W., ¾ of a mile from Cramond island, is Cramond town, at the entrance of the River Almond. A sand here, called the Drum Sand, dries a mile out, and stretches along the shore from Granton to Hound point. There are many dangers in this vicinity, and particularly about Mickry island.

CRANBERRY ISLAND, North America, Nova Scotia. There is a lighthouse on this island, in lat. 45° 20′ N., long. 60° 57′ W. This light is not far from Cape Canso, a modern tower, with fixed light, intended to facilitate the navigation of the Gut of Canso.

CRANE ISLAND, North America, River St. Lawrence. Lat. of S. end, 47° 2′ 30″ N., long. 70° 38′ 10″ W.

CRANE ISLAND, North America, Maryland, in the Potomac, 30 miles S.W. of Annapolis.

CRANFIELD POINT, Ireland, the N. point of Carlingford bay, on the E. coast.

CRANSTAIL, Isle of Man, a creek dependent on Ramsay.

CRATOWNESS CAPE, Scotland, E. coast, Kincardine co., 3 miles S. of Stonehaven.

CRAWFORD ISLAND, African coast, Isle de Los; the establishment there is in lat. 9° 27′ 24″ N., long. 13° 48′ W.

CRAYFORD NESS. See Thames.

CREDEN HEAD, Ireland, the W. point of entrance into Waterford, on the S.E. coast; a high point from which a shoal lies off S.S.E. into the sea. Clear of this there is a broad and clean channel, having 7 fathoms water at the mouth of the haven, and 6 fathoms further within the headlands.

CREETOWN, Scotland, a small port on the E. side of Wigton bay, and the N. side of the Solway Frith.

CREMAILLIERE COVE, Newfoundland, in lat. 51° 18′ 30″ N., long. 55° 38′ 18″ W., the entrance of the E. point.

CRERIN, LOCH, Scotland, W. coast, beyond Ruinafanert point. A well-sheltered place, with good anchorage. The shore runs strait to it, after entering the sound between Lismore island and the main.

CRETA ISLAND, Greece, on the W. side of the strait of Paros, off the S. point of Port Maria, E. of the Port of Naussa 2½ miles. Naussa is in lat. 37° 7′ N., long. 25° 15′ 20″ E.

CREUX, CAPE DE, Spain, E. coast, a cape of no great height, but irregular, with a watchtower on its summit, in lat. 42° 19′ N., long. 3° 20′ E. The mountains rise loftily beyond it, seen a great way off at sea. It is the W. entrance to the Gulf of Lyons. The tower on the cape bears from Cape Begu N. ¾ E., distant 23 miles.

CREVE CŒUR, African S. coast, a Dutch factory, near Accra.

CREYL, THE, Holland, a bank or sand in the Zuyder Zee, running off to a point in a S.W. direction from the Point of Henlopen and Staveren, at the end of which is the Creyl buoy, to the S. nearly from Frees Plat buoy. The Creyl sand is a danger here, which the pilots leave half a mile to the larboard.

CRIANAMIL ISLAND, W. Isles of Scotland, a small barren island in Otervore road, northward of the Isles of Wia, Flattay, Hellesay, and Gigay, it is in one part of the channel, dividing Barra from S. Uist, but is in no way to be recommended, though there is anchorage, in moderate weather, to the N. of it between Gigay and the Fuddia islands, in 4, 6, or 7 fathoms.

CRIBACH HEAD, South Wales, W. coast, 4½ miles E.S.E. ¼ E. from Cardigan island, and W. of Aberporth road.

CRICHAETH CASTLE, North Wales, Caernarvon co., situated on a lofty promontory facing the

N.W., near the bottom of Traeth Mawr bay. It is high, and seen far at sea serves for an excellent mark, there being a village and some remarkable objects E. of it.

CRIMEA, Black Sea, a peninsula, which was the ancient Taurica Chersonesus, and is now included in the Russian Government of Taurida. It is almost divided into two parts by the River Salgir, which runs in a N.E. direction into the sea of Azof, that bounds it on that side as the Sivache or Putrid sea bounds it on the N., and the Black sea and Strait of Kertche, or Jenkaleh, on the E., S., and W. The climate is mild but unwholesome in the S., and in the N. cold or scorching. The capital is Simpheripol, or Akmetched. It has several ports and commodious anchorages. The exports are salt, honey, wine, wax, butter, horses, hides, furs, and lamb-skins. The Gulf of Kerkinet on the N.W. contains several bays where vessels may anchor. The coast, stretching S.W. to Cape Tarkan, then retires, and between Koslof and Cape Chersonesus forms a deep bay near the S. headland, of which is the noble harbour of Sevastapol, or Sebastapol. To the S. yet further is Balaklava and its bay; still further to the S. and E., passing Cape Aitador and the anchorage of Jaltie, the coast indented and irregular goes by Tekia and Kaffa, to the Strait of Jenkaleh or Kertche. The town of Jenkaleh lies near the entrance of the strait from the Sea of Azof, within which, on the W. side, the only place of note is Arabat at the bottom of the gulf of that name, and the S.E. extremity of what is denominated the Tongue of Arabat, a long strip of land between the seas of Azof and the Sivache, at the W. end of which is the isthmus, which connects it with the North Steppe, and at the N. end of that is Perecop. The Crimea is divided into six districts. Pop. 207,000.

CRINAN LOCH, Scotland, W. of the Gulf of Coryvreckan, open to the W., and in winter not well sheltered.

CRIO, CAPE, Greek sea, a promontory from the main land of Anatolia, dividing the Gulf of Symi from that of Boudroun, or Cos, E.N.E. ¼ N., distant 7½ miles from the N.W. end of Nisari, in lat. 36° 40′ 46″ N., long. 27° 21′ 6″ E. At a distance it appears like an island, being connected with the main only by a very shallow neck of land, on both sides of which there were once artificial harbours. Within this cape are the extensive ruins of Cnidus, situated on the side of a mountain about 400 feet high, called by the native Greeks Phrianon. The peninsula of Cape Crio consists of lofty mountains, retiring upwards from the port, and presenting to the W. a perpendicular face of rock inaccessible.

CRIPPLE COVE, Newfoundland, a small creek round the S.E. point of the island westward.

CRIPPLEGATE BAY, West Indies, Island of Antigua, a small inlet W. of Fort bay, open to the N.W., on the E. of the point, near which is a very small island called Ship Stern, the N. limit of Deep bay.

CRIPPLE SAND, Holland, S. from the buoy of the Creyl about half a league. See Creyl. It is not a broad sand, but extends W. nearly half a league, having at its W. end only 2 fathoms, and near 3 to the S. of its W. end.

CROCODILLO POINT, S. side of Cuba island, on the S.W. point of the Isle of Pines.

CROIX, ST., or SANTA CRUZ, West Indies, an island belonging to Denmark about 5 leagues

S.E. of St. Thomas, and as far E. by S. of Crab island, on the E. end of Porto Rico. It is about 30 miles long, by 8 broad. It is well cultivated. The E. extremity of the island is in lat. 17° 45' 30" N., long. 64° 34' W. The capital is Christianstadt, the lat. of which, at Fort Christian, at the flagstaff, is 17° 45' 59" N., long. 64° 41' 58" W. A mile N. from the entrance of this harbour is the W. extremity of a reef called the Scotch reef, stretching with its shoals full 1½ miles E.N.E., making the approach to the harbour dangerous. Nine nautical miles full from St. Croix N.E. by E. ¼ E., and about 11 miles E. by N. from the E. point of Buck island, commences the E. extremity of an extensive bank or shoal, the N. end of which goes off from thence to the N.W., and soon after stretches out W., inclining to the S. of a W. direction towards Buck island shoals and reefs, with which it may be considered as connected. The N. edge is of coral, several miles in extent, on which 5½ fathoms is the best depth found, the more common being 6, 6½, and 7 fathoms. The sea has been seen to break on the whole line of the N. edge, and to the very extremity of the bank, in a very alarming manner, during a N. ground-swell in the winter months.

CROIX, ST., HARBOUR, Newfoundland, on the N.W. side, the N. point of which is only 4 leagues W. from Quirpon island at the N. extremity.

CROIX, ST., RIVER, North America, Nova Scotia, the limit between that country and New England. Its mouth is in the Bay of Fundy, on the N. side, in lat. 45° N., long. 67° 30" W. Off its entrance is the Great Manana island.

CROIX, POINT, ST., at the E. end of the Island of Groa, off the W. coast of France. This point is a mark for the Birvadaux rock, and others near, at 2½ leagues S.S.E. from it.

CROIZETTE CAPE, France, Bay of Marseilles, the E. cape of that bay. Several mountains are to be seen on approaching it to the E. and W.; 15 miles S.E. by E. ¼ E. from Cape Couronne.

CROISIC, France, W. coast, S.E. of the River Vilaine, a point N. of the entrance of the Loire river. There are two harbour lights shown here, the first near the shore, a quarter of a mile from the church; the second, 151 feet S. ¼ W. from the first; in a line they lead into the harbour from the N.W.; one is elevated 13 feet, the other light 32, visible 6 miles. This direction ranges near two rocks, half a mile from the beacon on Le Trehis rock. The entrance to Croisic is dangerous, the currents are strong, and a pilot is absolutely necessary. It is high water at 3ʰ 45ᵐ.

CROMACII HEAD, Scotland, on the E. side of Spey bay, and S. shore of Murray Frith, 2 miles W. from Scarnose. There are rocks off this point for a cable's length or more.

CROMARTY, Scotland, co. of Cromarty, a town W.N.W. from Cowsey Point, and N.E. by E. ½ E. about 2 leagues from Fort George, with a good harbour, a mile wide at the entrance, having from 30 to 22 fathoms water. Pop. 2,900. Close to the S. point there is a small rock, called the W. Sutler, and on the opposite side another, called the E. Sutler, the only dangers. The town stands on a pier, the S. side of the harbour, 2 miles within the W. Sutler. Upon Cromarty point is a lighthouse in lat. 57° 40' 58" N., long. 4° 2' 7" W., within the entrance of Cromarty Frith, a fixed red light, open from W.N.W. round to S.E. by E. ½ S., in a N. direction, visible 10 miles, 50 feet above

the sea. The stream tide runs off Cromarty entrance from 3½ and 4 springs, to 2 neaps.

CROMER, Norfolk, a town upon the verge of the German sea, visited as a bathing place; pop. 1,232. Here is a lighthouse called Cromer or Foulness light, revolving, exhibiting a bright flash every two minutes. The lantern is 274 feet above high water, and in clear weather is visible 7 leagues. Off Cromer, on what is called Foulness spit, a buoy is laid in 3 fathoms on the outer part. High water, springs, at 7ʰ. Lat. of the town, about 53° 5' N., long. 51' E. Cromer outer bank or knoll, lies 7½ leagues N.E. ½ E. from Cromer lighthouse, extending N.W. by W. about 2 miles, and three quarters of a mile broad. Cromer inner bank lies 10 miles W.S.W. from the outer, 13 miles N.E. from Cromer lighthouse, extending above 3 miles N.W. by W. and S.E. by E., a mile broad, with 4 fathoms on its middle part, increasing at the ends to 6 and 7 fathoms.

CROMPTON POINT, West Indies, Island of Dominica, on the N.E. side. Lat. 15° 32' N., long. 61° 24' W.

CROMWELL'S LEDGE, Newfoundland; lat. 50° 12' N., long. 53° 30' W.

CRONENBURGH, or CRONBERG CASTLE, Denmark, in the Isle of Zealand, at the mouth of the Sound, where vessels bound to the Baltic pay a tribute to the Danes. Lat. 56° 4' N., long. 12° 37' E.

CRONLIN ISLANDS, Scotland, W. coast, near Loch Kishorn; they are three in number, and lie out a mile from the main; small vessels might find a shelter between the two larger islands, and lie aground upon soft sand, or float with landfasts on each side, in 10 feet water; the creek will not hold more than four or five vessels. The channel dries at half ebb.

CRONSTADT, Russia, a fortress, town, and naval station, government of Petersburg, on the Island of Catline Ostrov, in the Russian tongue, and in that of Retauzari, in the Finnish language. This island is in the Gulf of Finland, about 5 miles long, and is 19½ miles W. of St. Petersburg. It is the chief station of the Russian fleet, and is defended toward the sea on the S. by the harbour fortifications, and round the rest of the line by bastion ramparts of earth, which are mounted with a vast number of cannon. On the N. is Fort Alexandrosky and the battery of St. John, built upon piles in the sea. The castle of Cronslot is erected upon a sand-bank, forming a cross fire with the guns of Cronstadt. There are here three harbours, one belonging to the merchant service and two to the Russian navy; the merchant harbour, which is the largest, will hold 600 sail; the naval harbour about 30 vessels, and therefore, for want of room, the men-of-war remain for the most part in the middle port. The nature of the water here is said to produce the dry rot more rapidly than elsewhere. The canal of Peter the Great, 238 fathoms long, 56 feet broad, and 25 deep, communicates with a large basin, 1,050 fathoms long, and 24 feet deep. It is entered from the sea, and communicates with dry docks, which extend 150 fathoms in the form of a cross, the middle of which is circular, and lined with granite. The dry docks are 40 feet deep, faced and paved with the same kind of stone. The water is pumped out by a steam-engine. A rampart of granite, built in the sea, encloses the mole. There is a foundry here for casting cannon, magazines, rope walks, and every accessory to a naval establishment of the first magnitude. The merchants' harbour is closed by a boom, and defended by batteries. The channel between Cronstadt and the coast of Ingria is narrowed by shoals to three-quarters of a mile, and the depth is only 4 fathoms. It is commanded by the castle of Cronslot on the Ingrian side, a round building with outworks, mounting 50 guns, and on the Cronstadt side by St. Peter's battery mounting 100. A sandy flat of shoal water, called Recs bank runs off from the S. shore full 3 miles from the land, having only 1, 1½, and 2 fathoms water, overlooked on the main by the royal residence of Raninbom or Orienbaum, built on an artificial terrace, 100 feet above the sea. The passage to the N. of Cronstadt has several shoals, and only 2½ fathoms water. It has been blocked up by the hulls of vessels sunk to prevent a passage. Off the W. end of the island, upon a rocky shoal, is the Tolbeacon light, upon a round tower 88 feet high, visible 14 miles; and on the opposite side, at the extremity of a bank jutting from the S. shore, there is most commonly a light-vessel called the London chest light-vessel, with three lights. The mouth of the Neva and space between Cronstadt and St. Petersburg is shallow, and the river has a bar with only from 7 to 10 feet water. The proper sea-port of the city, therefore, is Cronstadt, whence steamers ply continually to the capital, or the journey may be made overland, by Orienbaum, Peterhoff, and Strelna. On reaching Cronstadt, the visitor goes to the harbour-master's vessel, where his passport is examined, and he then enters the town, proceeding with an interpreter to the port admiral and military commandant, where his passport is registered, and he is at liberty afterwards to visit the capital. Cronstadt is, therefore, the advanced guard of the capital, too strong to be forced by line-of-battle ships. The streets of the town are of brick, or brick plastered, ill paved for the most part, but straight. It has seven churches, and a marine hospital. The inhabitants are principally persons connected with the sea, not more than 300 or 400 being free citizens out of a pop. of 40,000; seamen, soldiers, and labourers make up the rest. The Tolbeacon light at Cronstadt stands in lat. 60° 2′ 25″ N., long. 29° 32′ 25″ E.; the cathedral of Cronstadt 59° 59′ 43″ N., long. 29° 46′ E.

CROOKED HAVEN, Scotland, a bay on the N. coast of Banff, 2½ N.W. of Cullen.

CROOKED ISLANDS, West Indies, a little S. of E. from Long island, by which vessels take a windward passage to England. The N.E. breaker is in lat. 22° 43′ 30″ N., long. 73° 47′ W. Bird rock, off the N.W. point, lat. 22° 51′ N., long. 74° 22′ 15″ W. The extremity of the N.E. reef, lat. 22° 47′ N., long. 73° 49′ 45″ W.; the E. end of the E. reef, lat. 22° 18′ N., long. 72° 38′ 16″ W. See *Bahamas.*

CROOKHAVEN, Ireland. The entrance to this harbour is 8½ miles N.W. ½ N. from Cape Clear. It is a narrow, well-sheltered harbour, with good ground, and water for large vessels. Some rocks lie off the point, on the S. side of the entrance. The best anchorage is opposite the houses on the S. side, about 1¼ miles from the entrance. Ships drawing above 16 feet may anchor half a mile up from the mouth of the harbour, in 4 fathoms. There is a lighthouse upon Rock island point, at the N. side of the entrance to Crookhaven, having a fixed light 67 feet above the level of the sea, bearing from Cape Clear light N.W. ½ W. 8½ miles. It stands in lat. 51° 28′ 35″ N., long. 9° 42′ 31″ W.

CROQUE HARBOUR, Newfoundland; lat. 51° 2′ 30″ N., long. 55° 49′ 18″ W.

CROSS, or PORTCROSS, France, one of the Hyeres islands off Toulon; the Gabiniere rock S. of Portcross island is in lat. 42° 59′ 6″ N., long. 6° 22′ 40″ E.

CROSS BANK, England, River Thames, half a mile W. by S. from the two buoys of the Narrows, and almost abreast of the Hearn beacon, on the S. shore to the W. of Reculver church, whence it bears S.S.W. about.

CROSS ISLAND, North America, United States, near the coast of Maine, at the entrance into Machias bay, lat. 44° 30′ N., long. 67° 5′ W.

CROSS HAVEN RIVER, Ireland, on the W. side of Cork harbour, and upon the S. coast. It is round the point of Ram's head, near which is James's battery, over against Dog's Nose or John's fort, at the inner neck or entrance into the S. end of the long and spacious basin, of which, on the N. side, is Kuskine haven, and to the W. of it the garrison. Ships may run in here to wait for the tide, and be land-locked and safe from all winds.

CROSS ISLAND, Ireland, on the N.E. coast, and N. from Copeland island, near the entrance into the Bay of Carrickfergus. It is small, and near it is an islet, called the Mews; between which two and Copeland island there is a channel of near a mile in breadth, with 8 fathoms. They are 3 leagues from Carrickfergus at E. by S. ½ S.

CROSS ISLAND, North America, Nova Scotia, off Lunenburg harbour: there is a lighthouse here on the S.E. point of the island, at the entrance of Lunenburg bay, in lat. 44° 23′ N., long. 64° 7′ W. It is on a tower, painted red, with two lights placed vertically, 30 feet apart. The lower light is fixed, and the upper so eclipsed as to show minute flashes, changing abruptly from dark to light. Cross island is thickly wooded.

CROSS ISLAND, White sea, lat. 66° 25′ N., long. 40° 3′ E. It lies on the passage to Archangel, just S. of the Arctic circle. It is 9 leagues S.W. by S. from Ponnoy road, round the coast of Lapland, and 20 leagues N.E. from the Cat's Noss, or Grey Point. There is good anchorage all the way from Ponnoy to Cat's Noss, in 8 or 9 fathoms, where ships may stop for the tides, though they run strong there. A point appears between them, on which are three crosses, that has often been mistaken for Cross island, but is much larger, and has two crosses at the N. end, and three more at the S. end, by which it may be known. Ships may either sail behind Cross island, or anchor there in 6 or 7 fathoms; but the best place is to bring the great Cross at E.N.E.

CROSS SAND, off Yarmouth, 1½ mile E. of the Scroby, 6 miles long, and three-quarters of a mile broad at its broadest part. On this sand and its branches three buoys are placed, called the Cross Sand, S. of the Cross Sand Middle, and the N. cross buoys.

CROSS LODGE SAND, England, close to the middle of the land side of the Brakes, near the N. end of the Downs. There are 3 fathoms between the Cross Ledge and the Querns. The length of this sand is near a mile, W. ½ S.

CROSS-SAVAGH, LOCH, W. Isles of Scotland, 2½ miles E. of Stohenish, a large bay, with good ground, open to the S. and S.E., where a small vessel or two might be secure riding behind a rock at the head of the lake.

CROSSBY CHANNEL. See *Liverpool.*

CROTONA, or CORTRONE, S. coast of Italy, a port, distant 4 miles from Cape Nau, N.W. by W., pop. 5,000, who are poor, only exporting a little corn and cheese. The harbour is neglected, and vessels that go there must lie in the open road. There are the remains of two piers; but the shore from Cape Nau to Crotona has many rocks scattered about. From Crotona to Point Alice the land runs N. ¾ E. 19 miles. Lat. 39° 9′ N.

CROTOY, France, once a port, a league from the sea, on the N. side of the River Somme. Its trade is now removed to St. Valery.

CROUCH RIVER, England, Essex co. It runs parallel with the Thames more than half the length of the county, and falls into the sea at the end of Foulness island.

CROW HEAD, Ireland, S.W. coast, lat. 51° 32′ N., long. 10° 2′ W.

CROW ISLAND, North America, St. Lawrence river. Lat. of N.E. point, called Kamouraska, 47° 35′ 17″ N., long. 69° 55′ 48″ W.

CROW ROCK, South Wales, a mile from Lenwy, or St. Patrick's point, about 4 miles S.E. by S. from the entrance of Milford haven. It is only dry at half-tide.

CROW SOUND, Cornwall. See *Scilly Isles.*

CROW TAING, Scotland, a cape on the N.W. coast of Ronaldsha island.

CROWL BAY, West Indies, Antigua, S. of Willoughby bay, on the W. side, where the land trends S. from that bay to the S. point of the Isle of Antigua. Ships may lie sheltered in this bay, though exposed in some degree to the S.E. currents, it being open to their influence at that part of the island.

CROWN, or CRONSLOT CASTLE, Russia, Gulf of Finland, a small island in the Neva, 5 leagues from St. Petersburg. Here is a station of the Russian navy. Lat. 59° 56′ 42″ N., long. 29° 43′ 30″ E. A sand, called the S.E. Sand, runs off from the Point of Cronslot, bounding the S. channel. See *Cronstadt.*

CRUANACARRA, or CRUANAKILLY, Ireland, an islet rock, near the W. coast, lat. 53° 16′ N.

CRUMP'S ISLAND, West Indies, Island of Antigua, on the N.E. side, of some size, but irregular in form, separating Mercer's creek on the S. from Harris's bay. Lat. 17° 14′ N., long. 61° 25′ W.

CRUCES POINT, France, Bay of Biscay, a point at the entrance of Port Passages, in the S.E. angle of the bay, near the boundary of France and Spain.

CRUDEN, WARD OF, and CRUDEN SCARS, Scotland. The Ward of Cruden is a fishing town, at the E. end of a bay about a mile over Cruden Sands. The Scars are rocks extending half a mile from the shore, 6 miles S.W. ¼ W. from Buchanness, on the E. coast. Near the Scars there are from 5 to 6 fathoms water, which deepens to 20 and 21 three miles off. From the Ward of Cruden to Buchanness the shore presents high rugged cliffs, with 12 and 14 fathoms close to them, and 30 a little distance from the shore. Slanes Castle, a large house, looking remarkable at sea, is seen a little E. of the Ward, from which the rock called the Buss, lies a little to the S.W., about a cable's length from the shore.

CRUEYSFIORD, or CUEYSFIORD, a bay of Norway, 20 miles S.W. of Bergen.

CRUIT ISLAND, Ireland, a mere islet near the N.W. coast, 10 miles S. of Bloody Farland's point. Lat. 55° 2′ N., long. 8° 19′ W.

CRUMSTONE, a flat rock above water, a mile S.E. by E. from the S. part of Staples island, and a mile S. from the Longstone lighthouse, one of the Farn islands, off the coast of Northumberland.

CRUZ BAY, Africa, on the W. coast, S.E. by E. 7 or 8 leagues from Cape Geer, having the Moor's castle on a high land, 5 or 6 miles N., called also Aghadeer. The River Suze is to the S. of the town of Cruz or Aghadeer. A range of rocks runs off about a gun-shot, but the rest of the coast is clear. To anchor, a ship must run into the bay till the Moor's castle bears N., and the European warehouses N.E.; a ship will then be southward of these rocks, in 7 or 8 fathoms. When Cape Geer bears N.W. by W., that is the best station in the road; and vessels in summer may come so near the shore as from 6 to 7 fathoms. To come in from Cape Geer a ship must run along by the land of the cape till abreast of the castle nearly, because the N. winds blow the trades here; and if a ship gets too far from the shore she must work up again from the N.E. and N.N.E. to recover the shore. A vessel must come no nearer by night than 12 or 14 fathoms.

CRUZ, CAPE, Cuba island, West Indies, on the S. side, nearly N. from Point Negril, on the W. point of the Island of Jamaica. Lat. 19° 50′ N., long. 77° 45′ W. It is high land, and steep toward the sea, and a shoal runs out from it a good way to seaward.

CRUZ, SANTA, Canary islands, the name of the capital of Palma.

CRUZ, SANTA. See *Madeira Island.*

CRUZ, VERA. See *Vera Cruz.*

CRYLL, Scotland, on the E. coast, a small haven between Fifeness and Ellness, to the S.W. by W.; dry at low water; a shallow place, only fit for small craft.

CUAGUA, South America, an island between that of Margarita and the main, lying N. of the Gulf of Cariaco, with the intervening peninsula between, and on the S.E. the island of Coche, surrounded by rocks and shoals. See *Coche.*

CUB, or CUBB'S ISLANDS, North and South, Hudson's bay, North America; the N. in lat. 54° 25′, long. 80° 50′ W.; the S. in lat. 53° 42′, long. 80° 30′ W.

CUBA, ISLAND OF, West Indies, a large island subject to Spain, 620 miles in length by 65 in breadth, lying between lat. 19° 42′, and 23° 26′ N., and 74° and 85° W. long. It was discovered by Columbus in 1492; after which the Spaniards extirpated the original inhabitants. Cuba possesses many convenient harbours. The town of St. Jago is considered the capital, in lat. 19° 59′ 30″ N., long. 76° 11′ W., at the S.E. part of the island. The Havanna on the N.W. side is much more considerable in respect to trade and influence. Its lighthouse stands on the Morro, at the E. part of the harbour, where there is a revolving light, in lat. 23° 9′ 24″ N., long. 82° 22′ W. This island possesses upwards of 156 rivers, and is traversed longitudinally by a chain of high mountains, closed with dense forests of valuable timber. There are also wealthy mines in the interior, and several medicinal springs. The staple produce is sugar, but it produces all the valuable articles of export which are common to the West India islands in general. Before the disturbed state of the mother country upwards of 300,000 arobas of sugar were annually raised in the island, and the pop., white and coloured, was estimated at 350,000.

The harbour of the Havanna, capable of well accommodating 1,000 sail of ships, has a general depth of 6 fathoms. The entrance is narrow, 1½ mile long, and hulks have been sunk in it to render an entrance more difficult. It is defended by the Morro castle on the E., and the Punta fort on the W. The Morro is too high to be cannonaded by ships-of-the-line. It consists of two bastions towards the sea, and two towards the land, with a covered way and a deep ditch cut in the rock. The Punta fort, on the contrary, is low, well casemated, and the works very formidable. The city lies on the W. side of the port, on an island formed by two branches of the River Lagida. The principal bays visited by shipping are Nuevitas, 7 leagues long. It is within Savinal Cay, and has an entrance on the E. A poor village, called Villa del Princesse, stands at the mouth of the river in this bay. St. Juan de Remedios, Matanzas, St. Carlos, Ratabona, and Samana bay, are places visited more particularly by shipping. The headlands and capes of this island are very numerous, Cape Antonio is on the W., and that of Vera Cruz on the S. There are also numerous islands, as the Caymans, the Isle of Pines, and others, while nearly all are surrounded by dangers to navigation, more or less requiring the close attention of the mariner. Cuba is 19 leagues W. from St. Domingo island, or Hayti, as it is more commonly called. It is 25 leagues N. of Jamaica, and 41 S. of the coast of Florida.

CUBAIMARON BAY, St. Vincent's island, S. coast. Lat 13° 6' N., long. 61° 11' W.

CUCKHOLDS' HEAD, North America, Newfoundland, a bare round hill on the E. coast, having at sea the appearance of a haycock. It is N. of St. John's about 1 mile.

CUCKOLD'S POINT, Barbadoes, E. coast. Lat. 13° 19' N., long. 59° 42' W.

CUCKMERE HAVEN, Sussex, 4 miles W. of Beachy head; it has but 6 feet water at low ebbs, and 10 at high water. It is frequently stopped up in storms. From this haven towards Beachy head the shore is steep, and the soundings are regular. Between this place and Beachy head six caverns, with entrances 3 feet wide, leading to apartments 8 feet square, have been cut in the cliffs, and a place called Darby cave has been repaired, in order to afford a refuge for shipwrecked mariners beyond the reach of the waves.

CUDD'S CHANNEL, coast of Kent, another name for Ramsgate channel.

CUDDEN POINT, Cornwall, on the E. side of Mount's bay, a point of land a league E. of St. Michael's mount.

CUEVAS, West Indies, Isle of Trinidad, a bay on the N. shore, having a good anchorage and water.

CUL DE COHE, Martinique, West Indies, a bay on the N. part of Cul de Sac Royal, also another on the W. side.

CUL DE SAC DES ANGLOIS, Martinique, West Indies, on the S. coast, a little S. of Cape Ferre.

CUL DE SAC FRANCAIS, Martinique, a bay in lat. 14° 36' N., long. 60° 53' W.

CUL DE SAC GRAND, West Indies, Island of Guadaloupe, on the N. coast, lat. 16° 20" N., long. 61° 53' W.

CUL DE SAC MARIN, West Indies, Island of Martinique, at the S. end. This opening is N. by E. from the N. end of St. Lucia island, lat. 14° 31' N., long. 60° 45' W. CUL DE SAC ROBERT,

a bay on the E. of the same island, lat. 14° 34' N., long. 60° 59' W. CUL DE SAC VACHE, in the same island, lat. 14° 31' N., long. 61° 47' W. CUL DE SAC PETIT, Guadaloupe island, 7 miles S. of the Grand Cul de Sac.

CULEBRAS CAPE, America, on the N. coast of the Isthmus of Darien. Lat. 9° 36' N., long. 78° 52' W.

CULLERA, Spain, E. coast, town and cape on the N. bank of the Xucar, half a mile from the sea, at the scope of a mountain ridge, terminating in Cape Cullera, which is elevated, and has a tower upon its summit, in lat. 39° 9' N., long. 11' 17" E. Cullera is between Denia and Valencia, and is sometimes denominated Coulibre.

CULLERCOATS, Northumberland co., 2 miles N. of the bar of Tynemouth, where vessels may lie dry at the ebb. The passage to it is between rocks, but they are beaconed and marked. It is within the Borough of Tynemouth. Pop. 600.

CULLODEN ROCK, Scotland, 1 mile from Tarbet Ness lighthouse, which lies W. by S. from it. An eight feet buoy is laid down in three-quarters fathoms at low water, spring tides, in a line with Tarbert Ness lighthouse flag-staff, bearing W.S.W. ¾ W.; and the Duke of Sutherland's monument, in a line with the corner of a wood farther W. from Dunrobin castle.

CULMORE CASTLE, the W. point of the entrance of the river of Londonderry, in the N. of Ireland.

CULVER CLIFF, Isle of Wight, Hants co.; it is sometimes called Swan cliff, at the S.E. corner of the island. It is high, white, and steep, about 2 miles from Bambridge point, at the E. end of the island. It furnishes an excellent mark for ships approaching that part of it, either for St. Helen's, Spithead, or Portsmouth harbour. It is about 4 or 5 leagues W. ½ N. from the Owers sand.

CULVER SAND, coast of Somersetshire, a dangerous and extensive flat, lying N. of the channel to Bridgewater, a narrow ridge of which dries for 3 miles, having long spits at each end. The W. end lies E.S.E. ½ E. from the Foreland, distant 19 miles. The sand extends in the direction of the stream E. by S. for 6 miles. It breaks at half tide; the flood tide from Nash point sets directly across towards Bridgewater bay, and the ebb from Bridgewater, also sets directly across it. Quantock hill, which has a tower and beacon, when it bears S., shews that a vessel is abreast of its W. limit; and when the Flatholm is on with the Steepholm N. by E. ½ E., a vessel will be 1½ miles to the E. of its E. part. There are buoys upon this sand.

CUMA, Italy, Naples, a point so named near Baia, on the coast of the Bay of Naples.

CUMANA, a province and seaport of Colombia, South America. The town is situated in the department of the Orinoco, in lat. 10° 27' 42" N., long. 64° 11' W., near the entrance of the Gulf of Curiaco, about a mile from the sea shore on an arid, hot, sandy plain. It is liable to earthquakes. Its principal import is iron. The inhabitants are fishermen, or employed in rearing cattle. Pop. 20,000.

CUMBERLAND BAY, STRAITS and ISLAND, North America, nearly under the Arctic Circle, being in lat. 66° 40' N., and long. 65° 20' W.

CUMBERLAND BAY, West Indies, Island of St. Vincent, on the W. coast. lat. 13° 14' N., long. 61° 20' W.

CUMBERLAND FORT, New Brunswick, North

America, a fort at the head of the Bay of Fundy, in lat. 45° 49' N., long. 64° 10' 18" W. It is on the E. side of its N. branch, and will accommodate 400 men.

CUMBERLAND HARBOUR, or PORT DEL GUANTANAMO, on the S.E. part of the Island of Cuba, one of the finest in the world, capable of sheltering any number of ships. It is 20 leagues E. of St. Jago de Cuba. The entrance is in lat. 19° 55' 10" N., long. 75° 20' 25" W.

CUMBERLAND HOUSE, North America, one of the Hudson's Bay Company's stations, in New South Wales, 158 miles E.N.E. of Hudson's House, on the S. side of Pine island lake, lat. 53° 56' 41" N., long. 102° 13' W.

CUMBERLAND ISLAND, North America, United States, on the coast of Camden co., Georgia, between Prince William sound, at the S. end, and the mouth of Great Salilla river, at its N. end, 20 miles S. of the town of Frederica. It has a lighthouse at its S. end, on a tower 80 feet above the sea, a revolving light, which appears full twice in every minute and a half. This lighthouse is in lat. 30° 45' N., long. 81° 37' W.

CUMBERLAND TOWN, New Brunswick, in Cumberland co. This last comprehends the territory on the basin called Chebecton, and the rivers which fall into it, as the Lac, Missiquash, Napan, Au Lac, Macon, Memramcook, Petcoudia, Chapodiè, and Herbert; some of these are shoal rivers, only navigable for boats 4 or 5 miles; the Herbert is navigable for 12 miles.

CUMINO, Mediterranean, one of the Maltese islands, between Malta and Gozo. Its E. point is in lat. 36° 0' 15" N., and long. 14° 19' 40" E. Its length from N.E. to S.W. is nearly 2 miles, and its breadth 1. At its S. point is a tower upon an eminence, and to the W. the little Island of Cuminotto, with rocks around it. On each side of Cumino there is a channel, taking its name from the island nearest to it, that on the W. being called the Gozo channel, and that on the E. the Passage of Malta. Both are good and safe in mid-channel, with from 12 to 30 fathoms all through, and a bottom of fine sand. Ball's, or St. Paul's bank, about 3½ miles in length, and 1 in breadth, lies in the fair way of vessels, passing between Cumino and Malta. There are from 14 to 7 fathoms over it, the E. part having the deepest water and best bottom; as towards the N. it is more shallow and rocky, particularly when there is less water than 9 fathoms. The W. part is dangerous, and with E. and N.W. winds the sea breaks much.

CUNI ISLAND, Ireland, off the W. coast, in Sligo bay.

CUOLAGH BAY, Ireland, on the S. side of the entrance into Kenmare river, at the S.W. part of the kingdom.

CUPID'S COVE, North America, E. coast of Newfoundland, 4 miles S.S.W. from Sheep's cove, within the limits of Port Grave. It is small, but has good anchorage for one or two ships at a time.

CURACOA, ISLAND OF, South America, to W. of Buen Ayre 27 miles, lying nearly N.W. and S.E., 35 miles long and 6 broad. It belongs to the Dutch, and produces coffee, cotton, and sugar; it is stored with the commodities of Europe and the East, being a depôt for merchandize. The Fort is in lat. 12° 6' 18" N., long. 68° 55' 42" W. The Bay of St. Anne is in lat. 12° 6' 20" N., long. 68° 37' 13" W. at the entrance. The island is moderately high, and the coasts are clean, there being no danger in any part at a cable's length distance. There are

III.

many bays and harbours in this island, the principal of which is that of St. Anne, on the W. coast, 14 miles from Point Canou. A vessel making the harbour of St. Anne ought to make Port Canou, so as to run down the coast from 1 to 2 miles distant, taking good care not to get to leeward of the harbour's mouth, as the current sets strongly to the W. The entrance to the harbour is very narrow, formed by low tongues of sand, which on the inside form several lagoons. On the E. point is the fort named Fort Amsterdam, and the principal town. On an islet off the W. point is a battery, which, with Fort Amsterdam on the E., defends the mouth of the harbour. The catholics here reside separately from the Jews and protestants, on the W. side of the town. The channel into the bay extends N.N.E. three-quarters of a mile long and a cable's length wide, while between the forts is scarcely half a cable's length. The town wharfs and warehouses are on the banks of the channel, where the vessels anchor and careen. To enter, keep close to the coast to windward, and run down, but not less than a cable's length from the shore, as there are rocks and a reef extending out about one-third of a cable's length, and when a ship is abreast of the batteries on the point at Fort Amsterdam, she must luff up towards the battery on the islet at the W. point; and run through the mid-channel. S.E. from the E. point of Curaçoa, at the distance of 4 miles, there is a low sandy islet, called LITTLE CURACOA, which, though clean, is dangerous, from its lowness, in dark thick weather. The N. end is in lat. 12° N., long. 68° 37' 13" W. This island is a great mart for all kinds of trading, under the Dutch protection, and was once noted for its contrabandists.

CURISHAFF BAY, or Gulf of Courland, Russia, on the E. of the Baltic sea, separated from it by the Curisch Nerung, extending 60 miles from Memel to Labiau, of unequal breadth, but narrowest to the N.; 7 leagues N. by E. from Konigsberg. Lat. 55° N., long. 22° 15' E.

CURLEW HARBOUR, North America, Labrador, nearly S.W. of Gannet islands, on the main. It has a green round island at its entrance, and deep water within, while there is little danger in entering.

CURRENT ISLAND, Sicily, distant 2 leagues from Cape Passaro, the most S. point of the island. It is a sandstone rock, and is reported to have a shoal about 3½ miles from its outermost point, small, with 13 feet water, which bears from Current island E. ½ N. Current island is divided from the main by a ridge of very shallow water, over which boats cannot safely pass. On the N.E. of the island there is a sort of natural mole, where Maltese trading boats sometimes run for security. Cape Passaro bears from Current island N.E. by E. ½ E. 4 miles.

CURRENTS, OCEANIC; certain progressive motions of the waters of the sea in particular places, either superficial or as deep as the soundings where they occur, by which a vessel is carried forward more rapidly, or retarded in her course, according to the circumstances connected with the direction of the stream. There is no single word in the English language to express the motion of a limited portion of any fluid situated amidst a body of the same nature, but of a bulk immeasureably greater, in place of moving amid substances foreign to itself, as is the case with streams and currents running over land. Thus a word seems wanting of a less general character to express the currents

K

of the ocean, than that to which the paucity of language obliges recourse to be had. The existence of currents in the ocean is undoubted, and currents, too, which arise from dissimilar causes. Their effects have been so constantly experienced, that while the art of navigation is little enlightened in regard to their causes and mode of action, at least no further so than conjecture is able to lay them open, the reiterated experiences of the most able seamen have established their existence, and the general directions of those which have been most obvious in the track of navigators. There are not wanting those who, through prejudice, ignorance, or both in union, cavil at facts and attempt to account for undeniable consequences by any but the most obvious reasons. Vessels out in reckonings carefully kept, discovered by observation to be in positions wholly unexpected, cannot but attribute to acknowledged causes that indisputable fact. The existence of oceanic currents, though their effects had been experienced long before, seems not to have been fully established, and an examination instituted into their limits and mode of action, until comparatively recent times. Sets of the sea were noticed, it is true, for they could not escape observation, but a grand system of currents from primary causes was not until lately regarded as part of the great oceanic system. Secondary currents from the action of tides, and the impulsive effect of rivers as affecting bounded localities near their place of meeting with the great body of waters where they pass out of being, have been proved from general observation among practical seamen upon every coast. Of late years the greater oceanic phenomena have attracted much regard, both from seamen and philosophical observers, though not more than they merit. Efforts have been made not only to trace the action of this class of currents, but to systematise their effects, and look into their causes. The attempt to trace the causes has as yet been followed by no demonstrative proof, but continual experience of their effects every day adds to the materials for systematising this important branch of oceanic geography, thus leading the way to the knowledge so much desired. The continued attempts to ascertain the direction of currents by bottles and floating substances, have no doubt their value, but they must be inconclusive, because floating substances are always acted upon more or less by the wind, and impelled to the verge of the stream, and out of it altogether, even if originally cast into the centre of the current, unless the wind and current continue in a line of perfect coincidence, which for any time together cannot be imagined possible. The existence of currents in the ocean being undeniable, their causes, uses, and limits remain for discovery. Primary currents are those to which this more immediately applies, as tidal and secondary streams of that character are obviously owing to causes which are acknowledged to be the configurations of the shores near which they are found, or to similar agencies. The causes of oceanic currents belong rather to philosophical than nautical investigation, and it must be admitted that none have been yet advanced which go beyond the limits of plausibility. It is possible that the unseen, silent, and yet all-powerful agent, which seems under some one of its modifications to direct the course of all the operations of nature—that subtle agent, as recognisable in the growth of the humblest plant as in the terrors of the lightning, may have some important

influence through its attractive and repulsive powers in causing actions in the waters of the sea, essential, perhaps, to its existing ends, going deeper, too, into its abysses than the superficial wave can do, agitating it, and by shifting deep waters of high temperatures from one latitude to another, causing salutary and necessary circulations and alterations of heat and cold, and keeping up that system of change and motion which is an essential part of that of the universe itself. One of the most remarkable oceanic currents, the Gulf Stream, seems to afford certain well-founded arguments towards this conclusion. It is impossible, unless there be a strong substratum, by many degrees colder than the superficial water constantly passing into the Gulf of Mexico, for example, that such a continued stream can flow out at so high a temperature as the Gulf Current. The tendency would rather be the reverse, the less rarified fluid pushing back that which is most heated, and occupying its place. Now the warmer fluid flowing out over the surface of the colder, must have a considerable effect on a large atmospheric space. It may even act in attracting icebergs southwards. Mr. Redfield and Colonel Reid, in their ingenious, and there is reason to think, well-founded system in relation to West Indian tempests and their causes, exhibit undeniably a coincidence of the course of hurricanes (always accompanied or preceded by electrical phenomena as they are) with the Gulf Stream, the most obvious and remarkable of all the currents of the ocean with which we have an acquaintance, and therefore that from which those conclusions can be most easily drawn, and for which lesser oceanic movements of the same nature are too confined to afford presumptive proof, the accompanying signs being too indefinable and transient for observation. If something of this nature be not the reason of the deviation of portions of the ocean from a natural state of rest into one of motion, through indirect agency, time and laborious investigation will alone give the hope of a discovery of the latent cause. The idea of some, ill acquainted with the laws of motion governing the universal system, that the earth's rotation is connected with these phenomena of the sea, is specious, but contrary to fact. The movement of every mass on the surface of the earth is subordinate to the motion of the primary body, of which motion no inferior material action can be independent that forms an integral part of that body itself. To produce such an effect there must be an external and independent cause. Passing over tidal currents, which are of admitted origin, however varied in their results, it will be proper to notice a few of the more remarkable primary phenomena of this kind, which have been made matter of nautical observation, and the effects of which have been demonstrated beyond reasonable doubt. The most remarkable current near the English shore is that denominated "Rennell's Current," which entering the Bay of Biscay by Cape Finisterre, sweeps along the N. coast of Spain and France, and sets out W. and N.W. across the English channel entrance, and so away towards Cape Clear. Vessels had long found themselves singularly out of their reckoning between the parallels of 49° and 50° for which they could no way account. In 1793 Major Rennell published his remarks on the current of Scilly, whence the name of Rennell's Current. The instances of vessels thus thrown out of their reckoning, in a way for which they could not account, are numerous. Nor is it at the entrance of the

English channel alone that the existence of this current has been shown by its effects. Vessels have not only found themselves carried N. of Scilly, for the reason of which they could not account, but others in the Bay of Biscay expecting to be lost, have found themselves carried off shore during the night almost out of sight of land, as Captain King, R.N., stated he experienced. It is remarked that along the N. coast of Spain, the water is very deep, and free from mud, but after sweeping round by Cape Ortegal, into and out of the bay northerly, it deposits along on the N.W. side, great quantities of mud, before it sets out into and across the entrance of the English channel, with a rapidity more or less, as it is more or less retarded or accelerated by secondary causes. Thus while in some cases it obstructs, in others it accelerates navigation, as indeed almost all currents may be made to do under certain circumstances. Some voyages have been made in which, through their aid, a vast saving has arisen in point of time. Hence is a knowledge of currents, and all that concerns them, of such vast importance to the seaman. The breadth of the current which thus runs to the westward of Scilly is very considerable, apparently depending upon the prevalence of W. and S.W. winds. Another noted current passes along the coast of Africa, and is distinguished as the Guinea current. This runs to the E. along the coast of Africa, into the Bights of Benin and Biafra, whence it sets out to the W. in long. 7¾ W., 50 leagues S. of Cape Palmas. The outer edge of this current sets to the E., and continues to between 7° and 8° long. S. of Cape Three Points. It then takes a course more N. in the direction of the Bights of Benin and Biafra, about 2° N. of the line. Here it blends with the South African current, which coming N. and N.W. enters the Bight. Hence S. of the equator, the united currents set to the S.W., W.N.W., and N.W. in one stream, and help very considerably the passage of vessels bound from Fernando Po to Sierra Leone. The velocity varies from 15 to 40 miles in 24 hours. Currents are found in all climates on the oceanic surface. Thus there is the Greenland current, a powerful stream setting between Iceland and Greenland. There are currents from Hudson's and Davis's Straits, which are felt along the N. shores of the American continent, and in the Gulf of St. Lawrence. Between Cape Finisterre and the Canaries, a current exists setting S.E. from Cape Finisterre, then E., and then again S.E. to 25°, and beyond Madeira W., with a velocity of from 12 to 20 miles in 24 hours. A current is found to the N. and S. of Cape Verde, and along the African coast from Cape Bojador to St. Cyprian's bay. The currents described on the E. side of the Atlantic do not reach the S.W. of the Azores; they seem to terminate in the Sargasso sea, a place where they probably concentrate and balance with others not yet ascertained. One of the most remarkable of ocean currents sets from the E.S.E. across the equator into the Caribbean sea, and a current from the Indian ocean, passing round the Cape of Good Hope, and uniting with the equatorial current N.W. of St. Helena, thence (without doubt the same current) passes in to the W. of the Caribbee islands, and along the N. coast of the continent of South America. Another branch proceeds more to the N., and even to the N.N.E. From the W. of the Caribbean sea, a current enters the Mexican, from the S. or S.W. by the Yucatan channel, and from thence into the Gulf of Mexico, whence proceeds the Gulf Stream,

passing out through the Gulf of Florida, along the American coast on the W., and the Bahama banks on the E., at a very high temperature. This extraordinary current, thus entering the N. Atlantic, is on many accounts the most singular yet known. Besides acquiring a temperature of 84°, and even in the early part of its course 87° of Fahrenheit, it has a velocity varying from 1½ mile to 5 miles an hour. Its narrowest part is in the Gulf of Florida, where it is 36 or 37 miles broad only, running with great rapidity. Off Cape Hatteras, northward, 700 miles further, it is found to be 75 miles broad, and more N. it expands to an indefinite extent. It is in the latter part of August or early in September, when it may be supposed that the surface of the waters of the torrid ocean have received their greatest heat, that the Gulf Stream runs with most velocity. It has been known to proceed 100 miles in twenty four hours in the narrows among the Bahamas; more N. it slackens at that season to 70 miles. In October its rapidity lessens, but it is by no means uniform, varying according to circumstances. Vast in extent, it most displays its singularities and strength on its N. and W. borders. Near the Cuba shore it is weakest. As it flows to the N., and expands over the colder waters beneath, it seems to extend with itself the range of atmospherical disturbance when under perhaps certain peculiar governing circumstances. The notion of its movement being affected by the winds, other than its surface, as other portions of the sea are affected, seems utterly without foundation. A wind meeting the current may raise a rougher and more troubled surface, and pushing tidal waves into various channels, raise the waters, but it can have no share in the formation or tendency of the stream itself, though it may cause temporary disturbances and frightful agitations, subsiding with the cause. Hurricanes sometimes happen when E.N.E. and N. winds prevail at the full and change of the moon. See *Gulf Stream.* There are marked currents on the coast of Newfoundland and the shores of the St. Lawrence. No pains then need be taken to establish the existence of currents generally, whatever differences may prevail respecting such as have been designated more particularly. They must be admitted an essential part of the oceanic system, remaining for scientific investigation, assisted by the fruits of close observation. An investigation indeed of the utmost importance, since it is stated that there is only a difference of one day's sail between New York and Liverpool, and New York and Havre, while in time there is a distance of nearly a week. This may be solved by a knowledge of the intervening currents. American seamen say, that sailing from the Havannah to New Orleans, some vessels stem a current of 3 miles an hour, when they may have 1 or 2 miles in their favour. Currents are not confined to the greater oceanic surfaces, they prevail in comparatively small seas, where they are perhaps ascribed to causes often right, and as frequently erroneous. That currents should prevail near the outlets of internal seas or straits, is apparently a natural consequence, dependent upon their depth, and the extent of their superficies, in connection with the rivers of which they are the outlet, for their extent and their rate of velocity. Such a current as that found near the opening of the Bosphorus in the Euxine, for example, is clearly owing to secondary causes, which can have no analogy with

K 2

those in the Atlantic or Pacific, on the bulk of which the influence of the largest river could impart no impulsive power worth regarding, more particularly as the specific gravity of the different waters is also on the side of the opposing magnitude of the oceanic body. Differences of atmospherical temperature, as already said, elevating the rarified upper surface, and producing motion by the continual flow of the colder stratum of water beneath, in an opposite direction to that of the surface, may be one great motive power of what have been here called primary currents, the causes of which are not traceable to drifts, tides, or almost visible agency in straits, or peculiar conformations of circumambient land, but must be looked for in the natural laws which govern the material world. Currents of inland seas in existence under circumstances not traceable to the influx of rivers or pressure in straits, like those of the Bosphorus or Gibraltar, exist in some shape everywhere. That eminent observer, and lover of science, Admiral Beaufort, found them in existence on the coast of Karamania, and Captain Smyth in other parts of the Mediterranean, where rivers could not have caused them, and the tidal influence was small. It is probable, therefore, that in all climates where large bodies of water are found, differences of atmospherical and aquatic temperature may cause in a circumscribed way, and in miniature, the same action observed upon the vast expanse of the greatest oceanic superficies, and cause upper and under currents. Whatever be the cause of these mysterious streams within the waters of the sea, there can be no mistake in surmising, that when science has rendered their limits and courses familiar, and explored the laws which govern them, a great advantage will be gained by the navigator, through availing himself of the aid they will afford him in reaching his destined object, to a degree perhaps as yet little expected.

CURRITUCK ISLAND, SOUND, INLET, and COUNTY, North America, United States, North Carolina. The island is 30 miles long, and 2 broad, and crosses the sound, which goes 50 miles inland, and is from 1 to 10 broad, with no great depth of water. The Sound divides the county into two parts, and comprehends the entire sea-coast of North Carolina, from the Virginian line to a point 20 miles S.W. of Cape Hatteras, the lighthouse of which is in lat. 35° 14′ N., long. 75° 30′ W., including Roanoke island. The inlet is an entrance into the sound, between two islands.

CURTO, CAPE, N.W. coast of Africa, the S.W. limit of the River Belta.

CURVEL or BERGANTIN. See *Virgin Islands.*

CURZOLA, Gulf of Venice, a large island S. of Lesina, in a direction nearly E. and W., or almost parallel with Lesina. Its length is 25 miles, and breadth 4. There are good forests of useful timber upon it, adapted for ship building; there are also excellent vineyards and fisheries, but very little corn. It has 16 towns or villages, and a population of about 8,000. The town stands near the N.E. end of the island, and is walled. It contains a dock, a port with a mole, and a roadstead. It is defended by the Fort of San Blas, which stands in lat. 42° 57′ 25″ N., long. 17° 7′ 59″ E. This island is distant from the shore of Sabbioncello three-quarters of a mile, with a safe channel between, and a depth of water from 16 to 20 fathoms, except at the E. end of Curzola, where there are many rocks, but all above water. A ship passing

through this channel, E. or W., should keep nearest to the Sabbioncello shore, where there are 9 and 8 fathoms water, and it may anchor opposite the town of Sabbioncello in from 18 to 8 fathoms. There are some places of good anchorage on the S. side. The S.W. point of Curzola is Cape Kaneira, passing that, the Valle Grande opens, at the E. end of which stands the small church of St. Giovanni di Blatta, in lat. 42° 58′ 5″ N., and long. 16° 40′ 34″ E. There are several small islands in this bay, with sufficient water to sail round them.

CUSCO BAY, North America, New England, it extends from Cape Elizabeth to Small point, 25 miles in breadth, and about 14 inland. It is a beautiful bay, full of small islands. On the shore is the town of Brunswick, in lat. 43° 40′ N., long. 70° 10′ W.

CUTOCHE, CATOCHE, COTOCHE, or CATOUCHE, CAPE, North America, the N. point of Yucatan, in lat. 21° 34′ 30″ N., long. 87° 7′ W. It is almost hid among mangroves and islands, that continually change their appearance, so that it is seldom seen from a vessel, and Contoy island, 16 miles distant to the S.W. is mistaken for it. There are two islets off this cape that do not extend more than a mile from it, and with Jolva's island, form two mouths, which are called Jonjon and Nueva, only fit for canoe navigation. From this cape W. the coast bends somewhat to the S., for a distance of 18 miles to the W. end of Jolva's island, forming the Bocas or mouths of Conic. This coast is foul; a rocky bank, with a little water on it, extends 2½ miles out, and on the meridian of Cape Catoche, there are several shoals or ridges, which extend as far N. as 21° 45′ N., having from 2½ to 3¾ fathoms upon them; the outer one has 3½, called the Green Ridge. Along this coast are vigias, or look-out towers of wood, named after the places where they are situated, as the vigia of Santa Clara, or of Silau. In these towers there are watchmen placed to discover vessels. Point Piedras is the next point or cape of note on the peninsula of Yucatan, although several others intervene. The old mode of spelling this cape is Cotoche, and it is so found in former maps and in Spanish dictionaries. Catoche is now generally adopted.

CUTLER, THE, a rocky shoal off the coast of Suffolk. The lights of Orfordness in one, lead on its outer edge ; a black buoy is placed over its S.W. end, in 4½ fathoms.

CUTTEAU BAY, Newfoundland, about 6 leagues from the Burgeo islands. It has only water for small vessels.

CUTTER HARBOUR, North America, Labrador ; about a mile W. ¾ S. from the S. point of Pocklington island. There is good anchorage for small vessels in this harbour.

CUTWELL HARBOUR, Newfoundland. Lat. of E. point 49° 37′ N., long. 55° 41′ 48″ W.

CUXHAVEN, Hanover, a town at the mouth of the Elbe, well fortified ; it is in the province of Bremen, the old station of the English packets for Hamburgh, from which it is distant W.S.W. 56 miles. Cuxhaven is in lat. 53° 53′ N., long. 8° 44′ E. The stream of the flood tide from the Texel towards the Elbe, sets E. ; at Cuxhaven, S.S.E. The flood generally runs 6 hours, the ebb 6ʰ 40ᵐ, but at Cuxhaven the ebb begins late, and continues 6ʰ 45ᵐ, and the flood 5ʰ 40ᵐ. In the road the current does not wholly cease, the flood running in three-quarters or four-fifths of an hour, after the falling of the water on the N. shore. The velocity of the current is greater between Cuxhaven and the mouth of the river, than outwards at the sea. In the channel at mid ebb, the current where strongest, runs from 3 to 4 miles an hour, and with flood 2 and 3 miles. The tide rises on full and change 11 feet; on quarter days 8½. High water full and change, at 1ʰ.

CYCLADES ISLANDS, Greek Archipelago; the islands that surround the Isle of Delos were first thus named, being those which contributed to the support of the Temple of Delos; by degrees the term got to embrace a great number of the small isles of the Ægean sea.

CYPRIAN'S or ST. CYPRIAN'S RIVER, N. coast of Spain, about 6 leagues N.W. from Ribadeo; it is small, and useless to shipping, but off this coast, among other rocks, are the Farrallones of St. Cyprian, rocks behind which in a moment of urgent necessity, a vessel may shelter herself in winds from the N.W., round by S. to E., over a bottom of sand, riding well in 4 fathoms. Beyond these rocks, passing the river at the foot of Montsancho, a good mark of the coast, and Point Roncadoira, and some other rocks; the Naro point is seen on the opposite point Socastro, the W. boundary of the entrance to the haven of Vivero; near Socastro is the Island Graviera.

CYPRIAN BAY, W. coast of Africa, a bay with a shallow bottom, on which the sea breaks with great violence. The E. part is formed by a cliff 150 feet high, having a rounded form seawards, and a flat top, resembling a fortification. The W. side, a steep cliff, extends 2½ miles to the W., then turns S.W. abruptly, and forms Cape Barbas, in 22° 19′ N. lat., and 16° 39′ W. long.

CYPRUS, ISLAND OF, called Kupriss by the Turks, in the Mediterranean, between lat. 34° 32′ and 35° 41′ N., long. 32° 16′ and 34° 38′ E. It lies E.N.E., and W.S.W., 41 leagues in length, and 11 in breadth. Vessels from the W., bound to Cyprus, fall in with it about Cape Salizano, the N.W. point of the island, in lat. 35° 6′ 20″ N., long. 32° 16′ 30″ E., 23 miles from the town of Baffa, in the bay of that name, which runs S. by E. to Cape Blanco, a distance of 15 miles. From Cape Blanco to Cape Gatte is 6 leagues, that cape being the S. point of Cyprus, and projecting far into the sea. Three leagues N. of Cape Gatte is the town of Limasol, which has a commodious harbour. The island has gone much to decay, but it is still celebrated for its Muscadine wine. Larnaca, 5 miles from Cape Chiti, has an open roadstead, but with good ground. Famagousta is the most important town in the island, built upon a rock, and nearly 2 miles in circumference, encompassed with strong walls, and surrounded by a broad ditch. There is good anchorage before the town in 10, 9, and 7 fathoms. There are sunken rocks in the neighbourhood of this island. One is named Kephalatis Attalias, laid down as in lat. 35° 45′ N., long. 30° 51′ E., and from the town of Rhodes E. by S. ½ S. distant 45 leagues. Its existence is very much doubted. Another rock, above water, lies W. from Cape Salizano 26½ leagues, in lat. 35° 40″ N., long. 30° 40′ E., and in the direct way of vessels sailing from the W. for the Island of Cyprus. Some consider this rock to be doubtful. A third rock was said to lie 5½ leagues to the S. of Cape Chiti, and E. of Cape Gatte, distant 11 leagues. The existence of this rock has been ascertained by Sir John Franklin; the sea breaks over it.

DADAYA and MOLI, ISLANDS, Mediterranean, among the Balearic Islands, having havens of the same name. See *Balearic Isles.*

DÆDALUS MOUNT, coast of Anatolia; on its W. shore a useful sea-mark.

DÆDALUS ROCK, Portugal, off Cape St. Vincent; so named because the Dædalus transport is said to have struck upon it in 1813. The spot is about 13 leagues S.W. from the Cape, but the true position has not been ascertained. It was reported as observed by a schooner from Plymouth in 1839, but there still seems a doubt of its existence.

DAGERORT, Dago, Gulf of Finland, a light-tower E. of the town of Dagerort, 538 feet above the sea, visible 30 miles. It carries a fixed light. Lat. 58° 55′ N., long. 22° 11′ 54″ E.

DAGO HAVEN, Dago or Dagden Island, at the entrance of the Gulf of Finland, on the E. side; an insignificant place only fit for small craft. It is in the Russian Government of Revel. Dagerort light is nearly W. of it, upon the opposite or W. coast of the island.

DAILO, PORT, a mere cove in the Island of Negropont, N. of Cape Doro, open to the S.E.

DAINS ISLAND, Ireland, W. coast, in the Bay of Galway, between which and Eddy island, E. of Blackhead, there is on the S. side a good road.

DAJOU or TAGOU POINT, W. coast of Africa, E. by N. 5 leagues from Cormantine. Seamen sometimes call it Tagu, and the point E. of it Rough point.

DALE ROAD, TOWN, BAY and POINT, South Wales, at the entrance into Milford Haven. From this point Milford or Man-of-War's road is distant 4 miles E.S.E. Dale town lies N. of St. Anne's point and lighthouses. Dale road lies in sight of the town; that and the bay and point being within St. Anne's point. It is necessary to distinguish West Dale bay, without St. Anne's lighthouses, from Dale bay within, or the consequences might be disastrous, and also a bay, W. of Dale bay, from that anchorage. There are lights upon St. Anne's point, being two lighthouses, painted white. The one low, 15 feet high, is 159 feet above the sea level; the other, 44 feet high, is 192 feet above the sea, and bears from the low light N. by W. ¼ W. by compass, distant 263 yards; visible 4 or 5 leagues. When these two lights are in one, or one is under the other, bearing N. by W. ¼ W., the upper light never being brought to the N. of the lower one, Lenny head may be rounded safely, and the passage in is open, the course in being N.E. by E. until Dale road be made.

DALKEY ISLAND, Ireland, E. coast of Dalkey point, the S. limit of the Bay of Dublin, in lat. 53° 18′ N., long. 6° 5′ W. The narrow passage between the island and the point is called Dalkey Sound, and has from 7 to 9 fathoms water, on any side of which there is good anchorage. The passage is narrow here, and the current strong. The little Island of Muggel lies off from Dalkey, having a small narrow channel between, with 5 and 6 fathoms water, but the ground is bad. Muggel or Mughal is often called Little Dalkey.

DALLAS, France, a small fishing place between Dieppe and Fecamp.

DALRYMPLE POINT, Dominica island, 2 miles S. of Charlotte town.

DAM, a town near the Baltic sea, on the River Oder, 10 miles from Stettin; lat. 53° 31′ N., long. 14° 50′ E.

DAM, a town on the River Danester, a league from the sea, and 5 leagues S.W. of Embden, in lat. 53° 22′ N., long. 6° 48′ E.

DAMALA, Greece, on the W. shore of the Gulf of Athens.

DAMANHOUR, Egypt, a town on the canal from Alexandria to Ramanieh, E. of Lake Mareotis.

DAMARISCOTTA RIVER, America, United States, Maine. The E. side is formed by Hern Island; high, and covered with fir-trees. Varnum's point, on the W. side, is high and wooded, and the shores on both sides bold. It is rather an arm of the sea than a river. It receives the waters of the Damariscotta pond, and is navigable for large vessels 16 miles.

DAME MARIA CAPE, or CAPE DONNA MARIA, Hayti, lat. 18° 36′ N., long. 74° 27′ W. There is the false cape, Donna Maria, as well as the true, by the first of which the true cape is soon discovered. It is very like Beachy head, in the English Channel. A ship will get ground in from 18 to 15 fathoms, and may run along by it for the distance of a quarter of a league, in from 12 to 8 fathoms weedy water. To enter the bay of Donna Maria, a distance of half a mile off must be kept, to avoid a reef, which stretches out W. a cable and a half from the point. A vessel must keep the same distance from the coast of half a mile, until past the cape to the S. of Cape Donna Maria, or the False Cape, which is foul also. Then the vessel must haul her wind, and steer S.E., keeping the lead going, and anchor W.N.W. of a large white tapion or hummock, on which there is a battery, and within a musket-shot of which 5 fathoms water will be found. There is anchorage all over this bay; at a mile from the shore there are from 4 to 6 fathoms, and at two miles from 6 to 10; and there is shelter from all winds between the N. and S., passing by the E.; but ships feel the swell lying in 9 or 10 fathoms when it blows fresh outside.

DAMGARTEN BAY and TOWN, Baltic Sea, 6 leagues W. of Stralsund, upon which the town, with the same name, is situated, in lat. 54° 20′ N., long. 12° 10′ E.

DAMIAN, CASTLE OF, Spain, Harbour of Ribadeo, is the mark of the anchoring place, in 5 fathoms.

DAMIETTA, a considerable town of Egypt, on the E. branch of the Nile, 8 miles above its junction with the sea in lat. 31° 25′ N., long. 31° 47′ E. It is situated on a narrow neck of land, formed between the Lake Menzaleh or Monzaleh and the River Nile. The town is built in a circular form along the right bank of the river, opposite to the village of Selanieh. It was once walled, but its walls are now dilapidated. It wants a secure harbour, even trading vessels being compelled to lie in the road before the mouth of the river, exposed to every wind. The ground is good to a distance for anchorage; but both E. and W. the bottom is hard. On approaching the town, the clumps of trees carry the appearance of islands until the land itself is seen. At 3 leagues distant, in 11 or 10 fathoms, the masts of the vessels in the road will be descried before the lowland is seen. The water along the coast is shallow, the soundings regular, and anchorage may be found anywhere, free from danger, in 7, 6, or 5 fathoms. Fort Lesbeh, standing at the entrance, is seen long before the line of land. At the mouth of the river there is a bar, both shallow and dangerous, but native boats await passengers and goods, and take them safely over to the town. There are two castles not far from the bar, one on each side of the river, and higher up is a fort, built by the French, under the batteries of which is the custom-house, where boats stop and luggage is searched. It is 24 leagues from Damietta to Rosetta, the land all movable and sandy. The W. half is occupied by the Lake of Bourlos, communicating with the sea, about long. 31° E. There are regular soundings all the way from Damietta to Rosetta.

DAMNABLE BAY, Newfoundland, to the N. of Salvage bay, between Cape Bonavista and North Cape Freels. To sail in from Gull island, Bonavista, a vessel must steer W.N.W. ¾ N. about 7 leagues, and round the Shag rocks, thence N. of the Baker's Loaf, steer W. ⅛ S. from Ship island, 5½ miles, which reaches the entrance of the harbour, that, on account of the narrowness of its entrance, is only fit for small vessels.

DANE'S HILL, Africa, S. coast, about 1 mile E. from Cape Coast castle.

DANIA RIVER, Africa, W. coast, between the Rio Pongo and Sierra Leone; the Isles de Los lie S.W. a few leagues from its entrance; it is S. of the River Dembia.

DANT, or DAUNT ROCK, S.W. coast of Ireland, three-quarters of a mile S.S.E. ¾ E. from the end of Robert head, upon which there is a signal tower. This rock is S.W. ⅛ S. 4½ miles from Roche lighthouse, and is marked by a black buoy with a white head, having "Daunt Rock" upon it. The rock lies in the fair way nearly of vessels passing between Cork and Kinsale. Cuskinny house, in Cork harbour, kept in view to the E. of the point of land under Fort Camden, leads clear of it. Robert Head tower in a line with Robert head will clear it, about a quarter of a mile to the S.

DANTZICK COVE, GREAT and LITTLE, Newfoundland, N. by E. from Cape May.

DANTZICK, West Prussia, about 4 miles from the sea, in lat. 54° 21′ 6″ N., long. 18° 39′ 42″ E. This city stands on the W. branch of the Vistula river, and its entrance is defended by the two forts of Weichselmunde and the Westerchantze, and communicates by canal with the Mattlan, a river that traverses the place, into which any vessel drawing 8 feet of water can pass. The chief export from this city is corn, for storing which it possesses large magazines. It has an arsenal, and 4 docks for merchant-vessels. The pop. is 45,000. Vessels making for Dantzick from the sound, after coming round Bornholm, proceed on a S.E. by E. course between Bornholm and Eastholm, and thence towards Reserhooft, or Rixhooft, or Rosehead, the most N. point of the shore of Prussian Pomerania, on the W. side of the Gulf of Dantzick. There are two lighthouse here which burn from sunset to sunrise, lit with Argand lamps and reflectors, and rising 61 and 77 feet above the sea, so that in clear weather they are seen from a ship 10 and 14 miles. Hela light, at the S.E. end of the promontory, which is a continuation of the coast from Reserhooft, has a revolving light, which shows a half-minute light, elevated 130 feet above the sea level, and about 4 cables' length N.E. ¼ N. from the extreme point of the land. These lights burn all the year. From Hela the course is S.W. ¼ S. about 12 miles, the water shoaling all the way to 5 fathoms near Dantzick. The best anchorage is with Dantzick light-towers bearing S. or S. by W., running into 6, 5, or 4 fathoms. There are two lights at the harbour of Dantzick. The lesser is fixed in a small iron lighthouse, on the summit of the Eastern Harbour Mole, and, with the large light, kept burning every night from sunset to sunrise. This new light is situated N. by compass, 1,647 yards distant from the great lighttower, 44 feet above the sea, and may be seen in all points of the compass, from W.S.W. to S.E., and from sea, in clear weather, at the distance of 10 English miles. Ships leaving Dantzick roads in the night, having arrived as far as Old Weichselmunde (the mouth of the Old Vistula), must bring the higher, or S.W. light not more W. than S.W., and the light of the E. Mole not more N. than W., in order to avoid the shoals of the Old Vistula, which extend to a great distance. The light on the E. mole bearing S. by E., or S.S.E., with the soundings of 5 fathoms water, offers safe anchorage in the roads. Both lights, which, observed in a S. direction appear one, are at a considerable distance from each other, and the great tower is W. of the one at the Mole.

DANUBE, MOUTHS OF THE RIVER, Black sea; this river enters the sea by four different channels, Kilia to the N., Soulina, St. George, and Portitcheh to the S.; separated from each other by low islands covered with reeds and trees, the shores in the Black sea bordered by shoals of 2 or 3 miles in extent. The greater part of the vessels that enter it proceed by the Soulina mouth, that being the deepest. For strangers who are sailing up the first time, it is best to take a departure from Serpent's island, from whence a ship should steer W. by S., the distance being 20 miles. In clear weather the mountains ahead, forming part of the Beche-tepeh, or Five Hills, are seen first, situated on the N. side of the river. A wooden tower, in ruins, is then seen on the port side, and to the N.W. a large building. The tower must be brought W. without nearing the land too much, and then W.N.W., on which a course must be held till the buoys and shoals are seen, and a pilot comes off, a thing sometimes much delayed. There are 9 and 10 feet water on the bar of this channel in general, but in spring and autumn about 12 feet. The water deepens gradually afterwards, and near Isaktcha and Keni reaches 78 feet. Toultche, a Turkish town, lies on the W. shore, and from thence ships bound for Ismail run into a N. direction, and double the point of the island Tchutal, which divides the river into two branches. The branch that runs to the N.E. must be taken, and arriving at the quarantine station the vessel is moored to stakes placed in the shore. Ismael is 70 miles from the mouth of the Danube, and 12 from Toultche. W. of Ismael 35 miles is the town of Reni, below the confluence of the Pruth and Danube. To proceed thither, a vessel must keep along the S. side of the Danube from Toultche as far as Isaktcha, whence numerous windings are passed, running among islets on the S. side of the river. This town is Russian, and exports much agricultural produce. Galatz, higher up, in Moldavia, between the mouths of the Pruth and Sireth, 10 miles above Reni, exports similar commodities. The Kilia channel is abandoned as a passage up on account of its shallowness. Kilia town is about 17 miles from the sea. The lat. of this mouth is 45° 27′ N., long. 29° 42′ E. It is not recommended that strangers should run from the sea into the Soulina mouth in rough weather without a pilot, since the bar and shoals are apt to shift, and what was deep in the spring may become shallow in the autumn. Even the buoys have been known to shift during a stiff gale; the depth of water on the bar varying at different times from 9

to 12 feet. The land round is low, so that the entrance cannot be discerned, except when it is quite close, or not more than 2 or 3 miles from the shore, and because there is no visible mark to guide into the entrance, for the houses spoken of above are low. In moderate weather a pilot boat crosses the bar to bring in vessels, but in rough weather remains inside, waving a flag, seldom seen till a vessel is out of danger. The freshes of this great river even to the month of July, contribute to increase the difficulties of the navigation at the entrance. Vessels during their prevalence have been 3 or 4 weeks getting up to Galatz and Brailow, that make the passage in a period very much shorter at other times.

DARAITH'S ROCK; this is ranked as one of those dangers, the position of which has not been fixed. M. Bellin, in his account of it in 1742, treats it as a certain danger, seen in August, 1700, by M. Daraith, who approached it within 1½ league, sailed round, and took an altitude within view of it. The rock is described as being three-quarters of a league broad, and 1½ long. The long. is considered very uncertain. It is given as in long. 54° 53′ W., and lat. 40° 50′ N.

DARDANELLES, or THE HELLESPONT, Turkey, Europe and Asia. These celebrated straits, at the N. of the Greek archipelago, commence, on the Asiatic side, a little N.E. of Cape Janissary, or in lat. 39° 59′ 30″, long. 26° 12′ 30″ E., and at what is called the Castle of Asia, on the same side, in lat. 40° 9′ N., long. 26° 23′ 12″ E., and are about 36 miles long. On the European side stands the castle of Europe, mounted with 70 cannon and 4 mortars. From castle to castle here is about 1½ mile. The cape on the side of Europe is called Cape Hellas, or Cape Greco in modern parlance. Within the castles the strait widens in some parts to nearly 4 miles. At Point Barbeirre, or Kehpiz Bouroun, it narrows again, then widens, and soon narrows once more so near that the two sides are not more than 400 fathoms asunder. The strait then widens a little opposite Maida or Meita, a Turkish town, again narrows at Point Nagara, and again widening a little between Sestos and Abydos, continues nearly the same width to Point Galata, on the European side, opposite Lamsaqui, on the N.W. of which, but on the European side, is the town of Gallipoli, where the sides of the straits approximate for the last time before entering the Sea of Marmora. At Gallipoli there is a lighthouse. The defences of this celebrated strait on the side of Europe are, an old battery of 15 guns near the castle of Europe, and that castle itself of 70 guns and 4 mortars. A fort 2½ or 3 miles within the W. entrance, having 12 guns; the Dardanelle of Europe, mounting 64 guns, 16 of which throw stone shot. A new battery; a second with 35 40-pounders; one with 30 60-pounders; a battery near Maida mounting 30; and a new battery at Sestos mounting 50 guns. On the Asiatic side is the castle at the entrance of the strait, having 80 guns, of which 16 throw stone shot, with 4 mortars. The Dardanelle of Asia, having 120 guns, 18 of which carry stone shot, and a strong battery near, having 25 heavy guns; a second with 30 guns; a new battery on the point of the Bay of Abydos, called Point Nagara, having 44 guns, and a new fort higher up mounted with 84 cannon. These constitute a formidable defence. The navigation of these straits to the Sea of Marmora is in general commenced, when possible, with a fair wind to stem the current; but experienced seamen advise entering with such a wind as will enable a vessel to weather Imbro, after which, almost invariably, a wind will be found coming from the N.W., enabling a vessel to work into the eddy on the Asiatic side, whence she can reach the anchorage of the White cliffs; there she may wait a change of wind which will enable her to pass between the castles of the Dardanelles. This done, she will afterwards keep the Asiatic shore on board on account of the eddy, and if not able to pass Point Nagara through an unfavourable wind she can anchor off the Kiosk; thus proceeding by degrees and saving time and expense. The Asiatic side of these straits presents a very fine landscape, thickly peopled, and well cultivated. S. winds sometimes blow at the entrance of the straits, but extend only a short way up, when a calm will intervene. In such cases, oftentimes at some distance upwards, the wind will be met blowing down; in such cases, between the two winds, a vessel should take in her studding sails and prepare to anchor or make fast to the shore. The chief obstacle in these straits is that the wind blows too often 10 months out of 12 in the direction of the current, and in the summer months being almost continually from the N. The wind here is most variable about the equinoxes. In summer the nights are generally calm, and the S. wind seldom blows for more than 2 or 3 days together.

DARDANELLE OF ASIA, or SULTANI KALESSI, a castle of 120 guns, on the Asiatic side of the strait of the Dardanelles, N. of Point Barbierre, having opposite the *Killid Bahar* or *Dardanelle of Europe*, mounting 64 guns. These castles must not be confounded with those of Asia and Europe, at the entrance of the Dardanelles, from the Archipelago sea on the S., being situated one third of the way through the strait. Between these two castles there are 20 and 22 fathoms water. They are S. of Maida and Point Nagara, as well as of the ancient Sestos and Abydos, and less than 800 yards asunder.

DAR EL BEIDA, N.W. coast of Africa, near Azamor, in lat. 33° 34′ 40″ N., long. 7° 30′ W. It is a small walled town on the sea shore, within a point that projects half a mile N.N.E. true, inside which is a cove three-quarters of a mile deep, and sheltered from E. winds. The principal export is corn. There is water here in plenty, and there are palm trees and gardens around the town, which has a pop. of about 700. There is a British consular agent here. The landing-place is behind a reef of rocks, one-third of a mile off the town.

DARGONEERS, or DRAGONERAS more correctly, Greek archipelago; 2 rocky islets on the E. side of Cerigo island.

DARIEN, GULF OF, South America; Point Caribana on the E., and Cape Tiburon on the W., forming the entrance. The Bay of Uraba applies to the bay between Point Uraba on the S., and Sandy point on the N., upon the E. side of the entrance of the deep Gulf of Darien. From Cape Caribana to Cape Tiburon is 29 miles, bearing N. 84° W. The Cerro de Aguila, or Hill of the Eagle, within Point Caribana, is in lat. 8° 37′ 50″ N., long. 76° 56′ 30″ W. That of Sandy point, or Punta Arenas, is in lat. 8° 33′ N., long. 76° 56′ 15″ W., being more within the bay. Cape Tiburon is in lat. 8° 41′ 15″ N., long. 77° 22′ 45″ W. The Culata, or bottom of the Gulf, is in lat. 7° 57′ N. The Bay of Uraba on the E. side, sometimes called North Darien, and in fact the whole bay in the E. and S. parts, offers secure anchorages at all seasons, as far as to the Bay of Candelaria, in the S. part, towards the bottom. Within the Punta de Arenas, which has two points, is the Aguila lagoon, extending E. 5¾ miles, and 3 miles broad. From Sandy point S. there are several rivers that enter the bay N. of the point Uraba; of these the principal is the Rio Salado, 5½ miles S. The land is low, with hillocks, but the bottom clean all the way to the hill and point of Cayman, 14 miles S. 14° E.; the shores swampy on both sides. Numerous small streams fall into the gulf all the way round the bottom to where the River Atrato, by several mouths, enters the bay on the W. side, carrying its main stream W. of the Bay of Candelaria. Caution is required in navigating near this bay, the lead must be in constant use, and a vessel must not go into less than 17 fathoms at the entrance, nor beyond 12 within. A portion of the bay shoals from 13 to 5 fathoms, and from that to getting aground. The only reason for entering is to communicate with the interior by the River Atrato, which though it has many mouths has only 8 navigable for boats, and the best of these, the Little Faysan branch, is in the S. part of this bay. On the bar of this branch there are 3 feet water only, and the tides rise but 5 throughout the Gulf of Darien.

DARIEN, SOUND, North America, United States, Georgia, called also Doboy; vessels making the sound bring the beacons on Wolf's island to bear W. ¼ S., and run to the buoy on the bar in this course and pass it on either side; they continue the same course until near the inner buoy to escape an 8 feet knowl midway within the heads of the shoals. The bar has 12 feet at low water. The N. breaker is to be kept on the starboard as well as the knowl, but the inner buoy is to be passed on the larboard, taking care the flood does not set the vessel on the N. breaker. When abreast of the inner buoy she must run from it 1¼ mile N.W. by N. into excellent anchorage and in sight of the lighthouse on Sapello island, the S. point of which is revolving, 74 feet high. This being made, steer for the beacons on Wolf island, bring them in a line, keep the lead going, the lighthouse being W. ¼ N. true, and cross the bar in 13 feet water. Neap ebbs here are 7 feet. Lat. of Doboy bar by computation 31° 20′ N., long. 81° 22′ W.

DARS HEAD, Pomerania, S.E. by E. from the Trindelen shoal, distant 15 miles, these two points being considered the extremities of the Danish belts. Dars Head is covered with trees, and the land is low; near it is erected a lofty seamark; at a distance it appears like a church. To the E. is the low Island Zingst, which continues flat nearly to the entrance of the strait leading to Stralsund and Griefswald. E. of Dars head, a flat of 3¼ fathoms stretches off full 4 miles; a shoal, with from 11 to 14 feet, lies E. by N., 3½ miles from the head.

DARTMOUTH BAY, America, Nova Scotia, by the Bay of Malaguash or Lunenburg, between Oven and Rose points. There are some settlements about its shores and on an island in the bottom of this bay. Off Lunenburg the lights on Cross island, S.E. point, are two, in lat. 44° 22′ N., long. 64° 6′ W., the upper revolving; lower fixed.

DARTMOUTH, Devonshire co., N.E. ½ N. 7 miles from the Start point, and 4¾ miles from Berry head, in Torbay, S.W. by W. The entrance to this harbour and town is between two high points; and it is a safe harbour for vessels against all winds. There is a castle here upon St. Peter's point, on which a light of a deep red colour is shown, elevated 49 feet above high water, visible 7 miles off, between N.W. ¼ N. and N. by E. It is lit all the year, except in June, July, and August. If vessels cannot at once sail in, they may anchor securely in the Range, in from 7 to 10 fathoms, if the wind should not blow in, and if it should do so, it will lead up the harbour. When the wind is from S.W. to E.S.E. it blows in, but from N.W. to N.E. it blows out. From all other quarters it blows in sudden flaws, and hence it is inconvenient to enter or leave the harbour without a leading wind. There are some sunk rocks on the E. side of the Range, extending 150 fathoms from the shore. It is high water at 6^h 5^m full and change, springs rise 19 feet, neaps 11. This town may be known by the square steeple of Stoke Fleming Church, on the hill, half a mile to the W. of the entrance, and by the Mewstone. The town is in lat. 50° 21′ 24″ N., long. 3° 33′ 12″ W. The harbour is capacious and perfectly landlocked. The number of vessels belonging to the port amounts to nearly 400. The coasting trade is extensive. This town is seated on the River Dart, the estuary of which forms the harbour; the pop. is about 5,000.

DASSAU, Duchy of Mecklenburg, a port on the Trave, of no great moment as respects size or commercial importance.

DATTOLO ISLETS, Mediterranean, a cluster of islets which once perhaps composed a single large island. Of these the most N. lie 2½ miles to the E. of Panaria; they are called Basiluzza and Lisca Nera, the former being the largest of the entire group, in shape like a gunner's quoin. There are on this islet vestiges of ancient buildings. Dattolo, about 1½ mile from the E. point of Panaria, is a remarkably steep white rock of lava, decomposing, in the holes of which the inhabitants place beehives. The channel between Basiluzza and Panarelli is more than a mile in width, and has from 30 to 35 fathoms upon it. Lisca Bianca, Tila Navi, and Bottaro are three islets on one bank. A lump of lava to the S. of Tila Navi, has been named by Captain Smyth St. Anne's Reef, having over it from 2½ to 5 fathoms, while all around it there are 12. A league to the S. of the three islets last mentioned, the same officer discovered a bank of sand, shells, and lava, which he named Exmouth Bank, lat. 38° 31′ 45″ N., long. 15° 13′ E. It has over it from 15 to 40 fathoms.

DAUME, France, S. coast, one of the rocky islets in the Bay of Marseilles.

DAUPHIN, FORT or PORT, West Indies, Island of Hayti, in Mancenille bay, once the boundary place of the French part of the island, which only extended to Massacre river, a little E. of this port. It has been sometimes called Bayaha.

DAUPHIN PORT, America, on the E. side of Cape Breton, at the S. extremity of the Gulf of St. Lawrence.

DAUPHIN, PORT, Greek Archipelago, a port on the N. side of the Island of Scio.

DAUPHIN ISLAND, North America, Louisiana, forming the W. point of the entrance to Mobile bay; Mobile point, on the W., being distant 3½ miles. There is good anchorage between Dauphin and Pelican islands. This island is 7 miles long, and is next to Massacre island, the W. point of which last is in lat. 30° 15′ N., long. 88° 32′ W.

DAVID'S, ST., HEAD, South Wales, N. of St. Bride's bay, having Pen y Maen point and Ramsay island between, with an irregular W. coast. This point is N.W. of the city of St. David's, which last is in lat. 51° 52′ 56″ N., long. 5° 14′ 53″ W. From St. David's head to Strumble head the course is E.N.E. ¾ E., distant 11 miles.

DAVIS'S INLET, North America, E. coast, lat. 55° 45′ N., long. 64° 48′ W.

DAWLISH, Devonshire co., 2¾ miles from Teignmouth, N.N.E. It is chiefly noted as an agreeable watering place; pop. 3,200.

DAWPOOL, and DAWPOOL DEEPS, Cheshire on the River Dee, approximating to Parkgate.

DAWRUS HEAD, N. coast of Ireland, 8½ miles E. by N. of Glen head; Roanninish island lies about 2½ miles N.E. ½ N. from Dawrus.

DAWSON'S COVE, America, Newfoundland, on the N.W. side of Connaigre bay, and N.N.E. 4 miles from Connaigre head.

DEAD MAN'S CHEST, THE, or CAXA DE LOS MUERTOS, West Indies, S. side of Porto Rico, E. ½ S. from Cape Roxo, an island shaped like a wedge, the form of the coffin, in these islands. Lat. 17° 50′ N., long. 66° 3′ W.

DEADMAN or DODMAN, Cornwall, on the S. coast, a high bluff point 379 feet high, in lat. 50° 13′ N., long. 4° 48′ W. It has a town and village upon it, the land appearing double. From it Pennar point bears W. ¾ S., distant 4¾ miles, and St. Anthony's head W. ½ S. 9¼ miles, and Black head W.S.W. 17 miles.

DEAD MAN'S HARBOUR, America, Labrador, a league N.N.W. from Square island harbour, which last lies 3¼ miles N.N.W. from Cape St. Michael.

DEADMAN ISLET, America, Gulf of St. Lawrence, N. 52° W. about 7¾ miles from the W. cape of the Magdalens. It is only about 300 fathoms long, and half that breadth, lying in an E.S.E. direction. It is 170 feet high, with steep sloping sides, but at a distance looking like a body laid out for burial. On the W. side it is bold, and a vessel may pass it within 2 cables' length in safety.

DEAD ISLANDS, Newfoundland, W.N.W. ¾ W. about 4 leagues from Rose Blanche point; between these and the main is the passage to Dead Islands harbour, a good anchorage for shipping, but dangerous of access to strangers, about 15 miles from Cape Ray, which lies W.N.W. from them, in lat. 47° 37′ N., long. 59° 20′ W.

DEAL BANK and TOWN, Kent, a shoal lying off the town of Deal, about half a mile from the shore, with no more than 12 feet on any part of it. There is a buoy of a red colour placed at its E. extremity, in 6 fathoms. The town stands on the shore, and is a place at which shipping homeward bound send letters and passengers ashore, or await orders. There is a bold beach before the place, defended by a long rampart of pebbles thrown up by the sea. S. of the town is a castle, having a ditch and drawbridge. There is a naval storehouse and an extensive naval hospital at this place. The pilots and boatmen here are excellent and intrepid seamen, active in affording assistance to vessels in distress. Pop. 7,600.

DEAL SOUND, one of the harbours of Pomona, in the Orkneys. See *Orkneys*.

DEANE HARBOUR, or POPE'S HARBOUR, America, Nova Scotia, on the W. side of Gerrard's isles.

DEBA, RIVER, Spain, N. coast, between St.

Sebastian and Cape Machicaco. Its E. point is high and steep, having a hermitage on its summit. There are 4, 5, and 6 fathoms within this river, but its entrance is so filled up that it is only fit for small craft.

DEBUCKO or REBUCKO, Africa, Rio Nunez, one of three settlements which exist some way up this river, the entrance of which is in lat., at Sand island, 10° 36' 37" N., long. 14° 42' W., according to Captain Belcher. Debucko settlement is in lat. 10° 57' N., long. 14° 21' 23" W. Captain Owen gives the entrance of the river at the S. point as in lat. 10° 35' N., long. 14° 41' 30" W. Of the settlements, Walkeria is the lowest; Debucko is 10 miles higher, having the settlement of Cassasez between itself and Walkeria.

DEE RIVER, Scotland, E. coast, a rapid stream, descending from the Grampian hills. Its mouth, confined between piers, forms the port of New Aberdeen.

DEEP SOUND, Norway, a haven between Lexe and Hitteren, visited by the coasters that are bound N. behind Lexe.

DEER HARBOUR, and ISLAND, America, Newfoundland, lying in the Bay of Bulls; it is extensive, and has good anchorage, but is barred with many shoals. Deer island is close to this bay. There is a harbour of the same name 2¾ leagues from the most N. of the Battle islands, Labrador, lat. about 52° 15' 44" N.

DEER ISLAND, Ireland, W. coast, Galway Bay, round Black head. There is a ledge E. of it, extending a quarter of a mile, and three rocks in a line with the island and Durus point, dry at half ebb.

DEGERHAMN, Sweden, on the Oland side of Calmar sound, standing on a cliff, and having a small boat harbour of 4 feet water only.

DEGERO ISLAND, situated on the N. side of the Gulf of Finland. Lat 59° 58' N.

DEGRAT COVE, America, Newfoundland, on the E. side of Quirpon island, and to the N. of Cape Degrat. Bauld cape, after Captain Bayfield, is in lat. 51° 39' N., long. 55° 26' W.; it is the N. extremity of Quirpon island.

DELAWARE BAY and RIVER, America, United States, between the states of New Jersey, Delaware, and Pennsylvania. Its entrance is between Cape Henlopen and Cape May. The lighthouse on the last is in lat. 38° 55' N., long. 75° 2' W. It bears N.E. by N. from Cape Henlopen lighthouse, about 12¼ miles. Cape Henlopen is in lat. 38° 47' N., long. 75° 6' W. The lighthouse is an octagon, built of stone, 115 feet high, and nearly at the same elevation above the sea at its foundation. It carries a fixed light. E. of this lighthouse is the Hen and Chickens shoal. The lighthouse on Cape May is 80 feet high, revolves every three minutes, and bears from the Brandywine light-vessel, distant about 2¼ miles, the last lying N.W. by N. from it; Cape Henlopen light is distant about 12 miles. This vessel carries two lights, and is only anchored off the Brandywine shoals from March 10 to September 10. Vessels approaching the Delaware by Cape May get the light to bear W.N.W., in 4 or 5 fathoms, and then run on, making a safe entrance into the Delaware, clear of all shoals, with vessels drawing only 12 or 13 feet water. The main ship-channel and deepest water is on the side of Cape Henlopen, on the N.W. of the Hen and Chickens shoal. Here there are 16, 15, 13, 12, and 15 fathoms to a good distance within the Capes of one of the finest rivers

estuaries in the world, the river, as high as Philadelphia, having a depth of 6 fathoms. The bay, too, is well-lighted, having the following light-houses besides those on the capes at the entrance, viz., one at Mispillion point, on a dwelling-house, for vessels bound up Mispillion creek, drawing less than 6 feet water. One at Mahon's ditch, S. of Bombay Hook island, on a dwelling-house, for vessels drawing 10 feet or less. One on the N. side of the bay, on Cohansy point, on the larboard side at the entrance of the Creek of Cohansy. One on the starboard side at the entrance of Christiana creek, 4 miles above Newcastle, leading up to Wilmington, Delaware. Within the Delaware capes the tides set W.N.W. the first quarter flood, N.N.W. the second to last quarter, E.S.E. first quarter ebb, S.S.E. the second to the last quarter.

DELAWARE SHOAL, West Indies, E. of the Island of Trinidad. This is said to have been found by Captain Ross, U.S. ship Delaware, in 1839, in lat. 10° 38', where he struck soundings in 37 fathoms, shells and sandy bottom. Sailing S., he passed, at 3 P.M., over a rocky bank, having 5, 7, and 10 fathoms; the bottom was seen clearly, and it was inferred that the shallow part must be in 10° 37' N. lat.; the longitude by chronometer was 60°.3' W. At 3¼ P.M. 70 fathoms were found.

DELEN CHANNEL, Sweden; the space between the Isles of Aland and the coast of Finland, is named the Waltus Kiflet; it is filled with numerous islands, through which two channels lead, one of which is called the Delen.

DELFT HAVEN, Holland, within the River Maas, on the N. side of that river, between Scheidam and Rotterdam. The canals run along the streets, and are planted on each side. It was once celebrated for its potteries, and gave birth to Grotius. It is 8 miles from Rotterdam, and has a pop. of 10,000. To go up to it, vessels, after passing Scheidam, keep the soundings on the N. shore, and run up before the haven, where they come to an anchor. The town is in lat. 52° 16' N., long 4° 15' E.

DELFZYL, Holland, in the province of Gronigen, at the mouth of the Fivel or Damster Diep, which joins the Dollart, a strong fortress, in advance of Dam, a handsome open town, a league above it, on the same arm of the sea.

DELLIS, or TEDELLES, N. coast of Africa, E. of Cape Bingut, lat. 36° 56' N., long. 3° 55' E. On the point is a fixed harbour light.

DELMINA, ELMINA, or ST. GEORGE DEL MINA, S. coast of Africa, 2 leagues E. of Commenda, lat. 5° 4' 48" N., lat. 1° 20' 12" W. There is a castle here, mounting heavy cannon. The town is one of the best upon the coast, the houses of stone. It is 10 leagues from Chama, and lies W. by S. from Anamaboe.

DEL NORTE RIVER, or RIO BRAVO DEL NORTE, North America, Gulf of Mexico, Texas. The mouth is in lat. 25° 56' N., long. 97° 12' W. Few vessels enter this river by the central mouth, as it has a bar with 6 feet water; but the trade to Matamoras is carried on through the Bar of Santiago, which is deeper, having 7 fathoms. Intending to make the bar, vessels should keep S. of lat. 26°, because, falling to the S. of the bar, it is easily made from a N. current, which runs there two or three knots an hour. In September and March they should keep as near as possible in lat. 26° The water has the same cast as that approaching the Mississippi. A large house is seen on the land, between the river and the Bar of Santiago, at the entrance of a creek called the Boca Chica, which is 5 miles S. of the Bar of Santiago. Except with an E. wind, the current here is northerly. Signals, necessary to be understood before a pilot boards, are made by flags from the shore.

DELOS ISLAND, Greek Archipelago, now called Rhenea island, 16½ miles from the S.W. point of Syra. It is divided by a narrow channel from a smaller island, which is the proper Delos. Both are called Sedilis by the Turks, and are now little better than two uninhabited rocks, on the larger of which the natives of the Island of Miconia pasture their sheep. On the smaller island are magnificent vestiges of the Temples of Apollo and Diana. The strait between these islands is an excellent harbour; the best entrance to which is to the S., with 35 and 40 fathoms water. The centre of Rhenea is in lat. 37° 25′ N., long. 25° 15′ E.

DELPHI, MOUNT, Greece, lat. 38° 37′ N., long. 23° 51′ E. Elevation 5,725 feet, W. of Cape Kili, on the N.E. shore, and inland.

DELUTE HARBOUR, America, United States, Maine, on the W. side of Campo Bello. Moose island lies on the opposite side of the channel.

DEMAS, CAPE, N. coast of Africa, Tunis, N. by W. ¾ W., 6½ miles from Cape Africa. The Balta islets lie off the N. coast of this cape, whence a sandy reef runs out 11 miles, having at its end the Islets of Kuriat, forming the S. point of the Bay or Gulf of Hammamat.

DEMBIA RIVER, Africa, W. coast, a little N. of the Isles de Los, S.E. from Cape Verga, and distant 8 leagues from the mouth of the Rio Pongo. It appears to be one of the mouths of the Cakungee river.

DEMERARA RIVER and HARBOUR, America, British Guyana; the entrance is in lat. 6°49′18″ N., long. 58° 11′ 30″ W., at the lighthouse, Georgetown, on the E. bank. This river has a bar of 9 and 11 feet at low, and 18 and 19 at high water; it is half a league broad at the entrance. The harbour is formed by Point Corrobano (on which is a fixed light, 100 feet high) being its E. point. The pilotage here is so heavy, that unless particularly required, it is advisable to anchor outside, 4 or 5 miles from the shore; Corrobano point bearing W.S.W. is the best anchorage, 6 or 7 miles distant. Vessels frequently fall in with the coast to the leeward of Demerara, which occasions much anxiety and loss of time in gaining the port, particularly as the sands at the mouth of the Essequibo run out so far from the land. To know the windward or leeward of Demerara, when to the northward of the latitude of 7° 12′ N., if land is made to the westward, a ship is assured of being to leeward, and must haul on a bowline directly to the S.E.; if land be not seen in the parallel of 7° 12′, then a ship must be to the windward. If to leeward of Demerara, she must ascertain which way the current is running; if it is flood, the Essequibo stream will drift her further to leeward; in that case, anchor immediately, and wait for the ebb; this, after heavy rains, will run long and strongly towards the N. Then weighing at the first of the ebb, and taking, by turning to windward in the stream, the advantage of a whole tide, and a vessel may always gain sufficient offing to fetch Demerara bar, on the larboard tack; then she must make a long stretch to the E. with the first of the following ebb.

DEMETRI, ST., CAPE, Mediterranean, the N.W. end of Gozo, in lat. 36° 3′ N., and long. 14° 10′ E.

DENARES, ISLET OF, Cyprus island, off Cape St. Andrea, the N.E. point of the island, in lat. 35° 42′ N., long. 34° 39′ E.

DEN CAPE, Newfoundland, the S. point of White bay, in the Machigonis river, on the E. coast, having to the E. the Islands of St. Barbe, off the mouth of Green bay; lat. 50° 17′ N., long. 56° 15′ W.

DENIA, HARBOUR and TOWN OF, E. coast of Spain, N. of Cape St. Antonio, beyond the Point of Saida. The haven is formed by two banks of mud; that on the N.W. has 2, 3, and 4 feet upon it; that on the S.E. from 2 to 11 feet. Outside there is the N. bank, called El Caballo, which has 12 feet over it, and a second, called the Ladrona, with 16 feet; and between them there are from 55 to 22 feet of water. There are from 11 to 18 between El Caballo and the entrance of the port, and between the Ladrona and the S.E. bank from 11 to 21. The channel has from 22 to 16 feet water; but to enter it a rock must be passed on the port side, having only 11 feet. Up the channel to the town further there is from 9 to 5. There is an old castle in the centre of the town. Lat. 38° 45′ N., long. 33′ E.

DENNIS HARBOUR, America, United States, Massachusets, E. of Hyannes light on Point Gammon, S. side of Cape Cod. This light is·70 feet above the sea, and fixed. Bass river lies E. of it, and near that is this harbour.

DENNIS NESS. See *Orkneys.*

DENNY ISLAND, in the Severn, E. of Goldcliffe point, on the coast of Monmouthshire, and N.W. of King's road, at the mouth of the lower Avon river, leading up to Bristol.

DEONG RIVER, S. coast of Africa, called also the Jong. It is the most central of the three rivers, the Bagroo, Deong, and Boom Kittam; the last meets it at right angles at the S.E. end of Sherbro island. This and the other rivers were once extensive slave marts, but the trade has been rooted out. The Forks of the Boom Kittam, where the Deong or Jong unites with it, are in lat. 7° 28′ N., long. 12° 30′ W.

DEPANO, or PORT DEPANO, Mediterranean, an anchorage reached through the Ducato passage, or, as more commonly called, the Santa Maura channel, the entrance being among the islands a few miles from Cape Ducato, the first of which is Arkudi, E. ¼ S., distant 7¼ miles from Cape Ducato, a dependent isle upon Santa Maura, and secondly, Maganisi. The passages most used in visiting this anchorage are those between Maganisi and Kalamo, and Maganisi and Santa Maura.

DEPTFORD, Kent co., a town on the S. side of the River Thames, W. of Greenwich, from which it is divided by the Ravensborne river, over which were two bridges. It was anciently called Deptford Strond, or West Greenwich, and divided into Upper and Lower Deptford. It has a dockyard belonging to the navy, and the principal victualling department of the service is established there. There are two hospitals at Deptford, one of which was incorporated in the time of Henry VIII. for pilots and decayed masters of vessels or their widows. Pop. 19,795 in 1831.

DERBY HAVEN, Isle of Man, on the S.E. end, N. of Castletown, formed by St. Michael's island, and joined to the mainland by a causeway 100 yards long.

DEREE ROCKS, at the W. end of the Minquiers, 11 miles S. by W. ¼ W. from the S.W.

point of Jersey island, and 11 miles N.E. ½ E. from Cape Frehel. The W. rocks are always above water.

DERG LOCH, Ireland, a part of the River Shannon, that expanding in that place is so named; the Shannon afterwards contracts again, and flows in a broad, fine channel to Limerick and the sea, newly and finally expanding into a grand estuary.

DERNA, Tripoli, N. coast of Africa, a town 29 miles distant from Cape Razatin, a place of some extent, giving its own name to the district around. The Folfelli islands lie before the town, inside which vessels most commonly anchor. From this town the coast continues W.N.W. to Point Tourba, and then W.N.W. ½ W. to Cape Heilal.

DEROUTE CHANNEL, English Channel, between Serk and Jersey islands. After a vessel has passed through the Race, as if going to the Great Russel, Serk is to be left on the starboard about a league, to avoid the Blanchard, a sunken rock, which is dry at low water, and bears E.S.E. from the windmill on Serk island.

DERTSWICK, Holland, a village on the E. side of the Zuyder Zee, within the Texel island, on the N. of Henlopen.

DERWENT RIVER, Cumberland; it forms at Workington, the best haven on the coast, admitting vessels of 400 tons from the sea.

DESART ISLAND, America, Maine, near Penobscot bay, lat. 44° 3' N., long. 68° W. Near this island are some rocks bearing the same name.

DESERTAS, THE, islands so called, two in number, lying S.E. of Madeira; the most N. is in lat. 32° 35' N., and long. 16° 32' 30" W.; the S. island is in lat. 32° 23, N., long. 16° 27' 30" W., after Capt. Didal. The extremity of the S. island is given by Lieut. Raper as in lat. 32° 38' N., long. 16° 31' W. See *Madeira*.

DESIRADA, DESIRADE, or DESEADA, ISLAND, West Indies, a small rocky island, destitute of wood and good water. It was the first land which Columbus made on his second expedition, and was called Desiderada, or the Desired, by its discoverer, after his long voyage. Cotton is cultivated here in a small quantity. There is an anchorage off the S.W. side, 1½ mile from some houses standing near the shore. The depth is 5 or 6 fathoms. The ground rock, Petite Terre, bears S.E., on which is a light E. 108 feet, and Point Chateau W. by S.; the last is the most E. point of Guadaloupe. Lat. 16° 14' N., long. 61° 12' W. The lat. of the N.E. point is 16° 20' N., long. 61° 3' 45" W. The same point, according to Borda, is in lat. 16° 20' 30" N. Captain Monteath places the centre in long. 61° 9' 7" W.

DESOLATION, CAPE, America, W.N.W. from Cape Farewell, the S. point of Greenland, and near the entrance into Davis's straits. It is nearly on a parallel with the S.E. entrance of Hudson's straits, to the W. of it. Lat. 60° 48' N., long. 48° 10' W.

DESPAIR, CAPE, America, Gulf of St. Lawrence, the N.E. point of Chaleur bay; it has red sandstone cliffs.

DETROIT RIVER, North America, Upper Canada, connecting the great inland seas of Lake Erie and Lake Huron. Expanding in its course into the little Lake of St. Clair, it carries the superfluous waters of the Huron lake, 1,000 miles in circumference, as far in its progress seaward as to its outlet in another vast body of inland water, still at a great distance from the ocean.

DEURLOO CHANNEL. coast of Flanders, bounded by the Raen and Elboog to the S.W., and the Rassen to the N.E. In time of peace this channel is buoyed; in war the buoys are taken up. It is pointed out by six white buoys on the S. side, and five black on the N., marked "D." for Deurloo, and numbered from the inner ones outward. It is the most difficult to take of all the channels of the Scheldt; therefore not to be entered without an experienced pilot.

DEVIL'S BAY, America, S. coast of Newfoundland, 4½ miles N.W. from Hare bay, and a league N.E. from Hare's Ears point; it is a narrow inlet, with no anchorage except close to the head.

DEVIL'S BAY, West Indies, Jamaica, round Pedro point, on the S.W. point of the island, to the N., facing the W., where the coast suddenly turns to N. and W., and renders the knowledge of it indisputable. Lat. 18° 5' N., long. 77° 45' W.

DEVIL'S CREEK, America, on the coast of Labrador, within the Gulf of St. Lawrence, S.W. of Francis river, and between that and Wolf creek. Lat. 51° 33' N., long. 55° 50' W.

DEVIL'S EYE, Gulf of Finland, a dangerous shoal, lying between Ragnild's Grounds and Kokskar. There are only 7 feet on the shoalest part. It bears from the Kokscar light W. ¾ N., distant 4 miles; from the Revel Stone S.E. by E. 5¼ miles; and from the Nargen light E. by N. 12 miles. The Nargen revolving light is in lat. 59° 36' 22" N., long. 24° 31' 4" E.

DEVIL'S HILL, Africa, W. coast, 2 leagues N.E. by E. from Tagu or Rough point. It lies E. of Tantumquerry, near Winebah, with Barracoe point E. ½ N. and W. ½ S. 4 leagues distant. At 4 leagues to the N.N.W. it appears shaped like a haycock.

DEVIL'S HORN, or HORN REEF, Denmark, a long narrow bank in the shape of a horn, which runs out sharp into the sea for 7 or 8 leagues, between the S. and W. Zyt, on the W. coast of Jutland, about S.E. from Numet island, in lat. 55° 40' N. The depth of water upon it varies from 2½ to 3 fathoms, and near the land to 3½ fathoms. It is therefore dangerous for large ships, rendered still more so by its steepness, as there may be 22 fathoms at one cast of the lead, 15 at the next, and only 3 at the third, when the vessel is on the sand. The high hill of Blawen creek is 4 leagues to the N. of it, looking black and steep, and to the N. of that there are white sand hills. The coast here trends E.S.E. one way to Grawe Deep, and the other way N. to the Western Zyt. Some affirm this bank to be a part of the Great Dogger bank, to the E. end of which it very closely approximates, though there is a narrow channel, with 16 and 17 fathoms between.

DEVIL'S HOLE, West Indies, Antigua, at the E. end of the island, S. of Green island, and on the N. side of the peninsula.

DEVIL'S HOLE, West Indies, a creek on the S. coast of Jamaica, 4 miles directly E. from Great Point Pedro.

DEVIL'S ISLAND, African coast, in the River Mellacoree, below which there are factories established, S. of the Isles de Los. Valuable timber is obtained by vessels from this river.

DEVIL'S ISLAND, West Indies, off the S. point of the crooked peninsula of Martinique; it extends to the W., and there forms Port Royal bay.

DEVIL'S ISLET, Greece, E. coast of the Gulf of Napoli, where there are many scattered islands.

DEVIL'S POINT, West Indies, Jamaica, 5 leagues E. of Great Pedro point, on the S. side of the island; on each side the coast forms a bay, called Coquar Plump bay and Long bay, and to the W. it trends N.W. about a league, having red cliffs. There are high hills within land, a little way from Devil's point.

DEVIL'S POINT, Plymouth, a projecting point of limestone rock, on the N. side of the passage from the Sound into the branch of that celebrated port, called Hamoaze, in which the dockyard is situated. The channel narrows here considerably between Devil's point and Mount Edgecombe, the battery of which is a little higher up on the opposite side. It is just round the rocky point of Firestone bay, which adjoins Mill bay, and faces Redding point, which lies nearly due S. of it, and the Vanguard shoal a little S.E. There is an old tower at this point which is whitened, and serves for a mark in some of the passages from the Sound. The depth of water between Devil's point and Mount Edgecombe is 19¼ fathoms. The narrowness of the channel makes it important to proceed up this passage a little before high water, and to come down a little after, unless the wind be fair, the current setting with remarkable strength at spring tides round this point. See *Plymouth.*

DEVIL'S ROCKS; these are rocks of which the existence is not yet certainly ascertained. They are stated to lie in lat. 46° 35′ N., and long. 13° 7′ W. It was mentioned by M. Bellin, in his memoir of 1742, that in lat. 46° 55′ N., W.S.W. of Ushant 110 leagues, a rock had been discovered level with the surface of the water, by a Captain Brignon, of St. Malo, in 1737. These rocks, in lat. 46° 35′ N., and long. 13° 10′ W., it is supposed may be the above danger. They are marked in most charts. They were seen by Captain Thomas, of Havre, in 1764, off the larboard bow, 3 feet above water, and about 40 feet in diameter, in lat. 46° 24′ N. In 1819, the master of the Brothick, Captain Peter, from Liverpool to Rio, observed a rock at noon, 10 feet from the starboard quarter, and 2 feet above water, in lat. 46° 35′ N., but he made the long. 13° 2′ W. The water broke upon it. Captain Scott, of the Voast, in 1829, saw such a rock. Captain Henderson, of the Fortescue, observed a rock 2 feet above, in lat. 46° 33′ N., and long. 13° 2′ W., he thought he saw more than one head. This rock was seen by Captain Swinson, in the Fortitude, in lat. 46° 35′ N., long. 13° 8′ W. He thought that the water, in fine weather, would not have broken upon these rocks. Lieutenant Sprigg, of the Brisk, in 1842, being 35 miles distant from their supposed site, shaped his course for them. He saw a change of colour in the water to a greenish black near that spot, extending in a N.N.W. and S.S.E. direction, for 1½ mile. There was no doubt of a shoal there, but a heavy swell prevailed at the time. It disappeared from the deck astern in about 15 minutes, the vessel going 8 knots. Lieut. Sprigg, by dead reckoning, placed the spot in lat. 46° 12′ N., long. 15° 3′ 30″ W. In 1820, in lat. 46° 9′ 30″ N., long. 12° 50′ W., similar indications were observed, and the same impress of the truth of the existence of a shoal on that spot, was produced on Captain Livingston, in the Friends, who passed the place in broad day, at 2ʰ 20ᵐ P.M.

DIABOLOS MORROS, America, a point of land on the E. coast of Yucatan, N. from the Gulf of Honduras, and of the island of Cozumel. The coast between them all the way is flat and low,

and covered with islands, at a small distance from the shore. It consists of a haven among hills, denominated from the bucaneers, who used to land there for provisions, and to divide their plunder.

DIA, Candia island, another name for Standia isle, of the port of Candia.

DIAMOND GROUND, Gulf of Finland, three shallow patches, small, and of a triangular form, marked by beacons. These are distant from the Seascar light E.S.E. ½ S. 12½ miles. On the S. patch there is only 4 feet water, and this, the most dangerous, is properly the Diamond Stone. These shoals are within the range of the Seascar light, when it is clear weather, during which, if the light be not seen, they are passed. They are somewhat nearer to Dolgoinoss than to Seascar island, being not far S.E. of the mid-channel.

DIAMOND ISLAND, sometimes called ROUND ISLAND, West Indies, one of the Grenadilloes, between Grenada and Curaçoa.

DIAMOND POINT, or MUCARAS, West Indies, S. bank of the Bahamas, in lat. 22° 10′ N., long. 77° 20′ W., after Capt. Barnett's survey.

DIAMOND ROCK, Island off Martinique, 10 miles to the W. of Point Salines. It is sometimes called the Devil's island, and by the French the Isle de Barque. It is 600 feet high, and about a mile in circumference, looking like a ninepin with its top broken off, the S., S.W., and E. sides are inaccessible, rising perpendicularly out of the sea. The landing place is on the W. side, but lined with breakers. To the N.E. of this rock is Great Diamond Cove, where Admiral Rodney anchored in 1762.

DIAMOND ROCK, Spain, a sunk rock in the Bay of Cadiz, generally avoided by steering between that and the Puercos, over which the sea always breaks, and between which and the Diamond there is a channel more than a mile wide.

DIAMOND ROCK, West Indies, a little W. of Acken's Kays, 5 leagues N.E. by N. from the E. end of Ash island, and 3 leagues E. northerly from the entrance of Port Louis, Hayti.

DIAMOND SHOAL or REEF, Antigua island, West Indies. It is about a mile in diameter, with from 1 to 9 feet water. Between the Diamond and a dangerous reef a mile N. of Boon's point is a channel with 5 or 6 fathoms, on bringing the Leeward Sister and the Fort Flagstaff in one.

DIANA LAKE, Island of Corsica, in the territory of Aleria, now a lagoon, but supposed once to have formed the port of that city.

DIANA REEF, West Indies, in the Crooked Island passage, discovered by the Diana packet, in 1805. The centre lies in lat. 22° 31′ N., long. 74° 47′ 30″ W. It is sometimes called the Monkey Bank. The reports of the water observed over it in different places varied greatly, from 20 fathoms to 7, and in one place to 4 feet only. This reef was supposed to lie about 9 leagues N.W. ½ N. from Castle island, but Commander Owen, R.N., having surveyed it, fixed its centre in lat. 22° 31′ N., long. 74° 48′ W.

DIANA'S SHOAL, Turkey, at the entrance of the Dardanelles, opposite the lighthouse, on the S.E. coast of the Sea of Marmora.

DICK'S COVE, DIXCOVE, or **DICKY'S COVE,** Africa, on the S. coast, E. of Cape Three Points, in lat. 4° 47′ 45″ N., long. 1° 56′ 40″ W. There is here only a creek fit for small craft. English ships used to visit it principally to load with rice. There is a good landing place. A changeable current exists on the coast, but there is clear oozy ground in 15 fathoms, where ships may anchor.

DICKENSON'S BAY, West Indies, Island of Antigua, near the N.W. angle, bounded by Wetherell's point to the N.E., and Corbizon's point to the S.W. Off this bay there are two islands contiguous, nearly parallel to the coast, called the Great and Little Sisters, of which one has its S. end nearly W. from Corbizon's point. Those islands are sufficient to distinguish the bay, and to afford shelter for small ships within them, as there is not depth for large ones.

DICTATOR'S SHOAL, Denmark, Great Belt passage, a knoll of 3 or 4 fathoms water, off Luus harbour, at the S. end of Samsœ isle, out of the way of the course of vessels through the Belt.

DIEGO, ST., CASTLE. See *Corunna*.

DIEGOS ISLANDS, West Indies, Trinidad, W. side, S. and S.E. of Prince's point, Chaguaramas bay, close to the Coloras islets, two in number, and about a quarter of a mile each in extent; there are 9 and 12 fathoms between them, and from 20 to 25 between them and Prince's point.

DIELETTE, PORT, France, department of La Manche, 8 miles S. of the Nez de Jobourg, and 2 N.E. of Cape Flamanville. The entrance is between two ledges of rocks, that dry to nearly one-third from the shore, marked by two beacons, one on each side, with a channel of 70 fathoms in breadth, on fine sand. The harbour is formed by two piers, that on the W. side is 360 feet long, well protecting vessels from W. winds. Half a mile from the entrance W. by N., is the bank called the Coucous, having only 3 feet water over it. Half a mile W. of Port Dielette, the coast rises into cliffs, and forms Cape Flamanville.

DIEMEN, Holland, a long mark used in sailing over the Pampus, coming out of the city of Amsterdam, towards the sea. It lies on the S. side of the channel which goes up to the city, and is the steeple of Diemen church, kept over the point of Tyoort to the N.E. from it, and on the N. side of the channel.

DIEPPE, France, department of the Seine Inferieure, a port having a pop. of 17,500, situated on the S. side of the English channel, at the junction of the little river Arques, increased by the waters of the Bethune, with the sea. Havre deprives it of all commercial importance, but it is a regular passage port to and from England, by way of Brighton; it is much visited as a watering place, possesses a considerable fishery of herrings, of which 36,000 barrels are annually taken; has manufacturers of lace, and works in bone and ivory. It is situated in lat. at the jetty, 49° 55' 42" N., long. 1° 5' 12" E., or by the French observations, in lat. 49° 55' 34" N., long. W. from Paris, 1° 15' 31" W. The coast from Berneval trends W. ¼ S. 4½ miles from Berneval towards this port. At N.E. by N., about 10 miles from its entrance, are in the channel what are called the Ridens of Dieppe, a bank having 4 fathoms water, with patches detached towards the W., having 6, 7, and 10 fathoms upon them. S. E. of the Ridens there are others with 5 and 9 fathoms generally, with a bottom of sand and shells, while around are from 10 to 15 fathoms, with a similar bottom. Two miles S.E. of the Ridens, there are 6½ fathoms, part of the ground connected with the Bassurelle of the River Somme. N. by W. 7½ miles from Dieppe, are situated the Little Ecamias shoals, with 7 and 9 fathoms N.N.E. from Cape d'Ailly, and on the same bearing 4½ miles from the cape, are the

Great Ecamias, with a similar depth of water. Ships waiting for a tide to run into Dieppe, anchor in good ground in 6 or 7 fathoms, in a good road, 14 miles W. of Treport. This road is well sheltered from E. and S. winds, but exposed to those from the W.N.W. to N.E. This road is W. of Dieppe, and opposite a small church on the cliff, with a remarkable steeple. The Port of Dieppe lies in a valley between two cliffs, and on coming in from the sea, shows two steeples and a castle on the W. of the town. On the N.E. side is a suburb called Le Pollet, and two stone piers on each side of the entrance. The harbour is difficult of access, by reason of the rapidity of the current, on entering and coming out it is pretty much the same. The harbour is dry at low water. The quays are secure and well sheltered from the winds. The time for entering is shown by a fixed light on a tower, 35 yards from the W. pier head, 39 feet above high water level, visible 3 leagues off. It burns only while there are 10 feet water in the harbour, English measure. On the E. pier, 33 feet from its head, there are 3 lanterns on a mast, the first, 23 feet above high water, burns all night, visible 2 leagues; the second is a tide light, 8 feet above the permanent light, lit 2½ hours before, and remaining 2 hours after, the time of high water, the third is a tide light midway between the two others, lighted two hours before high water, and extinguished at high water. The two last are never lit when the weather is such as to forbid access to the port. The mast which supports the lights is vertical when the vessel keeps a proper course, but is inclined when she does not, to the side towards which she should steer. The tide flows at Dieppe at 11ʰ 11ᵐ; springs rise 32, neaps 21 feet.

DIEU ISLAND, coast of France, 5 leagues S.W. of Noirmoutier, a granite rock, a league and a half in extent. There is no good haven in, nor any commendable anchorage around it.

DIGGS' CAPE, to the S.W. from Cape Walsingham, on the E. point of the entrance to the channel from Hudson's straits to Hudson's bay, in North America. It is in lat. about 62° 45' N., long. 79° W.

DIGGS' ISLAND, at the N.W. end of Hudson's straits, in the fair way, 120 leagues from Resolution island, at the entrance of the bay; it lies S. of Nottingham island, and W. of Cape Diggs, nearly in the parallel with the N. end of Mansel island.

DILDO HARBOUR, America, Newfoundland, 2½ miles from New Harbour, with good anchorage in from 10 to 20 fathoms; both harbours lie between Baccalou island and Cape Bonavista.

DIMARSKAR ISLETS, Baltic sea, at the entrance of the Gulf of Finland, S. of Rosala island.

DIMITRI, ST., Corfu island, an islet in the road N.W. of the town of Corfu, one of three that shelter the port.

DINA ISLAND, coast of Italy, a small isle in the Gulf of Policastro.

DINAN BAY, France, W. coast, on the S. side of the peninsula of Camaret, and in the bight of the S. coast, as Camaret road is in the bight of the N. In this bay there are from 4, to 8, 10, and 12 fathoms. Off the S.W. point are several small islands, called the Castles of Dinan, the town of that name standing on the point. From the S. point of Dinan bay the coast stretches S. about 4 miles, as far as the Bec de la Chevre, on the W. point of the extensive Bay of Douarnenez.

DINAS HEAD, Pembroke co., Wales, a projecting point of the coast, having Newport and the opening at Nevern river on the E., and Fishguard bay on the W. and S.W. From Dinas head the coast trends nearly N. by E. 3 leagues, to the mouth of the Tivy or Cardigan river.

DINGLE BAY, Ireland, W. coast, lying E. by S., and W. by N., with steep shores on both sides. Dunmore head, in this bay, is in lat. 52° 6' N., long. 10° 29' W. Only two rocks, one called Crow Rock, half a mile W. of Dingle harbour, and half a mile from the shore, covered at spring tides, and with deep water all round ; the second E. of Kaynalass point, drying only at springs, need be avoided. In moderate weather a vessel may anchor a mile from the shore, in any part of the bay.

DINGLE HARBOUR, Ireland, W. coast, in the bay of the same name. It is only fit for small vessels, that must lie aground at low water, upon soft mud.

DINMOR BANK, North Wales, off Beaumaris bay; it is a small narrow bank, with only 8 or 9 feet over it. It lies N.N.W. one-fifth of a mile from the Menai lighthouse, and W. by N. ½ N., three-quarters of a mile from Puffin island.

DIRKROOM'S SAND-HILL, a mark for finding the Land Deep or channel along the W. coast of the Helder on the coast of the Low Countries. In coming in from the sea, with a free wind, a ship must keep the great beacon upon Huisdown over this sand-hill, and run in directly as far as the outermost buoy of the Land Deep. Dirkroom's sand-hill is the second hill S. from Huisdown, and N. of Blenk sand-hill.

DISCORD CAPE, Greenland, E. coast, S.W. from Whale island, in lat. 60° 53' N., long. 42° 26' W.

DISKEN SHOAL, or as denominated in Denmark, the Disken Ground, lies off Cronborg or Cronenburg Castle, on the S.; the N. edge is distant more than a mile from the shore, extending S. ¼ E. about 2 miles, its breadth being about three-quarters of a mile. It lies in the channel direct, and has 4 or 5 fathoms, hard ground not good for holding. There are 10 and 12 fathoms on the Danish side, and from 12 to 16 on the Swedish.

DIVELAND, or DUVIVELAND, Holland, an island of Zealand, to the E. from Schonen, separated from that island only by a narrow channel. It is about 3 leagues long, by 2 broad.

DIVETTE RIVER, France, N. coast, the stream that flows into the sea at Cherbourg.

DIVI, or IVI, CAPE, Africa, N. coast, E. of Cape Ferrat and W. of Teddert bay. Lat. 36° 6' N., long. 12' E.

DIVIS HILL, Ireland, N. coast, a conspicuous object as a mark from seaward, lat. 54° 37' N., long. 6° 1' W. It attains an elevation of 1,800 feet.

DJEBAIL, a small town on the sea-shore of Palestine, between Cape Madonna and Beirout bay, near which there are Roman ruins; 5 miles S. of it is a little river called Nahr Abraham, which descends from Mount Libanus.

DJEMERMA POINT, Africa, N. coast, the W. point of the Gulf of the Arabs. Lat. 31° 57' N.

DJIEBA, Africa, N. coast, W. of the Gulf of Buschaifa, a small town opposite the rocks of Ishailoo, lat. 31° 30' N., long. 25° 37' E.

DJIMOVO POINT, Greece, in the Morea, on the E. side of the Gulf of Koron.

DNIEPER, GULF and RIVER, Black sea; the entrance to the gulf is between the point of Otchakov on one side, and Kilbouroun on the other, distant about 2 miles on the N. side of the Euxine. The entrance is to the E., and, in a general sense, the gulf may be said to be common both to the rivers Dnieper and Boug, the last falling in from the N., and the former from the N.E. The entrance is low and sandy. From the Point of Kilbouroun, a shoal-spit runs out about a mile to the N. and W. of this point and a long sand-bank stretches away to the W., to within 8 miles of Odessa. On its S. side the depths diminish gradually from 7 to 6 and 4½ fathoms, while, off the N. edge, there are 8 or 10. It may be crossed in 4½ fathoms, with the Isle Berezane bearing N. by E. There is a light-vessel at the entrance of this gulf, with three fixed lights, vertical, in lat. 46° 25' N. In proceeding towards Otchakov, the S. point of Berezane island is to be rounded at the distance of more than a mile, or not in less than 20 feet water, though, to the S., there are 8 or 9 fathoms. The two N. beacons on Berezane island brought into one, and the two on the N. side of Kilbouroun into the S.E. by E., and a vessel will be opposite the Gulf of the Dnieper in 45 feet water. The main channel passed, and the Point of Otchakov brought to bear W. by N., and then an E. course taken, and there will be found from 4 to 3 fathoms water. Within the gulf the N. shore may be approached much easier than the S. About 32 miles from Otchakov, passing the mouth of the Boug, and the town of Gloubok, and 14 miles from Kherson, appears the principal mouth of the Dnieper, pointed out by the reddish colour of Cape Kizime, which gives a name to that outlet of the river; there are 7 feet water here, and the shoals are marked. Numerous islets lie further up, covered with reeds, forming a difficult navigation. The broadest channel is to the port side, and gives from 25 to 37 feet water. The Dnieper is commonly frozen over for two months in the year, from the 1st to the 15th of December until the 10th or 20th of February. Opposite Kherson this river is about 2 miles broad and 50 feet deep. Even when frozen a S. wind, in winter, will sometimes break up the ice.

DNIESTER RIVER, Black sea; the S. mouth is 18 leagues N. by E. ½ E. from Fidonisi, and 17 leagues N.E. ¼ N. from the Kilia pass; the N. mouth is 2 miles from the S., between which is a narrow and low islet. The coast between the Danube and Dnieister is low, bordered with a sandy beach, with trees that may be seen 5 or 6 miles off before the shore is visible ; and there are some extensive lagoons inland. Two sandy points, one running S.W. by S., the other N.E. by N., with a low islet from the two passages, of which the N. is called the Passage of Otchakov, and the S. that of Tsarigrade, or Constantinople. The first is most used from being the deepest; for while on the bar of the N. there are only 4½ feet, the S. has 7. The Gulf of the Dneister is 22 miles in extent from its entrance to the mouth of the river. In the broadest part, N. of Akerman, it is 6 miles across, but between that town and Starvigoroditche it is only 2½ miles across. Inside the bar of the river there is a depth of 40 feet. Akerman is 8 miles from the Constantinople passage, and Ovidiopol 10½ miles from the same. The Constantinople passage is in lat. 46° 7' N., long. 30° 30' E.

DOBBS, CAPE, on the W. shore of the N. part of Hudson's bay, or the Welcome sea. It is the S. entrance of Wager bay, in lat. 65° 10' N., long. 86° 25' W.

DOCE LEGUAS KAYS, West Indies, Cuba, S. side, from Cape Cruz, extending W.N.W. 20 leagues. These are the Cayos de las Doce Leguas, or Twelve League Kays. Here kays and reefs rise from the bank in considerable numbers, and ships may run along them within 3 miles during the day, in 7 fathoms. They have two openings, the Boca Grande and Boca de Caballones, the last nearest to Cape Cruz.

DOCKING SAND, Norfolk coast, N. of Hasborough sand. The N. end lies E. by S. ½ S. 3½ miles from the S. end of the Inner Dowsing, and W. 2 miles from the red buoy on the N. end of Race's shoal, having a chequered black and white buoy upon its N. extremity. A part of this shoal is very dangerous, as the lead gives no warning; for there are 9 and 11 fathoms close to the bank. The chequered buoy lies in 9 fathoms, with Hunstanton lighthouse S.W. 18 miles, and Ingoldsmel church W. by N. 13 miles. The S.W. point of this sand lies three-quarters of a mile from the N. point of Burnham flats, between which is a swashway of 5 or 6 fathoms. There are only 7 or 8 feet of water on the N. part of the sand.

DOCKUM, Holland, a town of Friesland, on the River Ee, sometimes called Dockum river. It has a good port, well situated for trade, on the W. coast of an inlet of the sea, on the very N. part of the Low Countries, between Friesland and Groningen. The passage is S. of Schiermonikoog island, a small island off the coast to the E. of Ameland. It communicates with the interior by means of canals. Lat. 53° 18′ N., long. 5° 41′ E.

DODO, or DODA, RIVER, Africa, S.W. coast, N.N.W. of Cape Formosa about 13 leagues.

DOGGER BANK, on the E. coast of England, and between that country and Holland, a very extensive sand, between the Well bank and the Fisher's gat. Its W. point is nearly 12 leagues E. of Bridlington, in Yorkshire, and it is the shoalest, having but 9 fathoms water. It extends nearly E.N.E. and N.E. by E. to within 12 leagues of the coast of Jutland, being, in some places, 20 leagues broad, running out narrow and pointed towards the E. end. Shoalest towards England and its middle part, it has, in some places, 25, 26, and 27 fathoms over it. On its N. side a long bank extends from it with deeper water, called the Great Fisher, or Long Bank, the N. and S. ends of which are distinguished on the W. side by the names of the North and South Flats. This bank is a noted station for the cod fishery, principally carried on by the English and Dutch, where boats repair in considerable numbers. On the S. side are what are called the Great Silver Pits, being narrow and irregular ground, 10, 12, and 15 fathoms deeper than on the adjoining bank to the N. The Well bank lies S. of this irregular ground, between the Dogger bank and the Dutch coast.

DOGGER and BOAT SHOAL, Durham co., N. by W. of Hartlepool, distant about 5 miles.

DOG ISLAND, America, an island in Mobile bay, belonging to a chain of which Massacre and Horn islands are part, Cat island being the last of the chain, having on its W. point a fixed light.

DOG ISLAND, West Indies, one of the islets near Anguilla, on the N.W. by W. of the W. point of that island, due N. of Passage and Prickly Pear islands. Lat. about 18° 26′ N.

DOG KAY, Bahamas, the N. of the Cat kays, generally denominated the Gun kay, on which there stands a lighthouse, revolving in 1ᵐ 9ˢ. Lat.

25° 34′ 30″ N., long. 79° 18′ 24″ W. See *Bahamas*. It would appear that of the present Cat kays, so called, one was formerly known as Dog kay, a second as Wolf kay, and a third as Cat kay.

DOG and PRICKLY PEAR ISLAND, West Indies; 14 miles N.W. ½ N. from Great Anguilla kay lies Prickly Pear island, whence a rocky reef stretches off E. 14 or 15 miles, several parts are always above water. Near its W. part is Passage island, between which and Dog island there is a channel 3½ miles to the W., with deep water through which vessels pass. The islands are all low and uninhabited, except Dog island, which is the largest and has inhabitants. From the lowness of these islands they cannot be seen any considerable distance off. Detached from the N. part of Dog island is an islet called the North Dog, and to the W. a larger one called Hat kay, with a rocky reef off its S. point.

DOG REEF, or THE EL PERRO REEF, Spain, S. side of the entrance of the Guadalquiver; it is a dangerous rocky shelf, running out from Point Chipiona N.W. about three miles. This point is in lat. 36° 44′ 18″ N., long. 6° 26′ W.

DOG ROCK, America, Nova Scotia, S., 1½ mile from South-West, or Holderness island. There are several sunk rocks near it. It is sometimes called the Horse-Shoe rock.

DOG ROCKS, West Indies, W. of the Great Bahama bank, on Salt Kay bank, N. of the Anguilla islands. The N. Western of the Dog islands is in lat. 24° 4′ N., long. 79° 52′ W.

DOG'S HEAD SAND, between Foulness and the Humber, covered at quarter flood, lying S.W. ½ S., 4 miles from which a bar of shoal water extends across the entrance, and joins the end of the Outer knock. New buoys are placed on the bar, the southern buoy is red, and lies in 6 feet water. N. by W. ¼ W. from this red buoy, is a black beacon-buoy, with a ball; S.W. by S. from the beacon-buoy is another black buoy; and farther in, on the side of the Dog's head, is a red buoy. The latter lies nearly S.S.W. from the second black buoy, distant half a mile; and from the outer red buoy of the Dog's head S.W. ¾ W., almost a mile. These black buoys, in entering, must be left on the starboard side, and the red buoys on the port or larboard. This is the common channel into Boston Deeps.

DOG'S ISLAND, or CHAON. See *Cape de Verde Islands*.

DOG'S ISLAND POINT, called also CHARLES' POINT, Africa, on the N. side of the River Gambia, 8 miles higher up, on the N. side, than Barra point. The island is opposite in the river way to the point.

DOG'S POINT, West Indies, Tortuga island, on the N. side of Hayti, the N.E. point.

DOIG'S ROCK, Scotland, in the fair way between Inchcallen and Inverkeithing, or St. David's road, S. by E. ¼ E. from St. David's head, and 3 cables' length from the shore. There are only 3 feet on it at low water.

DOL, France, department of the Ille and Vilaine, on the River Cardequin, 2 leagues from its mouth, a small port and town, with a pop. of 3,700. Its exports are principally wheat, cyder, and flax, coastways.

DOLCE GOLFO, *anglice* FRESHWATER GULF, America, in the S.W. bight of the Bay of Honduras, and W. of Port Omoa. The inhabitants of Guatimala at only 8 leagues from the Pacific Ocean, carried on formerly their commercial relations with Spain through this gulf, the distance being shortened by making use of it, though not less than 40 leagues from the Gulf of Mexico.

DOLGELLY, Wales, a town of Merionethshire, 10 miles from Barmouth, at the head of a boat navigation which comes up to that town by means of the rivers, which form a junction just below it, named theWinion and the Maw. Cader Idris, or the Chair of Idris, a mountain 2,914 feet high, a conspicuous sea-mark from Cardigan bay, rises with a craggy double-crested summit immediately over the town. It is situated inland from the sea-shore 8 or 9 miles, but its height renders it sufficiently striking to be of service.

DOLGOI ISLAND, Black sea, low, and of small superficies, S.E. of Otchakov, in the Gulf of Kilbouroun, within which there is room for small vessels, secure from N.W. and N. winds, in from 18 to 9 feet water, 2 miles from Dolgoi island. The revolving light on Tendra island, 92 feet high, in lat. 46° 18′ 54″ N., long. 31° 29′ 30″ E., is about 17 miles S.W. from it.

DOLGOI NOSS, Gulf of Finland, in the channel to St. Petersburgh, it constitutes,and is often styled there, the S.W. cape. There is a wooden tower on this point, 50 feet high and 171 above the sea, erected as a landmark. Lat. 59° 56′ N., long. 28° 57′ E.

DOLLABARAT'S SHOAL, Azores, S.S.E. of the Formigas. It was shown in a chart of the Atlantic in 1766, its existence long disputed, it was afterwards omitted in many charts. After conflicting statements, but evidence on the whole that led to a belief of the existence of this danger, Captain Vidal fixed its position at S. 44° E. true, from the Formigas distant 3½ miles, in lat. 37° 13′ 30″ N. He declared it a most fearful and insidious danger, only showing itself in a high swell of the sea, and the least depth on it 11 feet at low water.

DOLLART BAY, a large gulf, on the N. side of which is situated the city of Embden, at the extreme part of Groningen. It divides East Friesland, in Germany, on the N.E., from the Low Countries to the S.W.

DOMBOCARRO, Africa, W. coast, a settlement belonging to the slave-dealers, recently destroyed by the Wanderer sloop-of-war. It was on the Gallinas or Galhinas river, the nearest slave-mart to the bar of that river.

DOMESNESS, the S. point at the entrance into the Gulf of Riga, whence a reef runs off a good way direct from the point, which must have a berth of 4 miles in passing. The reef or bank goes off in a N.E. ¼ E. direction. There are 8 and 15 fathoms at the end of the reef, shoaling suddenly to 6 and 4. There are double lights here in lat 57° 46′ N., long. 22° 32′ E.

DOMESNESS POINT, coast of Norway, 4 leagues from the small island of Swan Holm, lying near the shore, having a channel round it, through which vessels may pass.

DOMINGO, ST., CAYO DE, S. edge of the Great Bahama bank, and the most S. kay on the bank. It is a cable's length long, dry, and in its middle forms a small hill, covered with the Indian fig bush, which appears like a vessel upset, and may be seen at the distance of 3 leagues; a breaker extends from its S.S.W. side to the distance of a cable's length. In 1835 the crew of the Thunder erected a beacon of stones upon it, 15 feet high, about its centre: most part of it is only

5 feet above water. It is in lat. 21° 42′ N., long. 75° 44′ 45″ W., and is 15 feet where highest.

DOMINGO, ST., RIVER, Africa, W. coast, E. of Cape Roxo, sometimes called the Cacheo river. It falls into the sea opposite the Cacheo bank. The entrance is in lat. about 12° 11′ N., 16° 30′ W.

DOMINGO, ST., or HAYTI, ISLAND OF, in the West Indies. It was discovered by Columbus in 1492, who named it Hispaniola, giving the name of St. Domingo to a city which he founded there two years afterwards. Subsequently the island became divided into two parts, one French, the other Spanish, and more recently the whole became an independent republic, under the name of the *Republic of Hayti*. This island is 114 leagues long, from Cape Enganno at the E., in lat. 18° 35′ N., long. 68° 18′ W., to Cape Tiburon W., in lat. 18° 22′ N., long. 74° 28′ W. It is 45 leagues broad where widest, from Point Isabella or Isabelica on the N. to Point Beata on the S. The former is in lat. 19° 59′ N., long. 71° 1′ W.; the latter in lat. 17° 35′ N., long. 71° 27′ W. The surface of the land is greatly diversified, the most elevated spots being 6,000 feet above the level of the sea. The climate is hot and particularly unhealthy to Europeans. The pop. in 1823 was calculated, mulattos and negroes, to be about 820,000, and the whites at 30,000. The ports, bays, and harbours in this island are numerous, and some of them of great extent and beauty.

DOMINGO, CITY OF, Island of Hayti or St. Domingo, on the S. side. It stands on the W. bank of the River Ozama, having on the same side a large fort, and W. of the fort a fine and extensive savannah. The harbour is peculiarly commodious, and ships lie close to the shore to take in their lading, over planks from their sides to the wharf. The city itself is built upon a rocky point. It has a spacious cathedral, footways paved with brick, and is surrounded by a wall. To sail into the harbour, a vessel must steer directly to the castle with a steeple, and approach it within a mile. There will be found 15 fathoms water nearly opposite a point of considerable size on the starboard side, and a little within it a small fort on the larboard; here a vessel may run in without hazard. From the entrance of St. Domingo harbour the W. point of the bay, named Point Kissao, or Catalina point, bears S.W. ¼ W. 9 leagues. The city is in lat. 18° 29′ N., long. 69° 59′ W.

DOMINGO, ST., BAY OF, Hayti, or St. Domingo island. The great Bay of St. Domingo has Point Nissao on the W., and to clear it for that anchorage a vessel must steer S. by W. or S.S.W., and having run on either course a distance of 14 miles, she will be to the S. of it, or she may run 6 miles on the first course and then steer on upon the second. The coast from Nissao point trends about S.W. and W.S.W. to Salinas; it is so clear that it may be run along at less than 2 miles distance. From the Point of Salinas it bends to the N., and forms the Bay of Ocoa or Neiva, in which there are several good harbours and anchorages.

DOMINICA, West Indies, an island of the windward group, lat., Roseau town, 15° 18′ 24″ N., long. 61° 24′ 42″ W., and the road, lat. 15° 18′ 30″ N., long. 61° 25′ 15″ W. It can hardly be said that there is a harbour in this island, but there are excellent anchorages, perfectly safe, along the N.W. side, which is all bold.

III.

L

Vessels are, of course, exposed to the W. winds, but only in the winter months. In the Road of Roseau, the capital of the island, and on the S.W. side, ships may anchor in from 15 to 25 fathoms, good holding ground; and 3 or 4 miles from the N. end lies a noble bay and anchorage called Prince Rupert's bay. Scot's head, or Point Cachacrou, the S. point of the island, is a high rock, appearing at a distance like an island, from which the town of Roseau is distant 6 miles to the N. Merchant-ships generally anchor in the bay to the N. of the town. The course and distance from Roseau to Rupert's bay are N.N.W. 17 miles. Woodbridge bay is on the same side of the island as Roseau; St. David's, and Rosalie bay, are on the S. side, Hillsborough bay on the E. side, and Colebrook bay on the W.; but the N.W. side is to be preferred in all respects. The entire island is about 29 miles long by 16 broad, and contains about 186,436 acres. It was discovered by Columbus, on Sunday, November, 3, 1493, and named from that circumstance. The climate is moister than in the other islands. The maximum temperature is 88° Fahr., the minimum 69°, mean of hottest months 81°, of the coldest, 74°. The prevailing winds are northerly and easterly. The pop. 20,000. The value of exports about 140,000*l.*, of imports 35,000*l.* The revenue from 6,000*l.* to 7,000*l.* The expenditure about 28,000*l.* annually.

DOMMEL, Brabant, a river, which, joining the Meuse, forms the island of the same name, on uniting with the Wahal.

DON, RIVER and GULF OF, Sea of Azof, on the N.E. side. The Gulf of Azof is first entered between Cape Obriv on the S.E. and Point Bielosaria on the N.W. From Cape Obriv two banks run out, one called the Helena bank, 15 miles to the W., the other the Dolgoy bank, 14 miles to the N.W. Proceeding N.E., through the Gulf of Azof, towards its extremity, three spits extend, one from the S.E. and two from the N., which lock the S.E. between them, on the extremity of which there is a buoy, as well as on the most W. of the others, called Kossa Zolotoi, and following the crooked channel thus described, on rounding the third point, the gulf of the Don is entered. The river flows into the gulf by several mouths. On the S. branch or mouth is situated the town of Azof, an insignificant place. The N. branch runs nearly W. 4 leagues, then turning to the S. it forms a small bay, on the W. cape of which stands the fortress and town of Taganrog, 14 miles from the Don. The different mouths of this river are separated from each other by sand-bars, and small islets. Two or three of the branches might receive small craft, but they vary in depth at particular times, and those who visit them are continually obliged to trace out a new channel. Coasters have made most use of the Kalancha mouth, but even that may have changed. It is reported that the depth of water about this gulf has diminished 3 feet since Taganrog was founded.

DON, Scotland, E. coast, a river, on which stands Old Aberdeen, having a good salmon fishery. Small vessels enter its mouth.

DONAGHADEE, Ireland, E. coast, Down co., a small haven where the Scotch packet boats run from Port Patrick. Pop. about 3,500. It lies W.S.W. of Port Patrick, and has a pier haven, with 10 feet water at springs, and 8 at neaps. The pier-head is in lat 54° 38' 38" N., and long. 5° 32' 25" W.

DONAGHMORE BAY, Ireland, on the E. coast, between Wicklow and Wexford. It is formed by a small point to the S.E., and is the mark for passing within the banks of this coast through the channel, between the middle and S. grounds. When the Castle of Donaghmore is brought to bear W. by N., or W.N.W., a vessel may run right in, and come to an anchor just within the point, which is low ground. If the castle can be seen a little open of the point it is the sure mark for the channel, where there are 15 and 16 fathoms shoaling gradually to the shore.

DONAHA SHOAL, Ireland, W. coast; it bears from Carrigaholt town W.N.W. ¼ N., distant 2 miles, and has only 3 fathoms on it at low spring tides. It is about 3 cables' length from E. to W., and 1¾ from N. to S.

DONAN'S or DONEAN'S BAY, Ireland, S.W. coast, between Three Castle head and Sheep's head, best known as Dunmanus bay.

DONAT'S, ST., CLIFF or BLUFF, South Wales, coast of Glamorganshire, upon which is a castle in ruins, that, with its detached watch-tower, forms a remarkable sea-mark. It stands on the main, opposite the shore between Barry island and Dunraven castle.

DONCAL, THE, BANK, Spain, N. coast, at the entrance of the port of Santona; it has only 2 or 3 fathoms upon it, and lies S.E. ¼ E. from Passage point, distant 1 mile.

DONEGA, CAPE, White sea, N.W. of the mouth of the Dwina. Off this cape is the Island of Rovestra, or Gishginsh, having a lighthouse 140 feet above the sea, with a fixed light, on the E. side of the Bay of Onega, in lat. 65° 12' 17" N., long. 36° 51' 30" E. It is high water at the cape at 6ʰ.

DONEGAL BAY, Ireland, W. coast, N. of Sligo bay, between Gessigo point and Teelin head, distant from each other about 18 miles. Teelin head stands in lat. 54° 41' N., long. 8° 46' W.; it is very lofty. This bay is above 20 miles deep, and contains several harbours. A lighthouse is erecting on Rathlin O'Birne island, W. of Tillen or Teelin head.

DONEGAL HARBOUR, Ireland, W. coast, on the E. side of the Hassens, or Green islands, 2 miles below the town of Donegal, the lat. of the middle of which town is 54° 39' N., long. 8° 6' W. This harbour is only fit for small vessels, which may ride half a cable's length from the shore, in 2 and 3 fathoms water. The channel begins 2½ miles below the Hassens, and may be navigated there by vessels drawing 16 feet, at high water. High water here 6ʰ 26ᵐ. Springs rise 12 feet, neaps 6.

DONNA MARIA BAY and CAPE, Hayti; in lat. 18° 36' N., long. 74° 27' W. This cape is best known to the English as Cape Dame Maria. See *Dame Maria.*

DONNA NOOK BEACON, Lincolnshire; this is a beacon on a point of land N. ¼ E. 3½ miles from Saltfleet. It is of a conical form, red, 50 feet high, and visible 3 leagues; it is therefore a remarkable object on this part of the coast.

DOORLOY, or DOORLOG CHANNEL, Holland, one of the four passages for great ships, between the Isle of Walcheren and the Flemish coast, often known by the general name of the Weilings. See also *Deurloo.*

DORAN ISLAND, France, N. coast, with the Isle of Zearn, lying nearly 2 miles N. of the Tuscees and St. Maude islets, to the N.W. of Brehat, on the coast of France. Outside these, and to the E. the coast is foul and rocky.

DORCHESTER CAPE, North America, up the inlet on the N.E. from Cape Charles, towards a channel, which, communicating with Davis's straits, forms an island, of which it is the W. angle, on the E. side of Hudson's straits.

DORDOGNE RIVER, France, rising in the department of the Puy de Dôme, and falling into the Gironde at Libourne, near Bordeaux, whence they form a common channel to the sea, in the Bay of Biscay, at the tower of Cordouan. Its total course is about 140 leagues.

DORMET ISLAND, America, Nova Scotia, an islet at the entrance of Tatmagouche bay, on the S. coast, and on the S. side of the channel, between that and St. John's island on the N.

DORNBUSH POINT, Prussia, the most W. point of the Island of Rugen, in the Baltic, to the S. of which about 2 leagues, the entrance goes in to Straelsond or Sound. It is high steep land, yet there is good anchorage near it for E.N.E. and S.E. winds, in 3 or 4 fathoms. Whitmond point is 4 leagues distant from Dornbush.

DORNERSDORF, a town near the Gulf of Lubec, on the W. side of the Trave, which finds an outlet in that gulf.

DORNOCK, FRITH OF, and TOWN, E. coast of Scotland. This frith is sometimes called the Frith of Tain, which stands on its S. side, with Dornock on the N. This last is a town of 2,500 inhabitants, the only one in Sutherlandshire worthy of mention. Brora haven, at the mouth of the little river of that name, is situated on the N. of the entrance of the frith.

DORO, CAPE and STRAIT, Greek Archipelago, the S.E. point of the Island of Negropont, lat. 38° 9′ 30″ N., long. 24° 36′ E. The elevation of the land here is great, rising to 4,606 feet. Lieutenant Raper gives the lat. of the extremity at 38° 10′ N., long. 24° 35′ E., and of the summit 38° 3′ N., and 24° 28′ E. respectively. The Strait of Doro is sometimes called the Bocca Siloto. It is about 6 miles wide, and is the channel most used by vessels going to the northward. Vessels unable to get through the Doro channel, find it most advisable to run for Zea or Port Raphti, where shelter may be had until the passage can be made with success. Cape Doro, in Negropont, was the ancient Caparenm; Andros island forms the S. side of the Strait of Doro.

DORSET CAPE, North America, a point of land to the W. of Cape Cook, near the E. entrance of Hudson's straits, S.E. from Cape Charles.

DORT, or DORTRECHT, Holland, an island in the Blèsbosch, having a pop. of 20,000. It has a considerable trade in corn, timber, and wines; the timber is floated down the Rhine, and cut there by numerous saw-mills.

DORTACH, or D'ORTACH, British channel, the passage between the D'Ortach rock and the Caskets. It is not considered so safe as the Passage du Singe, and is only used in cases of necessity; in a calm the different settings of the tides between the Caskets and Guernsey make it dangerous. It is a rule going through to keep well to the E. towards Ortach, to avoid the sunken rock, called Le Quest. See *Alderney.*

DOUARNENEZ BAY, France, S.S.E. ¼ E. from Ushant about 9 or 10 leagues, and beyond the Point de Chévre. It is a large and commodious bay, the entrance wide, ground clean, and sounding so regular that no leading mark is required, only avoiding the dangers laid down in the chart. The best ground is towards the N. shore, clear sand, with a depth of from 9 to 15 fathoms. The S. shore of the bay runs in a W. direction to Point le Van, a distance of nearly 16 miles, with land high, steep, and having projecting points with rocks off them, dangerous to approach too near.

DOUBLE-HEADED SHOT KAYS, West Indies, Gulf of Florida, on the N.W. edge of Salt Kay bank, a narrow ridge of detached barren rocks. A fixed light has been erected here 160 feet above the sea, visible from 14 to 20 miles in all directions, except S.S.W. ¼ W., where it is hidden by Water Kay, at the distance of 9 miles.

DOUGLAS BAY, Isle of Man, 2¼ leagues N.E. from Derby haven, on the E. side of the island. There is good anchorage here in 20 fathoms.

DOUGLAS HARBOUR, North America, within Cape Dobbs, on the N. side of the large inlet of Wager bay. There are several small islands before it. The lat. is about 66° 32′ N., long. 91° 30′ W.

DOUGLAS ROCKS and PASSAGE, West Indies; these rocks lie S.W. ½ S. two-thirds of a mile from the W. end of Booby island, having two beacons erected on them. They are surrounded with a reef, and on the N.E. extremity of the reef is placed a black buoy, bearing N. by E. ¼ E., 179 fathoms from the S.E. beacon on the rocks. Opposite to it is the end of a reef which extends from Booby island one-third of a mile, leaving a passage of 160 fathoms in breadth, with 4½ fathoms at very low tides; this is called Douglas Passage, and is that which is used by ships-of-war to and from the New or Cochrane's anchorage. To the S. of Douglas rocks is another passage, of nearly the same breadth, called the South Passage; it has only 2½ fathoms in the shoalest part, and lies between a dangerous sunken rock, at the distance of a large quarter of a mile, S. by W. ¼ W. from the high beacon and the Douglas reef. Some islets and a reef extend from the N.E. end of Rose island to within a cable's length of the above dangerous rock, in which space is a depth of 2 fathoms, but the tides run through with great velocity, and render it very hazardous. Immediately within these passages the water deepens to 4½ fathoms, and at the distance of half a mile S.S.E. from the Douglas beacons, has a depth of 5 and 5½ fathoms.

DOUGLAS TOWN, America, Nova Scotia, a small town of fishermen and farmers, on the S. side of the entrance of the River of St. John's. The water is deep in the outer part of the bay, from 30 to 60 fathoms over mud, but the depth decreases to the anchorage.

DOURO, RIVER, Portugal; the N. point of the entrance at the castle of S. Joaõ de Foz is in lat. 41° 8′ 48″ N., long. 8° 37′ W. A dangerous shifting bar crosses the entrance, always rendering a pilot necessary. If no pilot will venture off, no attempt should be made to sail in. A tower stands, exhibiting a revolving light, 185 feet high, at the chapel of N.S. de Luz. From the castle of that name, a ledge of rocks extends to the S.W., some of which are always above water, and without this ledge is another, called the Filgueira, always visible, and to be left on the starboard when entering. To the S. of these is a sunk rock, called the N. Lage, and further S. about 15 fathoms, the S. Lage. After the bar is passed, there is a rock with a pillar on it, called the Cruz, another named the Agulha, and to the S., a long rocky ridge, called the Breakwater. A low sandy point stretches out on the S. side of the entrance, called the Cabedillo, going N. to within a cable's length of the N. shore, and this forms the shifting and dangerous bar. Vessels for the River Douro, should be well assured of their lat., for with S.W. and N.W. winds, a heavy sea sets along this coast. The bar is liable to alterations from gales of wind, and freshets or freshes of the river, and no stranger should attempt to enter without a pilot. The pilots first enter with the chapel of St. Catherine in a line with that of St. Michael, called also Ango Dome, bearing E. by S.; or the latter on with the bar mark, a white tower half-way up the hill behind it, and thence act according to circumstances. Freshets most frequently take place in the spring, from the melting of the snow on the mountains. The rise of the water at such times is often 40 feet, and the rapidity of the stream breaks vessels from their moorings, when it is impossible to afford them the least assistance. As no dependence can be placed on the anchors in such times, vessels are secured by a cable to trees on the bank of the river, or to stone pillars, erected for the purpose. They have time for preparation, as the approach of these freshets is communicated from the interior several days before their arrival, during which time the river gradually rises. The ordinary rise of springs is from 10 to 12 feet, and that of neaps from 6 to 8 feet, at the entrance of this river.

DOUVRES, or THE ROCHES DOUVRES, N.N.E., about 5 miles from Rock Gautier, on the N. coast of France, and 16 miles N.E. ¼ E. from Brehat island, is the centre of these rocks, the heads of several of which are always above water, and others dry, more or less, at spring tides, having shoals scattered about within a circumference of 9 miles. These rocks bear S.W. ¾ W., 23 miles from St. Martin's, the S.E. point of Guernsey, and W. by N. 24 miles from the S.W. end of Jersey. The S.W. end of the Douvres is in lat. 49° 6' N., long. 2° 51' W. The channel between the Roches Douvres and the Gautier and Barnouic is about 2½ miles wide.

DOVE, CAPE, America, coast of Nova Scotia, 6 leagues S.W. from Prospect harbour, and S.W. of Chebucto bay, in which is the town of Halifax.

DOVE CRAIG, Scotland, above Queen's Ferry, in the Frith of Forth, near the N. shore, a small island with a heap of stones upon it, a quarter of a mile from the land. On the N. side it dries at low water; between that and the shore, and on the S. side, it is flat for a cable's length off.

DOVE SAND, coast of Cheshire, within the Hoyle sand, off the N. entrance into the River Dee, below Chester. There is a fair channel between through which ships may pass. The Dove sand comes off the point of land, and stretches both ways along the shore on the N. side of the Dee river, to the entrance of the Mersey, below Liverpool.

DOVER BAY, America, Nova Scotia, on the S. coast, near Cape Canso, a wild indentation, at its head full of rocks and islands.

DOVER, England, Kent co., seated upon a small stream, which falls into the harbour. It is one of the Cinque ports, and a borough returning members to parliament. The lat. of the castle is 51° 7' 48" N., long. 1° 19' 30" E. The harbour will receive vessels of 300 or 400 tons, its entrance is narrow, between two piers, the channel N.N.W. and S.S.E. The W. pier projects further into the sea than the E.; on it are two flagstaffs. Upon the highest staff a red flag is hoisted while there are 10 feet water and above in the harbour. Two bright gas lights are lit at night while there is the same depth of water. The lights are of a red colour, that they may not be mistaken for any others near, and they stand nearly N. and S. of each other, at the distance of 15 feet. The lights do not indicate the channel, as the sand is constantly shifting. The upper light, in clear weather, is visible 12 miles, and may be seen seaward from E.N.E. to W.N.W. There is also a red light on the N. pier head in addition, at the entrance of the harbour, while the tide is in the same state as above mentioned. Dover lies W. ¾ S., or nearly from the South Foreland, distant 3 miles. In coming from the W. making Dover, a vessel should run for the jetty, W. of the pier heads, then steer towards the castle until the gates of the basin appear, and then haul quickly round for the harbour. A pilot is necessary to enter after a S. or S.W. gale, as under such circumstances a dangerous bar across the entrance may be raised. The harbour and town are defended by a castle and fortified heights. The castle occupies a commanding eminence towards the harbour and sea, presenting a precipitous cliff. It has subterraneous works, and casemates which are capable of accommodating 2,000 men. Dover is a principal port of communication with the continent, by means of packet-vessels that run daily to and from Calais, which lies S.E. ¼ S., distant 23 miles. The pop. of Dover is about 12,000.

DOVER ROAD, off the harbour of Dover, has good anchoring to stop a tide; but generally so great a swell, that ships roll much. This swell is supposed to be owing to a counter-current setting athwart the harbour's mouth from the last quarter flood to the end of the first quarter ebb. In the harbour of Dover the water rises on spring tides nearly 20 feet, and on the bar at half-flood 10 feet. The marks for anchoring in Dover road are St. James's church, known by its flat steeple, bearing N.N.W., and the South Foreland, E.N.E., in from 9 to 5 fathoms, or the white way, to the N.W. of the castle, directly over the hill, or between the hill and St. James's church.

DOVER PORT, America, Nova Scotia, between Halifax and Cape Sable, at the W. entrance to Blind bay, formed by Taylor's and the adjacent islands. It is the Port Durham of old charts.

DOVER, STRAITS OF, the narrow sea between the coast of Kent, on the English side, in lat. 51° 7' 48" N., and long. 1° 19' 30" E., and Cape Grisnez, on the French shore, in the Pas de Calais, in lat. 50° 52' N., long. 1° 35' E. The distance is 18½ miles between these two points, and the depth in the middle of this celebrated strait is from 18 to 24 fathoms. The French denominate this strait the "Pas de Calais." It is observable that the N. sea tides push through this strait, and those of the channel meet them some way within the strait W. of Dungeness, so that where the seas meet near that point the tide rises 7 feet higher than it does either E. or W. of the place of their meeting. The flood tides run 3 hours longer in the Strait than they do E. or W. of it, owing most probably to the contractions, and consequent obstacles presented by the shores on both sides. The stream runs E. in the Strait of Dover 6¾ hours, and W. 5½. This is caused by the meeting of the contracted waters in the two channels. At Dover the tides at full and change rise 15 feet springs, and 9 at neaps, while at Dungeness springs rise 24, neaps 15. At Hastings, more W., they begin to lessen again, being there, at springs, 22 feet, neaps 14. Thus clearly showing the meeting of the tides to be between the straits E. and Hastings W. It also proves that the N. sea tide is the prevalent tide, although it passes through a strait, and meets the influx of the channel, which is so much more capacious in the bulk of its waters, and from its wedgelike form might be supposed to press forward its tides by the weight of water contracted in the funnel-shaped passage upwards. Thus there would be produced a different result, in place of the channel waters being counterpoised so far within the straits, by the stream from the E. It has been thought that the tide, running into the strait from the North sea, proceeds in its natural course to the W., while that coming up channel, having passed along the coast of Ireland, comes by inflection to the E., and consequently with diminished power. When the flood to the E. has flowed about 3 hours, the water becomes higher than the level of that in the strait, runs in and continues to set to the W. for 5½ hours, then it turns, and runs E. about 6½. Four hours before the flood has done running at Dungeness, the ebb tide W. of the strait begins to run down the channel, so that the flood of Dungeness may be said to join the channel ebb in the last four hours, as during that time they both run W. At that period the ebb to the E. of the strait has run for the space of 3 hours, and the water fallen below the level of that at Dungeness, causes the ebb at this last place, to begin its course through the strait to the E. Then when the ebb at Dungeness has run through the strait E. for 2 hours, it is joined by the channel flood from the W., and they continue to run through the strait together for about 4½ hours longer. Such seems to be the solution of this phenomenon in the tides at the Straits of Dover. It may be observed that the similarity of the land on both sides, and the shallowness of the soundings, with the same ground all the way, is strong presumptive proof of the land between the two coasts having been once connected. It may be added, that it is the state of the tide currents here which render the passage from Dover to Calais so much more favourable than the reverse, across the strait, Dover being made with much more facility from Boulogne. Cape Grisnez, the nearest point to the English coast, is nearer Boulogne than Calais, it lying 4 leagues W. by S. from Calais, and the entrance of Boulogne being S.S.W. only 3 leagues. Dover was one of the five or cinque ports, when they were first privileged by Edward the Confessor, and confirmed by William I. It sent twenty-one ships to Calais with Edward III.

DOVEY or DYFFI, Wales, a river separating Cardigan and Merioneth, forming a good haven at Aberdovey, in which vessels not drawing more than 9 feet, may lie afloat close to the town, or aground on fine sand, out of the stream of tide which runs out 4 miles an hour.

DOWLAS or DOULUS HEAD, W. coast of Ireland, at the S. point of Dingle bay, and N. of Valentia island, in lat. 51° 57′ 6″ N., long. 10° 19′ W.

DOWNIE POINT, E. coast of Scotland, the S. limit of Stonehaven bay. The coast from Todhead, about 7 miles N.E., consists of high rugged cliffs, very steep, with 14 fathoms water close to the rocks, and 24 at 2 miles offing.

DOWNPATRICK HEAD, near the Bay of Killala, on the N. coast of Mayo, Ireland, in lat. 54° 20′ N., long. 9° 21′ W.

DOWNPATRICK, Loch Strangford, E. coast of Ireland, the chief town of the county of Down; vessels of 50 tons go up to the quay. It stands in a creek in the S.W. part of the loch, and possesses a little trade.

DOWNS, THE, Kent co., a roadstead, extending between the N. and S. Foreland, within the Goodwin sands. What are called the SMALL DOWNS, is the space between the South Brake and the shore, extending about 2 miles from Sandown castle towards Ramsgate. There is good anchoring here in from 6 to 2½ fathoms. The best marks for anchoring are, Bullock sand-hill, 2 miles N. of Sandown castle, between the two churches, or with Deal Mill and Sandown castle in one, or St. Margaret's church, on with the small mill N. of Deal. The marks for anchoring in the Downs are the S. Foreland high lighthouse, on with the middle of Old Stairs bay, and Upper Deal Mill, on with Deal castle, in 7, 8, or 9 fathoms, clay, and good holding ground. Ships moor with the best bower S., so as to have an open hawse with S. winds.

DOWSINGS, OUTER and INNER, and INNER DOWSING OVERFALLS, N. of Hasborough Sand, off the coast of Norfolk. The S. end of the outer lies 21 miles N. by E. ¾ E. from Foulness, and its shoalest part near the S. end, extends 3 miles N.N.E. and S.S.W., with from 5 to 7 fathoms water. There is a small patch with only 4½ fathoms; lat. 53° 17′ 30″ and long. 1° 16′ E. A bank runs to the N., having 14 fathoms on its W. side, and 11 and 12 on its E., uniting the S. shoals with the dangerous ones on the end of the Outer Dowsing. The Inner Dowsing's N.E. end lies 37 miles N.N.W. ½ W. from Foulness, and 14½ miles N.W. ½ W. from the Dudgeon light-vessel, 22 miles S. by E. ½ E. from the Spurn, and 10 E.S.E. from Trusthorpe church; it extends thence 6 miles S. by W. ½ W., and is about half a mile broad. The least water is 4 feet. A black buoy, with a staff and ball, is placed on the N.E. end of this sand, in 3¾ fathoms at low water. The Inner Dowsing Overfalls lie a mile N.W. by W, from the black beacon buoy on the N. end of the Inner Dowsing, consisting of four or five small and dangerous patches, about three-quarters of a mile in extent every way. They have only 12 feet upon them in some parts, with from 6 to 8 fathoms close to them all round. Trusthorpe church bears from their centre N.W. by W. ½ W. 9 miles. There are 8 and 9 fathoms between these patches and the Inner Dowsing; a shoal also lies three-quarters of a mile E. of the S. end of the Inner Dowsing, with only 3 and 3½ fathoms. It extends N.N.E. and S.S.W. 1¾ mile, and is one-third of a mile broad. The tide sets variously on the N. side of the Outer Dowsing. The first quarter will run S.W. by W., half flood S.E. by S., and near the latter part of the flood E.N.E., the ebb the contrary way.

DOXYA, Sea of Marmora, between the Gulf of Nicomedia and the channel of Constantinople, one of the Prince's islands, which are nine in number.

DOYLE REEF, America, Gulf of St. Lawrence, S. ½ E. 6¾ miles from East point, consisting of sharp rocks about 300 fathoms long, and 50 wide. The least water is 3 fathoms, and there are 12 and 13 around it. The only mark for it is the N. cape of the Magdalen isles, open two-thirds of its breadth to the N.E. of N.E. cape. On the reef the angle between these marks and the W. point of Coffin island is 24° 27′.

DRACO, Denmark, a small place on the W. side of the channel from Elsineur to Copenhagen. The buoys off Draco guide along the ground, and are necessary to be regarded in the passage up the Sound, as in coming along that channel there are only 26 or 27 feet of water, they must be used as guides in connection with the directions by which every vessel passing through that intricate navigation will steer its course.

DRAG, or DRAGSTONE, THE, Plymouth, off Penlee point. See *Plymouth.*

DRAGONERA, Mediterranean, lat. 39° 35′ N., long. 2° 16′ E. See *Balearic Isles* and *Dargoneers.*

DRAGON'S MOUTHS, or BOCAS DEL DRAGO, S. America, Island of Trinidad, the name of the N. entrance of the Bocas or mouths leading into the Gulf of Paria, between the island and the main upon the W. side. See *Bocas de Paria.*

DRAGSMARK, Sweden, a town on the N. side of the N. channel between the Island of Ouroust and the main.

DRAGSVIC, or DRAGSICK, Gulf of Finland, on the N.E. side, opposite the N.E. of the peninsula of Hango head, a small town upon the shore of a narrow arm of the sea.

DRAGUET POINT, N. coast of France, nearly S. of Cape Barfleur. It is foul and rocky, and requires to have a berth of a league at least. There is still, however, a sunken rock 5 miles from the coast, which must be carefully avoided.

DRAGUT POINT, island of Malta, forming the N. side of Marsamusceit harbour, in that island.

DRAKE'S CHANNEL, SIR FRANCIS, West Indies, this is among the Virgin isles, and in the centre, where nature has formed an extensive and secure basin for vessels, among islands with shores peculiarly rocky; this basin is 15 miles long by 3½ broad. See *Virgin Islands.*

DRAKE'S ISLAND, Plymouth Sound, placed almost in the middle part. It is east of Mount Edgcumbe, to which it sends off a reef of rocks, a part dry at low water connecting it with Redding point to the westward. There is one narrow passage through this reef, locally termed "the Bridge," which is found by keeping the old white tower at Devil's point on with the tower of the chapel at Devonport. This passage is extremely narrow, and is only to be ventured through at the turn of tide, by vessels of moderate burden. This island, named from the great navigator, who is so justly the glory of Plymouth, from having been the scourge of the boasted Spanish Armada, is often called St. Nicholas' island, perhaps from a prior custom, but in that case one more honoured in the breach than the observance. The island commands the sound, and is strongly fortified with heavy artillery. It is just of such an elevation above the level of the water, that while most effective against an enemy afloat, it is itself secure from the horizontal fire of ships of the line. All vessels going up Hamoaze, must pass the island between that and the main, and it is therefore a most formidable position, and with the citadel batteries, fully adequate to resist every attempt of an enemy to enter either of the inner harbours. The W. end of the breakwater is within a nautical mile and a half of its S. side, in lat. 50° 20′ 6″ N., long. 4° 9′ 30″ W. See *Plymouth.*

DRAKE'S ISLAND, America, an islet off the W. point of the harbour of Porto Bello, on the N. coast of S. America.

DRAKER REEF, Denmark, in the way from Elsineur, towards Falsterhorn, in the passage to the Baltic. The fifth buoy of the grounds lies upon it.

DRAM, Gulf of Christiania, Norway, a river on which are situated the towns of Tangen, Stromsæ, and Bregnæs, whence they received the general name of Drammen. The principal export is timber.

DRASTON HAVEN, Wales, a small port on the E. side of St. Bride's bay, at the entrance of a little river only adapted for boats.

DRAPANOS, or DREPANO, CAPE, Greek Archipelago, the S.W. point of the Gulf of Monte Santo, which shows on the promontory, that forms its E. side, and near its S. extremity the celebrated Mount Athos, now called Monte Santo, 6,349 feet above the level of the sea, in lat. 40° 10′ N., long. 24° 20′ E. Cape Drapano is in lat. 39° 56′ N., long. 23° 58′ E.

DREADFUL SAND, mouth of the Humber, a name once given to the shoals or sands near the Spurn, which are now called the Stone-banks and the Binks.

DRIG ROCK, coast of Cumberland, about a mile from the mouth of the Calder river. It is a sunk rock, and has sometimes not been seen more than twice in the space of seven years. It lies N.W. of Ravenglas, along the coast, as does also the opening of a stream called Port Drig, to the N.W. of which is the Calder.

DRINO, Albania, a river on the E. side of the Gulf of Venice: it is S.E. by E., 10 miles from Bojano, and falls into the sea by several channels or mouths. It is navigated by rafts nearly 100 miles inland, as it flows through extensive forests. Near its E. bank is the town of Alessio, and to the N. of the mouths are the port and anchorage of St. Giovanni de Medua, adapted only for small vessels.

DRINO, or LO DRINO, Albania, a bay in the Gulf of Venice, the N. point of which, Dulcigno, is in lat. 41° 53′ 45″ N., long. 19° 11′ 40″ E. The S. point is Cape Rodoni, distant 20 miles. This includes the first half of the gulf, the second extends 15 miles to Cape Pali. The River Drino falls into the N. half of this gulf. The headlands bear from each other from Dulcigno to Cape Rodoni S.E. by S., and from Cape Rodoni to Cape Pali S. by W. The coast is a long, sandy beach, with regular soundings, decreasing towards the land. The Mole of Dulcigno, on the N., is on one side of the main entrance to the River Bojano, navigable as far as Scutari, the capital of North Albania, with a pop. of 12,000. Cape Rodoni is in lat. 41° 37′ 40″ N., long. 19° 28′ 10″ E. There is anchorage all the way to Cape Pali, which last lies in lat. 41° 23′ N., long. 19° 24′ 30″ E. Cape Rodoni must have a good berth in passing from the sunken rocks about its extremity.

DROBAK, Norway, a small town on the E. side of the Gulf of Christania.

DROGDEN CHANNEL, or THE GASPAR, Denmark, near Copenhagen. It has on the W. side the Isle of Amak or Amag, extending in a S. direction to Drogoe, and thence W.S.W. ¼ W. to Alland's Hage. Amak is surrounded by a sandy flat, with only 7 and 8 feet of water; on the edge of this flat, marking the boundary of deep water, is the Kna buoy No. II., in 3 fathoms. On the E. side is the Saltholms island. Dragoe town is on the S.E. point of the Isle of Amag. All the dangers and channels are buoyed, and the buoys shifted according to circumstances, so that a pilot is necessary.

DROGHEDA, Ireland, E. coast; the centre in lat. 53° 42′ 50″ N., long. 6° 22′ W. This is a seaport of the co. Louth, seated on the River Boyne, 5 miles from the sea. It is a borough, and stands near the spot where the battle of the Boyne was fought, in 1690. Pop. about 18,000.

DROINAIA, Russia, near Cape Kherson, an inlet marked by several white cliffs, where there is good anchorage, from 6 to 2 fathoms, on avoiding two shoals which are found on entering, on which account the starboard shore should have a good berth. It lies between Sevastapol and Cape Kherson, with other anchorages very similar in character.

DROME, France, a river falling into the sea in the department of Calvados, having, at its mouth, Port en Bessin, a small haven occupied principally by fishermen.

DROMIA ISLAND, Greece, at the entrance of the Gulf of Salonica, another name for the Island of Cilidroni, the ancient Halonesus.

DRONTHEIM, or TRONDHEIM, Norway, a large town and capital of a province, on the S. bank of an arm of the sea. It exports principally iron, timber, and fish. The S. channel, to the entrance, is called the Grib Hoelen,' in lat. 63° 15′ 30″ N. Proceeding up this channel there are four fixed lights to be passed, which are lit from August 15 to April 30. The Tyrhong light, at the E. end of Eddo island, 35 feet high, to be left on the port side; one at Tonningen, 24 miles E. of Eddo, 35 feet high, to be left also on the port side going up; one at Agdaness point, 20 miles E. of Tonningen, 113 feet high, to be left on the starboard side, on rounding it, to the S., running towards Drontheim, and a fourth, which is the harbour-light of Monkholmen, 43 feet high, 21 miles S.E. of Agdaness point, which serves as the guide to the harbour. The N. entrance of Drontheim is the Ramsoe fiord, lat. 63° 30′ N.; running in to the E. of the Island of Smoelen, an island surrounded by rocks, above and under water, particularly at its S.W. and N.W. parts. It must always have a wide berth, both in entering the Grib hoelen and the Ramsoe fiord. As in entering all the Norway ports, a pilot is absolutely needful here. A light is established on the Island of Prœstoe, Gulf of Folden, in the Province of Drontheim, in lat. 64° 27′ 26″ N., long. 11° 8′ E., 33 feet above the sea, and visible 10 miles. Drontheim is considered the third city in Norway. It stands at the mouth of the River Nid. The buildings are generally of wood. Several hundred vessels annually enter the port. The pop. is about 9,000. There is little or no tide apparently visible. With strong W. winds the current sets continually N. and N.E., and what flood there is N.E., and the ebb S.W.

DROOG, Holland, the names given to the guides, or buoys, the second and third in the Fly, or Stortelmbeck channel, into the Zuyder Zee. The second buoy has the name of Droog drie, and the third of the innermost Droog drie; the first lies in 11 feet water, the second in 16 feet.

DROSSELBURG, Denmark, an island in the Great belt, one of those belonging to the government of Zealand.

DRUID'S REEF, a vigia, said to have been observed in lat. 41° 19′ N., long. 41° 25′ W. This was reported, in 1831, by Captain Treadwell, of the Druid of London, who stated that he passed it, at not more than 30 yards' distance, in calm weather. The reef had the appearance of from 7 to 10 sugar-loaf heads, and its length from E.N.E. to W.S.W. was estimated at from 10 to 14 feet. The reef had been inserted in a Spanish chart as having been seen in 1803, in lat. 41° 24′ N., long. 41° 20′ W. Some suppose it a rock said to have been seen by Desmaires, a pilot, in 1683, who reported it the height of a sloop out of the water. Bellin gives it in lat. 42° N., long. 41° 10′ W. The Spanish chart gives a vigia, seen in 1798, in lat. 43° 30′ N., long. 37° 35′ W., but this can hardly be the same, without great error of observation.

DRUMLAND SAND, Scotland, on the N. side of the River Tay, on the E. coast, off the church and village of Moray Frith: ships do not go into less than 3½ fathoms of water on that side to avoid it.

DRUM POINT, America, United States, on the N. side of the Patuxent river, on the starboard hand running up, a low, sandy bold point, with some small bushes upon it. When a ship has doubled it, and is in 2½ and 3 fathoms water, she is secure from all winds.

DRURO CAPE, Greece, on the S.W. side of Negropont, N. of the Petali islands.

DRY BANK, West Indies, a shoal so called, situated in lat. 18° 38′ N., long. 83° 40′ W. Most probably an older name for the Misteriosa bank.

DRY HARBOUR, Jamaica, in Rio Bueno bay, exposed to all winds between N. and N.N.W. The anchorage is very indifferent. The harbour is formed by two reefs. Dry harbour is 3 miles more E., and is a tolerable place for small vessels; it has a narrow channel, and 16 feet water.

DRYŒ, Denmark, a small island between Funen and Ærœ, of little importance either in trade or productiveness.

DRY POINT, or SICCA PUNTA, South America, in the Gulf of Trieste, on the N. coast, E.S.E. of the Island of Curaçoa. This bay is in lat. 10° 45′ N.

DUAN AGLAN, Sea of Marmora, a conspicuous hill on the W. shore to vessels passing through the Straits of the Dardanelles, N. of Point Playtar.

DUARTE ISLETS, America, Colombian coast, 3 miles S. 69° W. from Boquerones point, N. of Porto Bello, four in number, with reefs close to them. Lat. about 9° 40′ N.

DUBLIN BAY, Ireland, E. coast, in lat., from the city observatory, 53° 23′ 12″ N., long. 6° 20′ 30″ W. The Bay of Dublin is about 7 miles broad, from Brah or Bray head to Dalkey island. There is good anchorage in the bay, at the entrance, when the wind does not blow hard from the E. or S.E., for with these winds there rolls in a very heavy sea, which forces vessels to seek safety, with N.E. winds, in Kingstown harbour, on the S. side of the bay, inclosed by two large piers, affording effectual shelter. On the head of the E. pier, 2,800 feet long, there is a lighthouse showing a white and red revolving light, alternately attaining their greatest brilliancy at intervals of 30 seconds. It is 40 feet above the mean level of the sea. A fixed light is, at the same time, shown from the W. pier head. This E. pier bears from the Kish bank floating-lights W. by N. ¼ N., distant 6¾ miles, lat. 53° 19′ N., long. 5° 58′ W., and from Poolbeg light, S., distant 2 2-5ths miles, visible 9 miles. The entrance is to the N.E., with 4 fathoms water, and within from 4 to 2¼.

DUBLIN HARBOUR, and CITY, Ireland, E. coast. This harbour lies between two banks, called the North and South banks, and is sheltered on the S. by a wall and piles carried E. 3 miles into the bay. At the end of the S. wall there is a lighthouse called the Poolbeg, having a fixed light exhibited all night, and a lower light kept from half flood to half ebb, a ball being hoisted in the day, at the same period of the tide, to direct vessels into the harbour. There is a bar across the entrance, on which there are 10 feet at low water, and between the bar and lighthouse from 12 to 24 feet. A red buoy is moored a mile E.S.E. ½ E. from the lighthouse. The passage is a cable's length to the S. of this buoy, which will carry clear, with 10 feet at low tide. Over the bar the depth of water increases to 14, 18, and 20 feet. The tide flows at the lighthouse, on full and change days, at 10^h 47^m, and rises 12 feet on springs. Vessels from the N.E., bound for Dublin, give the S.E. of Howth a berth to avoid a bank called the Rosebeg. The city stands on the bay, which is of a semicircular form, and about 6 miles in diameter, into which the River Liffey runs, after dividing the city into equal parts, its course being nearly W. and E. From the point where the Liffey enters the bay it is embanked on both sides with freestone, forming a range of five quays throughout the city. The river is crossed by seven stone bridges. Two canals also communicate with the river, connecting it with the interior country. The city is in the form of a parallelogram, around which a road called the Circular is carried. The public buildings are very fine, particularly the college and the courts of justice. The pop. is 210,000. The trade is considerable, the imports various; the exports principally the productions of the soil, as the larger part of the manufactures exported are sent from the other ports. Dublin is 60 miles W. of Holyhead. It is high water at Dublin bar, on full and change, at 10^h 30^m.

DUBMILL SWAFE, England and Scotland, a sand in Solway Frith, below Carlisle, about half a mile in width. When Dubmill house and Skiddaw hill are in a line, a vessel is on the tail of the Swafe, and may turn down the Solway.

DUBON POINT, Island of Jersey, on the S. side; the W. point of Elizabeth bay, on the coast of that island, from whence it trends N.W. by W. to St. Brelaud's bay.

DUBREUIL'S VIGIA, said to exist in lat. 14° 50′ N., long. 29° 40′ W. It is supposed to be an imaginary danger.

DUCALLA ROCK, Ireland, W. coast, about 1 mile E.S.E. ½ E. from Roonharrick island, dry at low water. This rock is avoided on the S. side by keeping the Point of Rimore on with the sharpest hill summit, E. of the head of the bay in which it appears.

DUCATO CAPE, Leucadia, or Santa Maura island, one of the Ionian isles. This cape is given in lat. 38° 33′ 30″ N., long. 20° 32′ 45″ E., but according to Lieut. Raper, lat. 38° 24′ N., long. 20° 33′ E. To the N. of the cape about 2 miles, is a large overhanging rock, 200 feet high, with a number of pointed rocks beneath. This is denominated "Sappho's Leap;" and a ridge of limestone rocks, extending S. to Cape Ducato, is called the "Leucadian Promontory."

DUCK HARBOUR, America, W. side of Stony island, Labrador.

DUCK ISLAND, West Indies, between which and the island of St. Thomas there is a channel half a league wide, in the course from Ram head, in St. John's, to St. Thomas harbour. The course is W.N.W. to this island, Bird's Key being left on the S. There is a sunken rock at the entrance of this channel, in the fair way, with only 5 feet water.

DUCK KAY, West Indies, W. coast of Florida, the most E. of the Vaccas or Cows' kays, 5 miles S.W. of Viper's kay. There is fresh water among the rocks at the W. end. The Vaccas are in lat. 24° 42′ N., long. 81° 5′ W.

DUCKSBURY POINT, North America, United States, the N.E. angle or point of States island, on the W. side of the Channel of New York, in the part called the Narrows, opening to Yellow Hook, upon the point of Long island, on the E. side.

DUDDON SANDS, Lancaster and Cumberland cos., formed at the estuary of the River Duddon. Some part of these sands is considered dangerous from the uncertainty of the tides; but they are scarcely visited by the smallest class of vessels, and are in consequence of trivial comparative importance to the mariner.

DUDGEON SHOAL, coast of Norfolk, N. of Hasborough sand. This shoal lies N.N.W. ½ W. and S.S.E. ½ E., nearly 3 miles in length, and a mile broad. There are 9 or 10 feet on the shoalest part, which is S.E. about a mile from where the vessel is stationed. On the other parts there are 3 or 4 fathoms. A light-vessel lies 24 miles N. by W. from Cromer lighthouse, with one light, moored a little to the left of the Dudgeon, in lat. 53° 15′ N., long. 57′ E., and a gong is struck on board her in fogs. It is high water here at 6^h, but it continues to run until $7\frac{1}{2}^h$; the flood S., the ebb N.

DUE UDDE CAPE, the S. point of Bornholm island, in the Baltic, in long. 15° 5′ E.

DUHEERTACH ROCKS, Scotland, W. side, N. of the Sound of Icolmkill, 6 miles W. by S. ½ S. from the Torrin rocks, that lie S.W. from Dorrel island. The Duheertach is a remarkable rock above water, seen 3 or 4 leagues distant. A ledge of rocks runs out half a mile from its W. side, the end and part of the centre of which is only covered at high water. From this rock, Bennimore hill, in Jura, bears S.S.E. ¾ E.; the W. end of Isla, S., and Bentinish hill, in Tiri, N. ¼ W., or nearly. The flood runs N., but is not even perceived with neap tides.

DUINKER'S GAT, Holland, Texel, a channel separating the N. from the S. Haaks, narrow, and used only by the fishermen of the country, though it has 3½ fathoms water near the middle of the passage. It is circuitous, and has no buoys or beacons.

DUIVELAND, Holland, one of the islands forming the eastern division of the province of Zealand. It contains no town, but has several large villages.

DULAS ROCKS, North Wales; S. ¾ E., 2¼ miles from Point Lynas, is situated this dangerous cluster. On the central rock, always above water, is a beacon-tower, 42 feet above the sea-level at half tide.

DULCE RIVER, or DOLCE RIO, America, Gulf of Honduras; lat. of entrance, according to Captain R. Owen, 15° 49′ 45″ N., long. 88° 46′ 32″ W.

DULED RIVER, Pembrokeshire, Wales. It passes Haverfordwest, and falling into Milford haven, forms a principal back-water of that fine harbour.

DUMBAR SAND, Cornwall, at the entrance to Padstow harbour, on the E. side of the Camel river, the estuary of which forms that harbour. It is dry at low ebb.

DUMBARTON, Scotland, W. coast, situated on the Leven, which, issuing from Loch Lomond, falls into the Clyde. This town has a pop. of 2,500, and possesses some coasting vessels. It has a castle on a double-headed rock, on one side washed by the Clyde, on the other by the Leven.

DUMET, coast of France, an island opposite the mouth of the River Villaine.

DUMFRIES, Scotland, on the E. bank of the Nith, 9 miles from its mouth. It is a town of considerable trade, having a pop. of 12,000. It has some vessels employed in foreign trade, with a number of coasters. The Nith falls below Dumfries into the Solway Frith.

DUMPTON STAIRS, Kent, connected with Broadstairs, and principally visited for sea-bathing.

DUNAMUNDE, in the Gulf of Livonia, an island on the W. bank of the Dwina, near its outlet, on which is a strong fort commanding the passage of that river to Riga.

DUNAVARRE FORT, Ireland, N. coast, a ruin upon Fair head, the N.E. point of the mainland of Ireland.

DUNBAR, Scotland, E. coast, 13½ miles from St. Abb's head, N.W. ½ N., a town on a low rocky point, having a pier harbour in solid rock. On the W. side are the Staple rocks, some of which are always visible, having deep water close to them. Ships may anchor off in 7 or 8 fathoms. The usual roadstead is abreast of the piers, in 10 or 12 fathoms, sandy ground. The entrance between the piers is narrow, and the passage in not free from danger if there is a fresh breeze.

DUNBEG BAY, Ireland, in the space between Loop head and Arran island.

DUNCANNON FORT, Ireland, S. coast, on the E. point of the entrance of Waterford, on the same side as the Hook lighthouse, commanding the entrance to the port.

DUNCANSBY HEAD, Scotland, the N.E. point of that kingdom, 10½ miles N.N.E. ½ E. from Ross head, to which it is very similar, formed of perpendicular cliffs; but is easily distinguished by having over it Duncansby castle, otherwise John O'Groat's house, on a high rock visible 16 miles off. This head is in lat. 58° 39′ N., long. 3° 1′ W.

DUNDALK, BAY and TOWN, Ireland, E. coast. Dunany point, the point of Dundalk bay, is 4 miles N. by E. ½ E. from Clogher head, off which, about a mile E. of high-water mark, lies the outer edge of a rocky reef, with not more than 5 or 6 feet over it at low springs; it should therefore have a good berth, as broken patches run off E. 2¼ miles. To pass to the E. of all these shoals, Clogher head should not be brought to the S. of S.W. by S. The distillery chimney at Dundalk, bearing N.N.W., leads clear of them to the N.E. On the N. side of the bay, bearing N.E. ½ E., 8 miles from Dunany point, is Cooley point, off which a shoal extends 1¼ mile S.W., on which is the Castle rock, nearly uncovered at low springs. Shoal water extends to Ballaggan point, full 1¾ mile from the shore. Great caution is required in passing between these points; and vessels should not approach nearer than 7 fathoms. The town of Dundalk is situate on the N.W. corner of the bay, 2 miles within Soldier's point, on the S. side of the entrance to the harbour. This place has a considerable trade in corn and manufactures of linen and muslin. The pop. is 10,000; lat. 54° 2′ N., long. 6° 20′ W.

DUNDEDY HEAD, Ireland, S. coast, 4 miles S. of Ross. With offshore winds and moderate weather, vessels anchor on its W. side, towards Ross harbour, half a mile from the shore.

DUNDEE, Scotland, E. coast, a port on the River Tay, upon the N. side. It is the largest town in Forfar, having a pop. of nearly 50,000, and considerable manufactures of linen, canvass, and cordage. It is W.S.W. of Forfar. There is a bar at the entrance of the Tay. Buttoness must bear N.N.W. ¼ W., the light on May island being shut in, and the course must be thus until the lighthouses can be both seen. They must be brought on with each other, bearing N.N.W. ¼ W., and a vessel may then run in, with that direction, safely over the bar, close to the fair way buoy, and into the proper channel, until the Terry lights are in one. This will lead through the best water in 6, 3, 7, and 5 fathoms, to the S. of the Horse buoy. Then steering towards Broughty castle, a vessel runs midway to her anchorage in Dundee. The Frith of Tay is regularly buoyed. The bar should never be taken at a spring ebb. The ground in the river is all sand or gravel. It is best to wait for the flood tide to pass the bar, if the weather be bad. Off Red head the tide sets strongly. In the night, or in thick weather, it is not advisable to come nearer the coast than 26 fathoms; there are 20 only 1½ mile from the shore. The lat. of Dundee is 56° 27′ 24″ N., long. 2° 58′ W.

DUNDRUM, Ireland, E. coast, a village upon a bay and harbour, esteemed dangerous for shipping, as it is crossed by a bar, upon which the sea breaks, except in the finest weather. It is visited as a bathing-place, and several kinds of sea-fish are caught there of superior excellence. On its E. side, St. John's point, is an intermittent light, 62 feet high, lat. 53° 13′ 30″ N., long 5° 40′ W.

DUNGARVON HARBOUR, Ireland, S. coast, 22 miles W. by N. ½ N. from Hook point. There are only 9 feet water at the quay, and a foot or two more a little distance off, at spring tides. Small vessels alone may find good shelter here.

DUNGENESS, Kent co., a low point, 7 leagues W.S.W. ½ W. from the South Foreland. It has a lighthouse with a fixed light, 92 feet high, lat. 50° 55′ N., long. 58′ E., visible 6 leagues distant. The lighthouse itself is painted red, for the purpose of serving as a mark by day. The point of this Ness may be rounded, if needful, in 12 and 9 fathoms; 11 fathoms are found at the distance of three-quarters of a cable's length from it; and at 1½ cable's length there are 15 fathoms water. The water rises 24 feet at spring tides; the strongest tide runs in 13 fathoms.

DUNGENESS SHOAL, sometimes called STEPHENSON'S SHOAL. This shoal has 4 fathoms on its E. and W. ends, 6 or 7 round it, and between it and the shore from 6 to 2 fathoms. It extends 2 miles W. by S., and is from half to three-quarters of a mile broad, its broadest part being to the W. Lydd church-tower and Post mill in one leads to the W. of the sand, and Shakspeare's cliff open to the S. of Dungeness, to the S. of the shoal. To the W. of Dungeness, the North sea and channel tides meet.

DUNIER ISLANDS, GREAT and LITTLE, America, Newfoundland, between Cape Freels and the Strait of Belleisle.

DUNKERNEY or DONKERNEY BAY, Ireland, a small bay on the N. shore of Kilmore sound, at the S.W. extremity of Ireland, bearing N.N.W. from the Bay of Kilrush, on the N. shore of the same sound. It is an excellent road for ships of any draught of water.

DUNKIRK, France, a town in the department Du Nord. Pop. 24,520. It stands E. by N. of Calais 20 miles, and E. ¾ S. from Gravelines. It is readily known from the sea by its steeple, the highest on that part of the coast, seen 5 or 6 leagues off. The Stadt house has a small spire, visible 4 leagues. This town is a place of considerable commerce in fish, corn, and colonial productions. It is approached by a canal 1¼ mile in length, the port and basin being within the town. The road is situated at the outer extreme of the canal, and is formed by a sand-bank running parallel with the shore. The houses are neat, and there are extensive barracks, but the place has the disadvantage of possessing little good water. There is a lighthouse here on the pier-head, containing a revolving light 1,531 yards N.W from Heuguenar tower, and 193 feet above the sea. It stands in lat. 51° 3′ N., long. 2° 22′ E. To a vessel distant 4 or 5 leagues the light appears to revolve, being eclipsed every minute. Within that distance there is always a faint light visible. There is a harbour light upon the W. jetty-head, 23 feet above high water mark, visible 8 or 9 miles. The entrance to the harbour is between the jetties, dry at low water; they have beacons on their extremities; the course is S. by E. ½ E. High water 11ʰ 55ᵐ, (establishment of the port 11ʰ 48ᵐ). Lat. of town 51° 2′ 10″ N., long. 2° 22′ 5″ E.; long. from Paris, 2′ 23″ W.

DUNKIRK ROAD; this roadstead is bounded by the Snouw, Brack, Hils, and Traepageer banks, and the banks which border the shore, its length being about 12 miles. The distance from the W. end of the Snouw to Dunkirk is 7½ miles, the road running E. by S. ½ S. and W. by N. ½ N., then continuing 4½ miles further, as far as the Zuydcoote channel. Its breadth from N. to S. is not more than half a mile, with a depth of 4 fathoms, The soundings are from 7 to 8½, with a bottom of mud and sand, which holds well, but the only shelter is that of the surrounding banks, always under water.

DUNKLIN'S CLIFFS, Jamaica, on the N. side of the island, W. of Brasiletta kay, and N. of Bread-nut hills. They are between Britton's bay on the W. and Platform bay on the E. Lat. 19°, long. 77° 38′ W.

DUNLEARY, Ireland, E. coast, on the S. shore of Dublin bay, 2½ miles from Dalkey island. It has a pier haven for vessels of light burden, with a depth of 12 feet at springs, and 9 at neaps.

DUNMANUS BAY, Ireland, W. coast, and N.E. of Three Castle head. It has deep water and clean ground nearly up to Manin island, but it is exposed to W. winds, and seldom visited, except by vessels of a small class that can ride in one of its creeks, called Dunmanus creek, on the W. side of which there is anchorage in 3 or 4 fathoms.

DUNMORE HEAD, Ireland, W. coast, lat. 52° 6′ 3″ N., long. 10° 29′ W.

DUNMORE HARBOUR, Ireland, S. coast, on the S.W. side of Dunmore or Whitehouse bay, distant 1⅝ mile from Creden head. The packets from Milford haven resort to this port. At the end of the pier a lighthouse stands, having two faces, one seen from the sea W. to W.S.W. ¼ S., of a red colour; the other seen from the harbour, clear and bright, but not visible beyond Creden head. This light is 44 feet above high water, on a white building. The harbour is shallow, there being within the pier-head only one spot with 14 feet, 9 and 12 being the common depth. Lat. 52° 10′ N., long. 6° 58′ W. High water 5ʰ 18ᵐ.

DUNNAFF HEAD, Ireland, N. coast, the E. point of the entrance to Lough Swilly, Fannet, with its fixed light 90 feet high, forms the left, about 3½ miles distant, in lat. 55° 16′ 24″ N., long. 7° 38′ W. High water 5¼ʰ, full and change.

DUNNET HEAD, Scotland, about 11 miles from Duncansby head, the two points bearing from each other N.W. by W. and S.E. by E. The head is steep and rocky, and upon it is a lighthouse with a fixed light, built of stone, 45 feet high and 346 above the level of the sea; in lat. 58° 40′ N., long. 3° 21′ W. 6 miles W. ½ N. from the lighthouse is a little island called Holburn island, the land between them bending inward and forming Thurso bay and Scrabster road, within both of which anchorage may occasionally be found.

DUNNOSE, S.E. side of the Isle of Wight, a promontory at the S.W. limit of Sandown bay. Lat. 50° 37′ 8″ N., long. 1° 11′ 50″ W. It lies from Beachy head W.N.W. ½ W. 55 miles. Vessels are accustomed to come no nearer to this point than 18 fathoms, if bound down the channel. There are 22 fathoms at 4 leagues off the headland, with mixed soundings. The Owers sand lies E.N.E. of Dunnose, or from the elbow of the Owers to Dunnose is W.N.W. ½ W. 19 miles.

DUNNOTTAR CASTLE, E. coast of Scotland, on a high perpendicular cliff overlooking the sea, and almost surrounded by it; it is a very conspicuous mark at sea.

DUNSKERRY CASTLE, Scotland, W. coast, 1 mile S. of Port Patrick, on the edge of a precipice.

DUNSTANBURGH CASTLE, Northumberland, on the shore between Coquet and Farn islands, the ruins of a castle built by Edward I.

DUNSTER, England, Somerset co., W. coast, a town of two good streets, nearly a mile from the shore, on the Bristol channel, and surrounded by hills, except towards the sea.

DUNTULIM BAY, between Hulim island and Skye, W. coast of Scotland. The anchorage here is securely sheltered, the ground clean, and the water sufficient for large vessels in depth, but too narrow except for small ones. This bay may be entered from the N. or S., but the N. entrance is preferable. There is a rock in it above water lying nearest to Skye island.

DUNWICH, Suffolk co., a mean village of only 184 inhabitants, though formerly a considerable town, having sent 6 ships and 102 mariners to the siege of Calais under Edward I. The sea has obliterated all traces of its harbour.

DURAZZO, Gulf of Venice, S. by E. ¼ E. from Cape Pali, and distant about 5½ miles. The citadel is now in ruins, and the pop. fallen to 5,000. Durazzo stands on a point 7 miles from Cape Laghi, on the opposite side of the bay to Cape Pali, in lat. 41° 17′ 30″ N., long. 19° 26′ E.

DURELL'S LEDGE, called sometimes the SNAP ROCK, America, Newfoundland, about 7 leagues N.W. by N. from Funk island, over which the sea constantly breaks; N. by W. from this ledge, 3 leagues, is Cromwell ledge, supposed to bear E.S.E. ¾ E. 10or 15 miles from Little Naga island, between Cape Freels and the Strait of Belle Isle. It is said to lie 7 leagues N.W. by N. from Funk island.

DURGERDAM, Holland, a place on the N. shore of the channel from the Zuyder Zee to Amsterdam, at the second point of land, a mark for sailing towards that city.

DURLESTONE HEAD, Dorset co., a point about 4 miles from St. Alban's head, this last bearing from it W. by N.

DURNESS BAY, Scotland, Sutherlandshire, N. coast, W. of Tongue bay, with Loch Eribol, a spacious inlet between, on the W. shore of which is Port Ruspin, a small dry harbour.

DURSEY ISLAND, Ireland, W. coast. Lat. 51° 35' 5" N., long. 10° 14' 10" W. From the W. entrance of Beerhaven to Blackball head is 5 miles W. by N., from thence to Crow head W.N.W 4½ miles; the S.W. end of Dursey island lies from Crow head N.W. by N. 2½ miles. The Bull, Cow, and Calf rocks lie W. of Dursey island, four in number, always above water, having 36 fathoms close to them. 6 miles W. by S. from the W. point of Dursey island, and 20 miles N.W. ¼ W. from Mizen head, this last being in lat. 51° 27' N., long. 9° 50' W. The Leek bank, extends N.W. and S.E. for some miles, with 32 and 45 fathoms over, and 65 around it. Ballydonagon bay, with deep water, lies between Dursey island and Cod's head. With E. winds and moderate weather, vessels may lie between Dursey island and Crow head.

DURTICK SAND, Northumberland co., in the haven of Tynemouth, a little above Shields, on the N. side of the river. A vessel to avoid it must edge over to the S. side until the ship has passed Crooked reach, where the river trends to the northward.

. DUSAGHTY ROCK, Ireland, N. coast, half a mile from Moytog head, dry at half ebb.

DUTCH CANAL, THE GREAT, Holland, extending from the Helder to Amsterdam, to afford a passage for large vessels from the city to the sea. The city has 40 feet water fronting its port; but the bar in the Zuyder Zee has but 7; hence all ships of considerable burden unload part of their cargoes with lighters before they can enter. Ordinary means of improving the access to the port being ineffectual, it was determined to cut a canal to the Helder. The distance between the two points is 41 miles, but the length of the canal is 50½ The breadth of the surface of the canal 124 feet, the breadth at bottom 36 feet, and the depth 20 feet; its level is the high tide of the sea, from which it receives water. It has two tide-locks, one at each end; there are two sluices besides, with a flood-gate. The locks and sluices are double, or two in the breadth of the canal. These are built of brick, with bands of limestone at intervals, protecting it from abrasion. A broad towing-path lies on each side. The canal is wide enough to admit one frigate passing another. It takes its course from the river Ye; running N. to Pumerend, then W. to Alkmaar lake, then N. again by Alkmaar to a point within 2 miles of the coast near Petten, and thence in a direction parallel to the coast as far as the Helder, where it joins the harbour of Neiuwe Diep, where there are two fixed lights, one red, 20 and 24 feet high. The time spent in tracking vessels from the Helder to Amsterdam is commonly 18 hours. Opposite the Helder there are 100 feet water at high tides; and at the shallowest part of the bar, to the W., 27 feet. There are 40 feet water at the port; above and below it not more than 12 or 10 feet.

DUTCH KAY. See *Cape Verde Islands.*

DUVARDELET ISLAND, N. coast of France, about a league along shore, within Herqui point. The Neiras islets, three in number, lie about a mile N.N.W. from it, and the Lejon and Bouvillon rocks, are in the same direction. This island is a mark for the Lejon rock, W. due nearly from Cape Frehel, and 5 miles N.W. by W. from Herqui point.

DUXBURY, a town of the United States of North America, in Plymouth co., Massachusets bay. It is 2 miles from the mouth of a river that falls into the bay near Gurnet point. The pop. does not exceed 2,500.

DUYO, Spain, Bay of Corcubion, a small shoal S. of Caldebarcos, having 2 fathoms over it.

DWAL, or DVALDEN, at the entrance of the Gulf of Livonia, a long reef which stretches out from the N. side, S.W. by S. nearly 9 miles. It is distant 6 miles S.S.W. of the lighthouse on the Oesel side.

DWALE GROUND, E. coast of Jutland, on the passage from the Scaw to the Belts, 7 miles S. of Sæbye, and 15 from Hirtsholmen, 4 miles from the shore. Vessels drawing little water often pass between that and the main in 4 fathoms, with a bottom of fine sand. There are 2, 3, and 3½ fathoms on this shoal, which has a passage 3 miles wide, with from 9 to 14 fathoms between itself and the shoals of Lessoe. W.S.W. of the Dwale 3 miles, and close to the shore, is Rimmen sand, or Asser bank, parallel with the land extending from abreast of the Dwale 9 miles, and having within a small harbour, over a sort of bar, for vessels drawing no more than 10 feet water.

DWINA, Russia, a river rising in the government of Vologda, it falls into the White sea at Archangel, in lat. 64° 32' N. In the N.W channel, below the mouth of this river, at the Island of Mudoska, opposite the town of that name, there is a lighthouse 140 feet high, in lat. 64° 55' N., long. 40° 17' E. The Dwina divides itself into many channels below Archangel. At the bar the rise of the tide is 3 feet.

DWINA, or DUNA, RIVER. See *Riga.*

DYCK BANK, coast of Belgium, a long narrow bank in 3 divisions, called the W. Dyck, the Middle Dyck, and the E. Dyck, or Clif bank. The W. end of the W. Dyck lies N.E. from Calais, distant 6 miles, with from 2½ to 7 fathoms on it, the deepest water being at the W. end. The Middle Dyck, properly the Dyck, is separated from the W. Dyck by a narrow channel, having from 4 to 5 fathoms. Its W. end is N.W. ¼ N. from Dunkirk, distant 9¼ miles, and N.N.E. ¾ E. from Gravelines, distant 6¼ miles. Its greatest breadth is about a mile.

DYER'S BAY, America, United States, on the coast of Maine.

DYER'S CAPE, North America, the N.E. point of Exeter sound, on the W. side of Davis's straits. Lat. about 66° N.

DYER'S SOUND, North America, S. side of Cumberland straits, leading through Metainco into the Straits of Frobisher, at the W. end of those islands.

DYET ROCKS, E. of Bermuda; the existence of these rocks is disputed. They are shown in charts as existing 100 leagues E. of the Bermudas. They were first reported to exist by M. Bellin. They were since supposed to have been seen by the commander of the Francis Freeling packet. The belief of their existence was revived by Mr. Robert Dyet, of the Catherine Green, of London, in 1845, who stated on May 17th of that year, he passed within 30 or 40 feet of two sunken rocks, having only 6 or 8 feet of water over them, the sea very smooth, in lat. 32° 46′ N., at noon, long. 60° 6′ W. Mr. Dyet suspected they were the rocks marked doubtful in lat. 32° 30′ N., long. 59° W.

DYMCHURCH WALL, Kent co., a dyke between Hythe and New Romney, nearly 3 miles long, preserving Romney marsh from the action of the sea. The repairs cost 4,000l. per annum, and the superabundant water is let off by sluices, the level of the sea at low water only being below that of the marsh.

DYMON, in the FERRO, or FÆROERNE, or FÆR ISLANDS, a rocky islet S. of Sandoe isle. See *Ferro Isles.*

DYNE, a town of Sweden, situated on a fiord or arm of the sea called the Dynekile, the centre of the last being in lat. 59° N.

DYSART, Scotland, a parish and burgh in Fifeshire, on the N. shore of the Frith of Forth, pop. about 7,000. It had once considerable salt manufactories, and many merchant-ships were built there for the Baltic trade.

E AGLE COVE, America, coast of Labrador, S. of Hawke island; a place affording a good roadstead for large ships, in 30 and 40 fathoms water, as well as anchorage for smaller vessels in 7 or 8 fathoms higher up the bay.

EAGLE HARBOUR, America, Long island, at the W. end, to the E. of Ha-Ha bay, formed by a cluster of islands, and capable of sheltering a great number of vessels, the ground holding well in from 20 to 10 fathoms water. To find this anchorage a vessel should make for the Great Mecatina island, and then shape her course for the Fox islands. The Great Mecatina point is in lat. 50° 44′ 12″ N. The Fox islands lie S.S.E. ¼ S., a good mile from the W. entrance of the harbour. It may be known by a great bay seen E. of it, without any islands, while to the W. there are a great many. In the W. passage to this harbour there are 2¼ fathoms, but the channel is narrow, and fit only for small vessels. All this part of the coast is dangerous for vessels, being full of low islands and sunken rocks.

EAGLE, MOUNT, West Indies, the highest land in the Island of Santa Cruz, 1,162 feet above the level of the sea.

EAGLE'S ROCK, N. coast of Madeira, called by the natives *Penha d'Aguia*, a vast insulated mass of rock of a cubical configuration, rising black and large out of the sea to the height of 1000 feet. See *Madeira.*

EARN PORT, Isle of Man, on the W. side of the island, about 4 miles from the Calf of Man, and at the S.W. point.

EARTH HOLMS, Baltic sea, 5 leagues E.S.E. ¾ E. from the N. end of Bornholm island. They consist of a cluster of bare rocks, which form a harbour, with water tolerably deep. Only small vessels visit them. On the largest there is a round castellated tower. There is a revolving light placed on these rocks, 92 feet high.

EASTHONIA or EASTLAND, N.E. in the Baltic, and the S. coast of the Gulf of Finland. From Wrangen island to the road of Narva, S.W. from Petersburgh, it is barren of interest for the seaman, except about the 3 small islands in Kock harbour, an extent of 45 leagues at E.S.E. and S.E.

EAST BANK, off the coast of Lundy island, Bristol channel. This bank lies nearly a mile E.N.E. of that island, trending on that bearing two-thirds of a mile. It is about a quarter of a mile in breadth, with 6½ to 8 fathoms over it, and 10 near by, sloping gradually into deep water.

EAST BAY, Newfoundland, between Cape Ray and the Strait of Belle Isle, not far from Red island, which last is in lat. 48° 34′ N., long. 59° 16′ 16″ W. There is anchorage here in from 8 to 12 fathoms.

EASTBOROUGH HEAD, Sussex coast, between 2 and 3 leagues S.W. by S. of Arundel, a bank which dries at low water for nearly a cable's length. The whole sand is more than a league long. [This description, found in old accounts of this coast, does not appear in modern sailing directions under the present name.]

EASTBOURNE, Sussex co., about 5 miles N.E. of Beachy head. The coast is foul and rocky a mile from the shore, but there is good ground farther off. S. of the town there is anchorage in 6 and 7 fathoms, sheltered from W. winds, but E. the ground is not good. This town is visited by strangers for sea bathing; it possesses a chalybeate spring. Pop. 2,726.

EAST CAPE, Anticosti, Gulf of St. Lawrence, America. This is a perpendicular cliff of limestone rising to 100 feet above the sea. It is the S.E. termination of a ridge trending W. inland. At Heath point, which is low land, there is a lighthouse 100 feet high, but it was not lighted in 1847; it bears from the E. point S.W. ¼ S. 3¼ miles, and between those two points is Wreck bay, dangerous, and offering no anchorage. Off S.E. from the E. cape a reef extends more than one-third of a mile. The E. cape or point is in lat. 49° 8′ N., long. 61° 43′ W.

EAST END BAY, West Indies. See *Virgin Isles.*

EAST GAT, Flemish coast, the passage to Flushing leads through this gat and Zoutland channel, by steering with W. Kapelle S.E. by S., which will lead clear to the W. of the S. Stone bank, and passing this bank, steering S.E. by E. ¼ E., it will lead to the N. of Caloo, up to the outer or red buoy.

EAST HAVEN, E. coast of Scotland, a fishing town, with a creek through the rocks, about a mile from West Haven. The coast here is all rocky and foul.

EAST HARBOUR, West Indies, Jamaica, near the N.E. end of the island, one of the two harbours at Port Antonio.

EAST HEAD, Africa, on the Old Calebar river, upon the S.W. coast. When this head bears E.N.E. Tom Shot's point is seen. This is the W. point of entrance, bearing 4 leagues N.N.W. from the head. Shot's point is in lat. 4° 36′ N., long. 8° 19′ W.

EAST HEAD, Denmark, coast of Jutland, in that part called Jutt's reef, 4 leagues E. of the Holms, between which, 2 leagues from the head and 5 from the land, there is a dangerous stony bank, where several ships have been lost. There is but 2 fathoms of water upon it at half tide. It is about 17 leagues W.S.W. from the point of Jutland called the Scaw.

EAST MAIN, Hudson's bay, North America, the E. coast of the Bay of New South Wales, which is yet more barren and unsusceptible of cultivation than the W. coast, and is lined with innumerable rocky islets.

EAST PASS, America, S. coast, Gulf of Mexico, the only channel for large ships to Apalachicola, leading to within 12 or 13 miles of the town, having 15 or 16 feet water, and no bar. The West pass has only 12 feet on the bar, and very little more is found beyond it.

EAST POINT, America, the extreme E. point of the Island of St. John's, in the Gulf of Nova Scotia, to the W. of the Island of Cape Breton, and N. by W. from the N. entrance of the Gut or Channel of Canso, often called the Straits.

EAST REEF, West Indies, Island of Hayti or St. Domingo, on the N.E. coast, in lat. 19° 20' N., long. 68° 59' W.

EAST RIVER, North America, United States, the sound or channel between Long island and New York island, extending E. between that island and the mainland of Connecticut.

EAST SAND HEAD, Island of Bonavista, one of the Cape de Verde islands. See *Cape de Verde.*

EAST and WEST HAVENS, Scotland; two fishing creeks on the S. shore of St. Andrew's bay.

EASTER BOOM'S GAT, or CHANNEL, coast of Holland, one of the passages near the Scheveling island. The Schorr ground, or St. Peter's ground, is near the strand of the Scheveling. The buoy of this gat, in coming in from the N. or E. is little more than a mile from the N. coast of Scheveling island, and about 4 miles or 1½ league from the W. end of that island.

EASTER DALE, Norway, a town upon Kyn sound.

EASTER POINT, Holland, a mark for sailing along the coast of Holland within the Texel island.

EASTERN BAVENT, Suffolk, the remains of a village washed away by the sea, near where the Point of Easterness, formerly the E. point of England, has wholly disappeared.

EASTERN CHANNEL, or SEDLEY'S CHANNEL. See *Plymouth.*

EASTERN GAT, or the OUTER DEEP, Denmark, leading from Elsineur road to the grounds.

EASTERN PIGEON, Mosquito shore, a low small kay covered with bushes, in lat. 15° 43' N., long. 82° 55' W. There is also the West Pigeon, which bears from the Eastern N.W. by W. ¼ W., distant 3 miles, and is similar to that in appearance. These kays are about 7 miles from the Cocorocuma kays, which lie W. ¼ S. of the Eastern Pigeon.

EASTON NESS, or EAST NESS, Suffolk coast, N.E. from Southwold, being the N. cape of Southwold bay. It is a mark for the Bernard shoal, thwart of Covehithe to the N., and not far from the land: the Ness, bearing W.N.W., is S. of the shoal.

EATON'S NECK LIGHT, America, United States, Long Island sound, S. side, E. of Huntingdon bay, in lat. 40° 57' 6" N., long. 73° 24' 12" W.

EASTWARE, Kent co., between Dover and the S. Foreland, a place frequented by fishermen, off the shore of which, and of St. Margaret, shell-fish of a superior quality are taken.

EBBWY RIVER, Monmouth co., flowing into the Usk, below Newport, in the Severn sea, where these two rivers form what is called Newport haven.

EBLETOFT, Denmark, a small town on the E. shore of a large bay, formed by two peninsulas on the coast of Jutland.

EBRO RIVER, Spain, on the E. side; it rises in the mountains of Asturia, and flowing E. by Calahorra, Tudela, Saragoza, and Tortosa, falls into the Mediterranean. This river is only navigable at its mouth for small vessels. The entrance is divided into two channels by the Isle of Buda, which is, like all the ground near, low and flat. The most prominent part of the island is Cape Tortosa, in lat. 40° 43' 25" N. The City of Tortosa is far inland. The Bay of Fangal, formed by the silt and mud of the Ebro, is all low land. The points of this bay are 3 miles apart, and lie about N.W. and S.E. from each other. N.W. winds are particularly troublesome here, blowing with great violence. The water is shallow, the river having only 16 feet at the entrance, and 5 further in. It is wholly unfit for vessels with any tolerable draught of water.

ECATHERINODAR, Russia, a town of the Cossacks, on the Black sea, in the government of the Taurida, on the left bank of the Kuban or Couban river. It is the chief place of the Tchernomors Cossacks, and where they hold their tribunals.

ECKE SOUND, Norway, on the W. side, N. of the Naze.

ECLAT BANKS, France, on the W. coast, S.W. by W. of Cape la Heve. The two roads at the entrance of the River Seine are separated by the bank of Eclat, and that called the high ground of the road. These banks are buoyed.

ECNOMUS, Sicily, S. coast, a hill, W. of the Port of Alicata, the last being in lat. 37° 4' 3" N.

ECREHOU ROCKS, British channel, near the Island of Jersey.

ECREVIERE BANK, France, near the Road of Granville.

EDA, Orkney islands. See *Orkneys.*

EDAM, Holland, a town on the Zuyder Zee, about 6 leagues S.S.W. from Euckhuysen and nearly 3 leagues from Hoorn. It has a good port, formed by the River Ey, on the banks of which it is situated, and this, with the dam which is thrown up against the inundations of the river, gives the town its name. It stands in lat. 52° 32' N., and long. 4° 58' E.

EDDISTO INLET, America, United States, South Carolina. The course here, from Charleston bar, is S.W. by W. ¾ W., and the distance 5 leagues, which will take a vessel clear of the shoals that lie off Stono inlet, lying further off than any that are in the way to Eddisto. Stono inlet is 2 leagues from the S. channel of Charleston. The course from thence is W.S.W. to Eddisto inlet, and the distance 11 miles. The soundings are regular between, and shoal gradually from the offing towards the shore. The bar of North Eddisto, and the shoals which are near it, lie 4 or 5 miles from the land. At the bar and shoals there are 9 and 10 feet at low water. South Eddisto is W.S.W. 3 leagues from North Eddisto. This inlet is in lat. 32° 32' N., long. 80° 10' W.

EDDLE SHOAL, coast of Essex, one of the shoals which lie off the entrance to Maldon, some of which are dry, and others have a little water over them, it lies to the N. of the sand called the Knowl. A ship must bring the high land above Maldon on the W., a sail's breadth open with Bradwell point, or the point at S.W., which is the E. limit of Burnham water. This will take a vessel clear between the Eddle and knowl, keeping on with it till the steeple on the N. shore is open of Red Cliff End.

EDD OE ISLAND, Norway, lying in the

Drontheim's Leed or the channel, which vessels take to that city.

EDDY ISLAND, Ireland, in the bottom of the bay of Galway, in the S.E. angle and W. from the smaller island called Darus island. There is a good road between these two on the S. side, by which ships may go in.

EDDYSTONE LIGHTHOUSE, coast of Cornwall, 8¼ miles distant from Rame head, S.W. ¾ S. It is a fixed light, and is lit up with lamps and reflectors. It is 72 feet high, built upon a rock, one-third part consisting of granite stones, dovetailed into each other, as well as into the rock at the foundation, as it is exposed to a tremendous sea during gales of wind, when, lofty as the building is, the waves frequently bury two-thirds of its column in their angry waters, and the spray has been known to break the thick glass of the lantern. The rocks on which this wonderful work of art stands, slope away to the S., so that the waves meeting the rock roll upward to the highest part, which is but a few feet above high-water mark. These rocks lie nearly in the fair way from the Start to the Lizard, and are therefore an object of the utmost importance. It is on the summit of the largest rock that the lighthouse has been erected, for all the rocks but one are covered with the flood tide, but some appear at the ebb. The first lighthouse was erected in 1696, and resisted many violent storms, but was blown down on the 27th of November, 1703; when the projector, who happened to be in it, and all his attendants, perished. The corporation of the Trinity House afterwards erected another in 1709, and to support the expense laid a duty on all passing vessels. This was burnt down in 1755, and rebuilt by Mr. Smeaton, within four years afterwards. The building, as thus constructed, consists of four rooms, one over the other, and at the top a gallery and lantern. The stone floors are flat above, but concave beneath, and each is additionally secured from pressing against the sides of the building by a chain let into the walls. Portland stone and granite are united together by a strong cement, and let into each other horizontally by dovetails. The architect discovered that Portland stone was likely to be destroyed by a marine animal, and as the working of granite was expensive and laborious, the external part only was constructed with this stone, and the internal part with the other. To form a broad base, and a strong bulk of matter to resist the waves, the foundation is one entire solid mass of stones to the height of 35 feet, engrafted into each other, and united by every additional means. About a quarter of a mile from the lighthouse to the N.E. is a rock under water, which only shows at the low ebbs of a high spring tide, and then only about the size of a butt. There are 10 fathoms all round it, within a ship's length. The rocks received their name from the variety of contrary currents or sets of the tide which are found to prevail near them. From their situation in relation to the Atlantic ocean and the Bay of Biscay, they are exposed to the swells of both, from all the S. western points of the compass, and the heavy seas which roll in from thence come upon these rocks, and there break with a fury inconceivable to those who have not witnessed it. A ground swell will also sometimes meet the slope of the rocks, and break upon them frightfully, when the surface of the sea is perfectly smooth, and without the slightest breeze. This obstruction, even at these times, is so great as to impede and put a stop to everything which may be carrying on upon the rock, and to prevent landing upon it, though for the weather persons might go upon the sea in the smallest ship's-boat. The W. side of the main rock, where it rises but 6 or 7 feet above water at the ebb of a spring tide, is bold and steep; but it is foul for half a mile off to the S.S.E., and there are several other rocks, which show themselves above water at the ebb. It lies from the Bolt head W.N.W. about 6 leagues, and from the Lizard E. ¾ N. 38 miles. In the fair way of this rock is 40 fathoms, and between it and Rame head 35 or 36 fathoms, so that a ship in coming up channel in the night, if she cannot see the light, must keep without the depth of 40 fathoms, as 35 fathoms is a sure mark for being within it. The lat. of the Eddystone is 50° 11′ N., and long. 4° 16′ W. It is high water there at spring tides at half-past 5 o'clock. Generally speaking, the flood tide sets within the Eddystone, from the Rame head to the Start, at E.S.E., and ebbs to W.N.W.

EDEN ISLAND, America, United States, on the coast of South Carolina, to the N.E. of the S. of Eddisto channel or river.

EDEN RIVER, Scotland, E. coast, about half a mile from St. Andrew's. Its mouth has not a good entrance, nor does it admit anything better than small barges.

EDEN RIVER, Cumberland co., falling into the Solway frith, below Carlisle. After entering the frith, the channel continually shifts from the effect of land floods at low ebbs, and at times renders the sands thus formed dangerous to venture upon.

EDENTON, a town of North Carolina, in the United States of America, upon a bay of its own name, on the N. side, and within Albemarle sound. It stands at the mouth of the Chowan river, and has much business in the shipping line. Lat. 35° 58′ N., long. 76° 40′ W.

EDER, Africa, W. coast, a small town on the edge of a cliff, about 14 miles S. of Cape Blanco, and S. of the ruins of Woladia.

EDGARTOWN, HARBOUR and TOWN, America, United States, between Martha's Vineyard and Cape Poge, Massachusets. The town itself is about 18 miles from Falmouth. A fixed light is erected at the entrance of the harbour on a pier running from the W. side 1,000 feet into the sea, and 50 feet above high water level. There are numerous shoals about this harbour, which, when passed, there is good anchorage within, in 4½ and 5 fathoms, holding ground. The light is fixed, 40 feet high, in lat. 41° 23′ N., long. 70° 31′ W.

EDGECUMBE, MOUNT, Cornwall and Devon cos., on the W. side of Plymouth sound, and connected with Drake's island by a ridge of rocks, called the bridge. It is celebrated for some of the finest scenery in England. See *Plymouth.*

EDINBOROUGH KAY, America, Mosquito shore, in lat. 14° 48′ N., long. 82° 43′ W. Two miles to the E. of this kay is Edinborough reef, curving to the N. by W. for nearly 3 miles. Between the reef and kay there are 2 fathoms water; N. of the kay and W. of the reef, 4 and 7 fathoms, and to the E. of the reef 15, 16, and 13 fathoms. Between this kay and reef to the N., and the Martinez reef to the S. is the Edinborough channel; it is 10 miles wide; the middle is in lat. 14° 42′ N., having 15, 14, and 13 fathoms through it.

EDINBURGH. See *Leith.*

EDKO, a lake of Egypt, E. of Lake Mediah.

EDMUND'S, ST., POINT, coast of Norfolk; from this point the coast trends one way almost S. towards Lynn, and on the other N.E. towards Burnham and Wells, and from the coast of Lincolnshire W. by N.

EEGHOLM, Denmark, an island off the coast of Zealand, in the Great Belt.

EEL POINT, coast of Wales, the cliffy point near the N.W. end of Caldy island; a reef of rocks extends N.N.E. to about half way over the E. entrance of Caldy sound from this point.

EGEAN SEA, Greece, the ancient name for the sea of the present Archipelago, between continental Greece on the W., and Asia Minor on the E. See *Ægean*.

EGG CHANNEL, West Indies, among the Bocas de Paria, sometimes called the Umbrella channel. See *Bocas de Paria*.

EGG HARBOUR, GREAT, America, United States; on sailing from Sandy Hook light-tower, as soon as a vessel is to the E. of the bar, she will steer S. till past Barnegat, if it be night, but if it be daylight she may approach the breaker to within 5½ fathoms. When Barnegat is passed, she must steer S.W. by S. 10 or 11 leagues, which will carry her up with Great Egg harbour. It has a shoal bank a league from the shore, with not more than 6 feet water upon it. Great Egg harbour is at the S. end of Barnegat, a long, narrow island, parallel with the coast. Lat. of the entrance 39° 19′ N., long. 74° 37′ W.

EGG HARBOUR, LITTLE, America, United States, on the coast of Jersey. During winter it often happens that vessels are prevented from entering the Delaware, or Sandy Hook, by the violence of N.W. winds, and are sometimes driven from the coast into the gulf. In order, therefore, to make a safe harbour, Blunt, in his "American Coast Pilot," advises the running through the Sod channel, then to keep within 30 or 40 yards of Small point, in 2½ fathoms; to pass the point, haul round gradually, giving the breakers a little berth, or to steer in for the beach when opposite Tucker's house, until the vessel is in 4 fathoms, then she must steer W.S.W., which course will take her through the same channel. There are five buoys placed at the entrance of Little Egg harbour, from which Absecum lies 5½ miles S.W.

EGG ISLAND, America, off the coast of Labrador, lat. 49° 38′ 21″ N., long. 67° 3′ 10″ W.

EGG ISLAND, America, United States, a small island so named in Delaware bay, about W. by N. or N.N.W. from Cape May.

EGG ISLAND, West Indies, on the edge of the Bank of Bahama; the extremity of Egg island reef is in lat. 25° 34′ N., long. 76° 55′ 30″ W.

EGG, or OVO, ISLAND, in lat. 35° 37′ 30″ N., long. 25° 34′ 30″ E., having deep water all round it. This island is N. from Cape St. John, in Candia, about 22 miles.

EGG, or OVO, ISLAND, S.E. coast of Greece, in lat. 36° 4′ 45″ N., long. 23° 0′ 30″ E. It is a high rock, very easily seen from a vessel, and avoided. It is 500 feet high, and lies due S. of the Port of Kapsali, in Cerigo.

EGGER-OE ISLAND, Norway, S. coast, an island having a haven on the S., running into the main land; it lies S. of Stavanger.

EGGERSUND HARBOUR, S. coast of Norway, W. of the Naze, near Lundervüg. It has two entrances, the S. running in between the E. side of Egger-Oe and the main, and the N. channel passing to the N. of Egger-Oe island.

EGINA, or ATHENS, Gulf of Greece, E. coast, between Cape Skylli, the S.W. extremity, in lat. 37° 27′ N., and Cape Colonna, in lat. 37° 19′ 15″ N.

EGINA ISLAND, Greece, E.N.E. from Cape Estimo. See *Ægina Island*.

EGLWYS DINAS, Wales, a fishing village on the W. shore of Newport bay, in Pembrokeshire.

EGMONT OP ZEE, between the Maas and Texel, Dutch coast, 7¾ miles from Wyck op Zee. It has a bluff square steeple, and two light-towers standing close to it. Two leagues N. of Egmont op Zee is Camperdown or Schoorldown, the highest land on the Dutch coast, a large sandy down, in lat. 52° 41′ 30″ N. It appears in hummocks from afar, and may be seen in clear weather from 14 fathoms water, distant 20 miles.

EGMONT CAPE and BAY, America, in Prince Edward's island. The cape is in lat. 46° 53′ N., long. 60° 22′ W.

EGMONT HARBOUR, or CALAVINE, Grenada, West Indies, two leagues E. of Salines point. It is very deep, and so large that sixty sail of the line might lie there at anchor in safety. The entrance is about half a mile in breadth. Inside it forms two harbours, or an outer and inner port. The depth is generally 7 fathoms good ground. Vessels may lie here alongside the warehouses.

EGMONT ISLAND, America, West coast of Florida, near Espirito Santo bay. This island has a beacon upon it 80 feet high, for the purpose of directing vessels into Tampa bay.

EGRIPOS, Greece, the name given by the Turks to the ancient Euboea, now the Island of Negropont, separated from the coast of Greece by the Strait of Euripus.

EGRY LIMAN, coast of Anatolia, 2 leagues to the S. of Cape Karabouroun, a narrow and compact harbour, discovered and explored by that zealous and scientific officer, Admiral Beaufort. It is in shape like a dock; nearly a mile from N. to S.; the inner half a shoal of mud, but the outer part having from 7 to 10 fathoms, mud and sand. The entrance, its discoverer states, is not easy to find, unless close in with the land, which is very high, but it may be recognised when a small peak on the highest part of the Karabouroun range of mountains, bears E. by S.; a patch of grey barren rock, with some cultivated ground in its vicinity, may be observed about one-third up the mountain, which brought under the peak thus mentioned will be nearly in one with the extremity of the peninsula that forms the entrance. Vessels may lie sheltered here from all winds but those from N. by E. to N.W. This port is in lat. 38° 34′ N.

EGYPT, African coast, commencing at El Arisch, which separates it from Palestine, and extending as far as the Lybian Desert, W. of the Nile, or from long. about 29° E., to 33° 48′. The coast is low and sandy, having neither bay nor harbour of considerable moment. From El Arisch W. 22½ leagues, the site of the ancient Pelusium presents itself, under the modern name of Tineh. It was once seated on an important branch of the Nile, at its exit into the Mediterranean, but time has made it little better than a muddy channel, traversing a sandy waste, and except when the Nile rises, has no connection with its waters. There are yet some remains of the ancient Pelusium extant, but they consist of a few columns only, and heaps of rubbish. There is a castle at Tineh, but that is also in decay. W. from Tineh the Lake Monzaleh receives some of the lesser channels of the great river. It extends W. as far as Damietta, where a main channel of the Nile disembogues its waters. From Damietta a desert sandy flat shore extends to Lake Bourlos, the receptacle of some small channels from the Nile. The second great branch takes its course into the Mediterranean beyond, though very near the W. end of that lake. At the mouth of this second great river branch, stands the town of Rosetta. These two larger river branches are generally called the Eastern and the Western Nile. A bay renowned for the battle between Nelson and the French, called Aboukir bay, extends from Rosetta to the castle of Aboukir, presenting a beach of arid sand, with low sandy hillocks along the shore. Around the point of Aboukir bay the coast runs W. to the pharos and ports of Alexandria, terminating at Cape Marabut, from whence to the W., commencing with what is called the Gulf of the Arabs, the shore is of the same desert character as that from El Arisch to Tineh.

EIG, Scotland, one of the Hebrides islands, S. of Skye, about 6 miles long, and 2 broad. It has numerous basaltic columns, and the coasts are formed of a very porous lava.

EIGHT STONES, N. of the Madeira islands, an extensive and dangerous reef, said to have been discovered in 1732, by a Captain Vobonne, of London. He reported the sight of eight rocks, even with the surface of the water, situated between 34° 30', and 34° 45' N., near the meridian of 16° 40' W. This report was the cause of much alarm to navigators, and naturally excited attention, but there is good reason to believe that there is no foundation for these dangers, or at all events not in the position thus assigned to them. Numerous vessels, some sent out for the express purpose of naval surveying, have been again and again over the spot, and unable to detect these rocks. From 1828 to 1836, no less than twelve vessels of the royal navy passed over and about the place pointed out. The latest of these, the Etna, sounded near the meridian of 17°, but could find no bottom at 200 fathoms; she had before gone over the supposed centre of the danger, and could discover none at 70 fathoms. Captain Fitzroy, in 1832, searched in vain for these rocks, and found nothing to indicate rock or shoal, or even shallow water. He met with a deep blue sea, uniform in colour, in which he could get no soundings. He felt certain that if there had been a rock or shoal within 7 miles of him, at any hour of daylight, he must have discovered it. He hardly thought the existence of such rocks possible, and remain unobserved. It seems that there had been traditional accounts at Madeira, that a large ship had been lost on the N.W. Baxio, an extensive shoal, now marked as the Falcon rocks in the charts. The aged inhabitants report, too, that the sea is not now what it once was there, for though it still breaks on these banks with great fury, they assert that rocks were formerly to be seen upon it. They must have consequently subsided or fallen since, if they ever really existed.

EIL LOCH, Scotland, W. coast, the entrance is opposite to the ruins of Fort William. A vessel may enter this loch with the flood tide, and sailing between the two first islands at Camusnagaul, anchor above a mile N. of the entrance, because on both sides the ground is too rocky there.

EKESIO, Sweden, an island town of Smaland, in East Gothland, 15 leagues N.W. from Calmar, in lat. 57° 28' N., long. 15° 12' E.

EKHOLM, THE, Gulf of Finland, a lighthouse so called, built upon the Island of Ekholm. It is a fixed light, 75 feet above the sea, visible 15 miles, in lat. 59° 41' 8" N., long. 25° 47' 58" E. The Island of Ekholm is small and unimportant, bearing from the Koksear light E. by S., distant 8 leagues.

EKNAES, Gulf of Finland, a small town in a bay, E. of Hango head, there are several islands in the bay, with only 7 or 8 feet of water between them.

EKRENFORD BAY, Denmark, coast of Sleswick, on the Baltic, 3 leagues E. from Sleswick. The town stands on the bay, and has a good harbour for ships, in lat. 54° 50' N., long. 10° E.

EL ARAISCHE, W. coast of Africa. See *Araiche*.

EL ARISH or ARISCH. See *Arisch*.

ELBA, ISLAND OF, Mediterranean, off the coast of Italy, about 15 miles long, and 10 broad, in the widest part, but at its western end not more than 2½ miles across. The country is mountainous, the climate healthy, the population full 14,000. It produces corn and fruit in abundance, and possesses mines of iron and copper. Elba has two chief harbours, Porto Ferrajo on the N. side, and Porto Longone on the S. The island is immediately opposite to Piombino, in Tuscany. Cape Vita, the N. point, is in lat. 42° 52' 12" N., long. 10° 25' 10" E. Porto Ferrajo has a light on the W. entrance, in 42° 49' N. lat., and 10° 20' E. long. It bears from Piombino W.S.W. ⅜ S., distant 6 miles, and from Cape Troya W.N.W. ¼ W. 16 miles. Several islets are nearly close to its shores, in different directions; such are those of Topi, distant about half a mile, with 7 fathoms between, Palmajola, Cervoli, and others. There is a knoll of 3½ fathoms to the N. of Cape Vita, about 1 mile distant, with 9, 10, 11, and 14 fathoms around it. Porto Ferrajo is the capital of the island.

ELBE RIVER; this important river of Europe rises in the mountains of Bohemia. It is navigable by ships as far as Hamburgh. It lies E. of the River Weser. Vessels making the entrance, pass between the Vogel sand and the Schaarhorn reef, and steer S.E. by S. for the red buoy; with a flood and S. winds, the course is S.S.E., and with an ebb and N.E. winds, S.E. In running in the fairway for the red buoy, a ship will have 20, 17, 15, 14, 13, and 12 fathoms, soft clay ground, of a bluish colour, and at the red buoy, which lies in 10 fathoms, fine yellow sand. If, in the course from Helgoland to the Elbe, a hard sandy bottom be found, of a reddish colour, a ship is to the N., and out of the fairway. Vessels coming from the west, and acquainted with the Weser and Elbe, do not sail to Helgoland, particularly with southerly winds, but being arrived between Wangeroog, on which is an intermittent light 64 feet high, and Helgoland, and having the one or the other of these islands in sight, they steer, with an E. course, directly for the Elbe. The islands Wangeroog and Helgoland bear N.N.E. and S.S.W. from each other, distant nearly 8 leagues. When midway between these islands, the direct course for the red buoy is E.S.E., distant 5 leagues. Allowance must be made for the wind and tide, the course with flood being S.E. by E., and with ebb E. by S., somewhat more S. or E., according to the wind, a vessel will have 17, 15, 14, 13, and 12 fathoms, with soft bluish ground. When standing towards the south shore, and coming into 10 fathoms or less, the bottom is hard fine white sand.

The sandy shore between the rivers Jahde, Weser, and Elbe, is very dangerous, because it is steep-to, from 10 to 9 and 7 fathoms, and then dry. If it should be dark or thick weather, a vessel must not approach nearer than 13 fathoms, and then, if it be flood-tide, anchor. With an ebb she might perhaps, keep under weigh until daylight, or until the weather becomes clear. Great attention to the winds and tides is necessary, observing that the flood sets N. and E., and the ebb W. and S., and when near the entrance of the Jahde and Weser, in 12 fathoms, the flood sets into these rivers; but the ebb sets always to seaward. These currents are also stronger the nearer to those rivers, or to the passages between the sands. It is a rule that vessels should run into the Elbe and Weser with the tide, and always in the daytime. The best guide for the entrance is the signal-vessel, which is stationed at the mouth of the river, a mile N.W. by N. from the red buoy, in 11 fathoms at low water, and 13 at high, having the great tower of Neuwerk, the Schaarhorn beacon, and red buoy in a line; and there moored with iron chains, she does not leave her station in stormy weather, except when forced by the ice, in the winter season.

ELBING, Prussia Royal, a large city in the district of Marienberg, on a bay of the Baltic sea, known as the Frisch-haf, 30 miles E.S.E. of Dantzick. It has a pop. of 17,000, and enjoys a considerable trade. Lat. 54° 20′ N., long. 19° 45′ E.

ELBINNY COVE, Torbay, a small hollow on the S. side of the bay, that kept open with a rocky point about half a mile E. of the Cove, serves as a mark for the distance of the anchorage of line-of-battle ships from the S. shore of the bay.

ELBOOG SAND, Flemish coast, a hard narrow bank, drying at low water; the part of its S. end which dries, lies W. by N. from the pier of Flushing, distant 1¼ mile. From this it extends N.W. more than 3 miles, and from thence a bank of shallow water, which is only passable by small vessels, connects it to the Raen. The N. side of the Elboog forms the S. boundary of the Deurloo channel.

ELBOW BANK, coast of Suffolk, W. ¾ S., about 1½ mile from the W. end of Copperas bank, having over it 3½, and at its S. end 4½ fathoms. It bears nearly S.E. from the Tripod, distant 1 mile.

ELBOW BUOY. See *Portsmouth*.

ELBOW REEF, West Indies, Little Bahama bank. Lat. 26° 33′ N., long. 76° 50′ W., after M. de Mayne, but according to the American statements, in lat. 26° 34′, long. 76° 52′ W.

ELBOW SHOAL, coast of Kent, 2½ miles S.E. by E. from the N. Foreland lighthouse, having a white buoy upon it, to be left 2 or 3 cables' length to starboard.

ELBURG, Holland, a town of Guelderland, on the E. shore of the Zuyder Zee. Lat. 52° 30′ N., long. 5° 50′ E.

EL CAMBRE, Grand Canary island, the highest mountain peak it exhibits rising to an elevation of 6,648 feet above the level of the sea.

ELEFSIS, LEFSINA, or LEPSINA, Greece, the ancient Eleusis, once celebrated for its mysteries in honour of Ceres. It is now a miserable Albanian hamlet, in the Bay of Eleusis.

ELENA, ST., POINT, Spain, S. coast, the E. boundary of the Llanos or Plains of Almeria, and in fact the S.W. limit of the bay of that name. It lies N.E. ¾ E., distant 2½ miles from Cape Savinal,

III.

and is a low point, having the tower of Los Carrillos between.

ELENNA, ST., or HELENA, POINT, and CAPE, Greek island of Scio. The two are 1¼ mile from each other, with a clear, white sandy bottom between. Cape St. Elenna is a low, rugged point, gradually rising to a round hill, upon which is an old watch-tower. A shoal stretches out from this cape E. ¾ S. about 900 yards, running along as far as Point Elenna. To avoid it, Cape Proseaux, on the S., is to be kept open of Point St. Elenna, about a sail's breadth, which will take a vessel clear in 10 and 11 fathoms gravel and coarse sand.

ELEPHANT SHOAL, Denmark, passage of the Great Belt; in the middle of the passage between Reisoe and Romsoe. It is about a mile in length and three-quarters of a mile broad, and on its centre there are but 12 feet water, while around it there are 4 fathoms. Further off the water deepens on all sides to 6, 10, 11, and 14 fathoms. A mill on Fyen island, touching the N. part of Romsoe, will lead directly on the N. part of the Elephant shoal. Kierteminde steeple, touching the S. point of Stavres head, leads on to its S. part. A remarkably long white house on Zealand, bearing S. 40° E., and being open of the S. point of Musholm, will clear it to the S.

ELEUSIS, BAY OF, Greece, on the N. side of the Island of Colouris, or Koulouri.

ELEUTHERA ISLAND, West Indies, one of the Bahamas, the lat. of which, at Pigeon kay, is 25° 11′ N., long. 76° 15′ W. Vessels from the E. should first make the coast of Eleuthera, in a track between the parallels of 25° 30′, and 25° 40′, not exceeding the parallel of Harbour island, if bound thither.

ELFSBORG, and NEW ELFSBORG, Sweden, the names of the fortresses that defend Gothenburg on the approach from the sea. The first a citadel below the town, and New Elfsborg, on a rocky island, in the middle of the channel.

EL GAZIE, African coast, a low, sandy beach, supposed to be about 400 miles N. of Senegal, upon which the American ship Charles, John Horton, master, was wrecked in 1810, and in which was also wrecked John Adams, a seaman, who published a narrative of his three years' slavery among the Arabs, from whom he was ransomed by the British consul at Mogador.

ELGIN, a town of Scotland, standing on the Lossie river, and having its port 5 miles distance, where that river enters the sea. Its export is principally corn.

ELGOE ISLAND, and ELGOE FIORD, Sweden; this island is N.E. of Koe-Oe and Marstrand, and the Fiord leads up to it, nearly on a parallel with the Paternoster rocks, in lat. 57° 55′ 30″ N.

ELIAS, ST., CAPE, Island of Sardinia, in the Gulf of Cagliari, having on the E. side the Bay of Quarta.

ELIAS, ST., MOUNT, Coast of Anatolia, a mountain seen a great way off, having a remarkably dark appearance, and rising suddenly to a great height, to the W. of Cape Karabouroun, a remarkable seamark.

ELIAS, ST., MOUNT, Greece, Bay of Egina, on the Island of Poros, a mountain nearly in the centre of the island, N.E. from the town. This mountain, according to Captain Copeland, stands in lat. 37° 31′ 12″ N., long. 23° 27′ 50″ E.

ELIAS, ST., MOUNT, Greece, Island of Milo,

M

at the S.E. end. It is one of the highest in the Archipelago, elevated 2,530 feet above the sea. The summit is in lat. 36° 40' N., long. 24° 23' E., after Lieutenant Raper.

ELIAS, ST., MOUNT, Greece, S. end of Negropont, near the Strait of Doro. See *Doro*.

ELIAS, ST., MOUNT, Greece, Island of Paros, in lat. 37° 3' N., long. 25° 12' E.

ELIAS, ST., MOUNT, Greece, Island of Psara, or Ipsara, its centre rises to a great elevation. It is in lat. 38° 35' 34" N., long. 26° 36' E. Like all the other mountains of the same name in these islands, it is an important object seen from the sea.

ELIAS, ST., MOUNT, Greece, Island of Santorin; the highest point, in lat. 36° 21' 56" N., long. 25° 28' 33" E.

ELING, a hamlet on Southampton water, Hants co.; it is also a parish in the Hundred of Redbridge. It has a considerable trade in corn and timber. Shipbuilding has been carried on there to a considerable extent, there being slips and docks for building vessels, some of which have been constructed for the West India trade.

ELIZABETH CAPE, America, the S.E. point of the Island of Good Fortune, on the N.E. side of Hudson's straits, having Cumberland straits and islands on the N.E. The coast trends N. from it, on the E. side. Lat. 62° 35' N., long. 64° W.

ELIZABETH CAPE, America, Maine, about 8 miles S.W. from Long island, and S. of the entrance of Portland harbour. It has two lights, one fixed, the other revolving, 300 yards apart, bearing from each other S.W. ½ W., and N.E. ½ E. The W. light revolves once in two minutes. The E. is a fixed light. Lat. 43° 33' 36" N., long. 70° 15' W., N.E. light. Lieutenant Raper gives the cape as in lat. 43° 33' N., long. 70° 10' W.

ELIZABETH CASTLE, Island of Jersey; it is connected with the island by a ridge above half a mile in length, which is dry at low water. Ships sail over it at high water, when the castle becomes insulated.

ELIZABETH ISLANDS, America, coast of New England, a short distance from Cape Cod. They are four or five in number, running off S.W. from Falmouth, the S.W. point of Barnstaple peninsula.

ELIZABETH TOWN, America, United States, situated on a bay 15 miles from New York. Vessels of 30 tons come up to the town, and those of 300 to the mouth of the river.

ELKTON, America, United States, Maryland province, near the head of the Chesapeake; it has a considerable trade in grain.

ELLE, Scotland, one of the small towns and havens within the Frith of Forth, between Fife ness and Elle ness, a space of about 3 leagues. It is the best of the small havens in that part, all of which are dry at low water. It has only 9 or 10 feet at high water, neap tides, and but 12 and 13 at springs.

ELLE NESS, Scotland, 3 leagues W.S.W. from Fife ness; from whence to Kinghorn ness is 4 or 5 leagues, also to the W.S.W. and S.W. by W., and between them is a tolerably deep bay.

ELLELOS, TOWN, Sweden, W. coast, at the head of Ellelos fiord, on the S. shore of the Island of Oroust.

ELLEN'S CUT, America, Yucatan coast, between Gladden and English kay; the channel lies N. 13 miles of Gladden kay, in lat. 16° 47' N. It has 2½ fathoms water.

ELLIOT'S KAY, West Indies, one of the Florida kays, near Kay Biscayno, the S. end of which is in lat. 25° 41' N., long. 80° 3' W., and had a fixed light on it 70 feet high, in November, 1847.

ELLIS BAY, ANTICOSTI, the only anchorage that is tolerably sheltered in the whole island, so that vessels, whose draught of water is not too great for 3 fathoms, may lie there safely during the three finest summer months, but should moor, with an open hawse, to the S. Larger vessels, whose object is a few hours' stay only, may moor further out, but the best anchorage is far up the bay, the best berth in a line between Cape Henry and the White cliff, bearing W.S.W. ½ W., and E.N.E. ½ E. respectively, from each other. A vessel will then be in 3 fathoms over a muddy bottom, distant 300 fathoms from the flats on either side, and half a mile from those at the head of the bay. The extremities of the reefs, off Capes Henry and Eagle, will then bear S.W. by S., and S. ½ E. Thus three points of the compass are left open, but in a direction from which heavy winds are of rare occurrence, and then the sea, at the anchorage, is much less than might be expected from its heaviness between the reefs, which are limestone and dry at low water. There is no anchorage from S.W. point to Ellis bay. The bay can easily be made out at sea, Cape Henry being a bluff point, and the land at the head of the bay being low. Cape Henry, the S.E. extremity, according to Captain Bayfield, is in lat. 49° 47' 50" N., long. 64° 25' 44" W. The line of breakers, seen on either side, presents a formidable appearance to those who see them for the first time, but there is no danger, if the directions imparted by those acquainted with the bay are attended to.

EL MANSORIA, W. coast of Africa. Lat. 33° 46' 10" N., long. 7° 20' W. An Arab village, upon an iron-bound part of the coast. It has a mosque with a tower, about 80 feet high, and stands 180 feet above the ocean level, visible from the deck of a frigate at the distance of 6 leagues.

EL MEDIAT, MEHEDIA, or MAHADIAH, Africa, on the N.W. coast. It is surrounded by the sea and well fortified, with a good harbour. It stands in lat. 34° 18' N., long. 6° 36' W., from the observation of Captain Washington, R.N., 1830.

ELME, ST., ROAD, France, S. coast; this road lies on the S. side of the isthmus, by which the peninsula, that runs out nearly E. to Cape Sepet, is united to the mainland, on the S. side of Toulon roads and harbour. The course from Cape Sicie, a high and steep cape, with the chapel of Notre Dame on the summit, in lat. 43° 3' N., long. 5° 51' E., to Cape Sepet, is nearly E.N.E ½ E. about 4 miles, all low land, and nearly midway lies the Road of St. Elme. The shore is low and sandy, but vessels may anchor there in case of necessity. Some have mistaken this road for the entrance to Toulon, because that city is seen as a vessel passes, and ships at anchor in it. This road lies, however, a league W. of Cape Sepet, which is the entrance to the Bay of Toulon.

ELMINA, DELMINA, or ST. GEORGE DE LA MINA, Africa, S. coast, the principal establishment of the Dutch, on the coast of Africa. Lat. 5° 4' 48" N., long. 1° 20' 12" W. See *Delmina*.

ELMO, ST., LIGHTHOUSE and CASTLE OF, Island of Malta, 1¾ mile from the Bay of St. Julian, in lat. 35° 54' 15" N., long. 14° 31' 10" E.

ELOY POINT, West Indies, Island of Grenada,

a mile S. of Molenier's point, and near St. George's bay.

EL ROQUE, PORT OF, Colombia, to the E. of the PIRATE'S KAY; it is in lat. 11° 56′ N., long. 66° 41′ W. There is here a space about 2 miles long, and half a mile wide, at the narrowest part, between reefs, where ships may anchor in from 12 to 14 fathoms, having five entrances all clean, the reefs steep-to, and consisting of coral rock. The bottom is hard, and fish abound. It is high water here, at full and change of the moon, at 4ʰ 30ᵐ, rise 3 feet.

ELSHOLDMEN, Sweden, a small town opposite Walho island, in the Alands haf, lat. 59° 58′ N.

ELSIMBORG, Sweden, a port and town 7 miles E. of Elsineur, in lat. 56° N., long. 13° 20′ E.

ELSINEUR, ELSINORE, or HELSINGFORE, Denmark, the second town in importance belonging to Zealand, at the entrance of the Sound, connecting the Cattegat sea with the Baltic. It is opposite to Helsingborg, in Sweden, and about 8 miles N. of Copenhagen. The Danes here exact a toll from all vessels proceeding up the Baltic. This town is in lat. 56° 2′ 54″ N., long. 12° 16′ 42″ E. There is a reef runs off N.W. from the point of land upon which stands the Castle of Cronenborg, but by keeping in 7 and 8 fathoms, a vessel will pass clear of it. It is commonly known by the name of the Lap or Lapsand. When round Cronenborg castle a vessel should edge in nearest to the starboard shore, to avoid a sandbank near the middle, between the two shores, at N. by W. from the Island of Huen. There may be a sufficient depth of water, but the shoalings may occasion an alarm, which is thus avoided. After this all is clear as far as Copenhagen, giving the shore a moderate berth in order to keep clear, and 6 or 7 fathoms will bring a vessel to the opening of the Harbour of Copenhagen, where there is a buoy to mark out the point of a sand running out N. from Amack or Amag island, on the port side. Here every vessel which passed through the Sound was once expected to lower her topsails. There is anchorage at Elsineur in 9 and 10 fathoms.

ELSINGVOS, Sweden, a channel on the coast, 16 miles E. by N. from the channel of Luys, and about 11 leagues W.N.W. from Pelting sound.

ELY'S BAY, West Indies, Island of Antigua. On the N. side of the island, edging off to the S.E. from a point of land called Beggar's point, its N.W. limit.

EMANUEL HEAD, Northumberland. From this head Berwick-upon-Tweed is N.N.W. ¾ W., 8 miles distant; Whapness N. by W. ¼ W., 5 leagues; and St. Abb's head N. by W. ¼ W., 19 miles. Vessels taking a course N. from this head, by keeping it in one with Budle signal-house, S. ¾ W., will clear all the shoals in Berwick bay; and when Berwick lighthouse bears N.W. by W. ¼ W., may steer directly towards it.

EMBDEN, a town of Hanover, and the capital of the province of East Friesland, situated on Dollart bay, at the mouth of the Ems river, in lat. 53° 22′ 6″ N., long. 7° 12′ 42″ E. Vessels of considerable tonnage are unable to come up to the city. It has considerable trade, and a pop. of about 12,000. The bay is divided into two channels by the Island of Nessa, within which, upon the main, the town stands. The W. channel is the widest; but both are entered from the S. either from the German ocean, coming from the W., or from Der Dollart, to the E. of Groningen, at S.S.W. There are several islands at the mouth of

the Ems; the most N. is that within which is Embden road, before mentioned. The tide flows here at S. and S. by W., and at N. and N. by E., that is from about 12ʰ to a little before 1ʰ, on days of full and change.

EMBROS ISLAND, in the Archipelago of Greece, sometimes called Imbros. It is W.N.W. 10 miles from the entrance into the Dardanelles, and N.W. of Cape Janissary, off the Gulf of Coridia, to the S.W. On the N.W. the Island of Samothraki lies 13 miles distant.

EMENEH, CAPE, on the W. coast of the Black sea, N. of Bourgas; lat. about 42° 43′ N., long. 27° 54′ E.

EMMICK BAY, Cornwall. This bay is situated just round the Deadman or Dodman point. It has clear sandy ground in 6, 7, or 8 fathoms, where in E. winds a ship may securely anchor, with the Deadman point brought to bear S.E.

EMPERORS, THE, France, S. coast, two remarkable rocks, S. of Riou.

EMPEROR'S POINT, Spain, E. coast, a name once given by the English to Cape Artemus; by the French called Cape St. Martin. It is a high, steep headland, and forms the S. point of the Bay of Valencia.

EMS RIVER, between the Texel and Elbe. This river has three entrances, called the West Ems, the East Ems, and the Homme Gat; the two last separated at their commencement by the Juister reef. The West Ems has upon its W. side the Island of Rottum, the Huiebert Plaat, and other sands, and on the E. side the Borcum reef and island, the Randzel, and others. The entrance to the West Ems is between two banks, called the Geldzak Plaat on the N. and the Huibert's Plaat on the S. side. The outer buoy is black, and lies in lat. 55° 36′ 30″ N., about 13¼ miles E. and S. from the red buoy at the entrance of the Vriesche gat. Within this channel there are the middleground shoals, called the Meenwen Staart, 4½ miles long, extending S.E. by S.; the Ems Horn, 3 miles S.E. of the Meenwen Staart; and those called the Hond and the Paap, 5 miles in length, in a direction S. by E. The Eastern Ems is entered by bringing the great beacon upon Borcum in one with the light-tower, until the vessel is in 8 and 7 fathoms, where the first black buoy is found in lat. 53° 40′ 45″ N., the light-tower bearing S. and E., fixed light 142 feet high, distant 6 miles. The Homme gat passage into the Eastern Ems lies between the Juister reef, on the W. side, and the Schaape sand on the E., and is nearly a mile wide. On the N. side of this sand there is a red buoy, lying at some distance E. from the entrance, with the two beacons on the E. end of Borcum in a line, distant 6½ miles. Vessels bound into this gat should not approach the Juister reef nearer than from 8 to 6 fathoms, until they bring the Borcum light-tower to bear S.W. by W. This river should never be entered without a pilot. In bad weather the buoys shift, and the tides run with great velocity. To the E. of the Ems lie the islands Juist, Norderney, Baltrum, Langer Oog, Spiker Oog, and Wangeroog, all low and sandy, fronting the main, having channels both between each other and the shore. These channels open a line of communication all the way from the Texel to the Ems, and from the Ems to the Jahde and Weser; but they are only navigable by small craft, and known exclusively to the natives.

EMSWORTH, Hants co., a small and thriving

M 2

village, accessible to small craft through Langston and Chichester harbours.

ENCKHUYSEN, Holland, a fortified city near the mouth of its harbour, on the Zuyder Zee, 10 miles E. from Hoorn. It was formerly a good port, but the sands have accumulated, and large vessels can no longer enter. It lies on the starboard side of the channel called the North and South Gat, in regard to their situation from this city, and nearly due N. from the Ton on the Cripple. It is high water here, at spring tides, at 12ʰ. Lat. 52° 39′ N., long. 5° 20′ E. It is 27 miles N.E. of Amsterdam.

ENCKHUYSEN SAND, Holland, lying nearly E. and W. in the Zuyder Zee. The N. side is nearly in a line W. from the Island of Urck, towards the north buoys of the North Gat, from the Ton of the Cripple to the city of Enckhuysen. The S. side extends a long way S. of E., and is called the Hout Rib, running nearly at an equal distance by the main body from the N. side, and turning in like manner to the E.

ENDELAVE, Denmark, an island between Samsoe and Jutland.

ENDYMION ROCK, West Indies, in the Turk's island passage, 8 or 9 miles S.W. by S. from Sand Kay, in lat. 21° 7′ N., long. 71″ 18′ W. The Square Handkerchief, at its S.W. part, is 7 leagues E.S.E. from this rock.

ENGANO CAPE, Hayti, the E. point of the island, low, but visible at the distance of 10 leagues. It lies in lat. 18° 35′ N., long. 63° 18′ W. The land of Porto Rico can be seen from it in clear weather. A shoal extends 3 miles from this cape to the N.E., and having but little water on it, should have a good berth. This cape, bearing W. by S., 6 leagues, appears with a head like two wedges.

ENGERHOLM, Sweden, a town on the River Ronne, which falls into Skelder bay; it has a little trade, and a pop. of about 1,000.

ENGLE'E HARBOUR, Newfoundland, on the N. side of Canada bay, near an islet of considerable size, in lat. 50° 42′ 30″ N. A low white point, forming the N. entrance of the bay, being passed, a vessel must keep the shore until abreast of the next point, which makes the harbour, hauling round it to the S.E., giving the point a good berth, and anchoring in from 15 to 7 fathoms.

ENGLISH BAR, France, W. coast, a sandy bank, called by the French La Barre Anglaise, running off nearly a league from Point Terre Negrée towards Point de la Coubre, the two points of a large bay, affording anchorage in 3, 4, and 5 fathoms, on the N. coast of the Gironde, and almost N. from the tower of Cordouan.

ENGLISH CAPE, Newfoundland, on the S. coast, at W.N.W. from Cape Pine, forming the opening of the narrow and deep bay of St. Mary's. Within this point, nearer to the bay, is a harbour on the starboard or E. side of the bay. Lat. 46° 49′ N., long. 53° 45′ W. St. Mary's Cape forms the W. cape of the entrance.

ENGLISH COVE, Island of Minorca, in the harbour of Port Mahon, a little below the town, where the harbour is about a mile over. It is the watering-place of the shipping.

ENGLISH CREEK, West Indies, a narrow cove on the S.E. part of the Island of Martinique, a little round a headland from Point Salines, the most S. cape of the island.

ENGLISH GROUNDS, in the Severn river, or Bristol channel, E. of the Holms, a large shoal that stretches out from the E. land, with a

long tail shoaling towards the W. To avoid it, in coming from the W., as soon as a vessel is past the Flat Holm, she must sail N.E. or N.E. by E. for a league, by keeping the Flat Holm at S.W. or S.W. by W. When at this distance, a ship must bear away E. and E. by N. towards Portished point, in 14 or 15 fathoms water, looking out for a small island lying near the E. shore, which will then be nearly athwart of the ship. There are two mills on this island, and another on the main above; and when the last is in a right line with the W. end of the island, a ship will be past the tail of the English Grounds.

ENGLISH HARBOUR, America, Newfoundland, between Chapeau Rouge and Cape Ray. It is a little to the W. of Grand Pierre. To the W. of English harbour is the little Bay de L'Eau. Both are small, and fit only for boats.

ENGLISH HARBOUR, America, Newfoundland, at the S.E. entrance of Salmon cove, a clean bay, with from 4 to 5 fathoms water. It lies W. of the Horsechops about 3 miles, these last being in lat. 48° 21′ 30″ N., long. 53° 14′ 18″ W.

ENGLISH HARBOUR, America, on the N.W. shore of the River St. Lawrence, from 7 to 8 leagues W. by S. from Trinity point. Lat. 49° 29′ N., long. 67° W.

ENGLISH HARBOUR, Hayti, or St. Domingo, West Indies, at the E. part of the island. The coast from this cape turns in W. to the Gulf of Samana. This harbour is entered from the N., off the E. part of the entrance of the Gulf of Samana. There is a bank of sand, which must be avoided by keeping either the N. or S. shore well on board. A vessel must not open the harbour more than to have it at S. by E. before she gets within the E. point of land. It was on a rock near the entrance of this harbour, in 1782, that the Scipion, 74 guns, was lost, when running into the harbour after engaging the London, 90. Lat. about 19° 8′ N., long. 69° 5′ W.

ENGLISH HARBOUR, West Indies, Island of Antigua, lat. 17° N., long. 61° 45′ 42″ W., a temporary light, is said to be hoisted on Fort Barclay, lying close under the W. part of the E. highest land, sheltered from all winds, where ships-of-war commonly lie during the hurricanes. See *Antigua.*

ENGLISH ISLAND, Greek Archipelago, Gulf of Smyrna, 2 miles within Cape Kanlubouroun, and about 2 leagues from the Vourla islands. It is also called Sahib and Sagleosa; it has some resemblance to Portland bill.

ENGLISH ISLANDS, West Indies, some islets lying off the S. point of English creek, near the S. end of the Island of Martinique.

ENGLISH ISLET, W. coast of Africa, in the centre of the River of Senegal, abreast of which there is a signal-post.

ENGLISH KAY, America, Yucatan coast, N.W. by N. 12 miles from Kay Bokel, in lat. 17° 19′ N., long. 88° 2′ W. It is covered with trees, has three houses upon it, and a flag-staff, 70 feet high, on which, at night, a light is placed (1846), and where a jack is hoisted when ships pass by day. It is 13 miles from Belize, the capital of British Yucatan, and situated on the S. side of the channel. The N. side is formed of Goff's kay and the Sand Bore.

ENGLISHMAN'S BAY, Island of Tobago, upon the N.E. side, near the middle of the island.

ENGLISHMAN'S HEAD, West Indies, Island of Guadaloupe. Lat. 16° 31′ N., long. 61° 50′ W.

ENGLISH POINT, West Indies, Grand Turk island, in the windward passage, off the S. end of the island.

ENGLISH POLE, Flemish coast, one of the shoals or banks, having several channels through them, by which means vessels may sail between Flanders and the Isle of Walcheren. The inner bank, or French pole, is along the Island of Cadsand; the English pole lies off the W. end of the inner bank, between a sand without, called the Raen, and the coast of Flanders.

ENGLISH ROAD, BONAVISTA, forming a bay on which the town is situated. It is 5 miles in extent from N.E. to S.W. See *Cape de Verde Islands*.

ENHALLOW SOUND. See *Orkneys*.

ENION POINT. See *Porth Enion*.

ENIS BOSINE, Ireland, W. coast, N. of Dog's Head point, and S. from Achill head. The smaller island of Enis Turk lies from it E.N.E. Lat. about 53° 33′ N., long. 10° W. This word, a prefix to many Irish names, seems to mean "island," as Enis Turk, or Turk's island—Enis Bosine, or Bosine island. Sometimes it is written Inis or Innis.

ENISHERKIN, Ireland, S.W. coast; an islet between Cape Clear and the main, N. by E. from the cape, and W. from Baltimore haven.

ENISKERRY, or INNISKERRY ISLANDS, Ireland, W. coast, S. of Galway bay, and nearly N.E. from Loop head, the N. limit of the River Shannon. Although there is a passage within the most S. of them from the S.W., nearly along shore, yet to the N. a ledge of rocks runs off from the point of Trumore castle. No passage can consequently be found N. within these islands. The castle and middle of the E. side of the island are the utmost limits of the rocks, brought to bear due E. and W.

ENISMURRY, Ireland, an island on the N.W. coast, in Donegal bay, 3 leagues W. from Ballyshannon haven, about half a league in length; at its S. end is a small island or rock, above water. The S. end is very foul, having a ledge of rocks, stretching a great way into the sea. There is good anchorage on the S.E. side of the island.

ENISTERHUL, INNISTRAHUL, or MUNSTER HULL, Ireland, at the northern extremity, nearly N.W. from Coledagh head, where there is a revolving light every two minutes, 167 feet high, lat. 55° 26′ N., long. 7° 14′ W. This island is 4 leagues W.N.W. from Lough Foyle, with a passage between, but dangerous to venture through by those not well acquainted with it, on account of shoals and foul ground. There are some small black rocks not much above water, scattered about the island and the point on the main opposite to it.

ENIS TURK ISLAND, Ireland, W. coast, to the S. of Achill island.

ENNIS TUISKAN ROCK, Ireland, W. coast, 4½ miles N.E. by E. ½ E. from the most W. Tiraught rock, and W. ¼ S., 4 miles from Sybel head.

ENOS, GULF OF, Greece, E.N.E. ¼ E. from Lagos 15½ leagues. This gulf has a very narrow entrance, but the inside expands and runs in between 14 and 15 miles, in a direction E.N.E. ½ N. The town stands upon a point of land on the S. shore, and is reported to carry on a considerable traffic with the adjacent country. In the N. part of this gulf the River Maritza enters by several streams, and forms a channel which runs to the N.

of the gulf. A considerable fishery is carried on at its entrance. The coast runs S. nearly 3½ leagues to Cape Paxi, or Grimea, in lat. 40° 35′ N., long. 26° 7′ E.

ENRAGE, CAPE, America, within Chignecto cape, the W. extremity of the bay of that name, about 12 or 13 miles within which, on the N. shore, running out seaward, is Cape Enragé, or the Angry cape, on which there is a lighthouse, with a fixed light 138 feet high, in lat. 45° 36′ N., long. 64° 29′ W.

ENRAGE, CAPE, West Indies, Island of Martinique, a projecting point of land on the W. side, a little curved, and S. of St. Pierre, whence the coast trends S.E. easterly as far as Cape Negro. There is a shallow bay between Cape Enragé and Point Negro.

ENRAGEE POINT, America, Newfoundland, separated from Cape Ray, in lat. 47° 36′ 56″ N., long. 59° 20′ 10″ W., by a sandy bay, in which vessels may anchor with winds from N.N.W. to E.

ENSENADA, or COVE OF UNARE, South America, coast of Cumana. This cove affords good anchorage and shelter from the sea breezes. In order to enter it, keep a mile from the N.E. point, around which a rocky shoal extends the same way, and as soon as the point is rounded, a vessel may anchor in 5 fathoms on a sandy bottom. There is water in this cove, and upon a small hill E. of the rivulet which brings it, there is an Indian village, named San Juan de Unare. There is a reef off the S.W. point of the cove also, stretching out half a mile, with some islands about it. Passing outside these, about 2 cable lengths, all danger is avoided.

ENSENADA DE LA RADA, South America, coast of Cartagena, E. of Port Cispata.

ENSTER, Scotland, one of the smaller havens on the N. coast of the Frith of Forth, between Fife ness and Elle ness, the second from the E., the first being Cryll. This harbour has from 7 to 8 feet water on neaps, and from 10 to 11 at spring tides.

ENTRY ISLAND, America, Gulf of St. Lawrence, the highest of the Magdalen islands, its summit being 580 feet above the sea. Its cliffs are red and lofty, rising at the N.E. point to 350 feet, and the S. point to 400 feet perpendicular. Off the N.E. point there is a high rock, about a cable's length from the cliff, and on its N. side a remarkable tower rock of red sandstone, joined to the island, and seen from the S.W. over the low N.W. point, as well as from the N.E. The E. point of this island is in lat. 47° 17′ N., long. 61° 45′ W.

EPEA, Africa, S.W. coast, in the kingdom of Ardra, a portion of territory so named, 15 leagues E. of Whydah, and 5 from the sea, the distance between being occupied by lagoons, navigable for boats, 40 leagues to the N.

EPEES DE TREGUIER, ROCKS, France, N.W. coast, on the E. side of the River Tréguier, 4 miles to the W. of the Heaux de Brehat.

EPELS ROCKS, or EPELS SCAREN, N. coast of the Gulf of Finland, a cluster of sunken rocks, 5 leagues E. of a great rock in the sea, called Putsfagre, 9 leagues W. southerly from the Goe Scars, and 16 from Rhodel, at the mouth of the sound leading into Wyborg.

EPHESUS, Anatolia. In the bottom of the Bay of Scala-nuevo, N. of the town of that name, stands what remains of Ephesus, upon the River Cayster, now called the River Kuchuk Meinder, about 2 miles from the sea, extending towards which the ruins are scattered; an extensive marsh has formed between them and the shore. The stone embankments that once confined the river are yet seen in many places, but the river has raised a bar at the entrance, and the beach is become an unwholesome fen. A little way above Ephesus is a bridge of seven arches, and much fine scenery, including grand precipices. The ruins of the celebrated temple of Diana are sought for in vain, but the area of the Stadium and Theatre are yet to be traced. The inhabitants are a few miserable Turkish peasants, or haply, the wandering Turcomans. From the Cayster, or the more modern river named Kuchuk Meinder, where it enters the sea, N.W. by W. about 5 miles, is Ghiaoor Kioy bay. There are soundings all the way on a bottom of sand and mud, with a depth from 4 to 30 fathoms. Within the external bay there is another, on the E. side of which are the extensive ruins of Colophon and Claros. There is a good road here across the land to Smyrna. Ephesus is 2 leagues N. of Scala-nuevo; this last place is in lat. 37° 51' 40" N., long. 27° 16' 50" E.

EQUATOR, or EQUINOCTIAL LINE, an imaginary circle circumscribing the earth at an equal distance from the poles, in consequence the line of swiftest motion on the superficies of the globe. It is at right angles with the meridian line which intersects the poles, and as on these last the points of N. and S. are coincident in their direction, so those of E. and W. bear a similar relation to the Equinoctial. It is the centre of the imaginary lines which define the frigid, temperate, and torrid zones, is intersected by the line of the ecliptic, or sun's path; and from it, in parallels N. and S. at certain distances, the old geographers measured the climates, making the first parallel N. and S. of the Equinoctial, that in which the longest day was 12¹ 30ᵐ, the second, in which the longest day was 13 hours long, and so on, up to the polar day of six months' duration. It is from this line that the parallels of lat. commence N. and S., which are of such importance in navigation. It is divided by mathematicians and astronomers into 360° for convenience, and the illustration of that beautiful system which enables the human mind to acquire the most sublime and positive truths regarding the frame of the universe, which could be obtained by no other means. To the navigator, whose object is of a practical nature, and who seeks for a guide to pass in a right direction over the trackless surface of the ocean, only a portion of that science, which the calculation and experience of ages has brought to its existing perfection, is at all necessary, but in that portion the line thus circumscribing the earth, and furnishing, at the same time, data for ascertaining the motions of the heavenly bodies that pilot his vessel over unfathomable depths, is an object of the utmost importance. Not only does it furnish his parallels of lat. but it is intersected by his lines of long. and a knowledge of his position at all times depends upon calculations, which are, to a great extent, based upon this line. But not only is it thus important in matters relative to ascertaining maritime positions, in modes it would be superfluous to detail, it defines a belt of the earth N. and S. of it, which includes the sites of atmospherical phenomena of a peculiar character. The winds, currents, temperature, calms, as well as celestial aspects, are attended there by certain peculiarities, with which it is essential the seaman should

be familiar. In passages to the East Indies or the S. hemisphere, vessels are hastened or retarded to a degree of vital importance according to the mode and place of crossing the Equinoctial. In one place calms that try the mariner's patience to the utmost are encountered, on another part of this line the passage shall be made to the full content of the commander, in speed and comfort, as his knowledge may enable him to avail himself of those particular accidents which will aid him in carrying out his intentions. Thus most modern navigators shape their course for the S. to the westward of the Canaries and Madeira islands, keeping to the W. of them all, for the sake of steadier winds. But in passing too far to the W. it is necessary to be careful of not getting so near the Coast of Brazil as not to be able to weather it. The N.E. trade wind is sometimes lost in 11° and 12° N., when a continuance of it is hoped for, sufficient at least to lead into variable breezes or shifts, of which advantage may be taken to stand upon the tack which will gain most S. Near the Cape de Verde isles the trade wind has often been found to fail, a great reason for keeping clear of the islands, and their light, eddy winds, and thus escaping the dreaded calms near the line. The line, when possible, should be passed between the meridians of 18° and 23°. On losing the N.E. trades these are the longitudes to be kept, avoiding long E. or W. tacks with a dead S. wind. Care must be had of the currents about the line, particularly of the Equinoctial setting N. On proceeding homewards, the equator should be crossed between 18° and 25°, but when the sun is N. of the line 21° and 23° are preferable, owing to the light, varying winds extending from the African coast after Midsummer to the close of September. If S. winds become light, then a N. or N. by W. course may be kept to reach the N.E. trade wind as soon as may be, but if light breezes prevail far to the N., the Cape Verde islands should be passed at the distance of between 40 and 50 leagues. In crossing the N.E. trade, a vessel should do all possible to gain the N. speedily. Going S. a ship should even cross in 20° or 21°, if able to do so, in order to have the advantage of the wind directly free the moment the S.E. trade sets in, but this is a contingency dependent upon rare circumstances as to possibility. The passing the line to the W., as far as is thus recommended, is much accelerated by the use of the chronometer, and thus modern navigators have an advantage which those of past days did not possess. On the S. of the Equinoctial the mariner seems to enter upon a new world, in which the aspect of the heavens is peculiarly favourable for his objects, and the observations based upon the imaginary line that thus girds the globe become, for his objects, the foundation of his calculations, and the actual base of his operations under a firmament, the signs of which are new and beautiful objects of vision as well as silent oracles to consult upon the locality where he moves, and the yet unseen, perhaps unknown, shores to which his bold prow is ultimately directed. Thus this imaginary line points out the existent path, and the first studies of the stars by Chaldean shepherds have ripened into a science that has made the infinitely remote and measureless heavens the oracle of the seaman in moving a fragile and solitary point over an almost unbounded waste of waters.

EQUILLES D'ENTRETAT, France, W. coast,

rocks so named, situated off Cape de Caux, or Antifer, near the shore, and extending a league from W. by S. to E. by N. They are bold-to on the N. side.

ER, ISLES D', France, N.W. coast, on the W. side of the River Tréguier, about 4 miles W. of the Heaux de Brehat: they are mere rocks.

ERCOLE, PORT, Italy, in the Duchy of Tuscany, on the W. side of an open bay, which reaches from Point Stella to Point Ansidonia, open to all winds from the S. quarter. It is a small place shut in between two high points, upon the W. of which is the town and fort of Stella, and on the E. the Fort St. Philip, of considerable strength. All the W. part of the port is filled with sand-banks continually on the increase. At the end of the port there are a few houses, and water may be obtained for shipping.

EREKLI or RAKLIA, coast of the Sea of Marmora, the ancient Heradia, a small town where the Turks had once a considerable manufactory of gunpowder.

ERICEIRA BAY and TOWN, Portugal, 14 miles from the Town of Vimeiro, and 25 miles S. of Cape Corvoeiro. From Vimeiro to Ericeira the whole way is a flat sandy beach. About a league inland is the village of Mafra, where, situated on a hill, is the celebrated convent and church of Mafra, being the object most distinguishable on the whole coast, a magnificent structure of white marble. It stands in lat 38° 55′ 54″ N., long. 9° 20′ W. Ericeira is in lat. 38° 57′ 24″ N., long. 9° 24′ W., according to Captain Owen.

ERICHO, a small decayed town at the head of the Gulf of Valona, in the Adriatic.

ERICK'S FRITH, Greenland coast, lat. about 63° 30′ N., from whence the coast turns more S. towards the W., as far as Whale island, about lat. 62°; and to Cape Discord, in lat. 60° 53′ N., long. 42° 26′ W.

ERIE, LAKE, America, United States and Canada, 300 miles long, and about 90 wide. In fair weather there is anchorage all over it. This is one of the vast chain of lakes that traverse the interior of the North American continent, communicating, through the St. Lawrence, with the Atlantic, in one or two cases interrupted by falls, shoals, and rapids, but in such cases having the navigation supplied by artificial channels. The N. shores are rocky, as, indeed, are numerous islands at the extremity of the lake. The S. shore is a fine sandy beach, so low that in storms the water frequently inundates a large extent of country. It has several harbours for vessels of moderate burthen.

ERIESUA, Mediterranean, one of the Lipari islands. See *Lipari Islands.*

ERISKAY ISLAND, Scotland, W. isles, S. of the Isle of South Uist, forming the N. side of the road of Otervore. It has on its E. side a small creek named Acherystvore, to enter which a leading wind is required, care being taken to avoid two sunk rocks, between which there is one above water. One of these lies a cable's length N. of Mealnacabel head, on the larboard side of the entrance, and dries at low spring tides. W. of this is another, at about 2 cables' length distance, right in midway, that dries at half ebb. To avoid them Mealnacabel head must have a berth of a cable's length, and one-third distance must be kept from the S. shore. This is a safe creek for small vessels when once in. Off the S. end of the island there are three islets, and to the E. the

Gelig rock, always above water. Half way between Gelig and Acherystore creek there is a shoal, over which the sea breaks in blowing weather, it has 6 feet water over it, and the leading mark to it is Gelig rock. A vessel may ride safely too in a creek at the N.E. end of this island.

ERITH, Kent, a town on the River Thames, on the S. side, 5 miles from Woolwich. The church and town are conspicuous marks sailing up or down the river; the former is said to have stood upon the same spot from the later Saxon times.

ERK ISLAND, Baltic sea, the entrance into the Haaft of Aland or Aaland, near the S.W. of Finland, and about 9 leagues W. of the Island of Uttoy.

ERME RIVER, Devonshire co., falling into Bigbury bay, between the Bolt head and Plymouth sound to the W.

ERNE LAKE or LOCH, Scotland; a considerable branch of the Frith of Clyde, on the N.W. part, is thus named; it extends in that direction to Kilreny, and then turns N.E. to Inverary.

ERQUI CAPE, France, N.W. coast, a point of land bearing W. ¼ N. from Cape Frehel, distant 7 miles. N.N.E. of this cape there is a cluster of rocks, the principal of which, called the Robinet, is always above water. They are cleared N. by keeping the light on Cape Frehel S.E. by E. ¼ E. Inside these rocks is the channel of Erqui, which has from 2¾ to 3½ fathoms upon it. To the S. of the cape is the road of Erqui, where small vessels may ride in 2½ fathoms, protected from N.E. and E. winds.

ERYTHRÆ, BAY OF, coast of Anatolia; it is now called Ritra; to the S.E. are the ruins of the once celebrated town of Erythræ, N. of Chesmeh.

ESAMEAUX PASSAGE, France, entrance of the Gironde river; the channel between the Mauvaise sand, S. of Point la Coubre a league, and the Cordouan tower. It has 4 fathoms water.

ESCALONA, anciently ESCALON, Palestine, on the shore of the Mediterranean, 15 leagues S. of Jaffa; it exhibits little more than ruins of ancient buildings; it is some miles N. of Razza or Gaza.

ESCARCEO POINT, or PUNTA DEL ESCARCEO, South America, N. of the Gulf of Cariaco, opposite the Island of Cuagua, between Margaritta island and the mainland of Cumana.

ESCARPADO ROXO POINT, South America, forming one of the entrances of the Gulf of Santa Fe, New Barcelona, in the province of Cumana upon the Main.

ESCARROS TOWER, E. coast of Spain, a ruin that serves for a mark on going into the Harbour of Denia.

ESCATARI ISLAND, America, Cape Breton island, 5 leagues N. of Louisburg, in the S. part of the Gulf of St. Lawrence.

ESCALETA TOWER, E. coast of Spain, near Cape Santa Pola, which cape includes the space from the tower of Escaleta to Point Argibe. The cape is in lat. 38° 12′ N., and long. 30′ W. It is but a short distance from this cape to Alicante.

ESCOLLOS, Spain, E. coast, one of three rocky islets in the Bay of Cadaques, about 400 yards from the Cucurucu, a triangular island in that bay, in lat. 42° 17′ 15″ N.

ESCOMBRERA BAY, Spain, S.E. coast, to the E. of Cartagena, capable of receiving vessels of any size, and sheltered from N.E., S.E., and N.W. winds. At its entrance there are 14 and 15 fathoms, but on proceeding the depth decreases to

5 and 6 fathoms. In the middle the bay is sandy, with patches of sea-weed. It runs about a mile inward, which is nearly the breadth at the entrance. The usual anchorage is under the S. shore, in 7 fathoms. The land is high all the way from the S. point of Escombrera island, towards Cape Aqua, a distance of about 3 miles E.S.E. ¾ E., that cape appearing with 3 peaks.

ESCONDIDO HARBOUR, or PUERTO ES-CONDIDO, Cuba, 3 miles W. from Point Malano. The port is well described as "hidden" because within a mile of it it is not easy to be descried. It forms an anchorage secure in all winds. Its E. point lies in lat. 19° 55′ 30″ N., long. 75° 12′ 20″ W. It has various bays in the interior, fit for all kinds of vessels, but its entrance between the outer points is only a cable's length wide, and each point sending out a reef; the channel is not more than 90 yards broad. It has no windings, and the whole length of the strait is not more than a cable and a half. To enter it, a vessel must steer N. 43° W. It may always be done with a free wind, even if the breeze is at N.E. This place has no commerce, and is rarely visited.

ESCONDIDO, or HIDDEN HARBOUR, Hayti, nearly N.W. from Point Salinas. The entrance is above half a mile in width, but near the S. point it is clean, and the water is so deep that at half a cable's length from it there are 5½ and 6 fathoms. A reef stretches out from the N. point.

ESCONDIDO HARBOUR, West Indies, Trinidad island; to the N. of Punta San Carlos, or Escondido, is a little harbour thus named. Distant S.S.E. ¼ E. from the point, 4-10ths of a mile, is Prince's point, and from thence to the E. end Chaguaramas peninsula bears E.N.E. 1-10th of a mile.

ESCONDIDO HARBOUR, South America, N. coast, in Cumana, W. of the city of that name, in lat. 10° 40′ N., long. 63° 29′ W.

ESCONDIDO POINT, America, Yucatan, in lat. 18° 56′ N., long. 91° 12′ W.

ESCONDIDO POINT and KAY, Hayti, N. coast, near the port and town of Samana, towards the E. end of the island.

ESCOSES POINT and PORT, South America, in lat. 8° 52′ N., long. 77° 43′ W. from Greenwich, and 71° 25′ 30″ W. long. from Cadiz. This point is the S E. headland of a bay named Carolina, the isle called De Oro, or Santa Catalina, forming the N.W. The two points bearing from each other N.W. and S.E., distant 4 miles. This bay runs in 1¾ mile. In the S.E. part is Puerto Escoses or Escondido, Scotch, or Hidden Harbour, extending 3 miles S.E., and affording excellent shelter. It has several shoals in it, and anchorage in 5, 6, 7, and 8 fathoms, over a bottom of sand.

ESCRAVOS RIVER, coast of Africa, 4½ leagues from the Bight of Benin. It has only 2 fathoms water at the entrance, and is about a quarter of a mile in breadth. It was formerly a noted slave-mart, but is now abandoned. It joins the river Forçados, about 3 leagues from the entrance, leading to the town of Warree, and is easily distinguished by the Island of Forçados, in the centre of its entrance, and another called Poloma, close to the shore, on the S. side. The town of Poloma is 6 miles up the river. It has two passages, one on the N. side and the other between two islands. The last is the best, having 2 fathoms water. The anchorage outside the entrance has 6 fathoms.

ESCRIBANOS HARBOUR, America, Darien; from Cocos point on this coast, that of Escribanos bears S. 80° W. about 1½ mile, and thus a bay is formed, in the centre of which is Escribanos harbour, going in to the S. from its entrance about half a mile. It is shallow, having only from 1 to 1½ fathoms water.

ESCRIBANOS SHOALS, America, N.E. of the mouth of the harbour of the same name. They are two in number, with little water over them, and lie near each other. On the reef nearest to the coast there is an islet somewhat less than 2 miles from Cocos point. This reef extends a mile from W.S.W. to E.N.E. The second reef, W.N.W. from the islet or rock, is nearly a mile from E. to W. Both are steep-to, with 3 and 4 fathoms water. On the bank there are from 8 to 12. ESCRIBANOS BANK lies nearly N.W. by W. from the shoal of that name, distant 5½ miles. It extends N. 56° W., and S. 56° E. nearly 2 miles, and has from 5 to 8 fathoms water on a rocky bottom. The N.W. part lies N. 32° W., distant 8¼ miles from Escribanos point.

ESCUDO ISLAND, America, Darien; the course to this island from Chagres is W. by S., by compass, 92 miles. It lies 10 miles from the main, in lat. 9° 6′ N., long. 81° 32′ W. It is low, covered with cocoa-nut trees, and is only 1¼ mile in length, with a reef of rocks at each end. Point Valiente (eastern entrance to Chiriqui Lagoon,) bears from this island W.N.W. 22 miles, and may be seen from it in clear weather. To the E. are two small kays covered with trees, and to the W. about 3 miles are the Tyger kays. Between the island and the main is a passage about a mile broad, having from 7 to 16 fathoms water, in lat. 9° 12′ N., long. 81° 57′ W.

ESHIBREKY LOCH, Scotland, W. coast, 3 miles N.E. from Muligrach, in the promontory of Ru More, a place where the ground holds well, with sufficient water for any vessel, but as the bay is narrow it would be proper to lay one anchor on shore. Any vessel may also lie on the S.E. side of the island, at the entrance of the harbour, in good sandy ground.

ESK, Scotland, a small river, at the mouth of which Musselburgh is situated, having its name from the muscle-banks before the town. The haven is small. The pop. about 5,000.

ESK RIVER, Scotland, called the S. Esk; at its estuary is situated the town of Montrose. It forms a basin at the town 250 yards broad, accessible to vessels of 400 tons.

ESK RIVER, Scotland and England, a stream that forming a part of the boundary between England and Scotland falls into the sea at Solway Frith, to the S.W.

ESKE ADALIA. See *Adalia Eske.*

ESKE RIVER, Yorkshire, upon which is situated the important seaport of Whitby, which it divides into two unequal parts, connected by a drawbridge. The climate at the mouth of the river is almost as cold and ungenial as that of Orkney. The depth of water in this estuary is 12 feet at neap tides, at common springs 12, and at the equinoctial springs from 23 to 24. In the vicinity of the entrance of this river there have been found the skeletons of various animals unknown to its own N. lat., even that of the crocodile.

ESKEFIORD, Iceland island, a small port, which has the grant of a city, with five others in the island, to which grant considerable privileges were annexed by the crown of Denmark in 1787.

ESKIFOROS, or TARKHAN, CAPE, Black sea, S. by E. ¼ E., distant about 3 miles from Cape Karamroune, between which the coast bends in and forms a small haven before the valley of Karadji to the N.E. Vessels cast anchor there in 5½ fathoms on mud and gravel, but exposed to W. winds. There is a lighthouse on this cape, which exhibits a bright fixed light, 108 feet high, visible in clear weather at the distance of 17 miles. Lat. 45° 20' 42" N., long. 32° 29' 23" E. To the W. of the cape is a reef of rocks, 5 miles long and 2½ broad, extending from the shore. At the distance of 4 miles from the land it breaks into two parts, with a channel between them, so that small vessels knowing the navigation may pass through. Near the edges of the reef, and about the cape there are from 20 to 22 fathoms water. From Cape Eskiforos the coast turns to the E. by S. ½ S., 7 miles, to Cape Ouret.

ESKIRKS, ESQUIRKS, more properly the SKERKI, ROCKS, two reefs of volcanic origin, N.E. and S.W. of each other, in the Mediterranean sea, surrounded by a bank of soundings, in lat. 37° 49' N., long. 10° 56' E. It was upon this reef that the Athenienne man-of-war was lost.

ESMIRALDA BAY and ISLAND, Colombia, between the Gulf of Paria and Cartagena, W. of Morro Blanco and Esmeralda Morro. It is a large bay, with irregular soundings, over a bank in its centre, extending about a mile S. and S.W. or nearly, then S.W. by S. to within one-third of a mile from the coast, whence it turns E., and runs parallel with the shore, at about that distance from it. On anchoring in this bay, vessels may keep as close as is necessary to the N. and W. of Esmiralda island, and moor under shelter of it at the distance of 2 cables' length, in 5½ to 6½ fathoms, sandy mud. From this bay the coast runs W. about 5 miles to the point and Morro of Manzanilla all the way bounded at the distance of one-third of a mile by a bank which extends from the Esmiralda island.

ESPADA POINT, West Indies, the S.E. point of the Island of Hayti, or St. Domingo; it is the point nearest to the Mona passage, through which the current sets N. 11½ leagues, W.N.W. from that island. The Espada is a low point encompassed with a reef and a white shoal, distant from Cape Engaño, the E. point of the island, 5 leagues, in lat. 18° 20' N., long. 68° 30' W.

ESPADA PUNTA, or SWORD'S POINT, South America, at the entrance of the Gulf of Maracaibo, with Point Macolla on the opposite or E. side, forming the headlands to that extensive bay, at the distance of 50 miles from each other. Lat. 12° 4' N., long. 71° 9' 50" W.

ESPAGNE, PORT D', PUERTO D'ESPANO, or PORT OF SPAIN, West Indies, Island of Trinidad. The Gulf of Paria being entered through one of the W. bocas or passages in Chaguaramas bay, a vessel must steer a course so as to gain the W. part of the coast of Trinidad, sailing along 2½ or 3 miles, as far as Brea point, from which Port d'Espagne is distant 8 leagues, and when steering N.N.E., the town will soon be seen. If the wind will not permit such a course, a vessel beating up must never approach the shore nearer than 3 miles, for the water becomes shallow along the coast. There are some strict regulations to be attended to respecting this harbour, with which a vessel must be acquainted. If unknown, it is sufficient

that no boat must go on board before the harbour-master has visited the vessel arriving, from whom the regulations may be obtained. Some of these regulations savour far more of a port under Spanish than British sway. This town is embosomed in hills, and is one of the finest in the West Indies. The buildings are of hewn stone, and their very form is prescribed by law. The streets are broad, and shaded with trees, in parallel lines from the land towards the sea; thus every breeze that blows is enjoyed, and there is a fine public walk. The town is divided into barrios or quarters, each of which is placed under magistrates responsible for civic duties. There are barracks for the reception of 600 men. The harbour is the second best in the island, upon one of the most extensive bays in world. Fortified heights overlook both the bay and the town. A fine stone quay runs out several hundred yards into the sea, and has a battery on its extremity. In fact the whole of the W. coast near this port is a series of bays, where vessels may at all times anchor securely. There is a fixed harbour light on the pier, 51 feet high. The time of high water here at full and change is 5ʰ 20ᵐ. Port d'Espagne, at Fort St. David, stands in lat. 10° 38' 42" N., long. 61° 36' W.

ESPALAMACA, Fayal, the N. point of the Bay of Orta.

ESPALMADOR ISLET, POINT, and CASTLE, Mediterranean, close to the Island of Formentera. The Bay of Espalmador is the space between the castle or tower and what is denominated the Trocados.

ESPANO CAPE, or CAPE DEL SPANO, Island of Corsica, the N.E. point of that island, about 6 leagues from Point Renella. It is the N.E. limit of the Bay of Calvi.

ESPARTO, Mediterranean, a small island lying S. of Conejera, distant about a mile, near the Island of Iviza.

ESPECIA, more correctly SPEZIA, Mediterranean, a small port of Genoa, S.E. from that city, and midway between Genoa and Leghorn. A fixed light stands on Tino island, 384 feet high. See *Spezia.*

ESPERANCE RIVER and BAY, West Indies, Island of St. Lucia, falling into the bay of the same name. The bay is at the N. end of the island, between Cape Marquis on the E., and the N. cape of the island on the W., and nearly due S. from Diamond point, the most S. point of Martinique. Lat. 14° 5' N., long. 61° W.

ESPERONE, CAPE, Island of Corsica, near the Strait of Bonifacio.

ESPICHEL, CAPE, Portugal, S. of Lisbon, 15 miles, to which, within about 2 miles, a low sandy beach extends. The land then rises and turns E. towards St. Ubes. Cape Espichel mounts from the sea almost perpendicularly, to a moderate height. The summit is irregular, whitish on the N. side, and reddish towards the S. It has a chapel and lighthouse near its extremity, the last carrying a fixed bad light, elevated 623 feet above the sea, visible 12 miles To the E. of this cape, Mont St. Louis is seen lying inland, and a little to the N. of this is a hill of less magnitude, in figure like a haycock. These are good marks coming from the sea, to know the land by, and may be descried before the cape itself is seen. The lighthouse stands in lat. 38° 24' N., long. 9° 13' W., after Captain Owen.

ESPIRITO SANTO, PASS and BAY (the first leading into the bay), America, coast of the

Gulf of Mexico, between the N.E. of San Josef island, and the S.W. end of Matagorda. Lat. 28° 5' N., long. 96° 51' W.

ESPOIR, CAPE D', America, S.W. of Bonaventura island, on the W. coast of the Gulf of St. Lawrence, so named by the French.

ESPONJA, Mediterranean, near the Island of Iviza, a mere islet rock.

ESPRIT ISLAND, or ST. ESPRIT ISLAND, America, on the coast of Cape Breton, at the S. part of the Gulf of St. Lawrence. An islet off the S. entrance of Rigaud haven.

ESQUIMAUX, or ESKIMAUX COUNTRY, America, lying between 50° and 64° N. lat., and between 59° and 80° W. long. It is bounded by Hudson's straits on the N., separating it from Greenland; the Atlantic on the E.; on the S.E., by the Gulf and River St. Lawrence, which separates it from Newfoundland and Nova Scotia, and by Hudson's bay on the W. The inhabitants are of the same race, apparently, as those who are found roving over the inclement shores of Greenland and the part of the American continent nearer the pole, and who have been accustomed to exchange their utensils and skins for articles of utility or trinkets of European manufacture. They subsist, for the most part, upon fish, and reside in huts of snow during the winter season, made with considerable skill, sometimes, indeed, excavated in the earth, with a subterraneous passage of some length leading into them, and invariably opening towards the S. At first they were represented as a savage and ferocious people, but subsequent voyages and observations of travellers have given ground for qualifying this opinion in a considerable degree. They are a distinct race from the N. American Indians who visit the Hudson's bay factories, both in their mode of living, in manners, habits, and physiognomy, all perhaps modified by the necessities and privations of a climate more inclement than any nation of hunters can be supposed to inhabit.

ESQUIMAUX BAY, America, S. coast of Labrador, within the Straits of Belle Isle, opening into the Gulf of St. Lawrence. It lies on the N. shore of the gulf, W. from Bradore harbour, and E. from Shecatica bay and St. Augustin's river. Lat. about 51° 25' N., long. 57° 45' W.

ESQUIMAUX BAY, America, on the E. coast, its S. point being in lat. about 53° 49' N., and long. 56° 42' W., as seen in old maps. It is probable that this is the Sandwich bay of modern charts.

ESQUIMAUX, CAPE, America, on the W. coast of that part of Hudson's bay called "The Welcome," to the N. of Seal river and Button's bay. Lat. about 60° 40' N., long. 95° W.

ESQUIMAUX ISLANDS, and BAY, America, Labrador, on the St. Lawrence. The course and distance to the outer Esquimaux island from Point Belle Amour are W. by S. 10 or 11 miles, N.N.E. About 4 miles from this island there is a good anchorage between two high islands, but only for small vessels. Within these islands lies the River Esquimaux. The Esquimaux islands are at the entrance of the River St. Lawrence. From the river to the Dog island is a chain of small islands and rocks, the E. of which have the name of Esquimaux, the middle ones, of the Old Fort island, and the W. ones, of the Dog islands. Within these there are many good bays and places of shelter for vessels, but the entrances are narrow, difficult, and dangerous, so that none but those

well acquainted with them should attempt their navigation.

ESQUIMAUX RIVER, America, Labrador, falling into the bay of the same name.

ESQUIMINAC, CAPE, America, New Brunswick, lat. 47° 4' N., long. 64° 51' W. It bears a fixed light, 70 feet high.

ESSA NESS. See *Shetland Isles.*

ESSEQUIBO RIVER, South America, Guyana; this is a very large river, its mouth being 15 miles wide, but full of shoals and islands, obstructing the passage, and rendering it difficult to enter. Although the islands and shoals form a passage deep enough for every class of vessels, yet much use and practical knowledge are required to enter them. These islands are numerous, low, and bushy, the greater proportion of them from 1 league to 2 and 3 in length, very narrow, and lying N. and S. The extremity of the Leguan bank in this river is in lat. 7° 0' 20" N., long. 58° 18' W. There are two principal channels for entering, the E. and W. channel. The E. is the best, having from 15 to 35 fathoms water. The islands at the entrance passed, another cluster of them is seen, which should be passed on the E. side, the channels between them being so deep as to have from 40 to 70 fathoms. About 10 leagues up from the entrance there is a fort in the middle of the river, the town or village being situated on the W. side in front of the Fort. This river is the boundary between British and Colombian Guyana. At the W. end of the Leguan bank is the Island of the same name; pointing up the river, and parallel with it is Wakenaam island, 11 miles long, having at its E. end Kiwakewaraba island. Between the two shoals, which are both dry, the Wakenaam bank lies athwart the channel, having 24 and 18 fathoms in the passage at each end. The ship channel on the S. of Leguan island has good anchorage within the island, and from 30 to 60 fathoms water. Tiger island is the most W., narrow, 3 or 4 miles in length, separated from the main by a channel, and having from 13 to 26 feet water; a spit of sand runs out from its N. end full 4 miles. The ship channel on the E. is rather difficult of access from the various sand-banks which lie off from it. About 60 miles from the entrance the river divides into three parts, the most N. being the Cayoni, and having a cataract and islands near its place of junction; the centre river is the Mazarani, and the third the Essequibo. All these rivers are studded with islands. The influence of the tide is felt as high up the river as the lower cataracts, 30 leagues from Tiger island.

ESTACIA TOWER, E. coast of Spain, inside the island of Grossa, upon a projecting point of land, where a small bight is formed, within which coasters seek shelter. The point has several rocks running out from it. N.W. by N., 2½ miles from this tower, is the Torre de la Encanizada, near the entrance of Mar Menor, a shallow place, only adapted to receive boats.

ESTACIO, a small rocky isle off the E. end of Esparto, among the Balearic isles.

ESTANGUES, ANCHORAGE OF, America, by the Paraguana peninsula. It is very good for vessels of the largest class, even within a cable's length of the beach, and with room for 20 sail to anchor in security. The marks for this anchorage are the Pan, or mountain of Santa Anna, seen in clear weather 8 or 9 leagues off. This, when bearing E. ¼ N., leads to the anchorage. The spot may also be known by there being a long tongue of sand, with some huts at the extremity, occupied by fishermen. There is no good water, nor can any supplies of moment be obtained.

ESTAPONA, SIERRA DE, Spain, S.E. coast, bearing from the Rock of Gibraltar N.N.E. ½ N., and S.S.W. ½ S., in lat. 36° 28′ N. This mountain is a landmark for passing along the strait coming from the E., and should always be sighted if possible. Point Doncella is the S.W. extremity of the low shore of Estapona, the N.E. point of which is called Marmoles, the points being 2 miles asunder. There is also a town of the same name on the shore. Between the points is a good road for small vessels in 4 and 5 fathoms, and further out there is anchorage for large ships in 14 and 18 fathoms, opposite the town, and sheltered from N.W. winds. There is a shoal, reported to be 3 miles S.E. by E. from Estapona town, called in the French charts Ouvreil Oeil. The direct course and distance from Europa point to Estapona are N.N.E. ½ E. about 7 leagues. The Sierra or high land of Estapona is 5 miles N.W ½ N. from the town. All the way to Europa point vessels may anchor in any part a mile from the shore, with W. and N. W. winds, or by taking advantage of the tides, ply to windward until they gain the Bay of Gibraltar.

ESTEDELLA, TOWER OF, Mediterranean, upon Point Negra, in the Island of Majorca.

ESTEIRAS, BIGHT OF, W. coast of Africa, where the land trends away to the S.E. There is anchorage in this bight in 6 or 7 fathoms, and a sandy bottom. From hence the coast runs, thickly wooded, as far S. as Cape Clara, distant from Cape Esteiras 8 leagues.

ESTEIRAS, or CAPE ESTERIAS, W. coast of Africa, near Corisco bay, lat. as given by Captain Owen, R.N., 37′ 48″ N., long. 9° 18′ W. It is on the S. side of Danger river, between which and the N. side is a bank of breakers in mid-channel, with a passage of 3 or 4 fathoms on each side, leading to the river between the two islands of Corisco. A vessel should steer N.E. from the breakers in the direction of the little island until the E. side of the great island bears N.N.W., to clear the bank, which lies 3 miles E. of the S. point. Then steer for the N. side, bordering on its shore, to avoid a bank, which extends 3 miles from the small island. When the river is thus open, run for the bar, keeping Musquito island a little open on the starboard bow. The tide flows here, full and change, at 5¼ʰ.

ESTELLAS, Coast of Portugal, a high rock running out with a cluster of small ones, to the S.W. of the Great Berling. Near the N.E. side of these is another high rock, called the Farilhoen da Velha.

ESTEREL, CAPE, France, S. coast, above 1½ mile N.E. ¾ E. from Ribaud. It is close to a rock above water, is moderately high, and marks the commencement of the Bay of Hieres.

ESTIMO, CAPE, Greece, 4½ miles W.N.W. ½ W. from Paros island, off which is the small island of Petali.

ESTRATA, ST., or HAGIOS STRATI, Greek Archipelago, S.W. from Cape Stata, 15 leagues distant; it is a small triangular island about 5½ miles in length and 4 in breadth, of a moderate height, rising in the centre; its summit is in lat. 39° 31′ N., and long. 25° 2′ E. On the W. side there is a village and a roadstead, in which there is occasional anchorage, sheltered from N. winds. On its S.E. side there are two low islets, and off

its S. point a rock under water. To the N.E. of this island 5 leagues, the Guerriere rock is said to lie, having only 4 feet water.

ESTRELLA CASTLE, West Indies, Island of Cuba, in the Harbour of St. Jago de la Vega, on the narrowest part of that harbour. To avoid getting on shore near this castle, when a ship is abreast of the Morro point, it should begin to keep away, but this should not be done too suddenly. The Morro castle here is in lat. 19° 57′ 29″ N., long. 76° 3′ 45″ W., according to the Spanish astronomers.

ESTRIDGES RIVER, West Indies, near the N. point of the Island of St. Kitt's, E. from Sandy bay.

ESTUDIOS BANK, West Indies, St. Domingo, in the Bay of St. Domingo, stretching along its whole front, half a mile out to sea, and having 5, 6, and 7 fathoms water. Vessels sometimes anchor upon this bank, not without risk, during the season of the S. winds which throw in a heavy swell, from which there is no shelter, with a wild and rocky coast. There is safe anchorage within the River Ozama in this bay, but it has a bar of rock which will not admit of vessels drawing more than 13 feet water passing over it, and during the S. winds even then they are in danger of striking. To anchor on the bank, it is necessary to coast the windward land from Cape Caucedo, at the distance of from 3 cables' length to half a mile. It is clean and deep, as only on the E. point of the river is there any shoal water. A little shoal stretches out thence almost 2 cables' length, and to keep clear of it a vessel must not haul to the N. in the smallest degree, until the W. point of the river bear N.

ESWEEK MOUTH, N. of Calford Voe. See *Shetland Isles.*

ETANG HARBOUR, America, New Brunswick, to the S. of Magagadarve, running in N.E. of Campo Bello. Several islands lie before it. Bliss island, at the entrance, is situated in 45° 2′ 30″ N. lat., and 66° 53′ W. long. There are three entrances into this harbour, so that vessels may come in and go out at all times. The W. entrance leads into La Tete harbour, where there is anchorage in from 10 to 5 fathoms, but there is no passage for vessels round Payne's island at the N. end. The channels between Bliss and Payne's islands are considered to be the best, because they will admit of vessels working through them. The E. passage requires a leading wind. A pilot is necessary on account of the intricacies of the channel, but one is always to be obtained; water is also procurable, and the bay is extensive, secure, and has everywhere good anchorage.

ETAPLES, France, department of the Pas de Calais, 5 leagues W.N.W. from Montreuil. Pop. 1,700. It is situated upon the little River Canche, near its entrance into the sea, and possesses a small port which is much used for the herring fishery. On the Point de l'Ornel here, on the N. side of the Canche, there is a fixed harbour-light, 32 ft. 10 in. high, and on the S. side, at Touquet, there are two others, 42 ft. 8 in. distant from each other, and 52 ft. 6 in. high, visible 2½ leagues.

ETATES or ETETES, BAY OF, France, N.W. coast, near the town of St. Malo.

ETIVE, LOCH, Scotland, W. coast; N.E. of Keraray island lies Ilan Beach island, at the entrance of this loch. It is an excellent harbour when a vessel is within it, but the entrance is narrow, shallow, and rocky, while the stream is rapid. The entrance between Lidiack point and Ilan Beach dries at low water almost half-way over, the remaining part having only 6 feet, is therefore adapted for small vessels alone. There are two rocks in the channel, the S. of which dries at low water, and the N. at half ebb. The channel lies on the S. side of the rocks, about a vessel's length from the shore. The stream of tide in the Narrow runs 8 miles an hour. The Castle of Dunstaffanage stands upon a point of this loch. Spring tides run on the E. side of Ilan Beach about 4 miles an hour, and between Dunstaffanage and Ilan Beach about 3 miles.

ETNA, Sicily, in the E. part of the island, a celebrated volcanic mountain 10,874 feet above the level of the sea, and a most conspicuous mark for vessels at a great distance.

EU RIVER, France, E. coast of the English channel, between St. Vallery on the N.E. and Dieppe on the S.W., not far from St. Remy. It is a small stream half-way between the Somme and Dieppe. Treport stands on its S. side, a tide haven, with a fixed tide-light on the mole, 26 feet 3 inches high, visible 3 leagues, into which vessels may go at high water, lying dry at the ebb. It is sometimes called the Brele or Bresle river. Near it, and not far from Treport, the King of France has a country residence, at which he received a visit from the English Queen Victoria. The inhabitants of this little port, in number about 2,500, are principally occupied with their fisheries.

EUBŒAN SEA, Mediterranean, Greek Archipelago, another name for the Gulf of Athens, derived from Eubœa.

EUDOSAGUA POINT, Mediterranean, the N. point of the Bay of Polenza, in Majorca.

EUFEMIA, ST., GULF OF, W. coast of Italy, bounded on the N. by Cape Suvero, and on the S. by Cape Vaticano. Cape Suvero has a tower upon it, and is in lat. 39° 2' 55" N., and long. 16° 8' 40" E. The Cape Vaticano is in lat. 38° 36' 30" N., long. 15° 51' E. These capes bear from each other N.N.E. ½ E., and S.S.W. ½ W., distant 29 miles. The gulf or bay runs to the E., having a sandy beach of low land next the sea. About 5 miles from Cape Suvero is the town of St. Eufemia, nearly 2¼ miles from the shore; a little river runs up to the town, at the mouth of which there is a fort.

EUGENIA ISLAND, Gulf of Venice, near the coast, and N.W. of Mount St. Angelo, and not far from the small towns of Civita a Mare and Termoli.

EUNIEH, CAPE, and TOWN, Black sea; the town is built on the E. declivity of the cape. The pop. almost wholly Greek. It is a place where shipbuilding has made great progress in later times, and the vessels built carry a good character. Rope and canvas are also manufactured here. The road in summer is good, though N. winds make it dangerous. At a quarter of a mile from the shore there are 4 and 5 fathoms water, but close in not more than 1 and 2. About 17 miles E. from Cape Eunieh Cape Iasoun advances slopingly into the sea. These two capes form the extremity of a bay 8 miles in depth, at the bottom of which stands the town of Fatsah, 8 or 9 miles S.W. by W. from Iasoun. Eunieh is in Trebizonde, at the S. side of the Black sea, in lat. about 40° 51' N., long. 37° 17' E.

EUPHATORIA, or KOSLOV, TOWN and BAY, Black sea, N. side. Passing Sevastapol the coast takes a N. direction, and for 13 miles is bordered by red cliffs. After the land about Cape Kherson has been seen, and a run of 35 miles N. by E. made, over a depth of 50, 45, 30, 19, and 11 fathoms water, with a muddy bottom, the road of Euphatoria will appear on approaching the point on which the Lazaretto stands, and must be kept at a distance of 3 cables' length, on account of a spit of sand which extends out 350 fathoms. The wind generally blows strong in these roads, but the Lazaretto point and shoal break the swell of the sea, and shelter the vessels which lie close in shore in 3 fathoms water. There are 4 and 6 fathoms further out. Vessels generally buoy their cables here. The articles of export are generally wheat, barley, hides, lamb skins, and salt. Lat. about 45° 9' N., long. 33° 18' E.

EURIPUS, Greece, the ancient name of the Talanta channel and channel of Negropont, between the island of that name and the continent of Greece. The island being joined to the continent by the bridge, the upper half has received one name and the lower another, but this appears to have been a denomination adopted only in modern times. The tides during the first 8 days of the moon, and from the 14th to the 20th inclusive, have been observed to be very regular in their ebb and flow here, but at other periods have been observed to flow several times, and in like manner to ebb, during 24 hours. Numerous attempts have been made in ancient as well as modern times to account for this singularity.

EUROPA POINT, Spain, on the Mediterranean, in lat. 36° 6' 30" N., and long. 5° 22' W., being the S. point of the rock of Gibraltar. There is a powerful lighthouse upon this point showing a fixed light of great power, at an elevation of 150 feet above the level of the sea. It is visible from Sandy bay, on the Algeziras coast, to the mouth of the river Palmones. There is anchorage even S. of this point, but the water deepens from that within the bay. A pinnacle rock lies about 300 yards N.E. of Europa. This rock is of small dimensions, and there is water deep enough for a man-of-war to pass between the rock and the point. In moderate weather a vessel may anchor on a bottom of sand in 10 fathoms water, a considerable distance off. Point del Carnero, on the S.W., is the opposite point of the bay. Point Santa Catalina is the S.E. point of the straits, distant about 14 miles, but Point Leona, under Ape's hill, on the African side, is nearer to Europa point, not being beyond 12 miles distant.

EUROPE, CONTINENT OF, the smallest of the four great divisions into which the surface of the earth has been quartered by geographical writers. Though least in extent it is the most important in the history of human civilization— the power and influence acquired by its intellectual superiority have been exerted and are acknowledged in the remotest regions of the globe. Although Africa may claim through Egypt the merit of laying the foundation of the sciences and arts, it was Europe that refined and carried them to perfection in times when the empires of Greece and Rome were in their glory. Nor have the intervening ages since those great nations passed away, witnessed the diminution of the superiority of this continent over the other three. The sun of civilization has continued to shine brightly upon the fields of Europe, while, by her children, vast territories on the earth's surface have been subdued, or, before uninhabited by civilized man, have been peopled with sturdy and persevering races. The fleets of Europe have rode in and subjugated every sea; her vessels have explored the remotest shores, and mapped, so as to be familiar to every eye, the limits of arctic shores and the burning confines of torrid countries. Yet is Europe, as already observed, the smallest of the quarters into which the globe is divided. Its length, from Lisbon on the S.W. to the Uralian mountains in the E., is about 3,300 miles, while from the North Cape to Cape Matapan, the S. extremity of Greece, the distance is about 2,350. But the land superficies of this continent are much less than such longitudinal dimensions might lead to be supposed. In the N. part the White sea, the Gulfs of Bothnia and Finland, the Baltic, Cattegat and Sleeve, with part of the Black sea and of the Mediterranean in the S., (in the latter case, drawing a line from E. to W., or from Cape Matapan to Gibraltar,) exhibit a vast space occupied by water, and show the land to be much less than might have been supposed from the sum of its extent from N.E. to S.W. Several attempts have been made to estimate the square miles of land in this continent, but a sight of the map must show how very erroneous such an estimate is likely to prove, from the difficulty of ascertaining the extent of a surface so irregular and indented. Any estimate, therefore, can be only comparative, and 2,500,000 square miles attributed to its superficies, is perhaps as near the truth as it is possible to arrive. The modern boundaries of Europe seem to have been unknown to the ancients, the name itself originating in a small spot near the Hellespont. A full third had not been explored until later days. The Mediterranean sea washes Europe along its whole S. extent, if the addition on the E. of a small portion of the seas of Marmora, of the Euxine, and sea of Azof be made. W. of the Straits of Gibraltar it is bounded by the North Atlantic, which contains its remoter islands, as the Azores and Iceland. The N. boundary is the Arctic ocean, with the Island of Spitzbergen. The Uralian mountains and the River Cara, flowing into the sea of Karskey, may be admitted as the N.E. boundary, and S. of the Uralian line, the division has remained in the confused distinctions of some of the Russian petty governments as far as the Volga and Don, into the Euxine. The N. cape in Lapland stands in lat. 71° 10′ N., long. 25° 10′ E. ; Cape Matapan, the S. extremity, is in lat. 36° 21′ N., long. 22° 28′ E. The most W. point of the mainland of this continent is the rock of Lisbon, in long. 9° 30′ W. Among the islands, however, Dunmore head, in Ireland, is more W., standing in long. 10° 29′, and the Skelligs, off the same coast, in 10° 32′ W. Including the islands, Europe exhibits a great extent of sea coast, and owing to its position with regard to the ocean, over a considerable portion of its shores experiences a great mildness of temperature. Almost the whole of Europe is N. of the 40th degree of lat., while 10° and 12° more N. the climate is far more temperate and genial than on the coast of America to the S. of 40°, where the excesses of heat and cold are not only very great, but changes from one extreme to the other are sudden and pernicious. The aspect of the European coast offers much diversity, under nearly similar local circumstances. In sailing N. along the dreary coast of Norway, the eye rests only upon chains of mountains broken into a variety of strange forms. Vast insulated masses of rock line the coast, sometimes in long reefs, within which there is a smooth navigation, but generally honey-combed, and, as it were, eaten deeply into by the ceaseless action of a cold tempestuous ocean for countless ages. Rocky summits are observed towering at a distance of 20 leagues, serving as marks for seamen. Seals and shrieking sea-birds occupy the lower islets and rocks, the loftier being visited alone by the tempest. Rarely here and there on this coast of desolation, a Lapland fisherman is now and then seen, as if to tell that the most inclement climes were no proof against the pliant constitutional temperament of man, which accommodates itself alike to the rigours of an arctic winter and the burning atmosphere of the tropics. The rocks of Norway generally consist of gneiss stratified and dipping towards the E. In some few spots on the coast tracts of morass are found in peculiar situations, resting upon broken shells and marine plants, 30 feet above the present level of the northern ocean. Some of the mountains near the coast have glaciers, one of which, in 1756, fell into the Langefiord; and though that fiord or sea-arm is 12 miles long and 2 broad, it raised the waters so suddenly that they swept away houses 200 paces from the shore. Beyond 65° N. lat., fruit trees disappear, and the fiord of 65° is the limit where the oyster disappears from the ocean. Turning S. along the coast, if the climate amend a little, the character of the coast is the same still, deeply penetrated by fiords or inlets that run 15 and 20 leagues inland, with depths of from 300 to 400 fathoms at 10 leagues from their entrance. In these fiords, and among their rocky islets, vessels find a safe shelter. In the S., in the constant internal navigation about Bergen, where the fiords are too deep for anchorage, iron rings are fixed in the rocks 2 fathoms above the water-level, for attaching vessels. Even at the N. extremity of Finmark these fiords do not freeze. Much of the same character, but modified by local circumstances, are the S. shores of Norway. These peculiarities extend as far as the western limits of Sweden, beyond the Christiania fiord. The shores of that country are high, consisting of rocks and promontories, rugged and desert for the most part, and partaking of the same character along the Scagerack and Cattegat. The coast of Schonen is flat, and the roads for shipping exposed. All the way from Blekingen to the Gulf of Bothnia the Swedish shore is lined with islands and rocks, broken into gulfs and bays, having safe but useless ports, and mostly somewhat difficult of access, serving only for the refuge of coasting vessels. The shores of the Gulf of Bothnia are rugged and broken, and on the Swedish side a ridge of mountains lines them. The N. and E. coasts of Jutland are low, and the N. extremity composed of barren sands, sometimes elevated into hillocks of a few hundred feet high. The Lenifiord nearly divides the peninsula of Jutland, being separated from the sea on the W. of it by the sandy isthmus of Aggerland, only 3 or 4 miles broad. From Holstein E. the German shores are low and sandy, and the beaches so heaped up by the waves as to form causeways or dikes, of which the most considerable is that of Dobberan, in Mecklenburg. Along the coast of Pomerania the action of the water produces somewhat similar effects as at the peninsula of Dars, with its basin. The Isle of Rugen and others speak the same action of the Baltic waves. Beyond Pomerania is the Gulf of Dantzick, and on the Prussian coast the lagoons of Frisch Haf and Curisch Haf appear, consisting of the fresh water drained from the rivers of Poland, and each having but a single outlet. The whole coast of Prussia is low, and covered with sand-banks; it is the same with that of Courland, which has only the ports of Liebau and Windau. The hill of Domberg, a noted seamark, is almost the only elevation it possesses. The Gulf of Livonia or Riga is large, the coast composed of sand, gravel, and calcareous strata, but it is higher than that of Courland. The icy Gulf of Finland is filled with barren rocks and islands, dreary as might be expected in the inclement atmosphere and long winter climate of 60° N. The Baltic receives 240 rivers. The coast of the North sea, from the Scaw to the mouth of the Scheldt, on the E. side, is for the most part low and encumbered with sand-banks. A large proportion of the land seems to have been heaped-up sand out of the ocean depths, and covered with a thin vegetative soil adapted for pasture. Much, however, is waste, or has been made useful only by unremitting industry, as the well-known provinces of Holland at once proclaim, where the sea is barred out by dykes, which alone prevent the keel of the fisherman from navigating over fertile pasturages. The great extent of the coasts of the British isles is naturally varied in character, but offers little that is not observable on the shores bordering upon great seas in general. The promontory of Cornwall and the headland of Cape Wrath, alike the N. and W. extremities of the island, equally exhibit rocky shores, that resist the encroachments of the stormy seas which assail them. The promontory of Cornwall, pushing into the Atlantic, consists of walls of impenetrable granite. The coast from the Land's End to the S. Foreland possesses a superior mildness of climate to the rest of the country. The E. coast, from the mouth of the Thames to the N., is for a large space encumbered with sands, but as the N. part of the coast widens and Scotland is approached, these diminish, and rocky shores and better harbours are found. The force of the Atlantic waves is not experienced so far up the British channel as it is on the N.W. side of the island. The long unbroken waves and miniature rollers, called locally the ground-swell, are not experienced much higher than the Start point. The whole S. coast possesses excellent harbours, and, compared with the French side of the channel, is invaluable for navigation; it offers in few places any great elevations close to the sea, though some headlands, in both Devon and Cornwall, rise to 600 feet. Blown sands are found on the N.W. of this part of the island, but not on the S. It has been observed that a great proportion of the English coast in the North sea, especially near the mouths of rivers, is troubled with accumulations of sand and banks of considerable extent. In many places from the Straits of Dover northwards, these are observable, but not in this respect equalling the large accumulations of the same character which trouble the coast from Calais to the N. of Jutland, rendering the navigation perplexed, intricate, and dangerous. The coast of France in the British channel is the reverse of the English, the ports are few, rocks and dangers abound, and from Brest to Cherbourg there is no shelter for vessels of burthen, nor again from Cherbourg to the Scheldt. The shores of the Bay of Biscay are proverbially treacherous, but they possess several safe harbours. The N. coast of

Spain offers a totally different character from that of France northwards. Here steep rocky shores are lashed by deep waters, so that anchorage, unless close up to them, is impracticable. A depth of 200 and 250 fathoms is not uncommon very close in, suggesting some very peculiar cause, for which even the current inwards along that coast and out to the N.E. of Biscay cannot satisfactorily account. From Cape Finisterre to Cape St. Vincent there is a varied coast, sometimes bold, at others low and sandy. The Straits of Gibraltar exhibit rock shores of marble and a strong current between, respecting which last there has been much difference of opinion. The current from the Atlantic is generally observed to set into the Mediterranean; but that there are under-currents passing out seems to be generally admitted. Vessels wrecked or sunk have been found thrown up at distances and in places that proved they must have been carried to where they were found by some means not superficially visible. The current, too, is regularly affected by the tides. The shores here on both sides are not equally lofty; but, the straits passed, they expand, and a noble inland sea offers every variety of coast, under a warm and genial climate, being, as it were, the common point at which all the continents of the old world meet within a circumference of great extent, embracing the site of empires the most important in the earlier extant history of the human race. The population of Europe has not been clearly ascertained; an approximation to the truth has been made nearer to it perhaps in regard to this quarter of the globe, than has been attainable in relation to any of the other three, but too uncertain to be worthy of record as a total to place reliance upon. The number of inland seas in Europe, and their extent, is evident at the first glance over a chart of this quarter of the earth. The Mediterranean, 2,000 miles long, has been already mentioned, but there are its important branches and feeders. Scarcely escaping a junction with the Red sea at Suez, it passes up through the Greek Archipelago to meet the waters of the Black sea and Sea of Azof, with all their tributary rivers. It sends a long arm up the Adriatic as far as Trieste, along the classic shores of Italy and Greece. Next to the Mediterranean comes the Baltic in the N., with its gulfs of Bothnia and Finland, its island-strewed sound, and singular outlet in the Cattegat, already alluded to. It was called by the Germans the Eastern Sea, from whence came the word "Easterlings" in English history. Then there is the White sea on the N., the voyage to which is so circuitous round the N. Cape, supposed to have been known even in Saxon times. N. of all is the Arctic sea, the great N. reservoir of ice, that expands and contracts (as telescopes show to be the case in the planet Mars) during the winter and summer seasons. This vast region of cold, is that where the herring propagates its millions for the fisheries, and the whalers annually pursue their dangerous avocation. In the extent of naval power among the European nations, England is unquestionably foremost. The next nation is France; but all follow, at a vast distance, the island navy in number, strength, and the experience only conferred by practical skill. In Europe, naval science first developed itself, and here originated those progressive improvements which have rendered the ocean as safe a highway as the land. The Phœnicians traded with the inhabitants of Cornwall for tin, and their best mariners were their

Carthaginian colonists in Spain. In later times the Venetians and Genoese in the south of Europe were the principal navigators of the world. It was Europe that dispatched Columbus to seek out a new world, to be peopled by her children; and from her Drake sailed to circumnavigate the entire planet. Here, too, the invention of steam has introduced a new era into the science. To Europe the world is now, as it was 2,000 years ago, indebted for all that contributes to extend the sphere of knowledge, unite the common family of man in brotherhood, spread through commerce the intercourse that softens and humanises the most savage races, and bringing to its own door the product of distant countries, rewards labour, and increases comfort by enlarging the variety of those luxuries of which the expanded minds of its inhabitants continually demand new creations.

EUROTAS, Greece, a river of the Morea, the name of which is changed to that of the Iris, or Basilipotamos.

EUSTACE, or EUSTACIA ISLAND, America, at the inlet of the harbour of St. Augustine, on the Gulf of Florida, which, in union with a long point of land, may be said to form its entrance. It has also been called Metanzas, or Slaughter island, and Anastatia more commonly still; a long narrow island, principally consisting of sand and bushes. It is 12 miles in circuit. N. point, lat. 29° 50' N., long. 81° 25' W.

EUSTATIUS, or EUSTATIA, ST., ISLAND OF, West Indies, appearing from the ocean like a huge pyramid rising out of the sea, when seen at a distance, but on a nearer view, taking the resemblance of two mountains, the E. being higher than the other, hollow at the top, the crater of a volcano. These mountains are cultivated to the very summit; the pop. is reckoned about 18,000, of whom 12,000 are Africans; this includes the Island of Saba, 15 miles off, which is smaller. The exports are the common exports of the West India islands, but it produces in abundance all kinds of provisions, including hogs, poultry, and rabbits, in great plenty. The town of St. Eustatius is divided into the Upper and Lower town, the last close to the edge of the sea, the former on higher ground, and protected by a fort. Both the upper and lower towns consist of a single street. Landing is very difficult here on account of the surf, especially if the wind is from the S.E. The anchorage is before the town in 12 fathoms, about three-quarters of a mile from the shore, bringing the church to bear E.N.E. or N.E by E., and the W. end of the bay N.W. by N. The anchorage, however, is not good, though it is much used, the ground being rock and coarse sand. Vessels may anchor further out, but at the hazard of a squall, which will require instant putting to sea. This island is in lat. 17° 29' 30" N., long. 63° 4' 30" W.

EUXINE, another name for the Black sea, derived from Pontus Euxinus, the name given to it by the ancients.

EVORT, LOCH, Scotland, Western isles, in North Uist, N. of Loch Rueval, towards which the coast is free of all danger. This loch extends 6 miles inland, the anchorage is at the broadest part about a mile within, well sheltered, and the ground good, the depth fit for vessels of any size. There are two rocks to be avoided, one on the N., the other on the S. side. The first dries two hours after high water. The best anchorage is on the N. side of the harbour.

EX, RIVER, Devonshire co., falling into the sea at Exmouth. The tides run up to Topsham, where all sea-borne vessels that can pass the bar at its mouth discharge their cargoes. The bar has not more than 6 feet at low water. The city of Exeter stands upon the banks of this river, about 4 miles above Topsham, which may be called its port.

EXCHANGE BAY, E. end of Antigua, where the coast begins to trend S.W. It is N.E. from Hudson's point, the E. extremity of Willoughby bay. Smith's island lies before the opening, and is to be left on the starboard in going in. There are but 2½ fathoms between the island and the main, and within the bay itself not more than 2.

EXETER, America, United States, a town of New Hampshire, on the N. side of the Exeter river, having a good harbour. It was once noted for its shipbuilding.

EXETER BAY, America, United States, one of the numerous bays of New England, which has been little noted, on account of the superiority of many others which it possesses.

EXETER SOUND, America, N. off Cape Walsingham, and N.E. from the Cape of God's Mercy, the most S. point of Cumberland's island. The island itself lies on the N. side of Cumberland's straits.

EXMOUTH BAR and TOWN, Devonshire co. This bar is 12 leagues N.W. by W. from the Bill of Portland, and 20 miles W. ¼ N. from Lyme Regis. It is a dangerous bar, and should not be attempted except by those who are well acquainted with it. The entrance is narrow and intricate, having rocks on the E. side, and sands on the W., with no more than 7 feet over the bar at low, and 16 or 17 at high water. Vessels that have passed the bar commonly anchor off Starcross, where they remain afloat in 10 and 12 feet water. Such ships as go up to Topsham lie there alongside the quay, and their cargoes are sent up to Exeter in lighters. Outside the bar, about a mile from the W. shore, there is good anchorage in 6 or 7 fathoms. Strait point bearing E. by N., or E.N.E., and Longstone rock N.W., or N.W. by W. The outermost sand here is called the Pole, and is covered at high water. The common channel is E. of this sand; the passage narrow and not easy to find, is buoyed. At the entrance is a red buoy in 4 fathoms water; to the W. of this buoy a vessel may stop in 22 feet, until the tide serves for entering. Vessels in coming to the entrance from the W., to clear the reef which runs off Strait point, bring the obelisk on Haldon hill a sail's breadth open of the high land of Oakam. It is high water here at 6ʰ 25ᵐ, full and change. The least depth over the bar at low water spring tides is 6 feet, and with neaps 8. The tide rises with springs 14, and with neaps 8 feet. The town is on the E. side of the mouth of the Ex, 10 miles S.S.E. of Exeter. It is visited in the bathing season by strangers. The pop. is not considerable, nor is it a place of any commercial importance.

EXUMA ISLANDS, West Indies, two in number, called the Great and Little Exumas. They lie to the W. of Long island 22 miles. See *Bahamas*. The Great island has a port of entry N.W. of Exuma; 2 leagues on the E. edge of the Great bank there are a number of small islands called Exuma kays. Others succeed them more N. for 26 leagues, the most N. called Ship Channel kay, E. by S. 10 leagues from New Providence, while between the bank which surrounds the Cat islands and the E. edge of the Great Bahama bank, is situated Exuma sound, a channel 6 leagues wide. The beacon at Exuma is in lat. 23° 32′ N., long. 75° 49′ W. The Galliot cut on the bank, lat. 23° 55′N., long. 76° 15′ W.

EYA, Piedmont, a small town on the sea-shore, between Nice and Monaco.

EYDER RIVER, N. coast of Europe, N.E. of the Elbe. It is buoyed, but the sands continually shift, and so frequently that a pilot is always necessary to enter. The outer black buoy lies E.S.E. ½ E. from Heligoland lighthouse, distant 22 miles; N.E. by E. from the red buoy of the Elbe, distant 13 miles; and E.N.E. from the Schlussel, or outer buoy of the Weser, distant 29 miles. A light-vessel with a single mast is stationed off the entrance of the Eyder, between February and November. She is painted red with a white streak, and in the daytime has a small Danish flag flying 60 feet above the water. In the night a lamp is burned at the height of 34 feet. When the vessel bears E. by S. she may be safely approached. She has on board pilots to conduct vessels to Husum, Tonningen, and the Elbe. When the weather is thick, a cannon is discharged, or a bell tolled; as signals to vessels that they are taking or may take a wrong course. The channel of this river is kept regularly buoyed.

EYEMOUTH, Scotland, a town of Berwick-shire; it is sometimes called Aymouth. It is small, having a tide haven, which is subject to freshets. There are some rocks at the entrance, and to go in, a vessel must keep in the midway of one of the two channels which they form, as may be best suited to the state of the wind. When well within them, keep close to the beacon, fixed on the port side rocks, and so into the harbour. There is much corn shipped at this port, and great quantities of fish are cured here. Two harbour lights are displayed for the benefit of the fishermen during the herring season, one on a pole 26 feet high, seen at the distance of 6 miles, and a smaller and less brilliant light on the pier head, showing the entrance into the harbour, and from its position relatively to the other light, the best passage into Eyemouth bay.

EYERLAND ISLAND, North Holland, to the N. of the Texel, from which the sea separates it at every high tide. It is small and flat. It is a mark for the buoy on the Burgh sand, to the S.E., within the Texel island.

EYNORT, LOCH, Scotland, W. coast, S.W. in the Isle of Skye, and 5 miles N.W. of Loch Brittle. Between these the coast is free from rocks and every kind of danger, having deep water close to the shore. Opposite to Loch Eynort there is a rock always above water, called Duskere rock, about half a mile from the land; and there is a small rock at the S. point of the entrance, but above water, and close to the shore. At the N. point of the entrance there is also a ledge extending half a cable's length S., which must be carefully avoided. A ship may lie in this loch almost land-locked, if not drawing more than 12 or 15 feet water, below the Green point of Glen Eynort, in 3 and 3½ fathoms, and large vessels may anchor half a mile lower in 6 and 7 fathoms, off Crakenish.

EYNORT, LOCH, Scotland, situated behind the Isle of Scalpa, about N.E. of Loch Eynort on the opposite or N.E. side of the Isle of Skye. It is well sheltered, its access safe, the ground good, and the depth moderate. It is, however, exposed at times to squalls that blow down from the mountains in stormy weather. On the port side of the entrance there is a sand-bank, reaching nearly one-third of the way over, on which there is not above 1½ fathom, and at its upper end only 6 feet. Large ships may anchor anywhere on the port or larboard side, and small vessels above this bank on the S. side. The harbour, as it is called, of Kylescoog is in the channel between the Islands Skye and Scalpa.

EYNORT, LOCH, Scotland, W. isles, in South Uist, a good harbour for vessels of all sizes, and of easy access, well sheltered, with good ground, and moderate depth. The upper part, above the Narrows, is called Loch Arinaban. The best anchorage for large vessels is about the middle, in the broadest part, in from 6 to 12 fathoms. Ships going higher up to Loch Arinaban must take the flood tide, as the spring tides in the Narrows run 3 miles an hour. A rock must also be avoided on the port side, which dries at low water.

EYRGOS, or PYRGOS, Greece, near the Gulf of Koron, E. of Point Stupar, 2½ miles inland.

EYSDIL HARBOUR, W. coast of Scotland, about a mile S. of Inish island, and 3 miles N. of Black harbour, on the coast of Argyleshire. It is difficult to enter, and only adapted for vessels drawing under 8 feet water. It is shallow, and little used. The place of anchorage is a little N. of a small point in the middle of the harbour, about half a cable's length from the shore. It possesses valuable slate quarries.

EYSORT, LOCH, Scotland, W. coast, S. of the Island Oronsa, on the coast of the Isle of Skye. It is customary for vessels to stop near the N. side of the mouth of Loch Eysort, in 20 and in 8 fathoms, on the E. side of Borereg rocks, in 6 fathoms water, or going up to Island Heast, on the E. side of that island, in 7 fathoms.

FAABORG, or FABORG, Denmark, Isle of Funen or Fyen, a town on the S.W. part of the island, having a pop. of 2,000. It is 17 miles S. of Odensee, and is situated in the centre of a bay, before which there are two islands.

FACHEUX, America, Newfoundland, between Cape Ray and Chapeau Rouge, a bay W.N.W. from Bonne bay about 4 miles. The entrance is the easternmost of two bays, the second called Dragon bay. It is easily seen from the sea, and runs in about 2 leagues N.N.E. It is one-third of a mile broad at the entrance, with deep water on most parts of it. On the W. side there are three coves, where ships may anchor in from 10 to 20 fathoms. Little Hole haven, where small craft may shelter, lies a mile to the W.

FACHOLA, COVE OF, Island of Corsica, between Capes Feno and Bonifacio. It has several islets before it, and affords good shelter for small craft and even for ships.

FACTORY ISLAND, Africa, W. coast, the same with Idolos island, one of the Isles de Los.

FÆRDER ISLAND, Norway, on the W. side of the entrance to Christiania fiord. It is in form high and conical, having a lighthouse on the summit, with a fixed light, 216 feet above the level of the sea, visible 7 leagues. Lat. 59° 3′ N., long. 10° 37′ E.

FAIR AND FALSE BAY, America, Newfoundland, between Cape Bonavista and Cape Freels. This is a bay so full of small islands and rocks, that though it may contain good anchorages, they have not been described. One cluster of islets extends full 20 miles off the mouth of this bay, and between them are innumerable channels with deep water, and navigable. There is also a channel running from Fair and False bay and Morris island to the N., leading to Bloody bay, which there turns W.; but no description given of this bay and its vicinity can be of the slightest use to the seaman, from the intricacies with which it abounds.

FAIR HEAD, Ireland, N. coast, S.E. ¾ E., 11½ miles from Bengore head. Lat. 55° 13′ 30″ N., long. 6° 9′ 30″ W. The promontory of Fair head exhibits pillars of greater length and coarser texture than those of the Giant's causeway: the basalt of the latter is very compact, and the angles of the pillars have preserved their sharpness, although exposed to the sea no doubt for thousands of years. Mr. Hamilton, who examined these singular structures, is of opinion that the pillars are magnetic, and affirms that in the semicircular bays about Bengore and Fair heads, the compass becomes considerably deranged. This should be recollected by vessels in the vicinity, when in Ballicastle bay or Church bay, Rachlin island. See *Bengore Head*.

FAIR ISLE. See *Orkney Isles*.

FAIRWEATHER ISLAND, America, United States, Connecticut, off Black Rock harbour. To enter this harbour, which is safe and easy of access, no direct course can be given, in consequence of its depending upon the distance a vessel may be from the light at the time she may make the island. This light on the Black rock stands in lat. 41° 8′ 24″ N., long. 73° 13′ 30″ W.; it is fixed 35 feet high. Vessels coming from the W. to the harbour, in order to avoid the reef called the Cows, bring the light to bear N. by W., and run for it until about three-quarters of a mile distant, then, if necessary, stretch in to the W. in a fine channel having good ground. On approaching the light, which is on the E. side of the harbour, the water gradually shoals to 2 fathoms. The mouth of the harbour, although not very wide, is not difficult to enter, for the light bearing E. brings a ship completely into the harbour. The island on which the lighthouse stands, and the reef called the Cows, on the S. and W. side, form the harbour of Black rock. On the E. rock of the reef there stands a spindle, half a league from the light, which bears due N. from it, and the light is 44 rods from the S. point of the island at low water. From this point, the island pushes off a single rock for 30 rods, on which there are 8 feet at high water, making in all about 75 rods. The light bears from this point N. by E. ¼ E. When a ship has passed this rocky point, the harbour is fairly open to the N. in any point from N. to W.N.W., and with the lead the light may be run for in safety.

FAIRY BANK, one of the banks between Ostend and the Hoak or Hook of Holland. It is 3 miles to the W. of the S. end of the West Hinder, extending N.E. ¼ E. about 8 miles, and about half or three-quarters of a mile broad, with from 4 to 9 fathoms water upon it. Its shallowest part, with 4½ fathoms, is 2 miles in length, and nearly in the centre of the bank, between lat. 53° 23′ and 53° 25′ N. Thirty-two miles from the North foreland, in the channel between this bank and the West Hinder, there are from 14 to 19 fathoms, except near the N. end. In the mid-channel there is a narrow shoal of 9 fathoms, about 2 miles in length, running parallel with the banks, having from 16 to 19 fathoms close to it.

FALAISE DES FONDS, France, N. coast, at the entrance of the River Seine; a point of land to the W. of Honfleur.

III.

FALCO, CAPE, Spain, E. coast, a small headland W. of Cape Norfeo, near the N. entrance of the Bay of Rosas.

FALCON, CAPE, Africa, N. coast, known also as RAS HAL HAISHFAR, in lat. 35° 47′ N., long. 48′ W. It lies W. of Oran, being the W. point of that bay. There are several rocks about Cape Falcon, one of which, above water, is called Avintar islet; these will require a berth in passing. W. ¼ S., 4 miles from cape, is the Islet of Plane.

FALCON CAPE, Balearic isles, Island of Iviza, the S.E. point.

FALCON ROCKS, Madeira islands. They were first described as a bank on which a mariner of Honfleur had grounded, N.E. of Porto Santo, and as being a ledge on which a Dutch ship was lost. The officers of the sloop-of-war Falcon, in 1803, found a ledge of rocks to the N. of Porto Santo, and the least water upon it 4½ fathoms. The rocks were steep-to, and lay at the distance of 8 miles from the N. point of the island, with the N.E. point of Porto Santo bearing S.S.E.; the Ilheo da Fonte, or N. rock, S. by W.; and the W. point of the island S.S.W. This is supposed to be the reef first described as that on which a Dutch ship was lost. It extends E. and W. true nearly 1½ mile, terminating in a reef to the W.

FALCONE, CAPE, Island of Sardinia, the N.W. point of that island, S. of the Island of Asinara, between which and Cape Falcone there are two passages, formed by Viana or Piana island. The N. passage has only 4 or 5 feet water, but the S. has 10 or 12 feet, upon a bank which may be sailed over. Lat. 41° 2′ N., long. 8° 10′ E.

FALCONERA, CAPE, Minorca, on the N. side of the island, E. from Cape Grosso.

FALCONERA, PORT, Gulf of Venice, in the N. part; a small port at the entrance of the River Tagliamento.

FALCONERI, a Greek island, S.E. of the Gulf of Nauplia or Argos, and W. of Anti-Milo. It is high and uninhabited. Lat. 36° 50′ 40″ N., long. 23° 53′ E.

FALKENBORG, Sweden, a town on the shore of the Cattegat. It lies a little S.E. of the Morups Tange lighthouse, at the mouth of the Athoan river, 12 miles N.W. of Halmstadt.

FALKNER'S, or FAULKNER'S ISLAND, America, United States, on the N. side of Long island sound, S.E. of Sachem's head. There is a fixed light, 93 feet high, upon this island, which is nearly W. from Black island, within towards the centre of the sound, in lat. 41° 13′ N., long. 72° 40′ W.

FALLEN CITY, or OLD JERUSALEM, West Indies, a remarkable cluster of islands to the S. of Virgin Gorda. S. of them is a large bluff rock, called Round rock, next to which is Ginger island. See *Virgin Islands*.

FALMOUTH, America, Nova Scotia, in Hants co., on the S.E. side of the Basin of Mines, opposite Windsor, and 28 miles N.W. of Halifax.

FALMOUTH, America, United States, on the S.W. point of Cape Cod, between Buzzard bay and Vineyard sound. It has several good harbours, of which Wood's hole is the best, having 3 to 6 fathoms water. It has a number of vessels employed in the fishery, and in the coasting trade.

FALMOUTH, America, United States, a township in Maine, upon Casco bay. It contains 2,600 inhabitants, and was incorporated in 1718. It is 120 miles N.N.E. of Boston.

FALMOUTH HARBOUR and BAY, Antigua, on the S. side, W. of English harbour, separated

N

from it only by a narrow isthmus. It is more spacious than English harbour, more open, but with less depth, though sufficient for most ships. A vessel to sail into this harbour should run close in towards the W. point, called Proctor's point, and she will then pass clear of a ledge of sunk rocks called the Bishops, which lie towards the middle, within the entrance, and terminate a shoal, extending from the E. point, on which there is a redoubt for the defence of the harbour. Beyond these rocks there is good anchorage, in from 3 to 6 fathoms water. The Harbour of Falmouth is about 1½ mile to leeward of English harbour, where water is not to be obtained, but at Falmouth harbour it may be procured, though somewhat brackish.

FALMOUTH HARBOUR and TOWN, Jamaica, or as sometimes called MARTHA BRAE HARBOUR. It lies 6 leagues E. of Montego bay, and is a bar-harbour. The channel, at the entrance, is very narrow, and has not more than 16 or 17 feet of water, and can only be attempted with a pilot, owing to its intricacy. The town stands on the W. side of the harbour; it has throughout a regular depth of from 5 to 10 fathoms.

FALMOUTH PORT, England, on the S. coast of Cornwall, the most secure in the channel, situated in lat. 50° 8′ 48″ N., and in long. 5° 2′ 42″ W., at Pendennis flagstaff. Falmouth stands upon the S.W. side of the harbour, strictly so denominated, within Trefusis point on the N. and Bar point on the S. opening into Carrick roadstead, one of the finest in the kingdom. On the N.E. side of the harbour are the villages of Flushing and Little Falmouth, over against the N.W. end of the town. The castle called Pendennis, built by Henry VIII., and since strongly fortified, stands on a hill promontory about a mile off, and is held by government of the heirs of Killigrews at a rent of 13*l.* 6*s.* 8*d.* out of the Exchequer. The entrance into the harbour and the roadstead on the W. side of the Black rock are commanded by Pendennis, the Castle of St. Mawes commanding it on the E. side. The town is of modern date, having been built by the Killigrews in the reign of Charles II. Its inhabitants have ever been adventurous seamen. In their small packets they fought several gallant actions during the last war. The Merchants' Hospital of London for the support of disabled seamen and the widows and children of those lost in the merchant service, have a branch here. Falmouth is the only bonding port for tobacco, except Plymouth, in the counties of Cornwall and Devon. It exports most of the commodities of the county, principally tin, fish, and copper. The pop. of the town and parish is 7,695. The old town of Falmouth is marked by its original narrow streets and inconvenient turnings, but the vicinity is stocked with pleasant houses. The distance from London is 270 miles. The trade of Falmouth has been diminished, in consequence of the removal of the post-office packets, of which it was for more than a century the convenient rendezvous, as they sailed from thence for Spain, Portugal, the West Indies, and America. With these countries Falmouth drove a more thriving trade formerly, as well as with Newfoundland. The road out of which the harbour opens westward is extensive, and took its name from the Carracks or large ships of the old time making it an anchorage. It is so winding in its creeks and bays that it was observed in King Charles's reign that 100 sail might anchor in them, and not one ship see the mast of another. During the last French war, Admiral Cornwallis, keeping the sea off Brest in the winter, made it his favourite point of shelter, when compelled to run during a gale with 20 sail of the line among hundreds of transports and merchantmen, all riding in perfect safety. In approaching the road from the S., after coming round the Lizard, the Manacle rocks are the first dangers. These lie S. and S. by W. from the entrance of the harbour just off St. Keverne, (offshore 1 mile) where the coast begins to trend N. 6 miles S.S.W. ½ W. from Pendennis point. On nearing the road at the entrance between the two points of Pendennis and St. Mawes, rather nearer to the W. shore, there is a rock called the Black rock, on which is a beacon. Ships may sail in on either side, but the E. side is best, since she may borrow with the lead towards St. Mawes into 5 or 6 fathoms at low water. The Bay of Falmouth and the outer roads may be entered at any time, since the lighthouse on St. Anthony's point has been erected. The light is 65 feet above the level of the sea at high water, visible from S. 40° E. (S.E. ½ E.) round seaward, and up Falmouth harbour. It exhibits a rapid succession of bright flashes every 20 seconds, visible 4 leagues off. There is good anchorage in the outer road, Pendennis castle N. by W., St. Anthony's point N.E. by E. ½ E.; a third of a league off there is good ground at 10 fathoms, not going more W. than bringing Pendennis point a little open W. of Mylor point. When the Black rock is covered with water, the largest ships may go in on the W. side. There is only 16 feet at low water on a shoal running out S.E. by E. ½ E. from the Black rock, about a cable's length. There is a flat from St. Anthony's point to Turnaware point on the E. side, and another on the W. from Pendennis point to Pill creek; the channel is in the middle. That on the E. side is called St. Mawes' bank; it extends from the shore nearly half a mile, and forms a point three-quarters of a mile N. ¾ W. from St. Mawes' castle, and S. ¼ E. about half a mile from Penarrow point. The mark of its N. edge is Budock church over the sloping land of Trefusis point; the W. flat extends across the entrance to Falmouth harbour, and is called Falmouth bank. Abreast of the land of St. Mawes the two banks approach within 130 fathoms, with from 12 to 16 fathoms in midchannel. On the W. side of the Road between Trefusis and Pendennis is the entrance to Falmouth harbour, within which are 3, 2½, and 1 fathom—it is here that the packets usually lie. To sail in on the E. side, Mylor point must be brought on with Killyganoon house N. ½ E. until Budock church bears W. ½ N. about a sail's breadth open S. of Falmouth church, over the slope of Trefusis point. This last mark leads through the Cross road into St. Just Pool. To sail in by the W. channel let Feock house be one with Mylor point bearing N. by E. ½ E. This will lead over Falmouth flat in 5 fathoms at half tide. Frigates and line-of-battle ships may pass here at half tide, but it is not safe at low water. Coming round from the E., and passing the Zone or St. Anthony's point, a good berth should be kept until the Black rock come at N.E. or N.N.E.; when S., or S. of W. from the point, a ship may alter her course and pass the Shag rocks, near which stands the lighthouse, upon the starboard, lying about W.N.W. from the point. This light is not seen until the point is rounded from the E. The road may be entered at all times by day or night with a leading wind. Vessels off the Lizard at night should steer E.N.E., keeping the Lizard lights in sight until St. Anthony's light bears N.N.E., then steer directly for it, until just within 2 cables' length from the light. Next steer N. by W.; this will take to the eastward of the Black rock, up to the 3 fathoms or Falmouth bank. When St. Mawes' castle bears E.S.E., or the centre of Falmouth town lights W.N.W., the anchor may be dropped. High water, at full and change, 5ʰ 15ᵐ.

FALSE BOAT ROCK, Ireland, in Tralee bay, W. coast, near the entrance.

FALSE CAPE, America, United States, Virginia, 7 leagues from Cape Henry, S. towards Carrituck. After getting to the N. of False cape a vessel may again keep in 7, 8, and 9 fathoms, until up with Cape Henry; 9 and 10 fathoms having been near enough before to False cape.

FALSE CAPE, West Indies, Hayti, on the S. side of the island, a little W. of Cape Beata, and one of the double points of that cape thus named.

FALSE CAPE, Sierra Leone, W. coast of Africa. The coast at the foot of Cape Sierra Leone forms a sandy bay, bordered with trees, extending 3 miles S. of that cape, and ending in a rocky point. At three-quarters of a mile further there is another point more projecting and conspicuous, named False Cape, and bearing from Cape Sierra Leone S. by W, ¼ W., distant 4 miles. The extremity of False Cape is in lat. 8° 25′ 48″ N., long. 13° 17′ 48″ W., according to Captain Owen.

FALSTER ISLAND, Denmark, an island near the entrance of the Baltic, a mile E. of Laaland, from which it is only separated by a narrow strait. Lat. 55° 10′ N., long. 12° 14′ E.

FALSTERBO POINT, Sweden, the S. limit of the Sound in Sweden, on which there is a conspicuous lighthouse, in lat. 55° 23′ 6″ N., long. 12° 51′ E. It exhibits a fixed light of the second class, 74 feet above the sea. It is only lit during the winter months. Reefs extend from this point more than a league seaward, surrounded by shallow water. S.S.W. ½ W., distant 5 miles from the lighthouse point, there are but 19 feet water. The point must therefore have a wide berth, keeping in 7 fathoms, until Falsterbo church bear N.E. ¼ N., and when the church comes N.N.E. a ship is in the Baltic sea. In returning from the Baltic, Falsterbo light being brought to bear S. by E., a ship will be in the fair way for the Dragoe black buoy. There is also a light-vessel with two fixed lights, moored in 6½ fathoms water a mile from the outermost point of the reef, with the Falsterbo light bearing N.E., distant 6 miles, and the light on Steven head bearing W.N.W. ¼ W., compass bearings. The lights are 50 feet above the level of the sea, and may be seen 8 or 10 miles. There are five pilots stationed in this light-ship, round which the largest vessels may sail to procure one; the flag on board is pulled down if all happen to be employed.

FALULO BANK, W. coast of Africa, 5 leagues in length and 3 miles from the shore, on nearly the same meridian; its W. point bearing S.E. from the Cape, on the E. side of which, and N. of the bank, is a bay, and still more E. of the bay is a cape of the same name, but W. of the N. bank, and of the river Cacheo or St. Domingo.

FALULO BAY, Africa, on the W. coast, between Cape Roxo and Cacheo river.

FALULO, BREAKERS OF, W. coast of Africa. S. by E. 17½ miles from Cape Roxo, to the S.W.

of the River Cacheo. These breakers are divided into two groups, and extend in a true E.S.E. and W.N.W. direction for 3 miles. They are very steep-to, and close to them are from 6 to 3 fathoms water. They are remarkable for never ceasing to break under any circumstances. The breakers of Falulo, at the W. point, are in lat. 12° 5′ N.; long. 16° 38′ 30″ W.

FAMAGOUSTA, PORT and TOWN, Island of Cyprus, 13 miles from Cape Grego. The town is built upon a rock, and is nearly 2 miles in circuit. It is surrounded with strong walls, flanked by 12 enormous towers, and having a ditch cut out of the rock, very deep, and 20 paces wide. There is a castle with bastions inside the walls. The town has only two gates, one on the land and one on the sea-side. The harbour is narrow, and closed by a chain, unloaded vessels alone being permitted to enter, to be hove down or refitted. The River Pedio, descending from Nicosia, enters the Gulf of Famagousta a little N. of the town. There is anchorage in the road before the town in 10, 9, and 8 fathoms water, and vessels resort here to winter in security. Cape Grego, 13 miles S. of Famagousta, is in lat. 31° 58′ N., long. 34° 6′ E.

FANAES POINT, N.N.E. of Baxio point, in the Island of Flores, one of the Azores.

FANCIULLA ISLAND, Italy, Gulf of Tarento, a mere islet on the N.E. side, close to the shore, having behind it the Torre delli Pali. Several small rocks lie about this islet.

FANGAL, PORT, Spain, on the E. coast, 2 leagues from Cape Tortosa, nearly N.W. Both this port and that of La Sofa are denominated by seamen "the Alfaques," but these comprehend in reality the low flat shore from Rapita to Ampulla, through the midst of which the Ebro discharges its waters into the sea, and not the ports only which are situated in the space of low land formed by the sand and mud of the river, of which Port Fangal is one. This port, or rather bay, has its points 3 miles apart, lying about N.W. and S.E. from each other. N.W. winds blow in with great violence, but a sandy point in the S. part shelters vessels from an E. wind. The water is shallow and narrow. There are but 16 feet at the entrance, and some way in only 5. The place is undefended and uninhabited. The land shifts that forms the bay, consisting only of sand and mud, low, and covered with pools and marshes. Port Fangal is in lat. 40° 48′ N., long. 48′ E.

FANNET POINT, Ireland, N. coast. Lat. of the lighthouse 55° 16′ 24″ N., long. 7° 38′ W. The light-tower is 90 feet high, and stands on the W. side of Lough Swilly; it is a fixed light, red to seaward, and bright towards the Lough, visible 14 miles. This point is low.

FANO, Italy, Gulf of Venice, a town surrounded by a lofty wall, with towers and bastions towards the sea. There is a lighthouse here, in lat. 43° 51′ N., long. 13° 2′ E. It is about 6 leagues N. of Sinigaglia.

FANO, Mediterranean, a small island N.W. from the Island of Corfu, towards the straits of Otranto, and nearly due E. of Cape Santa Maria de Leuca. Lat. 39° 50′ N., long. 19° 20′ E.

FANOE, Denmark, an island in the jurisdiction of Sleswick, upon that coast, about 7 miles in length.

FAO, Portugal, a town near Villa del Conde; opposite to it are two ledges of rock running out 1½ mile, even with the edge of the water, and

N 2

called the Cavallos de Fao. A vessel should come no nearer the shore here than 14 fathoms.

FARALLONES OF ST. CYPRIAN, Spain, N. coast, rocks near the entrance of St. Cyprian river, about 10 miles W. of the harbour of Ribadeo. In case of urgent necessity, a vessel might find shelter behind these rocks in winds from the N.W. round by S. to E., riding over a bottom of mud, holding well in 4 fathoms water.

FAREWELL, CAPE, Greenland, the S. point, called also Statenhuk, is in lat. 59° 49' 12" N., long. 43° 53' 40" W. In some charts this name has been applied to an island 45 leagues W.N.W. from the above point.

FARILHOEN DA VELHA, coast of Portugal, near the Great Berling, on its N.E. side, this last being in lat. 39° 25' N., long. 9° 30' W.

FARILHOEN ROCK, THE GREAT, Portugal coast, 5 miles N. from the Great Berling, a broad round, ragged rock, nearly as high as the Berlings, with a number of rocks about it both under and above water. Half a mile from the Great Farilhoen is the N.E. rock of the same name. There are sunken rocks on the W. side, on which the sea breaks at low water.

FARILLON, or FARELLON, a Spanish term applied to rocky islets generally, often occurring in the description of similar islets or rocks upon the coast of Spain.

FARINA, CAPE AND PORT, Africa, the N.W point of the Gulf of Tunis, distant from Cape Carthage N. by W. 19 miles. Lat. 37° 10' 55" N., long. 10° 14' 30" E. At 2 miles E. of the point is a small low island called Piana; it may be approached on every side to the depth of 4 fathoms, and in the channel between that and the cape, are 7, 9, and 10 fathoms. Abreast of the little rocky isle of Pila, N.N.E. of the N.E. point of the cape, there is a spot with only 4 fathoms. In making the cape care must be taken of the Cane rocks, or the Dogs, as well as in approaching the Island of Piana. Port Farina is 4 miles within the cape, a capacious but shallow inlet, not admitting large vessels, but the road before it is equal to Tunis.

FARLEY'S BAY, West Indies, Island of Antigua, in the N.E. part of the island, within a creek called Mercer's creek, formed by several islands, as those of Codrington, Crump, Pelican, and others, besides reefs of rock beyond them, but the channel in is difficult.

FARMER LEDGE, Ireland, rocks so called on the port side upon entering the harbour of Kinsale, and to be carefully avoided, equally with the Bulman rock upon the starboard.

FARN, or FERN ISLANDS, England, Durham co., two groups of islands 17 in number, in the parish of Holy Island. That nearest the shore is called House island, and is 1 mile 63 chains distant from the main, while the remotest is 7 miles distant. House island is noted as the place where the Romish Saint Cuthbert spent his last years. There is a chasm in the rocks here, through which in stormy weather the sea is forced with great violence upwards, and forms a fine *jet d'eau* 60 feet high. The passage between this island and the main is dangerous, from a cluster of rocks that lie in the middle of it. There are two lighthouses on the larger island, bearing E. by S. from Bamborough Castle, distant 2 miles. This, the highest island of the group, on the S.W. steep and cliffy, but sloping to the N.E. It is about 100 paces in diameter. The high lighthouse is about 80 feet from the S.W. cliff, coloured white, and the lantern red. This light revolves, is seen all around the horizon, and shows the full face of the reflector every 30 seconds, the centre of which is 82 feet above the level of high water. The low light stands near the N.W. part of the island, and can only be seen in a N. direction, it bears from the high light N. by W. ¼ W., having also a red lantern and a fixed light 38 feet above high water. The lights in one bearing S. by E., will lead between the Goldstone and Plough Seat, but directly across the Megstone. E.S.E. from this island, are two rocky islets called the Wide Opens or Little Farns, and E. of these are two black rocks called the Scare-crows or Start Cars: these are always above water, and on the S. side are steep-to, with 9 and 7 fathoms close to them. At low water they are all dry, at high water portions here and there only appear. The vicinity of these islands is very dangerous, and many lives and much property have been lost here. This has been occasioned too much by want of attention to the lead. It is recommended, therefore, by the Trinity Board, that the deep sea lead be kept going upon all that part of the coast between Coquet island, and St. Abb's head, and if a vessel be found in less than 30 fathoms, to haul out into that depth of water without delay. Great caution is given to mariners not to involve themselves near, much less among these islands, on any account whatever. See also *Staple's Island*.

FARO DI MESSINA, or STRAIT OF MESSINO, between Italy and Sicily. Upon the Faro point on the N. there is a lighthouse standing in lat. 38° 16' N., long. 15° 41' E. This point was anciently called Cape Pelorus, and is low and sandy, having upon it an old fortified tower, with two martello towers and two batteries besides, the whole commanded by a redoubt called the Telegraph redoubt, on a hill about half a mile W. of the lighthouse. The village of Faro stands S.W. of the lighthouse, small and dirty. From the Faro point the N. coast of Sicily runs W.N.W. about 5½ miles to a point called Arena Bianca, from whence it turns more to the W. for 2 miles to Cape Rasaculmo. This cape is a flat promontory of middling height, having upon its outer point the remains of a strong Saracenic tower, and near it a turret with a telegraph upon it. The Arena Bianca may be known by a fishing village near, called Aqua Ladrone: there is excellent anchorage near here in from 20 to 15 fathoms, stiff mud. From Cape Rasaculmo the land runs W.S.W., 14½ miles in the direction of Milazzo, a town situated at the S. part of a peninsula running out N. nearly 2½ miles, and having a lighthouse at its termination, in lat. 38° 16' N., long. 15° 13' 30" E. Returning E. and entering the strait, the land goes circularly in towards Messina, which bears from the Faro or Pelorus light S.W., distant 7 miles, the shore being a sandy beach, and there being forts, and several towns and villages within the space. Messina, a large ancient and beautiful city, has a lighthouse in lat. 38° 11' 30" N., long. 15° 34' 42" E. (See *Messina*.) The passage through these straits is best made by attending to the currents, which do not seem to be well known. The difficulty is to work through the strait to the N. The different currents in these straits always take the same course at certain times of the tide, and the tides flow regularly and change with the moon, the flood running 5 and the ebb 7 hours. On full and change times, the tides in the stream off the Faro lighthouse flow about 5, and at Messina about half past 7 o'clock. The difference is 2½ hours, which in fine weather is equal to half the flood. When the wind blows hard from the W., the W.N.W., or N.W., or even N.N.W., the tides are greatly affected, as is usual in all similar straits where a tide rises and falls, nor does there seem to be any other obstacles to navigation than might be expected from the form of the land on either side, and those different tendencies of the stream that are thus locally produced, combined with the power of the winds and the tides in counteraction. As is customary on such occasions, advantage is taken here of eddies and counter currents, particularly between the Faro point and Messina. Care is had, in going N., to avoid the stream of the Faro tide, but none save the experienced mariner, who has made himself well acquainted with the set of the current, under every phase of wind and tide, can venture to navigate in a situation where no self-confidence can compensate for the absence of that degree of local knowledge which, though of a confined character in all which is foreign to itself, is in such cases the one thing needful. In passing the Faro of Messina a seaman who is well acquainted with the strait is absolutely necessary, to prevent accident when the breeze might drop, or from the anxiety natural to such positions occasion the mischief it seeks to remedy. Only to make the harbour of Messina is oftentimes by no means an easy task, and should not be ventured upon by a stranger, especially as the wind is never to be depended upon there, though to enter that port itself requires no pilot. The dangers of this strait, the Scylla and Charybdis of the ancients, have long ceased to be apprehended. Captain Smyth describes all that remains of this Sicilian whirlpool as a spot of agitated water, from 70 to 90 fathoms deep, circling in rapid eddies; these appear to be occasioned by the meeting of the harbour and lateral currents with the main current, the last being forced over in this direction by the opposite point of Pezzo. The same distinguished officer states that no danger of moment is to be apprehended from it.

FARO, Portugal, a city in the S. part of that kingdom. It is 12¼ miles from Point Albufera, the coast forming here a circular light. The lat. of St. Antonio of the Hill here, is, according to Captain Owen, 36° 59' 24" N., long. 7° 50' 30" W. This city stands near the shore, at the entrance of the bar that divides the islands, constituting Cape Santa Maria, from the mainland. The channel leads to the E., and afterwards bends a little N. towards the great bar, on which there are 13 feet water, and here vessels generally enter. There is another channel, but it has only 9 feet water. In making for Cape Santa Maria, an inland hill appears at 16 leagues distance, this is called Monte Figo; it bears N.E. from the Cape, and is not like any other elevation of the land about that part. It is 2,000 feet above the level of the sea, in lat. 37° 9' 42" N., long. 7° 42' 12" W. In coming from the W., it is not discovered until a vessel has passed Sagres, and cannot be seen at the distance of Cape St. Vincent.

FARO POINT, Spain, N. coast, the E. entrance to the harbour of Vivero.

FAR-OUT HEAD, N. coast of Scotland, the E. point of Durness Kyle. There are rocks about this head, but the middle of the Kyle is free of all danger, and has 10, 9, and 8 fathoms of water upon it, but is too open to the N.E. for vessels to anchor

there in security. In the N.W. part of the entrance is Garran island, with rocks around it.

FARRAN POINT, W. coast of Africa, an elevated point, with a house upon it, which has sometimes been mistaken for Cape Sierra Leone, although nearly 2 leagues to the E. from that cape; and vessels have thus touched on the Middle ground.

FAS, or FAZ, PEAK OF, N.W. coast of Africa; it serves for a mark for Old Marmora, from which it bears due E. It stands in lat. 34° 6' 3" N., long. 4° 58' 15" W., after Don Juan Leblich.

FASALIS, CAPE, Spain, the S. point of the Bay of Bayona, off which are the rocks called the Lobos reef.

FASANA TOWN, Istria, Gulf of Venice, 11 miles S. of Rovigno, situated in a species of bay. There is good anchorage before this town in 10, 9, and 8 fathoms over a bottom of mud; midway between Fasana and Brioni, the anchorage is not to be recommended. Vessels may sail through the Fasana channel with good water, until they get near the islands Girolamo and Koseda; at the S. part of the passage, a little N. of these islands, there is a bank, called the Rangon, in 3 and 4 fathoms water.

FASOURO RIVER, N. coast of Spain, small and useless to shipping, falling into the Bay of Biscay a few miles W. of Ribadeo.

FASTNET, ROCK, Ireland, on the S. coast, S.W. from Cape Clear; it lies in lat. 51° 24' N., long. 9° 36' 15" W. It stands high and solitary, having at first view somewhat the appearance of a sail. This rock kept in sight to the E. of the S.E. end of Cape Clear island, will lead within half a mile of the entrance of Baltimore harbour. It is nearly 5 miles W. ¼ S. from Cape Clear lighthouse, and rises 98 feet above the level of the sea. The bottom, W., S., and N.E. of it is shoal and rocky. On the N.E. there is a flat rock, a quarter of a mile from the Fastnet, having only 9 feet over it at low water.

FATATENDA FACTORY, Africa, on the River Gambia, S.W. of Fooliconda, about lat. 13° 5' N., and nearly as far up as the river has been ascended.

FATHER ISLAND, South America, the largest of the Remire islands off the coast near the Aprouak river.

FATHER POINT, America, River St. Lawrence, remarkable as the place of residence of most of the pilots of the St. Lawrence. Lat. 43° 33' 30" N., long. 68° 30' 40" W., according to Captain Bayfield.

FATSAH, TOWN OF, Black sea, in the bottom of a bay formed by Capes Jasoun and Eunieh. The sea is very deep here, and the River Balama runs into it, on the W. bank of which is the town of Fatsah.

FATTIKOO, a town inland upon the Gambia.

FAVIGNANA ISLAND, one of the Ægadean islands, off the coast of Sicily. See *Ægadean Islands*.

FAVORITE COVE, America, Nova Scotia, in the West channel, about the middle of Cape Sable island, a haven where small vessels may run in and find anchorage in 2 fathoms, behind a small islet in mid-channel at its entrance, having a passage on either side, but that to the E. is best, having deeper water.

FAYAL. See *Azores*.

FEAR, CAPE, America, United States, North Carolina, lat. 33° 48' N., long. 78° W. The

American observations give lat. 33° 48' N., long. 77° 57' W. One lighthouse of this cape stands near the W. bend of the Frying Pan shoal, 17 miles from the land, on the S.W. end of Smith's island, and is called Bald Head lighthouse. It is painted black, in order to distinguish it from the lighthouse on Federal point, from which it bears S.W. by S. distant 8¾ miles. It stands a mile from the sea, and exhibits a fixed light, at 110 feet. There is a second lighthouse belonging to Cape Fear, erected on what is called the Inlet of Cape Fear, at Federal point, upon the N. side. It is a white tower, with a fixed light at 44 feet above the sea. From this light the bar extends E.S.E. 1 mile. The whole coast of North Carolina has a large proportion of barred ports, with innumerable shoals within, so that even pilots who live on the spot find it difficult to take ships in without accident. The shoals about this Cape and Cape Hatteras require the utmost caution when navigating in their vicinity. The currents on the coast are governed by the wind, and are liable to very sudden changes.

FEATHARD, HARBOUR, Ireland, on the S. coast, a small dry harbour, 10 miles N.N.W. of Saltee island; there are 8 and 9 feet water between the pier heads at high water, spring tides, and 6 or 7 with neaps. It is on the N. side of Ingard point, from which a rocky ledge extends E. about 2 cables' length.

FECAMP, France, N. coast, W. of Dieppe, in the department of the Seine Inferieure, a small port situated on a river of the same name, with a pop. of 7,850. The town may be known from the sea by its standing in a large valley, and having a church upon the edge of the shore, N.E. of the town. The entrance of the harbour lies S.E. by E. and N.W. by W., lat. 49° 46' 5" N., long. 0° 22' 19" E. The entrance is obstructed by a bar of shingle, which dries at low water, and stretches along the coast S.W. of the harbour for above half a mile. There are some rocks, called the Carpenters, a quarter of a mile from Port Fagnet, where a fixed light is placed. The access to this port is easy, at the proper time of the tide, except with gales from the S.W. or W. It is high water at Fecamp at 10ʰ 47ᵐ full and change. Spring tides rise 27 feet, neaps 19. There are two roads before Fecamp, named the Great and Little Road. The Great lies opposite Criquebœuf, 2 miles from the shore, and has 13 fathoms, on clay, mixed with sand, and very good. The Little Road lies over against the W. side of the harbour, and has from 10 to 7 fathoms. A fixed light is shown on a tower upon the Monte de la Vierge, on the left of the entrance to the harbour, 56 feet above the ground, and 427 above the level of high water at equinoctial tides; visible when clear weather 7 leagues. There is also a tide light here, 262 feet from the end of the N. jetty, at the foot of the Monte de la Vierge, 29 feet above high water, and may be seen 3 leagues off; this light is varied by a flash once in three minutes, and is lighted when there is 10 feet water in the entrance to the harbour.

FEDDERWARDER CHANNEL, one of the channels of the River Weser.

FELES BANKS, coast of France, near Granville, and, with the Basses de Taillepied, forming one ridge of dangerous ground. The E. tail of the Félés bank has not more than 1½ fathom upon it at low water. It lies S. ¾ W. 5½ miles from Cape Carteret, and extends thence, N.W. ¼ W., 4 miles to the W. tail, which last lies S.W. ¾ W., 6 miles

from the cape, and has 1½ and 2¾ fathoms upon it.

FELICUDI ISLAND. See *Lipari Islands.*

FELIU, ST., or ST. PHILIOU, BAY and TOWN, Spain, E. coast, sometimes called St. Felice da Guizols, N.N.E. of Cape Tosa, and immediately beyond Bosquet. It is a good bay for summer anchorage, but open to E.S.E. and S. winds, which blow with great strength there in autumn and winter. It is about half a mile wide, and has a good depth of water. The bottom is sand and sea-weed; the place is much visited, being well supplied with necessaries. There are 10 fathoms water between the points at the opening of the bay. The small town of St. Feliu is opposite to the entrance, situated on a plain; once large and well fortified. Its ramparts are now levelled with the ground. The existence of a rock is reported on the W. side, with only 3 or 4 feet of water upon it. The anchorage is good in the midst of the bay, after passing this rock, rather inclining to the E. side, in 7, 8, or 9 fathoms, or even near the E. shore, in 5 fathoms. The Balellas rocky islands lie off the E. shore of St. Feliu, having a clean passage between them and the land.

FELIXTOW LEDGE, connected with the Wadgate, English coast, between Hollesley bay and Harwich, on the N. side of the channel, and stretching out a mile from Felixtow cliff, with 3 and 2 fathoms on its outer parts. The Wadgate is a rocky patch of 3 fathoms, a little S.W. of the Felixtow ledge. Burnthouse cliff just open of Naze cliff, bearing S.W. by W. ½ W., will lead S. of these dangers, which ought not to be neared in less than 4 fathoms.

FELOOPS SHORE, Africa, the coast of a country inhabited by the Feloops or Floops. It lies between Capes St. Mary and Roxo, and N. of the rivers St. Peter and Casamanza, the last river being in lat. 12° 35' 20" N.

FEMEREN ISLAND, Denmark. It is 10 miles long by 5 wide, separated from the coast of Holstein by a narrow channel, called Femeren sound. It is S. of the W. end of Laaland island, opposite to which is the light, on the N.E. side. It stands on the Oldenburg huk, in the vicinity of the dangerous reef of the Ruttgard; 96 feet above the sea-level there is a light placed, visible, when the eye is elevated 10 feet above the water, over the whole island, at the distance of 15 miles, until a vessel gets within 2 or 3 miles of the shore. Within this distance, or any bearing S. 70° W. and S. 15° E., the light is hid by the Hill of Catharienhof. The light revolves, appearing for 10 and disappearing for 20 seconds. This light bears from the light on Giedser Odde, the S. point of Falster, W. 9° N., distant 25 miles, and from that on the S. end of Langeland, S. 33° E., distant 23 miles.

FEMME HARBOUR, America, Newfoundland, between Chapeau Rouge and Cape Ray. It is half a league W. of New harbour, and has above 20 fathoms water. There is an islet at its entrance, with rocks about it.

FENO CAPE, Island of Corsica, S. by E. from Point Ventilegna, distant about 3 miles, forming the N.W. point of the entrance to the Straits of Bonifacio, the coast between that and Point Ventilegna making the bay of this name. Cape Feno is foul, and from thence towards Bonifacio there are many rocks, both above and below water, the course being E. by S., about 2½ miles.

FENO FIORD, coast of Norway, a wide channel, N. of Feye fiord.

FERMOSE HARBOUR, America, Newfoundland, sometimes called FERMOUSE HARBOUR, between Cape Race and St. John's harbour. It has a narrow entrance, but there is no danger in entering. On the N. side is a small cove, where a fishery is carried on, but the anchorage is indifferent. Further in is Admiral's cove, where merchant-ships ride land-locked in 7 and 8 fathoms, and a mile further in still is Vice-Admiral's cove. Large ships anchor on its S. side, in 12 and 15 fathoms, muddy ground.

FERNANDO, ST., RIVER OF, America, Gulf of Mexico, between the mouth of the River Santander and that of the River Bravo.

FEROLLE BAY, North America, Newfoundland, between Cape Ray and the Strait of Belle Isle. It is no more than a little cove, lying to the E. of the point of the same name, is all over flat, and has not more than 2 or 3 fathoms in any part.

FEROLLE, OLD, ISLAND and HARBOUR, America, Newfoundland, between Cape Ray and the Strait of Belle Isle. Dog island is 3 miles E. of Point Ferolle, and Old Ferolle is 5 miles E. of Dog island. The Island of Ferolle lies parallel with the shore, and forms the harbour of Old Ferolle, which is safe and good. The best entrance is at the S.W. end of the island, passing to the S. of a small island at the entrance, which is bold-to.

FEROLLE POINT, America, Newfoundland, between Cape Ray and the Strait of Belle Isle, N.E. by E. from Point Ritch, 22 miles distant. It is joined to the main by a neck of land, which divides the Bay of St. John from New Ferolle bay. Lat. 51° 2′ 24″ N., long. 57° 9′ W.

FERRAJO, PORTO. See *Porto Ferrajo*.

FERRAT, CAPE, or RAS AL MISHAF, N. coast of Africa. It may easily be known by a rock which lies off its point, having a resemblance to the steeple of a church. Lat. 35° 55′ N., long. 23′ W. The shore all about it is steep-to, and there are 20 fathoms within a quarter of a mile of the land. From the cape the land turns W.S.W., 5 miles, to Point Abujah. Between this and Cape Falcon is the Bay of Oran.

FERRATO, CAPE, Island of Sardinia, E. side, N. of Cape Carbonera, the S.E. extremity of the island. The tower of Mount Ferro stands on the summit of this cape.

FERRE, CAPE, West Indies, Island of Martinique, on the N.E. side. It must have a wide berth, on account of the shoals of Vauclain, the outermost of which is distant from Cape Ferré nearly 4 miles. To avoid them, the Carvel or Caravel rock should be brought well open of the land before advancing to the N.

FERREIRA REEF. See *Constante Reef*.

FERRIER SAND, England, on the coast between Foulness and the Humber. It lies in a direction S.W. by S., and joins the dry sands which encumber the entrance to Lynn. On its E. edge there is a black buoy with staff and ball, bearing from the buoy of the Sunk, S.S.W. ½ W., distant 2 miles, and from Hunstanton lighthouse, W. by N., distant 2¾ miles.

FERRO, FEROE, FARO, FAER, or FÆROE-RNE ISLANDS, a group of islands subject to Denmark, nearly due N. of the W. isles of Scotland, and N.W. of the Orkneys about 77 leagues. These islands are 25 in number, and 17 are habitable. They lie between lat. 61° and 63° N., and long. 5° and 8° W. Some of them possess harbours

deeply indented, many are lofty and faced with high precipices. The chief island is Stromoe. The others most noted, beginning on the N., are Fugloe, Wideroe, Kalsoe, Oosteroe, Svinoe, Boroe, Myggenes, Waagoe, Kolter, Kaalsoe, Sandoe, Skuge, Dymon, Suderoe, and Monk. There are numerous rocks and rocky islets besides amid the group. The productions are of little moment. The longitude was formerly reckoned from the Island of Ferro by several European nations.

FERRO, CAPE, Island of Sardinia, on the N.W. part, off which lie the Caprera, Madalena, and other islands or islets, often denominated the Archipelago of the Magdalen islands, between which there are numerous anchorages and passages.

FERRO, CAPE, more correctly, RAS HADEED, and the ISLAND OF FERRO, Africa, on the N. coast. The cape is the N.E. point of the Gulf of Storah; W. from which, distant 32 miles, is the large headland called Ras Sebbah Rous, the Seven Capes, or Cape Bugiaroni, in lat. 37° 7′ N. Cape Ferro is in lat. about 37° 6′ N., long. 7° 11′ 12″ E.

FERRO ISLAND. See *Canary Islands*.

FERRO, MOUNT, Spain, a hill which serves as a seamark to the entrance of Vigo by the N. passages.

FERROL HARBOUR, Spain, N. coast. This harbour is enclosed with hills on all sides, but the channel is clear, and there is no difficulty in sailing into it. The best winds to enter are from the S.W. round by the W. to the N. Any of these will take a vessel S. of Cape Priorino, passing it at the distance of half a mile. If the wind blows from the S.W., a ship may keep nearer to the S. shore; and if the wind be N., tack towards the N. shore, avoiding the ledges which run from St. Philip's and Palma castles. From the round point near, a small reef projects, and off Bispon point there is a rock, therefore the land should have a good berth. The tides run strong in this harbour with equinoctial gales; it is therefore advisable to come in and go out about an hour before either high or low water, in order to head the current. In such gales the tides rise 15 feet, which is 2 more than at ordinary spring tides. Should the wind be unfavourable, the Bay of Carino may be reached, in which there are from 8 to 14 fathoms water, sheltered from the N.W., N., and N.E., but the harbour must be kept open, that with a S.W. wind a vessel may run in, those winds being there exceedingly dangerous. If the wind be adverse or too powerful to make this bay, a ship must run for the Groyn or Corunna. Cape Caitelada is the S. point of the entrance to Ferrol, and the N. point of the entrance to Ares and Betanzas. In making the harbour, either of Ferrol or Corunna, it is requisite to be careful, and to keep off at night. The harbour of Ferrol is one of the best in Europe. It runs 10 miles inland, and is from a quarter to half a mile broad, with a depth to Ferrol for the largest ships, 5 miles from the entrance, and for frigates 2 miles further up. Both the shores are lined with forts, and the arsenal, formed by piers, may be closed with a boom. Here the Spaniards possess the docks and magazines necessary for the use of a large fleet; barracks for 6,000 workmen, and a marine school. The town formerly contained about 8,000 inhabitants. Foreign merchant-vessels and merchandise were never permitted to enter it, and consequently it has never been a place of trade, and has of late years been greatly neglected. The Mole of Ferrol is in lat. 43° 29′ 30″ N., long. 8° 12′ 42″ W.

FERRYLAND HEAD and HARBOUR, America, E. coast of Newfoundland, between Cape Race and St. John's harbour; this harbour is entered between Ferryland head and Bois island, the entrance being little more than half a cable's length wide. The head has two rocks near it, which being passed, and a vessel within Bois island, it becomes wider, with good anchorage. These rocks are called Hare's Ears.

FETEIRA COVE, S. coast of Fayal, between Points Santa Catalina and Guia.

FETEIRA, PUNTA DE LA, in St. Mary's island, one of the Azores.

FETLAR, an islet of Shetland, with a small sand attached, called by the Shetlanders the Tresta. Whales have oftentimes been killed on this sand.

FEYE FIORD, Norway, one of the passages to Bergen, 7 miles N. of the Feye Oosen passage, a wide and extensive channel, having its entrance N. of Holmengraa.

FEYE OOSEN, Norway, W. coast, one of the two great N. passages to Bergen, situated in lat. 60° 44′, running in between Flissa and Feye-Oe, 1¾ of a mile wide, and clear of all danger, except the Klevesk, on the S. side of the entrance, which must have a good berth.

FIARSCAR, Sweden, in Wingoe sound, leading up to Gothenburg, a rocky island on the E. of the entrance.

FICO, POINT and TOWER OF, Italy, E. of Monte Circello. This point must not be approached on account of some rocky banks which extend from it. A square tower stands upon the summit, and about a mile N.E. of it is a low sandy point and a village named St. Felice, where, in a case of necessity, anchorage may be found in 4, 5, 6, or 7 fathoms water, sandy ground.

FIDALLAH POINT, N.W. coast of Africa, lat. 33° 44′ N., long. 7° 23′ 32″ W.

FIDDLER'S REACH, best known as ST. CLEMENT'S REACH, River Thames, lying from N.E. by E. to E., running from Greenhithe and turning round Broad Ness to the southward.

FIDONISI, or SERPENT'S ISLAND, or ULAN ADASSI, an island of the Black sea, off the mouths of the Danube, in lat. 45° 15′ N., long. 30° 11′ E. It bears a revolving light. There are 18 and 19 fathoms of water on approaching this island from the S. and W. Vessels generally find the best course to Odessa to be to the E. of Ulan Adassi, or Fidonisi island.

FIDRA ROCK, between St. Abb's head and the Frith of Forth, on the coast of Haddingtonshire. It is a black rugged rock, having a hole through it, N.W. by W. from Lamb island, 1 mile distant, and 2 miles distant from Craig Leith. All around it the ground is foul and rocky, and a reef called the Bridge, between the rock and the shore, does not even permit a passage for boats, being dry at low water.

FIELDS' VIGIA, said to exist in lat. 37° 31′ N., long. by account, 66° W. Captain Fields was master of the schooner Little Mary, and stated, that bound from Antigua to St. John, New Brunswick, in 1833, on the 1st of April, he tried the temperature of the water at 4 A.M., and found himself in the Gulf stream. At 6 A.M. the water was still warm. At 8 it was very cold, and of a dark, muddy colour. At about a mile to the W.

the sea appeared as if breaking, or rather rolling, over a shoal, and he saw quantities of small fish arise and porpoises. At 10 found the water warm and of the blue ocean colour, and continued thus until 4 P.M. of the 2nd. Lat. of the shoal at noon, corrected, 37° 31′ N., long., by account, 66° W. On making the Bay of Fundy on the 6th, the reckoning was only 18 miles of long. to the W. of the vessel, and Bermuda having been sighted on the passage, the long. of the shoal, Captain Fields considered, could not be far from the truth.

FIENO CAPE, Island of Corsica, 5 miles distant from Cape Sanguinario, the course to which is S. by E.

FIERRO POINT, America, the S. point of Atchafalaya bay, on the mainland of Louisiana.

FIFE NESS, Scotland, the N.E. entrance to the Frith of Forth. This headland lies in lat. 56° 17′ N., long. 2° 35′ W. High water full and change 11ʰ 10ᵐ. It is steep-to on its S. side; but W.S.W. from it is a high stone called Kilmeny craig, which forms a very remarkable object. The shore is foul a full cable's length distance, and N.E. ½ E. 1 mile from Fife ness is a dangerous ledge of rocks called the North Carr, stretching out in that direction for a full mile. It dries at the last quarter ebb, and the outer rock appears of the size of a boat, having 12 fathoms close to it, with a little reef running out towards the N. To clear the North Carr southwards, the Kilmeny craig should be kept in sight and open of the land. To go E. of it, Trarapene law, a hill on the S. side, should be kept its apparent breadth E. of the Bass rock. In the night it must not be approached nearer than 16 or 15 fathoms. This rock bears from the Bell rock S.W. by W. ¾ W. distant 10¼ miles, and the Isle of May lighthouses in one will clear the N. Carrs also N.N.E. 6½ miles. A beacon, 25 feet high, is erected on the North Carr.

FIGALO CAPE, N. coast of Africa, W. of Cape Falcon, and the E. point of the Gulf of Tremezen, the W. being formed by the Islands of Karaka.

FIGARI CAPE, Island of Sardinia, the S. point of the Gulf of Congianus.

FIGARONI ISLANDS, Island of Corsica, called also the FINOCCHIAROLA ISLANDS, distant 17 miles from Bastia, the coast rounding N.N.E. ½ N. nearly 7 miles to Cape Sagro, and then 10 N. by W. to the Figaroni islands. They are three in number, and to the S. very near the islands is Figaroni road, where is anchorage in 6, 7, and 8 fathoms, sheltered from N., N.W., and W. winds.

FIGTREE BAY, West Indies, Island of Jamaica, towards the N.E. part of the island, a little E. of N. from Kingston on the opposite side of the island, and between Palmeto point on the E., and Dry river on the W.

FIGUARI, PORT and POINT, Island of Corsica; both are between the rocks called the Monks and Cape Feno, the most south-western cape of the island. Both are unimportant. The Briccia rocks lie about 1½ mile somewhat northerly of Figuari point.

FILES KIRT, S. coast of Ireland, three islets so called, to the S. of Swiny head, on the W. side of the entrance to the harbour of Waterford.

FILEY BRIG, and BAY, England, Yorkshire co. The brig is a bold, rocky promontory 8¼ miles from Flamborough head, advancing into the sea, forming a sort of hook. Behind, or to the S. of this hook, small coasting vessels sometimes ride, sheltered from the N.W., but open to all other winds. Filey bay is clean, and shoals gradually up to the beach. It is high water here, full and change, at 4ʰ 20ᵐ. Springs rise 18, neaps 10 feet.

FILLETTES, France, W. coast; these are rocks so called in the mid-channel of the Bay of Brest. They dry 6 feet at low water, and with the Goudran always covered, and the Mingan never, but at spring tides, renders it necessary for vessels sailing into Brest Goulet to coast along on either side of the passage, when there is nothing to fear. The Fillettes bear from Bertheaume castle S.E. by S. 4 miles distant, and from Point Capucins N. ¼ W., distant three-quarters of a mile. The Mingan lies nearly a mile further up the Goulet, and a beacon upon it, and in a line between these are the Goudron rocky banks.

FIN, PORT and CAPE, Greek Archipelago, on the E. side of the Island of Scio: Cape Fin lies on the S. part of the port, and is sometimes called Fin head, between which and St. George's islet which lies from it S.S.W. ¼ S., to N.N.E. ¼ N., there is a channel 150 yards broad, and 2-5ths of a mile long. There are three separate coves or harbours in this port, one of which is adapted for vessels of any burthen.

FINDHORN, Scotland, a considerable fishing-town of Morayshire, on the E. coast of Scotland. It possesses a small tide-haven, about 4 leagues S.E. by E. from Cromarty.

FINISTER, CAPE, from the Spanish Finisterra, in Gallicia, the most W. headland of Spain, expressed often by the French word "Finisterre," though given in all old English maps with the abridgement of the *ra* as above. Cape Finister is only half a league S. of what is called the Navé of Finister, a high mountain, at the distance of 5½ miles to the S.S.W. from Cape Toriñana. This elevated land has a flat summit, and at one-third of its height above the sea there appears a short point with hummocks upon it, and at its base a high but small island. Vessels in N.E. and E. winds may anchor in the bay between Cape Toriñana and the Navé of Finister, off a little stream in from 6 to 8 fathoms, sandy bottom; the ground is foul in the deeper water. Cape Finister is easily recognised from seaward by the bight between the cape and the Navé, with a low beach, and the land less elevated behind it, resembling no other points on the coast near. The lat. of Cape Finister is 42° 54′ N., long. 9° 20′ W. The French mode of spelling has been adopted in the Admiralty charts, and may therefore be considered as sanctioned.

FINISTERRE, CAPE, France, department of Finisterre, the most W. land of Britany, nearest on the mainland to the Lizard point, on the English side of the channel. This point, though laid down in many French maps, is always omitted in the English.

FINISTIRES, or FINISTERRES ROCKS, France, on the W. coast, near Conquet haven, to be avoided by running close along shore until the haven is entered. Point Conquet, at the entrance, is 2 miles from St. Matthew's point, at the entrance of Brest roads. Between the Finistires and the rocks called the Monks there is good anchorage.

FINLAND, GULF OF, Baltic sea, stretching E. as far as St. Petersburgh. Its N. boundary commences at Hango head, and its S. limit at Dagerort, about 19 leagues apart. In some places, in the E. part, it is full 20 leagues across, but approaching the channel leading to St. Petersburgh it becomes narrow; its length, from Hango head to St. Petersburgh, is computed at 73 leagues. Numerous islands and shoals are scattered about this gulf, which hitherto have made its navigation hazardous; but these are now rendered safe and conspicuous by lighthouses, beacons, and flags. The latter are commonly either white or red, and are supported by high wooden crosses; the greatest attention is paid to keep them in repair, and galliots are constantly employed to superintend them during the time the gulf is navigable. The white flags are placed on the N. end of the respective shoals, the red flags are on their S. extremities, and the lights continue to burn until the winter sets in, which happens generally at the latter end of October, or the beginning of November. Thus the Port of Riga (in the Baltic) is commonly closed with ice in October or November, and opened again in March or April. The ports to the E. close in November or December, and open in February or March. Narva, Petersburgh, Wyborg, and Frederick's haven close in October, or early in November, and open again in April, except Petersburgh, which is never free of ice until May. The whole of the N. and S. shores of Finland are included in the Russian territory, and the S. coast is under the governments of Revel and Petersburgh, the N. coast, under that of Carhelia and Nyborg. Worms road, Habsal, Rogerwick, or Port Baltic, Revel, Kolkawick, Paponwick, Monkwick, Kasperwick, Kunda, and Narva harbours, are under the government of Revel.

FINO, PORT and TOWN, or PORTO FINO CAPE, or POINT CHRAPA, Italy, bearing E.S.E. from the Mole of Genoa, distant above 4 leagues, a large high headland, looking round and steep on all sides. It is easily known by some towers, and a square fort on its summit. A little chapel also is a remarkable object here, between two rocks, like an entrenchment behind a breach. It is situated in lat. 44° 18′ N., long. 9° 14′ E. Sailing near the headland, and rounding it, a vessel enters the Gulf of Rapallo, commonly so called, when the town and harbour of Porto Fino are immediately discovered, the last not being seen until a vessel is opposite its mouth. There is a quay here, upon which pillars are placed for the purpose of vessels mooring by them. A vessel may anchor with contrary winds outside the port in 30 fathoms water, bringing the hill upon which the fort is built opposite the little chapel. Porto Fino is at the S.W. entrance of the Gulf of Rapallo.

FINNIS BAY, Scotland, W. isles; E. by N. from Rowal, in Lewis island, and distant 3 miles, is Finnis bay. It is well sheltered, and capable of accommodating at anchor 12 or 13 sail of large vessels. The anchorage is anywhere above the small island a little way up the harbour, beyond the two islands at its entrance. There is a rock half a mile from the head of the harbour, near the E. side, drying at half ebb; it has a perch upon it, and shoals half a cable's length from where it dries at low water.

FINOW CANAL, Prussia, an artificial communication between the rivers Oder and Elbe. It is fit for small vessels alone, as it only admits a draught of 6½ feet. It is 6 leagues long, and has several locks.

FIORENZO, ST., GULF OF, at the N.W. end of the Island of Corsica; it is broad and deep, having several watch-towers on the E. coast, and one on Cape Martello, upon the W. side, near which is anchorage, 3 cables' length from the land, in 12 or 13 fathoms, or in 8 or 9 mooring on shore. Cape Martello must not be too nearly approached, on account of some rocks which lie off from it. Small vessels may anchor before the town of St. Fiorenzo, but there is a rocky ledge here above water, about 250 fathoms from the coast. In entering or sailing out of this gulf, the mid-channel should be kept, or if nearer either shore, rather the W. than the E. The town of St. Fiorenzo stands in an unhealthy position, in lat. 42° 41′ 3″ N., long. 9° 17′ 42″ E. It is well fortified.

FISGARD BAY, South Wales, 3 miles E. of Strumble head. This is a bay in which good anchorage may be found with any wind, except from N. to E.; the ground holds well. Within Anglas point on the W. side, at the distance of three-quarters of a mile, there is a remarkable rock, always above water, called the Cow, with several lesser rocks about it to the extent of a cable's length. The best anchorage is to the S. of the Cow rock, in 4 or 5 fathoms, the rock bearing N. by E. ¼ E., distant a quarter of a mile. A vessel drawing only 10 or 11 feet water can go up to the quay of Fisgard at high water, spring tides, and one drawing 6 or 7 feet at neaps. There is always a sufficient depth of water in this bay for large ships at all times, and it is without rocks or danger. The ground is sand mixed with clay without any stones.

FISHERMAN'S CLIFFS, coast of Africa, near Cintra Bay, the same with the cliffs of Rio Ouro, lat. 23° 26′ N., long. 16° 5′ W.

FISHERMAN'S HARBOUR, North America, on the S. coast of Nova Scotia. It is near Country harbour, and requires great care in entering it between Cape Mocodame and the Black ledge, in order to avoid the Bull rock, which dries at low water, but at high water is covered, the sea only breaking upon it when the weather is bad.

FISHERROUGH, Scotland, one of the small towns between Guillaness and Musselburg, about 5 miles from Edinburgh, and not far from the mouth of the Forth. It lies in a large bay, where there is good anchorage everywhere in 6, 7, and 8 fathoms, and good riding in S.W. or S.E. winds. Fisherrough has a small harbour, where vessels may lie and deliver their cargoes in summer.

FISHER'S ISLAND, America, United States, 5 miles from the coast of New England, and a little E. of S. from the mouth of the River Thames. It is about 8 miles long by 1 wide. Here, too, is Fisher's Island sound, which is considered perfectly safe for a vessel in the hands of those acquainted with it, and to be preferred, if bound E. on the flood, or W. with an ebb tide, to going through the Race, but it is not to be attempted by strangers without a leading wind, and great attention to the lead, though possessing the best directions. This island is in lat. about 41° 13′ N., long. 71° 53′ W.

FISHING ISLANDS, North America, Labrador; N.N.E. ¾ N. distant 3 miles from St. Francis island is the position of the most N. These islands are three in number. The two northern are connected by a beach, which, with the main, forms what is called Fishing Ship harbour, where vessels may ride land-locked, secure in all winds, in from 5 to 14 fathoms water, the entrance being to the S. of the Southern Fishing island; the best passage is between the two western islands, that entrance bearing from Hare island N. by W. There is no danger in this channel, and vessels may sail directly through it, in nearly a N. by W. direction up to the head of the harbour, and anchor in 12 fathoms, with good room to moor. There are other passages in, but none preferable to this.

FISHING RIP SHOAL, America, coast of the United States. Upon this rip or shoal, one of those to the N. and E. of Nantucket, there are from 5 to 7 fathoms water. It is about 12 miles long from N. to S., and very narrow. The N. point is 24 miles from Sancoty head. There is a good channel of 12 to 22 fathoms, uneven bottom, 12 miles wide between this and the Great Rip.

FISKENOSET, or FISKER-NAER, Greenland, an establishment of the Danes, where they send missionaries. It is at the head of Fisker Fiord, on the W. coast of Greenland. Lat. about 63° 30′ W.

FITFUL HEAD, Shetland isles, 2 miles S.S.E. from the Island of Colsa. See *Shetland Isles.*

FITCHER, CAPE, coast of Portugal; the old name of Cape Espichel among mariners. See *Cape Espichel.*

FIUME, called also ST. VEIT, Istria, in the Gulf of Venice, a considerable port, defended by a castle, and having a pop. of 6,000. The W. point of the river is in lat. 45° 19′ N., long. 14° 26′ E. It carries on a considerable trade in corn, wood, tobacco, fruits, and salted provisions from Hungary. There is a mole before the town, but the harbour is not convenient, for being surrounded by high mountains the wind reacts from them, and several islands lying before it, the S.W. wind prevents vessels from getting out; and in winter, when the bora blows hard, it is requisite for ships at anchor to have moorings against that part of the mole where the river passes. The place is only fit for vessels of a small draft of water. The whole coast of Fiume is encumbered with islands, having channels of deep water and good harbours inside them.

FIUMICINO, Italy, the N. channel of the River Tiber, at its entrance, so denominated; this entrance is pointed out by a light on the Clementine tower, 28 miles more to the N. than that on the tower of Cape Anzo. It is in lat. 41° 45′ 49″ N., long. 12° 11′ 30″ E., and is S.E. from Cape Linaro, distant 22 miles. The other entrance is called the Fiumara mouth. See *Tiber.* The light at Fiumicino is visible 11 nautical miles, enlightening the sector comprised between Capes Linaro and Anzo.

FIUMENICA POINT, Italy, Gulf of Tarento, the next point within Point Alice to the N.W.

FIVE FATHOM BANK, America, United States, off the Delaware river. Vessels from the N., or having fallen N. of Cape Henlopen, should be careful not to approach nearer than 12 fathoms until they have got into the lat. of the cape, to avoid this shoal. A light-vessel is moored upon it, having 2 masts, with a lantern on each, in 7¼ fathoms water, Cape May lighthouse bearing W. 20° 30′ N., distant 15¼ miles. The centre of the shoalest ground, on which there is found 12 feet water, bears N. 23° E. from the lighted ship, distant 2¾ miles. It is three-quarters of a mile long, half a mile broad, and very bold at its E. edge, as there are 12 fathoms half a mile E. of the shoal water. Vessels from the N. should not run for the light-ship while bearing from it between N. 14° E. and N. 41° E. There are 5 fathoms water three-quarters of a mile S.E. from the light-vessel.

FIVE HEADS, a vigia, reported to exist in a French chart of 1766, as a rocky shoal, in lat. 44° 10′ N., long. 19° 25′ W. It was given on the authority of M. Van Keulen by Bellin, but there is no other ground to credit the existence of such a danger.

FIVE ISLANDS HARBOUR, West Indies, Island of Antigua, so denominated from the number of Islands which lie off the S. point, but are connected with the point by a sandy reef. The entrance to the harbour is three-quarters of a mile wide, and the water gradually decreases in depth inwards. The N. point of this harbour is named Pelican or Fullerton's point, and has rocks about it.

FIVE ISLANDS INLET, America, United States, in Long island sound, lat. 40° 37′ 10″ N., long. 73° 12′ 35″ W., at the lighthouse. Blunt's "American Pilot" makes no mention of Five Island Inlet, but instead gives Fire Island Inlet, which has a bar with 7 feet water. The light on the island revolving, elevated 89 feet 3 inches above the sea, and bearing N. 77° 35′ E. from Sandy Hook light, 12 leagues distant.

FIVE ROCKS, England; these rocks lie off Hatherwood point, in the Needles passage, on the coast of the Isle of Wight. They are close to the land, out of the way of vessels, and have from 1½ to 3 fathoms water over them.

FIVEL, or DAMSTER DIEP, Holland, a river which joins the Dollart, at the mouth of which is Delfzyl, a strong fortress in the province of Groningen, with a tolerable trade.

FLADDA HUAN ISLAND, Scotland, W. coast, bearing from Ru-Borniskilag, N.E. ½ N., distant 7½ miles, and from Hulim island N. ¼ E., distant 4½ miles. There are rocks off both its N. and S. extremities, some above and some under water.

FLADSTRAND, or FREDERICKSHAVEN, Denmark, a seaport of Jutland, 30 miles N.N.E. of Alborg. It is easily distinguished by the high land of Bangsboe, which lies about a league to the S. On approaching the land the chapel N. of the town is seen, and the tower of the castle close to the shore. The Island of Hirtsholmen lies in front of the town, on which is a quadrangular tower showing a revolving light once in every 1½ minute. The lantern is 43 feet above the level of the sea, and the tower is whitened for a mark. The anchorage for large vessels here is in 5 and 6 fathoms, with the castle W.N.W. or N.W. by W. Small vessels find anchorage within the Hirtsholmen, and in the N. and S. roads. A new harbour too has been made here, and piers built affording shelter within to ships drawing 13 or 14 feet water. A lighthouse 23 feet above the sea level, with a fixed light, stands on the point of the S. pier. The pilots are established on Hirtsholmen island. Fladstrand is defended by several forts.

FLAG BAY, or WHITE FLAG BAY, West Indies, at the N.W. end of the Island of St. Kitts, between Boyer's bay to the N.E. and Fougeaux bay to the S. There are soundings near the shore in 3, 5, and 6 fathoms or more, and above 30 fathoms little more than a mile from land.

FLAMANDE, or FLEMISH BAY, West Indies, near the extreme W. part of the Island of Hayti, on the S. side of the great west peninsula of that island. The Isle Vache or Cow island, is

S. of Flemish bay, and the coast trends on the W. side of the bay to the S.W., and then turns short S. and S.E. to Point Abacou, which with Vache island forms a shelter from the N. and W.

FLAMBOROUGH HEAD, England, E. coast, Yorkshire co., in lat. 54° 7′ N., long. 5″ W. The land is high here, and so continues to Speeton cliffs, bold-to, and without danger. From the outside of this head to Scarborough is 14 miles N.N.W. There is a lighthouse upon this head, 400 yards from the extreme point, close to a bluff point on the S. side of Silex cove, the only landing place near the head. This is a revolving light having 3 faces and each 7 reflectors, and in order that it may be distinguished from Cromer and Tynemouth revolving lights, which show a face every minute, one appears illuminated every two minutes, and of these the colour of one is red; and the light from that face being diminished will not be visible as far as the others in hazy weather. Therefore when only two faces are seen the interval of time will be regularly 2 and 4 minutes alternately, which distinguishes it from any other light. Cromer light bears from Flamborough head light S. ¾ E., and is distant between 29 and 30 leagues: the Spurn head bears S. by W., nearly, distant between 11 and 12 leagues; and the Dudgeon light about S. by E., distant about 21 leagues.

FLAMANVILLE, CAPE, N. coast of France, E. a little northerly from the Island of Guernsey, and the S. limit of Vauville bay, about midway between Cape la Hogue and Cape Carteret, at S. ¼ E. It is sometimes called Gros Nez, the Great Ness or Point. The village is upon the headland. Its lat. is about 49° 32′ N, and long. 1° 48′ W.

FLANNAN ISLANDS, W. Isles of Scotland, W.N.W. ¼ N., 16 miles distant from Gallan head, and sometimes called the Seven Hunters. The largest appears to contain 80 acres, the second about 20; the rest are smaller and uninhabited.

FLATHOLM, Severn river, W. coast of England; it is the most N. of two islands known as the Holms, the second being called the Steepholm. It lies about W.S.W. ½ S. from King's Road, in the estuary of the Severn river, distant 5 leagues. A lighthouse stands upon this Holm, 156 feet high. It is a fixed light, in lat. 51° 22′ 36″ N., long. 3° 7′ W.

FLAT ISLAND, Mediterranean, Island of Minorca, in the harbour of Port Mahon, a little more than a mile from the entrance at Fort St. Philip. There is anchorage on the port side of the harbour at going in, with 14 or 15 fathoms water, the other side of the island is shoaler from 8 to 3 fathoms.

FLAT POINT, Ireland, on the N.E. coast, W. of Copeland island, which last is 2 miles due E. from the main, and open with the Bay of Carrickfergus.

FLAT POINT ISLAND, America, S.W. of Louisburgh, and half a league nearly W. from White point or the E. limits of Gabarrus bay in Cape Breton.

FLAT ROCK POINT, America, coast of Newfoundland, low dark land, the N. point of the harbour of Torbay, the coast running from it northerly to Red head, the S. point, a distance of 2 miles.

FLEES SOUND, Norway, between Stemmeshest and Drontheim, having an opening between the islands directly through to the sea, the passages among which are only to be made by pilots.

FLEET, or LARNE HARBOUR, Ireland, 2 leagues north of Carrickfergus, at the N.E. part of the coast. Coming from the N. it is known by two flat towers, one of which is somewhat higher than the other, seen on the N. side of the bay. At the S. point is a small round island called the Knee, by which it may be known, and ships may sail in along the mid-channel. Vessels coming from the S. as soon as they get a sight of the Knee island, must avoid a rock known by the name of Hunter's rock, on which there are only 2 fathoms at high water. It is 8 leagues from hence N.W., northerly to Fair Foreland, and all the way a bold shore with nothing to fear. The Maidens, dangerous rocks, some of them a little above water, but many more covered, lie N. by E. from Carrickfergus bay, 2½ leagues N.E. from Old Fleet harbour. This harbour will only admit vessels drawing 10 feet water, but it is quite land-locked.

FLEKKER-OE, Norway, S. coast, near Christiansand, forming between itself and the main a roadstead several miles in length, with good anchorage in 8 or 9 fathoms, much visited by shipping. The Haven of Flekker-Oe, at the entrance, is divided by the Island Flekker-Oe into the East and West gats, the latter of which lies 3½ miles to the E. of Hellis-Oe. Upon a small island in the bay is the fortress or castle, which is very remarkable from before the entrance, when not hidden by Flekker-Oe. This harbour is capable of containing a number of ships, made fast by rings on the shore. The depth is from 14 to 18 fathoms; but far off, the bottom is in several places rocky; and, in some parts, apparently clean, the cables will frequently be found damaged. Ships-of-war and other heavy ships should lie to the S. of the castle, where there is some sea, if the wind blow directly in through the opening. As there are two entrances, ships may sail from this place with winds from W.S.W., round to N. and E.S.E. With W. winds they may also go readily from hence, within the ridges, for Christiansand. At the S.W. end of Flekke-Oe is Grundvigkil creek, wherein ships, not drawing more than 10 or 12 feet water, may stand into 3 or 4 fathoms, sandy ground.

FLEMISH BANKS, off the coast of the Low Countries, the W. portion of which is best known by the name of the Dunkirk Banks. These banks are as follow:—the Sandetie bank, Outer Ruytingen, Inner Ruytingen, the Bergues, Dyck, Inner Ratel, Outer Ratel, the Breedt bank, the Smal bank, and the long narrow sand, bounding the Road of Dunkirk to the N., divided into the Snouw, the Braek bank, the Hils bank, and the Traepegeer.

FLEMSDEN, or FLEEMSDEN, Norway, one of the five large islands on the coast N. of Broad sound. It is sometimes called Roof island, and has a hummock at the W. end, which shows like the roof of a house. Within the E. point of this island there is a bay which has a good road, land-locked from all winds.

FLENSBERG, Denmark, a town of Sleswick, on a bay of the Baltic, with a harbour for large vessels. It runs 4 miles inland, and is 18 miles N. of the city of Sleswick, in lat. about 54° 50' N., long. 9° 45' E.

FLEUR DE LYS HARBOUR, America, Newfoundland, between Cape Freels and the Strait of Belle isle. The E. point is in lat. 50° 6' 40" N.,

and long. 56° 10' W. This harbour lies to the E. of Partridge point, distant about a league, and derives its name from three remarkable hillocks, which are over it. It is small, safe, and has excellent anchorage in 4 fathoms water, within its N.E. arm.

FLIE, or FLY ISLAND, coast of Holland, at the entrance of the Zuyder Zee, the nearest island E. of the Texel island, whence it bears E.N.E. a short distance from that point of it called Eyerland. There is no channel to the W. of this island, between it and Eyerland, but to the E. between this and Schelling island is the Stortelmick pass, and to the N.E. is the Westernboom gat, or channel, for ships from the N.W., and to the E. of that is the Easternboom gat, for those from the N.E.

FLINT BAY, West Indies, on the S. side of the Island of Jamaica, N.W. from the most S. point, called the Pitch of Portland, on the W. side of Milk bay, and contiguous to that of Calabash.

FLINT ISLAND, America, coast of Cape Breton island, and N. of Cape Brule, lat. about 46° 8' N. of the Bay of Morienne. The French gave it the name of Island Platte, or Island de Pierre Fusil.

FLOGE CREEK, Norway, about a Norwegian mile from Drontheim. The winds and currents are considered very violent here, and the sea so deep as to forbid anchorage, having 300 fathoms close to the rocks.

FLORIDA, CAPE, America, S. of Cape Canaveral; a narrow island, separated from the main on the N. side by an opening called the White Inlet, or Boca Ratones, is thus denominated.

FLORIDA REEF, West Indies; the W. range of this reef lies S. of the W. extremity of a sand-bank to the W. and N.W. of Cayo Marques, which is in lat. 24° 31' 30" N., and the W. end of the general reef in lat. 24° 20' N., long. 82° 39' W., being the S. part of the whole. At this, the W. end, the reef is about 2½ miles broad, and the least water on it is 5 fathoms, with irregular soundings, to 7 and 8 fathoms. The water over it here is discoloured with white and brown patches of sand and coral rocks, and the bottom is plainly seen. The reef in general is steep, there being from 30 to 20 fathoms within a mile or two of it. There is a channel between the W. part of the Florida reef and the Tortugas, within 2 or 3 leagues of them, through which there is in the day-time a safe passage. There seems little doubt but that the natural causes which formed the Florida reef formed the Tortugas, the last being but an interrupted continuation of them. Following these reefs W. and N.W., next comes the Boca Grande E. of the Cayo Marques, 2 miles in breadth, with a channel running N. and S., in which the least water is 9 feet. Next is W. Kay, or Cayo Hueso, called also Thompson's island, to the E. of Kay Boca Grande, with mangrove islets about it, and white sandy beaches. Kay West, 5¼ miles in length, inhabited, is next Cayo Canaletto, having a harbour, visited by turtle-hunters and wreckers. There is a new lighthouse building for a fixed light on the S.W. point of Kay West, 83 feet above the sea. A passage is made this way into the Gulf of Mexico for vessels drawing only 9 or 10 feet of water, by passing through here, for which the United States have made the most convenient arrangements, by placing a light-vessel in lat. 24° 48' N., long. 81° 53' W. The S.W. point of Kay West lies in lat. 24° 32′ N., long. 81° 54′ 30″ W., and 28 leagues N. 13° E. true from the Havanna. Sand Kay, or Porpoise island, lies 2½ leagues S.S.W. ½ W. from the S.W. point of West Kay, or Cayo Hueso, in lat. 24° 25′ 15″ N. Here also there is a lighthouse, and a fixed light, elevated 83 feet above the sea. The deepest water within the reef here is 8 fathoms. At the E. end of this Kay, there is a canoe opening, called Boca Chica, but it leads only among mangrove islands. Next E. are the Cayos Samboes, about 6 miles from the Boca Chica. The islands turn now to the N. of E., and next come the Pine islands, so called from being covered with pine trees. Here is Looe Kay, so called from the loss of the British ship-of-war the Looe, on a small sandy island. It has now a white tower upon it, with a black pole and large ball, in lat. 24° 33′ N., long. 81° 23′ W. Bahia Honda, the Cayos Vacas, and Cayo Sombrero, nearly due S. of the mainland of Florida, or Cape Sable, are situated where the reef begins to sweep round more decidedly to the N. After the Vacas Kays, Viper Kay, or Cayo Bivoras, and Old Matacumbe follow in order, the last 3½ miles long, in a N.E. direction. The N. end is in lat. 24° 51′ N., long. 80° 42′ W. Indian Kay lies to the W. of Matacumbe. The whole reef from Cayo Sombrero is for the most part very broken ground, as far as the W. end of Matacumbe. There are also dangerous coral rocks in the channel between the reef and the S.W. part of the Cayos de Vacas, the largest of which, about 3½ miles N.E. ½ N. from Cayo Sombrero, and 1½ mile from the Cayos de Vacas, has only 4 feet water. From the Cayos de Vacas shoals, the channel continues to be 2 and 2½ miles broad, to the E. of Matacumbe; 4 fathoms being the deepest water, but 2½ and 3 the general depth along Kay Bivoras, at 2 or 3 miles distance. New Matacumbe lies N.E. of the Old about 3½ miles long, covered with thick tall trees. Next is Tavernier Kay, or Cayo Tabano, lat. 24° 59′ N., long. 80° 30′ W., a small island near a larger lying within it, off which is Cayo Rodrigues, a large mangrove island, in lat. 25° 3′ N. From this Kay, the patches of coral rock increase in number and size, forming double and triple reefs, with small channels between them, which are imperfectly known. Cayo Largo, 26 miles in length, of irregular form, and from 1½ to 6 miles in breadth, forms on its S.E. side a remarkable projection, called Sound point, opposite to which and to the Kay of Rodrigues, is the Great inlet of the Florida reef, the N. side of which is formed by the Carysfort reef. The indraught here makes a near approach very dangerous for large ships, especially with a light or on-shore wind. There is a floating light-vessel moored on the Carysfort reef, bearing E. by S. 7 miles from the highest land on Cayo Largo. Lat. 25° 12′ N., long. 80° 17′ W. From the N. end of this last kay there commences a range of islets and kays that terminate at what is called Cape Florida, although improperly. After Cayo Largo there is an inlet called Angel Fish creek, having Jenning's kay on the N. Angel Fish creek lies 6 leagues N. by E. ½ E. from Sound point. Jenning's kay is 2½ miles in length, and has 1½ mile to the N.E., the S.W. end of Elliot's kay, and between them an islet called Black Cesar's creek. N. of Elliot's island is a small island called Las Tetas, or the Paps, but Pownal kay by the English. The inlet between has the name of Saunder's cut, where a vessel of 4 feet draught may sail into a wide sound between these kays and the watering-places on the main. To the N. and N.E. of the Paps are seven rocks just above water called the Soldier's kays, or Mascaras. Next come the kays Pollock, Knox, Paradisos, and Lawrence. Oswald kays succeed, inaccessible to all but boats, and then follows Biscayne kay, or Cayo Biscayno, 5 miles beyond Oswald kays, and 5 leagues N.N.E. from the N. end of Elliot's kay. The edge of the bank on which these kays lie is very irregular, trending N.E. by N., next N. and N.N.E. Kay Biscayne has its S. end in lat. 25° 41′ N., long. 80° 3′ W. There is a light on this kay at 70 feet above the level of the sea, lighted in 1847. The Florida reef extends in a direction N.E. ½ E., and N. ½ E. from Carysfort reef to Fowey rocks, the N. extremity off which is in lat. 25° 43′ N, the S. extremity of Carysfort reef is in lat. 25° 9′ N.; and it is upon this reef that almost every vessel cast away meets her fate. It was so named from the British ship-of-war Carysfort having been run aground there in 1770. In lat. 25° 35′ N. the Fowey ship-of-war beat over the reef into 3 fathoms water, having lost all her anchors, and when within the reef drifted 5 leagues to the N., and was in danger of drifting out into the stream. The Fowey rocks lie at the N. end of the reef, and are partly dry. The E. edge of these rocks lies about 4½ miles from the E. of Kay Biscayne. They have many bad bars within them. Kay Biscayne has a bank lying off its E. side. There are openings and outlets all over the reef, which are safe communications between the Hawke channel and Florida stream, having 18 feet water. By having a boat placed at any of these inlets it will point them out in such a mode as that they may be entered safely for shelter or for water. Great inlet is in lat. 25° 15′ N., long. 80° 11′ W. It has a knoll of dry rocks on the S.E. point of the reef on the edge of the channel, by which it may be known easily. Spencer's inlet, lat. 24° 50′ N., is opposite Old Matacumbe, and above 6 miles wide. There are no visible marks; the eye and chart must guide here. The Martyr's islands, lying S. of Cape Sable, are divided into the High and Low, or High and Drowned islands. The High are grounded upon rocks grey, white, or black in colour. The Low, or Mangrove islands, are upon coral rocks covered with a rich wet soil. There is a passage from Biscayne kay, on the W. side, being the N. entrance between the Florida kays on the W. and N. side and the Florida reefs on the E. and S. side, called the Hawke channel. The coast for 4 or 5 leagues N. of the kay has a formidable appearance, from the breaking of the sea, but it has nowhere less than 3 fathoms until abreast of the S. end of the kay, where there are only 11 feet upon a bank, which should have a good berth. In this way, by making a careful ocular observation, and the chart in hand, the passage may be made S. and W. round to Cayo Marques, within the outer reef; but to return, to the N. course of the reef, from the Fowey rocks, at the N. end of the general reef. Cape Florida, already mentioned, is formed by what is called Narrow island, separated from the main on the N. side by an opening called the White inlet, or Boca Ratones, admitting only small craft. 18 miles more N. is the entrance of Middle river, which is shallow. 10 miles more N. is New inlet. 6 miles N. of New inlet is Rio Seco, or Dry River inlet, across which is a narrow sand-bar. 8 leagues more N., in lat. 26° 49', is Grenville inlet, with only 5 feet water. It is known by a high mount of sand and rocks, called Cooper's hill, to the S. of it, as well as by the Rocky spring, a high ledge of rocks, out of which a stream of fresh water flows to the sea. It is S. of the mount, in lat. 26° 43', and is 6 miles S. of the inlet. The coast here abounds in cabbage trees, sea grapes, and cocoa plants. Venison and other game are obtainable. Here the gulf stream approaches very near the beach, and the colour of the water changes approaching the coast. 3 leagues N. of Grenville inlet, on the beach, are several high black rocks, and to the N.W. a remarkable hill, with numerous white spots upon it, called the Bald mount, or Bleach yard. There is a small reef here half a mile from the shore, just under water. At the back of the sand-hills, in a meadow a mile S. of the rocks, are two wells of excellent water. The coast is bold-to, with regular soundings. Hillsborough inlet, in lat. 26° 23' N., Admiralty chart, has a shifting bar, sometimes dry, at others with 8 or 10 feet water. A vessel must be moored here, as the tide runs strong, rising 5 feet. The fishing here is excellent, and oysters abound. 8 leagues N.W. ¾ N. from Hillsborough inlet is a range of small hills called the Tortolas or Hummocks, off which are some sunken rocks. The coast here bends gradually to the N., terminating its direction at Cape Canaveral, the shore for 6 miles N. of Hillsborough inlet being flat, and covered with palm trees. This cape is in lat. 28° 28' N., long. 80° 32' W., surrounded by shoals, the outermost breakers being 2 miles from the cape. Vessels must here take a very large offing, or incur great danger. After Dry inlet, or Rio Seco, the bank takes a N. by W. ¼ W. direction, increasing in breadth to the N., so that, in lat. 29°, it extends nearly 20 leagues from the shore. Then it changes to N. ¼ W., and again N. increasing in breadth. In lat. 30° N. it extends 22 leagues from the coast, from whence it bends N.W., until off Cape Canaveral, it does not extend more than 9 leagues from the land, and 2 miles only from the Cape shoals, whence it takes a direction more W. The depths on it are various, from 3 to 90 fathoms, and it is the same with the quality of the ground. Cape Canaveral is N. of the N. opening of the Gulf of Florida, on the Bahama side, as the Matanilla bank, the N. of the Bahama banks, between which runs the gulf stream, is on a parallel with the Tortolas, on the side of Florida. The Matanilla reef or bank, the N.W. end, is in lat. 27° 25' N., long. 79° 9' W., according to Capt. Barnett, 1846, the most W. part, 79° 13' W. The singular conformation of the Florida banks and reefs on the W., and of the Bahamas on the E., yet requires deep study, in order to explain their nature, intricacies, and peculiar phenomena, considered in relation to the wonderful current which flows between them out of the Mexican gulf. While some imagine the work of abrasion of the shores to be constantly going on upon both sides of the gulf, it is possible that the entire tendency may be the reverse, as coral is so plentifully dispersed there on all sides, and is generally rather a mark of the increase of land in the ocean depths than the reverse: this, however, is a difficult and complicated question, although of deep interest to the mariner, and would require a long period of time and a close watching of the increase or decrease of the ground over a vast superficial extent before it could be known. In a few generations more of the human race, the world will perhaps be made acquainted with the real state of the fact, by the recorded labours of hydrographers, and the general increase or decrease of the land which is at present above the ocean level. The productive hand of nature is equally active with its destructive power, and the Florida stream may be as active an agent for the one as the other. The rivers in the Gulf of Mexico extend the land at their mouths, bringing down whole forests and embedding a good part of them in the ocean, thus laying the foundations of new land. The operations of nature, however, are slow. The observations of man, confined to a brief space of time, cannot be effective in deciding upon changes that thousands of years may be occupied in completing. As the case stands, the Florida gulf and stream are among the most curious and surprising objects connected with the ocean.

FLORIDA STRAIT, America, the space or strait included between the meridian of the Dry Tortugas and the parallel of Cape Canaveral. This is easily discovered from the inspection of a map or chart to be the course of a stream confined between land on both sides, with an accelerated current in consequence, flowing for the most part to the N. In fact, the gulf stream, the current which thus sweeps round to the N. and N.E. from the Gulf of Florida, it is of the utmost importance for mariners to comprehend in all its bearings, because all who navigate the coasts of North America must experience more or less of its influence. This current or ocean-stream flows first to the E.N.E. as far as the meridian of the western part of the Double-Headed Shot kays, (on the Elbow kay is a fixed light, in lat. 23° 56' N., and long. 80° 27' W., 160 feet high,) by which kays the stream is diverted from E.N.E. to N. by E., the direction which it pursues on the parallel of Cape Florida; from thence to Cape Canaveral it runs N., a little inclining to the E. This is spoken of the main current, for small branches run off here and there in other directions, setting E.S.E. and S.E. from the Tortugas soundings (see *Gulf Stream.*) The main current then runs as stated above, confined on one side by the continent on which is the peninsula of Florida to Cape Sable, and next externally by the reefs, islets, and channels, which they enclose upon the Florida side, and on the other side of the strait or channel by the Bahama banks, islands, and channels, beginning at the Matanilla bank, on the N.W. and Square Handkerchief on the S.E.

FLORIDA STREAM. See *Gulf Stream.*

FLOTA, Scotland, one of the Orkneys. See *Orkney Isles.*

FLOTTE, PORT LA, France, 2 miles S.E. of St. Martin's, Isle of Ré, Bay of Biscay. A lighthouse has been constructed here at the extremity of the New Mole, visible 3 miles.

FLOWACK ISLANDS, coast of Norway, between the N. point of Stadland and Broad sound. Four leagues from the N. point, within these islands, is the island Swynoe, N.N.E.

FLOWER'S POINT, America, coast of Newfoundland, about 1½ league from the N. head of Catalina bay; 3½ miles from the point, at N. by W., is Bird island.

FLUSHING, Holland, Island of Walcheren, in Zealand, a town having a secure harbour, 8 miles S. of Middleburg, in lat. 51° 26' N., long. 3° 34' E. It is fortified, and defends the passage of the Scheldt and all the islands of Zealand. The port lies between two moles, raised to break off the force of the sea, and is entered by two canals, which form two basins in the town, into which loaded vessels may sail; they are dry at low water; but there is a basin for men-of-war, where they are kept afloat. There are several channels by which ships may sail in and out of Flushing, as the Bodkill, the Deurloo, and the Spleet. The time of high water on full and change is 1¼ᵇ. Flushing is 7 miles S.E. from West Cappel. Flushing may be known by its lofty spire and large square-built Stadthouse. Between Walcheren and Cadsand island is the entrance to the Hondt or West Scheldt, the principal branch of which runs up to Antwerp. The breadth of the river opposite Flushing is 2½ miles; but the sand-banks are so numerous that there is no navigation in safety without a pilot. There is a fixed light at Flushing, upon a wooden elevation, on the W. harbour bulwark, 49 feet above high-water mark, visible 10 or 12 miles, and seen around the horizon from E.S.E. through S. to N. by W. At Terneuse, on the opposite side of the river, 10 miles from Flushing, there is also a light established on the W. harbour dyke. The continental and proper name of this town is Vlissingen. Pop. about 10,000.

FLUSHING, Cornwall, a small town on that branch of Falmouth harbour which goes up to Penryn. It stands on the E. side of the harbour, and has a commodious quay and a small pier.

FOCEVERDE, FOGLIANO, and FOSSO DI CAPO D'OLMA, towers between Point Astura and Monte Circello, on the Italian coast, which are all more or less sea-marks.

FOCHE POINT and ISLAND, Africa, on the S.W. coast, the W. point of the entrance to the New Calebar and Bonny rivers. It is a steep bluff point, almost perpendicular, 5 leagues from Sombrero bluff, with Rough Corner, 7 miles across, forming the entrance to the above-mentioned rivers. The passage between these points is rendered exceedingly difficult for sailing vessels, by reason of the numerous rocks and shoals in the vicinity. Besides a bank called the Sombrero or Foche bank, there is a shoal of hard sand called the Baleur, connected with the former by a bar across the N. side, and a second on the S., upon which there are only from 2½ to 3 fathoms at low water, with from 3 to 5 between. This point stands upon an island of the same name.

FOGO CAPE, Newfoundland, E. coast, upon the island of that name. Lat. 49° 40' N., long. 54° 1' W. There are some dangerous rocks W. of this cape, called the Rocks of Fogo.

FOGO ISLANDS, America, Newfoundland, between Cape Freels and the Strait of Belle isle, to the N.W. of the Wadham isles. Great Fogo is a large island, 4 leagues long and 3 broad. The Indian isles lie off its S.W. point, and N.E. by N., 4 miles from the body of Great Fogo, are the Little Fogo islands, while numerous shoals and rocks are scattered about the vicinity. Fogo harbour is at the N.W. part of Fogo island, the E. extremity of which is in lat. 49° 44' 20" N., long. 54° 19' W. The S.E. extremity of Cape Fogo is in lat. 49° 39' 30" N., long. 54° 2' 48" W., according to the Admiralty surveys made by Captains Bullock, Bayfield, and their assistants. The Americans give Cape Fogo in lat. 49° 41' N., long. 54° W., according to Blunt.

FOGO ISLANDS, LITTLE, America, New-foundland. These islands lie nearly N.E., distant 4½ miles from Joe Batt's point, in the larger island. There are numerous rocks about them, both above and under water, making the coast near them exceedingly dangerous. A little E. of Fogo is the rock called the North-eastern rock, and in somewhat of the same direction as Cromwell's ledge, distant 10 or 11 miles, the position of this last not being well determined. N. of Little Fogo are the Turr rocks, and from thence in the direction of Great Fogo island there are the Storehouse rocks, the Seals' Nests, Gappy and Stone islands, the Jigger and Black rocks, with various others, all having deep water around them.

FOGO or FUEGO ISLAND. See *Cape Verde Islands.*

FOGO, ILHA DE. See *Cape Verde Islands.*

FOGNAN ROCK, W. coast of Norway, about lat. 63° 7' N., a sunk rock E.N.E. from Quitholman 2 miles. It may be passed on either side without danger.

FOGS. These are common appearances, both on sea and land; in the last case, however, differing very much according to local circumstances, and impregnated oftentimes with substances which the purity of a sea atmosphere forbids in those which ascend from the surface of the ocean. These fogs are at sea identical with clouds. They are minute aqueous particles, floating in the air at a greater or less elevation, in proportion to their own density, or to the rarity of the atmosphere. When at a high comparative elevation, they are considered to be clouds; but when they are low, they are called fogs. Dew and fog are different in their causes. Dew is the condensation of water in the form of vapour, upon a condensing surface, in a state of rest. Fogs and clouds are condensed in the air, from the particles floating there, and being so condensed that the air will no longer support them, they descend in rain. Fogs, or rather mists, which are suspended over "the surface of green fields oftentimes are *stratus* clouds; fogs which involve elevated objects are *cumulous* clouds." Light, shade, perspective, distance, and reflection, give to clouds their varied appearances. Clouds have been arranged by scientific observers in seven classes. 1. A *Cirrus*, feathery, or resembling a lock of hair. 2. *Cumulus*, or heaped up, conical, or paplike in form. 3. *Stratus*, in horizontal beds or sheets. 4. *Cirro-cumulus*, or small fleecy or rounded clouds. 5. *Cirro-stratus*, or a wavy and undulating stratus. 6. *Cumulo-stratus*, or cumulus and cirro-stratus mixed. 7. *Nimbus*, or a cumulus spreading out into cirrus, and raining underneath. The cirrus is generally the most elevated cloud, putting on the likeness sometimes of a gauze veil, or parallel threads. Its height varies considerably. Clouds are seldom found at a greater elevation than the summits of the highest mountains, and they frequently descend to 600 or 700 feet; when thus low, the earth may have no misty appearance lower. They hang round and obscure the summits of hills of this elevation in moist weather, putting on massy forms not devoid of grandeur from their slow rolling movements. From the evaporating nature of the atmosphere in very high regions, as well as from the extreme cold prevalent there, the clouds are seldom seen in distinct forms at their greatest elevations. Somewhat lower they are often at a temperature below freezing, while the lower atmosphere is above it, hence the phenomenon of snow storms and hail, by reversing this circumstance, for if the clouds be above the freezing point, and the air nearer the earth's surface be below it, there is no difficulty in comprehending how snow and hail arise. To be in a fog at sea is one of the most trying situations of the mariner when he is near land, for it requires the utmost attention to the action of the tides and currents, when the breeze is no more felt, and yet the motion of the vessel is perhaps stealthy towards destruction.

FOKIA HARBOUR, GREAT FOKIA ISLAND, and NEW, or LITTLE FOKIA, Gulf of Smyrna, on the Asiatic coast, and called in old maps Fogia. Great Fokia island is situated in the N. part of the Gulf of Smyrna, on the E. side, and forms with the bend of the coast at Raphia, the harbour of Raphia or Rapphina. In going in, St. George's islet may be passed on either side, the passage being bold between that and Great Fokia island, and having between 17 and 20 fathoms. There is a shoal on the E. side of the harbour. Between St. George's and Middle island there is a good passage, a quarter of a mile broad, having in it 12, 13, 14, and 16 fathoms. St. George's island is oval in form, and somewhat high. Great Fokia island is larger and higher. Fokia head here is a peninsula, a black cliff appearing like an island joined to the main by a beach, to the S.E. of which is a kind of bay, where there is anchorage with N.E. winds in 14 and 16 fathoms. New Fokia harbour lies between St. George's and Middle island, 2 miles from Fokia head, which bears from St. George's island S.S.E. ¼ E., distant 1¼ mile, having from 25 to 28 fathoms. This harbour is sometimes called Follery harbour. New or Little Fokia is 2 miles from Fokia head, a small safe harbour, 1 mile long and ⅛ of a mile broad.

FOKIA, or FOGIA, OLD, Anatolia, on the S. side, and on the W. shore of a peninsula in the Gulf of Tchanderli or Sandarlie, S.E. of Port Olivier in Mytelin island, also nearly due S. of Arguinoussi island.

FOLDEN, GULF OF, coast of Norway, on the Island of Praestoe, in this gulf, and in the province of Drontheim; there is here a fixed light elevated 37 feet above the level of the sea, visible 10 miles, lat. 64° 47' N., long. 11° 8' E., and lit between the 15th of August and 30th of April. Vessels are warned not to stand so much to the E. as to lose sight of this light.

FOLFOLA, or FILFOLA, ROCK, S. coast of the Island of Malta, in lat. 35° 46' 30" N., long. 14° 28' 30" E.

FOLIART, LOCH, Scotland, W. coast, round Dunvegan head, a fine bay for vessels to shelter in, having in it no rocks or shoals that are not at a distance from the shore. In the wide part of the loch the water is too deep for any anchorage, there being 50 fathoms water within 2 cables' length of the shore, but at the head of the loch in Dunvegan harbour, or within the small isles on its W. side, there is good anchorage for ships of any size.

FOLKSTONE, a town of Kent; it is a member of the Port of Dover, and has a harbour, from whence travellers make the Port of Boulogne, after a railroad journey from London. It is said to have had formerly five churches, of which only one is left, which serves as a sea-mark, and stands in lat. 51° 4' 45" N., long. 1° 11' 6" E. The pop. in 1831 was 3,638, including the entire parish.

FOND LE GRANGE BAY, Hayti island, West Indies, 4 miles W. from the Bay of Chouchou. It is 600 fathoms broad, and the W. point is named Palmista, distinguished by a chain of reefs, extending nearly a league to the W., almost

to Port d'Icaque. This a good roadstead, and in case of necessity a vessel of the line might ride in it, for throughout it there are not less than 6 fathoms water, within about a cable's length from the shore. The E. point should be approached on entering, and a vessel should anchor in the middle of the bay.

FONTANE, CAPE, Black sea, 6 leagues N.E. from the mouth of the Dniester, and 6¼ miles to the S. of Odessa lighthouse, stands in lat. 46° 22' N., long. 30° 44' E., elevated 203 feet.

FONTARABIA, Spain, on the N. coast, at the bottom of the Bay of Biscay, and only separated from France by the River Bidassoa. There is a fixed harbour light on Cape Figuera. It is a fortified town, opposite to the village of Andaye, on the French side of the river, the mouth of which is very shallow, and only admits of the entrance of small vessels. Lat. 43° 23' N., and long. 1° 46' W.

FOOLICONDA, Africa, on the Gambia river, the name of several native towns, and of one situated opposite the Devil's point, E. of the Badiboo river. There are 4 and 5 fathoms water in the channel opposite the town.

FOORD, GREAT and LITTLE, Lancaster co., England, two banks at the entrance of the River Wyre so named; they are both buoyed. The Little foord is never seen above water, even at the lowest tides.

FOOTABAR, or TAMARA, ISLAND. See Isles de Los.

FORA, Denmark, a small island in the jurisdiction, and on the coast of Sleswick.

FORBISHER'S, or FROBISHER'S, STRAITS, so called from their discoverer, in 1578, upon the E. side of what are now generally called Davis's straits, in lat. 62° 30' N., nearly E. from Resolution islands, at the mouth of Hudson's straits or passage into Hudson's bay.

FORCADOS, N.W. coast of Spain, a cluster of rocks about 6¾ miles from Cape Cé, and 2 from Caldebarcos. There is only a channel for boats between them and the land.

FORCAS, CAPE DE TRES, Africa, N. coast, lat. 35° 28' N., long. 2° 56' W. It is the extremity of a large promontory, and has a tower on its summit. The ground is foul near the cape, and there are several rocky islets about it. A large ship may round this cape at the distance of a gunshot with safety. N. ¾ W. from this cape, distant 10 leagues, is the little Island of Alboran.

FOREECAREAH RIVER, Africa, W. coast, lat. 9° 16' N., long. 13° 26' W., a river in the country of the Soozees, or the Mandingos, between the Isles de Los and Sierra Leone, N. of the River Mellacoree, or, as some charts express it, Malacary.

FORELAND, BLUFF, Africa, S.W. coast, 2 leagues from the River Cousa, with a small reef at its foot.

FORELAND, FAIR, Ireland, N. coast, opposite Rachlin island, on the main, best known as Fair head.

FORELAND LEDGE AND POINT, N. coast of Devonshire, a rocky shelf extending 2 miles E. and W., about a mile distant from a point called the Foreland, near Lynmouth, and abreast of it. There are only 19 feet on its S.E. part, but the general depth is 4¼ fathoms. Between this ledge and the Foreland there is a depth of 7 fathoms, and close to it on the outside there are from 10 to 14 fathoms. This point is a little E., southerly of Morte point. It is high bold land off the outer point, but on the W. side the shore is shallow. Lat. 51° 12′ N., long. 4° 12′ W.

FORELAND, NORTH, Isle of Thanet, Kent co., the N.E. limit of Kent, a noted promontory, and the most S. part of the port of London, the boundary extending across the mouth of the Thames to the point called the Naze on the Essex coast. All the towns and harbours within these limits on both shores are subject to the jurisdiction of the metropolitan port. This noted headland, according to Lieutenant Raper, stands in lat. 51° 22′ N., long. 1° 27′ E. In the trigonometrical survey, the lighthouse is given as in lat. 51° 22′ 31″ N., long. 1° 26′ 47″ E., the lighthouse being elevated 184 feet above the sea level, and carrying a fixed light. It is high water here at 11ʰ 15ᵐ, the rise 17 feet at full and change. Off this promontory in hard gales of wind the tide will often set almost round the compass. Commonly the first of the flood sets S. by E. or nearer to the S.; the middle of it a little to the W. of S., and the last to N.W. ½ N. The ebb commonly sets first to the N. by E. ½ E., the middle to E. by S. ½ S., and the last nearly towards the S.S.E. A headland thus situated is with its light a remarkable and serviceable mark, amid the anchorages, shoals, sands, currents, and various dangers within sight of which it rises out of the channel waters.

FORELAND POINT, the same with OLD HARRY POINT, on the coast of Hampshire, about 1½ mile from Poole bar, this last being N. by E. from it.

FORELAND, SOUTH, Kent co., S. a little W. of the North Foreland, in lat. (on the same authorities as those given for the North Foreland) 51° 8′ N., and long. 1° 22′ E., and at the high lighthouse 51° 8′ 24″ N., long. 1° 22′ 22″ E. There are two lighthouses on the South Foreland, at an elevation of 372 and 275 feet. Both lights are fixed, and when in one, W. by N., lead clear into the North sea, S. of the Goodwin sand. From the South Foreland to Calais, the distance is 7 leagues S.E.; to Dieppe, 25 leagues S. ¼ W.; and 6½ leagues W. by S. ¼ S. to Dungeness. Like the N. Foreland, this headland is a very important direction to vessels at sea, and for those anchoring in the Downs. When a ship is in the channel in 24 fathoms, both Calais cliff and the South Foreland may be seen from her, and Fairlight and this headland may be seen in 26 and 27 fathoms, between Winchelsea and the coast of Picardy. The situation of the South Foreland, too, is a great security to the Downs anchorage, as it breaks off the violence of the sea from the S.E. by S. and S.W., and in those winds renders the Downs a tolerably secure anchorage.

FORIA, Italy, a small town of Ischia island, near its S.W. point, and the entrance of the Bay of Naples.

FORLORN HOPE, South America, a point at the entrance of the harbour of Cayenne, between which and Malingre a vessel should come to an anchor, in order to obtain a pilot, and to be able to cross and avoid the shallows. The bottom is a clayey ground, with from 20 to 25 feet at low water.

FORMAGIER'S BAY, West Indies, a small bay in the Island of Tobago, having anchorage for vessels not exceeding 150 tons.

FORMBY CHANNEL, coast of Cheshire, the best channel into the port of Liverpool. It has from 3 to 5 fathoms at low water. There is here

a light-vessel in lat. 53° 32′ N., long. 3° 10′ W. It carries two fixed lights.

FORMENTARA ISLAND, Mediterranean, one of the Balearic isles.

FORMENTON, CAPE, the N.E. cape of the Island of Majorca, in lat. 39° 57′ N., and long. 3° 15′ E.

FORMICHE ROCKS, Italian coast, off the S.E. point of the Island of Ponza, half a mile from the shore.

FORMICHES, FORMIGUES, or ANTS' ROCKS, on the coast of Italy; 2¾ miles from the Africa rock, which last lies in lat. 42° 21′ 15″ N., long. 10° 8′ 20″ E., is the Africa shoal, on which there are 2½ fathoms water, and near it are 4, 6, and 7 fathoms, with 17, 19, and 28 fathoms between; these once had the above name of "the Ants," and are very dangerous, especially at night, and in day during a calm, lest the current should set vessels upon them. Giglio island bearing S.E. by E. will lead clear between Pianosa and Elba; and Giglio island bearing S.E. ¾ S. will lead clear of the Formiches, and between Palmalola and Cervoli.

FORMICOLA ISLAND, coast of Italy, Lower Calabria, between Belmonte and Amantea, an islet at a little distance from the shore, and close to Panicello, another rocky islet of the same coast.

FORMIGAS, or ANTS, West Indies, Jamaica coast, dangerous coral spots on a sand-bank about 3½ leagues long, and 2½ wide. The N.E. part of this bank bears W. ¼ N. 13 leagues from Navaza, and 16 leagues N.E. by N. from Morant point, Jamaica. Its S.W. extremity bears N.E. by N. ¼ N. 12½ leagues from Morant point. From the N.E. end of Jamaica the body of the shoal bears N.E. ¾ E., distant 43 miles. The E. part is the shoalest, having in some places no more than 14 feet water.

FORMOSA, CAPE, Africa, W. coast, Bight of Benin, in lat. 4° 15′ N., long. 6° 11′ E. Lieutenant Raper characterises this as no distinct cape, the land appearing to sweep round without any remarkable projection. The River Nun enters the sea on the coast to the W., and the River Bento, a branch of the same river, to the E. of the long. given to this cape.

FORMOSA, or WARANG ISLAND, Africa, W. coast, the most N. of the Bissago or Bijooga islands, 6 or 7 leagues long and 3 broad, about 13 leagues from the E. coast of Cape Roxo.

FORNELLES, PORT, or HAVEN OF, Island of Minorca, N.E. side, having water deep enough for vessels of all sizes. It is sheltered from every wind. The anchorage is abreast of the old castle at the W. part of the haven, where the town stands, in from 7 to 11 fathoms water, mooring E. and W.

FORNO, BAY OF, Mediterranean, Island of Gozo, near the middle of the N. coast. It only affords anchorage for small vessels.

FORT BAY, West Indies, Island of Antigua, another name for Deep Bay, on the W. side of Falmouth harbour.

FORTEAU BAY, America, entrance of the River St. Lawrence, 5 or 6 miles to the W. of Wolf's cove. It is about 3 miles broad, and runs in for nearly the same distance. On the W. side near the head of the bay there is good riding in from 10 to 16 fathoms, but exposed to the S. Off the E. point of the bay there is a rock which appears like a shallop under sail, and on the W. side of the bay there is a fall of water, which coming from the E. is easily seen. The point of

IV.

O

this bay, similarly named, is in lat. 51° 25' 36" N., long. 56° 59' W.

FORTH, RIVER and FRITH of, Scotland. This stream, after a course of 40 miles, meets the sea near Stirling, from whence it receives the name of the Frith of Forth, being, in fact, the estuary of this river. The headlands at the entrance from the ocean are on the N., Fifeness, in lat. 56° 17' N., and Dunbar on the S. in lat. 56°; being about 17 miles asunder. At the entrance, on the E. or N.E., is the Isle of May, about 3 miles in circumference, in which there are two lighthouses, in lat. 56° 11' N., long. 2° 33' W., on an elevation of 240 and 110 feet. These lights in one N.N.E. ¾ E. and S.S.W. ¾ W. clear the Carr rocks, and on the S. side, but more to the W., is the Bass rock, 400 feet high, and about a mile in circumference. Within this rock, opposite N. Berwick, the Frith contracts to about 7 miles in width, and further up, near Leith, 22 miles from the extreme headland, it is not more than 5 miles broad, while at Queen's Ferry, yet higher, it narrows to 1 mile, widens again to 3 miles, and at Alloa takes the character of a navigable river. Vessels coming from the North Sea for the Frith of Forth, in about the latitude of 56° 12', or nearly that of May island lighthouse, will first perceive the high land about St. Abb's head lofty and regular; and the Cheviot hills, which, in clear weather, may be seen 24 miles off, will easily be recognized, by their appearing above all other hills to the S. If making towards the coast of Fifeshire, the High Lomonds, Largo Law and Kelly Law, will first appear, making unequal and detached heights, of conical forms, like the tops of sugar-loaves, long before the low land between is visible. If coming from the S. the round hill near Dunbar will appear, making somewhat like the Bass rock. Some navigators have mistaken it for the Bass rock; but the North Berwick Law, seen to the N. of it, will always distinguish it from the Bass. If going within the Bass, between it and the main, the South Carrs must have a good berth; and when passed, keep a moderate distance from the shore, and go either inside or outside of Craig Leith; if the former, keep close to it, and stand out between it and Lamb isle. The depths are various, and the ground near the shore rocky. The mark to go between the Bass rock and the shore is Fidra, between Craig Leith and Lamb island. It is more customary, however, and safer to sail up the Frith outside to the N. of the Bass. In this case, a vessel must steer from St. Abb's head nearly N.W. by N. until the Bass is passed, and then W.N.W. to Inch Keith, on which is a revolving light, 220 feet high. The course to Inch Keith from the midway between the Bass and May island is W. by N. and from the S. end of May island W. ½ N. about 21½ miles. To sail from St. Abb's head to the Bass rock, keep the East Lomond in the Bass rock, and, in the night, keep within the stream of 20 fathoms midway between May island and Fifeness. The course to Inch Keith is W. ½ S. If close in with Fifeness, then a vessel must steer W.S.W. ½ W. 9 miles, or to abreast Elie Ness, and from thence W. towards Inch Keith.

FORT ISLANDS, or OLD FORT ISLANDS, America, a group of islets in Esquimaux bay, on the N. coast of the Gulf of St. Lawrence.

FORT ISLAND, South America, E. coast, an island at the entrance of the River Essequibo.

FORT ROYAL BAY, West Indies, Island of Martinique, on the W. side, thus affording a shelter from the prevalent winds. It is nearly 5¼ miles between Point Negro and Cape Soloman, its W. limits, and it is 7 miles deep, but narrows so towards the middle as to be not more than 2 miles wide. Fort Royal, the capital of the island, is situated on the N. side of the bay, and 1¼ mile E.N.E. of Point Negro. The pop. is about 4,000. At its E. extremity, near the careening place, there is a fine parade called the Savannah, forming the glacis of Fort St. Louis, upon which there is a fixed light on the S.W. part, 130 feet high: which is in lat., by Lieut. Raper, 14° 35' 54" N., long. 61° 4' 12" W. The anchorages in this bay may be found suitable for all circumstances, in one part or another. There is the German anchorage, which is safe and good during the E. winds from November to July. In the next three months, a shelter is found at the Careenage, the Three Islets, or Cohe du Lamentin. The first of the three anchorages has its channel between banks of Madreporic rocks and gravel, which extend on one side 4 cables' length S. of St. Louis Fort. The anchorage of the Three Islets is one of the most important in the whole bay. It is very safe during the rainy season, but contains a number of banks, which occasion a diminution of the space for anchorage. The best is on the middle of a line drawn from the easternmost of three small islands to the top of the Great Islet. In going further S. banks are met with, reaching almost to the Great Islet. The entrance to the third harbour, or Cohe du Lamentin, is at Point Salle. This harbour is a bay, extending 1 6-10th of a mile N.N.W. and S.S.E. Its greatest width perpendicular to its length is 1½ mile wide, and the entrance 7-10th of a mile wide. There are gravel and rocks in this bay, which lessen the space, but there is excellent holding ground. There are also several coves and islets, and, S. of Pigeon island, a roadstead for small vessels, in the W. part of which is good anchorage. Although this harbour is one of the best in the West India islands, it is impeded by numerous shoals, and even rocks, so that a stranger cannot prudently attempt to sail among them.

FORT ROYAL ISLAND, France, a rocky islet, near the harbour of St. Malo, N. by E. from Great Bay Fort; N.E. of this islet there is a cluster of rocks.

FORT ROYAL TOWN, West Indies, Island of Grenada; it lies at the bottom of a spacious bay, capable of containing 25 sail of the line, in perfect security. It is on the S.W. side of the island, and is the seat of government. Lieut. Raper gives the lat. of St. George's Fort at 12° 2' 54" N., long. 61° 48' W.

FORTUNE BAY, America, Newfoundland, between Chapeau Rouge and Cape Ray; Fortune Head is E.N.E. from Dantzick Cove and Point. The entrance to the bay is between Point May and Pass island. This bay is from 22 to 23 leagues deep, and has numerous harbours, bays, and islands, among them Brunet island, the Little Brunets, the Plata islands, and Sogona island. Point May is the S. extremity of Fortune bay, and the S.W. extremity of that part of Newfoundland known by a great black rock nearly joining the pitch of the point, and something higher than the land, which makes it look like a black hummock upon the point. A quarter of a mile from this black rock there are three sunken ones.

FORTUNE HARBOUR, Newfoundland, between Cape Freels and Belle Isle. Fortune harbour is also between the Bay of Exploits and New Bay, but the entrance is extremely narrow and dangerous, on account of the high land about it, from which all winds baffle, except when blowing in.

FORTUNE ISLAND, or ISLAND OF GOOD FORTUNE, on the W. side of Davis's straits, N. America, bounded on the W. by White Bear bay, and on the N.E. by the Cumberland strait and islands.

FORTUNE ISLAND, or LONG KAY, West Indies, windward passage. This island is 10 miles in length, and lies about N.N.E. and S.S.W., and is about 1½ mile in breadth. The S. point is in lat. 22° 32′ N., long. 74° 23′ W.

FOSSAND, Norway, N.E. of Jomfruland, and S.W. of Skeen, an inconsiderable place at the head of a small fiord.

FOSSE ROAD, West Indies, Island of Hayti, at the W. end at the bottom of the bight formed by the trending of the N. coast of the peninsula from thence to the W. of Cape Dame Marie. It lies within Gonave island at E.S.E. nearly, and from St. Mark's bay and cape at S.E.

FOTU MONTÉ, Africa, S.W. coast, a small place between Commenda and the Castle of Minas, or St. George; ships cannot approach it, on account of the reefs and rocks along the coast.

FOUCHEE COVE, Newfoundland, between Cape Freels and the strait of Belle Isle. It is very little visited, and there is no anchorage but at its inner end, where there is a cove on the N. side. On the whole, it is a place almost useless to shipping.

FOUDREY. See *Piel of Fouldrey*.

FOUGEUX BAY, West Indies, at the W. extremity of the Island of St. Kitts, between White Flag bay on the N. and Sandy point, or Beltate bay on the S. There is a rocky and sandy bank running out near a mile from this last point; and some rocks close to the shore on the S. side of the mouth of Fig Tree river, without which is anchorage in 5 fathoms, and more northerly in 10, within a mile of the shore; nearer White Flag bay there are but 5 and 6 fathoms.

FOUL BAY, West Indies, on the E. side of the Island of Barbadoes, to the S. of which is a continued ledge of rocks, of which one part running to the N.E. is called the Cobler's Rocks; these rocks run up in the shape of a horn for 3 or 4 miles, about 2 from the nearest part of the coast.

FOUL ISLAND. See *Orkneys*.

FOULHOLM SAND, England, at the entrance of the River Humber. N.W. by N. distant 3½ miles from the white buoy of the Sunk Spit, begins the dry part of the Foulholm sand. It joins the Paull to the N., and is extensive and dangerous. There is a channel within it leading to Stone Creek, towards the town of Paull or Paghill, only to be used by small coasters. The usual passage is to the W. of these sands. A sandy flat of shallow water lies off the S. part of the Foulholm, upon which there is a white buoy, and, nearly 2 miles beyond that, a chequered buoy, black and white.

FOULNESS, on the coast of Norfolk, 4 miles E. of Cromer. It is sometimes called Cromer light. It stands in lat. 52° 55′ 36″ N., long. 1° 19′ E. It is a revolving two-minute light, 274 feet above the sea level. See *Cromer*.

FOULNESS, or FOWLNESS, ISLAND, England, Essex co., on the S. side of the channel into Burnham Water.

FOULNESS SAND, sometimes called the MAPLIN SAND, England, an extensive flat running off from the N. shore of the Thames from Leigh and Southend eastwards as far as the entrance of the river Crouch. It mostly dries, and is covered at about 1½ flood. Near Shoebury it extends more than a mile from the shore. Off Foulness island it is 4½ miles broad, on which is a fixed light on piles (red) 36 feet high, and a bell is sounded in foggy weather. At Crouch point, the entrance of the river, its breadth is almost 6 miles. Its E. edge is steep, and duly buoyed and marked.

FOULNEY ISLAND, England, Lancashire coast, contiguous to the Pile or Piel of Fouldrey. See *Piel of Fouldrey*.

FOURAS CASTLE, France, within the entrance of the Charente river, from the sea, on the port side, and nearer the sea than Fort Lupin, which is besides upon the starboard bank.

FOUR BANK, France, W. coast, a sand-bank in the way of the entrance to Croisic, 3½ miles W.N.W. of Point Croisac or Crozic. It is full 2¾ miles long, and its S. part bears from the Cardinals S.E. by S. about 8 miles. A portion is dry every tide, and at its S. part there is one spot with only 3 feet water, called the Gouëvas. The Semaphore de la Romaine open to the right of Guerand will lead a vessel directly upon it. Between the bank and Point Croisic there are from 12 to 7 fathoms. A round stone tower stands upon the N. dry rocks of the Foul bank, and exhibits a revolving light 56 feet above the level of the sea, eclipsed every minute and visible 5 leagues off, but the light does not entirely disappear within a distance of 7 or 8 miles. It bears W.N.W. 3½ miles from Point Croisic, and E.S.E. ¼ E. 18 miles from the S.E. end of Belle Isle.

FOURCHU CAPE, America, Nova Scotia, near Yarmouth, the Forked Cape, as named by the English from the island which forms it, having two prongs projected to the S. The inlet between these must not be mistaken for Yarmouth harbour, which lies E. of both. This cape is a remarkable object, rocky, high and barren; bearing from Jebogue head N.N.W. ¾ N., distant 4½ miles, lat. 43° 47′ N., long. 66° 10′ W. The light on this head is revolving in 1m 15s, and 140 feet high.

FOUR PASSAGE, France; this is a passage into Brest harbour, thus called from a large black rock never covered with water named the Four or Oven, hence the Passage du Four. It lies a mile from the N.W. coast, and E. ¾ S. 3½ leagues from Ushant lighthouse, being the chief mark to pass by the Isle of Ushant towards Brest. The course from the Four to St. Matthew's point, at the entrance of Brest, is S.S.W., but no vessel can sail in that direction on account of the numerous rocks which obstruct the passage. St. Matthew's point is readily known by the lighthouse upon it, which exhibits an intermittent or flashing light 177 feet above high water, visible 6 leagues. The flashes succeed each other every half-minute, but the eclipses do not appear total when within the distance of 3 leagues. A good berth must always be given to the Four in passing, to avoid the Boureau bank, a shoal which lies a mile to the W. of it. Trizien Mill, in ruins, open to the W. of a house called *Maison de Remar*, will lead clear to the W. The marks for the centre are, Landuneves and the S. point of the Isle of Jock in one, and the lighthouse of St. Matthew seen open 17° to the right of Point de Corseu. W. by S. ¼ S. distant 2¾ miles from the rock Four lies a shoal called the Basse Muer; the marks for it are the Landuneves windmill and Point Melgrone in one, and the lighthouse of St. Matthew seen between the peninsula of Kermorvan and the summit of Kermorvan. There are several rocks S. of the Four. The Platresses are 1½ league N. ¾ W. from Point Conquet; they lie nearly midway between the land and the Plateau de Helle. There are two rocks between the Platresses and the land, and W. of the Platresses, about half-way between them and the islands S. of Ushant, is a high rock called La Helle, appearing like a ship under sail, from which to the islands there is no safe passage on account of the rocks scattered all over it. The islands extending from the Passage du Fromvuer to Point St. Matthew are Le Bannec, which is the N., Le Balanec, Le Molene, Le Trielen, Le Quéménes, and Le Beniguet, which is the S.E. island: these have passages between them, but too difficult for a stranger to navigate; they are all surrounded with smaller islets, rocks, and dangers, and extend S. to the lat. of 48° 17′ 30″ N.

FOURNI ISLANDS, Greek Archipelago, a cluster of islands and islets 5 miles from Nicaria and 3½ from the Isle of Samos; they are said to be uninhabited. The middle island lies nearly N. and S. and is 3 leagues long; the island named Menas lies on its E. side, and on its W. is a similar island with rocks about its N. end.

FOURNIGUES of ESCAMPEBARION, Cape and Rocks, France, S. coast, between point St. Margaret and the bay or road of Hyeres. The rocks are 1¼ mile off the extreme point of the cape, above and under water; from the great swell often prevailing near they require much attention when they are approached.

FOWEY ROCKS. See *Florida Reefs*.

FOWEY TOWN and HARBOUR, Cornwall, 7½ miles to the W. of Looe island, and about 3 leagues E.N.E. ½ E. from the Deadman point. It may easily be known by the narrowness of its entrance, and the high land on each side: the entrance is about a cable's length across. On the west side stands St. Catherine's castle, and on the east side are the ruins of St. Saviour, an old church. In the best of the channel there are not more than 3 fathoms at low water; opposite the town there are 3 fathoms, but near to Polruan, on the east shore, only 12 and 15 feet. As this harbour lies about N.E. by E. and S.W. by W. it has a better outlet to the W. than many others on the S. coast of England, and in cases of necessity vessels may enter it without cable or anchor, and run on shore abreast of the town on soft mud, floating again when the tide rises. It is high water here on the change and full of the moon, at $5^h\ 14^m$, and the water rises on spring-tides nearly 17 feet, and neaps 8 feet. The ebb tide runs one hour after it is low water by the shore, and is always stronger than the flood. There is a good anchorage without the harbour in from 5 to 10 fathoms. With the tower of Fowey Church over St. Saviour's point, and the three points to the eastward open, is a good anchorage in 7 or 8 fathoms, fine sandy bottom; if in deeper water than 10 fathoms, there is foul ground. The danger to be avoided going into Fowey is a ledge of rocks called the Canness; it lies a quarter of a mile S. by W. ½ W. from Predmouth point, and may be seen at half-ebb. Between these rocks and the point are only 10 feet water, and the ground foul. To go clear of the Canness, Deadman's point must be open, without the Gwineas or Gull Rock, or Gorran Rock, a sail's breadth within Deadman point, until the pinnacles of Fowey church come in sight. A beacon has been placed upon Gribben head, to the W. of the harbour of Fowey, 85 feet high, standing upon an elevation of 257 feet above the level of the sea. This town sent 47 vessels to the siege of Calais under Edward III., being more than any other town in England, London not excepted.

FOX COVE, America, Newfoundland, at the bottom of Little Mortier bay, 1½ mile from Mortier island, between Cape Race and Cape Chapeau Rouge. There is fair anchorage here for one ship to moor in 9 fathoms water, two points open to the sea from S.S.E. to S.E.

FOX HARBOUR, America, Labrador, fit only for small craft, about 1¼ mile N.N.W. ¾ N. from Cape St. Louis.

FOX ISLAND, America, Nova Scotia, on the S. side of Chedabucto bay; it is small and lies near the shore. The anchorage of this island is one of the greatest mackerel fisheries in North America during September and October. On sailing in, a vessel must pass to the W. of Fox island, giving it a berth of a quarter of a mile, as there are rocks both under and above water. The anchorage is from 4 to 10 fathoms, with the W. end of the island bearing from E.N.E. to N.N.E., keeping about midway between the island and the main. Lat. 45° 21′ N., long. 61° 7′ W.

FOX ISLAND, America, Newfoundland, between Cape Race and Cape Chapeau Rouge. It is a small round island 3 miles from Point Latina, N.E. ¼ N. and N.W. by W. 3 miles from Ship Harbour point.

FOX ISLAND HARBOUR, America, Newfoundland, formed by the island of the same name, and situated about half a league W. of Old Mosquito harbour. There are rocky islets and rocks between, but there is a commodious harbour for vessels to anchor in 8, 9, and 10 fathoms water. A ship may go in on either side of the island.

FOX ISLAND PASSAGE, America, United States, Maine, in Penobscot bay. On Brown's head here, which forms the W. side of the southern Fox island, there is a small lighthouse with a fixed light 2 rods from the shore, and 80 feet above high-water mark.

FOX POINT and BAY, America, St. Lawrence, Anticosti island, 4 miles to the S.E. of and lower than Table head. The bay is not quite 2 miles to the S. of the point. It is about a mile wide, and deep inland, with a sandy beach at its head, where a stream issues from a small lake. Boats at high water may enter through the outlet of the lake. The house of M. Godin on the N.W. of this bay was the scene of the horrible sufferings of the passengers and crew of the Granicus, wrecked on this coast in Nov. 1828, who all perished from famine before the next spring.

FOY ROCK, France, N. coast, between Rochefort point and the Groin or Groyne of Cancale. Off this last point, a little E., is the Isle des Landes, and S.W. the Point de Barboult.

FOYLE, or FEOYLE, Lough, Ireland, N. coast, a large harbour, where vessels of any size may lie in every weather. The Tuns bank lies on the S. side of the entrance, having a channel between itself and the land three-quarters of a mile wide. There are two lighthouses on the W. side of the entrance, 460 feet apart, on two white towers three-quarters of a mile S. of Innishowen head, having fixed lights 67 feet high, in lat. 55° 14′ N., long. 6° 56′ W. There are some dangers to be avoided in reaching the anchorage, but there are two pilot stations, one close to the lighthouses, and one 8 miles off Innishowen head, at Turmore, and the pilots are fined if they do not board. The best anchorage for large vessels is 3 or 4 miles above the entrance, on the N. side of the bay, half or or three-quarters of a mile from the shore. Vessels not drawing above 12 feet anchor off Quigley's point, in 2½ or 3 fathoms.

FOYNE'S ISLAND and HARBOUR, Ireland, W. coast, in the River Shannon. The harbour is S. of the island, but can be safely made by a stranger. The entrance is about three quarters of a cable's length wide, having 3½ fathoms water in some parts, and inside 9 fathoms.

FOZ, BAY OF, France, S. coast, near the Rhone, E. of the most easterly mouth of which Point de Tignes forms the S.W. limit, having Martigues for the eastern point of this bay, which lies between that haven and the Rhone.

FOZ, ST. JOAO DE, coast of Portugal, a castle at the mouth of the river Douro, together with a town of the same name in lat. 41° 8′ 43″ N. See *Douro*.

FRAIL ROCKS, Ireland, near the Great Saltee Rock; they are said to be 1 mile N.E. by E. ½ E. from Coningmore and from the S.W. point of the Great Saltee, S.E. by E. 1 mile. They are two rocks that appear at half ebb. It has been supposed that the two Brandy rocks, formerly so named, off the S.E. side of the Great Saltee, are probably identical with the modern Frail rocks. To pass clear of them on the S. side, a vessel must keep the castle of Balliteigue open with the E. end of the Little Saltee, and to go clear of them on the S. having Coningmore rock N.W. by W. ½ W.; and on the W. side keep Balliteigue castle just open of the W. end of the Little Saltee, or bring the S. end of the Great Saltee N.N.E. ½ E.

FRAMULA, Italy, a village on the sea-shore, between Porto Fino and Cape dell Mesco.

FRANCIS, PUERTO DE, ST., West Indies, Isle of Pines, 2½ miles from Point Pedernales, and near Cayo Francis, an anchorage and watering place. It is a small roadstead, the bank of which extends half a mile from the sandy beach, with a depth of 5 fathoms on sand. It is much visited by vessels for timber. The cape is in lat. 21° 37′ N., long. 83° 12′ W.

FRANCIS, ST., CAPE, America, Newfoundland, on the E. coast, the S.E. limit of Conception bay, in lat. 47° 48′ 4″ N., long. 52° 51′ W. A whitish point of low land. The opposite point is Baccalou island, which may be considered the other boundary of Conception bay, the course and distance being N.N.E. ¼ N. 5¾ leagues. From Cape St. Francis to Belle Isle is 4 leagues S.W. by W On this vicinity the following remarks have been made as recently as 1847 :—" The whole of the land," says the *Newfoundland Times*, " in and about the neighbourhood of Conception Bay, very probably the whole island, is rising out of the ocean at a rate which promises at no very distant day materially to affect, if not to render useless, many of the best harbours we have now on the coast. At Port de Grave a series of observations has been made, which undeniably proves the rapid displacement of the sea-level in the vicinity. Several large flat rocks, over which schooners might pass some 30 or 40 years ago with the greatest facility, are now approaching the surface, the water being

scarcely navigable for a skiff. At a place called the Cosh, at the head of Bay Roberts, upwards of a mile from the sea-shore, and at several feet above its level, covered with five or six feet of vegetable mould, there is a perfect beach, the stones being rounded, of a moderate size, and in all respects similar to those now found in the adjacent land washes." The last observation is not novel, because such results are found in most parts of the world, the effect of similar upheavings of the land, but the changes alluded to in the first part of the above extract should be carefully recorded, with a view to future precaution on the part of the navigator.

FRANCIS, ST., CAPE and HARBOUR, America, Labrador. This cape is in lat. 52° 37′ N., long. 55° 31′ 18″ W. Half a mile W.S.W. from the cape is the harbour of St. Francis. This harbour is snug and secure, but very small, generally filled with the vessels that use it during the fishery. Sealing Bight, 1½ mile W., is more commodious.

FRANCIS, ST., LAKE, America, British provinces of Canada, an extension of the river St. Lawrence, between Kingston and Montreal, through which the line passes dividing Upper and Lower Canada.

FRANCISCO, PUNTA DE, West Indies, Island of Porto Rico, 7½ miles S.W. by W. from the village of Aguadilla.

FRANCISCO, ST., RIVER OF, America, 4½ miles to the N.E. of Campeachy, in the bay of that name.

FRANCOIS BAY, America, New Brunswick, nearly half a league from the Bay of Chaleur. It is a small inlet, running in N.W. ¼ W. about a mile, and at the entrance a quarter of a mile broad. It has 17 fathoms water, just within 50 and 60, and at the head from 30 to 20, and anchorage.

FRANCOIS, OLD CAPE, West Indies, Island of Hayti, on the N. coast, at a remarkable projecting part of the island, in lat. 19° 40′ 30″ N., long. 69° 55′ W., according to the Spanish observations. The land trends away W. southerly along the N. side of the island, and S. nearly towards the N.E. and E. part.

FRANCOIS, PORT, Hayti island, N. coast; from Port Picolet, on the N. side of the island, the coast trends W. to Honorat point, which is the N. point of Port François, whence a reef stretches out a cable's length to the N.W. At its extremity are 3 fathoms water. The anchorage off Port François is in a small bay about 2 cables' length in extent, but between the points there is good shelter from the breezes. A vessel to enter must run along the edge of the reef of Honorat, which is on the N. side, and after having gone about 2 cables' length to the S.S.E., anchor in 8 or 10 fathoms, clayey sand, about S.W. by W. from the fort. Lat. 19° 49′ N., long. 72° 15′ W.

FRANGEROLA POINT, Spain, S. coast, S.W. of Malaga 13 miles. This point is rocky, but the beach on the N.E. is clean. The ancient castle of Frangerola is seen here seated upon an eminence, and having five or six lofty towers. It is about 1½ mile N.N.E. ¼ E. from the Cala de Burra tower. All the way from this tower to Frangerola point is rocky, but from thence to Torre Blanca, about 2½ miles, the beach is low, clean, and sandy, and the anchorage before it good for vessels of all sizes, and safe from all but N.E. winds. The French charts place two small banks S. of Frangerola point, of which the Spanish surveys take no notice, and therefore their existence must remain undecided.

FRANK'S ISLAND, America, Gulf of Florida,

at the mouth of the N.E. pass of the Mississippi river. A lighthouse has been erected here 78 feet above the level of the sea, having a steady fixed light, seen in clear weather 6 or 7 leagues. A large bell is also kept tolling during fogs, when the light cannot be seen at 4 miles distance.

FRANK'S LANDING, West Indies, Island of Anegada, a spot of ground where the land rises gradually from N. to S. on the S.E. side of the island, about 60 feet above the level of the sea, being the highest land in the island.

FRASCA, CAPE, Island of Sardinia, on the W. side, being the S. point of the gulf of Oristano.

FRASERBURGH LIGHT, Kinnaird's head, Scotland, E. coast, the name of the lighthouse upon this head, situated in lat. 57° 42', long. 2° W. It is a fixed light, 120 feet high.

FRATELLI, or the BROTHERS' ISLAND, Greek Archipelago, an uninhabited islet E. of Lero, or Leros.

FRAYLE, or EL FRAYLE, E. coast of Spain, an insulated rock near the port De los Aguilas, N.E. of Moxacar.

FRAYLECITO, ISLET and POINT OF, Africa, between Cape Spartel and Tangier bay, about midway, lying close into the shore; the distance from Cape Spartel to the W. point of Tangier bay is 5¼ miles, Cape Spartel being in lat. 35° 47' N., long. 5° 55' W. The shore is high and clear, though close in it is bordered with rocks, and is without any beach as far as Fraylecito point.

FRAYLE, PUNTA DEL, the S.E. point of the mainland of Spain, W. of Cabrita point, between which is a bay with anchorage on the W. side. On the E. side is Palamos or Pigeon island, and half a mile S. of this last is the Pearl rock, having only 12 feet, and between that and the island 6, 7, 8, and 10 fathoms water. Point del Frayle has a watchtower upon the slope of a hill, with a small island off its extremity, said to resemble a friar in appearance. There are several rocky islets about this point, and on the E. side a sandy cove, only fit for small craft, sheltered from the W. and N., and having a castle for the defence of the anchorage.

FRAYLES, or FRIAR'S ISLET, off the Island of Terceira, in the Azores; it has two pyramidical peaks, and a shoal extends off the S.E. side of it.

FRAYLES, THE, coast of South America, a group of islets, of which the most S. is the largest. They are all clean except the most N., which is surrounded by a reef extending from it 2 cables' length. The Frayles, or Los Frayles, lie about 8 miles E.N.E. of Margarita island.

FRAZER'S BANK, Ireland, Bay of Dublin, commencing about half a mile from Dalkey island, more than a mile in length, and nearly a quarter of a mile in breadth, lying N.N.E. and S.S.W., with from 3 to 5 fathoms water generally; but one-third of a mile from the N. end, there are not more than 16 feet. In E. and S. gales, the sea breaks heavily over it. Lat. 53° 16' N., long. 6° 5' W.

FREDERICA, North America, United States, Glyn co., Georgia, on the W. side of St. Simon's island, at the mouth of a branch of the River Alatamaha, which washes the side of this pleasant island, and forms a bay before the town, where there is safe and commodious anchorage for vessels of any burthen. The town, which has never much increased, was founded by General Oglethorpe, in 1735.

FREDERICA, Denmark, a town on the shore of the Little Belt, opposite the N.W. part of Funen island, where merchant-ships pay toll for the Little Belt passage.

FREDERICK BAY, S.W. coast of Africa, the name of a bay W. of Cormantine about 2 leagues; it has a tolerable road, lying between the Dutch forts of Nassau and Maurice.

FREDERICK'S HARBOUR, or FREDERIK'S HAAB, W. coast of Greenland, a station of the Danes, at the head of a bay, in lat. about 62° 30' N.

FREDERICKSHALL, Norway, a seaport in the province of Smaalhenes, in Swine sound or bay. The harbour is commodious. The streets are wide; the houses of wood. On the opposite side of the harbour stands the strong and almost impregnable Fortress of Frederickstein, in lat. 59° 12' N., long. 11° E. To enter this harbour, a pilot can be had at Fœrder island. The middle of the town is in lat. 59° 7' N., long. 11° 26' E.

FREDERICKSTADT, Norway, a port near the entrance of Christiania fiord, on the E. side, upon the Glommen river. It has a considerable trade in timber. It is 15 miles N.W. of Frederickshall. Lat. 59° 12' N., long. 11° E.

FREDERICKSTAED, or WEST END BAY, an extensive and beautiful bay in the Island of Santa Cruz, having a town of the same name upon it. The lat. at the flagstaff of the fort in this bay is 17° 43' N., long. 64° 52' W. The bay affords excellent anchorage, except when the wind has westing, when, like all open anchorages, it becomes dangerous.

FREDERICKSTED, FORT OF, West Indies, Island of Santa Cruz.

FREDERICKSOE, Denmark, a small island off the coast of Bornholm.

FREDERICKTON, America, New Brunswick, a town 90 miles up the river of St. John, to which there is a sloop navigation.

FREDERICK'S-VAEN, properly STAVAERN, a small town near Christiansand, in Norway, on the S. coast. It is protected by a fort on an island close by, and has a port capable of receiving large vessels, as well as being a common place for their construction.

FREDERICKSVORN, Norway, on the S. coast, and on the W. of the entrance to Christiania fiord, a town on the S. of Laurvig.

FREDERIK HALL, Sweden, on the E. coast of the Scagerac, and S.E. from Frederikstad, at the mouth of the River Tiste, having a safe and commodious harbour. It is in lat. about 59° 7' N., long. 11° 2' E.

FREDERIKSBORG, Africa, Gold coast, a Danish fort near and to the W. of Cape Three Points, 21 leagues from Cape Coast castle, and in lat. 4° 46' N., long. 2° 7' W., built in 1734.

FREDERIKSORT, Holstein, near the mouth of the Eyder, and S. of the Bulkhost light.

FREDERIK WARK, Denmark, a town on the S.W. coast of Zealand island, principally important as possessing the chief cannon foundry of the Danish government.

FREE BANK, coast of Flanders, the same as the Broad bank. The S. end lies N.E. by E. from Calais cliff, distant 4 or 5 leagues. At the S. end there is a shoal, with only 1½ fathom water; everywhere else it has 3 and 4 fathoms; and at the N. end 6, 7, and 8, from which end Nieuport bears E. by S.

FREELS, CAPE, America, Newfoundland; lat. at Gull island 48° 19' 6" N., long. 53° 28' 46". This cape is formed of three points, called the

South Bill, the North Bill, and the Middle Bill, the last Cape Freels, more properly. There are many shoals and dangers about these points; hence a good berth should be given them. There is high land over them visible at a considerable distance, and called Cape Ridge. There is another cape of this name on the S. part of Newfoundland, near St. Mary's bay.

FREEMAN'S BAY, West Indies, on the E. side, within the entrance of English harbour, in the Island of Antigua.

FREEPORT, TOWN, America, United States, in the province of Maine, situated at the head of Casco bay, 140 miles N. by E. of Boston.

FREETOWN, BAY and TOWN OF, Africa, W. coast. The town stands on a hill-side, protected by a fort; and on the summit of the hill above are barracks. The general aspect of the country is that of prosperity and salubrity. Thus deceptive to the vision are the haunts of pestilence. Within Cape Sierra Leone, the trend of the coast is nearly due E. 6 miles, broken by inlets there called bays. The first is called Cape bay; the second Pirate's bay; the third Whiteman's bay: the fourth St. George's or Freetown bay, where the town of that name stands in lat. 8° 29′ 42″ N., long. 13° 14′ 18″ W. The N. battery is given by Lieut. Raper as in lat. 8° 29′ 54″ N., long. 13° 14′ 30″ W., well determined. To anchor a large vessel off Freetown, bring Fort Thornton to bear S. by W., and King Tom's point W. by N., where she will be a quarter of a mile from the shore in 12 and 14 fathoms, muddy bottom, mooring with an open hawse to the N.

FREES, or FREIZE PLAT BUOY, Holland, in the Zuyder Zee, nearly W.S.W. from Hinlopen, on the coast of Friesland, and almost due N. from the Creyl buoy.

FREHEL, CAPE, France, about 4 leagues N.W. ¾ W. from St. Malo. It is a high cape, and has a remarkable lighthouse, bearing a revolving light, the eclipses of which succeed each other every half minute. It stands on the point of the cape, at an elevation of 260 feet above high water, and may be seen 7 leagues off. Half a mile W. from the cape there is a high rock, called Amas de Frehel, which serves as a mark; and half a mile S.E. ½ E. from the cape is the rock Tendrée, from whence a shoal runs out S.E. for three-quarters of a mile, having from 1 foot to 2½ fathoms upon it. Between this cape and St. Malo are several creeks, off which there is anchorage.

FREIDERA SHOAL, Spain, Cadiz bay, having seldom less than 2½ fathoms; but whenever any swell arises, the sea breaks over it.

FREJUS, GULF OF, and TOWN, France, S. coast, between Capes Alexandre and St. Egou, distant 3 miles, the coast running high. The town so named stands in the middle of the gulf, at a little distance from the sea-side, in lat. 43° 26′ 15″ N., long. 6° 44′ 20″ E. A small river falls into the sea to the E. of it; and to the W. is another river, called Argens. Off the shore there are two islands, named the Lions; and from the outermost of these to Cape Egou is about 2 miles. There is good anchorage in the Bay of Frejus, in from 11 to 7 fathoms; but it is open to S. winds. The pop. of the town is 2,300. There are numerous antiquities about this town, principally Roman—A gate, said to have been built by Cæsar; the ruins of an aqueduct, temple, and amphitheatre, with an ancient lighthouse and walls. Here Napoleon disembarked on his return from Egypt. The trade of Frejus is unimportant.

FRENCH COVE, America, Nova Scotia. This cove is easy of access, and is in fact a natural dock, extensive, having plenty of water, and well sheltered. There is a shoal, with 10 feet water over it, opposite the entrance of the cove, at the distance of two miles. But the island and shore are bold, and no vessel need go so far out into the bay as to approach this danger.

FRENCH KAYS, called also the PLANAS, West Indies. These lie between Mariguana and the Crooked islands. Lat. 22° 35′ N., long. 73° 33′ W. They are described as very little known, and to the pilot of the Eagle, a French bark, the world was first indebted for a description of them. This bark anchored in the white water on the W. side of the largest of the French islands, about 2 cables' length off, sheltered by a reef, which runs N. and N.W. 2 miles. The islands are two in number. Rocks extend about two miles to the N. of the W. island. The largest island is about 5 miles E. and W., and half that N. and S. The E. and N. sides are surrounded by a reef 1 mile off shore. The anchorage is about three-quarters of a mile from the S. point, near which is a landing-place, and by digging two or three feet in the sand, good fresh water is to be had on the W. island. The islands are low, and nearly level, though risings appear at a little distance. The ground is sand and rock, with some bushes, only fit for firing. The second island is smaller, lies nearly N. and S. 2 miles to the W. of the first. The passage between is very narrow, owing to the reefs on both sides. It is not a mile wide, and is only fit for small craft.

FRENCHMAN'S COVE, America, Newfoundland, between Chapeau Rouge and Cape Ray, and between the Grand bank and Point Enragée, a place where small vessels may run in and anchor in 4 or 5 fathoms water, tolerably sheltered from sea winds.

FRENCHMAN'S COVE, America, W. coast of Newfoundland, opposite to the S.E. end of Harbour island, on the S. side of the bay, in which there is good anchorage in from 20 to 12 fathoms.

FRENCHMAN'S ISLAND, or ISLE OF QUAILS. See Verde, Cape.

FRENCHMAN'S ISLAND, West Indies, one of the Virgin islands.

FRENCHMAN'S RIVER, North America, S. part of Hudson's bay, in the S.E. angle of James bay. Lat. about 51° N., long. 79° W.

FRENCHMAN'S SHOAL and INNER FRENCHMAN, called also the LIME-BURNER and CARRICKHAVRANK, Ireland, N. coast, E. of Sheephaven, having 7 feet on it. The first is 2½ miles N. by E. ½ E. from Melmore point, a shoal on which the sea breaks with W. winds, and in calms before a gale from that quarter. It is small. The second consists of two shoals, one half a mile N.E. by E., and the other one-third of a mile N.W. ½ W. from Ballyorskey or Mulroy point. There are generally breakers upon them. The Tory island light bears from the Frenchman's shoal W.N.W. distant 5 leagues, and the Fannet light S.E. ¼ E., distant 5¾ miles.

FRENCH SHOAL, a vigia, laid down in lat. 4° 5′ N., long. 20° 34′ W. This has been searched for very carefully both by officers of the British and foreign naval services, and no trace of real danger discovered. The British man-of-war Leven, Capt. Bartholomew, R.N., and Capt. Wilkes, of the American navy, so well explored the reputed place of this shoal, that its existence cannot be credited.

FRENEY ROAD, France, N. coast, on the E. side, and a little S. from Point de Late, or Cape Frehel, and nearly S.W. by S. 31½ miles from St. Aubin, near the S.W. point of the Island of Jersey.

FRESHWATER BAY, America, Newfoundland, on the S. coast, between Signal hill and Castle hill, having a pebbly beach around it, and with 6 or 7 fathoms water, over good ground. Caution is necessary in taking this anchorage, to avoid the Gibraltar rock and Green point, and to use the lead.

FRESHWATER BAY, Isle of St. Nicholas, one of the Cape Verde Islands. See *Verde, Cape*.

FRESHWATER BAY, Isle of Wight, near the S. W. extremity of the island.

FRESHWATER BAY, West Indies, Island of Hayti, to the leeward of Port au Prince.

FRESWICK CASTLE, Scotland, in Caithness, upon Sinclair bay, N. of Noss head, seated upon a bold promontory.

FREZENDORF, Pomerania, a port with some trade on the left bank of the Peene.

FRIAR'S BAY, West Indies, Island of St. Kitt's.

FRIAR'S HEAD BAY, West Indies, Island of Antigua, E. end, a small bay S. of Great Deep Bay, and W. a little S. from York island; so called from the S. point, which has the name of the Friar's head.

FRIDERICKS-HAMN, or FRIDERICKS-HAMM, Russia, in the Gulf of Finland, a large and well fortified town, near an islet, 50 miles W. of Wyborg. It carries on a good trade, and exports similar articles to Wyborg, more especially planks and tallow. The Russians keep in this town a garrison of 6,000 men.

FRIESTON, or FIRESTONE, HILL, Ilha de Sal, or Salt island. See *Verde, Cape*.

FRIGATE ISLAND, West Indies, one of the small rocky islands between St. Vincent's and Grenada, called the Grenadines or Grenadillas.

FRISCH HAF, or FRESHWATER SEA, Prussia, a species of lagoon 16 or 17 leagues long, separated from the Baltic by a tongue of land called the Frisch-Nehrung. This land commences near Dantzick, and extends E. 17 leagues, with a breadth of 1 to 3. The portion near Dantzick is well cultivated, the rest is barren. The communication of the basin within with the Baltic has several times changed. The present passage near Pillau was formed in 1500. It is 1,900 or 2,000 fathoms wide, with a depth of water of from 13 to 16 feet.

FRISCO RIVER, Africa, Ivory coast, or Rio de Lagos, to the E. of St. Andrew's river and bay, in lat. 5° 1' 8" N., long. 5° 32' 5" W., according to Captain Vidal, R.N.

FRISKLEYER, Denmark, a village of small moment upon the E. coast of Schonen.

FROE ISLANDS, W. coast, part of those numerous clusters of which the shores of Norway present so extraordinary an assemblage; they lie in the Froe Havet, or broad sea entrance on the N. to Drontheim.

FROG'S POINT, America, United States, West Chester co., New York, on the coast of Long Island sound, 9 miles from Haerlem heights.

FRONTIGNAC FORT, America, a fort in Canada, at the head of a bay on the N.W. side of the outlet of Lake Ontario, about a league from the mouth of the lake, and 300 miles from Quebec. It was built by the French, and afterwards de-

stroyed by them, again repaired, and finally taken from them by the English, in whose possession it remained after the peace of 1763. It may now be said to have merged in Kingston.

FRONTON DE ALMANZA, Africa, E. of Cruzes point, on the S. side of the Straits of Gibraltar, and nearly N. of Ape's hill; its direction is E. by S. a broad, abrupt, lofty precipice. The land bends in between Cruzes and Almanza, and forms a small sandy cove.

FROUSAC or FROWSAC, CAPE and CHANNEL, America. This cape is the N.W. point of the Strait or Gut of Canso, and the channel is the strait itself, thus formerly named.

FROUSEIRAS POINT, Spain, N. coast, E. of Cape Prior and of Mount Campelo; it has a watch-tower on its summit, appearing with a sharp peaked top.

FROYEN ISLAND, on one side of the Froy fiord, leading N.E. to Drontheim, having Hitteren island between itself and the mainland of Norway.

FROY FIORDEN, Norway, W. coast, one of the passages or channels leading to Drontheim, within Froyen island.

FRUCHILAN ISLAND, Ireland, on one side of the entrance to Ballinakill harbour, which must be kept at a cable's length on entering; the anchorage is near the E. end of this island.

FRY, CAPE, America, in Hudson's bay, the N.E. limit or point of Chesterfield inlet, projecting to the E. in lat. about 64° N., long. 88° W.

FRYING-PAN SHOAL, North Carolina, United States of America, a shoal near Cape Fear. At the entrance of the river is Bald head, with a fixed light, 110 feet high. It is needful on this coast to keep a degree S. of the lat. of the place intended to be made, until a vessel may reckon to be on the edge of the Gulf stream. It is not wise to sail, if it can be possibly avoided, to the N. of 33° 20', or, at the highest, 33° 25' N., until 10 fathoms water are obtained, and in this depth a vessel will be within the S. or outer end of the Frying-pan shoal, which lies in lat. 33° 34' N. In approaching 33° 20' lat., the first soundings will be 30 and 35 fathoms near the edge of the Gulf stream, fine grey sand, with black spots; next 17 fathoms, a long flat in this depth, and westward there will be little shoaling for 5 or 6 leagues. In 14 fathoms, the water will shoal gradually quicker. The land will be seen from 10 fathoms, if clear weather, and a vessel may then be certain of being within the Frying-pan from the outside of this shoal. To the W. of N.W. no land can be seen when without the shoals.

FUCHAN LOCH, to the N.E. of Inish island, on the W. side of Scotland, off the coast of Argyleshire. The entrance is nearly dry at low water spring tides, yet the stream runs through it. Small vessels might enter this loch at half-flood. The ground is good, but the soundings are unequal; the anchorage is off Ardintellan house.

FUEGO, ILHA DE. See *Verde, Cape*.

FUENTE DE LAS YEGUAS, Spain, a small hill, serving as a sea-mark, near Camariñas, and seen on approaching the bay of that name.

FUERTE ISLAND, America, coast of Colombia, the N.E. end S. 84½° W., 21½ miles from Port Mestizos, and from the N.E. part of Point Piedras, N. 57° W., 6½ miles. This island is 1½ mile in length from N. to S., and a little less from E. to W., high in the middle, and covered with trees, in which palms appear beyond the others. It is to be approached only at Arenas point, on the S., it being surrounded by reefs and sunken rocks, some of which show themselves above water. Outside the reefs on the banks, and even upon them, there are from 2 to 4 fathoms water, over rock and coarse sand. Besides these banks surrounding the island, there are two others smaller, one with 4½ fathoms water on sand, lies S.S.W. from the island, a mile distant, and the other, with 5½ water on sand and gravel, lies S. 28° E. a long mile from Arenas point. In the channel between the island and the main there are from 6 to 14 fathoms water. This island may be seen from the deck of a low ship at the distance of 20 miles. Lat. 9° 23′ 30″ N., long. 76° 11′ 15″ W.

FUERTEVENTURA ISLAND. See *Canary Islands*.

FULEHUK ROCKS, coast of Norway, N.E. ¾ N. from Fœrder island, distant 7 miles. There is a lighthouse on one of these rocks, exhibiting a revolving light eclipsed every half minute, 43 feet high, to distinguish it from the light on Fœrder island. It is visible 8 miles. Near the lighthouse, in foggy weather, when it cannot be seen at the distance of half a league, a bell is struck 10 or 12 times; at night, every quarter of an hour, and during the day, every half hour. A course N. by E. from this light will carry a vessel up the fiord in mid-channel.

FUGLOE SKERRY. See *Shetland Isles*.

FULA, or THULA, ISLAND. See *Shetland*.

FULL-MOON SHOAL, America, United States, off Cape Hatteras, one of the formidable shoals which lie about this dangerous cape. There is reason to believe that this shoal has much diminished in extent since the coast of America has become known to Europeans.

FULO, or FOWL, ISLAND. See *Shetland Isles*.

FUMENDOSA, Island of Sardinia, one of the principal rivers entering the sea upon the S.E. coast.

FUNCHAL. See *Madeira Islands*.

FUNDY, BAY OF, America, S.W. coast of Nova Scotia. This is a bay in which currents, fogs, and changes of weather are likely to confuse the mariner whose judgment is soundest, taken together with the numerous rocks, islands, and its exposure to the Atlantic seas. It is fortunate that it abounds with secure harbours, which those versed in the navigation well understand. Those who are comparative strangers will have much less to fear if they will frequently use the deep-sea lead, and keep anchors and cables ready for immediate use, and not as they are stowed in the middle of the Atlantic. "Though," says Blunt, "the loss of vessels in these parts fully justifies a perilous apprehension, there are few obstacles which a moderate share of skill and resolution will not overcome, and we fear it is more to the want of these qualifications than to the dangers of the navigation, that losses have ever occurred." In this bay, a knowledge of the tides is more than commonly necessary to the navigator. The entire bay runs about 200 miles in a N.E. direction, bounded easterly by Nova Scotia, and separated only at its head by a narrow isthmus at the end of Verte bay from the Strait of Northumberland, that divides Prince Edward's island from the main. On the N.W. side it has Passamaquody bay, between New Brunswick and Maine, with a number of bays and islands, which contract its width. From St. John's, in New Brunswick, to the Gulf of Annapolis, in Nova

Scotia, eastwards, it is 25 miles across the bay. Above this it preserves a tolerably equal breadth, until separated by a peninsula, the W. point of which is called Cape Chignecto, dividing the head of the bay into two arms. In this bay the tides rise with astonishing rapidity, even above 40 feet; they even rise 30 feet in the Gulf of Annapolis. This rapid and great rise is rendered more dangerous from islands, rocks, and shoals. Beginning at Cape Sable, and proceeding W., there are the Blonde rock, the Seal (on this last there is a lighthouse painted white, having a fixed light, in lat. 43° 23′ 50″ N., long. 66° W., on the S. part), the Mud islands, Tusket islands, Gannet rock, Green island, and others. The Seal island S. point bears from Cape Sable nearly W.N.W. ¼ W. distant 16 miles, about 2 miles long; 2 7-10 miles from the S. part is a rock on which the Blonde was wrecked in 1777, with 7, 9, and 10 fathoms around it. A mile W. of this rock are dangerous overfalls. Off the W. of Seal island a mile, are two rocky islets, called the Devil's Limb and Limb's Limb, the last only seen at half tide. The Mud islands are sometimes called the North Seals, 5 or 6 in number, the largest N.E. by N. 3¼ miles from the South Seal island, 1¼ mile long, 2 miles N.W. by N. from the largest Mud island, is a ridge called the Soldier's ledge. Tusket Bald islands lie N. of Mud islands, to the S.W. of the entrance of Tusket river; some are considerable in size, but the passages among them are full of shoals and dangers. Between Tusket and Mud islands is the Acteon rocky shoal. Pubnico harbour is 12 or 13 miles from the S. end of Seal island, easy of access. Near its entrance is St. John's island, under the N. side of which there is good shelter. To the W. of the Mud and Tusket islands, going N., the Gannet rock lies N.W. by W., about 6 miles distant from the S. Bald island, and S. ½ W. 10 miles from Cape Fourchu, 36 feet above the surface of the water, and white with birds' dung. Two miles S.W. of the Gannet is the Opossum's ledge, visible at half tide, which has endangered many lives, from having been once represented as 4 miles W. by N. from the Gannet; the lat. 43° 40′ 40″ N., long. 66° 9′ W., after Des Barres. Green island is N.N.E. ¾ N. 3½ miles from the Gannet rock light. A reef runs out from this island to the S.W. nearly three-quarters of a mile. There are round it 5 and 6 fathoms, and between it and the Gannet from 12 to 17. Cape Fourchu, or Yarmouth, has a revolving light on the W. part, in lat. 43° 47′ N., long. 66° 8′ 50″ W., 140 feet above the sea level. From this cape the mainland extends N. by W. to Cape Mary, and from Cape St. Mary to the lighthouse on Bryer's island N.N.W. 13 miles. Almost opposite to Cape Fourchu is the Lurcher's rocky shoal, and between that and Cape St. Mary the Trinity ledge. From Seal islands up to Cape St. Mary, the soundings extend 20 and 30 leagues off the land. W. of Bryer's island light, and near the Manan ledges are 60, 80, and 100 fathoms, at 3 or 4 miles distance, therefore the lead must be continually going. Bryer's island, at the S.W. entrance of St. Mary's bay, is 4 miles long and 1½ broad, and has on its W. side a lighthouse, coloured white, 92 feet high. It bears a fixed light, about half a mile inshore, in lat. 44° 15′ N., long. 66° 23′ W., 92 feet high. There are reefs stretching off from this island from the S. part 3½ miles, and other dangers around it. The N. part of this island should not be approached nearer than 4 or 5 miles. Long island is separated by a narrow channel from Bryer's island. In St. Mary's bay, there are flats on the S. shore for three-quarters of a mile, but on the N. it is deep water. About two-thirds of the bay in mid-channel there is a rocky bank with 4, 5, and 6 fathoms over it. Grand passage runs in between Bryer's and Long island. Petit passage lies at the further extremity of Long island. Annapolis gut or channel is best made by keeping 1½ mile from the land, in 50, 40, but not less than 30 fathoms, all the way to Point Prim; when 1½ mile from the lighthouse, the same depth will be found. The Gut of Annapolis, about the middle, is in lat. 44° 43' N., long. 65° 44' W. The lighthouse on the S. side of the entrance stands in lat. 44° 41' N., long. 65° 47' W.; it is a fixed light, 76 feet above the water. From the Gut of Annapolis the coast continues straight to Cape Split, and does not much vary to the entrance of the Basin of Minas or Mines. Haute island lies at the entrance of the Mines or Minas channel, not quite 1½ mile long and half broad, bearing from Cape Chignecto, distant S.W. 4 miles, (a light is to be erected on Haute island) the channel on either side good. Capes Chignecto and D'Or are high lands with steep cliffs of rocks or red earth, which last prevails to the head of Chignecto bay, where there are extensive flats left dry at low water. Here the tides rush in with a bore very rapidly, at the equinoxes having been known to rise 60 and 70 feet perpendicular. The Bay of Chignecto runs up E.N.E., and may be considered the N.E. branch of the Bay of Fundy, divided from the Mines or Minas channel by the peninsula of which Cape Chignecto is the W. extremity. 12 or 13 miles within this cape, on the port or N. shore, a headland runs to seaward, having a lighthouse and fixed light on the pitch of the cape, in lat. 45° 36' N., long. 64° 47' W., 151 feet above the water. This headland is Cape Enragé. 11 miles further on is the entrance to Cumberland basin, and by the River Missequash to Verte bay, in the Gulf of St. Lawrence. The River Missequash forms the limit between the provinces of Nova Scotia and New Brunswick. The N. coast of the Bay of Fundy from Cape Enragé to Quaco, even as far as St. John's, is thinly inhabited; the weather humid, the winds boisterous, and the sunshine evanescent. On a small rock off Quaco head, W. ¾ S. from St. Martin's head, there is a revolving minute-light, in lat. 45° 19' 30" N., long. 65° 32' W. Quaco ledge is a dangerous shoal, 12 miles S.E. ½ E. from Quaco, and W. by N., 11 miles from Haute head, extending W. by N. and S.E. by E. about 3½ miles, and half a mile broad, showing at half tide. The night tides in the bay are generally highest, and are so at St. John's during the summer; but it is the contrary there in winter. To clear the Quaco ledge, S. Cape D'Or must be kept on with the S. side of Haute island. The entrance to St. John's harbour is 11 leagues from the Gut of Annapolis. It has a fixed light on Partridge island, in lat. 45° 14' N., long. 66° 3' 30" W., where a bell is rung in foggy weather; it is 119 feet above the water. There is a succession of islands and points on this side the bay; among them are Musquash harbour, Point Lepreau, Mace's or Mason's bay, Beaver, Etang, and St. Andrew's harbours; the Wolf and Manan islands, Long Island bay, Wood island, Manan ledges, the Gannet and Western Seal islands, or Machias Seal islands, on which

are two lighthouses. The E. is in lat. 44° 30' N., long. 67° 6' W. These lights are fixed, 48 feet above the water, and distant from each other 140 feet. From the most W. of these lighthouses, according to Blunt, the following are the bearings:—To the S. of the Murr ledges E.S.E.; to Gannet Rock lighthouse E. by S. ¼ S. about 12 miles; to N.E. rock N.E. by N. 1½ mile; to the S. head of Grand Manan E. by N. ½ N.; to West Quoddy lighthouse N.N.E.; to Little River head N. by W.; to Libby Island lighthouse N.W. by W.; to the S. part of Kent island (on the chart three isles) E. The intercourse, continually increasing, of England with the settlements about this bay, renders a knowledge of its troublesome navigation more and more necessary. The difficulties here are seldom encountered in common navigation, being peculiar to the locality, such as the dense fogs, high tides setting with extraordinary rapidity, and difficulty of anchorage from the deep water. Between Manan and Bryer islands the greatest caution is requisite, from the flood tide setting upon rocks and shoals, during thick fogs. The prevalent winds along the coast are from W.S.W. to S.W. nearly, and very steady, except during the summer months, when they become more S., but with little intermission of the fog. The last only disperses with a N.W. wind. Anchorage at dark seems continually necessary, thus proceeding from one place of anchorage to another at night or when a fog comes on, which, with a S.W. wind, is very sudden in enveloping everything. At the entrance of the bay, though rapid, the tides are regular. The flood sets from Cape Sable to the N.W. through the Seal, Mud, and Tusket Bald islands, at the rate of 2 or 3 miles an hour, and increases in the channels among the islands to 4 or 5 miles. Thence, in the direction of the mainland, it flows past Cape St. Mary, and next N.N.W. towards Bryer's island. Up St. Mary's bay it flows slowly, but doubles its strength on the E. shore, increases in rapidity as the bay contracts, rushes with a bore or head into the Basin of Mines, and up Chignecto bay, and there rises sometimes to the astonishing height of 75 feet. An undertide below 5 fathoms exists in this bay, ascertained by the lead being carried the same way as the upper current down to that depth; but when lowered deeper, being drifted in an opposite direction. The positions above are from the recent surveys of Captain W. F. Owen, R.N., received December 17, 1847.

FUNEN ISLAND, or FYEN, Denmark, the second island in magnitude, at the entrance into the Baltic. It is about 150 miles in circumference, separated from Jutland by the Little Belt strait, and from Zealand by the Great Belt. The coast is much indented. There is a fixed light at Assens, in lat. 55° 17' N., long. 10° 2' E.

FUNK ISLAND, America, Newfoundland, between Cape Freels and the Strait of Belle Isle. This island is distant 27 miles N. 54° E. from Cape Freels. It is a barren island, visible only at 10 or 12 miles distance, but easily distinguished by the vast number of birds that hover over it. Escape, or E. point, on Funk island, is in lat. 49° 44' 21" N., long. 53° 7' W., according to Messrs. Holbrook and Bullock, Admiralty surveyors, in 1826. About 200 yards N. of this island there is a large rock above water, and distant from this rock N.W. by W. 180 yards, are still larger rocks, the resort of sea birds. Between these rocks there are 18, 27, and 42 fathoms, with a clear passage, but between the E. rock and Funk island, 120 fathoms distant, there is a sunken rock, with 10 feet water only, over which the sea generally breaks. Near this dangerous rock there are 14 and 16 fathoms. Off the W. point of Funk island there are rocks, and at its E. end a sort of creek, with 5 fathoms water. A ledge of rocks is reported to lie S.W. of Funk island, about 7 miles distant.

FURNA, or OVEN BAY. See *Verde, Cape.*

FURNESS, North Lancashire, a peninsula formed by Morcombe bay and sands on the E., and the River Duddon on the W. The Island of Walney, and six or seven others, lie off its W. extremity; on one called the Pile or Piel of Fouldrey, there are extensive ruins of a castle. There is a revolving light, every five minutes, at the S. extremity of Walney island, 70 feet high, in lat. 54° 3' N., long. 3° 11' W.

FURRUNACH ROCK, Ireland, near Ardbear harbour, on the W. coast. It is always above water.

FURZE ISLAND, Ireland, in Dunmanus bay, S. coast. It is almost connected with Horse island by a rocky ledge, covered an hour before high water.

FUSCALDO, Italy, a town on the sea coast, between Cirella island and Cape Suvero, 14½ miles S. of that island.

FYNE, LOCH, Scotland, W. coast, county of Argyle, an extensive lake, having a connection with the sea. It is nearly 40 miles in length, and is more correctly an arm of the sea than an inland lake. The Isle of Arran lies opposite the entrance. The sides are finely indented with bays, and the water deep. At certain seasons it is a great resort of vessels in the herring fishery, which enter the loch by incredible numbers.

GAA SAND, Scotland, a sand-bank on the N. side of the River Tay, extending 1¼ mile from Buddon Ness, on which are two fixed lights, 85 and 65 feet high, and part drying; it is well buoyed.

GABARUS BAY, America, Cape Breton; this bay lies from between the entrance of Louisburg and Guion island, called also Portland island; the course is S. by W., and the distance 3 leagues to where lies Gabarus bay, which is spacious and has a depth of from 20 to 30 fathoms. Off the S. point of this bay, called Cape Portland, are the Cormorants, a number of islets and rocks which are very dangerous. Four leagues to the W. of Gabarus bay is the Forked harbour, a narrow winding inlet, into which small vessels may run and be landlocked. Five miles S.W. of this appears a remarkable white cliff called Cape Blancherotte.

GABBARD, INNER and OUTER, coast of Suffolk; two shoals, the first of which is 6 miles in length, lying N.E. ½ N. and S.W. ½ S., having on its shallowest part 2 fathoms at low water for about 2 miles in the central part, on the other parts having from 4 to 10 fathoms. There is a black buoy on the central part with Orfordness high light N.N.W. ½ W., distant about 15 miles; the S. buoy of the Shipwash W.N.W. ¾ W. 11 miles; the Sunk light-vessel W. ½ S. about 17 miles; and the Galloper light-vessel S. by W. ¾ W. 8 miles. The Inner Gabbard is steep-to, and at half a mile distance on each side there are from 13 to 16 fathoms. The Outer Gabbard is about 4 miles to the E. of the Inner, a narrow shoal of above 3½ miles in length, divided by a swashway. This shoal lies N.N.E. and S.S.W. and has from

9 to 3 fathoms upon it; the N. end bearing S.E. ¼ E., distant 19 miles from Orfordness high lighthouse. There is a buoy striped red and white upon this shoal in the least water, about a mile from its N.N.E. end with Orfordness lighthouses bearing N.W., distant 18¾ miles. This shoal is steep-to, there being from 15 to 17 fathoms within half a mile of it all around. The buoys of the Inner and Outer Gabbard bear from each other E. and W., distant 5¼ miles. The tide makes a constant ripple over these shoals except at slack water, by which their situation is easily known if hazy weather prevent the buoys from being seen.

GABIA ISLAND, Mediterranean, coast of Italy, off the N. end of the Island of Ponza, with a channel between only adapted for boats. The Island of Zannone lies 2½ miles N.E. ¾ E. from the Island of Gabia.

GABILEH TOWN, coast of Syria, distant 5 leagues from Cape Ziaret, S.E. ½ S., standing at the foot of an inland mountain.

GABLES OF SOTTEVILLE ROCKS, France, N. coast, W. of St. Valery en Caux, 1¼ mile to the E. of Veules, near the coast, and off the church of Sotteville.

GABON, or GABOON, RIVER, Africa, W. coast, N. of Cape Lopez. The entrance is in lat. 18' 30" N., long. 9° 17' E.

GABRIEL or GAVRION PORT, Greece, Island of Andros, a haven on the W. side of that island, about 7½ miles from Cape Guardia, in lat. 37° 53' N., long. 24° 45' E. It is encumbered with rocks, the entrance being bad and not to be recommended for safety.

GADA, PUNTA DEL, Island of Flores, Azores, the N. point.

GAETA, PORT and TOWN, Italy, kingdom of Naples, an ancient town situated on a peninsula, and strongly fortified, having a pop. of 15,000. The citadel commands the city, and is built upon an eminence. This town stands in lat. 41° 12' 18" N., long. 13° 34' 12" E. On approaching the port from the W. it may be recognised by the Monte Trinita, or Mount Trinity, on the W. of which is the sandy cove of La Serpa. The mount is remarkable in appearance, being divided from top to bottom, and having towards the sea an oratory at the foot of the mountain, and a convent near the place where the mountain is cleft. Gaeta is built on a steep point near the sea, forming the port entrance to the bay. Near its east extremity there are a lighthouse and a bastion, and there are other fortifications further in. Ships generally anchor before Gaeta, but not in the middle of the bay, owing to the great depth of the water. Anchoring near the town, the force of the sea is avoided, the most troublesome being with winds from the N.E. to the E. The lighthouse in the S. of the upper city is a fixed light, in lat. 41° 12' 25" N., long. 13° 34' 45" E.

GAGLAN ISLET, Gulf of Venice, a mere rock off the S. coast of the Island of Veglia, in the Quarnerola strait, on the coast of Croatia.

GAGLIANO, VILLAGE and RIVER, Italy, in Otranto, 3 miles N. of Monte Lunga point, and about 4¾ from the Cape Donna Maria de Leuca.

GAHOLA, Gulf of Venice, a small island belonging to Croatia.

GAILLAC, France, department of the Tarn, a town with a pop. of 7500. It is built on a fertile plain and produces excellent wine, which is exported by the River Tarn, upon the right bank of which it stands, and which becomes navigable at that place. It is 5 leagues W.S.W. of Albi.

GAINSBOROUGH, England, Lincoln co., on the River Trent, which falls into the Humber about 20 miles below the town. It has a pop. of 8000, and through the Humber and Trent receives vessels at its quays of 200 tons, where, by means of canals, it distributes their cargoes over the N. of the kingdom, and communicates with Liverpool and London. The tide runs up the Trent at this town with very extraordinary rapidity.

GAIRMOUTH, Scotland, on the river Spey, above the Moray or Murray Frith, where there is a good haven for small vessels, which take on board timber floated down from Strathspey for the purpose of export, or of ship-building; vessels of 500 tons burthen have been constructed here.

GAITHRONISI, or GAIDARO ISLAND, called also PROVENSALA, Greece, at the N.E. entrance of the Gulf of Egina, 3 miles W. from Cape Colonna, between which there is good anchorage in 12, 10, and 8 fathoms water.

GALAM, Africa, W. coast, on the left bank of the river Senegal, 255 miles above St. Louis, a settlement of the River Traders.

GALATA, a suburb of Constantinople, on the N. side of the port by the water, the principal seat of trade, inhabited by merchants and mariners. It stands upon sloping ground, having 20 and 23 fathoms water in the harbour, which terminates it southwards.

GALATA CAPE, Black sea, the S. point of the Bay of Varna, in lat. 43° 10′ N., long. 27° 58′ E. The common anchoring place is between the town and Cape Galata to the S.E., in about 8 fathoms water, open from the E. to S.S.E. A small reef runs out from Cape Galata, round which there are 5 and 6 fathoms water. The N. point of this bay is Cape Sanganlik.

GALATZ, a town of Moldavia, on the River Danube, between the mouth of the Pruth and Sireth, 10 miles above Reni. It has a considerable commerce with the towns on the Black sea, particularly with Odessa. An E.S.E. wind will take a vessel up to this town without towing.

GALEA POINT, Spain, N. coast, W.N.W. from Cape Villano 5 miles, the E. point of entrance to the harbour of Bilbao, remarkable for its white colour.

GALEOTA POINT, West Indies, Island of Trinidad, the S.E. point, 59 miles E. of Point Icaque. Lat. 10° 9′ 30″ N., long. 60° 58′ 20″ W., Admiralty chart, after Don J. F. Fidalgo, in the "Direcion Hidrografico." A dangerous rock is said to exist 6 or 7 leagues from the land here, its site not ascertained.

GALERA DE ZAMBA, anchorage, S. America, Columbia. The point of Galera is so low that when there is a fresh breeze the sea washes over a good part of it. To the W., W.N.W., and N.W. of its western extremity, and at the distance of 2 miles from its projecting part, there are four small banks of different sizes, with 5½ fathoms water, on black sand. Between these banks, and between them and the coast, the depths are 7, 8, 9, and 10 fathoms. The Galera point of Zamba projects 8 miles into the sea, and forms on its S. part an anchorage sheltered from the breezes. In taking this anchorage, great care is necessary on account of the banks on it, and of the Isle de Arenas, or Sandy island, which lies in the middle of the Bay of Galera de Zamba; and every vessel entering must make great use of the hand-lead and pay it close attention.

GALERA POINT, South America, facing the W., upon the N. side of the Island of Margarita, from whence to Point Maria Libre it is S.W. ½ S. 3½ miles.

GALERA POINT, Spain, N. coast, low and surrounded by rocks, which have deep water close to them, forming the points, a mile asunder, of the entrance to the Bay of Corcubion.

GALERA POINT, West Indies, Island of Trinidad, on the N. side, being the N.E. point, 46 miles from Chaguarimas to the W. Lat. 10° 51′ N., long. 60° 54′ W., according to Lieut. Raper. A bank of 16 fathoms, with deep water round, was found by Capt. Columbine, R.N., to be S.E. ¾ S. 7 leagues from this point.

GALERA ROCK, Italy, Gulf of St. Eufemia, between Capes Suvero and Vaticano, above water and very near the shore, 12 miles from the latter cape.

GALERA, rocky shoal, Spain, in the harbour of Cadiz, extending N. and S. nearly 400 fathoms, and distant from the Diamond shoal 412 fathoms. It has 9 feet water upon the shallowest part.

GALINA POINT, Island of Jamaica, 3¼ miles E. of Ora Cabeza, a low point rising gradually to a hill within, very remarkable when seen from the E.

GALIPOLI, or GALLIPOLI, Turkey, in the passage of the Dardanelles, and the principal town in the Strait on the European side, built upon a peninsula and having two harbours, with an old castle commanding the anchorage. A mole here bears a fixed light in lat. 40° 24′ N., long. 26° 39′ 40″ E., at the W. entrance of the strait from the Sea of Marmora, answering to the E. fixed light on the Asiatic side. There is a bay to the W. of Galipoli, where there is anchorage with any wind inside a shoal of 2 fathoms water, discovered by H.M.S. Rifleman in 1829. The road of this port is in front of the arsenal, and has a depth of from 25 to 10 fathoms. The bay behind, where the old lighthouse stood, is avoided on account of a rocky bank there of considerable extent. In sailing by Galipoli, the ledge of rocks at the back of the town being passed, the European side should be kept to avoid the current. The course into the Sea of Marmora, on leaving Galipoli, is E.N.E.

GALISANO, CAPE, Spain, N. coast, near the Island of Santa Marina, the E. point of the entrance into Santander port.

GALISSIONIERE'S ROCK, a vigia, said to exist about lat. 12° 20′ N., long. 54° 49′ W. This rock is mentioned by M. Galissionière and some other navigators; and a danger nearly in the same situation had before that been named the Island of Fonseca. It is said to have been once seen by the Rainbow, a ship of the Royal Navy, and it is vaguely reported as having been seen by some other vessel again, in 1822.

GALITA, FORT OF, Sicily, near Palermo, on the opposite side of the cove to the citadel.

GALITA ISLAND, N. coast of Africa, N.W. by N. from Cape Serrat, distant 7 leagues. It is about 3 miles long, 1 broad, and of moderate height. Off the N.E. point there are some small high rocks, called the Sandfort and the Dog, and there is a sunken rock to the N., with a depth of 3 fathoms over it. Half a league from the S.W. end are two other rocks above water, called Galitona and Agulia. There is a channel between these rocky islets and Galita, with from 10 to 20 fathoms water. All the shores are bold, except the N.W. part. Vessels from the N.E. or

N.W., that seek anchorage in the S. road of Galita, must give a good berth to the rocks off the N.E. point of the island, and having passed them in 10 and 14 fathoms, go on in a similar depth for the E. end of Galita, which may be neared to a cable's length, then rounding this point, turn in to the W. for the roadstead about the middle of the island. There are two creeks here; near the western is a patch of grey land, not far from the water, and opposite that is the best anchorage in 10 and 11 fathoms water, sandy ground, close inshore, and sheltered from N., N.W., and N.E. winds. In N. gales the sea runs round the island, and causes vessels to roll very deep. With a fresh wind, the sails of a vessel should be reduced before passing under the lee of the island, or the masts may be endangered by sudden gusts from the peak at the E. end. In such cases there is anchorage under the E. point of the island in 20 and 22 fathoms. The summit of Galita rises to 1562 feet, in lat. 37° 31′ N., long. 8° 55′ E.

GALLEGA and GALLEGUILLA REEFS, Mexico, at the entrance of the harbour of Vera Cruz; the last lies farthest out, and the castle of San Juan de Ulua, or Uloa, stands upon the inner edge, on the N.W. part of which is a revolving light, 80 feet high. These banks are on the E. side of the entrance, with the reef of Punta Gorda on the W., lying nearly two miles asunder, and between these, just N. and S., is the fair way of the port.

GALLEON'S BANK, lat. 15° 56′ N., long, 49° 50′ W., a bank or reef discovered in July, 1730, by Longueville, the pilot of the San Fernando, commanded by N. de Navarro, admiral of the Spanish galleons. By the statement of M. Longueville, there is no doubt of the existence of this danger. The San Fernando struck and passed over it without damage. Other vessels in the same fleet struck with more or less violence, but without actual injury. This statement is in the Depôt de la Marine, in Paris. In the same depôt there is the memorandum of the existence of a bank in 15° N. lat., and 228 leagues E. of Martinique, upon which there is said to be 40 fathoms water, and a bottom of fine sand, over which some one named Joachim Voette is said to have passed and sounded.

GALLINAS POINT, South America, province of Guajira, the most N. point of that province, N.E. of Bahia Honda.

GALLINAS, or GALLINHAS, RIVER, coast of Africa, the entrance being in lat. 7° 0′ 1″ N., long. 11° 38′ 5″ W., according to Capt. Vidal, R.N. The bar is only passable for large boats, or small coasting craft, and is impassable during the rains, or else extremely dangerous to attempt. On passing the bar, a spacious sheet of water is discovered widening in every direction for 3 miles, studded with islands, which were not long ago occupied by slave-dealers, affording favourable situations for trading factories. Thence the river runs in three branches to the N.W., N., and N.E. The first branch joins the Boom Kittam river, and affords a direct inland water communication with Sierra Leone, but in the dry season too shallow for canoes to pass. The second branch runs past the town of Ghindamar, 9 miles from the sea, and is navigable about 5 leagues for large canoes. The third branch runs close inside the sea-beach to the S.E. about 4 miles, and then turns suddenly to the N.E. at a place called Sulimane, whence it is navigable for large canoes about 7 miles. This branch is the S.E. boundary of the Gallinas territory.

GALLINHA ISLAND, west coast of Africa, on the Rio Grande or Bolola river, in lat., the W. point, 11° 27′ 30″ N., long. 15° 45′ W.

GALLION'S REACH. See *Thames*.

GALLIPOLI, PORT AND TOWN OF, Italy, on the S. coast, and on the W. side of the Gulf of Taranto. There are islands in the port, upon the most W. of which, named St. Andrea, there is a tower, in lat. 40° 2′ N., long. 17° 56′ E. There are 5 and 6 fathoms water between these islands, and also between them and Gallipoli 9, 10, and 12 fathoms. The town is built upon a peninsula and well fortified, having a pop. of 8000 or 9000. The principal exports are oil and cotton. There is a good road to the E., where a vessel may be sheltered in S.W., S.E., and N.E. winds. To go into the road before Gallipoli a vessel should sail outside the islands and anchor before the city in 10, 11, and 12 fathoms, within gunshot of the castles. An E. wind blows off the shore here, and a S. one directly off the city. The town must not be approached nearer than gunshot, on account of a rocky bank under water. From the Island of St. Andrea a bank extends 2 miles to the W., on which the depth is from 6 to 12 fathoms, and near the edge suddenly drops into 50 and 54 fathoms. The middle of the gulf is very deep, but there are soundings along the shores, the bottom generally clay. Gallipoli bears from Point Alice N.E. distant 52 miles; from Cape Nau N.N.E. ¾ E. distant 68 miles; and from Cape St. Vito E.S.E. ¾ S. distant 44 miles.

GALLO CAPE, Greece, on the mainland, in lat. 36° 41′ 20″ N., long. 21° 54′ 56″ E., about E.N.E. ¾ E. from the S.E. point of the Island Cabrera, distant 6½ miles, and by a channel a mile wide from the Island of Venetico. The land from Cape Gallo runs N.E. ½ N. to the point of Koron. Due E. from Cape Gallo, in lat. 36° 47′ N., and long. 22° 1′ E. There is said to be a shoal near with 12 fathoms water, marked upon most charts, but considered to rest upon very doubtful authority.

GALLO, CAPE DI, Sicily, near to and N. of Palermo, in lat. 38° 14′ N., long. 13° 19′ E. Cape Orso lies W. by S. of it, distant 10 miles, and Cape St. Vito, the N.W. point of Sicily, W. ½ S. 26 miles.

GALLO, or COCK'S POINT, West Indies, Island of Trinidad, W. of Cedar bay, and N. of Icaque point, at the S.W. part of the island.

GALLO, POINT DE, Spain, Bay of Gibraltar, E.S.E. ¼ S. from Fort Mirador three-quarters of a mile.

GALLOPER SHOAL, coast of Suffolk, a dangerous bank, having not more than 8 feet in some places at low water. It extends 5 miles N.E. and S.W., has 7 fathoms on each end, and is not a mile across at the broadest or middle part. This part lies S. by E. ½ E. from the high light at Orfordness distant 23 miles; E.S.E. ½ E. from the buoy on the Long Sand head, distant 13 miles; E. by N. 13 miles from the Kentish Knock light-vessel; and N.E. by E. ½ E. 31 miles from the North Foreland lighthouse. There is a light-vessel at the S. end of this sand, bearing two fixed lights, to clear this sand and others. The lights are placed horizontally on separate masts. The tide flows here full and change till three-quarters after 11, running nearly three knots. The channel between the Galloper and the Long Sand head is about 12 miles wide, and has from 20 to 17 fathoms in it, shoaling gradually towards the Long Sand. On the E. side of the Galloper are 12, 16, 18, and 20 fathoms; near the

S. end 14 fathoms; three-quarters of a mile outside the N. end are 14 fathoms, and close to this end 9, 8, and 7 fathoms. A new shoal extends S.W. $\frac{1}{2}$ S. 5 miles, from 9 fathoms at each end, on which are some patches, with only 4 and 4$\frac{1}{2}$ fathoms. The N. lies S.S.W., about 3$\frac{1}{2}$ miles from the buoy of the Galloper. S.W. by S. from this spot is another, with similar depth, three-quarters of a mile distant. The soundings at about half a mile from the E. side are from 19 to 20 fathoms, and on the W. side rather more. The distance between the N. end of the shoal and the S. end of the Galloper is above 2 miles, with 16 to 20 fathoms. These shoals are called, in Captain Hewett's survey, the Four Mile Knolls, and are situated on and connected with the North Falls, and continue in a S.W. $\frac{3}{4}$ S. direction, having 7, 9, and 10 fathoms, to the lat. of 51° 35′, where they deepen to 13, 14, 12, and 17 fathoms, and turn in a S.W. by W. direction, to the lat. of 51° 28′, where commences what may be termed the South Falls; these stretch down to the lat. 51° 13′, and are scarcely more than a mile across in any part, the shoalest water being 4$\frac{1}{2}$ and 5 fathoms, in the lat. from 51° 22′ to 51° 17′, about 6$\frac{1}{2}$ miles to the E. of the Goodwin light-vessel, between which there are from the Falls 24, 28, 30, then 13, 12, and 10 fathoms to the light-vessel.

GALLOPING ANDREW SHOAL, America, Newfoundland, between Cape Race and Cape Chapeau Rouge. On the mainland, within Pardy's island, are two singular white marks in the rocks. The N. of these brought on with the N. part of Pardy's island and Iron island, N.E. $\frac{1}{4}$ N., will lead upon the Galloping Andrews, having upon it 5 fathoms water.

GALLOWAY, PENINSULA OF, Scotland, the S. point of which is called the Mull of Galloway, which is the most S. point of Scotland. There is here a three-minute intermittent light, 325 feet high, lat. 54° 38′ N., long. 4° 51′ W. The principal port is Port Patrick, a neat town of 1,000 inhabitants, with a small haven from whence packets sail daily for the N. of Ireland.

GALLY BOY'S HARBOUR, Newfoundland, between Cape Race and Cape Chapeau Rouge. It lies on the E. side of La Poile bay, opposite Tooth's head, and is small, snug, and convenient for ships bound to the W. The N. point is high and steep, with a white spot in the cliff, and near its S. point are some hillocks close to the shore. To sail in or out, a vessel must keep the N. side on board, and anchor as soon as she is within the inner S. point in 9 or 10 fathoms, good ground and sheltered from all winds. A mile to the N. of Gally Boy's harbour, between two sandy coves on the E. side of the bay, and nearly two cables' length from the shore, is a sunken rock that just uncovers at low water.

GALTANS' HARBOUR, America, Newfoundland, between Chapeau Rouge and Cape Ray. There are four harbours on the S. side of Long island, the most E. of which is called Galtans; it is small, and lies near the S.E. point of the island. In the harbour there are from 15 to 24 fathoms.

GALVESTON, PORT AND ISLAND, America, Gulf of Mexico, State of Texas. This harbour may be approached with less apprehension of danger than any other in the United States of America, to which it now belongs. The entrance of the bay is between Point Bolivar on the N. and the E. end of the St. Louis or Galveston island on the S.

lat. of the island, E. end, 29° 19′ N., long. 94° 48′ W. There are four channels between them, by which ships may enter. Galveston island is about 32 miles long, and trends N.E. and S.W. It is low, but it cannot be mistaken. There are three single trees in the middle. At the W. end there is a wide pass, with a small island nearly in the middle of it, and back about 7 miles is a long wood called Oyster and Chocolate Dye wood. From the S.W. coast of the island the coast continues S.W. distant 10 miles, to the mouth of the Rio Brazos, with 3 and 4 fathoms water at 2$\frac{1}{2}$ and 3 miles from the shore. Passing up to the mouth of the river, a long house is seen on the N.E. point, called Michaels; and to the N.W. of the mouth the woodlands of Brazos, distant 7 or 8 miles from the beach. The bar at the entrance has from 4$\frac{1}{2}$ to 5 and 6 feet water on it, but like all the other entrances is subject to changes. At the port of Velasco, at the mouth of the Brazos river, there are always experienced pilots, and a vessel drawing 6 feet water can enter the river without difficulty. To return to Galveston : it has been considered difficult to make on account of its being low land. The increase of houses, however, has been so great, and many of them are so lofty, that from the masthead of a vessel they may be seen at the distance of 20 miles. Vessels of heavy burthen should not approach the bar nearer than six fathoms, and then by making the signal for a pilot they will be instantly attended to. Vessels drawing 8 feet or less may approach as near as 4 fathoms in order to obtain a pilot. Galveston harbour is perfectly safe, but vessels should never attempt to enter here at night without a pilot, nor approach nearer than 5 fathoms. The lat. of the bar is 29° 15′ N., long. 94° 49′ W. The anchorage of Galveston island at the E. end is 29° 18′ N., long. 94° 48′ W. High water full and change 8 A.M., rise 3 feet.

GALWAY, BAY and TOWN, Ireland, W. coast, the N. entrance to which is Gulin head, the S. being the point of Durus. There is good holding in Galway road, but when the wind is between S. and W.S.W., a great swell sets in. The anchorage for small vessels is on the E. side of Mutton island, 2 cables' length from the shore, when the middle of the island is on with Black head. Ships drawing 10 or 12 feet must ride more to the S.E., so as to have Black head fairly open of Mutton island, the steeple of Galway bearing N. Vessels drawing 10 or 12 feet may go into the harbour of Galway at high water, taking the last of the flood, steering between the two perches and lying between the two quays. Ships coming in from sea, and bound to Galway, after making the three-minute revolving light on Great Arran island, 498 feet high, in lat. 53° 7′ N., long. 9° 42′ W., may take either the N. or S. channel, both being free of danger. The N. shore, from Casleh bay to Barna creek, a distance of 5 leagues, is all clear. The S. side is also free from danger; and there is anchorage for a large ship about two miles S. E. of Black head, in 6 to 10 fathoms, sandy bottom. After passing Black head, by bringing Mutton island fixed light, which is 33 feet high, to bear E. by N., and proceeding in that direction, a vessel will pass to the S. of the Black rock, and carry not less than 5 fathoms up to the lighthouse, to which a berth of a good cable's length must be given; and as soon as the lighthouse bears N.W., within half a mile, the anchor may be dropped. The town of Galway is situated inland, at a considerable distance from the port, from which goods are conveyed by lighters. It was formerly a walled town, and contained many ancient buildings, most of which have been newly built. The collegiate church is an old gothic building. It carries on a considerable traffic, having an exchange, and several linen manufactories. Its fisheries are productive and extensive; and large quantities of kelp are burned in the bay.

GAMBIA RIVER, Africa, W. coast. Cape St. Mary, the S. point of entrance, lies in lat. 13° 30′ N., long. 16° 41′ W. Between this cape and Banyan point, to the E., there are extensive banks in the river, running out 6 miles from the land, called St. Mary's or Banyan shoals; they only break when the breeze is fresh. The N. side of the entrance forms a spit in the direction of Broken islands, between the distance is scarcely 5 miles. Three leagues S. by E. from these islands is Barra point, which, with Banyan point, forms the Narrows, the passage being only 2½ miles across, and here the tide runs with wonderful rapidity, especially at the ebb, for nearly 9 hours. There are rocks off Barra point about a mile out. From thence to Dog's island, on the same shore, is 9 miles, and it is 14 from Dog's island to James' island and Old Fort. In sailing into the Gambia river, a vessel should keep towards the Bird islands in 5 or 6 fathoms, as far as Broken islands, and come no nearer to St. Mary's shoals than 6 fathoms, and when Banyan point bears S., steer for Barra point, thus clear of the banks, but approach no nearer than 6 fathoms, there being on both sides but 5 fathoms, and the next cast but 3. The flood sets directly over Barra point, and the ebb over the shoals, which, in light winds, renders it necessary to have one anchor always ready as a precaution. There are numerous new and old establishments, English and French, on this river, which is navigable for vessels of burthen as far as the port of Cassan, or 70 leagues upwards. The tide flows at full and change here at 10ʰ 30ᵐ, rise 7 and 8 feet. The winds in this part are generally N.W., except during the winter in December, January, February and March, when they are between N.E. and E.S.E.

GAMBLE SAND, Durham, Hartlepool, at the entrance, between the Tees river and Sunderland. It is avoided by keeping close to the pier.

GAMLA CARLEBEG, Finland, Gulf of Bothnia, a small seaport, accessible only to vessels that draw 10 feet water. The principal exports are tallow and timber.

GANDI RIVER and PORT, coast of Colombia, South America, province of Darien, S. of Cape Tiburon, at the W. entrance of the Gulf of Darien, and about 12 miles S. of that point.

GANDO, PORT OF, Grand Canary island. See *Canaries*.

GANDOROULL ISLAND, Greek Archipelago, E. of Khios island, one of the islands forming with that of Spalmatori and others, the Port Spalmatori.

GANELLY ISLAND. See *Scilly Isles*.

GANNET ISLES, Labrador, a cluster of islands lying from 7 to 11 miles off the mainland. The outermost island bears from the Wolf rock N.N.W. ¾ W., distant 10 leagues, on the coast between York point and Sandwich harbour.

GANNET STONE, coast of Devon, near the N.E. point of Lundy island. Moderately sized vessels bring up here in 10 fathoms sand and mud, with the Gannet stone just shut in with Tibbet point, bearing N. ¼ E.; the farm-house appear-

ing just over the land W. by S.; and Rat island S.S.W., distant half a mile; in this spot they are able to clear either end of the island upon a change of wind. On this island are two lights (vertical) in one tower, revolving and fixed, 86 and 40 feet high.

GANNY or GRANNY COVE, America, Newfoundland, S. of the Bay of Bulls, on the coast from St. John's harbour to Baccalou island.

GAPEAU, France, S. coast, an anchorage in the Bay of Hyeres.

GARAFFA, TOWN OF, Italy, E. of Cape Spartivento on the S. coast, and 7 miles N.E. of Bruzzano point; it is an inconsiderable place.

GARAJAO CAPE, or BRAZEN HEAD, Funchal bay, Madeira.

GARCIA, SAN, POINT, Spain, Bay of Gibraltar, N.N.E. ½ N. from Carnero or Cabrita point, and 1¾ of a mile distant.

GARDENSTONE, Scotland, a small fishing place on the E. coast, in Banff co.

GARDIOLA, Spain, E. coast, a small round hill, with a tower upon it, between Monsia and La Rapita, serving as a mark of direction for that part of the low coast of Spain called the Alfaques of Tortosa.

GARDNER'S ISLAND and BAY, Long island, in the United States of America, in New York. The entrance of Gardner's bay is formed by the N. end of Gardner's island and the S. end of Plumb island. The passage from Gardner's bay through the sound towards New York is between the W. end of Plumb island and Oyster pond, in which there are from 4 to 20 fathoms water. The N.E. end of Gardner's island is 4 leagues W.N.W. from Montock point.

GARE, SOUTH and NORTH, Durham co., at the entrance into the River Tees, the channel in being between these two banks, running from the bar inwards nearly S.W. by S. for 3 miles. This channel is regularly marked by beacons and buoys.

GARET RIVER, Africa, N. coast, S.W. by W. ¼ W. 17 miles from Cape Tres Forcas.

GARFANTA POINT, Spain, N. coast, a low point with rocks round it, N.W. ¾ N. from Port Santona.

GARGALO CAPE, Mediterranean, Island of Corsica, 14 miles from Cape Rivellata S.W. by S. It is a large point, rugged at the summit, and has an island near it about 2 miles in circumference, with a channel between them, which will admit galleys. Off Cape Gargalo there are two detached rocks.

GARGANO, PROMONTORY OF, or MONTE GARGANO, Italy, Gulf of Venice, off one point of which is a rock under water, called the Testi di Gargano. There are several points round this obtuse headland, as Point Felice, upon which is a tower to the W. of the Testi di Gargano; Porto Novo, and, 2¼ miles beyond Porto Novo, Cape Viesti, having a town and fort upon it, and a watch-tower on its N. point, near which is a rocky island with a passage for boats with one fathom between that and Point Croce, whence the island is named. With W. and N.W. winds there is a good anchorage under Cape Viesti, which is in lat. 41° 52′ N., long. 16° 10′ E. This is the extreme point of the promontory of Monte Gargano. The town here has a pop. of 5000, but its trade is very limited.

GARIA BAY, Newfoundland, between Cape Ray and Cape Rouge, and also between Little Ireland and La Moine bay; this bay is only adapted for small vessels. It has several coves, and before them are islets and sunken rocks scattered along shore, which in bad weather discover themselves. Sailing into the bay in coasting along the shore, a white headland will be seen, which is the S. point of an island lying under the land of the E. point of the bay, and a little W. of two green hillocks on the main. A vessel should bring this white point or headland N.N.E., and steer directly towards it, keeping between that and several islands that lie to the W.S.W.; from the white point the course into the bay is N. by W.; a vessel should keep towards the E. point, which is low.

GARIGLIANO TOWER and RIVER, Italy, in Naples, E. by N. from Gaeta 7 miles.

GARNET BAY, W. coast of Africa, the Angra dos Ruivos of the Portuguese. 2 leagues to the S. of it are seven small hills, called the Seven Capes, an excellent landmark, in lat. 24° 41′ 12″ N., long. 15° 0′ W., according to Capt. W. F. Owen, R.N.

GAROUPE CAPE, France, S. coast, about 1¼ mile from Antibes. There is a tower upon this headland, near the chapel of Notre Dame de·la Garde, 79 feet high, on which there is a fixed light 338 feet above the sea level, visible 18 miles, in lat. 43° 33′ 51″ N., long. 7° 8′ 10″ E. In coming from the E., vessels see at the same time the lights of Villa Franca, Antibes, and Garoupe, but this last is only visible to vessels from the S. or S. W. till round the peninsula.

GARRACHICO, Canary islands, in the seventeenth century was the principal town of Teneriffe, ruined by an earthquake and volcanic eruption in 1706, and at present a village without a port, that which once existed having been filled up with lava.

GARRAF, LAND OF, Spain, E. coast; the land 7 miles E. of the Point of Christoval, with a watch-tower, is thus named, S.W. of the River Llobregat, 8 miles.

GARRAWAY, Africa, S.W. coast, and N.W. of Cape Palmas, a town well known by three hills between itself and Grand Sestros or Sesters.

GARRIMORE POINT, and MOUNT GARIMORE, French Guyana, South America, 15 leagues N.W. of Cape Orange. This is the E. point of the River Aprouague or Appouaque, of which the navigation is little known.

GARROTE HARBOUR and ISLAND, South America, about 4 miles W. from Point Manzanilla, Darien. The harbour is formed on the E. by Great Garrote island, and on the W. by the Pelado islet. The mouth is not more than three-tenths of a mile between the reefs, to the W. of Great Garrote island and Pelado islet. Its direction is from N. to S., and afterwards to S.E., with a depth of from 6½ fathoms in the interior of the harbour, to 12 and 18 in the mouth, on clay. It is sheltered from the seas and winds of the N.E. breeze. The Hill of Garrote, of moderate height, with a peaked top, is seven-tenths of a mile from the coast. S. by E. ½ E., at the distance of 3½ miles from the little harbour of Garrote is the mountain of Capiro or Campana, high, and continually covered with clouds.

GARSAY ISLAND, one of the Orkney isles. See *Orkney*.

GARS HEAD, Prussia, in Pomerania, 9 leagues N.E., and N.E. by E. from Coslin, on the S. coast of the Baltic sea. It lies W.S.W. 12 leagues from Kolberg.

GARVEN PORT, Cornwall, to the E. of Port Isaac, on the N.W. coast, and with a similar depth

of water, or sufficient for a vessel of 200 tons at high tide. It is principally used by fishing-vessels. At spring-tides there is a depth of 22 or 23 feet; at neaps from 11 to 12. This port is in the district of Padstow.

GARVENY ISLANDS, Ireland, N. coast, on the E. side of Malin head, and about 1½ mile from the shore. There are several of these islets. At a quarter of a mile E.N.E. from Stanoff, which is the N. round rocky isle of the Garvenys, are Doherty's rocks, the E. ends of which dry with spring tides, and the other parts at half ebb.

GASKINAAN SOUND, Ireland, S.W. coast. The passage between Cape Clear island and the Skerkin islands, so called from two rocks that lie in the middle, having a channel on each side.

GASPAR CHICO ISLANDS, Gulf of Paria, Trinidad island, West Indies, W. of Port d'Espagne, and S. of the Careenage. Lat. about 10° 38′ N.

GASPAR GRANDE ISLAND, Trinidad, West Indies; S. 4° W., 1 mile from Taitron's point, is the W. end of Gaspar Grande, named Espolon, or the Cock's Spur. This island extends E. nearly 1¼ mile to Punta de la Reyna, the E. point, and is about half a mile broad. Its elevation is 337 feet; and its coasts form coves or little bays, in which small vessels may anchor.

GASPE BAY, America, Gulf of St. Lawrence, a most important place in a maritime point of view. It has a good outer roadstead off Douglas town, with a harbour at its head, capable of receiving and holding a numerous fleet in perfect safety, as well as possessing a dock, where the largest ships might be hove down and refitted. The course of this bay, from Flat island to the end of Sandy beach point, is N. by W. ½ W., rather more than 16 miles. From the Flower-pot rock (which fell down in 1846) to the same point, the course was N.W. ½ W., distant nearly 11½ miles. From Point Peter the land rises in undulations to a chain of mountains 5 miles inland, from the S.W. shore of the bay. These mountains reach an elevation of 1,500 feet above the level of the sea, and sweeping round Mal bay, terminate with the Percé mountains. The S.W. shore of Gaspé bay, from Point Petre to Douglas town, 12 miles, is a succession of precipitous headlands. The cliffs of bituminous shale and sandstone, are 200 feet above the sea in the highest parts. Shoal-water extends nearly a third of mile from the cliffs; all vessels should beware of this danger.

GASPE CAPE, America, Gulf of St. Lawrence, a lofty and remarkable headland of limestone, having on its N.E. side a magnificent range of cliffs, which rise from the sea to the height of 692 feet. Flower-pot rock lay off the S.E. extremity of this cape, and was also a very singular object, the base being worn so much away by the waves, that it seemed wonderful how it resisted the pressure of the ice. It disappeared in 1846 during a gale of wind, according to the report of Captain Bayfield, R.N. The N.E. side of this bay is thickly covered with the houses of the fishermen, for a distance of 5 miles within Cape Gaspé. The principal fishing establishments belong to Jersey merchants. There is anchorage, with good holding-ground, in not less than 17 fathoms, except within a quarter of a mile of the shore, abreast of St. George's cove, Grande Grève, and Little Gaspé. The word cove, however, is inappropriately applied to any part of the shore between Grande Grève and the cape; for though there are establishments for fishing at that spot, there are no coves. That side is bold and free from danger in every part, with the exception of the Seal rocks, which are the only detached dangers in the bay. The Seal rocks are 6¾ miles within Cape Gaspé, 1 mile S.E. by S. from Cape Brule, and half a mile offshore. The length of the reef, from 3 fathoms to 3 fathoms, in a direction parallel with the shore, is half a mile, and its breadth a quarter.

GAT, the general name for a channel in Holland, along the Belgian coasts, and to the N. and N.E. connected with some word explaining the locality, as East gat, the channel W.N.W. from Walcheren, between Caloo sand and the Queen's or the Western gat, between Eyerland and the N.E. end of the Texel island.

GAT, CAPE DE, Spain, S. coast, more correctly CAPE DE GATA, a Spanish term for the head of a vessel. It stands in lat. 36° 43′ N., long. 2° 13′ W., having upon it the tower of Testa del Cabo, it being the E. point of the Gulf of Almeria, distant from the W. point, or Point St. Elena, 21 miles, bearing from each other W. ¼ S. and E. ¼ N. About half a mile from the Testa tower is the castle of St. Francis de Paula, on the summit of a steep eminence, having an island and some rocks very near it. There are many cross-currents about this headland, which enhance the dangers on this part of the coast. Captain Toup Nicolas, of the Royal Navy, reported a danger unknown before off this cape, being a sunken rock full a mile from the land, within which Captain Nicolas passed almost close. The old building on Cape de Gat bore about E.N.E. by compass, and the white mark in the rock to the E. of the cape about N.E. by N. by compass, the variation just two points W.

GATEHOLM ISLAND, South Wales, a small island very near the shore, at the bottom of a shallow bay, between St. Bride's bay and the W. point of Milford haven, or West Dale bay.

GATEHOUSE, W. coast of Scotland, on the River Fleet, an inconsiderable place, to which coasters of 80 tons navigate from the sea.

GATHRONISI, or GADARO, ISLAND, Greek Archipelago, 4 leagues from Cape Monodendri. It is 5 miles long, and has several rocky islets attached to it, situated nearly between Cape Monodendri, the island Lipso, and the S. side of Samos, in lat. 37° 27′ N., long. 27° E., directly S. from the S. entrance of the Strait of Samos.

GATHRONISI, or GADARO, ISLETS, Greek Archipelago, E. of Tenedos, between the N.E. point of that island and Koum Bouroum, on the main. A shoal lies between the two Gadaro islets and Koum Bouroum, with 19 feet water.

GATO TOWN and CREEK, the last generally called REGGIO, on the S.W. coast of Africa, uniting with the River Benin, the point between being named Reggio point. In the rainy season the whole coast is inundated, almost close up to Gato.

GATTANDAR, Africa, W. coast, a small Moorish town of huts, upon a sand-hill, on the strand, opposite the town of St. Louis, on the Senegal river. Lat. about 16° 0′ 48″ N., long. 16° 33′ W.

GATTO, CAPE, Island of Cyprus, S. side, in lat. 34° 33′ N., long. 33° E.

GATTEVILLE, CAPE, France, near Cape Barfleur. It extends three-quarters of a mile from Barfleur to the N., a low rocky point, stretching 300 fathoms into the sea, in a line with the steeple of the church of Gatteville, on the summit of the

coast, which is very rocky. This point bears the light of Barfleur 236 feet above the sea, in lat. 49° 42′ N., long. 1° 16′ W. It is a revolving half-minute light, and with the light of Cape D'Ailly and the light of La Héve, is of great use to the vessels of Havre, in going to or returning from sea to the mouth of the Seine.

GAUDIA, Italy, a small town on the coast between Cirella and Cape Suvero, 26 miles from Diamante.

GAVANDE ROCKS, Channel islands, about a quarter of a mile from Brecqhou island.

GAY HEAD, America, United States, a peninsula in Martha's vineyard, about 3 or 4 miles long, by 2 broad. It has a lighthouse on the extremity, 172 feet high, the light revolving every four minutes, in lat. 41° 21′ N., long. 70° 50′ W.

GAYO, or GAJO, PORT OF, Greece, Island of Paxo, on the E. side, affording good anchorage for a number of small vessels. It has an inner harbour, formed by an island almost in contact with the other, and defended by a circular battery. There is a lighthouse on this island, called the Convent island, having a fixed light in a building 70 feet high, the lantern 107 feet above high water, in lat. 39° 11′ 30″ N., long. 20° 12′ 20″ E.

GAZA, RAZZEH, or RAZZA, Palestine, S.W. of Jerusalem, on the W. coast, in lat. 31° 28′ N., long. 34° 30′ 10″ E., formerly deemed the frontier town of the Land of Canaan towards Egypt. The current runs strongly to the N. along this coast and that of Syria.

GEAR BANK, coast of Flanders, near the land, between Nieuport and Ostend. Within and near the shore there is a channel of 8, 9, and 10 fathoms water.

GEDGES SHOAL, S. coast of Cornwall, on the N. shore of Helford haven, to be avoided by keeping rather nearest to the S. shore. It has only 6 feet water. On the S. side there is a steep point, and on the N. a low and flat one.

GEER, GHIR, or GHEER CAPE, Africa, W. coast, more correctly RAS AFERNI, situated, according to Lieut. Arlett, in lat. 30° 37′ N., long. 9° 52′ W. It rises boldly from the sea, 25 miles S. of Cape Tepelneh. The back land has an elevation of 2,895 feet above the sea. The country appears wooded, and villages and tents are seen here and there. On the W. side, this cape presents a bold, bluff slope, the highest part 1,235 feet above the sea. The depth of water diminishes gradually, and soundings are found 26 miles off. Lieut. Raper makes the long. of this cape 9° 54′ W., lat. 30° 37′ N.; the Chevalier Borda 30° 37′ N., long. 9° 53′ E.

GEFLE, GIAWLF, or GEVALIA, Gulf of Bothnia, on a river of the same name, half a mile from the sea. The town is well situated for commercial purposes. It is the capital of the province of Geflaborgs, in Sweden, upon an arm of the sea, which divides its site into two islands, 56 miles N.N.W. of Upsala. Its exports are principally bar-iron, timber, pitch, tar, and planks. Lat. 60° 40′ N., long. 17° E.

GEISSINGBOAN ROCK, Norway, W. coast, at the entrance of the Ramsoe fiord, approaching to Drontheim. It lies at the entrance; and a vessel should steer to the E. of it, in seeking to reach and open the Drontheim Leed.

GELLEN, STRAIT OF, Pomerania, between the Island of Rugen and the continent, being from 1¼ to 14 miles broad. The E. entrance is called the Bodden or Boden, and is navigable for

IV.   P

the largest vessels: but in the narrows the depth is only from 3 to 4 fathoms. They require to be constantly freed from the sand that accumulates in them, to defray the expense of which a toll is levied on all vessels passing through.

GELSA, TOWN and HARBOUR OF, Gulf of Venice, in the Island of Lesina, principally visited on account of its quarries of beautiful marble.

GENARO, POINT and TOWN OF, Gulf of Venice, a little N. of Cape Otranto, the S.W. limit of that gulf, on the E. coast of Italy.

GENERAL TURNER'S PENINSULA, W. coast of Africa; the tongue of land between the Boom Kittam river and the sea-shore, to which that river runs parallel 8 leagues in length, terminated by the Forks, in long. 12° 8′ 30″ W., lat. 7° 14′ 24″ N., according to Captain Vidal, R.N.

GENNIS, ST., HAVEN, a tide-harbour for small craft on the N. coast of Cornwall.

GENOA, GULF and PORT OF, in the Mediterranean, N.E. by E. ¾ E. from Savona, in the kingdom of Piedmont, standing at the foot of high mountains, on the N. part of the gulf of the same name. The city is enclosed on the land side by a double wall,—one about 6 miles in circumference, enclosing the city, and one embracing an outer circumference of 12 miles, including several hills. From the sea, the city rises in the form of an amphitheatre, the harbour in the centre. The background is formed by the Apennines, while on the left the distant Alps exhibit their snow-crowned summits. The buildings beyond the port are magnificent, particularly those of the suburb of San Pietro d'Arena. The port is formed by two moles, which rise from 16 to 18 feet above the sea, that on the E. being called the Old Mole, and that on the W. the New Mole. The heads of these moles are about 300 fathoms asunder. On the old mole are guns for the defence of the harbour. The lighthouse is a square white tower, on the new mole, the lantern of which is 370 feet above high water, the building being 100 feet high. The light is revolving, seen in clear weather 24 miles, the time elapsing between the flashes is 30 seconds. This tower is in lat. 44° 24′ 36″ N., long. 8° 53′ 5″ E. There is a fixed light on the W. mole head, and another of the same kind on the E. mole, both harbour lights. Signals are made from the great light-tower in the daytime to indicate the approach of vessels. A vessel once within the moles moors immediately with a cable or two upon them and an anchor out ahead. There is space enough for four or five sail to lie at their anchors, and moor with a cable each way. The usual anchorage for merchantmen is within the old mole, on the E. side, where the water is too shallow for large vessels. There is anchorage also outside the moles, 2 or 3 miles distant, in from 10 to 24 fathoms water, the lighthouse bearing about N.N.W., the bottom clay, and holding well. S.W. winds raise a great swell, but the N. wind is the most violent, and drives ships from their anchors. At the bottom of the port, about the centre of the city, there are two basins, one of which is designed to receive the galleys, and the other the smaller trading vessels. Both are closed with chains. From the sea this magnificent city is seen and recognised at a distance by the whiteness of its marble buildings, and merits its character of "the superb," bestowed upon it by the Italians, in times when it balanced the power and riches of Venice. The suburb of Pietro d'Arena, the first entered

from the sea, separated from the rest of the city by a hill on the left of the port, forms a promontory, which advances into the waves, and it is upon a point of rock here that the celebrated light-tower is built, called La Torre della Lanterna, at the foot of which, on the right hand, the new mole commences, on which are batteries for the defence of the port. The pop. of Genoa is about 76,000. No longer free, but a dependency of Sardinia, the trade of Genoa is greatly reduced. It has still, however, manufactures of velvet, gold and silver tissue, and paper, but its mariners are no longer renowned on the seas, and the Dorias and naval commanders of this renowned city are not now discoverable in rivalry with Venice, nor their descendants as competitors with the seamen of other nations in skilfulness and daring.

GENOVESE, PORT or COVE, Africa, N. coast between Capes Mavera and Bona. Vessels may anchor before this cove in from 15 to 7 fathoms, under the protection of a small fort, but wholly exposed to N.E. and E. gales, which force in a heavy sea.

GEORGE D'ARBORA, ISLAND OF, Greece, a small island between Zea and Hydra, and nearly facing the centre of the Gulf of Egina. It is about 2 miles in length from S.E. to N.W., narrow, rocky, and uncultivated. A few families occupy it, possessing scanty flocks of sheep and goats. There is no danger about its shores. The summit is in lat. 37° 28′ N., long. 23° 56′ E.

GEORGE, FORT, Scotland, on the Inverness side of the strait communicating between the two inner lakes of Murray Frith. Lat. 57° 35′ N., long. 4° 9′ W.

GEORGE'S BAY, or GRAND BAY, St. Nicholas, one of the Cape de Verde islands.

GEORGE'S BAY and TOWN, West Indies, Island of Grenada, called by the French La Grande bay; bounded by Cabrit or Goat's point to the S., and Molenier point to the N.; Point Eloy, more than a mile S. of Molenier point, has a rocky shoal off its extremity, extending a considerable way into the sea. There is also a bank of hard uneven ground, the centre of which lies nearly W. from Moncton's redoubt, distant half a mile from the entrance to the Careenage, and S.W. from Fort George; it is said this bank has several spots of coral upon it, and towards its middle has only 2¼ and 3 fathoms water, but in other parts 4, 5, and 6 fathoms, increasing in depth towards its edges, so that to the southward there are 8, 9, and 10 fathoms. There is also a patch of rocky ground, called the Anna shoal, of 3 fathoms, lying about three-quarters of a mile or less from St. George's fort, but this is most probably only part of the bank; an extensive coral reef also runs about a mile off Goat's point, in a direction towards the S.W. extremity of the island, or Point Salines, which must be avoided, and passed by large vessels on the outside. George town has one of the finest harbours in the West Indies, called the Careenage, where vessels may lie land-locked in from 4 to 10 fathoms, close to the wharfs. The town is well-built, and divided into two parts, one called the Bay town, the other the Careenage; there is also a lagoon within the Careenage, from 12 to 24 feet deep, but the entrance is nearly dry. In entering the last, vessels should keep, if possible, in mid-channel, in 10, 12, and 14 fathoms water, as towards the land on either side it shallows to 4, 3, and 2 fathoms. On the starboard side, still further in, there is a coral bank, which nearly dries. Ships do not go E. of the bank, but borrow a little towards St. George's point, and steer off the Careenage E.N.E. Small spots of 3 or 4 fathoms are on each side, but in the middle of the passage there are 8, 9, and 10 fathoms, until towards the further end, where it becomes shallower.

GEORGE'S, ST., BAY, America, 10 miles S. by W. of Antigonish harbour, Nova Scotia.

GEORGE'S, ST., Bermuda, one of the four signal stations in these islands, and the name of the island and principal town, in lat. 32° 22′ 23″ N., long. 64° 37′ 40″ W. Fort Cunningham stands in lat. 32° 22′ 44″ N., long. 64° 41′ W., after Captain Barnett, R.N., 1840.

GEORGE, ST., CAPE, America, N.W. of the N. entrance of the Gut of Canso, a remarkable promontory, at the distance of 11 leagues to the E. of Pictou harbour, in the Gulf of St. Lawrence. A course of 6 leagues thence to the S.E. will lead to the entrance of the gut, whence a ship may run along the Cape Breton shore.

GEORGE, ST., CAPE, Greece, S.E. point of the Morea, 3 leagues N.N.W. ¼ N., 3 leagues from Cape St. Angelo, which last is in lat. 36° 25′ N., long. 23° 12′ E.

GEORGE, ST., CAPE, Mediterranean, N.W. side of the Island of Rhodes.

GEORGE'S, ST., CHANNEL, that part of the sea on the W. coast of England from the entrance of the narrow sea between the S.E. point of Ireland and St. David's head in Wales, according to some charts. Others limit it to the sea between Hartland point in Devonshire, and St. Gowan's head in the Welsh coast, while others include the whole of the narrow sea between St. David's head and the Mull of Galloway, and denominate it St. George's channel and the Irish sea indiscriminately. Within Hartland point, however, as far as the coast of Wales, at St. Gowan's head, the sea is more correctly called the Bristol channel. From Hartland point westward is, in fact, the open Atlantic, the coasts of Cornwall and Devon on that side even beyond Hartland not being bounded W. by any land. St. George's channel, therefore, would seem correctly to be the sea between St. David's head and the coast of Ireland, running northwards along the E. coast of that island.

GEORGE, ST., CHANNEL OF, Black sea, one of the four mouths of the Danube, Kilia being the N., and those of Soulina, St. George, and Portitcheh following in S. succession.

GEORGE, ST., COVE OF, Island of Cyprus, 7 miles N. of Cape Grego, a place only adapted for the smallest vessels.

GEORGE, ST., DEL MINA. See *Delmina* and *Elmina*.

GEORGE'S, ST., ISLAND, one of the Azores islands. See *Azores*.

GEORGE'S ISLAND, America, Gulf of Mexico, opposite the River Apalachicola, on the W. coast of Florida, in lat. 29° 37′ N., long. 85° 9′ W. It has a lighthouse upon its N.W. point, upon which there is a fixed light, 65 feet high.

GEORGE'S, ST., ISLAND, Greek Archipelago, to the S. of Great Fokia island, E. of Scio, a rocky islet, only about two-fifths of a mile long and 150 yards broad.

GEORGE'S, ST., ISLAND, Islands of San Bernardo, off the coast of South America, in the province of Cartagena.

GEORGE, POINT ST., Ionian islands, on the S. coast of the Island of Cefalonia, and harbour of Argostoli; a shoal runs out from it to the S. for a full mile, having 3 fathoms water upon its further extremity.

GEORGE, ST., PORT OF, Gulf of Sto. Georgio, Venice, on the N.E. side of the Island of Lissa, capable of containing a squadron of ships of the line. Lat. of the church, 43° 3′ N., long. 16° 10′ E. It is accounted one of the best ports in Dalmatia.

GEORGE'S SHOAL, America, United States, about 100 miles S.E. of Cape Cod, in which space there are four shoals; no accurate accounts are given but of two, Cashe's ledge and George's bank (see *Cashe's Ledge*); this last, as well as Cashe's ledge, is dangerous. Of the four shoals, properly speaking, upon George's bank, all are included between latitudes 41° 34′ and 41° 50′ N., and longitudes 67° 20′ and 67° 55′ W. Between them there are from 2½ to 37 fathoms water. The bank is composed of a number of sandy spits, very narrow, so much so that the width of a narrow vessel will make several fathoms difference in the depth of water. The general range of the spits is from S.E. to N.W. As there are no rocks, they are consequently liable to change continually their positions and ranges.

GEORGETOWN, America, Prince Edward's island, on a peninsula between the Rivers Brudenell and Cardigan. There is an anchorage without Cardigan bay, in from 10 to 15 fathoms, where pilots are taken.

GEORGE TOWN, South America, River Demerara, on the E. bank of that river, 1¾ mile from the fort which defends the entrance; this town is in lat., at the lighthouse, 6° 49′ 18″ N., long. 58° 11′ 30″ W.

GEORGETOWN and HARBOUR, America, United States. The entrance is 65 miles from Cape Fear, in South Carolina, S.W. ¼ S. from the cape. A bank lies between, having 5 fathoms water over it; the N. end about 6 leagues S.W. by W. from Cape Fear, and extending thence S.W. ½ S. 12 leagues. Several inlets are passed in sailing towards the town, among them the N. inlet, 3 leagues from Georgetown lighthouse. This inlet is the N. boundary of North island, on which the lighthouse is situated. There is a high circular white tower and black lantern erected on the S. end of North island, which is on the E. and N. sides of the harbour, at the entrance of Winyaw bay. It shows a fixed light, 90 feet above the level of the sea at high water, bearing N. ½ W. from the bar, 6 miles distant. From the E. of Cape Roman shoal to the entrance of the bar, the course is N.N.E., and distance 15 miles; and from the S. part of Cape Roman shoal, it bears N. by E. ½ E., 20 miles distant. There are 5 feet water on the outer cape shoal at low water, and 9 at high. The principal entrance to the harbour is to the S. of the lighthouse. The town lies in lat. 33° 22′ N., long. 79° 14′ W. The lighthouse in lat. 33° 12′ N., long. 79° 4′ W.

GEORGE TOWN and ISLAND, America, United States, district of Columbia; the island lies at the junction of the Potomac and St. Mary's river, where it forms a bluff on the starboard hand, about 5 leagues up the river. From this island on the starboard hand, proceeding up the Potomac, N.W. ¼ W., 6 miles to Ragged point on the port side (off which a vessel must not go into less than 8 fathoms) she must continue N.W. ¼ W. until she drop into 5 fathoms, when hauling up W. by N., 8 miles will carry her above Blackstone's island, on the starboard hand; then W.N.W. to Cedar

P 2

point, till in sight of the light-boat stationed there. Then she must steer N.W. till the boat bears N. ¼ E., keeping soundings on the larboard hand in 4 and 5 fathoms. When up to Cedar point, a vessel must steer for Watkin's point, giving it a small berth; then up for Cedar point, also giving that a small berth, and thence the course to George town is the midway of the river.

GEORGIA, America, United States, one of the states of the union, between lat. 30° 37' and 35° N., and between long. 80° 8' and 91° 8' W. Bounded E. by the Atlantic, S. by Florida, W. by the Mississippi, N. and E. by South Carolina and Tenesse. It contains 24 counties. The principal rivers are Savannah, separating it from South Carolina; Ogeechee river, Turtle, Little Sitilla, Crooked river, and St. Mary's. The climate is mild, but near swamps and marshes it is unwholesome, while in the hills it is pure and salubrious. The thermometer in summer ranges from 76° to 90°, and in winter from 40° to 60°. The prevalent winds are S., W., and E.; in winter N.W. The E. is coolest in summer and warmest in winter. The whole seacoast is bordered with islands, the principal of which are Skidaway, Wassaw, Ossakaw, St. Catherine, Sapelo, Frederica, Jekyl, and Cumberland. On the N. point of the latter is a fixed light. The islands are surrounded by navigable creeks, between which and the mainland there is a great extent of salt marsh, fronting the whole state next the sea, not less on the average than 5 miles in breadth, everywhere intersected with creeks, admitting, through the whole, of an inland navigation between the islands and the mainland from the N.E. to the S.E. angles of the state. The E. sides of these islands are generally clean, hard, sandy beaches, exposed to the ocean waters. Between these islands are the entrances of the rivers from the interior, winding through low salt marshes, and opening into sounds, which form good harbours, from 3 to 8 miles in length, communicating with each other by parallel salt creeks. The pop. of this state in 1830 was 516,823. The registered tonnage of shipping, 10,611 tons. The value of exports, 8,803,830, the imports, 776,068 dollars.

GERAK POINT and COVE, W. coast of Asia Minor, 4½ miles to the W. of Sighajik, on the N. shore of that bay. There are some ruins near this cove. The W. promontory near is high land, formed by the mountain Corycus, which rises 2,250 feet above the level of the sea.

GERAKA POINT, Ionian islands, Zante island; from the S. point of the Bay of Zante the next in succession is Point Basiliko, and S. a little W. from that is Point Geraka, opening to the W., and with Point Kieri forming between them an extensive bay, in which are the islands of Peluso and of Moratonisi. Cape or Point Kieri is the most S. point of Zante, in lat. 37° 38' N., long. 20° 50' E. This bay has from 13 to 4 fathoms within it. Behind Moratonisi island there is a fair harbour, called Kieri bay.

GERBI, or ZERBI, an island on the E. side of Gulf of Cabes; the shores of this island are covered with the lotus, from which it derived its ancient name of Lotophagites.

GERELOTE SHOAL, Gulf of Finland, in Biorko sound; when the N. point of Biskopsoe bears S. by W. distant 1 mile, the N. end of this shoal will bear N.E. by E. ¾ E. nearly 1 mile distant; the end of the Gerelote shoal lies almost a mile from the nearest part of the shore.

GERLOCH, LOCH, Scotland, W. coast, 4 miles N. of Loch Terriden, a large well-sheltered harbour, with clean and good holding ground, and room for a squadron of the largest ships. The entrance is between Coupachar point and Longa island, distant from each other nearly 2 miles. There are no dangers in or near it, and when the wind does not blow hard from the W. or S.W. vessels may ride safely in any part, especially on the E. side of Longa island.

GERMAN ROCK. See *Plymouth*.

GERMAN'S, ST. Cornwall, a small town of about sixty houses situated upon the Lynher river, at the head of that branch of Plymouth harbour called Hamoaze.

GERTRUYDENBERG, Holland, at the S.E. extremity of the Biesbosch, on the River Dungen, a fortress of importance, with a town of 13,000 inhabitants.

GETARES, COVE OF, Spain, Bay of Gibraltar, between Carnero point and Point San Garcia.

GHARA ISLAND, N. coast of Africa, Bengasi, in the Gulf of Sydra, or Greater Syrtis, in lat. about 30° 46' N., long. 19° 55' E.

GHELENDJIK, or GHELINDJIK, Black sea, E. coast, a port and bay 10 miles from the Valley of Tchiangoti. This harbour is of an oval form, about 3 miles long from N.W. to S.E., and 1½ broad. Both points, on entering the bay, must have a good berth. The mouth is about a mile wide, with 8 or 9 fathoms in the middle. The whole S. part of the starboard side of the bay is occupied by a ledge of rocks. The N.W. side also affords good anchorage, but is uninhabited.

GHIA ISLAND, W. coast of Scotland. See *Gia Island*.

GHIAOOR KIOY, coast of Anatolia, not far from the site of ancient Ephesus, a small town near the sea-shore, and a small bay, on the E. side of which are the ruins of Colophon and Claros, in the Gulf of Scalanuova. There is good ground for anchorage here in from 20 to 16 fathoms from 3 to 5 cables' length from the land; small vessels may lie close in with from 10 to 2 fathoms water.

GHIUK, or GAIUK SOOYOO RIVER, on the coast of Karamania, E. of the low point or cape called Lissan el Kabeh. At its entrance this river is 66 yards wide, and has a depth of 3 fathoms. The points which form the river's mouth, are steep-to, and the water deep. Lat. 36° 20' N., long. 34° 8' E.

GIA ISLAND, Scotland, W. side, 12 miles N.E. by N. from Duninmore head, distant from the shore of Cantire 1¾ miles. Off its S. end lie the small islands of Gigilium and Cara, and half a mile to the W. of the point of Biachmore, there is a rock which dries at half-ebb, to avoid which on the W. side, a small rock above water, near the point of Rulenakerdock must be brought in a line with the highest hill near the Mull of Cantire.

GIAFRETTA RIVER, Sicily, the place on the shore where Mount Etna terminates eastwards the circumference of its base, which is 87 miles in circuit.

GIANT'S CAUSEWAY. See *Bengore* and *Fair Heads*.

GIANUTI ISLAND, coast of Italy, about 1¼ mile in diameter, with a small port on its S.E. side, where vessels may lie sheltered from N. and W. winds. This island is not very elevated. It is S.E. ¼ E. from Giglio, distant 8 miles, and from Civita Vecchia W.N.W. ¼ W. distant 28 miles; there is deep water all around it. There is also deep water between it and Giglio. The lat. of the middle is 42° 13′ N., long. 11° 9′ E.

GIBARA, or XIBARA HARBOUR, West Indies, Island of Cuba, 5 miles W. ½ S. from Jurum, on the N.E. side of the island. Jurum is known by three hills seen to the S., like islands. The first and most E. is named the Silla or Saddle of Gibara, the middle one is like a sugar-loaf, and to the W. of the third there are some hills of irregular height. From the port of Gibara the coast continues scarped and clear to Punta de Mangle, or Mangrove point.

GIBBS' HILL LIGHTHOUSE, Bermuda; an iron lighthouse stands here on the W. coast, 20 feet in diameter, and 120 high, constructed in London, and taken out and set up at the island. It is in lat. 32° 14′ 54″ N., long. 64° 52′ W., a revolving light, the period of revolution 54 seconds; it is 362 feet above the sea level.

GIBILEH, Syria, at the foot of Mount Casius, the base of which is washed by the sea, one of the highest of the Syrian mountains under Lebanon.

GIBRALTAR, ROCK, FORTRESS, and BAY, Spain, at the S. extremity. The Rock of Gibraltar is situated on the N. side of the straits of the same name, and was anciently styled one of the Pillars of Hercules, answering to Abyla on the other side of the straits in Africa, under the name of Calpe. Some suppose that the ancient Tartessus occupied the site of the present Gibraltar. The name is derived from the Moors, or rather in the Arabic language, from Gibel Tor. It consists of a lofty, steep, and in some parts inaccessible rock, hardly 3 miles in length, and from three-quarters to half a mile in breadth. It is joined to Spain by an isthmus on its N. side, about 1½ mile long, which connects it with the continent. The N. face is of rock, almost perpendicular; the E. side carries a front of fearful precipices towards the Mediterranean; the S. side is narrow, and almost inaccessible; the W. side, though rugged, and composed of numerous precipices, slopes down towards the sea, and even presents a small level spot, upon which the town is built. This mass of rock is secured from hostile assault by strong batteries, so as to be rendered completely impregnable. The town consists of one street about half a mile long, filled with shops; the rest of the town conforms more to the nature of the ground. From the N. end, what is called the Old Mole runs out 1,100 feet into the sea, N.W. by W. The New Mole, 1¼ mile S. of the old, projects 700 feet, having an angle formed by the shore, and affording shelter for vessels of burden in winter, the furthest out lying in 6 and 5 fathoms water. The rock which is the basis of all, and which rises there to 1,100 or 1,200 feet, is of fine limestone, or marble, lying in strata of 30, 40, and 50 feet in thickness. This rock has been excavated into caverns and galleries of communication between one part of the works and the other, completely sheltering the garrison from the fire of an enemy. Roads have also been excavated in the rock, and made passable for carriages. There are towers and barracks on the loftier elevations, adapted both for health and convenience. The garrison consists of about 5,000 men; the inhabitants of the town exceed that number. This celebrated rock, at Europa point, answers to the point on the W. side of the bay, S.E. of Algeziras, called the Point and Tower of San Garcia, which lie immediately W. of it, but the true W. point of the Bay of Gibraltar, or Algeziras, as it is denominated by the Spaniards, lies more to the S., and is the Point Carnero or Cabrita, having a small bay between the Point San Garcia and itself, called Getares bay or cove, in which, somewhat S. of the centre of the bay, is a lazaretto. A little N.W. of the Point San Garcia, near the Spanish shore, is the Isla Verde, or Green island, off the N.W. end of which, between that and the mole of Algeziras, a river runs into the sea, called the Rio de la Miel, which discharges in winter a considerable stream. The Point of San Garcia is 1¾ mile from Point Cabrita, which is consequently the span of the Bay of Getares. From Point San Garcia, which, as well as Carnero or Cabrita point, is encumbered with rocks, which are most of them visible from Point San Garcia to Point Rodea, further to the N., is about half a mile; half a mile from that point, or 1 mile from Point San Garcia N.N.W. is the island Verde or Algeziras already noticed, surrounded by rocks, and dangerous of approach. It is fortified. About half a mile N.N.W. ¼ W. of Isla Verde is the Mole of Algeziras, running out E., and half a cable's length from the mole is the rock of Galera, partly uncovered, from whence a reef runs out E. There is a shoal, too, a quarter of a mile N. ½ E. from the Galera, with only 2 fathoms water, and from the middle of Algeziras town a reef runs out E., and from the N. part of the town, where Fort Antonio is situated, a second reef runs out of larger rocks, going on N. round the bay. About three-quarters of a mile further is the Tower of Almirante, surrounded with rocks, and three-quarters of a mile further still, Polvera tower; these two towers in a line are a mark of foul ground. Half a mile further is the Point of Reconcillo, surrounded by rocks, and here a low sandy beach begins which goes round the bay to Gibraltar, except at two points, those of Mirador fort and Punta Mala. The River Palmones is deep, and large enough to afford a rendezvous for the vessels belonging to Algeziras in winter, as a heavy swell rolls into the bay from the N.E. to the south-eastward, rendering the bay anchorage not good. This river has its mouth encumbered with sand-banks, which run out a great distance. The tower of Entre Rios is a quarter of a mile E. of Palmones river, and about the same distance from the entrance of Rio Guadarranque, which is a stream of little moment. Three-fourths of a mile further on is Fort Mirador, circular, upon a small elevation. E. ¼ S. of Fort Mirador is the Point de Gallo, whence the shore forms a sort of bight in the direction of Point Mala. There are a few rocks about this point, which has a castle on its summit. N. from this castle one-third of a mile is the Hospital di la Sangre; and S.E. by E. from it 1¼ mile St. Philip, and the W. extremity of the ruined Spanish lines, which, stretching across the peninsula, terminate at Fort St. Barbara. The land hence is low all the way to where Gibraltar rock rises precipitously to a great height, its summit rugged, extending S. and suddenly dropping at Europa point into the E. extremity of the celebrated bay, on which point is an excellent fixed light, in lat. 36° 6′ N., long. 5° 21′ W., elevated 150 feet above the sea. There is deep water at the entrance and in the centre of the bay. Ships may lie at single anchor half a mile off Cabrita point, in from 18 to 20 fathoms, on sand and small gravel, and from thence as far out as 38 fathoms good holding ground, and thence N. as far as Point San Garcia, but the ground is not so good. Half a mile E. or

N. of the Isla Verde it continues bad as far as 15 fathoms, and thence into deeper water as far as 1¼ mile, or into 38 fathoms, where the ground holds well, and in summer is adapted to large vessels; further out the water is too deep. The best anchorage for all vessels is from off Algeziras towards Palmones river, and over towards the bridge of Mayorga. Except where the two towers of Almerante and Polvora come in one, the ground is good either mud or sand as far out as 23 fathoms. Anchorage may be taken all the way from the Bridge of Mayorga towards St. Phillip's castle, keeping clear of Point Mala, but the preferable anchorage is between the castle and Old Mole of Gibraltar, as less deep, and under the guns of the garrison, the smaller vessels furthest in. The space between the Old and New Mole affords good anchorage, and a choice of depth from half a mile off, not further out, the water getting too deep, nor going too far in on account of the rocks; this may be done all round the bay to the Europa, though the water deepens there. The tides in this bay rise 5 and 6 feet; it is high water at full and change at Gibraltar at $12^h. 15^m.$, and at Algeziras at $1^h. 34^m.$ Lat. of Gibraltar mole 36° 7′ 18″ N., long. 5° 21′ 12″ W.

GIBRALTAR, STRAITS OF, Europe and Africa, bounded on the S.W. entrance by Cape Spartel, on the N.W. of Africa and kingdom of Morocco, in lat. 35° 47′ N., and long. 5° 55′ W., being high land, rising to 1,043 feet, and bearing from the tower on Cape Trafalgar S. ¼ W. distant 22 miles, and on the N. by Cape Trafalgar, the point of which is in lat. 36° 9′ N., long. 6° 1′ W. From Cape Spartel on the S. side to Tangier cape eastward, it is 5 miles, the coast clear and high. From Cape Tangier to Cape Malabata, which has foul ground a mile offshore with 3½ fathoms on it, called the Almirante, it is 3 miles E. ¼ S., the Bay of Tangier between. From Tangier to Ceuta the shore is rugged, rocky, projecting and lofty for the most part. From Cape Malabata E. there is a cove at no great distance, called Baca cove, where there is anchorage in 9 fathoms; to this succeeds a small bay bounded on the E. by Point al Boassa; 1 mile to the N.W. is the Jasseur rock with only 11 feet on it. The next point, E. ½ S. from Cape Malabata 9½ miles, is Point Alcazar, with a reef of rocks about it and a cove, on the E. side of which there is a bay with anchorage. E. ¾ N., two miles from Point Alcazar, is Point Zannar, low, having a cove or small bay on each side of it. E.N.E. ¼ N. 3½ miles from Zannar, is Cires point. The Point of Cires is low, but rises to a cornered mountain called El Cuchilla, or the Gap of Cires. Nearly 1 mile E.S.E. ¼ E. from Point Cires, is the Point of Lanchones, with a flat beach, and anchorage in 15 fathoms. Half a mile from Point Lanchones is that of Cruzes, with rocks near to the former point. Cruzes point is high and steep-to, with several rocks at its base; but Lanchones point is not so high. Less than a mile from Cruzes point, in a S.E. by E. direction, is the Fronton de Almanza, a broad, high, and abrupt precipice. Two miles, E. ½ S. from Point Almanza, is Leona point; between them is the island Peregil, a high barren rock, having 18 fathoms water close to its outer side. Point Leona, the N. point of this part of Africa, is steep-to, of a moderate height, and distinguished by a large tower in ruins. Between Point Leona and Peregil, or Parsley islet, there is good anchorage in 8 fathoms water. Some small islets lie off its western point. 1¾ mile S.E. by E. from Point Leona is Point Torre Blanca, or White Tower point, high, abrupt, and of a dark colour, with the ruins of a tower built upon it, and islets at its base. About a quarter of a mile to the eastward of Torre Blanca is a shoal, described by Tofino thus:—"This shoal is about the size of the hull of a man-of-war, with from 3 to 4 fathoms over it, excepting where the rock is highest, and the depth only 1 fathom, occupying a space equal to a large long-boat. Between this rock and the shore there are from 8 to 9 fathoms, rocky bottom, the channel being only about a cable's length broad, and rendered narrow by some islets, projecting two-thirds of a cable's length from the land. On the outside of this shoal are 8 and 9 fathoms, also rocky ground. From the shoal the outermost islet of Torre Blanca appears in a line with Point Leona; the ruins of the tower with the upper point of Ape's hill S.W., and Point Bermeja on with the N.E. corner of the walls of Old Ceuta." Between Point Leona and Point Blanca is the cove or bay of Benzus, where vessels may anchor in 18 fathoms, sandy ground, and at the distance of 2 cables' length from the shore, protected from S. easterly and S. westerly winds. From this anchorage the islets of Torre Blanca will bear E.N.E. ½ E.; a remarkable pyramidical peaked hill S. by W. ¾ W.; and Point Leona will be in one with Tarifa, bearing N.W. ¼ N. About one mile S.S.E. ½ E. from Torre Blanca point is Point Bermeja, or Red point; this also is high, and has a tower upon it in ruins. The coast between them is all bordered with rocks, some above and some under water; but on the outside of these the water is very deep. Nearly three-quarters of a mile to the south-eastward of Bermeja point is Point Benites, with the two little flat islets of El Campo before it. The next point is that of Santa Catalina, with Small bay between, and Ceuta on the S. with the Great Bay of that name, where there is a revolving light, in lat. 35° 53′ N., long. 5° 19′ W. The N. coast of the strait commences at Cape Trafalgar, whence to the N.E. the land rises, forming the Altos or Heights of Meca, the tower of which is 3½ miles S.E. by E. ½ E. from Cape Trafalgar, with a patch of sand between called the Boqueron. To this succeed the cove and river of Barbata, and next Cape Camarinal, passing by 4½ miles S.E. from Meca tower, the tower and point of Sara. Hence to Cape Camarinal is 2½ miles further, and 4¼ miles more S.E. is Cape Plata. The interior from Meca point to Cape Plata is mountainous, and of considerable elevation. Paloma point is S.E. ¼ E. distant from Cape Plata, and 3¾ miles further onward are the tower and point of La Pena, with the cove of Valdebaqueros between. The River Puerca falls into a small creek here between the ledges of rock. Tarifa is about 4 miles from La Pena, having the chapel of Santa Catalina between itself and the sea, and S. the peninsula of Tarifa, a small point connected by a sandy isthmus with the main. It is low and level, and has upon it a lighthouse, a white tower with a revolving light, in lat. 36° N., long. 5° 37′ W., the period of revolution of the light $2^m 18^s$, elevated 124 feet above the sea. The Cabezos rocky shoal lies 5 miles N.W. by W. from the Island of Tarifa, 2¼ S. by W. ¼ W. from Point Paloma, and nearly W.S.W. from the tower of Pena distant 3 miles. Comoro point is E.S.E. of Santa Catalina chapel, with rocks about it, high, steep, and wide. The coast then runs on to Points Canales and Gualmesi, which have watchtowers. Further on, about 2¾ miles, is Point Acebuche, having a bend inland to where stands the castle of Tolmo. A mile further on is Point del Frayle, with a watchtower. There is an islet off this point, and several rocks around it, and E.N.E. ¼ N. of Point del Frayle is Point Cabrita, or Carnero, the W. point of the bay of Gibraltar, having an island called Pigeon island between, with the Pearl rock to the N., and another point one-third of a mile from Point del Frayle, called Point Secreta. It is seen that the shores on both sides are irregular, but the African is most rugged, its indentations and rocky projections being much smaller and more numerous than on the European side. Both are well fenced with rocky shores, against which the deep currents that pass in and out between them can effect little in the way of abrasion. In the strait, midway, the soundings are very deep. They increase from the westward. At the entrance they are from 150 to 180 fathoms. S. of Tarifa they deepen to 250, 300, and 440 fathoms. N. of Point Cires, W. and E. of N. they are from 500 to 700 fathoms, sand and shells. Between Europa point and Point Leona 720 fathoms are found in coral rock, and more E. 980, sand and shells. The tide on both sides of the strait would be regular, if it were not for the effect of the prevailing winds. From Cadiz the flood sets towards Cape Trafalgar, where it is high water at a quarter after 5, full and change. At Tarifa it is high water about a quarter after 11; at Gibraltar at 2 o'clock, the tide rising 5 feet. The flood again sets from Europa point towards Cape Carnero, and thence to Tarifa, so that it meets the tide from the west of Trafalgar, where it is low ebb at the time it is high water at Tarifa. The tide in this way runs as far as 2 miles off the coast; and at that distance from Tarifa it flows until 2 o'clock. The tide continues perceptible as far as Malaga, where it rises about 3 feet; but beyond this it gradually becomes imperceptible. On the southern coast of the Strait, from Cape Spartel to Ape's hill, another tide runs along shore to the westward. In the great cove, opposite to Tarifa, it is high water at 10 o'clock; but more to the W., and near Tangier, at 12; and in the offing it continues an hour longer. In the middle of the strait the current from the Atlantic generally sets in to the E. It is probable that under-currents restore the equilibrium. The depth in the straits fully admits this, without its being so perceptible as it otherwise would become to the attentive observer. That such under-currents do exist in other places is pretty well ascertained, and the difficulty of verifying the fact in the Straits of Gibraltar cannot be admitted as evidence of their non-existence. Vessels that have sunk are said to have been thrown upon the shore to which they must have been borne beneath the surface in an opposite direction to the superficial current; but however this may be, it has been tolerably well ascertained that the tide sets both outward and inward along each shore; that from the vicinity of Malaga the flood sets round the coast in a W. direction towards Gibraltar, and thence along the coast of Spain, until it meets that which comes from the W. off Cape Trafalgar; and that a similar current, and in a similar direction, sets along the African shore, until it passes Cape Spartel, and is lost in the Atlantic ocean. But although it has been asserted that the central stream invariably runs eastward, rarely standing still, there is reason to believe this is not strictly so, but that a current is sometimes found running outward to the west, occasioned perhaps by an extraordinary tide, or a surcharge of water in the

Mediterranean. Captain Barret, R.N., states that in 1820, when he was off Tangier bay, he felt the current set for above 3 hours decidedly to the W.; and that while the water at the surface was going apparently to the eastward, a westerly current had hold of the bottom of the ship. A London ship, when at anchor at Tarifa, found the tide running at the rate of 5½ knots; Captain Walker says he has been obliged to have a man at the wheel steering his vessel for the tide, when at anchor under Cape Spartel. A writer, speaking of the under-current, states that the Mediterranean sea is said to be rather salter than the waters of the Atlantic ocean. That it is not more so is ascribed to an under-current, salter than that of the ocean, which runs out of the strait, and unloads the waters of their excess of salt. It will thus appear that the current of the Strait of Gibraltar is in some measure regulated by the tides, and that its velocity will, according to the tide, vary from 3 to 6 and 7 miles an hour. To the W. of Tarifa, about the full and change of the moon, it sometimes sets to the W. quite across the strait; but E. of Tarifa it sets, more or less, to the E. Mr. I. Reiner, pilot, of Gibraltar, says that with W. winds the current in the middle of the strait between Europa point and Ceuta, will often equal the rapidity of 7 miles an hour, and that at such times every point in the bay will form an eddy or whirlpool at a considerable distance from the land. There has an attempt been made to account for the current eastwards by the consumption of the waters of the Mediterranean through evaporation, by no means a satisfactory solution of the difficulty. It would be sufficient to keep up the current reported to exist, that the waters of the Mediterranean should be warmer than those of the Atlantic, or the reverse. The effect would be at once an exchange of waters, either at the surface or at some depth below, a species of circulation such as occurs when cold air is admitted into a rarer and warmer atmosphere—the very same cause, in fact, that with a more subtle fluid than water brings about the trade winds, and forces the chilling N.E. blasts in spring from the Pole to the warmer regions of the S.W.

GIEDESBY HEAD, Denmark, the S. point of the Island of Falster, called in Danish the Giedsers Odde. There is a lighthouse upon this point, having a fixed light, at an elevation of 45 feet, and seen 11 miles distant in clear weather. Lat. 54° 34' N., long. 11° 57' E. This point recently has been better known by the name of the Trindelen point.

GIENS, PENINSULA OF, France, S. coast, connected by a sandy and rocky reef with two of the islands of Hyeres, named the Great and Little Ribaud.

GIESING BOGEN, Norway, W. coast, one of the passages or channels leading to Drontheim, among the rocks, islands, and intricacies of that remarkable coast.

GIGLIO ISLAND, Mediterranean, nearly due E. of Monte Christo island, off the coast of Italy, W. of Mount Argentero, in the Siennese. It is about 4 miles long by 1½ broad, is high land, has a town, fort, and a small port on its E. side, and on its N. side another called Port Campese. The N. point is in lat. 42° 22' N., long. 10° 57' E. The pop. 1,200.

GIGUERO, PUNTA DE, West Indies, Island of Porto Rico, S.W. from the Punta de Francisco, and 8 miles from the town of Aguadilla, S.W. by S.

GIJON, Spain, N. coast; this is a roadstead

where vessels of any description may ride securely in summer weather. A large ship should anchor with the little island Orrio in one with the town of Candas, and the hermitage of Santa Cruz in one with Point Otero, in 10 and 11 fathoms, fine sand. Smaller vessels may run for the Pier of Gijon, to go over the bar of which a pilot is necessary, as there are not more than 9 feet of water. Within the pier vessels lie dry at low tide. In the outer road attention must be paid to the wind, as should it blow on shore the sea becomes very heavy, and care should be had to get under sail directly, for if a vessel gets to leeward of the spot appointed for anchoring it will be impossible for her to clear the bar.

GILBERT'S RIVER, Labrador; between Fishing island and Granby island is the N. entrance, and the S. and widest between Denbigh island and the main. There is also a channel between Denbigh and Granby islands.

GILES, ST., or MELVILLE'S, ROCKS, West Indies, Island of Tobago, in lat. 11° 22' N. The most W. has a large opening in it, running in a N. and S. direction, and giving it a remarkable appearance. There is deep water close to these rocks, and a vessel may run up to them and coast along shore towards Man-of-War bay; the distance from these rocks to the N.E. point of the bay is not more than 3 miles.

GILETO, CAPE, Africa, N. coast, W. by S. ¾ S. from Cape Tenez 7 leagues; opposite to this cape is the little Island of Palomas or Columbi, with a clear passage between. Lat. 36° 25' N., long. 55' E.

GILLES, ST., PORT, France, W. coast, E.S.E. ¼ S. from the E. end of Isle Dieu or Dyeu, distant 4 leagues. The haven is only adapted for vessels of 100 or 120 tons. The town lies in a valley, and is hardly to be distinguished except by a great wood near it.

GILLINGHAM, England, Kent co., a village which gives its name to one of the reaches of the River Medway, and commands a noble view of the Thames and Medway rivers. Some of the finest vessels of the navy of England lie in ordinary on the river, which is an appendage to the dockyard of Chatham. Pop. about 4,000.

GILLON RIVER, Cornwall, S. coast, a little creek on the port side, entering Helford haven, in which small craft may lie safely sheltered.

GILMAN SAND, England, mouth of the Thames, a small narrow sand, 2 miles long, having on it from 6 to 9 and 13 feet water, divided from the middle ground by a passage of 3 fathoms, about three-quarters of a mile wide. There is a black and white buoy, with rings, on the N.E. part.

GILTAR POINT, South Wales, on the N. side of Caldy sound.

GILTEN CUP POINT, coast of Dorset, near Bridport, the highest land in that vicinity, under which, E. about a mile, a vessel may run on shore if in distress, at an hour ebb, with a chance of preserving life if not property, according to some, but in the opinion of others a dangerous experiment only allowable under the last necessity. It is urged that in a quarter of an hour, the tide ebbing here with remarkable rapidity, lives have been saved by such an alternative. The place alluded to is Seaton beach, just E. of Gilten Cup.

GIMBLET ROCK and SHOALS, North Wales, a rocky peninsula forming the S. side of the harbour of Pwllheli, 2½ miles from Pen y Chain point, Caernarvonshire. The shoals are four in number, lying S. by W. from the entrance of the harbour, the outer shoal being 1½ mile distant, the inner half a mile, with from 1 to 1¾ fathom water over them and deeper between.

GINGER ISLAND, one of the Virgin Isles.

GIORGIO, ST., ISLAND, coast of Anatolia, in the province of Makry, situated in the gloomy Bay of Kalamaki, bold all round, with a creek of deep water on its S. side, and on the N. a small bay with a sandy beach. Vessels passing between this island and the main must avoid a sunken rock, on which there are but 8 feet at low water, lying about a mile from the shore. Another rock is said to lie midway between this island and that of Kastelorizo.

GIORGIO, ST. MOUNT, Gulf of Venice, Island of Lagosta, at the bottom of the harbour, having a chapel upon its summit; this hill is 1,529 feet high. Lat. 42° 45' N., long. 16° 51' E.

GIORGIO, ST., TOWN, TOWER, and PORT, in the Island of Lesina, Gulf of Venice, at the E. extremity. The port is small and only adapted for small vessels.

GIOVENAZZA, a town on the E. coast of Italy, 8 miles N. from Point St. Cataldo, in the Gulf of Venice.

GIRAGLIA ISLAND, Mediterranean, 2½ miles from Cape Bianco, the N.W. point of Corsica, bearing from the lighthouse of Genoa S. by E. ¼ E. 85 miles distant, lying off Cape Corso. It has a watch-tower upon it, and there is a passage between this island and the cape, having a depth of 7, 8, and 9 fathoms, which should never be attempted except with a leading wind. Lat. 43° 2' N., long. 9° 24' E. A lighthouse is building here, and just finished.

GIRALATO BAY, Mediterranean, Island of Corsica, between Capes Sandola and Ozani, which last is the S. point of the bay.

GIRAM, CAPE, Island of Madeira, Bay of Funchal, a magnificent headland 1,600 feet above the level of the sea.

GIRDLE NESS, Scotland, the S. point of the Bay of Aberdeen. Off this ness at low water there is a small rocky shoal called the Girdle, which appears only at low spring tides; Findon ness, open of Grey ness, clears it to the E., and the first head light open of Short ness leads to the N. of it. There is a lighthouse on Girdle ness, a double light, both fixed, one over the other, like stars of the first magnitude at a certain distance, but far off appearing in an elongated form. The lanterns are 115 and 185 feet respectively above the level of the sea, visible from 19 to 16 miles, in lat. 57° 8' N., long. 2° 3' W., bearing W. from the N. pier of Aberdeen, S. by W. 1,220 yards.

GIRDLER SAND, or FLAT, coast of Kent, an extensive flat, on the N.W. part of which is a narrow patch about 2 miles long, dry at low water. It has a buoy at the W. end in 17 feet at low-water spring tides, and just S. of it a small knoll with only 15 feet. The buoy is to be left on the port side, lying with the E. buoy on the Shivering sand bearing N. ¾ W. Redding-street mill open to the left of North Down tower, the length of that tower bearing S.E. by S. and open to the E. of the N. Knoll buoy; Pan sand W., beacon buoy S. by E. easterly, on with the middle of Lower Hale-grove; Ash church open to the right of the three barns which stand next W. of the Reculvers, the width of these barns bearing S. ¼ E., and the buoy on the Nob N. ¼ E.

GIRGENTI, TOWN and PORT, Sicily. The

city is on a hill-side 1,200 feet above the level
of the sea, and so built that every house may be
seen. It stands 3 miles inland and to the N.E.
from the mole ; the pop. 15,000. To the E. of the
city is a lofty rocky mountain called Rupea
Athanea, below which towards the sea is the site
of the ancient Agrigentum, with its celebrated
ruined temples of Juno, Concord, and other
vestiges of ancient magnificence. The mole at
the landing-place of Girgenti was only constructed
in 1756, yet it has the appearance of being much
worn. This landing-place is called the Caricatore.
There are about 100 houses here, with magazines
for corn, and on the mole-head a lighthouse, in
lat. 37° 15' 10" N., long. 13° 31' 40" E. There
is another lighthouse in the town in lat. 37° 15' 30"
N., long. 13° 31' 20" E. The river of Girgenti,
once called Acragas, divided into two branches,
runs through the site of the ancient city. The
mouth of this river is much encumbered with
shoals.

GIRONDE RIVER, France, W. coast, the
entrance in lat. 45° 35' N., long. 1° 10' W. This
river is formed by the junction of the Garonne
and Dordogne at Bec d'Ambes, where both streams
uniting their names merge in the Gironde or
River of Bordeaux, which falls into the sea at the
Tower of Cordouan in the position above given.

GISBOROUGH, England, N.E. coast, 4 miles
S.E. of the Tees river, on the Yorkshire side, at
the mouth of a small stream where there is a bay
and harbour for ships.

GISHGINSH, or JIJGHISK, called also RO-
VESTRA ISLAND, in the White sea, at the E.
side of the Bay of Onega, in lat. 65° 12' 17" N.,
long. 36° 54' E., the position of the lighthouse at
the N. extremity of the island, which shows a fixed
light 148 feet above the level of the sea, visible
from all parts of the horizon 17 miles. The N.E.
side of this island is foul, as well as the S. part,
and vessels must not approach nearer than 2
miles to the former until the W. point of the
island be brought to bear S., when it may be
neared to a mile.

GIULIANA, PORT OF, in the Gulf of Venice,
on the mainland of Sabbioncello, an open roadstead with several rocks and shoals about it ; off
its S. point there is a rock above water, having no
passage between that and the shore. This roadstead is E. by S. from Cape Speo in Curzola.

GIVILIN ISLET, W. coast of Wales, on the
starboard side of the entrance into the Bay of
Aberdarod, within Bardsey island at the N.W.
point of the Great Bay of Cardigan, between that
on the W. and the two small Stidwell islands on
the E.

GIZZING BRIGGS, Scotland, in the co. of
Ross and Frith of Dornoch, the local name for a
bank which lines the shore of that Frith, and is so
called from the noise which the sea makes upon it.

GLAMORGAN CANAL, South Wales, Gla-
morgan co. ; the River Taff, which falls into the
sea at Cardiff, forms one outlet for the mining districts; and the produce of these mines finds its way
to market through the Glamorganshire canal and
its sea-lock, but inadequate to the demands for
accommodation consequent upon the great increase
of trade since the canal was opened. The lock of
the canal is about midway between Penarth head
and the town of Cardiff; a little within it stand
two conical glasshouses. This lock cannot take
vessels of more than 95 feet in length and 27
beam, but it carries 25 feet of water over its sill

in springs, though only 5 feet at the dead of the
neaps. The canal within retains 15 feet water at
the lock, and 8 feet close to the town, to which it
extends, with means of receiving 70 or 80 coasters.
The Bute docks have since contributed to afford
the aid in accommodation that is required additionally in this canal.

GLANDORE HARBOUR, Ireland, 4 miles to
the W. of Ross harbour. 1 mile to the N. of the
harbour's mouth, and nearly in mid-channel, are
four small rocks called the Dangers; the most S.
is dry at half-ebb, and the N. appears at 4 hours
ebb. On each side is a channel deep enough for
the largest vessels, that on the W. side is reckoned
the best, as there is a mark to lead through, it
being the S. extremity of the little island next to
the Dangers, on with the W. end of Adams isle.
Vessels may go safely to the head of this harbour,
which is shallow, having but 4 or 5 feet at low
water, upon soft ooze, and lie there safely aground.

GLASCARRICK POINT, E. coast of Ireland,
Wexford co.

GLASGORMAN'S BANK, Ireland, E. coast,
1½ mile E. by S. from Kilmichael point, extending
in a narrow ridge about 7 miles S.S.W., the depth
on it at low water being from 6 to 20 feet; the N.
half is the shallowest, and is avoided when Farrow
hill bears W. by S. It would appear that the
sands on this part of the coast have shifted and
continue to shift, forming shallows. An extensive
shoal is said to exist off Arklow bank, midway
between that and the shore, where until recently
deep water had been found. One shallow spot
thus formed is reported to lie N.E. off Blackwater
bank. Such statements are often made un-
supported, and there is as yet no accurate information upon the subject.

GLASGOW, Scotland, co. of Lanark, the second
town in Scotland, and the most populous, seated on
the Clyde river, upon the N. side. Pop. 202,426.
The Clyde is not navigable as far as this city for
vessels that draw more than 8 feet water. Glasgow has numerous manufactories of muslin, cotton,
coarse woollen fabrics, glass, and other articles,
and possesses a university of considerable celebrity.
Port Glasgow is 3 miles E. of Greenock ; and here
the largest vessels discharge their cargoes at the
quays, or into lighters, by which they are conveyed to Glasgow. The Clyde at Port Glasgow is
2 miles broad, but so filled with banks and shoals
as only to afford a channel of 200 yards wide, close
to the Port Glasgow shore. The lat. of Port
Glasgow is 55° 56' N., long. 4° 43' 16" W., by Capt.
Robinson, R.N., 1847.

GLASHEDA, or SEAL ISLAND, Ireland;
13½ miles from Fannet head, is Lough Strabogy,
on the N. coast, and N.W. from the entrance of
this Lough, half a mile from the shore, is Glasheda, surrounded with rocks.

GLAS POINT, N. Wales, Caernarvon bay,
N.E. ¼ E., 4 miles from Braich-y-Pwll, the coast
bordered with hills and cliffs.

GLASS, or SCALPA ISLAND, Scotland, W.
coast, a low island at the E. entrance to East Loch
Tarbot, and S. of Loch Seaford. On the E. end of
the island there is a lighthouse, having a fixed
light, 130 feet high, and seen in clear weather 5 or
6 leagues off. The S. harbour lies on the W. side
of Glass, in face of Ru Grebanish, sheltered from
all winds, with water for the largest ships, and
good anchorage; but it requires a leading wind
to enter or leave the passage between the two
islands, which are not two cables' length apart.

The lat. of the lighthouse is 57° 52′ N., long. 6° 33′ W.

GLENAN ISLANDS, France, S.W. coast, to the S. of Quimper, and S.W. of Concarneau, 9 miles from the entrance of the first, and 6 from Trevignon point. They are a group of small islands, surrounded by numerous rocks above and under water, some extending 2¼ miles to the S. of the main body. On the centre island there is a small fort called Cygogne. The principal islands are Penfret, on which is an intermittent light every four minutes, 118 feet high, lat. 47° 43′ N., long. 3° 57′ W., St. Nicholas, Loch, Drenec, and Bananec. Within these islands lie the Pourceaux bank and island, and the Moutons; and within these are other rocky patches and shallows. S.W. ½ S., distant 3½ miles from Cygogne fort, is the Jument bank, having only 2 feet water. The Basse Anero is another rock of the same group, lying with the N.E. part of Penfret island, bearing N. ½ E., distant 2½ miles, and the S. end of Loch island on with the Ruol and Four rock. All the space between these two and the Glenan islands is dangerous ground, full of rocky spots and shoals, with little water. On the N. point of the Isle of Penfret there is a tower, having a light of the third class, 118 feet above the level of high water. It is varied by a flash every four minutes, lasting from eight to ten seconds, seen at the distance of 5 leagues. To the E. of these islands are several rocky spots and shoals, called the Pignon, Basse Jaune, Spencer, and Valiant rocks. The Pignon is a small knoll of 4 fathoms water, lying nearly E.S.E. from Penfret island, distant 3 miles; and a line drawn in that direction from Cygogne fort will pass through the centre of the Pignon, Basse Jaune, and Valiant rocks. The Basse Jaune is a knoll with only from 2 to 6 feet water over it, and a rock that dries a foot at low water, while around it are 7, 10, and 12 fathoms. The Spencer rock lies S. of the Basse Jaune, and has near its N. point a knoll of 2 fathoms water. There is good anchorage to the E. and N.E. of the Isle of Penfret, on which sides it is steep-to, in from 15 to 20 fathoms, muddy bottom, at about 2 miles distance.

GLENARM BAY, Ireland, N.E. coast, and N.W. of Glenarm. It lies between Larne or Fleet harbour to the S.E., and Fair head to the N.W., Red bay and Cushenden bay being between that and Fair head.

GLENGAD HEAD, Ireland, N. coast, 8 miles S.E. of Malin head.

GLENGARIFF HARBOUR, Ireland, S.W. coast, on the N. side of Bantry bay, opposite Whiddy island. It is small, and the entrance narrow, with 8 fathoms water. There is a harbour without the island, on the E. side of which is the passage in. Abreast of the island the passage is a quarter of a mile wide, and has 6 fathoms water. The harbour is small, and the ground indifferent, being seldom used but by coasting vessels. Without the island the largest ships may ride at the mouth of the harbour, in 8 and 12 fathoms, good holding ground.

GLENLUCE BAY, Scotland, on the S. coast, to the E. round the S. Point or Mull of Galloway, a deep bay running far N. into the land, and so named from a town at the bottom, close to the shore. On the E. side the coast trends away nearly S.E, so that the opening is broad; the E. point of which is the limit to Wigtown bay, still further to the E.

GLENRAFON BAY, North Wales, between the South Stack lighthouse and Penrhos point, having the Penlas rock off its N. point.

GLORY KAY, America, E. edge of the main reefs of Yucatan; one of the numerous kays upon that part of the coast, between the Zapotilla kays and English kay, in lat. 17° 5′ N., long. 88° 1′ 30″ W.

GLOSHOLM ISLAND, Gulf of Finland. On this island, and upon the N. side, there is a lighthouse, standing in lat. 60° 11′ N., long. 25° 51′ E., 126 feet above the sea, having a revolving light, visible from N.N.W. to W.N.W. 12 miles. The light completes its revolution every three minutes, during which period it is three times visible, for twenty seconds each time, with forty seconds between each illumination. The lighthouse is opposite the S. point of the Island of Pellinghé, and N. by E. 30 miles from Ekholm lighthouse.

GLOSSA TOWER, Greece, in the Island of Scopolo, on the point of a steep hill rising from the W. shore.

GLOUBOK, a town on the estuary of the River Dnieper, entering the Black sea, on the N. shore, 4 miles E.S.E. from the principal branch of the river, issuing from amid numerous islets. This port will not admit the approach of vessels drawing more than 10 feet water.

GLOUCESTER, Africa, colony of Sierra Leone, one of the free towns inhabited by the liberated slaves.

GLOUCESTER, America, United States, on the N. side of Cape Ann harbour, Massachusetts. The lat. of the lighthouse on Thatcher's island, in the N.E. of the harbour, is 42° 38′ 18″ N., long. 70° 33′ W. It has two lights, 90 feet high.

GLOUCESTER, CITY, England, the capital of the county of the same name, seated on the E. side of the River Severn, where, by its dividing into two streams, it forms the Isle of Alney. Vessels formerly came up to the bridge, but the navigation being intricate and circuitous, a canal was cut from the city to Berkley, at the head of which is a fine basin, capable of receiving 100 sail. The pop. is about 12,000. Lat. 51° 46′ N., long. 2° 16′ W.

GLOVER'S REEF, America, Honduras, lat., by Captain Owen, R.N., of H.M.S. Blossom, of the N. point, 16° 55′ N., long. 87° 44′ W.; of the S. point, lat. 16° 41′ N., long. 87° 48′ W. This reef has two sand spots on the N. end, and lies nearly S. from Hat Key, distant 15 miles, trending thence S.S.W. ½ W. to the S. end of the reef, on which there are five islands or kays, which may easily be known from the S. four kays, as they are bold on the S. side.

GLUCKSTADT, Denmark, a town of Holstein, at the mouth of the Ryn, a good port, and strongly fortified. It is the chief town of the duchy, and numbers 5,000 inhabitants. It possesses a considerable foreign trade, and sends vessels to the whale fishery, in which it largely shares. This town stands in lat. 53° 48′ N., long. 9° 26′ 30″ E.

GLUTTON BANK, Essex co., off Harwich, N. of the Cod bank, having from 7 to 12 feet of water over it, with a red buoy on the E. part. The W. edge bears N. by E. ¼ E. from the buoy of the Altar; the E. edge, marked with a red buoy, lies in 9 feet, with Harwich church spire N.W. by W. ¾ W.; Harwich Cliff end, W. ¾ S.; and Felixstow Martello tower, E. by N.

GLUVIAS, ST., Cornwall, a village and parish, with the town of Penryn standing also in the parish, upon a branch of the harbour of Falmouth, to the quays of which vessels of 200 tons ascend to load and discharge their cargoes.

GNOOPEE ROCKS, West Indies, islets so named on the coast of the Swedish island of St. Bartholomew.

GOAT ISLAND, coast of Africa, the N. extremity of Prince's island, in the Bight of Biafra.

GOAT ISLAND, S.W. coast of Ireland, on one side of the passage into Long Island sound, the other side being Long Island itself. It is necessary to keep one-third nearer to Goat than Long Island in going in, until past a rocky ledge which runs out in a N.W. direction from Long Island. A ship is N. of this ledge when Coglan's tower at Crookhaven appears in one with the N. side of the high wedge-shaped rocks called the Green islands, the mid-channel must then be taken until reaching the anchorage.

GOAT ISLAND, West Indies, Island of Jamaica, S. side, in the Bay of Old harbour, a large island, which has two hills, one at the E., the other at the W. end, and to the N. of it a remarkable hummock, called Cudjoe hill. When this hummock comes on the W. extremity of the slant fall of the E. hill of Goat island, bearing about N. ¼ W., a vessel may haul up N.N.W. ¼ W. for Old harbour.

GOAT'S, or CABRAS, ISLETS, Azores, 1 cable's length from each other, in the bay S.E. of Angra.

GOAVE, GRANDE, and PETITE, Hayti island, two small settlements on the S. shore of the gulf, each having a good port. The lat. of Petite Goave is 18° 26′ 51″ N., long. 72° 53′ 39″ W., confirmed by Mr. Dunsterville, R.N. It is S. of Goave island.

GOCONAT ROCK, France, W. coast, in Camaret road, nearly 5 miles S.S.W. ¼ W. from Point St. Matthew.

GODREVY ISLAND, N.W. coast of Cornwall, on the E. point of St. Ives' bay, part of a ledge of rocks which stretches out 2 miles into the sea at half tide. These rocks are dangerous to ships that do not keep a good offing. The vessels bound to St. Ives' bay go in clear by keeping over to the W. point. The N. wind sends a great sea into this bay past Godrevy island.

GOD'S MERCIES, CAPE OF, Davis's straits; lat. 62° 28′ N., long. 70° 48′ W.

GOE SCARS, Finland gulf, 9 leagues E. northerly from Epel's Scars, to the W. from Wyborg sound, on the N. coast of the gulf.

GOEREE, or GOREE ISLAND, Holland, N.E. of Schouwen, appearing at a distance in white hummocks, those at the W. end being largest. Near the N. shore of the island is Goederede, or Goeree church, having a square steeple, which is one of the principal marks for entering the W. Gat. A light is exhibited here at night, fixed, 147¾ feet above high water mark, visible at the distance of 6 leagues, and illuminating a circle from the S.W. round to the N. and to the S.E. Two miles nearly W. of the church is a tall spire, called the Steen Baak, or Stone Beacon, another mark, with a fixed light, being removed in 1845 from the Houten Kaap; it is 100 feet high. Further W. is the Houten Kaap, in lat. 51° 49′ 30″ N., long. 3° 53′ 46″ E., visible 10 or 12 miles. The last is called the Goeree coast light. In the channels between this island and the Hook of Holland, the flood in common tides runs but little more than 4 hours, while the ebb runs nearly 8. With the flood the water continues to rise about 3 hours, and with the ebb to fall about 7, remaining the rest of the time at a sta-

tionary level. With the quarter moon the tide rises here about 5 feet 7 inches, but on the Brielle bar 6 feet. The tide turns with the sun before the banks both of Schouwen and Goeree.

GOES, Holland, the chief place of South Beveland island, a neat town with a harbour for small vessels.

GOESBEER POINT, Sweden, 3 leagues S.E. by E. from Ysted, from which point a shoal runs off nearly two leagues, which vessels run over in a depth of water of from 6 to 10 fathoms. In the last depth Bornholm island bears E. by S., or E.S.E., 4 leagues, and the point of Sandhammer N. or N. by E. 2 leagues.

GOFF'S KAY, America, one of the kays on the coast of Yucatan, on the N. side of English kay, 1 mile distant.

GOLA ISLANDS, Ireland, W. coast, off the shore of Donegal, to the S.W. of Runardallach or Farland point, and N.E. of the Owy and Croit islands, forming several passages and places of shelter for vessels in bad weather, or when detained by contrary winds. There are three entrances, the best is between Enis Irhir and the Ballyconnel rocks, to enter which, a vessel should steer between Umfin and Inishmaan islands S. ¼ W., on for Inish Shinney, and when within a cable's length of the small island to the N. of Inish Shinney, haul up W., and anchor under Gola, in 4 or 5 fathoms, avoiding a spit running out from the S.E. end of Gola, and thus ride landlocked. The S. passage is between Gola and Innis Fry.

GOLDCLIFF POINT, S. coast of Wales, on the N. side of the Severn estuary and N.W. by N. from the W. end of the sand called the Welsh Grounds, and W.N.W. from King's road.

GOLDEN HILLS or MATAS DE SANTIAGO, on the N.E. side of the River Belta, on the coast of Africa, 17 leagues E. by N. in the direction of the coast from Cape Bojador.

GOLDHILL, Africa, S. coast, a hill a little to the W. of the Commenda Forts, and near the mouth of St. John's or Boosempra river.

GOLD RIVER, or RIO DE OURO, W. coast of Africa, in lat., N. point, 23° 36′ 18″ N., long. 15° 58′ 30″ W., after Captain W. F. Owen, R.N.

GOLDSBOROUGH HARBOUR, America, United States, Maine, N.N.W. from Titmanan lighthouse, 2 leagues distant, leaving an island covered with trees on the starboard hand, and two on the larboard. The course is then N.N.W. 1½ mile, then N. ¼ E. 4 miles to Goldsborough point, where there is safe anchorage in all winds in 3 or 4 fathoms muddy bottom.

GOLDSTONE ROCK, coast of Durham, a dangerous rock a little more than a mile from the Plough Seat, small, visible at the last quarter's ebb, and having a black buoy near its W. side. A narrow reef runs off from it, called the Stiel, extending half a mile S.E. by E. ¼ E., drying at low spring ebbs. The mark for the Goldstone is, the lookout on the Heugh touching the S. side of Holy Island castle. The mark for the W. part of the Goldstone is the Megstone, E. edge touching the W. edge of Farn island. The marks for the E. part of the Stiel are the N. side of the Heugh, touching the S. side of Lindisfarn castle, and the the Megstone well open to the E. of Sunderland point.

GOLDTOP HEAD, coast of South Wales, St. Bride's bay, one of the two anchoring places in this bay, that on the S. side, the other being on the N., called Salvack. The anchorage is to the E. of Goldtop head, which is sometimes called Barrow head. Small vessels may lie secure here from all winds between W.S.W. and E. in 3 or 4 fathoms water, sandy ground, anchoring with Goldtop head bearing E. ½ N., distant a quarter of a mile.

GOLETTA, N. coast of Africa, the entrance of the inner port or lagoon before Tunis out of the bay or gulf, 5 or 6 miles from the city. Vessels entering the Goletta pay three Spanish dollars per day while they remain there. A castle defends the entrance. Fresh water is procured from a reservoir in the Goletta.

GOLFE TRISTE, South America, N. coast, in lat. 10° 50′ N., long. 68° W.

GOLFO, EL, Island of Ferro, in the Canaries, the principal village, on the E. side.

GOLO ISLAND, Gulf of Venice, one of the islands of the Quarnerola, near St. Gregorio and Arbe islands.

GOMBAUD'S ROCK, a vigia, which rests upon one solitary authority, a M. Gombaud, the commander of a merchant-vessel of Rochelle, as stated by M. Fleureu. Its existence is very doubtful.

GOMENA CAPE, coast of Dalmatia, the N.W. termination of the Peninsula of Sabbioncello, in lat. 43° 2′ 38″ N., long. 17° 1′ E.

GOMENIZZA, PORT OF, coast of Albania; at the entrance there is a brown-looking island on the port side, which must be passed at the distance of nearly half a mile, when a vessel must steer E., giving the port shore a good berth, because it is all shallow. Near the starboard side the land is steep-to, having 7 and 11 fathoms water. This course is to be kept till the bay turns S.E., when a vessel may anchor in from 10 to 14 fathoms, sheltered from all winds. The anchorage is 3 miles deep and 1½ broad.

GOMERA POINT, West Indies, Island of Hayti, N. coast, the N.E. entrance of Aguada bay.

GOMERA ISLAND, one of the Canary islands. See Canaries.

GOMORROS, BAY OF, Greece, in the island of Parga, on the S.W. side.

GONAVES, or GONAHIEVES, BAY, and GONAVE ISLAND, Island of Hayti; this is a large and fine bay, the anchorage excellent, and the entrance easy. Running along by the N. shore at the distance of 1½ or 2 miles, a vessel must steer about E. ½ N., and, when in, anchor in from 6 to 10 fathoms water. From the entrance under Gonaves point, which is low, 1 mile to the eastward of Point la Pierre, there is 15 and 12 fathoms, and a decreasing depth advancing into the bay. After doubling Gonaves point, leaving it on the port hand, Fort Castries, in ruins, appears on the top of a little hill, which must not be approached too near, as there is a kay lying about a mile S. of the point: there is also a reef extending N.E. from the S. point of the bay, a quarter of a mile, with 6 fathoms of water close to it. The best anchorage is with Fort Castries bearing from N. to N.N.W., in 5 or 6 fathoms. The N. coast and the N. end of Gonave island to the S., Platform point to the N., and the coast from Gonaves to Cape St. Mark to the E., forms the Gulf of Gonaves. Cape St. Mark on the N., and the N.E. point of Gonave island on the S. side, form the entrance of St. Mark's channel; the channel lies N.W. and S.E., distant 7 leagues. The length of this island is 10½ leagues. It lies E.S.E. and W.N.W.; its breadth is 3 leagues. The

N.E. point, called Galot point, is low, and bordered with a reef which runs out half a league from it E. and extends along the shore to the N. Vessels drawing 9 or 10 feet water may sail within this reef. The E. point is high and steep, and it has no white shoals off from it, but the white grounds off the Little Gonave are very dangerous. Little Gonave lies near the S.E. point of the great island. The currents here run strong and irregular, but generally N.N.E. between the two Gonaves. The W. part of the island is an iron-bound coast, and may be approached to within a moderate distance; but from the S.W. point to Point à Retoures there are several small detached reefs almost even with the water. The W. point of this island is in lat. 18° 56′ N., long. 73° 20′ W. The hummock on Petit Gonave is in lat. 18° 41′ N., long. 72° 42′ W., Spanish observations, confirmed generally by Mr. Dunsterville, R.N., as regards the points on the E. coast.

GOODWIN SANDS, off the coast of Kent, 4½ miles, extensive and dangerous, drying in several places, and divided in the middle by a narrow swashway, which runs nearly S.E. and N.W. Tradition states that these sands were part of the estate of Godwin, Earl of Kent, and were overwhelmed by the sea. The N. of these sands is 2¼ miles long and 2¾ broad. Its N.W. edge is steep-to and dries, having a large ridge running along it called the W. Dyke, terminating at the Trinity Swashway, which is not a quarter of a mile wide. On the S. side of this Swash is a spit of sand called the Fork, extending S.W. by W. ½ W. 1¾ mile, where there is a chequered black and white buoy; that part of the spit which is nearest the Swash dries, and is named the Bunt Head, with a buoy with black and white rings. From the N.E. part of the Dyke a large dry patch begins, and bends circularly to the E. These form the N. and E. edges of the Goodwin, called the N. Sand Head Barrow and E. Dyke. A knoll of considerable extent has grown up to the N. end of the Goodwin Sands, much in the way of vessels passing into and out of the Gull stream, having on it in some parts but 12 feet of water at low water spring tides; it is buoyed. The S. Goodwin is divided from the E. Dyke by a channel of half a mile with 4½ and 5 fathoms, called the Gulf or Goodwin Swash; this runs nearly N.W. into Trinity bay. At the N.E. part of the South Goodwin is a dry sand, called the Barrier, which forms the S.E. boundary of the Gulf; and S.W. by W., about three-quarters of a mile from the Barrier, is a long forked patch, which also dries, called the South and North Callipers; the Southern Calliper extends 2¾ miles, and forms the eastern edge of the South Goodwin, which is steep-to; the North Calliper is of equal length, and bends to the westward; the flat thence stretches to the South Sand head, increasing in depth from 3 feet to 4 fathoms: from the Gulf, which divides the North and South Goodwins, to the South Sand head, the distance is nearly 6 miles. Off the northern end of the Goodwin a light-vessel is placed; a light-vessel also rides N. of the Trinity swashway; with two fixed lights, and a light-vessel near the South Sand head. The Goodwin light-vessel shows three very bright fixed lights on separate masts, at 42 and 28 feet above the level of the sea; it lies N.E., about 2 miles from the nearest part of the North Sand head, that dries at low water, in 10 fathoms, with the North Foreland lighthouse N.W. by N. 6¼ miles; Ramsgate Pier lighthouse, N.W. by W.

¼ W. 6½ miles; and the South Foreland high light S.W. by W. ¼ W. 13½ miles. To distinguish these lights from the two Foreland lights, they are exhibited in such a manner that the middle light appears considerably higher than the two extreme lights, forming an erect triangle, so that they can never be mistaken; and in foggy or hazy weather a gong is constantly struck on board her to warn ships that they are near the North Sand head. The situation of this light-vessel renders it impossible for any ship to get upon the North Sand head, or any part of the Goodwin, if due attention be paid to the simple directions issued for the observation of all pilots and commanders of vessels. Several attempts have been made to erect safety beacons on these sands, and one has been placed upon them experimentally by Capt. Bullock, R.N. The edge of the sand has three buoys, with beacons, placed there in 1847. It is greatly to be desired that some of the efforts thus made should be successful. The standing safety beacon was placed upon the E. edge of these sands in 1844 by the Trinity Board, elevated 51 feet above the level of the sand, surmounted by a ball, 18 feet below which arose a gallery of refuge, easily accessible in case of need, tendering its assistance to those unfortunate enough to be cast upon these dangers. From this beacon by compass the North Foreland light bore N. by W., the South Foreland high lighthouse W. by S. ½ S., the North Sand head light-vessel N.N.E. ¼ E., and the Gull Stream light-vessel N.W. ¼ N.

GOOSE BAY, Newfoundland, between Cape Bonavista and Cape Freels (north), running in E. about 7½ miles; by keeping in mid-channel a vessel meets with no danger, and will have 47, 40, and 36 fathoms water until past Lubber's hole, when the depth decreases to 12, 13, 10 and 8 fathoms. There is a small island to the W. of Goose head, where a vessel may anchor in from 4 to 7 fathoms, or further S. in 5½. In most of the inlets here a vessel lies perfectly landlocked.

GOOSEBERRY ISLAND, coast of Newfoundland, lat. 48° 57′ N., long. 53° 33′ W.

GOOSEBERRY REEF, Swedish shore, 7 miles E. of Trelleborg, and 6½ leagues from Falsterbo; a ship, to clear this, should not go into less than 12 fathoms water.

GOOSE CAPE, Newfoundland, near Cape St. Anthony; the S.E. point lat. 51° 18′ N., long. 55° 36′ W.

GOOSE COVE and CREMALLIERE, America, Newfoundland, on the shore of Hare bay; these places lie on the N. shores of the bay. Cremalliere has good anchorage in 7 and 8 fathoms. Goose Cove, on the W. side of Goose Cape, is small but very secure, and has excellent anchorage in 4 and 5 fathoms water. Hare bay lies between Cape Freels and the Strait of Belle Isle.

GOOSE ISLAND, or EL PATO, West Indies, Trinidad, in the S. part of the Bocca Grande, or Dragon's mouth, just within the Gulf of Paria.

GORDA FORT, West Indies. See *Virgin Islands*.

GORDA KAY, America, a small insulated barren rock, 68 miles N.E. of Cape Gracios a Dios, being the first danger on the Great Mosquito bank from the N.E. There is a small detached breaker 5½ miles E. by S. of Gorda kay, called Farrel's breaker. These dangers are detached; the bank is clear for 30 miles within them, and

they may be avoided by not coming into less than 20 fathoms.

GORDO RIVER, Spain, S. coast, to the W. of Malaga, at the entrance of which are sandy flats to be avoided by all vessels making that port.

GORE ANCHORAGE, Kentish coast; the marks are Monkton beacon any where between the W. end of Lower Hale grove and the middle of Upper Hale grove, and St. Peter's church midway between the house and barns in Westgate bay about S.E. In this part there are 4½ and 5 fathoms water. In the Gore the tide runs till 12ʰ, but flows upon the shore till 11ʰ, full and change.

GOREE ISLAND, Bay of Goree, coast of Africa, a small island having a population of 5,000 or 6,000, colonized by the French. This island is a mere rock, about 400 fathoms in length from N. ¼ E. to S. ¼ W., and is about 167 fathoms in breadth, a volcanic production, consisting of sand and basalt, of the same character as Cape Manoel, from which it appears to have been separated. The S. part is about 500 feet above the level of the sea, the rest of the island is low, and the N. point only distinguished by buildings and defences. The landing-place is on the N.E. side of the island, in a small sandy bay. The lat., according to Capt. Owen, R.N., is 14° 39′ 50″ N., long. 17° 24′ 30″ W. In the fort there is a fixed light, seen 6 miles. The roadstead is to the N.E. of the island, sheltered from all winds from S.S.W. to E.N.E. and is perfectly safe from November 1 to July 1, or for eight months in the year. The island produces nothing, but has two springs of good water. All supplies are derived from the main.

GORE PATCH, England, mouth of the Thames, a shallow spot, having upon it only 6 feet water, and a buoy in 6 feet water, with the W. end of Cleave Wood in a line with the Preventive station on Birchington cliff S. by E. ¾ E.

GORE POINT, England, Norfolk co., on the N. side of that county N.E. of Lynn, and about E. of Boston. The space bounded by this point and the adjacent land to the S., and the Woolpack, Middle, and Sunk, to the N., is called the Bays. It is much visited by shipping, but too dangerous to be attempted by strangers to the navigation.

GORE SAND, coast of Somerset, at the N. entrance of the River Parret; it is marked by a black buoy, which must be left on the port hand, and lies in 3 fathoms, with Worle windmill its apparent length on the S. part of Brean down, E.N.E. ¾ E.; Burnham high light-tower, three times its apparent length open S. of the low light-tower, E.S.E.; and Flatholm, its apparent width open E. of Steepholm, the light-tower upon the former island bearing N.N.E. ¾ E.

GORGONA ISLAND, Italy, W. coast. Ships from the W. frequently make this island when bound to Leghorn. The summit stands in lat. 43° 26′ N., long. 9° 54′ E., and it furnishes an excellent land-mark for that road, the lighthouse bearing from it E.N.E., distant 18 miles. This island is high, about 2½ miles long, looking almost round, having two towers, an old monastic building, and several fishing huts upon it. At the N. end there is a rock under water close to the island, and deep water around. When this island is made, vessels intending to go to the S. of Malora should sail E.N.E. towards the lighthouse, and bringing it to bear due E. by compass, pass to the S. of Malora tower, distant 1¾ mile, and when the tower bears W.N.W. ½ N., haul up to the N.E. for the road.

GOROGLOI TOWN, on the N.E. shore of the River Dneister, on the W. side of the Black sea, and in that part of the river estuary called the Gulf of Dneister, about 6 miles where broadest, and 22 in length. Gorogloi lies N.W. of Akerman on the opposite shore. This town has a small inlet forming its haven, in which the water is deeper than in the gulf.

GORO, PORT or ROAD OF, Italy, E. coast, in the Gulf of Venice, formed by low land at the mouths of the River Po. The ground here is muddy and soft. To sail into it, a vessel must steer to the E. side in 4, 5, and 6 fathoms water, being the W. branch of the Po, which here runs into the sea by numerous channels; there are mud-banks off the shore, and the entire ground is composed of that substance, loose in consistence in many places, so that during the bora there is no anchorage to brave its fury found in this port.

GORRAN HAVEN, S. coast of Cornwall, N. of the Deadman.

GOSLINGS ROCKS, THE, South America, on the S.E. coast of the mainland of Paria, and W. of Goose or Pato island in the Boca Grande, W. of Trinidad island.

GOSPORT TOWN. See *Portsmouth.*

GOTHA or GOTA CANAL, Sweden, a navigation partly artificial, but for the most part natural, through the interior of Sweden to the E. coast of the kingdom. It commences at Gottenburg, and traverses the districts of Elsborg, Skaraborg, and Lin Kœping: it consists of several canals, the first of which, on the west, follows the entire course of the River Gotha into Lake Wener, which is crossed. The canal then recommences at the E. shore of that lake to the N.E. of Mariestad, and proceeds S.S.E. to Lake Wiken, near Tatorp. At the Loch of Hairstorp, on the last canal, the highest water of the great Canal of Gota commences. Farther on, this division is formed by Lake Wiken, by a small canal which joins the lake to Billstrommen, by this river itself, and by a canal which joins the Billstrommen to Lake Botten. Five canals have become necessary on the slope of the Baltic. The four first connect the Lakes Botten, Wetter, Boren, Boxen, and Aspangen; the fifth establishes the communication between this last lake and the Baltic, which it joins at 1¼ league below Soderkoping. The distance from the eastern shore of Lake Wener to the Baltic is about 120 English miles, of which more than 60 is occupied by lakes. The sluices are fifty-six in number. The average depth of the canal is 9 feet 7 inches, and the breadth at bottom 41 feet 6 inches. On the western shore of the Lake Wetter is a fortress, commenced in 1820, and intended to stop the communication between the eastern and western parts of the canal, if necessary.

GOTTENBURG or GOTEBERG, a port of Sweden. This city is situated on the mainland opposite the island of Hysingen, in the Cattegat. It is approached through Winga sound, where vessels sometimes anchor before proceeding up the river to the city, which lies on the S.E. bank. Lights have been established at the entrance of Gothenburg, one at Winga, at the mouth of the inlet, a fixed light of the third order, visible 12 miles from all parts of the horizon, in lat. 57° 38′ N., long. 11° 39′ E., and two auxiliary lights for the guidance of vessels up to Gotenburg, one on

Buskar and the other on Botto island, are kept lighted throughout the year. By following proper directions, therefore, a stranger may bring a vessel to anchor in Winga sound. The anchorage is with Buskar beacon S.W. or S.W. by W., and Little Denmark N.E. by N. If during the night, a vessel must keep 2 miles to the southward of the light on Winga, then steer E.S.E. about 1½ mile; or, as soon as the light on Buskar bears N.N.E., steer towards it, keeping it on the port bow, until within half a mile of the island; a N.E. course from thence will then lead to an anchorage in Winga sound; or a vessel may endeavour to pass the end of Buskar, at the distance of a quarter of a mile, then a N.E. ¼ N. course, three-quarters of a mile, will bring her up to the anchorage, in 17 or 18 fathoms, with Buskar light S.W., and Botto light E.S.E. If approaching this anchorage from the southward, the Winga light must be kept N.N.E., to avoid the Vanguard shoal; and when within 2 miles of it at daylight a pilot will be found. This city is the capital of the province of its own name, and stands at the mouth of the Gotha, being the best harbour for foreign traffic of any in the kingdom. It has a great trade in iron, timber, salt, and colonial produce. The pop. is 21,788. A communication exists, too, by inland navigation, with the Baltic on the E., avoiding the passage through the Belts and Sound.

GOTTLAND or GOTHLAND, Baltic sea, kingdom of Sweden. This is a large island, and produces and exports corn, timber, chalk, and sandstone; its fisheries are prosecuted with activity. The capital is Wisby, situated on the N.W. part of the island, fortified, having a small harbour. A breakwater has been built at Wisby, as a security for vessels in the port. The S. end of Gottland is 2 leagues broad, and has a bank running out 10 or 11 leagues to the southward, over which are from 7 to 15 fathoms; it has a round hillock, called Hoborg, near its S.W. end. This island is of a semi-circular shape, running north-easterly, including the island Faro, separated from it by a narrow channel, called the Sound of Faro; the breadth of its widest part is 2 miles. On the W. side of Gottland are the two Carlso islands, having passages between them, and to the N. of these is the small Island of Utholm, surrounded by a reef. Vessels bound to the Gulf of Finland always pass on the E. side of the Island of Gottland, in doing which it is requisite to give the shore a good berth, as several reefs stretch out to the E. a considerable way. There is a shoal between Gottland and Oland, in lat. 57° 33′ 30″ N., long. 17° 52′ E. Its length from N. to S. is 4½ cables' lengths; its breadth from E. to W. 2½ cables, with 6¼ fathoms in its shoalest part. A stone lighthouse, 58 feet high, stands on the S. point of Gottland, a mile N.E. of Hoborg, 57° 26′ 30″ N., 18° 45′ E. It shows a revolving light, visible at intervals of 1½ minute, and stands 170 feet above the level of the sea; seen from E. by N. magnetic bearings through S. to N. by E. at the distance of 16 miles.

GOTTSKA SANDO, Sweden, Baltic sea, some leagues from the N. end of Gottland, an island surrounded by banks, uninhabited,—the rendezvous of seals.

GOUBA, a Russian term, sometimes applied to designate a bay or gulf.

GOUBINIÈRRE ROCK, Channel Islands, one of the numerous rocks, as the Anfroques, Long

Pierre, and others, near the Great Russell channel, between Herm and Serk islands.

GOUDRON ROCKS, France, W. coast, sunken rocks in the Bay of Brest, never uncovered, and lying between the rocks called the Fillettes and the Mingan, on account of which vessels going in avoid the mid-channel and coast along on either side of the Goulet.

GOUGH'S ROCKS, lat. 40° 28' N., long. 30° W. These are rocks of uncertain existence. They were laid down in the chart of M. Rochette, in 1778, as rocks seen by Captains Gough and Birch. Captain Livingston reported : "Captain Beauford, of the brig Concord, of North Yarmouth, told me at Malaga, in 1820, that he saw Gough and Birch's rocks, when bound from Newfoundland for Lisbon; that one of them is about 12, and the other 3 feet above the water; and that they lie nearly in the long. assigned to them in the charts, but 5' more northward." Another report has stated that they were observed by Captain Harrison, in the brig Hope, from Sierra Leone to Cork, on the 17th April, 1830; lat. 40° 16' N., long. 33° W. At 11 A.M., two rocks were seen close under the lee quarter. In smooth water they would be even with the level of the sea; in the hollow of the sea, Captain H. could distinctly see 6 or 8 rocks down in the water.

GOULVEN BAY, France, W. coast, 3½ leagues from the Isle de Bas; it is a large sandy bay, much encumbered with rocks, where there is anchorage with offshore winds. The little port of Pontusval is 1½ mile to the W. of the Bay of Goulven.

GOURJEAN, GULF OF, France, S. coast, between the E. point of the Island of St. Margaret and Point Ilette, being 2¾ miles distant, the gulf or bay lying between. Both points must have a good berth on entering, because there are rocks lying off each. Cape Garoupe, about a mile from Point Ilette, and having a chapel on the top, and the Lerin islands, will easily point out this bay, in which there is good anchorage for all kinds of vessels.

GOURDON, Scotland, Kincardineshire, more correctly the port of Inverbervie, 2 miles further N., at the mouth of the Bervie, frequented only by small craft.

GOUZACOALOS RIVER, America, between Point Xicalango and the Bay of San Bernardo. It is known by its E. point forming a steep hill or morro, the W. point being very low. At (S. 34° W.) S.S.W. ¼ W. from the east point of the river, at the distance of 4 4-10 miles, there is seen on a height a vigia or look-out tower, with a house at its foot, which serves as a warehouse or magazine for gunpowder; and somewhat more to the E. a corps-de-garde, with a battery, which has a flag-staff at its east part, and which serves as a mark for the bar of the river. When this bears (S. 13° 30' W.) S. ¼ W., it will direct over the middle of the above bar, the least depth on which is 2¼ fathoms, increasing past it from 7 to 13 and 15 fathoms.

GOVEN'S or ST. GOVEN'S, HEAD, South Wales coast, 3 leagues W. from Caldy island, the S. point of Pembrokeshire. It has a white appearance, and is a perpendicular limestone cliff, 142 feet high, bold and safe. There is a sandy cove to the N.E. called Broad haven, about a mile distant.

GOVERNOR'S POINT, America, S.S.E. of the harbour of Cape Gracios a Dios about 38 miles. It is a long round turn of the land, beyond which the coast trends nearly S.S.W.

GOVINO PORT, Ionian islands, Corfu; W.N.W. ¼ W. from the citadel 4½ miles, is the entrance to this port, a large natural basin, where the galleys were formerly accustomed to refit, but the vicinity is marshy and unhealthy. The village lies on the S. part of the port, and the arsenal opposite to the entrance. The mouth of this harbour is encircled with sand-banks, so that vessels of less burthen than a brig can alone be admitted; the best anchorage for these is close to the dock-yard. Between Govino and Corfu town are the Govinas, or salt works. The N. point of the entrance to this port projects to seaward, and 1½ mile beyond it on the land is the town of Ipso, whence Mount Salvador appears stretching away to the N.E. 3,000 feet high. The W. coast of this island should ever be approached with caution, as there are patches of rocky ground off it, some of which extends 1½ mile from the shore.

GOZO ISLAND, Mediterranean, near Malta, in lat., at Cape Demetri, 36° 3' N., long. 14° 10' E. This is a much higher island than Malta, and surrounded with rocky cliffs, particularly steep to the W. and S., so that there is no port of moment in the whole island. The surface of this island is of the same character as that of Malta, the pasturage is good, and the inhabitants rear cattle for the Maltese market in considerable numbers. Corn, cotton, and grapes are also produced in plenty. There are six villages in the island, a castle in the centre, and in the E. part is Fort Chambray. About the middle of the N. coast is the Bay of Forno, which affords anchorage for small vessels, as well as that of Ramla to the E. of Forno. Port Miggiaro is on the S.E. side, where there are salt-works. Broken cliffs border the W. coast, less lofty than those on the S., and in some places rocks rise from the water to the height of 160 feet above the sea. Cape Demetri is at the N.W. end of the island.

GRAAEN ISLAND, Sweden, W. coast, a small island at the entrance to the port of Marstrand, upon which stands a white pyramid used as a sea-mark.

GRABUSA AGRIA, and GRABUSA, Island of Candia, islets off Cape Buso, which form a sheltered anchorage, sufficiently deep for the largest vessels. Only small ships can pass between Grabusa and the cape.

GRACIOS A DIOS BAY, Mosquito coast, on the N. of Cape Gracios a Dios, formed by a tongue of land extending S.W. 5½ miles, and offering a shelter against the winds from S.S.W. round by N. to S.S.E. The E. part of this tongue of land is Cape Gracios a Dios, and to the S.W. of it there is a piece of land which once constituted two islets, named St. Pio; its S. point, called Sandy point, is also the E. point of the bay. The depth of water at the entrance is 14 feet, and within it there are 9, and in all parts there is a soft clay and clean bottom.

GRACIOS A DIOS CAPE, American continent, 81 leagues from the harbour of San Juan de Nicaragua, properly named the Mosquito shore, being the N.E. extreme. It is low land, the high land terminating at San Juan; it is intersected with numerous rivers and lagoons, which send off a bank of soundings 43 leagues from the coast, and to the N. 32 leagues. There are many kays and banks upon this bank, some near and others distant from the coast, and on the outside of the bank there are other banks, and various islands. This cape is in lat. 14° 59' N., long. 83° 11' W.

GRACIOSA ISLAND. See *Azores*.

GRADAZ TOWN, coast of Dalmatia, in the Gulf of Venice, a small place 18 miles S.E. of Macarska. It stands on a point near the sea, but has no haven.

GRADO, PORT OF, Italy, in the states of Venice, a small town which may be easily known by its steeple, 5 miles from Busa, very near Monfalcone, the N. part of the Gulf of Trieste.

GRAEMSAY, one of the Orkneys. See *Orkney Islands*.

GRAHAM'S VOLCANIC SHOAL, Mediterranean; first seen to rise out of the sea on the 10th of July, 1831. In August it had attained a height of from 80 to 170 feet. From that time its eruption subsided, and in January, 1832, it had sunk down again and disappeared. At the end of that month its place was occupied by a shoal of from 2½ to 3 feet water. The bearings and distances were, from the S.E. point of Pantellaria W. ¾ N. 40 miles; from Granitola S. by E. 25 miles, and from Cape San Marco S.W. ¾ S. 24 miles. In December, 1837, the spot was visited by the Princess Charlotte ship-of-war, when 9 feet were found on the shoalest part. From subsequent accounts, it would seem that the shoal part had wholly disappeared about three years after the Princess Charlotte's examination of the spot, and there were in 1846, 37 fathoms on the site.

GRAIN COAST, Africa, a portion of the Windward coast, or that extending from Cape Mount to Assinee, which is divided into the three countries of the Grain coast, the Ivory coast, and Coast of Adau or Quaqua. The Grain coast may be said to be all that S.W. shore which extends from Cape Mount, in lat. 6° 44′ N., and long. 11° 23′ W., as far as Cape Palmas, in lat. 4° 22′ 9″ N., and long. 7° 44′ 16″ W. The Ivory coast, from Cape Palmas to about Cape Lahou, in lat. 5° 8′ 3″ N., long. 4° 57′ 40″ W., and the Quaqua country, from Cape Lahou to the River Assinee, in lat. 5° 3′ 5″ N., long. 3° 12′ 7″ W., best known, as well as the coast farther W. as far as the River Volta, or about Cape St. Paul, in long. 59′ E., as the Gold coast, being all on the S. coast of the African continent, N. of the line.

GRAMBOUSA ISLAND, coast of Anatolia, at the entrance of the Gulf of Adalia, E.N.E. about 4 miles from Cape Khelidonia; it is little more than a high, rugged rock, having a spring of good water on the N.E. side, which is much visited by coasting vessels.

GRAMMER'S ROCKS, between Cape Race and Cape Chapeau Rouge, Newfoundland, a cluster of low rocks just above water, lying E. by N. ½ N. 1½ mile from the N. end of Valen island. There is a passage between Great and Little Valen islands, but it is encumbered with rocks.

GRAND BASSAM, Africa, S. coast, a town 7½ leagues W. of the River Assinee, near the mouth of the River Costa.

GRAND BAY, America, Newfoundland, 2 miles W. of Port aux Basques; there are several small islands in and before it, the outermost of which are not above a quarter of a mile from the shore; the sea generally breaks upon them. This bay is only adapted for small vessels.

GRAND BRUIT HARBOUR, Newfoundland, 4 miles from the rocky island of Cinq Cerf, between Chapeau Rouge and Cape Ray, a small but commodious harbour, easily known by a very high and remarkable mountain over it, half a league within land; it is the highest ground on the coast. There are several small islands before this harbour, the largest of which is of moderate height, having upon it three green hillocks.

GRAND CAYMAN, West Indies. See *Cayman*.

GRAND and LITTLE CURROW, or CULLOH, Africa, S.W. coast, towns, rivers, and rocks, between Grand Bassa and Cape Palmas, S.E. of the River Sesters or Sestros, in lat. 5° 39′ N.

GRANDE DEL ORO, SAN AUGUSTIN, and PIEDRAS ISLET, between Cartagena and Cape Catoche, near Puerto Escoces, South America. The island Del Oro is lofty; one mile S. of it is a smaller island, called San Augustin, and on the same bearing, a little more than a cable's length from San Augustin, is Piedras islet, which no doubt takes its name from the many rocks with which it is surrounded.

GRAND ENTRY HARBOUR, America, Gulf of St. Lawrence, at the N.W. end of Coffin island; between that and the sand-bars, is the entrance of Grand Entry harbour, which has water enough within for large vessels. Its opening is very narrow, not exceeding 100 fathoms, and the narrow channel leading to it, between sandy shoals which are said to shift, extends 1¼ mile to the W. A native pilot must be employed, and the entrance made with a leading wind and fine weather. The depth is from 4 to 16 feet, with 28 inside the entrance.

GRANDFATHER'S COVE, otherwise L'ANCE L'UNION, America, Newfoundland, between Cape Freels and the Strait of Belleisle, an inlet about 2 miles inland, 1½ mile from Little Harbour Deep. It is open to S. winds, and may be known when near the shore by the N. point appearing like an island, bearing N.N.W. ½ N. 5 leagues from Cape Partridge. It is a very indifferent shelter for vessels.

GRAND HARBOUR, America, New Brunswick, on the W. side of Ross island, a shallow safe basin, in which vessels that have lost anchors and cables may enter and lie secure on the mud, a convenience of some moment when anchors and cables have been lost on the outer ledges. The entrance has 4, 5, 6, and 7 fathoms water, with a clay bottom, a narrow opening, and security from the sea.

GRAND LE PIERRE HARBOUR, America, Newfoundland, between Chapeau Rouge and Cape Ray, a good harbour, half a league from the head on the N. side of Fortune Bay. It is sheltered from all winds. There is no danger in entering, and within there is from 8 to 4 fathoms water.

GRAND MANAN ISLAND, America, situated at the N.W. entrance of the Bay of Fundy, in the province of New Brunswick, being a part of Charlotte county, 11½ miles in length and 6 in breadth. The N.W. part is distant from Passamaquoddy head 5¾ miles. On its W. side the cliffs rise to 600 feet of elevation, nearly perpendicular, and Dark cove is the only place that can afford shelter even for boats along the whole range, situated about 4 miles from the N. end of the island. Bradford's cove is of no moment. The soundings near the land from Bishop's to S.W. head run 3, 4, 5, and 6 fathoms, close to the shore, deepening to 13, 20, 21, and 22 half a mile off, and to 30, 40, and 50, at a mile distance. Lat. N.E. head, 44° 47′ N., long. 66° 47′ W.

GRAND PASSAGE, America, Nova Scotia, between Bryer's and Long Island; its S. entrance bears N. 29 miles from Cape Fourchu, and N. by

W. 12 miles from Cape St. Mary. It has a depth of from 14 to 30 fathoms.

GRAND SESTERS, or SESTROS, RIVER and TOWN, Africa, S.W. coast, N.W. of Cape Palmas, and S.E. of Little Sesters, having Garraway and Rock Town between the Cape and itself.

GRAND TURK ISLAND, West Indies, or Grand Kay, in lat. 21° 26′ N., long. 71° 8′ W., S. point. It is one of the islets known as Turk's Kays, on an extensive bank, and is considerable, compared to many in its vicinity. It gives the name to the Passage between itself and the Caicos bank. Salt Kay and the Endymion reef lie S. of it in lat. 27° 7′ N., long. 71° 19′ W., the Hawk's Nest anchorage on the W., and the dangerous Mouchoir Bank to the S.E., with a passage between. Nine leagues to the S.E. is the Silver Bank; N.W. part, lat. 20° 55′ N., long. 69° 57′ W.

GRANE ISLAND, England, coast of Kent, an islet in the River Thames, off the E. point of the peninsula formed by the rivers Thames and Medway, the channel from the latter downwards leaving it on the port side. It is separated from the main only by a boat passage called the Scrag. This island is a low marsh embanked from the sea.

GRANGE CHANNEL, Lancashire coast, the entrance into the River Kent, near the Piel or Pile of Fouldrey. It is dangerous, and ought never to be entered by strangers except in a case of urgent necessity, as it has no buoys or marks whatever. If forced in upon that part of the coast, a vessel can only pursue the channel, leading to Havershaw, or the Furness, running to Ulverston. The former passage is bounded by flats of sand on the E. of Lancaster, and on the W. by Cartmel's Wharf sands, dry at low water. The channel which leads to Ulverston is narrow, with Cartmel's Wharf sand to the E., and dangerous rocky ledges to the W. of the entrance. Having entered this channel for Ulverston by night, a tide light is seen on the pier head, a mere lamp, but sufficient to direct the course up. In the day the passage is readily discovered from the appearance of the water, as at four hours flood there is sufficient on the sands for a vessel drawing 12 feet to sail over.

GRANGE POINT, or THE GRANGE, West Indies, Island of Hayti, N. coast, W. of Cape Isabella 35 miles, known by the hill or mount of that name, seen at a great distance before the sea coast is perceived. It appears insulated, stands upon a low peninsula, and looks much like the roof of a barn, from which it takes its name of the Grange. The N.W. part of the Grange point is a rocky islet named the Frayle, and from its S.W. point 3 cables' length distance, there is another, named Cabras. These are the islets of Monte Christi, so called from a town of the same name on the strand. The Grange is 820 feet in height, and stands in lat. 19° 54′ N., long. 71° 39′ W., as observed by Mr. Dunsterville, R.N.

GRANITOLA CAPE, Sicily, 7 miles S. of Mazzara, a long, low, sandy headland, round which is a dangerous reef where many ships have been lost, as from the flatness of the shore it cannot be seen until a vessel is upon the shoals: the coast, too, is here rendered dangerous by currents. This is the S.W. point of Sicily, and lies in lat. 37° 34′ N., long. 12° 36′ E.

GRANNY ROCK, Ireland, S. coast, off Granny Island, Crookhaven.

GRANVILLE, FORT and TOWN, France, IV.

N. coast, situated on the summit of a steep rock, in the department of La Manche, 6 leagues N.W. of Avranches. It stands at the mouth of the little river Bosq, and is surrounded with a wall. The pop. is 7,030. The establishment of the port is 6ʰ 45ᵐ. The port is small on the S. side, formed by a mole running out S. by E. for nearly a quarter of a mile, the end turning a little E., and has a harbour light 28 feet high, but only visible at the distance of 1¼ league, to be left on the starboard on entering. At a quarter of a mile S. by W. ¼ W. from the end of the N. mole there is a beacon erected on the Loup rock. There is a smaller mole further within the port. Upon Point du Roc, or Cape Lihou, there is a lighthouse 154 feet above high water, showing a bright fixed light, visible 5 or 6 leagues. This point is surrounded by rocks and shallows, which require the utmost attention of those accustomed to the navigation, and render it very perilous for strangers to attempt a passage, even in the most favourable time of the tide.

GRAO, Spain, E. coast, a small town on the sea, 2 miles from Valencia, where vessels unload and load for that city; it is commonly called the Grao of Valentia. Vessels anchor off the Grao by bringing the steeple of a church north of that place on the beach and the tower of the cathedral of Valencia in a line. It should be borne in mind that the Barretta rocks half a mile from the beach, running N. full 1½ mile, must be avoided.

GRASS GROUNDS and GRASS GROUND BANK, Gulf of Finland, to the E. of Odensholm. The beacon of the Grass Ground bank lies 10 miles E. by S. ⅓ S. from the beacon of the New Ground. It is small, and has 2 fathoms over some part of it, but there are two spots or patches on it generally above water. Between the New and Grass grounds there are 30 and 40 fathoms water, and between the last and the W. Roge, 2 miles to the E., there are 26 and 27 fathoms.

GRASSHOLM, coast of Wales, 6 miles W.N.W. of Skomer island, between which there is a passage free from all dangers, 6 miles broad. It renders, with the Hats, Barrels, Smalls, and other rocks, the navigation about St. David's head very dangerous. Both the Grassholm and Skomer may be approached within half a mile, and that distance is necessary only to avoid the races, which extend from each about a quarter of a mile. The flood sets through about N.N.E., and the ebb in a contrary direction. Grassholm is 146 feet above the level of low water spring tides, and is generally the first land seen in coming from the W. towards Milford haven. It lies N.W. ¼ W. 11¾ miles from St. Ann's head. Between the Barrels and Grassholm the passage is 2 miles wide, with 20 fathoms water, and may be passed through with safety in daylight. The tides run here with great strength.

GRATO PORT, Greek islands, Isle of Scarpento, on the S.W. side; shallow, and only adapted for small vessels.

GRAVELINES, France, W. coast, department du Nord, 4 leagues W.S.W. from Dunkirk. Pop. 3,570. The entrance to Gravelines is 10 miles E. ⅔ S. from Calais. The church has a tall steeple. The place is distinguished by two windmills, one of which stands on the E., the other on the W. side. This town is about a mile from the coast, and viewed at a distance from the sea, appears like an island; the land on each side low and full of hummocks. A little to the E. of the town stands an old monastery. Gravelines harbour cannot be entered but when the tide is high. There are two

Q

beacons, which, being brought on with each other, will lead between the jetties. The marks for anchoring off Gravelines, to the westward of the banks, in the place called the Pit, are Gravelines steeple S., and Calais cliff W. ¼ S., in from 9 to 11 fathoms at low water, on coarse gravel. Gravelines lighthouse, from which a fixed light is exhibited, stands in lat. 51° 0' 18" N., and long. 2° 6' 48" E. of Greenwich, at the E. of the pier-heads. The building is 83 feet high; and the light, being 95 feet above the level of the sea, will be visible from a ship's deck at the distance of 5 leagues. There are also two harbour-lights, one a tide-light.

GRAVESEND, Kent county, a town on the River Thames, and the first port, 23 miles from London. The river before the town is a mile across to Tilbury fort on the Essex side, and here homeward-bound vessels have been accustomed to anchor, until visited by the officers of the customs. Here outward-bound ships, too, take their passengers on board. This town lies on the great road from London to Dover, and has a pop. of about 8,000. New piers and quays have been recently erected, and it has become the visiting point of numerous steamboats plying between its piers and the metropolis to an extent which is unequalled, under similar circumstances, in any other place, at home or abroad. The environs of the town are pleasant. The rendezvous of the shipping before the place, and its remarkable salubrity. give it a character of great liveliness during the summer season. A number of vessels employed in the fisheries belong to this port. At the E. end of the town there are batteries which cross their fire with Tilbury fort on the Essex side. Gravesend reach of the River Thames lies nearly E.S.E. and W.N.W., and has three shoals, two on the N. and one on the S. shore. The first lies close to the N. shore, abreast of the Barways and below the upper point. Another shoal, called the Oven, lies on the N. side, stretching out from Coal House point about half a cable's length, having 10 or 12 feet upon it. The thwart mark for this shoal is the E. end of East Tilbury church on with the Coal House. The shoal on the S. side begins just below the New Tavern, and extends about a mile downwards, stretching about a cable's length from the shore, with only 4 or 5 feet upon it. The mark for clearing the shoal is Gravesend church open of the blockhouse. The tides are rapid in this reach, and in mid-channel the water is deep. The ships which ride in the reach generally choose the mid-channel, or between that and the S. shore; and therefore vessels passing in the night generally keep well over to the N. bank. At a short distance E. of the Customhouse, between that and the entrance of the canal, is a stone pillar, and close to it are exhibited two lights perpendicularly, one above the other, the lower green, the upper red. These by night, and the pillar by day, denote the limits of the port of London for the collection of the corporation dues.

GRAVESZANDE, Holland, a town of S. Holland province, near the sea, 6 miles W. by S. of Delft. Its steeple was formerly a noted sea-mark, but has been pulled down.

GRAVINAL, POINT, Mediterranean, a headland in the island of Formentera, one of the Balearic isles.

GRAVOSA, PORT OF, coast of Dalmatia, Gulf of Venice, in the channel of Kalamota, being, with Slana, the best anchoring ground for small vessels in that channel.

GRAWE DIEP, Denmark, on the S. shore of Jutland. This bay receives the Starup.

GRAY'S REACH. See Thames.

GRAYSTON BAY, Kent co., E. coast, S.W. from Ramsgate pier towards Sandwich river.

GRAY'S THURROCK, Essex co., a village on the N. bank of the Thames, having a small creek and a pier for the use of passengers embarking or disembarking from steam-vessels and passage boats.

GREAT BAHAMA BANK. See Bahamas.

GREAT BANK OF NEWFOUNDLAND. See Newfoundland Island.

GREAT BARRYSWAY HARBOUR, America, Newfoundland, 4 miles W. of the Burgeo islands.

GREAT BAY DE L'EAU, Newfoundland, 4½ miles N. from St. John's head. There is good anchorage in this bay sheltered from all winds. The passage in is on the E. side of the island which lies at the entrance, as only very small vessels can enter to the westward.

GREAT BAY SHOAL, West Indies, Island of Jamaica, a shoal in the new channel of Port Royal, having upon it 9 feet water.

GREAT CHANCE HARBOUR, America, Newfoundland, a convenient place of anchorage between Cape Bonavista and North Cape Freels, the entrance to which lies W. ⅞ S., distant 10½ miles from West head. The Bacon Bone rock being avoided, a vessel may sail directly for the harbour, the course being W. ½ N. until abreast of Chance point. There is a sunken rock to be guarded against at the south part of the entrance, with only 16 feet water.

GREAT COURLAND BAY, West Indies, Island of Tobago, on the W. side of the island, the N. point of which, or Guano point, is 13 miles to the S.W. of the Sisters, and to this point a berth must be given on account of a rock called the Beer-barrel, which breaks at low water, lying just off from it. There is anchorage here in 6 fathoms. Little Courland bay lies to the S., and has good anchorage within the windward point, which is pretty bold. Vessels ride more smoothly here than in Great Courland bay.

GREAT FOX RIVER, Gulf of St. Lawrence, America, 12 miles N.N.W. nearly from Cape Rozier. It is a mere brook, entering a bay three-fourths of a mile wide, running in half a mile. There are reefs in this bay, which afford shelter to boats and to very small schooners, in from 2 to 2½ fathoms.

GREAT GALLOWS HARBOUR, America, Newfoundland, between Capes Race and Chapeau Rouge. From the S. point of Long island, N.N.W. about 2 miles, lies a small island, called Green island, and from Green island N.N.W. 2½ miles lies Great Gallows harbour island, high land. Vessels may pass on either side of this island into the harbour, which is a mile E.N.E. of the island. Little Gallows harbour, fit only for small vessels, lies close round to the E. of Great Gallows harbour,

GREAT HARBOUR DEEP, or ORANGE BAY, America, Newfoundland, between Cape Freels and the Strait of Belle Isle. It is known from any other inlet by the land at the entrance being much lower than any land on the N. side of White's bay, and by its bearing N. distant 5 leagues from Cape Partridge. It forms a large harbour, dividing into three branches about three miles within the entrance. In the N. branch the water is too deep for vessels to anchor until near its head, but the middle branch has a good bottom and safe anchorage in 6 or 7 fathoms water. Lat. 50° 14′ N., long. 56° 32′ W.

GREAT HEAD, West Indies, Island of Barbadoes, 3 miles N. of Speight's town, off which a rocky sand-bank stretches to the W. full half a mile.

GREAT HENEAGUA ISLAND, West Indies, or, as commonly denominated, GREAT INAGUA. It is low, and cannot be seen at any considerable distance, in consequence of the haze generally prevalent in the latitude in which it lies, the N.W. point being in 21° 7′ 30″ N., long. 73° 39′ 30″ W.; the Middle point, 21° 1′ 45″ N., long 73° 41′ W.; the S.W. point in lat. 20° 55′ N., long. 73° 39′ 3″ W.; the lantern-head, 82 feet high, 20° 56′ 30″ N., long. 73° 19′ 24″ W.; the N.E. point in 21° 20′ 30″ N., long. 72° 59′ 30″ W.—according to the Spanish observations, corrected by Captain R. Owen, R.N. In making the W. end there is no danger, and two sandy bays there afford good anchorage. From the trade-winds blowing over the island, the N. anchorage is the best. On standing in, the line of soundings is perceived by the colour of the water extending half a mile from the beach; the anchor may be let go as soon as a vessel is in soundings, in 6 or 7 fathoms. The entire island is about 15 miles long by 10 wide, the main part of it lying in lat. 21° 5′ N. A shoal, marked doubtful, is shown in some charts as lying off the S.W. end of Great Heneagua, or Inagua. Little Heneagua lies to the N. of the N.E. point of Great Hencagua; it is low and uneven, except a little mount or hummock at an equal distance from its N.E. and S.W. points. Not far from the shore, almost round the island, it is sandy, except at the S.E. point, where a ledge of rocks stretches off and breaks nearly 1¼ mile. On the S. side there is a white bottom, bordered with a reef, at the foot of which is a depth of 40 fathoms. Off the S.E. part, the bank extends 7 miles to the southward, on which is the Statira reef, 3 miles offshore. This island is in extent 7 miles from N. to S. and from E. to W., and is divided from the larger by a deep channel, 1½ league in breadth. The E. point of this island lies in lat. 21° 29′ N., long. 72° 54′ W.; the N. point is in lat. 21° 33′ N., long. 73° W.

GREAT and LITTLE ADVENTURE COVES, America, in Newfoundland, between Cape Bonavista and the N. Cape Freels; two snug small coves, about three-quarters of a mile above Sandy cove on the same side of Newman sound, only adapted for small vessels.

. GREAT and LITTLE BURIN HARBOURS, America, Newfoundland. S.W. ½ W. from Iron island, distant 1 league, is the S.E. point of Great Burin island, and W.N.W. 1½ mile from it is the N. part of Pardy's island. On the main, within these islands, lie the harbours of Great and Little Burin. Vessels bound for Burin can pass on either side of Iron island, the ledge of rocks called the Brandys are the only danger, and on these the sea generally breaks. They lie nearly a quarter of a mile to the S. of a low rock above water close under the land of Mortier West head.

. GREAT and LITTLE CAT ARM, Newfoundland; N.E. of Coney Arm head, distant 10 miles, is the Great Cat Arm, and 5 miles beyond is Little Cat Arm, the latter running up to the W. 4 miles. There are rocks off its N. point, to avoid which

the S. shore must be kept; the shelter afforded is under the head or further end of the arm.

GREAT and LITTLE CONEY ARMS, America, Newfoundland; 4 miles N.E. of Frenchman's cove is Coney Arm head, between Capes Bonavista and North Freels. It is the most remarkable land on the W. side of White bay, and bears W. distant 8½ leagues from Cape Partridge. The land here projects a mile and a half, forming a deep bight, called Great Coney Arm. It has no shelter. Little Coney Arm, to the W. of the head, is a convenient anchorage for small vessels.

GREAT and LITTLE CORN ISLANDS. See *Corn Islands.*

GREAT and LITTLE HARBOURS, Newfoundland, between Cape Chapeau Rouge and Cape Ray; 2 miles within the W. point of La Poile bay, and N. ½ W. 2 miles from Little Ireland, is Tweeds or Great Harbour, its S. point low, extending inwards W.N.W. one mile, about 1½ cables' length in its narrowest part. The anchorage near the head of the harbour is in 18 or 20 fathoms clear ground, sheltered from all winds. Half a mile N. of this is Little Harbour, the N. point of which is called Tooth's head. This harbour has 10 fathoms water half way up.

GREAT and LITTLE ROUND HARBOURS. America, Newfoundland, between Cape Freels and the Strait of Belle Isle. The first is a good harbour, without danger, the shores bold-to; the anchorage is between two inner points, in 4 and 5 fathoms water. Little Round harbour is a mere cove, unfit for shipping.

GREAT and LITTLE ST. JULIEN, HARBOUR and ISLAND, America, Newfoundland, between Cape Freels and the Strait of Belle Isle. The S.W. end of the Island of Great St. Julien is little separated from the main. The E. harbour has no danger for a vessel until she is within the entrance, the starboard shore being foul nearly one-third of a mile over. There is anchorage within in from 8 to 4 fathoms. Little St. Julien is near the larger harbour, but is not so easy of entrance as the larger and requires mooring head and stern.

GREAT and LITTLE ST. LAWRENCE HARBOURS, America, Newfoundland. The first harbour to the E. of Cape Chapeau Rouge. It requires care on entering with W. and S.W. winds not to approach too near the Hat mountain, to avoid the flaws off the high land. There is no danger but what is very near the shore. The course is first N.N.W. until the upper part of the harbour be open, then N. ½ W. The best anchorage for large ships is before a cove on the E. side of the harbour, in 13 fathoms water. Little St. Lawrence harbour is the first to the W. of Sanker head, a high hill like a sugar-loaf, 3 miles E. from Cape Chapeau Rouge. The W. shore must be kept close on board on entering, to escape a sunk rock. The anchorage is above the peninsula in 3 or 4 fathoms water. Ships may anchor in 12 fathoms, good ground, outside the peninsula, but open to S.S.E. winds.

GREAT and LITTLE SANDY HARBOURS, Newfoundland, the first W. ½ S. distant 4 miles from the S. end of Barren island. To this place there is a passage between Ship island and the main, with 7, 9, and 17 fathoms. The entrance is narrow and rocky. Little Sandy harbour lies a mile S. of the great harbour. It has 6 and 7 fathoms water, good ground, and may be known by the Bell island, which lies S.E. ¼ E. 1½ mile

Q 2

from its mouth, and N.E. by N. 13 miles from the W. point of Merasheen island.

GREAT JARVIS HARBOUR and ISLAND, America, Newfoundland, between Chapeau Rouge and Cape Ray, situated at the W. entrance into the Bay of Despair, which lies between Long island and Great Jarvis island, in the mouth of the harbour of the same name. This harbour has good anchorage, with from 16 to 20 fathoms water, secure from all winds. The passage in is on either side of the island, but the S. channel is safest, as in the N. there are several rocks.

GREATMAN'S BAY, Ireland, W. coast, to the W. of Galway, between Garomna island and the main. It is a place of good holding ground, well sheltered, and adapted for vessels drawing 12 or 14 feet water. There is a small rock here, dry at half ebb, nearly a cable's length south of Trebaan point. Within this harbour, nearly a mile above the mouth, there is a rocky ledge extending on the E. side a cable's length from the shore, the extremity of which dries at three-fourths ebb. To avoid this ledge keep in the mid-channel nearer the W. side. Half a mile above this ledge, on the W. side, there is a sandy shoal extending three-fourths across the channel. W. of the entrance of Greatman's bay, half a mile from the middle part of Garomna island, is the Englishman's rock, dry at three-fourths ebb. Gulin head kept in sight by Live island, will lead S. of it.

GREAT MIQUELON ROAD, ROCKS, and ISLAND, America, between Chapeau Rouge and Cape Ray, Newfoundland. Miquelon island is about 7 leagues in length from N. to S., and 6 miles in breadth in the widest part. The middle is high land, called the Highlands of Dunne. The shore is low, except at Cape Miquelon. The road, which is large, lies towards the N. end, between Cape Miquelon and Chapeau, a remarkable round mountain on the shore. All the rest is clear of danger. The best anchorage is in 6 and 7 fathoms, on sand, but it is exposed to E. winds. Miquelon rocks stretch off the E. point of the island, some above and some under water; and under the high land, 1¼ to the E., there are 12 fathoms close to them, with 18 and 20 fathoms a mile off. N.E. ½ E., about 4½ miles from these rocks, lies Miquelon bank, on which are 6 fathoms water.

GREAT PLUMB POINT, West Indies, Island of Jamaica, 8 miles W. by N. from Cow Bay point, W. of Yallah's point.

GREAT POND HARBOUR, America, Gulf of St. Lawrence, a small creek, only affording shelter to fishermen, 16 miles N.W. ½ N. from Great Fox river.

GREAT QUIRPON HARBOUR, America, Newfoundland, on Quirpon island, the N. extremity of which lies in lat. 51° 39′ N., long. 55° 27′ 50″ W. The entrance of the harbour, which is on the N.W. of the island, lies between the island itself and Grove's island. The anchorage within is good, but there are some shoals which must be left on the port side upon approaching the entrance.

GREAT RIP SHOAL, America, United States, coast of Massachusets. This shoal lies in shallow water, and is 12 miles in extent from N. to S., between Fishing Rip and Sancoty head. The N. end bears E. by N. ¼ N. 10 miles, and the S. end E.S.E. southerly 11 miles from Sancoty head, Nantucket island.

GREAT RIVER (Rio Grande), America, Mosquita coast, 9 miles distant from the most N. of the Man-of-War kays. To enter the Great River anchorage, a vessel should pass between the Man-of-War kays and the coast; in which channel, until arriving at the river, there is nothing to be feared.

GREAT ROUND SHOAL, America, United States, off the coast of Massachusets, partly dry at low water. It bears E.N.E. 9 miles from Nantucket light. There is a black buoy on the N. end in 14 feet water.

GREAT SALMON RIVER, Newfoundland; the entrance is E.N.E 5½ miles from the N. part of Little Colinet island. It is three-fourths of a mile wide, and runs E.N.E. 7 or 8 miles. Three miles up, on the S. shore, is an opening called Little Harbour, opposite which is a small cove, the best anchorage in the river, in 5 or 6 fathoms. The river is good throughout.

GREAT SOUTH HARBOUR, America, Newfoundland, Placentia bay. The Great harbour lies 1 mile to the N. of the Little. The entrance is between the middle point and the Isle of Bordeaux, 1½ mile wide, with from 20 to 30 fathoms water. There is no danger entering, and the anchorage, 1½ mile up, is good in 6 and 7 fathoms.

GREBASCIZZA, or SEBENICA VECCHIA, an inlet in the Gulf of Venice, in Dalmatia, on the N.E. side, and to the S. of the entrance to Sebenico, 2 leagues. It has the Island Plana at its entrance, which must be left on the starboard upon going in.

GRECO, POINT, Italy, Gulf of Venice, N. of Manfredonia about 15¾ miles, whence the shore turns N. to Porto di Campi, which has a tower on each side, and three islands off the entrance, behind which are 3 fathoms water, and outside 6, 7, and 8.

GREEN BAY, America, Newfoundland, between St. John's harbour and Baccalou island; this place affords anchorage at the entrance in 15 or 16 fathoms, but is dangerous to enter far being open to the E.

GREEN BAY ISLAND, America, in Darien, 8 miles S.E. of the Chiriqui Lagoon. It has anchorage in 10 fathoms.

GREEN CASTLE, Ireland, on the starboard point of entrance into the Lough of Londonderry, on the N. coast of Ireland, the opposite point to Magiligan, a mile distant from the castle.

GREEN GROUNDS, OUTER and INNER, South Wales coast, to the N.E. of the Mumble point; several patches of rocky ground formed in two groups. The channel between has from 4½ to 6 fathoms in it. The leading mark through is the S. bluff of Mumble heights on with the N. head of the lighthouse bearing W. by S.

GREEN HARBOUR, America, Nova Scotia, to the W. of Rugged island harbour, with an island on its W. side of entrance, running in full 3 miles. It is little visited, being open to S. winds.

GREENHOLM ISLAND. See *Orkney Isles.*

GREEN ISLAND, within Stronfa frith. See *Orkney Islands.*

GREEN ISLAND, at the N. end of Brassa. See *Shetland Isles.*

GREEN ISLAND, America, Nova Scotia, between Halifax and Cape Sable, lat 44° 27′ N., long. 64° 0′ 18″ W. It is W.N.W. ¾ W. from abreast of Sambro lighthouse. It is a small island. There is a second island with this name on the coast between Cape Sable and Bryers' island. It lies N.N.E. ¾ N., distant 3½ miles from the Gannet Rock light: a reef runs off from it to the S.W.

almost three-quarters of a mile, and round it are 6 and 5 fathoms water, and between it and the Gannet rock from 12 to 17. W. of this island 1½ mile there is a sunken ledge, that lies directly in the fair way of the channel to the little harbour of Jebogue, which is shoal and intricate.

GREEN ISLAND, Greek Archipelago, the most E. of the Spalmatori or Bird islands, about 12 miles from Scio.

GREEN ISLAND, Newfoundland, about three-quarters of a mile in circuit, low land, E.N.E. about 4 miles from St. Pierre, and nearly in the middle of the channel between that and Point May. On its S. side there are several rocks above and under water, extending 1½ mile to the W.S.W. Lat. 46° 52' N., long. 56° 10' W.

GREEN ISLAND, South America, near the land, and not far from the entrance of the Sinnamari river on the coast of French Guyana.

GREEN ISLAND, Spain, a little S.E. of the city of Algeziras in the Bay of Gibraltar.

GREEN ISLAND, West Indies, one of the Virgin islands.

GREENLAND, a vast country extending from Cape Farewell, in lat. 59° 49' 12" N., up to the Pole itself. But a part of this vast territory, or about 7°, is S. of the Arctic circle. The E. coast is for the most part an impenetrable barrier of ice, vast fields of which prevent communication with the shore. On the W. coast, the temperature is milder, running along and forming the E. side of Davis's straits, in the S. part of Baffin's bay. Here there have been some colonies formed by the Moravians, and a scanty population obtains a poor existence for civilized man, in a climate of great severity, producing few of the necessaries of life. The W. coast possesses some good harbours, and affords a contrast to the E. side, in this respect affording a striking example of the difference of temperature experienced in situations between shores having an eastern and a western aspect. The E. coast was said to have been once colonized near the S. extremity by the Danes, but the colonists are supposed to have all perished. The Moravian brethren principally hold the W. settlements, and devote themselves to the benevolent object of preaching to and civilizing the native inhabitants, who are of Esquimaux descent. They have a few cattle, which subsist in winter on the grass cut in summer and dried. The coast is foggy and the winter gloomy and severe; the whaler in summer visits these remote shores, which have nothing to attract the inhabitants of happier climates.

GREENLY ISLAND, America, Labrador; the N.E. point is in lat. 51° 23' N., long. 57° 13' W.

GREENOCK, Scotland, W. coast, a considerable seaport seated on the S. side of the River Clyde, where it takes the name of the Frith of Clyde, and is most convenient for the reception and departure of vessels of all classes. It has a pop. of 25,000. It is defended by a small fort, carries on a great foreign trade, and is largely concerned in the fisheries both of Newfoundland and the coasts and islands N. and W. as well as in the whale fishery. This town possesses dry docks, ship-building yards, extensive rope-walks, and every adjunct connected with an opulent river port. Lat. of the church spire 55° 56' 53" N., long. 4° 45' 18" W.

GREENORE POINT, Ireland, E. coast, N. of Carnsore point in Wexford, and the S. point of the Bay of Greenore, N.W. from the Tuskar rock.

GREENPOND TICKLE, America, Newfoundland. Greenpond is a square island, about a mile each way, between Cape Bonavista and North Cape Freels. Ships have sometimes run and anchored between Greenpond island and the main, but the channel is narrow, the water deep, and it lies open to the S.W. Greenpond Tickle, on the S.E., is of little importance, and can receive no ship drawing above 14 feet water.

GREENPORT, America, United States, New York, near the N.E. end of Long island, 12 miles from Gardner's point. There is much shipping belonging to this port. The harbour will admit the largest ships, well sheltered from storms, and is rarely obstructed by ice. A pilot is necessary to this port from Gardner's point.

GREENWICH, England, Kent co., a town seated on the River Thames, about 5 miles from London bridge towards the mouth of the river, opposite to the Isle of Dogs. There was once a royal palace at this place, together with a royal park, pleasantly situated. The site of the palace is now occupied by the hospital for superannuated seamen; to which purpose it was first applied in the year 1705, when 100 disabled seamen were entered on the foundation. The building of stone has been successively completed in four quadrangular parts. The principal front skirts the S. bank of the Thames. The extent of 863 feet is occupied by a terrace along the water side, and in the centre is the principal landing-place, by a double flight of steps. This landing-place affords the best view of the entire edifice. On the summit of the hill behind it, in the park, stands the celebrated Observatory of Greenwich, the latitude of which, the result of 720 observations of the Pole star, made during 18 months in 1825 and 1826, has been deduced as 51° 28' 38" 955, or lat. 51° 28' 39" N. The longitude, both in the English dominions and in the United States of America, commences here. It is from hence that the mean time is regulated for all nautical purposes, a ball being dropped by an apparatus from the summit of the Observatory daily at noon, for the regulation of timepieces and chronometers on board the different vessels in the river. The number of inhabitants in Greenwich is about 25,000. Of these, perhaps 3,400, including nurses and attendants, reside within the hospital.

GREENWICH, EAST AND WEST, America, United States, in Rhode island.

GREG NESS, Scotland, E. coast, S. of Girdle Ness, the last being the S. point of Aberdeen bay.

GREGO, CAPE, Island of Cyprus, S.E. side. Lat. 34° 58' N., long. 34° 6' E.

GREGO, PORT, Gulf of Venice, Island of Lesina, between ports Pellegrino and Lesina, a narrow but safe haven, with deep water, where ships ride fastened to the rocks.

GREGORIO, ST., ISLAND, Gulf of Venice, near the Isle of Arba, in the Quarnerola, among the islands off the S.W. coast of Croatia. The navigation here requires a pilot.

GREGORY'S SOUND, Ireland, W. coast, the passage that divides Inishmaan and Killeney or Arran island. There is a bay in the Island of Killeney, at the N.E. end, called sometimes Arran harbour, or Killeney bay. Here the ground is clean, and the water on the W. side of Straw island deep enough for the largest ships, but much exposed to E. and N.E. winds, which set in with a great swell; it is, therefore, an anchorage only adapted for moderate weather.

GREISSWALD. See *Griefswald*.

GRELOGH ROCKS, Ireland, W. coast, 4 miles S.S.W. from Crow head. These are a patch of rocky ground well known to the fishermen, but not considered dangerous. They bear from Sheep's head W.N.W. ¾ W. 11 miles.

GREMISHADEL LOCH, Scotland, W. coast, on the E. side of Lewis island. It requires a leading wind to enter, it being narrow; but once within, the ground is good and the depth sufficient. At its head there is shelter from all winds.

GRENADA, ISLAND OF, West Indies, about 17 miles in length and 12 in breadth, the most S. of the Antilles except Tobago, and the most beautiful of the British West Indies. St. George's fort in this island is in lat. 12° 2′ 54″ N., long. 61° 48′ W. It is about 70 miles distant from Tobago, and 75 distant from the South American continent. It was discovered by Columbus in his third voyage, in 1498. The interior is mountainous and picturesque, the N.W. coast consisting of successive conical hills heaped upon each other, rising in some places to the level of 3,000 feet above the sea. This island is very fertile, and has a pop. of about 5,000 whites and persons of colour, and 23,471 blacks. The island is divided into six parishes. The capital is St. George, which is defended by the forts and works of Richmond hill, Fort King George, Hospital hill, and Cardinal heights. This is the chief port and residence of the governor. The climate is hot, but tempered by sea breezes, and the seasons very regular. This island possesses 26 rivers. The E. part is the lowest, while the mountains may be seen in clear weather 8 leagues off at sea, and the island approached by shipping within two miles. It is much less subject to hurricanes than some of the other islands. The gross revenue is about 12,500l., and the expenditure 13,400l. to 12,600l. The shipping amounts to about 260 entered inwards in the year, and 256 outwards—a number that necessarily fluctuates. The dependencies, some upon Grenada and some upon St. Vincent's, are generally termed the Grenadines. These islands are Little Martinique, Union, Cannouan, Castle island, Atand, Mosquitos, Balesso, Carriacou, Becouya, Redondo, and some mere rocky islets. There are several bays and creeks on both sides of the island, affording good anchorage. Those most used are St. George's bay, the principal bay of the island on its W. side, Caliviny, or Egmont harbour, on the S. side, and Grenville bay on the E.

GRENADILLAS, or GRENADINES, West Indies, a chain of small islands and rocks, comprehended in the space between Grenada and St. Vincent. The jurisdiction of these islands is divided between Grenada and St. Vincent; Carriacou, Redondo, or Round island, the Diamond, Levora, &c., belong to Grenada; and Bequia, or Becouya, Young's island, Maillerean, Balesso, Cannouan, Mosquitos, Mayo, Union, Frigate island, Little Martinique, &c., appertain to the latter. The two principal islands are Carriacou and Becouya; these are inhabited, and have plantations of coffee, indigo, cotton, and some sugar; but are totally without rivers; fresh water is obtained from a few springs.

GRENÆ, Denmark, a small town of Jutland, on the E. shore of a large bay formed by two peninsulas.

GRENALLY, or GREENLEA, POINT, Somerset co., near the W. end of the Gore sand, at the entrance to Bridgwater river or the Parret.

GRENNEFIORD, Iceland, one of the six towns which in 1787 received the privileges and rank of cities from the Danish Government.

GRENNEL CHANNEL, America, Honduras, near the River Belize.

GRENSUND, Denmark, separating Falster island from Moen.

GRENVILLE BAY, Island of Grenada, West Indies, on the E. side of the island, and has a separate custom-house. This bay is open, safe, and commodious for those acquainted with its navigation, but to those who are not, it is difficult and dangerous, in fact impracticable, since the lead is no sure guide, and being once near the rocks, it is not possible to stem the current, which will drive the vessel that thus ventures in upon them. It is necessary also to guard against running to leeward of this bay, for should this be done by any vessel, she must go round Point Salines, turn to windward on the other side of the island, and make anew for her destination, since the best ship cannot make head against the S. current which sets along its E. shores. This bay is formed between Telescope point and Soubise island, two miles asunder; it is circular in form and surrounded with dangerous rocks: the channel is near Telescope point, and there are two beacons erected on the shore to the S. of the town at the head of the bay. These must be brought into a line bearing N. 74° W., and keeping them in this direction, a ship may run in under easy sail. No vessel drawing more than 13 feet water, can enter after passing the Narrows without being lightened. Vessels are advised to heave to a little to windward of the Grenville rock, which lies N.E. of Telescope point; here, on firing a gun, a pilot comes off.

GRENVILLE BAY, America, Prince Edward's island; this is also called New London or Richmond bay. It has from 8 to 10 feet water, but the bar which extends outwards half a mile is very difficult; the channel runs W.

GRENVILLE INLET, Florida reef, North America, in lat. 26° 47′ N., long. 80° 6′ W. See *Florida Reefs.*

GRENVILLE, TOWN and PORT, West Indies, in the Island of Grenada. This port lies about half way down the island on the E. side, generally called the bay. Those who make it must be cautious of not going to leeward of it; a pilot comes off upon the requisite signal being made, on a ship heaving to on the windward of the Green rock. The harbour is impracticable to a stranger.

GRESMA ISLAND. See *Orkney Isles.*

GREVALGA RIVER, America, Bay of Campeche; this river empties itself by two mouths, which enclose the Island of Tabasco. The W., or Tabasco mouth, is 2 miles wide, but crossed by a bar with only 12 feet water, within which for a considerable distance there are from 3 to 5 fathoms. In the rainy season the volume of water is so considerable as to freshen the sea outside the bar. The town of Tabasco, small but well-built, stands on the island.

GREYMISTER, Brassa sound. See *Shetland Isles.*

GRIBBIN HEAD, Cornwall, the E. point of the Bay of Tywardreth on the S. coast.

GRIB HOELEN, THE, Norway, W. coast, in lat. 63° 15′ 30″ N.; the southern entrance to the channel of Drontheim, which is so denominated.

GRIB TARREN ROCK, Norway, W. coast, lying 15 miles from the land, and called also the NATTERGALENE. Its position is to the W. of both channels up to Drontheim. On the S.E. part is a rock with only 9 feet water, while to the N.W. there are from 5 to 7 fathoms and deep all round. The shallowest rock lies N.N.W. from Grib Oerne distant 12 miles; and from the outermost, or N. point of the Smælin banks nearly W.S.W., distant 17 miles. Care is necessary to avoid this danger, which in stormy weather shows itself by breakers; a vessel may pass on either side, the water being unfathomable.

GRIEFSWALDE, Prussia, Pomerania, a port on the Riak, small but convenient, with a considerable trade. Two fixed lights (vertical) are on this island, in lat. 54° 15′ N., long. 14° E.

GRIFFIN COVE and RIVER, America, Gulf of St. Lawrence, 6½ miles N.N.W. nearly from Cape Rozier. A small bay here affords shelter to fishermen, but is of no moment to shipping except to afford supplies.

GRIGNANO POINT, Istria, a few miles N.W. of Trieste.

GRIGUET BAY, and CAMEL'S ISLAND HARBOUR, Newfoundland, near Belle Isle. The N. bay is insecure in spring and autumn, being perfectly exposed to the S. gales. The S.W. bay is therefore to be recommended. Camel's island harbour has too intricate a navigation to be attempted by a stranger; lat. of E. point 51° 32′ 30″ N., long. 55° 27′ W.

GRIMAUD, GULF OF, France, S. coast, sometimes called the Gulf of St. Tropez. The town is to the S.W. of the Castle of Grimaud, standing on a low point of land. It has a population of 4,000, employed in the fishery and coasting trade, a citadel and school of navigation, and is the place from whence Napoleon embarked for Elba in 1814. There is a small mole here, but vessels must be careful to avoid the rocks called the Sardinaux, almost level with the water and stretching half across the gulf, when navigating at this place, which is very near Cape St. Egou, the W. point of the Bay of Frejus.

GRIMSBY, NEW, Cornwall, W. of Trescow. See *Scilly Islands.*

GRIMSBY, TOWN and PORT, England, Lincoln co.; it had once a natural harbour that had become choked up with sand. It stands near the mouth of the River Humber on the S. side of that river. This port is subordinate to that of Hull; the pop. in 1831 was 4,225. The road here is preferable to that of Hawk in S. and S.W. winds. A vessel intending to anchor there, should steer about N.W., along the E. edge of the Bull sand, in 6 and 7 fathoms, until the low light becomes a fathom's length open of the high light, then bearing S.E. ¾ E., with the lights in this direction, a vessel will be carried between the Middle on one side, and the Clea Ness sand and Borcum on the other. Having passed the Clea Ness sand, leaving it on the port side, and got three-fourths of the way towards the black buoy on the S.E. end of the Borcum, a ship will be in Grimsby road, where large ships lie a little outside of the stream of the buoy, in 5 or 6 fathoms. Small vessels may lie in the Inner roads within the buoy, with Grimsby church tower bearing W. by S., or a little below, in from 10 to 15 feet, on a mixture of clay and mud. Regard must be paid to the time of the tide on anchoring, whether spring or neap. A wet dock, with flood-gates, sluices, and other requisite buildings, has been constructed here. The lock within the haven is 150 feet long, about ¼ of a mile distant from the

Humber, having a depth of 12 feet at the lowest neap-tides, with ordinary springs 20, and high springs 23 feet. The whole excavation is about 18 acres, affording ample room for 250 ships to load and discharge. In addition, there is a dry dock, a building-yard, and various private docks for bonding goods. Should a vessel, bound to Grimsby, arrive there during neap-tides, she will find safe anchorage in the roads until the following springs; and the same wind that will carry a vessel from the Humber, will enable her to leave the port. There are two fixed lights on the Spurn point, and a floating revolving light at the entrance of the Humber.

GRIMSTONE ROCKY SHOAL, Durham co., a quarter of a mile from Sunderland point. The sea commonly breaks over it, and it requires a wide berth in passing.

GRINDSTONE ISLAND, America, Gulf of St. Lawrence, near Amherst and Alright islands; its summit rises 550 feet above the sea at high water.

GRISLEHAMN, Sweden, a small port opposite the Isles of Aland.

GRISNEZ, CAPE, France, W. coast; 2 leagues W. of Calais is Cape Blancnez, and nearly 2 leagues W. by S. of the last is Cape Grisnez. A low dry sand extends all the way at low water along shore from Calais to Cape Grisnez, drying on an average one-third of a mile from the shore. On the N.E. and S.W. of Cape Blancnez, there are rocks, and the water continues shoal some distance from the dry sand, and on the parallel of the Bas Escalles, stretches out 2 miles from the coast, turning S.W. Cape Grisnez, forming the Ligne bank, a part of which in an E. by N. direction from Cape Grisnez dries; but to the S. of which there is a very narrow channel. The Rouge Riden and the Quenocs lie to the N. by W. from Blancnez, the least water on either being 1½ fathom. The Quenoc is about 1½ mile from Blancnez. On the N.W. part of the Ligne bank, called the Barriére, there is not more than 4½ feet. To the E., near the dry sand of the coast, are some rocks called the Guards; 16 fathoms is as near as any vessel should stand inshore until Cape Grisnez bear S. From thence to Boulogne the coast is clear at one-third of a mile from the land. There is a lighthouse upon Cape Grisnez, having a bright revolving light every half minute, elevated 46 feet from the base, and 193 feet above the sea at high water, visible 8 leagues off in clear weather. This revolving minute light stands in lat. 50° 52′ N., long. 1° 35′ E. From this cape to the entrance of Boulogne is S.S.W. about 3 leagues.

GROAIS ISLAND, America, Newfoundland, N.E. of Belle Isle, 7 miles in length, and 2¼ broad. The N. point lies in lat. 50° 58′ N., long. 55° 32′ W. There are several rocks under water off this end, and also off the N.W. part; otherwise this island is free from danger, and bold all round, having from 20 to 70 fathoms between its shore and the mainland.

GRŒSHOLMEN ISLET, Baltic sea, one of the rocks denominated the Eartholms, that enclose an excellent harbour, which is in lat. 55° 19′ N., long. 15° 12′ E., and has a light on the Island of Christiansoe, revolving every three minutes, 92 feet above the level of the sea, and sufficiently distinguishable from that on Bornholm, which is bright and fixed, and 280 feet high.

GROIX, or GROA, ISLE DE, W. coast of France, 4 miles long and 1½ broad, lying S.E. by

S. and N.W. by W., at the distance of a league from the mainland, in lat. 47° 39′ N., long. 3° 30′ W. It forms a good channel into Port Louis and l'Orient, with 16 or 17 fathoms water, free from danger. Still the E. and S.E. part of the island should have a good berth, as off them there are some rocks and shallow spots of water, called Basses Milit, nearly half a mile out. To the S. of these, half a mile distant, is the Point de la Croix, where there are a battery and signal post. Thence a spit runs out full three-quarters of a mile to the S. of the Point des Chats, ending in a rock called Les Chats, which sometimes shows above water. On the E. point of this island there is a lighthouse, upon the Fort de la Croix, 154 feet high, visible when clear at the distance of 10 miles. It is a fixed light, obscured in the direction of the Glenan isles by the heights of the W. part of Groix. On Point Penmen, the N.W. part of the island, there is a fixed light, 194 feet above the level of high water, visible at the distance of 7 leagues.

GRONINGEN, TOWN and PORT, Holland, the capital of the province of Groningen, seated on the Hunes, by which large vessels ascend to it from the sea; it communicates, by means of a canal, with Delfzuyl on the Ems; pop. 20,000.

GRONSKAR ROCK, Sweden, near the Sandhamn entrance to the harbour of Stockholm; there is a bright fixed light upon this rock, 103 feet high, seen 14 miles. There is a second light near this entrance, on the small woody Island of Korssö or Korosoe. It is a revolving light, making a revolution every four minutes, from which the Grönksär light is distant 3 miles E.S.E., in lat. 59° 17′ N., long. 19° 2′ E.

GRONSOE ISLAND, Sweden, Baltic sea, below Sodertelje, 3½ leagues E. of the Landsort lighthouse. Here is a custom-house, where all vessels are examined before they are allowed to proceed to Stockholm.

GROS, CAPE, E. coast of Spain, 6½ miles from Tarragona; between this cape and the town of Tamaris to the S., is a small beach, a stream, and the town of Altafaya. The Torre Dembarra stands on this cape, which is higher than any of the surrounding land, but the ground round it is foul and rocky. No vessel ventures to stop between Tarragona and Cape Gros, unless keeping open to all winds, and stopping only for a commercial purpose.

GROS DE RAZ, ROCKS, France, N. coast, near Cape La Hogue; the lighthouse stands upon this rock, elevated 157 feet above the sea at high water, and visible in clear weather 7 leagues. It is a bright fixed light, in lat. 49° 43′ N., long. 1° 57′ W.

GROS FIORD, or GRIMSTADT, HARBOUR, Norway, E. of Christiansand, S. coast, and 3 miles N.E. from Bior Oe. It has several entrances. Large vessels can enter from the S., and then, in moderate weather only and with a fair wind, from the ledges of rocks and the shoals lying near the passage. The other entrances are intricate and dangerous, but the harbour is good, having 11 and 12 fathoms water, with a clay bottom. It can only be left with a N. wind, except by steamers. On the E. side of the entrance is Hesnaesoa beacon, 31 feet high, and of a triangular form.

GROS HAF, coast of Pomerania, a basin formed by the Islands of Usedom and Wollin. The waters are generally fresh, but become brackish in northerly winds. It has three outlets into the Baltic, the Peene, the Swine, and the Divenow. The ports here constantly require to be freed from sand; the action of the sea is sometimes so violent as to wash away the strongest moles and dykes in a few hours.

GROS ISLET, BAY, and ISLAND, West Indies, Island of St. Lucia, principally visited by men-of-war, having excellent anchorage. Gros islet, or Pigeon island, is a small fortified island, separated from St. Lucia by a shallow channel, half a mile broad, and full of rocks.

GROS MORNE, West Indies, coast of Guadaloupe, a small island, within gunshot of the shore, between which and this island there is an extensive anchorage, running out a considerable distance from the land, but between the island and Englishman's head, and between both and the shore, there are only 12 feet water, the sea in strong winds breaking all across the passage.

GROSA ISLAND, Spain, S. coast, 4½ miles from the Point of Palos; it is high and in shape like a haycock, somewhat triangular, and 1½ mile from the Manga shore. The channel between affords anchorage in 4½ fathoms water, but is only fit for small vessels. N.E. by E. about 640 yards from Grosa island, there is a rocky islet, or "Farillon," in the Spanish tongue, and N.E. ¼ E. there is a sunken rock, with only 6 feet water over it.

GROSO, CAPE, Mediterranean, Island of Minorca, 2 miles from Point Negra.

GROSSA, or LUNGA ISLAND, Gulf of Venice, one of the outermost of the islands on the coast of Dalmatia, mountainous, long, and narrow, stretching S.E. ¼ S. from the lat. 44° 10′ to 43° 52′ N. It possesses 13 villages, the principal of which is Sale, on the E. shore, not far from the N. end of Port Tajer; there is here a little port, sheltered by the Island Labdara.

GROSSA ISLAND, Mediterranean, one of the islets near Formentera, and one of the Balearic group.

GROSSA POINT, Gulf of Venice, in the bay of Trieste, distant 4½ miles S.S.W. ¼ W. from the mole of St. Theresa, and forming between them the Bay of Muaja, the land bending E. Beyond Point Grossa another bay is formed, in which is the town of Capo d'Istria, seated upon a rock.

GROSS, CAPE, France, S. coast, N.E. of Point Garoupe, both having sunken rocks near them, and therefore to be passed at a proper distance. A course N. ½ W. from these capes will carry a vessel abreast of Point Negre, from whence to the mole of Antibes is distant about 1 mile.

GROSSO, CAPE, Greece, in the Gulf of Kalamata or Koron, on the N. side, and N.W. of Cape Matapan. It is steep-to, and has 150 fathoms near the land, 220 a mile out, and at 4 miles distance Captain Smyth could find no bottom at 300 fathoms.

GROSSO, CAPE, Sicily, E. coast, a bluff headland, with several rocks about it, situated a league from Scaletta, opposite to Cape del Armi, on the shore of Calabria. A line drawn from cape to cape, defines the S. boundary of the Faro or Cape of Messina. Just within Cape Grosso, on the declivity of Monte Scudero, is the old town of Ali, celebrated for its mineral waters; and 8 miles S.W. by S. from Cape Grosso is Point St. Alessio.

GROSSO LONGO, POINT, Sicily, E. coast, three-quarters of a mile to the S. of Cape Santa Croce, of which it is one termination. Cape Santa Croce is in lat. 37° 15′ N., long. 15° 14′ E.

GROSSO POINT, Ægadean islands, Mediterranean, the N. point of Levanso island.

GROTTA POINT, Sicily, 2 miles S. of Rosalini, on the S. coast of Sicily, to the W. of which is a cluster of flat rocks called the Porri, and to the S. of the point itself there is a reef. A little to the E. of this point is the shallow Bay of La Marza.

GROUND, GREAT MIDDLE, Cattegat, to the S. of the Little Middle Ground. S.E. of the Anholt light. The N. part is the parallel of 56° 38′ N., extending S. to the parallel of 56° 31′ N., being in breadth 5 miles. On a spot near the middle there are only 5½ fathoms, and around it 8, 9, 10, and 11 fathoms, and towards its N. edge 12, and at its S. end 14, 17, and 16 fathoms. The approach to this bank may be known, coming N. or S., as the soundings lessen considerably; and the lead brings up shells and small black pebbles; and on the E. side the soundings are very irregular. On the N.E. part of the middle ground, large black stones are found, and 13, 14, and 15 fathoms water. N.N.E. of the bank is the Rhode bank, with a bottom of coarse red sand, and from 15 to 18 fathoms; a little to the W. of the Rhode bank are 18, 19, 20, and 23 fathoms, fine sand and clay. Near the S. end of the Little Middle ground are 18 and 23 fathoms, mud; and a little to the N.W. of the Great Middle are 16 fathoms, fine sand.

GROUND, LITTLE MIDDLE, in the Cattegat, 12 miles to the S. of the Fladden bank, a dangerous and extensive shoal, running N. and S., and the N. end being in lat. of 57° N. From its shallowest part, which is 19 feet, the castle of Warberg bears E.N.E., 13 miles distant; Anholt lighthouse, nearly S.W. by W., distant about 16 miles; and Morup church, E. by S., distant about 14. It is from N. to S., about a quarter of a cable's length, and its breadth from E. to W. half a perch; there are 4 fathoms close to it, and this deepens immediately. It is formed of large stones in part, which point upwards in a conical form, the lead gliding down their sides into unequal ground. The sea, in storms, often breaks over it, and vessels are advised to avoid it by keeping the lead constantly going. The stony bank has from 4 to 7 fathoms over it, and vessels navigating near should be very cautious, and never venture into less water than 8 fathoms. On the N. of the Little Middle ground, approaching it, there is fine loose and hard sand, with coarse shingle. From the N.E. loose sand, chalky stones, and red sand, with shingles. From the E., grey sand, shingle, and shells, yellow sand, and chalky stones. From the S.E. coarse sand, shells and stones, stones and red sand. From the S. fine sand, sand with shells, coarse sand, coarse gravel with stones. From the W. coarse sand, shingle with large stones, brown sand with black spots, coarse sand with shingle. From the N.W. fine sand with stones, coarse sand, brown sand, fine sand with stones. There is another bank on the E. side of the Little Middle ground, of less extent, and between them a channel a mile broad, with 12 fathoms only, muddy ground, and is 7½ miles offshore. Further inward towards the Swedish shore there is a little bank, 2 miles long and 1 broad, over which there are from 16 to 18 fathoms, having a channel between that and the inner bank before mentioned, 1½ mile wide, with 24 and 26 fathoms in it, distant 5 miles from the shore, the water between being deep.

GROUNDS, THE SCOTCH, off Redding point, Plymouth; also the New grounds. For both, see *Plymouth.*

GROUVILLE BAY, Channel islands, on the E. side of the Island of Jersey, having good anchorage in 5 or 6 fathoms water, about a mile S.E. of Mount Orgueil castle.

GROWA, or GROOWAY TOWN, S. coast of Africa, N. of the Line, and near Cavally river, E. of Cape Palmas.

GROYNE, THE, Spain, N. coast, a name sometimes used to designate Coruña.

GRUI, POINT, Gulf of Venice, Dalmatian coast, the S.E. point of the Island of Maleda, from whence an E. ½ S. course leads to the port of Ragusa. This point is in lat. about 42° 40′ N., and 17° 44′ E.

GRUNDBOEN, an island in the Cattegat, off the coast of Sweden, within which and other islands the passage is sometimes made N.E. into the Cattegat, on leaving Malo sound with a S. wind.

GRUNE ROCK, Channel islands, one of the numerous sunken rocks in their vicinity, over which there is seldom less than 30 feet water. It lies 2½ miles S.S.W. ¼ W. from the River Puta, and a quarter of a mile W.N.W. ¾ W. from the N.W. point of Brecqhou island.

GRUTING VOE. See *Shetland Isles.*

GRYHAMN, Sweden, in Skelder bay, a town on the N. side of the bay, principally visited by coasting vessels.

GUACHIN, PUNTA DE, or GUARANA-CHE, South America, coast of Cumana, on the meridian of the Island of Cabagua, which lies N. of it. It is 6½ miles from Punta Gorda.

GUADALAVIAR RIVER, Spain, E. coast, the river on which Valencia is seated, 3 leagues from the mouth. Sailing-vessels cannot ascend this river, but anchor in the road instead, in what is called the Grao de Valencia.

GUADALETI, RIVER, Spain, falling into the sea at Port St. Mary's, in Cadiz harbour.

GUADAL-MACER, or the RIVER OF ALMANZORA, Spain, the entrance of which is defended by the castle of Montroy.

GUADAL-NARZA RIVER, near the Rio Verde, falling into the sea between Estapona and Marballa, on the E. coast of Spain.

GUADAL-NIDINA RIVER, Spain, falling into the Mediterranean at the port of Malaga.

GUADALOUPE, ISLAND OF, West Indies, one of the largest and most important of the French islands. It is very irregular in form, and may be considered as two islands, since it is divided by a branch of the sea, called Rivière Salée, or Salt River. This strait is from 15 to 50 fathoms wide, and is capable of being navigated in some parts by a vessel of 500 tons, but in others not of 50; its breadth is confined by mangrove trees, but the stream is clear and smooth as a mirror, and the length about 2 leagues. The N.E. island is named the Grande Terre, being about 9 leagues in length, and 3 leagues in breadth. This part of Guadaloupe is very deficient in fresh water, not having a single river, and but few springs. The W. division of the island, or Guadaloupe Proper, is 8 leagues long and 5 broad. A chain of mountains, extending N. and S., occupies the centre of this island; the mountains are of great height, and apparently of volcanic origin. Towards the S. part of this ridge is La Souffriere, which still emits a thin black smoke, mixed with frequent sparks of fire, often visible at night. This mountain was in a state of eruption in 1815; its summit forms a level plain, covered with calcined stones, the earth being perforated in various holes and openings, through which the smoke continually issues. It has other mountains and numerous streams. In one division of the island there are not less than 50 rivers. The W. part is divided by the mountains, and named Basse Terre or Cabes Terre. The pop. of Guadaloupe, with its dependent isles, such as the Saintes, Mariegalante, Deseada, &c., is calculated, all races, at 120,000. The exports are those of the West Indies in general. The climate is among the hottest of the islands, and it is not considered among the more healthy for Europeans. This island may be seen in clear weather at the distance of 30 leagues. The capital is St. Louis, or Point à Pitre, on the W. part of Grande Terre, at the S. entrance of the Rivière Salée, or Salt Channel, which separates it from Cabes Terre. The anchorage at Point à Pitre is sheltered, and vessels that remain at Guadaloupe winter in it for the hurricane season; it is necessary to have a pilot to take this anchorage. If a vessel is bound to it, care must be taken not to get to the W. of it, keeping to the S. and E. On the S.W. of Basse Terre stands the town of the same name, the most considerable in the island. The anchorage is a very unsheltered and inconvenient one, there being a constant swell. Its bottom at the edge is so steep, that at 2 cables' length from the shore there are from 80 to 100 fathoms water, nor is the ground good. These circumstances make it necessary to keep close to the shore, and let go the anchor in 20 and 30 fathoms, on clay, and hang to it without letting another go, to be ready to sail the moment winds from the S.E. quarter come on. From the anchorage at Basse Terre a ship may approach to the W. coast as near as the hill Gros Morne, which is the N.W. extremity of this part of the island. The principal commercial ports of the island are on the S. To make Point à Pitre, a ship should approach within two miles of the S. coast or Grande Terre, continuing that distance to the Bay of Fergeant, on which stands the town of St. Louis, where a pilot is obtained for Point à Pitre. There are two roadsteads on this coast, with towns in them, the first called St. Louis, and the second St. Anne. Between the last and Fort Louis there is a little town called Le Gosier, nearly N. from an island of the same name. To Fort Royal or Basseterre, a vessel should shape her course so as to approach Cabes Terre about Point St. Saveur. Then follow the road at the distance of a mile, and pass half a cable's length from Point de Vieux Fort, or Old Fort point, which is the most S. point of Petit Terre, and luff up at once when round it, to keep the same distance of half a cable's length from the coast, until opposite the town, where the anchor must be dropped. When sailing N. or S. to leeward of this island, a vessel should keep within 2 miles of the shore, to have the advantage of a light land breeze, which will be always sufficient to pass it before day. Getting further from the coast it is common to be becalmed for days together. If not able to keep within two miles, a vessel should stand off 8 or 10 leagues to avoid the calms.

GUADALQUIVER RIVER, Spain, W. coast. The course to this river from Cape St. Vincent is E.S.E. ½ E.; the distance, 40 leagues. It is 18 leagues S.E. ½ S. from the Guadiana to the Guadalquiver. On the N. side of the last river stands the castle of Jacinto, appearing among sand-hills, about three cables' length from the shore. From the point on which it stands there runs a dangerous rocky shelf for a mile W., in some parts drying, in others having 6 feet water. A mile and a half W. by S. from Jacinto tower is a rocky shoal of considerable extent, named Juan Pul, having 7 feet water over it, forming the narrowest part of the entrance of the bar of St. Lucar. W. ½ S. from Jacinto tower, 2 miles, is the shoal of Pabona, and in the same direction the Picacho. S.W. by W. nearly 4 miles, is Point Chipiona, the S. point of the Guadalquiver, low and flat, with a reef running out N.W. about a mile, called El Perro, or the Dog. W.N.W. of Chipiona about two miles is the Sabinal shoal. There are other dangers; all, except the Pabona, Juan Pul, and Cape shoals, are on the S. side of the river. To enter the Guadalquiver, bring the convent of St. Geronimo on with the four peaked hills of Gibalbin: these are remarkable, and not to be mistaken; on whichever peak the convent is brought, it will lead to the channel. Continue until the castle of Spirito Santo bears S.; then, if in 3¾ or 4½ fathoms, stand S.E., if in less water, E.S.E. a little E., so as to approach the shore between Jacinto tower and Point Malandar; which run close under, till S.S.W. from that point; then stand over towards the harbour of Bonanza, and anchor in 6 or 7 fathoms, sand and ooze, mooring according to the direction of the current, and taking care to avoid the small bank lying E.S.E. from Point Malandar.

GUADAMAR, Spain, at the mouth of the Rio Segerra, a small town, the principal export of which is salt.

GUADIANA RIVER, Spain, W. coast, bounded on the entrance on the W. side by Point St. Antonio, which is low and sandy, having a reef running out one mile; and on the E. side by Point Canelas, from which a sand runs out two miles; between these banks is a channel and a bar about a cable's length wide. On the W. side, a mile from Point Antonio, is the town of Villa Real, and 2 miles further, on the same side, Castro Marin, and between a branch of the river runs to the W. 2 miles up the river, on the E. side, is the Spanish town of Ayamonte. The river divides into two channels before reaching Ayamonte, leading to Higuereta and Tuta, which, when the tide is up, have water enough for coasting vessels. To sail over the bar of the Guadiana, the church of Cazerla should be brought on with a small hummock near the mountain of Montegordo, or the Blue Hill, and followed until two white mills on the E. side of the river come into one bearing, N. ¾ W.; then run up mid-channel until abreast of the Villa Real, and, if a large ship, anchor there; small vessels go higher up. At low water spring tides the bar has no more than 9 feet water. The tide rises 12 feet. It is high water, full and change, at 3ʰ.

GUADIERO POINT, Spain, S. coast, near a river of that name, between Cullera and Estapona.

GUAIGIMICO RIVER, West Indies, Island of Cuba, S. coast, W. of Trinidad or Guaraba river.

GUAJABA ISLAND, West Indies, Island of Cuba, N.W. of Savinal, and E. of Cayo Romano, 3 miles on the N.E. coast, a small island, recognized by four little mounts, which at the distance of 4 or 5 leagues appear like islets.

GUAJAIBON, PAN DE, or HILL OF GUAJAIBON, West Indies, Island of Cuba, S.E. of the banks of Isabella, and W. of Bahia Honda, in lat. 22° 49' N., long. 83° 23' 20" W.

GUALMASA, TOWER OF, S. coast of Spain, 4¾ miles from Estapona, whence a sandy point runs out with a rocky extremity nearly S.S.W.

GUAMA BANK, South America, N. coast, having 9 feet water over it, in lat. 8° 40′ N., long. 59° 57′ W.

GUAMA RIVER and KAYS, West Indies, Island of Cuba, S. coast, N. of the Isle of Pines.

GUANAHANI or CAT ISLAND, now ST. SALVADOR, one of the islands on the Great Bahama bank; the N.W. end 6½ leagues E. ½ S. from Powell's point in Eleuthera, and extending thence S.E. 14 leagues, with a breadth of from 3 to 13 miles.

GUANA ISLAND, West Indies, one of the Virgin islands.

GUANAJIVO POINT, West Indies, Island of Porto Rico, 5½ miles N. from Puerto Real de Cabo Roxo.

GUANA POINT, Tobago Island, West Indies, the N. point of Great Courland bay, 13 miles from the W. point of Man-of-War bay. It has a rock called the Beef Barrel close to its N. point, over which the sea breaks, more particularly at low water.

GUANA TOWN and RIVER, West Indies, Island of Cuba, due N. of Cape Corrientes, and N.E. from Cape Antonio.

GUANAQUILLA, PUNTA DE, West Indies, Island of Porto Rico, the N. point of a large bay named Del Boqueron, full of reefs and unfit for anchorage. A shoal lies 2 miles W. of this point.

GUANICA, PORT, West Indies, Island of Porto Rico, the best anchorage on the S. coast of the island. It is adapted for all classes of vessels, with a depth of from 6½ to 3 fathoms, on mud and gravel, the latter depth more than a mile from the entrance. The mouth of this harbour is in the middle of a large bay, formed by the point and cliff of Brea or Pitch point on the W. and that of Picua on the E. The last point has a reef running out from it, which must have a good berth. There are other dangers which are avoided with facility. The E. point is in lat. 17° 57′ 44″ N., long. 66° 53′ W.

GUANIGUANICO CHANNEL, West Indies, N.W. coast of Cuba, joining on the N.E. the Channel of Rosario; both are within the Colorados reefs, and run parallel with the mainland between the banks and kays. This channel commences on the W. with a bar of 3 and 4 fathoms water N.E. of Cape Antonio, and N. of the bank which extends N. of that cape.

GUANJA ISLAND, American coast; it was named Guanja by Columbus. See *Bonacca*.

GUANOS, PUNTA DE, West Indies, N. side of Cuba island, the W. entrance of Matanzas bay.

GUANTANAMO, Cuba. See *Cumberland Harbour*.

GUAPE BAY, West Indies, Island of Trinidad, S. of Brea point, in the Gulf of Paria.

GUARABA RIVER, West Indies, Island of Cuba, S. side, called also the River of Trinidad; to enter which, vessels steer along outside the bank, and run without fear to within musket-shot of the shore, which in this part is very clean. After running some distance, the bay of entrance will be seen formed by the Punta de los Ciriales to the S., and that of the River Canas to the N.

GUARACAYAL RIVER, South America, in Cumana, falling into the Gulf of Carriaco on the N. about the meridian of Point Galera in the Isle of Margarita.

GUARAPATURO POINT, South America, Columbia, W. of the Bay of Esmeralda, in Cumana, and E. of the Morro of Chacopala.

GUARDA, a town on the Spanish bank of the Minho, 2 leagues from its mouth, fortified, and having a pier haven for small vessels.

GUARDIA, CAPE, Greece, Island of Andros, and its most N. point, in lat. 38° 0′ N., long. 24° 47′ E.

GUARDIA, ISLANDS OF, Mediterranean, N. of Cape Tranquillo in the Island of Rhodes; they are two in number, mere rocky islets.

GUARDIANA ISLAND, Mediterranean, an islet off Cape Aji, in the Gulf of Argostoli, Island of Cefalonia. This island is low, and has upon it a convent of Greek priests, together with a lighthouse on the S.E. point, built of white stone, 100 feet high, the lantern 122 feet above the level of the sea, having a fixed light, in lat, 38° 8′ N., long. 20° 26′ 30″ E.

GUARDIAS VIEJAS, CASTLE OF, Spain, S. coast, standing where the Llanos, or plains of Almeria commence, a little inland from Point Moro, whence the coast runs E. 7½ miles to Point de Savinal.

GUARICO HARBOUR, West Indies, Island of Hayti, called also the City of the Cape, is no more than a bay formed at the W. and S. by the coast, and shut to the E. and N. by a part of the reefs that rise upon the White Bank, extending outward at this place more than a mile. Vessels bound to Cape Haytien should always make the Grange, because they can there shape a course free of danger to the cape. The land at the cape has nothing remarkable to distinguish it, unless when near enough to see the Hummock of Picolet, and the rock of that name to the N. of and very near the Hummock; the last cannot be seen at more than a league distance. Fort Picolet is built upon the point of that name, at the extremity of which the rock lies. The city of Haytien is under Mount Picolet. The harbour must not be entered without a pilot. Vessels enter the harbour about 12 o'clock when possible to do so, it being the best time. Point Picolet is passed on the starboard side at the distance of a short musket-shot. The town is 2 miles from Port Picolet. Lat., at the watering place, 19° 46′ 40″ N., long. 72° 10′ 42″ W., according to Mr. Dunsterville, Master, R.N.

GUARICO POINT, West Indies, called also Cape François, lat. 19° 40′ 30″ N., long. 69° 55′ W.

GUARICO, PUNTA DE, West Indies, Island of Cuba, N.E. side, and near the E. end, and N.W. from Cape Maize 43 miles.

GUATARO POINT and RIVER, West Indies, E. coast of the Isle of Trinidad. The river is sometimes called the Ortoire; its bar is too shallow even for the passage of canoes.

GUATEMALA, America, one of the states of Mexico, bounded on the N. by Chiapa, W. by Oaxaca, E. by Honduras and Nicaragua, and S. by the Pacific ocean. Its produce reaches the Gulf of Mexico by means of one or two river outlets into deep bays of the sea.

GUAVAS or GWAVAS LAKE, Cornwall S. coast, Mount's bay, on the W. side, where vessels ride safely with W. and S. winds avoiding the sunken rocks called the Low Lea and Carn Base, the former bearing E.S.E. and the last E. ¼ N. from St. Paul's church. These rocks lie nearly N. and S. from each other, distant a quarter of a mile. On the Low Lea there are 5 feet water only, and

6 feet on Carn Base, with 5 fathoms between. Low Lea rock is one-third of a mile from Penlea point. The mark for anchoring in Guavas lake is St. Clement's island, just shut in with Penlea point, in 6 fathoms water. Large vessels commonly anchor without the Low Lea and Carn Base rocks, between St. Michael's mount and St. Clement's island, in 12 or 13 fathoms.

GUAYAGUAGARE BAY, West Indies, Isle of Trinidad, on the S. side, between the rocky point of Blanquiares on the E. and Grand Calle point W. There is anchorage at the entrance of this bay in 4 and 5 fathoms, sandy ground.

GUAYAMA BAY, or YAMMA BAY, West Indies, Island of Porto Rico, S.W. of Cape de Mala Pasqua, the S.E. point of that island.

GUAYRA, South America, the principal port of Caracas, on the coast of Venezuela, although as a port by no means entitled to such a preference, being unsheltered from N. to E. and W. winds, and the continued E. breeze keeping the sea for ever in agitation. There is a considerable resort hither, and the town contains a pop. of 6,000. The road is in lat. 10° 37′ N., long. 66° 56′ W., after the Spanish surveys.

GUDEN RIVER, Denmark, the chief river of Jutland, rising near the centre of the peninsula, and after a course of 40 miles falling into the Gulf of Randers. Its navigation has been artificially improved.

GUENERON ISLET, France, W. coast, between the Bay of Dinan and Point la Chevre S. of Point Toulinguet, by Camaret road.

GUERNSEY ISLAND, English channel, one of the most considerable of the islands denominated the Channel islands, situated to the N. of the Island of Jersey, St. Pierre, the principal port of the island, standing in lat. 49° 27′ 12″ N., long. 2° 32′ W. This island is about 36 miles in circumference, lying off the N. coast of France, and has a pop. of 28,097. St. Pierre, or St. Peter's Port, lies on the E. side of the island. It has a commodious pier, and a castle, called Castle Cornet, stands on an islet W. of the town, which is upon uneven ground, the environs sprinkled with gardens and shrubberies, which from the sea have a very pleasant appearance. The harbour is below the town, and the sea at high tide comes up to the houses, which are built of a blue stone peculiar to the island. The entrance of the port is formed by two stone piers 35 feet high, the opening 100 feet wide at the top, and 68 feet at the surface of the water. It is well protected by Castle Cornet, which is also strengthened with batteries. Fort George is a second fortress, intended for the defence of the town, and is seated upon a commanding eminence on the S. a short distance off. The roadsteads are also upon the E. side of the island. They are approached through the channels called the Great and Little Russel. The S. coast of this island presents towards the sea an inaccessible rocky cliff, rising almost perpendicularly from the water to the height of 270 feet. The channel of approach, called the Little Russel, leading to and from Guernsey roads, runs between that island and the islands of Herm and Jethou, with the chain of rocks extending to Anfroque on the E. side, and the rocks called the Brayes at the N.E. point of Guernsey, with others which continue to line the E. coast of the island on the E. side. These rocks have many of them vegetation upon their surfaces, and between the larger there are small rocks innumerable, yet their limits are nearly parallel, so that there is a passage formed between them. This channel is about 4 miles long by 1 broad. The course from La Hogue to the Little Russel is more W. than that to the Great Russel, and is more difficult of access to strangers. In proceeding from the Race of Alderney, the Anfroques are left to the E., with care not to approach them nearer than 1¼ mile, in order to avoid foul ground running off to the N.E. from the Petite Anfroque. A sunk rock, called Les Boufresses, lies about three-quarters of a mile N.W. from the Grande Anfroque. Near the Russel rock there is a black buoy, with the word Guernsey upon it, in white letters on red. It is close to the S. head of this rock, its sunken head being distant from it about 18 feet. The danger extends in a N.E. direction 120 fathoms, therefore the buoy should not be approached nearer than 150 in that direction, but in all others may be approached within 60 fathoms. This is the most dangerous rock in the Little Russel channel, and lies N.N.W. from the Roustel. There are 8 and 9 fathoms at low water spring tides in the fair channel of the Little Russel. The first of the flood tide runs towards Herm island, and the beginning of the ebb sets to the N.N.W. The W. passage of the Little Russel lies between the Roustel and La Plat, or the Flat rock. On the last there is a beacon standing on the S.E. point of the rock. This beacon must not be approached by large vessels nearer than 2 cables' length, to avoid a dangerous sunk head, lying in an E.S.E. direction from it. In case of shipwreck, the beacon on the Flat rock has steps fixed to it in such a manner as to preserve the life of whoever may reach it. The Great Russel passage is the most convenient channel for large ships generally, and more especially for those from the eastward, it being above two miles broad, and of great depth. In sailing through this passage, the island of Serk may be approached without danger within a quarter of a mile. There are no rocks but what are seen until the W. point of Brecqhou island is neared, to which the Givaude rocks are almost close, being less than a quarter of a mile from the island. This passage is bounded on the N.W. by a considerable number of rocks, and on the S.E. by rocks extending from the N. point of Serk, the Isle of Brecqhou, and others. When the Race of Alderney is passed, steering S.W. by W. to the S. of the Schôle bank, a vessel must then steer W. by S. leaving the Isle of Herm to starboard. If the Deroute channel is taken, a vessel must steer S.W. through the Race, leave Serk on the starboard, keeping it at a league distance, to avoid the Blanchard, a sunken rock that bears S.E. by E. from the old mill on Serk island, at the distance of 2 miles. Herm, Jethou, and the adjacent islets divide the Great from the Little Russel channels, and shelter the anchorage in Guernsey roads from N.E. to S.E. The first danger in entering the Great Russel from the N. is the Bonne Grune, a mile E. by S. from the Grande Anfroque. This may be avoided by passing 1¼ mile to the E. of the Grande Anfroque, or by keeping the W. point of Brecqhou to the westward of S.W. by S., or the W. extreme of Little Serk, open to the E. of the E. end of Brecqhou isle, bearing S.S.W. ¼ W., until Fort le Marchant in Guernsey comes open to the N. of the Grande Anfroque; then steer towards the Noire Pute, which lies about a mile to the E. of Herm; keep it on the starboard bow, and pass it at not less than a quarter of a mile distance; then steer S.W. by W. ¼ W., until St. Martin's point comes open to the southward of the Goubinierre, a large rock lying half a mile S. from Jethou. Proceed with this mark on until Vale castle in Guernsey bears N.W. by N., or touches the S.W. end of Herm; thence steer W.S.W. for 2¼ miles, which will lead to an offing from the S. end of the bank. The rocks on the S.E. side of Herm will be avoided by keeping the Grande Anfroque N.E. ¼ E. its own apparent breadth open to the E. of the Selle Rocque on the S.E. side of Herm; then keep this mark on until St. Martin's Church-steeple is in one with the Martello tower in Fermain bay, bearing N.W. ¾ W., clear of all Herm rocks, the southernmost of which is called Les Tètes d'Aval; all then is right to Guernsey. The stream of the tide in Guernsey roads sets through both ways in a direction nearly straight, the flood running N.E. ¼ N., and the ebb S.W. ¼ S. The flood between Guernsey and the rocks projecting from the S.W. end of Jethou assume two different courses, one running through the Great Road directly, and so into the Little Russel, the other branching off from the first at nearly right angles, the first running to the E. towards Jethou, and through the above rocks into the Great Russel. The ebb runs directly the reverse, the stream from the Great Russel uniting with that in the road, after passing the rocks at Jethou, and both setting to the S.W. There are properly two roads belonging to Guernsey. The Great road, comprehending the space from St. Martin's point to within a mile of Brehou, the anchorage being abreast of Castle Cornet, or directly off the port, by bringing the church open to the N. of the castle, or to the S. of the lighthouse on the S. pier head, bearing W. ¼ N. in from 9 to 14 or 16 fathoms water, good ground. To the S. of the road there lies a shoal called the Great Bank, extending for nearly two-thirds of a mile E. of Castle Cornet to nearly abreast of St. Martin's point, being half a mile broad. Near the middle is a narrow ridge of half a mile in length N.E. and S.W.; on this the least water, after the survey of Captain White, is 3 fathoms. A fixed light is erected upon the round-house of the S. pier of St. Peter's port, elevated 40 feet above the sea. It may be seen from the N. through the Little Russel channel, from the E. through the Great Russel, and from the S. when round St. Martin's point. The round-house serves for a day mark. The Little road is off the opening between Castle Cornet and the rocks called La Blanche and Sardrette, on which beacons are fixed. To moor, drop one anchor near the rocks off Castle point, and the other near the Blanche rock. In coming to anchor in this road, veer away cable briskly, otherwise the anchor will be apt to come home; the ground is good, but the anchors are often choked with long weeds. Vessels drawing more than 14 feet cannot well lie in the Inner road. There are convenient warping buoys, and pilots always in attendance. To go within the piers, a pilot is advisable. In the months of the piers, great spring-tides rise 29 or 30 feet, neap-tides about 18 feet.

GUETARIA ISLAND, N. coast of Spain, 7½ miles from St. Sebastian's lighthouse, W.N.W. This island is high, and is joined to the mainland by a pier, which forms a harbour for small craft only, but vessels may anchor on the E. side of the pier in from 7 to 10 fathoms, sandy ground, sheltered from the S.W. to N.N.W. winds, but open

to all other points. The island has a chapel on the summit.

GUGH, or HUGH ISLAND. See *Scilly Islands*.

GUIA, CAPE SAN JUAN DE, South America, W. of Cape Aguja. Lat. 11° 22′ N., long. 74° 3′ W.

GUIA, Nᴬ SENᴬ DA, Portugal, a lighthouse so called near the Bay of Cascaes, and the entrance to the Tagus, in lat. 38° 41′ N., long. 9° 27′ 12″ W. It bears a fixed light on a stone pyramid, 96 feet high, and is visible at 10 miles distance.

GUINEA, Africa, a name given to the coast of that continent lying upon the Atlantic, divided into Upper and Lower Guinea, the first comprehending the coast for a vast extent, from Sierra Leone to Majumbo, and even Anziko, and the second including Congo, and the country S. This name is deservedly become obsolete, as being out of all reason general and indefinite.

GULDBORG SOUND, Denmark, separating Falster island from the E. end of Laaland.

GULF STREAM, America, Gulf of Mexico, the most remarkable of the currents of the ocean, apparently originating in a current which, setting into the Caribbean sea from the E., passes up into the Gulf of Mexico, and sweeps W. around that gulf, mainly attaining an outlet through the Gulf of Florida to the N. It widens its stream over the surface of ocean and divides N. and S. at the S. of Newfoundland, becoming at last blended with the ocean of which it forms so small a comparative surface, but a surface so well distinguished both by its motion and difference of temperature. The main outlet of this current is but about 36 miles wide between Cape Florida and the Great Bahama bank, and 700 miles more N. its breadth is only about doubled, so that it is a mere thread amid the mass of water amidst which it moves to the N. It might be expected that such a body when immediately clear of the Florida and Bahama shores or banks would at once spread over and be lost in the great ocean surface, and such would undoubtedly be the case were the level of the current materially different from that of the ocean. About lat. 40°, and between long. 60° and 70°, the Gulf Stream appears to divide, one portion running N.W. and N. of the Azores, the other setting S.E., continually expanding until its limits can no longer be traced. The difference of temperature, however, lightening the water which composes the stream, has no doubt some tendency to keep its particles together, by carrying it along the surface of a colder denser fluid beneath. During the greatest degree of summer heat, in the latter part of August or beginning of September, this stream runs the strongest and is at the highest temperature. In the Gulf of Mexico at the mouth of the Mississippi the water has been known at the end of the month of August to reach 90° of Fahr., and at midnight to stand at 86°. In the Strait of Florida the Gulf Stream will be 86° in lat. 24° 11′, and further N., between Capes Canaveral and Hatteras, from 83° at midnight to 84° at noon in lat. 33° 44′. Between Cape Hatteras and the Azores in 39° 37′ it will have fallen to 77° and 76°, and at 41°, or about the Azores, to 71° and 70°. Now the temperature of the Atlantic in about lat. 43° 58′ is 60½° of Fahrenheit in June month. The greatest velocity of this current or stream is between the Bemini isles and Cape Florida, where it has been found to run above 100 miles in 24 hours, in the month of

August, while at some distance westward it has not exceeded 70. In October the rapidity of the stream lessens, fluctuating according to the seasons (the temperature ?). The strength of the current on its W. and N. limits is much greater than on the E. and S., where it has a tendency to spread over the ocean in eddies, and is consequently weaker. Near the Cuba side of the Strait of Florida the stream is weakest and sets to the E. Along the Florida reefs and kays there is a counter current setting to the S.W. and W.; the winds have some effect upon the surface of this current but do not affect its volume or change the motion of the mass. N. winds between Cuba and Florida press it S. to the shore of the former island, S. winds react to the N. in a similar manner. When turned N. easterly winds press the surface toward the Florida side, and W. perhaps nearer the Bahamas, S. winds give it a tendency to spread out. When a gale opposes the course of the stream it overflows all the openings, creeks, reefs, and low coasts among the Martyr isles, and even vessels have been carried over the low kays and left high and dry on the shore. The waters have sometimes risen 30 feet, and run with violence 7 miles an hour against the tempest. At such times the Strait of Florida becomes the scene of strife and combustion fearful to witness. That there is some connection between this stream, its course and temperature, and the phenomena of the atmosphere, in all probability the electrical influence being increased or retarded by its action, is evident from the fact that the winds which prevail from September to March materially affect it, and that hurricanes occur at a certain crisis, it is said dependent upon the moon and her times of full or change, connected with boisterous northerly or S. winds when prevalent at such times. The course of the Gulf Stream and of several hurricanes has been shown by Mr. Redfield and Col. Reid to be remarkable in coincidence, moving in true circles or large curves as all bodies in motion have a tendency to do. It is not improbable that the electrical agency is covertly at work even in the propulsion of the Gulf Stream and in its singular cohesion for so long a distance amidst an analogous fluid without spreading out superficially. No small difference of level between the waters in the Gulf of Mexico and those in the ocean, could satisfactorily account for this remarkable quality of the Gulf Stream. From lat. 26° to 28° the stream generally sets rather N.-easterly. In the parallels, or rather between the parallels of 28° and 35° it has a N., N.N.E., and N.E. direction. It is bounded on the W. by the banks extending from the shores of America, nearly parallel to which it runs. The nearest edge of the stream to the coasts of Georgia and Carolina is about 40 miles, but it approaches the Florida coast much nearer. The average breadth of the stream is about 60 miles, less near Cape Canaveral and more near Cape Hatteras. The average rate of the stream may be about 3 miles an hour. A fleet of merchant-ships has been carried by the current through the whole strait without wind by the stream alone, after passing the Double-headed Shot kays at the rate of 2½ knots an hour, not a breath of wind stirring. The stream after passing the coast near Cape Hatteras ranges up to the parallels of 38° and 39° at the rate of 2½ knots, when the George and Nantucket bank seem to front and throw it off to the E.N.E. and E. by N. northerly. Passing Cape Hatteras the W. edge

of the stream ranges from N.N.E. to N.E., and from thence bends E. over the edge of George's bank, about the parallel 40° N. Its S. border, in a N.E. and E.N.E. direction, crosses the parallel of 35° N. in long. about 73° W. or 120 miles E. by S. of Cape Hatteras. In winter the temperature of this extraordinary stream is naturally much reduced. In February and April, in 39½° N., the temperature of 62° has been found. In 37° N. 66° has been met with, and 70° half a degree more to the S. In May near the centre from 60° to 65°. In June and July from 66° to 72°. In August from 77° to 80° in lat. 37° 20', long. 70°. In September near the N. edge 71° to 74°, and near the S. border 83°. In October the temperatures begin to fall, but no temperature can be given in accordance with a particular time. There is not only a variance in this respect, but also in different parts of the stream, so that colder water has been conjectured to mingle with the main stream in occasional veins from the W. Cold currents are commonly found on the soundings between the stream and the coast setting in a S. by W. direction, particularly in the winter months. Between the meridians of 70° and 65° the N. edge of the stream appears bounded by George's bank, after passing which it pursues a more N. course, or nearly N.E. by E. to the meridian of 65°, which in summer it crosses above the lat. of 41°, but in winter more to the S. The S. edge of the stream is then found about lat. 35°. In the spring, the temperature towards the N. is affected by the presence of ice. Thus, in the month of March, the temperature has been affected so as to make a difference of 6° in no great distance of latitude. It has been imagined that the difference between the temperature of the air and water in this extraordinary current has been the cause of storms. In a storm experienced in lat. 38° N., the air was at a temperature of 48°, the water of 74°. It is not impossible that a storm might be produced from so considerable a disturbance of the common atmospherical equilibrium, and the effort of nature to restore a balance nearer the usual state of things, just as clouds overcharged with electrical heat produce thunder-storms, to bring about the customary state of atmospherical freshness and purity. The guide to trace the stream readily is a good thermometer, and a comparison of its results with those obtained in the ocean. This requires the closest observation on the N.E. part, where the limit becomes less defined, and it insensibly blends with the great mass of ocean waters. The appearance of the water is an indication of its presence, as well as the temperature. It has a clear blue surface where it flows undisturbed in its usual course in the lower latitudes. Without the line formed by the ripple on its edge, the water appears like blue water in a state of ebullition; in some places it foams even in dead calms, and in situations that are unfathomable. The ripplings in fine weather are very perceptible on the outer edge, and the appearance of sea-weed is an indication of the edge of the stream, more commonly larger in quantity without the edge than within it. It is not true, as some have said, that its water does not sparkle at night like other sea water. In the stream the water is always much warmer than the air; out of the stream the reverse is the case, and the more as soundings are gained, and the shore approached. It is on all accounts most important for seamen to know how to take an advantage of the Gulf stream, and to avoid its disadvantages, and therefore the study of its peculiarities and a careful examination of the counter currents setting along its edges, are essential to good navigation as well as to the security of the navigator. The strength of the current, too, should be ascertained as often as possible, as it continually varies. Off the island of Cuba, in the meridian of the Havana, stripes of the main current are found setting E.S.E. and S.E. from the Tortuga soundings; and these again must not be confounded with the eddy current near the Colorados. Hence the necessity of care and watchfulness, the use of keeping in mid-channel, and the indispensable obligation of always knowing a vessel's position, sighting objects certainly known on every possible occasion.

GULLAND ROCK, Cornwall, N.W. coast, between Trevose head, W.S.W., and Padstow harbour, 2 miles distant from the harbour.

GULL ISLAND, America, coast of Newfoundland, situated three-fourths of a mile N.N.E. from the extremity of Cape Bonavista, in lat. 48° 42′ 20″ N., long. 53° 8′ W. It is small, and of moderate height, like a round hat, visible 4 or 5 leagues off in clear weather. N.N.E. ½ E. distant 3½ miles from Gull island, is the Old Harry rock, having over it but 13 feet water, and a bank, or reef, extending from it to the N.E. 3 miles.

GULL ROCK, Cornwall, S. coast, N.N.E. nearly from the Deadman, or Dodman, towards Fowey harbour, half a league from the land. Small vessels accustomed to the coast pass between this rock and the land.

GULL ROCK and MARIA'S LEDGE, America, Newfoundland, W.N.W. from Bauld cape, in the Island of Quirpon, distant 2½ miles N.N.E. from Cape Raven, always above water. Maria's ledge lies S.W. from the Gull rock, distant 2 miles, and N. by E. 1½ mile from Cape Raven.

GULL SAND, coast of Kent, a narrow ridge, about 1 mile long, lying N.W. by N. and S.E. by S. On the middle is a shoal part of 17 feet, where a white buoy is placed; the marks for this buoy are Jacob's Ladder at Ramsgate, midway between the two mills, W. of the west pier, bearing N.W. by W. ¾ W.; the flagstaff of Dover castle on with that of Deal castle; and the North Foreland lighthouse N. by W. ½ W. 4 miles. Ships cross over the E. end of this shoal in 4¾ fathoms, about half a mile to the E. of the buoy. Nearly parallel to the Gull is a narrow shoal, called the North bar, distant about three-quarters of a mile, the least water on which is 3¾ fathoms; nearly a mile to the southward of that is the Middle bar, with 4½ fathoms over it; and between these is a shoal called the Boot, with 5 fathoms over that. These lie in the fairway of the Gull stream, but the leading mark kept on will carry to the E. of them all.

GULL STREAM, Kentish coast, a channel in the Downs, formed by the S. end of the Brake and the Bunt head, in which there are 8 or 9 fathoms water. The nearer a vessel is to the Bunt head the deeper the water. The old marks to go through it were to bring a little house upon the N. end of the cliff to the S. of Old Stairs bay, and when Ash church is on with St. Clement's church at Sandwich, a ship will be to the S. end of the Brake. A vessel must steer through from N.E. ½ N. to N.E. by N. as the wind may permit. On its W. side a white buoy leads into the Downs in 4½ fathoms at low water neap tides. St. Lawrence church, about the middle of the houses on the N. cliff of Ramsgate, called Albion place, bore N.W. ¾ W.; the North Foreland N. by W.; the South Foreland upper light on with the N.W. cliff of Old Stairs bay; and the Pitch of the South Foreland about S.W. There is now a light-vessel moored in this stream, in 8 fathoms, off that part of the Goodwin called the Trinity Swash. In this vessel two fixed lights are placed horizontally, on separate masts, and elevated 14 feet above the water, are shown every night, from sunset to sunrise, for the purpose of guiding vessels in the night through the Gull stream to and from the anchorage in the Downs. "Gull" is painted on her sides, and a gong is sounded in foggy weather. The vessel in which these lights are exhibited is moored in 11 fathoms water, with the following marks and bearings:—St. Lawrence's mill, on the left part of Albion buildings, Ramsgate, and also on the new warehouses to the westward of the Ramsgate Trust Committee Room, N. by W. ¼ W.; the South Foreland high light a boat's length open to the westward of Old Parker's Cap, bearing S.W. by W., distant 9 miles; North Foreland light N. ¾ E., distant 6¼ miles; North Sand head (or Goodwin) light E. by N. ¼ N., distant 4½ miles; South Brake buoy, W.S.W.; Fork buoy, S.W. ¼ S.; Gull buoy, N.N.E. ¾ E.; North Brake buoy, N. ½ E., and a little to the left of the North Foreland light. The nearest distance from the light-vessel to the sand (in 2 fathoms) is from a quarter to one-third of a mile.

GULLSTONE, England, coast of Northumberland, a small rock, showing at half tide, bold and with deep water around it. Between this rock and the Staples is the broadest channel to Holy island.

GUN-BOAT SHOAL, America, United States, 4 miles from Portsmouth lighthouse, bearing S. 4° W., with not less than 3½ fathoms over it, and that only upon its shoalest part, which is very small. It runs E.N.E. and W.S.W. about 2 cables' length, and bears from Whale's Back light S. by W. ½ W. from Odiorne's point S. ½ E.

GUNFLEET SAND, England, mouth of the Thames, 3½ miles offshore, an extensive sand running from the Spitway in a E. ½ N. direction for 12 miles, with a breadth of 1 and 1½; a black beacon buoy is on its E. edge. It is one of the sands which bound the passage from the Nore and through the Swin and King's channel on the N. side. The N. edge of this sand forms the S. boundary of the Wallet, and its S. edge is the N. limit of the E. Swin, or King's channel.

GUN KAY, West Indies, Island of Jamaica, in the Port Royal eastern channel.

GUNNEL BAY, Cornwall, N.W. coast, near Pentire point, N. of the Gull on the same coast.

GUNNOT ROCK, Scotland, Frith of Forth, W. ½ N., 1½ mile from Inch Keith lighthouse, formed of two rocks that join each other, having over them 9 and 10 feet water; they have a white buoy at each end.

GUN REEF, coast of Northumberland, two patches of rock running out from the S.W. point of the Staples, and curving towards the N. end of the Brownsman. The seaweed along this reef may be seen at low ebbs, and two heads of rock at about two hours ebb. The mark for the W. part of the Gun reef is the plantation on Haffer Law hill open with the N.W. end of Sunderland sea-houses.

GUNVER HEAD, Cornwall, N.W. coast, opposite a little N.E. to Trevose head.

GUNWALLO COVE, Mount's bay, Cornwall,

to the N. of which there is anchorage in 7 fathoms, about a mile to the S. of the Loo bar, and N. of that the ground is good for anchoring in E. winds only.

GUPHONISI ISLANDS,· Greek Archipelago, three of the smallest of a cluster lying to the S. of Naxos island.

GURNARD'S HEAD, Cornwall, N.W. of Cape Cornwall 7 miles.

GURNET POINT, England, Isle of Wight, the W. point of Gurnet bay on the N. coast, which should have a good berth on account of the Gurnet rock, or ledge. There is a white buoy on the Gurnet in 3¼ fathoms, a mile and a quarter W. ¼ S. from Egypt point.

GUSSARD SHOAL, coast of Durham, about the size of a ship, ½ mile S. ¼ E. from the Goldstone, having 2 fathoms over it at low water. The marks are, the E. side of the Megstone, touching the W. side of Farn island, and the beacons of Old Law in a line with the S. part of Wingate Gap.

GUSTAVIA, TOWN AND PORT OF, West Indies, the capital of the Swedish island of St. Bartholomew. The anchorage before the harbour is extensive, and vessels may lie off and on the harbour's mouth that are bound thither, on obtaining a permit from the governor.

GUUL RIVER, Norway, near Drontheim, having its outlet in the long Drontheim fiord.

GUYAMA BAY, and RIVER, South America, called also WAYEENA, S. of the mouth of the River Orinoco and Sabinetta cape. The bay forms a deep bight; the appearance of the coast is remarkable, and it is a useful place to make to the S.S.E. for vessels bound to the Orinoco. Three singular hills, or hummocks, will be observed standing inland in a S.W. direction. Thence sail may be made along shore, keeping in 3, 4, and 5 fathoms water.

GUYANA, or GUIANA, South America, a country extending N.W. along the coast from the entrance of the Maranon or Amazon river, to the entrance of the Orinoco. This large tract of country is divided among several nations, and is dependent upon the different Governments of France, England, Portugal, Holland, and Colombia. The S. part of this territory belongs to the Portuguese, and is separated from the French portion of the territory, N., by the River Oyapok, terminating N. by Cape Orange and the sea. Along the coast numerous rivers find an outlet, the principal of which are besides the Maranon, the Anaurapara, Campanatubo, Amari or Arouari, Aricari, Comawini, Cassipour, Wassa, and Oyapok, running mostly E. or N.E. French Guyana extends along the coast from Oyapok bay to the River Maroni; the coast facing the N.E. is watered by several rivers, the principal of which are the Ouya or Oyak, the River of Cayenne, and the Aprouak or Approuague of the French. The S. port is in lat. 4° 20' N., long. 51° 22' W. The coast of Dutch Guiana has a N. aspect, and extends from the River Maroni to the River Corentyn, between which are the mouths of the Copenam and Surinam rivers, with their tributaries. British Guyana, the next N., extends along the coast from the Corentyn river to the limits of Colombian Guyana, which do not appear to be completely defined, but there are English settlements on the Poumaron, which would lead to the belief that Cape Nassau on that river is the proper boundary, in lat. 7° 40' N., long. 58° 50' W., though Cape Barima or Moco-

moco point, at the mouth of the Orinoco is asserted to be the termination of British Guyana, which would extend the limit to 8° 45' N. lat., and 60° 22' W. long. The extent of coast is about 200 miles, and includes in the more confined view of the boundaries the rivers partly or wholly of Corentyn, Berbice, Demerara, Essequibo, and Poumarouni. Colombian Guyana is of large extent, and the Orinoco river forms the most remarkable feature on that coast, the ship channel of which between Cape Sabinetta and Barima on the S., and Congrejos or Crab island on the N., has an entrance of 12 miles wide. The Delta of the Orinoco commences 100 miles inland, and there are not less than 50 outlets or channels communicating with the main branch of the river, of which seven are navigable for large vessels. Colombian Guyana extends N. to this river, the limit of the province of Cumana on the S. Along the sea-coast generally, the soil of Guyana is alluvial for 30 miles inland, terminating in a range of sand-hills, to which the sea once reached; much decayed vegetable matter is found at the mouths of the rivers. Hence the coasts vary continually, and sand or mud banks and shoals constantly alter their shapes and dimensions. The appearance of the shores is low, covered with bushes in many places and the approach is naturally shallow from the land deposits brought down by floods. Between high and low water mark, the shore is frequently covered with mangroves, and 500 yards within these commence low, swampy savannahs; the water on the whole coast is bad and brackish. It is always advisable to run down the shore from windward to leeward, as the various points of the coast have not been accurately settled as to position. All the rivers, it must be recollected, have bars and shallow water, and a river must never be entered without a pilot. The prevalent winds on this coast are from E.N.E. to N.E., or S.S.E. and S.E., and the current which always runs W.N.W., makes the lesser lat. to windward; hence it is easy to increase the N. lat. and difficult to decrease it; the current here alluded to, is the general tendency of the sea. The flux of the rivers is only found to prevail near the land, or for about 12 miles off the shore, to about 9 fathoms water, which may be considered the outer limit. Beyond and to the E. of that only, the general current prevails, and within that and the land no other current than that produced by the tides is felt. Here the flood acts towards the coast, and the ebb away from it; the tides flow at full and change at Cape North at 7ʰ; about the coast of Mayez at 6ʰ; at Cayenne at 5ʰ; off Surinam at 6ʰ; at the mouth of the Essequibo at 6¼ʰ; at the great mouth of the Orinoco at 6ʰ; and in the Gulf of Paria at 4¼ʰ. The most dangerous part of the coast is in the vicinity of Cayenne, and there only are those dangers found that require some care to prevent the loss of a vessel. These dangers consist of several rocky islets, such as the Constables, the Remue islands, the Devil's or Health islands, and others. No attempt should be made to pass between the Constables, except with a fair wind. Anchoring at the distance of 3 leagues, if needful; giving them a wide berth and going outside them, will obviate danger. Vessels from the Caribbean islands keep their port-tacks on board, until they get a convenient lat. to make the land, which should be always to the S. of the place to which they are bound. Seamen from Cayenne to the Orinoco are frequently unable to ascertain their exact position, and without observations for lat. are likely to run into serious mistakes. In consequence, it has been recommended to make the land about Mount Mayez; now, however, the newly established lights in British Guyana, will render more easy a knowledge of that part of the coast, while near Demerara, the trees burned and cleared away, and the houses seen through the openings, aid the seaman in ascertaining his correct position. The general shoalness of the water near the coast, and the lead continually going, will pretty well indicate the vicinity of a shore in many places difficult to recognize accurately. The British lights along this coast are, a fixed light, 100 feet above the water, at the E. entrance of the Demerara river, in lat. 6° 49' 20" N., long. 58° 11' 30" W. Secondly, a floating light painted red with two masts and a broad pendant by day, lying in 4 fathoms off Demerara river, N.N.E. ¾ E., about 12 miles from the entrance of the river, in lat. 6° 59' 30" N., long. 58° 5' W., a fixed light. Thirdly, a floating light, with a ball on the foremast, roofed over, with a jigger mast aft, off the River Berbice, in 2½ fathoms water, N. by E. ¾ E., about 12 miles from the entrance of the river, in lat. 6° 26' N., long. 57° 26' W.

GUYSBORO HARBOUR, or MILFORD HAVEN, America, Cape Breton island, is situated on the S. side of Chedabuctou bay, at the head, and is impeded by a bar, over which nothing larger can pass than a sloop-of-war. Within the bar vessels lie perfectly secure, though the tide sets in and out with great rapidity. The town is a place of little trade, but is protected by a battery. It is high water here at full and change at 8ʰ, rising 7 and 8 feet.

HAAFT, a term for a channel or sound on the German coast of the Baltic.

HABSAL, or HAPSAL, Gulf of Finland, a port only calculated for small vessels. Its entrances are through the Worms road, and to the S. of Worms island, or between Worms and Nickoe. There is also a passage to it from the Baltic, through the Sele sound, also from the Gulf of Livonia, by the Moon island channels.

HADDERSLABER, Denmark, in Sleswick, a town with a harbour for small vessels on an inlet in the Little Belt.

HADDOCK BANK, coast of Norfolk, 5 leagues to the N. of the N. part of the Ower, and 12 leagues N.E. from Cromer lighthouse at its S. end. From the S. end it extends 6 miles N.N.W. ¼ W., being nearly 2 miles wide, and having on its middle 5 fathoms, increasing each way to 6 or 7. Near it, on either side, are from 10 to 14 fathoms, and between it and the Ower, from 15 to 20. About 14 miles N. ¼ W. from this bank lies the N.N.E. hole, where the ground suddenly sinks to the depth of 40 fathoms, while it has around it only from 16 to 18.

HADID POINT and REEF, Africa, W. coast, N.E. of Mogador, distant 12 miles; a sandy spit that projects a mile beyond the general trend of the coast, and terminates in a reef of half a mile in length. Lat. 31° 44' N., long. 9° 38' W.

HAFFA EL BEIDA, Africa, N. coast, S.W. ½ W. 9 miles from Arzilla, hills rising between to the height of 700 feet, for a part of the distance. Haffa el Beida rises 300 feet above the sea, and in all directions puts on the form of a wedge, which serves to identify that part of the coast. The

IV.

strata here appears to lie at an angle of 70° with the horizon.

HAGUE, THE, or GRAVENHAGE, Holland, the capital of the province of Lower Holland, and it may be styled of Holland itself, one of the handsomest cities in Europe. It has a paved way across the sand-hills, with trees on each side, leading to Schevelling, on the sea shore, about 2 miles distant. Pop. about 80,000. Lat. 52° 5' N., long. 4° 18' E.

HA-HA BAY, America, Labrador, to the W. of Eagle harbour, at the end of Long island; it has several small islands at its opening, which form separate entrances. The best of these is between Seal point and Round island, leaving all the islands on the starboard side, and having no danger but what is visible. Ha-ha bay runs in to the N. about 7 miles, and has many islands at its head. Within them to the E. numerous anchorages exist from 9 to 12 fathoms.

HA-HA BAY, America, Newfoundland, on the E. side of Burnt cape, between Cape Freels and the Strait of Belle isle, running in S. about 2 miles, and open to the N. winds, having anchorage in 6 and 7 fathoms.

HA-HA HARBOUR, Newfoundland, between Cape Race and Chapeau Rouge; the entrance is on the S. side of the island before King's harbour, which last lies round the point of Wolf bay, and is one mile from Boat island, running in W.N.W. about a mile and a quarter broad, with from 20 to 10 fathoms water.

HALDIMAND, CAPE, Gulf of St. Lawrence, 2 miles N. of Douglas, a bluff point of cliff, and the S.E. termination of the range of hills which separates the harbour, basin, and S.W. arm from the valley of St. John's river.

HALDIMAND, or PORT LATONE, America, in Nova Scotia, a little W. of Negro harbour, and separated from it by a narrow peninsula; the extreme points bounding the entrance are Point Jeffrey, or Blanche point, to the E., and Point Baccaro to the W. There are clusters of rocks between, which render the harbour only fit for small craft.

HALF-MOON BAY, West Indies, Island of St. Kitts, on the N.E. side of the island.

HALF-MOON KAY, America, Yucatan; the lat. at the lighthouse 17° 12' 11" N., long. 87° 32' 24" W. This kay is low and bushy. The light is a fixed one, 70 feet above the sea level, and useful to those vessels bound to Balize by day as well as night, as it frequently happens that they get considerably to the S. of the fair way, owing to an indraught to the S.W. The lighthouse is on the S.E. part of the kay, and resembles a pyramid, 22 feet square at the base, diminishing to the lantern to 10½ feet; it is painted white. The light is not a good one, not being visible more than 2 leagues off, but it makes this passage to Belize to be greatly preferred. There is much valuable mahogany cut up the rivers of British Honduras, and all vessels intending to load must enter at the custom-house of Belize, and having taken a pilot there, proceed to the mouths of the rivers, where the mahogany is cut for the cargo awaiting shipment. If they are bound to rivers in the domains of the Central American government, which has granted leave to cut the timber in the State of Honduras, then the vessels must enter at Omoa, and return from thence to Belize loaded, before they can sail for England.

HALF-MOON KAY, West Indies, Island of

R

Jamaica, near Old harbour, two islets near the N.E. part of a bank lying N.E. and S.W. about a mile from Bare Bush kay bank, between Helshire point and Portland point. This last is in lat. 17° 43′ 50″ N., long. 77° 7′ 24″ W.

HALF-WAY ROCK, America, United States, coast of Maine, a high and black rock about 600 feet in diameter, elevated 16 feet above the level of the sea at high water. At the distance of 600 feet from this rock on the N.W., N., N.E., E., and S.E. sides, there are from 5 to 6 feet, gradually deepening to 25 fathoms within three-quarters of a mile. A reef extends off it W. by S. about one-eighth of a mile. It lies N. by compass from Cape Elizabeth, due N. from Fort Preble, and W. of Great Hog island, less than 2 miles. There is another rock of this name about E. by S. from Marble-head rock, near the harbour of that name on the coast of Massachusetts.

HALIFAX BAY, West Indies, Island of Tobago, to the N.W. of Richmond island; a good bay for vessels of 150 tons only. A shoal lies at the entrance.

HALIFAX, PORT and TOWN OF, America, Nova Scotia, the capital; according to Lieut. Raper, from the pillar in the dockyard, in lat. 44° 39′ 24″ N., long. 63° 37′ 30″ W.; by Admiral W. F. Owen, 1847, 63° 35′ 40″ W. This town was first settled in 1749. The streets are parallel, and the houses on the side of a hill, one above the other, have a very picturesque appearance. The harbour is defended by forts. At the N. extremity there is a dockyard, supplied with stores of all kinds, for the use of the Royal Navy. The principal fort or citadel stands on the summit of the hill on which the town is built, about 256 feet above the level of the sea. The direction of the fine harbour here is N. and S. nearly, and 16 miles in length, terminating in what is called Bedford basin, an extensive sheet of water, containing 10 square miles of anchorage. The coast about the entrance of the harbour of Halifax is rocky and rugged, particularly to the S., with patches of wood here and there. The land is rather low, and not visible 20 miles off, unless from the deck of a 74, except the mountains of La Have and Aspotogon, that are seen at 6 leagues distance. When the hills of Aspotogon bear N., having a long level appearance, a vessel is 6 leagues distant, and an E.N.E. course will carry her to Sambro′ lighthouse, on Sambro′ island. This is a fixed light, 132 feet above the sea, in the centre of the island, in lat. 44° 26′ 17″ N., long. 63° 35′ 16″ W. If a gun is fired here during a fog, it will be answered from this island. This island and lighthouse lie on the S.W. side of the entrance to the harbour. The colour of the shores indicates the side of the harbour—the E. being red and the W. white. S. by E. from the lighthouse, 2 miles, lies the Henercy rock, with only 8 feet water over it. One mile and a half W. are the W. ledges, or the Bull, Horses, and S.W. rock. The E. ledges are the Sisters or Black rocks, nearly E. from the lighthouse, distant 1 mile; and 1¼ mile N.E. by E. from the lighthouse is a rock, with 12 feet water on it, discovered by Captain W. F. Owen, in 1844. There are other dangers before arriving at Mauger′s beach, which has a perilous reef running out from it. The dangers are all marked. On Mauger′s beach is a tower called Sherbrook tower, on which there stands a lighthouse, on the E. side of the entrance, in lat. 44° 36′ 30″ N., long. 63° 35′ 40″ W. It is a fixed red light, 58 feet

above the level of the sea; and when the light bears N. ¼ W. it clears the Thrum Cap shoal. It is high water at the dockyard, Halifax, at 8ʰ; springs rise from 6½ to 9 feet. At Sambro′ island it is high water at 8ʰ 15ᵐ; and springs rise from 5 to 7 feet. There is a shoal between Mauger′s beach and George′s island. Reid′s rock, with 12 feet water, lies between Point Pleasant and Halifax, and M'Nabb′s island, on the E. side of the channel, 3 miles long and 1 broad, having a small island to the E. of it, called Carroll′s island. There is also a shoal off the N. part of M'Nabb′s island, "Ives," with a beacon buoy on the N.W. part. On the E. side of the harbour, opposite Halifax, is the rising settlement of Dartmouth. On account of vessels finding great difficulty in making Halifax from the E., particularly in winter, owing to the winds from W.S.W. to N.W., it is recommended that ships bound there in winter should shape a S. course, and run down their long. in lat. from 38° to 36°, by which they will have the best part of their passage in a temperate climate, until they approach the American coast, when they will be met by W. or even N.W. winds, which will enable them to make their passage good along shore with a free wind; to cross the Bay of Fundy, and that shut in, to keep the shore on board the whole way to Halifax.

HALLANDS VADERÆ ISLAND, Sweden, in Schouwen, about half a league long, with good anchorage between itself and the main, in 7 and 12 fathoms.

HALL BAY, America, Newfoundland, between Chapeau Rouge and Cape Ray, N.W. by W. one league from Richard′s bay. It runs N.N.E. about 5 miles, and is one-third of a mile wide, with deep water close to both shores, in all parts, except 1 league up on the W. side. There is good anchorage here in from 8 to 15 fathoms, with plenty of wood and water.

HALLESKIAR ISLAND, Sweden, on the W. coast of that country, in the approach to Gothenberg.

HALLIDAY FLATS, coast of Essex, one of the numerous shoals off the passage from Hollesley bay to Harwich, to be left on the port side. They extend in a direction N.W. by W. ¼ W. from the inner buoy of the ridge, having on the edge 13, 11, 12, and 13 feet water. These flats form the S. boundary of the rolling grounds, while the Andrew′s spit and beach end form the N.

HALLIFIORD, E. coast of Greenland, in lat. 64° 30′ N., long. 37° 25′ W.

HALLIMAN′S SCARS, Scotland, a part of Coversea Skerries, on which there is a beacon that bears from Craighead lighthouse E.N.E. ¼ E., consisting of a framework of iron, surmounted by a cylindrical cage and a cross, 48 feet above high water. There are steps leading from the rock to the cage, by which persons seeking a temporary refuge may ascend it. (See *Craighead*.) From hence to Cromarty the course is W.N.W. ¼ W., distant 22 miles.

HALLÖ ISLAND, in the Skagerac. The lighthouse here stands about a quarter of a mile S.E. of the Sälö beacon, in lat. 58° 28′ 30″ N., long. 11° 18′ E. The light is revolving, 116 feet above the level of the sea, the reflectors completing a revolution in six minutes, showing nine flashes of short duration with dark intervals. Thus this light is distinguished from that of Marstrand, which shows only four lights of longer duration in each revolution of eight minutes. The tower is stone, coloured white, 44 feet high from the ground to the light, built upon the S.E. side of the island.

HALMSTADT, TOWN and BAY, Sweden, the capital of the province of Holland, in the Cattegat. At the bar here there are only 3 and 4 feet water, and within the river 7 and 8 feet.

HALS, Denmark, a town on the E. coast of Jutland, on the N. side of the entrance into the Gulf of Lymfiord, 26 miles E. by N. of Aalborg. Vessels of too great a draught of water for the gulf discharge their cargoes here.

HALSA HAVEN, Sweden, at the bottom of Skalder bay, on the W. coast, of which bay Kullen point and light form the W. side of the entrance.

HALSKOV HEAD, on Funen, Denmark, in the passage of the Great Belt, with Knuds head seated on the Zealand side, and the Island of Sproe midway between. It has a fixed light in lat. 55° 20′ N., long. 11° 9′ E.

HALTON MIDDLE GROUND, England, River Humber. The S. end of this ground lies nearly three-quarters of a mile N.N.W. ¼ W. from the chequered buoy of the Foulholm, and runs in a N. direction a distance of nearly 4 miles. It has upon it 3, 2½, and 2 fathoms water.

HAMBLE RIVER, Hants, within the points of Southampton water, and a few miles E. of that town, due N. of Calshot castle. There are from 9 to 12 feet water within the entrance, but vessels go in and out at high water, the same depth continuing to the village on the W. shore of the river.

HAMBURGH, CITY, on the N. side of the River Elbe, one of the four free cities of Germany, 78 miles from the mouth of that river. Hamburgh is from 3 to 6 miles broad, much intersected with water communications, partly of the Alster as well as the Elbe, which join at this city. The site is somewhat elevated, except one particular quarter, which suffers from inundations at particular times. The houses are high, built of brick, and many of them elegant. There are two considerable suburbs between the Elbe and Alster. The latter forms a lake before it reaches the city, and there is a fine basin within it called the Binnen Alster, which is the harbour. The Elbe forms here two other harbours, one on the E. used by boats only, and one on the W. used by shipping, called the Niederbaram. Hamburgh has a few manufactures, but it is mainly enriched by its freedom, rendering it the entrepôt of merchandize of every description, and by the supply of Germany with imports, while, in return, vast quantities of produce from the interior are shipped here, above 2,000 vessels being thus annually engaged. The territory of Hamburgh consists of a small district round the city, extending above it and along the Elbe about 18 miles, and varying in breadth from 1 to 6 or 7; a detached portion at the S. entrance of the Elbe, containing the town of Cuxhaven, and six very small detached parcels of ground on the S. side of the Elbe, nearly opposite the city; the others in Denmark, in the Duchy of Holstein, in the direction of Lubec. The revenue of the city, from the customs and similar sources, is about 150,000l. per annum. Pop. 115,000, and of the whole territory 130,000. The lat. of the observatory here is 53° 32′ 54″ N., long. 9° 58′ 30″ E.

HAMILTON, TOWN OF, Bermuda islands, standing upon a harbour of the same name. See Bermudas.

HAMMAMET, GULF and TOWN OF, N. coast of Africa, the S. boundary of which is the Kuriat islands, and its N. Cape Tusihan, the bearing and distance being N. by W. ¼ W. about 40 miles. Several rocks lie about the N.E. point of the outer Kuriat, and between that and the middle or Conegliera island there is a channel with from 2¾ to 4 fathoms water, but to the S. of that island there is no passage except for boats. The centre of the bay is in lat. 36° N., and long. 10° 45′ E. The town is built upon a projecting piece of land, forming a small harbour on the N. side of the bay; there is good anchorage in the roadstead before it, except in E. winds. A sandy beach continues from the town to Cape Mahineer, which is distant 11 miles. The whole gulf has a clear sandy shore with regular soundings, generally mud and gravel.

HAMMOND'S KNOLL, Norfolk coast, a narrow ridge running in nearly a similar direction to Hasborough sand. The S. extremity of this sand is in 6 fathoms S.E. by E. ½ E. 4½ miles from the black or S. buoy on Hasborough sand, it extends thence N. by W. ¼ W. 5½ miles, where there is a depth of 7½ fathoms. Both ends sink gradually to seaward and into 14 fathoms water.

HAMNEWAY LOCH, Scotland, W. coast, 4 leagues to the S.W. of Gallen head, a well-sheltered place, fit for vessels in all weathers. There is no danger in entering except from the Daskere rocks that lie between Scarp and Mabistay island, and are 1¾ miles N. of the former island.

HAMOAZE, Devon and Cornwall, the W. branch of Plymouth harbour. See Plymouth.

HAMON, or JEAN HAMON'S ROCK, lat. 36° 54′ N., long. 19° 49′ W. This is one of those reported dangers resting upon the authority of one individual, Jean Hamon, the commander of the Trois Amis, of Bordeaux. On the 8th of January, 1733, he is said, on the authority of M. Bellin, to have approached it within three-quarters of a league and carefully observed it. He calculated the position by the course and distance run from the discovery of it until his arrival at the Rock of Lisbon, which he made to have been E.N.E. true about 165 leagues. There is good ground to believe that no such danger really exists.

HAMPTON HARBOUR, America, United States, near Newbury port in Massachusetts, 5 miles N. from the S. extremity of Salisbury point, the N. entrance of that port.

HAMPTON, LITTLE, Sussex co., a small port at the entrance of the River Arun, 4 miles S. by E. of Arundel. See Arundel.

HAMPTON ROAD, FLATS, and TOWN, America, United States, in Virginia, West of Cape Henry at the entrance of the Chesapeake. These roads are a branch of the Bay of the Chesapeake, at the mouth of the James' river. The flats extend along the N.W. side of the roads to the S.E. of the town, which is small, not reckoning more than 1,200 inhabitants. The roads are deep enough for the largest man-of-war, and it is an important naval station. There is a floating light at the entrance of these roads, which lies between Old Point Comfort on the N.W. and Willoughby sand point to the N.E. The Americans have strong fortifications on both points to command the entrance; that on Point Comfort is called Fort Monro, near which is a fixed light, and the works on the opposite side are named Fort Calhoun. The channel from Hampton roads to Norfolk is too intricate for strangers to attempt.

R 2

HAMRARNE, Baltic sea, the space between the Island of Bornholm and the mainland of Sweden.

HAMS, or N.W. BLUFF, Island of St. Croix, West Indies; it is bold-to along its N. part. The lat., according to Mr. Dunsterville, R.N., as communicated to him by Sir A. Lang in 1832, is 17° 46′ 24″ N., long. 64° 52′ 3″ W.

HAM SOUND, Orkney islands. See *Orkney*.

HANDFORT POINT, Dorset co., a projecting headland separating Studland bay on that coast, without Pool harbour, from Swanage bay; the most E. point of the Isle of Purbeck, and a league N. by E. from Peveril point and ledge.

HANGCLIFFE POINT, Shetland. See *Shetland Isles*.

HANGO or HANGAOUD LIGHT, Gulf of Finland, on the Island of Kanning, S. of Hango head, a revolving light, every three minutes, on the W. side, 107 feet above the sea, visible 15 miles, in lat. 59° 46′ 30″ N., long. 23° 1′ E.

HANN BAY, Africa, Island of Goree, a place from which the supply of water is brought for the garrison and shipping, but it is not fit to drink until it is filtered, and then only is good for culinary purposes. It is soft and issues amidst a soil of dark coloured mud. Hann bay is noted for supplying excellent fish. It falls back from Belair point, from which the place where the water is supplied is distant about a mile N.N.W. Lat. 14° 45′ N., long. 17° 28′ W.

HANNAH'S SHOAL, lat. 10° 7′ N., long. 27° 32′ W. This shoal was discovered by the master of the brig Hannah, Mr. T. Fanning, on the voyage from Rio Janeiro to Trieste, June 25, 1824. It appeared to extend 150 fathoms N.E. and S.W., with two branches or arms on the N.W. side, and one on the S.E. Soundings in 15 fathoms showed granulated coral on the S.W. part, but it was supposed much shoaler on the N.E. points, as the weed was plainly seen from the mast head on the surface of the water. The long. was deduced from lunar observations taken the day before, but as a strong W. equatorial current was felt, it cannot be depended on within 20 miles. The lat. is deemed correct.

HANO ISLAND, Baltic, S. coast of Sweden, N. of Bornholm island, and E. of Lister's head. The pilot's house stands in lat. 56° 1′ 2″ N., long. 14° 48′ 25″ E. Hano sound lies between the island and the main.

HANSE TOWNS, a powerful association of commercial towns once existing in Germany, that flourished from the 13th to the 17th century, under the name of the Hanseatic league. At one period they consisted of 64 towns, of which the port of Lubeck was the capital. Since 1630 the towns have been restricted to Frankfort and the ports of Hamburgh, Lubeck, and Bremen, which retain to this day the name of the Hanse towns.

HANSTHOLMEN LIGHTHOUSE, N.W. coast of Jutland, in lat. 57° 6′ 50″ N., long. 8° 36′ 10″ E. It exhibits a flashing light, in a lighthouse 57 feet in height; the light 212 feet above the level of the sea, and visible 6 leagues. It shows a flash of 15 seconds' duration every half minute, and is therefore easily to be distinguished from the fixed light on the Scaw, and the Norwegian light on Oxöe, which last is varied by flashes every fourth minute.

HANT'S HARBOUR, America, Newfoundland; W. by S. 7 miles from Salvage point is Hant's head, and a mile to the E. of the head is the harbour. Both are in Trinity bay, between Bacalhao island and Cape Bonavista. It is only adapted for very small vessels.

HAPSAL, a port of Livonia, on the E. part of the Baltic sea, lat. 59° 4′ N., long. 22° 47′ E.

HARBOUR BRITON, America, Newfoundland, to the W. of Little Barrysway, and N.N.E. ¼ E. 2 leagues from the Island of Sagona. The heads at the entrance are high, lying S.E. and N.W., distant 2 miles. Near the E. head is a rock above water. The only danger going in is from a ledge of rocks stretching 2 cables' length from the S. point of the S.W. arm, which is more than a mile within the W. head. The only place for ships-of-war to anchor in is above this ledge, before the entrance of the S.W. arm, in 16 and 18 fathoms, mooring E. and W. The bottom is good, and wood and water plenty.

HARBOUR BUFFET, and ISLAND, America, Newfoundland, on the S. coast. It lies upon the E. side of Long island; a tolerably good harbour, with a narrow entrance, but it has within 13 fathoms water. It may be known by the island of the same name, which lies E. ½ S. 1 mile from the entrance, as well as the other islands S. of its entrance.

HARBOUR, COVE, Cornwall, N. coast, on the W. shore of Padstow harbour, about a mile within the entrance, a place where vessels may lie afloat at low water, well sheltered.

HARBOUR DELUTE. See *Delute Harbour*.

HARBOUR FEMME, America, Newfoundland, between Cape Chapeau Rouge and Cape Ray, half a league to the W. of New Harbour; it is narrow, but has in it 20 and 23 fathoms water.

HARBOUR GRACE, Newfoundland, between St. John's harbour and Bacalhao island; the entrance is to the N. of the Island of Harbour Grace, where there is a fixed light in lat. 47° 42′ N., or to the S. and between them and the shore the channel is narrow, and the ground foul. The course in is about W. Almost in mid-channel is the Salvage rock, outside of which there is no danger, though there is a rock called the Long Harry, but it is visible. The way in is not difficult after passing this rock, and the water is 22 fathoms quite up to the harbour.

HARBOUR ISLAND, West Indies, one of the Bahamas, passed in sailing from the Crooked islands or Europe to New Providence.

HARBOUR ISLAND, Nova Scotia, S. coast; there is excellent anchorage within this island. The islands on the E. of the entrance to Fisherman's harbour are Green island, Goose island, and Harbour island; the William and Augustine isles of Des Barres.

HARBOUR LA CONTE, America, Newfoundland, between Chapeau Rouge and Cape Ray, is situated 1 mile to the W. of Brewer's Hole. There are some islands before it, the outer called Petticoat, and the inner Smock island, and there are two smaller between these, with some sunken rocks. There are dangers at the entrance of this harbour, but within there is anchorage at any depth from 6 to 16 fathoms, sand and mud.

HARBOUR MILLEE, America, Newfoundland, S. coast, to the E. of the E. point of L'Argent. Before this harbour there is a remarkable rock, which afar off appears like a shallop under sail. This harbour branches into two arms, one to the S.E., the other to the E., at the upper part of both the anchorage is good. Between this harbour and Point Enragée there are several bar harbours, with sandy beaches. The water all along the coast is very deep, and anchorage is safe everywhere, but it must be near the shore.

HARBOUR ROCK, Ireland, at the entrance of Cork harbour, having over it 2½ fathoms water. It is surrounded by a shoal of 3½ and 4 fathoms for nearly a cable's length, and lies 3 cables' length N.W. ½ N. from Roche point. It is near the middle of the entrance, and is to be the first thing avoided on going in. There are two buoys upon the shoal.

HARBURG, a town of Lunenburgh, on the S. shore of the river Elbe, opposite Hamburgh, equally well situated for commercial purposes.

HARDERWYCK, Holland, on the Zuyder Zee, a port town of Guelderland, in lat. 52° 15′ N., long. 5° 31′ E.

HARDING ROCK, Ireland, near Sheephaven, to the W. of Horn point, and within the Norway rock, about 2 cables' length from the land.

HARE HARBOUR and ISLAND, Labrador, between York point and Sandwich harbour; this harbour is only fit for small craft, the bottom being foul except towards the head, where there is anchorage in 3½ fathoms. Hare island lies before the entrance of the harbour, and is high land.

HARFLEUR, France, on the N. side of the Seine, about 3 miles E. of Havre de Grace, at the opening of the little river Lezarde, which forms a port into which small vessels go up with the tide.

HARLECH POINT, North Wales, the S. point of the Harbour of Traeth Bach and Traeth Mawr. The two points lie N.N.W. ¼ W., and S.S.E. ¼ E., 2¾ miles from each other, and from them the sands extend which form the bars of the harbours.

HARLINGEN, Holland, on the E. coast of the Zuyder Zee, nearly S. from the Island of Schelling. The harbour is much visited, but the sands prevent vessels of burden from entering it without unloading a portion of their cargoes. The passage to it is made by the Western Boom Gat, at the S.W. end of Schelling island. There is here a fixed light, 56 feet high, in lat. 53° 10′ N., and long. 5° 25′ E.

HARLOSH LOCH, Isle of Skye, Scotland, W. coast, 3 miles N. of Loch Bracadale; the entrance between Haverser and Harlosh island. There are two other passages in, between Haverser and Colbust to the S., and between Harlosh and Balimore point N. This Loch is situated by the point of Balimore and the Islands Harlosh and Haverser. The best anchorage is above Balimore House, in 5 and 6 fathoms.

HARMATTAN WIND. See *Africa*.

HARPSWELL SOUND, America, United States, on the coast of Maine; there is a stone erected on Little Mark island to serve as a direction for vessels running into this sound, as well as a conspicuous sea-mark for the mariner standing in from sea in any direction between Capes Elizabeth and Small point.

HARRINGTON BAY, Island of Cape Breton, called also Little Rastico: it will only admit small vessels; it has a communication with Great Rastico or Harris bay.

HARRIS ISLAND, one of the Hebrides or Western Isles of Scotland, and the S. division of Lewis island, from which it is separated only by a narrow channel. The S. end is in lat. 57° 42′ N., long. 6° 56′ W.

HARRIS SOUND, Scotland, W. coast, S. of Toe head; it is so full of rocks and shoals that no vessel should attempt to pass it. The only practicable channel is along the Main of Harris, and without an experienced pilot the passage should not be attempted.

HARSHALS, Denmark, a point of Jutland, 8 leagues W. of the Scaw, the shore low and sandy, bending to the N.E. A dangerous rocky shoal stretches out to the N.E. of this point, the outer extremity of which is 2¼ miles from the shore, hence this part of the coast must have a wide berth.

HART ISLAND, America, United States, in Long Island Sound, New York State; there is anchorage both on the E. and W. sides.

HARTLAND QUAY and POINT, N.W. coast of Devonshire. From Cape Cornwall this point lies E.N.E., distant 23 leagues, and from St. Martin's Scilly to Hartland point, the coast is E.N.E. ¾ E., distant 31 leagues. All the way from Pentire head to the entrance of the Camel river the coast is rocky and steep, and a ship may stand within half a mile of the shore, in from 6 to 11 fathoms without danger. This point, in lat. 51° 1′ 45″ N., and long. 4° 31′ W., is a high bluff land, having a reef of rocks, called the Tings, projecting N. by W. ½ W. to the distance of three-quarters of a mile, on which the sea breaks heavily, with 9 fathoms close to. The mark to clear these on the W. side, is Sharp's Nose, a high land, nearly 3 leagues to the S., well open of the coast, bearing S.W. by S.; and on the N. side, Gallantry Bower, lying to the W. of Clovelly, with a tuft of trees on it, kept open, and bearing S.E.; observing that the flood-tide always sets a vessel off in light winds, but the first of the ebb, when in shore, draws over it. Between Lundy island and Hartland point, the tides in the mid-channel set E. and W., at the same time that they flow and ebb on the shore, and at springs they run at the rate of 3 knots, neaps 2. This point may be easily known, as a feature in the coast, the connecting cliffs trending back abruptly each way at right angles. It is of a dark brown colour. Its summit is elevated 330 feet above the sea, towards which it slopes abruptly; the other cliffs are perpendicular. There is a gap which separates the extreme point from the headland.

HARTLEPOOL, TOWN and PORT, Durham co., in the N.E. division of Stockton Ward. As a port it is a dependent of Newcastle; the original harbour being closed up, and the present only fitted to admit vessels of light burden. The limits of the port extend from the Black shore in the Tees on the S., to the Blackhalls on the seashore northward. It has a pier which extends about 154 yards, and the harbour in its existing state as a receptacle for small vessels is artificially improved. The town stands on a promontory, and is nearly surrounded by the sea, lat. 54° 41′ 8″ N., long. 1° 10′ 7″ W. There is a lighthouse on the pier head, showing a red fixed light. Ships from the N. open this light when it bears W.N.W., and should not in the night approach nearer the shore than 14 fathoms water, at high tide, and when the light bears N.N.W., and on with a beacon in 7 fathoms, 2 miles distant, they may anchor. There are two fixed lights in one tower, vertical, 84 and 60 feet high, on the Heugh, 1847, the latter is only a tide light, and is red. In the day a red flag is hoisted at half flood, and continues to half ebb. At the "West Harbour," two red lights are on the quay, and a fixed green light on each pier head. In running for the Old harbour, when the pier light bears N.N.E. ½ E. about 120 fathoms, a vessel must steer in a N. ½ E. direction, until the two lights are brought in a line bearing N. by W., which is the direct course up the channel to the entrance of the inner harbour. These harbours are undergoing great additional improvements, which will not be completed in 1848.

HARTLEY HAVEN, Northumberland co., N.W. from Tynemouth, an artificial port, constructed for the purpose of shipping coal to London.

HARTWELL REEF, Cape de Verde islands, on the N.E. of Bonavista. It was upon this reef that the Hartwell Indiaman was lost in 1787. The Resolution, Captain Cook, too, was nearly driven upon it by a southerly current.

HARVEY REEF and SHALLOP ROCK, America, Prince Edward's island, Gulf of St. Lawrence, E. of Colvill bay. A bank of 12 feet lies 1¾ league to the N.E., with a channel of 3 to 7 fathoms between it and the island.

HARWICH, TOWN and PORT, England, Essex co., standing upon a peninsula, and bounded on the E. by the North Sea or German Ocean, and W. by the rivers Orwell and Stour, which uniting near the town, form a spacious harbour, capable of receiving vessels of considerable burden. The pop. is under 5,000. The number of vessels belonging to this port in 1829 was 96. The packets for Holland and Germany all sailed formerly from this port. The entrance is beset with numerous shoals and flats, which are all buoyed. Through these, different channels lead into the harbour. The entrance from the sea is between 2 and 3 miles wide at high water, but the principal channel by which vessels must enter the harbour, lies along the Suffolk side of the port, and is narrow and deep, commanded by the guns of Landguard fort. There are two lights at Harwich; the highest is of grey brick, 69 feet above the level of the water; the lower is white and 29 feet high. Both are fixed lights, and when in a line bear N.N.W. ¼ W.: they stand in lat. 51° 56′ 36″ N., long. 1° 17′ 30″ E. The passage from the Nore to Harwich, through the Swin and King's channels, is bounded by the Foulness, or Maplin, Whitaker, Buxey, and Gunfleet sands on the N. side, and by the Mouse, Barrows, Middle, Heaps, and Sunk on the S. side. The passages into Harwich, and towards Orfordness, are formed by the West rocks, Cork sand, Ledge, and Knot, Upper and Lower Rough, Shipwash, Baudsey, Kettle Bottom, Whiting, and Cutler, with several shoals besides within these. Between the West rocks and the Gunfleet there is a channel into the Wallet called Goldmer's Gatway, about 1½ mile wide, and having 5, 6, and 7 fathoms in it. There is also a channel between the West rocks and the Cork sand, with only 6 feet water, too dangerous to be attempted by any but those well acquainted with it. There are buoys and beacons on most of the shoals enumerated.

HASBOROUGH, Norfolk co., W. from Winterton ness, 3½ leagues E. by S. and E.S.E. from Cromer.

HASBOROUGH LIGHTHOUSES, coast of Norfolk. These are two in number, one higher than the other, both having fixed lights, 137 and 100 feet high, and used in sailing in and out of Hasborough gat, the leading mark being the two towers, bearing N.W. ¾ W.; or, night approaching, a vessel may run out through the Hasborough gat with safety, by bringing the two lights of Hasborough in one, bearing N.W. ¾ W., thence steering to the S.E., and keeping them on, which will lead a mile to the N.-Eastward of the Newarp light-vessel. The E. side of the Newarp bears S. by W. from the light-vessel; therefore, in rounding this light, and hauling to the S., a vessel must not bring that light to the E. of N. or N. by W. till after passing the light 3 or 4 miles, when she may steer a S.S.W. ½ W. or S.W. by S. course, which will take her outside the Cross and Holm sands. In rounding the light-vessel, with a half-flood in a ship's favour, and a commanding breeze, she may safely steer S.S.W. The flood, setting to the S.S.E., will keep all clear of the Newarp; but care is needful with an ebb-tide (which sets in a contrary direction) not to haul up too soon, as it will drift towards the sands, and without a favourable breeze, oblige a ship to anchor.

HASBOROUGH SAND and GAT, England, coast of Norfolk, the sand nearly parallel with the coast. It has a black buoy on its S. end, and one quartered black and white upon its N. lying in a N.N.W. direction for 9 miles, and is generally about a mile broad, from 4 fathoms on one side to 4 fathoms on the other, at its widest part, being in some places nearly dry at low water spring-tides. This sand is steep-to on both sides, having from 5 to 7 or 8 fathoms close to its edges, and at a quarter of a mile distance from 13 to 15 and 16 fathoms water, which renders the lead essential to ships standing in from sea. Near the S. end, on the eastern side, it is somewhat shallower and irregular; and N. by E. from the south black buoy, is a narrow ridge growing up, extending 3½ miles, over which are 4½, 6, and 8 fathoms. The Gat lies between Newarp and Sea heads to the S.W., and Winterton ridge, Hammond's knoll and Hasborough sand to the N.E.; it is 6 miles wide, and in depth from 10 to 20 fathoms. There is a light-vessel moored off the N. extremity of the Hasborough sand, exhibiting two lanterns with fixed lights on separate masts, 38 feet high, having the words "Hasbro' Gat" painted upon her sides. This vessel is moored in 13½ fathoms, with Cromer lighthouse W. by N., Hasborough high lighthouse S.W. ¼ S. 9 miles; and the N. buoy of Hasborough sand, E. by S. distant 1¾ mile; the lights may be seen 3 leagues distance.

HASSAN KALASSI, GULF OF, and HARBOUR. This gulf lies N. of that of Boudroum, and E. of the Island Cero, once called Mandeliah. It is separated from the Gulf of Boudroum by the high land of Aydyn. There is a little harbour near its S. entrance called Pasha Limany, and 5 miles to the S. of that is the harbour of Gumishlu, small, but having a depth of from 13 to 4 fathoms, sand and mud. The next place is Karabogla, an open port, having a little island at its entrance. It has a multitude of barren islands and rocks about it. Chatal is the largest of these islands, returning N. and E. to the Gulf of Hassan Kalassi from the Wreck rock 21 feet high, on the S. side, the gulf runs in E. by N. 6¼ leagues. At the head of the gulf is the harbour of Hassan Kalassi, or Isene port, the town standing on a narrow neck of land, the site of some ancient ruins. The N. shore of the gulf extends W.N.W. ½ W. to Cape Arbora or Monodendri, lat. 37° 21′ N., long. 27° 13′ E. full 23 miles; it is little used, and was scarcely known until Admiral Beaufort visited it in 1812, and at the same period made a survey of Boudroum to the S.

HASTE SHOAL, W. coast of Spain, a rocky bank N.W. ½ W. from Cape Roche, distant 5½ miles, having over it 3 fathoms water. There are from 10 to 14 fathoms near it, and the sea rolls heavy over it, and sometimes breaks.

HASTINGS, Sussex, a town on the sea-coast, between a high cliff towards the sea, and a high hill to the E. It is one of the Cinque Ports, but has no harbour, and only an indifferent road for small vessels. The chief trade of the town is confined, if the fishery be excepted, to the building of boats and small vessels. St. Leonards, a new and handsome town, has been erected recently to the W. of Hastings, and is visited much as a watering-place. It was to this town that William the Conqueror marched after his landing at Pevensey bay, and the remains of a castle, most probably erected about that time, are still in existence. This town is in lat. 50° 52′ N., long. 36′ E.; pop. about 11,000. There are two light-houses erected here, which are illuminated from March to September, principally intended for the use of the fishermen, to direct them in making the shore. One of these stands at the Hastings Stade, the other at Hill-street, 400 feet apart, bearing from each other N.N.W. and S.S.E. The upper light white, 60 feet above the level of the sea at high water, visible from S.E. seaward to S.W. for 9 miles; the lower light is red, 30 feet above the sea level. There are several rocky patches between Hastings and Pevensey.

HAT KAY and REEF, America, Honduras, W.S.W. of Half-moon Kay, wooded and in form somewhat resembling a coronet. A dangerous reef extends from this kay S. by E. for 3 miles, to clear which, when 2 miles S. of Half-moon kay, they must steer S.S.W. ¾ W. 10 miles. From the edge of this reef to Kay Bokel, the course is W. or W. ¼ N. 7 leagues.

HAT KAY, West Indies, between the islands of St. Bartholomew and St. Martin.

HATCHET KAY, America, coast of Honduras, one of the kays between Tobacco kay and Kay Glory.

HATS AND BARRELS ROCKS, coast of South Wales, lying between the Smalls and Grass-holm. Three of the Barrel rocks dry a little after half-tide, or the last quarter ebb. They lie W. ¾ N. from the centre of Grassholm, distant nearly 3 miles. Although the Barrels generally show themselves by a heavy breaking sea, they must be approached with caution; near high water, in light winds and no swell, with a lee tide, they can hardly be seen. The tide sets very strong over them, particularly at springs: this causes so great an eddy on the opposite side, that the indraught is to be carefully guarded against, particularly at night or in dark weather. They bear from the Smalls, S.E. by E., 4 miles. The Hats lie E.S.E. 2 miles from the Smalls, and N.W., 2 miles from the Barrels. They are always covered with 8 or 10 feet at low water, therefore must be approached with great caution. During the weather-tide, at the springs, a fearful sea breaks over them.

HATTERAS, CAPE, North America, United States, North Carolina, 35½ leagues S.S.E. from Cape Henry. The inlets of Currituck lie between, which are shoal, and New inlet, on which there are 5 feet water. About 6 leagues N. by E. from the cape lie the Wimble shoals, on which are 3 and 4½ fathoms water, extending N. by W. and S. by E. about 5 miles, being 3 miles wide; and 5 miles N. by W. from the N. end of this shoal, and 3 miles S.E. from the W. end of Hatteras island, there lie some small knolls, on which there are only 9 feet at low water. Cape Hatteras shoals

extend 8 miles in a direction S.E., with 5 and 6 fathoms on the extreme parts. The most dangerous shoal of all lies in lat. 35° 11′ N., the middle of which is 4 miles distant from the cape, and has barely 9 feet water. This is called the Diamond shoal, between which and the cape there is a good passage for small vessels in moderate weather, but it would always be safest to go round the shoal in 10, 12, and 15 fathoms. There is a lighthouse on Cape Hatteras, the light of which can be seen plainly in 9 and 10 fathoms water, on the outer part of the shoals, when only 10 feet above the level of the sea, but on board a large vessel it might be seen in 20 and 25 fathoms. This light is fixed and exhibited on a tower painted white, 95 feet high, in lat., according to Lieutenant Raper, 35° 14′ N., long. 75° 30′ W. The point now extends more than a mile from the lighthouse. The soundings from the cape are 2, 3, 4, 4½, 5, 6, and 7 fathoms, and then deepen to 9 on the S.S.E part. There is a current with a S. wind here, which runs N.N.E. 2 miles an hour, and with a north wind S.S.W. a mile an hour. W.S.W. from Cape Hatteras, 8 leagues distant, is Ocracock inlet, on the N. entrance is a revolving light, 75 feet high, on the bar of which there are 9 feet water. It changes, and should not be entered without a pilot. From Cape Hatteras to Cape Henry, the entrance of the Chesapeak, the ground is fine sand, and to the N. of Cape Henry coarse sand with some shoals among it. It is high water at full and change of the moon at 3ʰ 45ᵐ. The tide flows from 4 to 5 feet, being governed by the winds in the offing. In E. gales it runs several feet higher.

HAULBOWLINE LIGHTHOUSE, Ireland, E. coast, off Loch or Lough Carlingford, upon Haulbowline rock. See *Carlingford*.

HAULBOWLINE ROCK or ISLAND, Ireland, E. coast, at the entrance of Carlingford harbour. It must be passed on the N.E. side, giving it such a berth as not to come nearer than 6 fathoms, nor to the main than 2. There are two fixed lights (vertical) in one tower, one a tide light, 100 feet high. There is also an island of the same name in Cork harbour.

HAUTE-FOND, West Indies, Island of Hayti, 3 leagues N.N.E. of the Grange point, a white shoal, of not more than 2 cables' length each way. There is a small spot in this shoal with only 24 feet water, on which the Ville de Paris struck in 1731. Close to it is a depth of 6, then 10 and 15 fathoms, and suddenly no ground.

HAUTE ISLAND, North America, Bay of Fundy, at the entrance of Mines or Minas channel. It is not more than 1½ mile long, and half a mile broad, bearing from Cape Chignecto S.W., distant 4 miles. The channel on either side is good.

HAVANA, PORT OF, West Indies, Island of Cuba, 11 leagues E.N.E. ⅓ E. from Port Cavanas, in importance the principal harbour of Cuba, able to accommodate in security several hundred sail of the line. The entrance is more than three-quarters of a mile long and 1¼ cable's length wide, and strongly fortified throughout the whole extent. The mouth is defended by two castles, one called the Morro, on the E. point or hill, built in a triangular form, and strengthened with bastions and forty pieces of cannon, and another on the W. side, called La Punta, to the S.E. of the city. There is a lighthouse on the Morro castle, presenting a regular revolving light, 30 seconds between the revolutions, seen 22 miles in clear weather, and about 130 feet above high water. It stands in lat. 23° 9′ N., and long. 82° 22′

W. The entrance of this harbour lies in S. 59° E. true. This celebrated harbour is distinguished at a distance by the hills called the Tetas or Paps of Managua, called also the Maiden's Paps, which lie on the meridian of the harbour's mouth. The coast, both E. and W., is low and equal, and there only arises one little hill, which is surmounted by fortifications. At 6 leagues to the E. the Sierra de Jaruco, or Iron hills, are discovered of moderate height and insulated. Seven leagues to the W. the Tables of Mariel, and sometimes the Hill of Cabanas, may be descried. The "Nautical Magazine" has described this fine harbour as protected from all winds, and as being seldom visited by hurricanes. The entrance to the Havaña harbour lies N.W. and S.E. nearly, and is a quarter of a mile wide from the Morro castle to what is called the Punta, or opposite shore. It continues the same breadth, or a little wider, all the way up to the public wharf, where it is the narrowest. It is there less than one quarter of a mile, and the wharf, which is on the starboard hand going in, is about a mile from the Morro. Above the wharf the harbour opens out into a basin, where a vessel may choose her anchorage. Merchant-vessels with a cargo anchor abreast of the wharf, but if in ballast on the Casa Blanca side. The public wharf here has sufficient depth to haul the largest merchant-vessel bows on to discharge or load, and there is room for mooring upwards of 100 sail, large and small, in the same manner. A square-rigged vessel bound to Havaña, after having made the Morro, must wait the sea-breeze, which commonly sets in about 11 A.M., and takes off about 4 P.M., and most commonly from the N.E. It is best to haul close round the Morro, and keep on that side up to the guard-ship, if the wind permit; there are 4 to 5 fathoms alongside the rocks. There is a shoal a cable's length within the Morro, and extends half that distance offshore, on which is a flat buoy, 1845. About one-half the way there is a red buoy with a flag, on the starboard shore, placed on South Telmo shoal, and which stretches half-way across the harbour, and which must be left on the starboard side, within a sail's breadth of it. There is good anchorage all round the shoal, but a vessel coming in after passing the Morro and the red buoy, if taken aback, had better anchor immediately, and warp up, which is the general plan adopted, or she must keep all sail on, so as to be under command with her anchor ready, and a good range of cable. Should the wind be favourable, it is only necessary to keep about mid-channel, and before coming abreast of the wharf the harbour-master will go on board, and will give directions where to anchor. In other circumstances there is good anchorage all the way up from the red buoy, and a vessel obliged to come to must wait until the evening, and warp up; but the anchorage outside the red buoy is not good, being deep water, rocky, and bad holding ground. The guard-ship lies about two-thirds up the harbour from the Morro to the wharf, and a cable's length from the north shore. Between her and about four berths above the wharf all vessels moor head and stern with bower or stream anchor, and not more than three abreast from the Casa Blanca side, but higher up the harbour the vessels lie at single anchor. The depth of water from the Morro up to the harbour is from 10 to 6 fathoms; in the harbour from 8 to 6 and 4 fathoms, tough clay, mixed with shells. No directions are requisite for going out, as it is either done in the morning or evening,

when the wind is fair; and a vessel is clear as soon as she is past the Morro.

HAVERFORDWEST, South Wales, Pembroke co., a well-built town on the W. bank of the River Dongledy, one of the principal streams falling into Milford haven. It has several vessels belonging to its port, and stands in lat. 51° 49′ N., long. 4° 58′ W.

HAVERING ISLAND, Swedish coast, in the Baltic, at the entrance of the channel of New Koppen, 7 leagues S. by W. from Landvort point.

HAVRE DE CARTERET, France, W. coast, a haven S. of Cape Carteret, extending E. to the town, which is 1¼ mile from the cape. The entrance is encumbered with rocks and dry sand. The channel is marked by three beacons on the W. side.

HAVRE DE GRACE, America, United States, Maryland, on the Susquehannah river. Here is a lighthouse at Point Concord, having a fixed light, the navigation to which in passing from the port is alone fit for vessels drawing not over 8 feet. On the shoals there are only 3 feet water.

HAVRE DE GRACE, France, W. coast, a strong maritime port of the Seine Inferieure, having a pop. of 21,000. It is situated at the entrance of the River Seine, at the foot of an eminence called Ingouville, 2¼ miles S.S.E. from Cape la Hève, from whence to about a mile E. of the cape the land is low. Lat. of the Lower Jetty, 49° 29′ N., long. 6′ 12″ E., according to Lieut. Raper. The French make the long. of the town 2° 13′ 27″ W. from Paris, and 49° 24′ 14″ N. lat. The harbour, which lies within the walls of the town, has capacious docks and basins, in which 500 or 600 vessels might remain afloat, those in the harbour lying aground at low water. The entrance is formed by two stone piers, upon each of which there stands a tower. The longest jetty is upon the N. side; at 36 feet from the end of which is a superior harbour-light, fixed, and 39 feet high—the range 3 leagues. There is a bell attached to the tower, which is used in foggy weather. On the extremity of the S.E. jetty, which is 246 yards from the W., is a small light of a yellow colour, only visible about a mile, very useful to vessels entering during dark nights, and is 23 feet high. In this harbour the water does not perceptibly ebb till 3 hours after high tide. In consequence of this peculiarity fleets of 120 sail have left it in one tide, even with the wind against them. This effect is generally ascribed to the Seine, which, when the sea begins to ebb, crosses the pier-heads with such force as to prevent the water in the harbour from running out until the water without has fallen to a certain degree below it. It has been remarked that in this part, at about the full and change of the moon, the currents are so strong, and the winds so high, that ships which happen to be in the roads are in danger of being lost in the mouth of the river, or driven against the coast. All bound into the port should take a pilot. They attend commonly as far off as Cape Barfleur, unless the weather be so bad they cannot get off; in that case they go to the N. side of the entrance and make signals, by which a vessel may know when she may enter the harbour; in doing which it must be kept always open, so as to see all the ships within, between the two towers; then she must steer in, passing nearer to the great tower on the port than to the one on the starboard side. There are two roadsteads without the entrance of the river: the great road, a league and a half from the harbour, 3 miles in extent from N. to S., lying N.W. by W. from Cape la Hève; and the little road, lying to the S. of the cape, half a league from the harbour. The two roads are separated by banks called the High Grounds of the Road and L'Eclat Bank. The little road is within, and the great one without, these banks. In the outer road are 9 and 10 fathoms at low water. The inner one extends about three-quarters of a mile every way. Its bottom clay and good ground, but covered with pebbles. Ships only waiting a tide prefer the outer to the inner road. The best anchorage is a league W. of La Hève, in 8 fathoms at low water, the Castle of Orcher a little open of Ingouville, land to the N. of Havre. The flood tide sets here the first two hours S., the next two S.E.; the fifth hour E., and during the remainder of the tide from N.E to N.W. In any part within L'Eclat bank vessels may anchor for a tide in from 3 to 4 fathoms. There are buoys on the shallows known as the Bank of L'Eclat and the High Ground of the Road. A code of signals was established for this port as long ago as 1829, consisting of black balls denoting the depth of water in feet. These signals are made from a mast erected upon the N.W. jetty of the port, and may be distinguished with a common glass at a league or a league and a half distance from the harbour. It is high water full and change at Havre at 9h 57m. Spring tides rise 26¼ feet, neaps 19¼. The sands and depths of water at the mouth of the Seine river are very variable, and it is scarcely safe to place dependence on any plan or chart of this part of the coast. The only two permanent banks are those banks of stones known as the Bank of L'Eclat and the High Grounds of the Road, which form the boundary of the little road on the W. of the harbour. There is a large tree 3¾ mile S.E. by E. from the W. lighthouse of Honfleur, named the Homme de Bois by the pilots, seen from the sea, presenting the appearance of a great tower, and forming a useful mark. The ground in the centre of the little road is very good, but not sufficiently known, and the depth too shallow for large merchant-ships to attempt stopping there.

HAVRE DE ST. GERMAIN, France, a haven S. ¼ E. 11½ miles from the Havre de Carteret; it is 13 miles S.S.W. to the Havre de Regneville from St. Germain, and 8 leagues E. to Granville.

HAWKE CHANNEL, Florida Reefs. See *Florida Reefs*.

HAWKE ISLAND and BAY, America, Labrador, about 1 mile to the N. of Stoney island, within which is the bay running in W. 2 leagues and then dividing into two branches, one going W. by S. 6 miles, the other N.W. by W. 5 miles. Lat. of the island, S.E. point, 53° 4′ 20″ N., long. 55° 26′ W.

HAWK ROCK, Ireland, S. coast, off a projecting point called Poor head, S.W. by S. nearly a quarter of a mile, with 10 feet on it, and 3 miles E. from the entrance of Cork harbour.

HAWKER ROCK, coast of Northumberland, a rock near Staples island close to and one of the Blue Caps, some above and others under water, dark and with ragged heads.

HAWK'S BILL, West Indies, Island of Tobago, a point opposite and S. of Courland point.

HAWKE'S HARBOUR, Newfoundland, on the W. coast, to the E. of Ingornachoix bay. Vessels to enter this harbour commonly go to the S. of Keppel island. The starboard shore is foul, and

there are several shoals to be avoided, which is not difficult; a vessel may then anchor in 12 fathoms water or run within half a mile of the small island into the harbour and anchor there, a place convenient for wood and water. It is the best harbour for ships bound northward.

HAWXLEY POINT and ROCKS, Northumberland, on the coast between Blyth and Coquet island. The dangers here are buoyed.

HAYE'S ISLAND, America, Hudson's bay, formed by the rivers Nelson and Hayes, which, at first running together, separate and form an island, at the point of which is York fort, in lat. 57° 10′ N., long. 93° W.

HAYLE HARBOUR, Cornwall, in St. Ives bay, the channel leading to which goes up among the sands which form the entrance of the River Hayle, about 2 miles S.E. by S. from St. Ives. It is only adapted for vessels of moderate draught of water.

HAYLING, or HALING, ISLAND, Hants, N.E. from South Sea castle, near Portsmouth, between Chichester and Langston harbours.

HAYTI, ISLAND OF. See *Domingo*.

HAYTIEN CAPE, West Indies, Island of Hayti. See *François, Old Cape*.

HAZE, that appearance of density which the atmosphere puts on at times in most parts of the world during hot weather, rendering objects more or less indefinable, according to their distance, from their being seen through a veil of vapour. It is totally distinct from what is called mist or fog, which is the effect of humidity. On the coast of Africa, more or less at all times, the presence of haze is constant, and it is at its maximum during the prevalence of the dry wind called the Harmattan. Its appearance there is almost obliterated in the rainy season, and it is dispersed by the tornados, but returns with increased density when these storms cease. It is, therefore, requisite that the mariner should be on his guard against its deceptive effect, especially in judging of distances, which it usually lengthens beyond the truth, and may therefore lead those incautious of its effect into disaster.

HEAD HARBOUR, or DELAWARE RIVER, Nova Scotia, at the further end of Margaret's bay, on the N.E. extremity. Here is anchorage of the best description, where a large fleet might ride out undisturbed the most violent hurricane, in 8 and 6 fathoms. The surrounding land is lofty and broken.

HEAD HARBOUR, Newfoundland, on the W. coast, close to West bay; there is anchorage here in 8 fathoms.

HEAD HARBOUR, North America, belonging to New Brunswick, at the N.E. point of Campo Bello island; it is a small safe place, easy of access, with 6, 7, and 8 fathoms water, muddy bottom. A fixed light stands on the N. entrance, 64 feet high, in lat. 44° 58′ N., long. 66° 54′ W.

HEALTH, ISLES OF, or ISLES DE SALUT, South America, coast of French Guyana, three small islands, lying N.W. by N., 8 or 9 leagues distant from the Forlorn Hope and the entrance of the River Cayenne. These islands form a triangle, and make an excellent harbour. The best anchorage is to the E.S.E. of the most S. islet, where there is from 5 to 6 fathoms water, with hard clay ground, a musket-shot from the islet.

HEART'S CONTENT HARBOUR, America, Newfoundland, between Bacalhao island and Cape Bonavista. It is a good harbour, with water for any vessel, and excellent anchorage towards the N. shore in from 8 to 12 fathoms.

HEATH POINT, Anticosti island, Gulf of St. Lawrence, consisting of limestone about 10 feet high, with a superstratum of peat. This point disappears in the horizon at the distance of a few miles. The most dangerous reef on the island runs out from this point to the E.S.E. nearly 2 miles, at which distance it has 5 fathoms water. Within that space the reef is composed of large cubical blocks of limestone, with very irregular soundings from 2 to 5 fathoms. The rocky and irregular soundings from 5 to 7 fathoms extend 3 miles off Heath point, so that vessels should approach no nearer. There is a lighthouse on this point in lat. 49° 5′ 20″ N., long. 61° 44′ 17″ W., fitted for a fixed light, 100 feet above high water, but unfortunately it is not lighted (1847). It serves as a mark by day, appearing like a sail off the island, and is thus far useful in marking the extent of the low land to vessels either from the E. or W.

HEBBLES SAND, England, mouth of the Humber river; it is narrow, and stretches along the shore about three-quarters of a mile, leaving a passage between itself and the land for small craft. A white buoy is placed on its W. end in 4½ fathoms half-ebb, neap tides.

HEBRIDES, or WESTERN ISLANDS, a name given to a great number of islands lying off the coast of Scotland, and extending from the small Island of Sanda, in lat. 55° 17′ N., to the N. extremity of Lewis, in lat. 58° 32′ N. There are several large islands among them. They are allotted to three of the shires on the W. coast of Scotland. The chief of these islands are Lewis and its smaller isles, which, except the district of Harris, belong to Ross-shire. North and South Uist, Benbecula, St. Kilda, Bara, Skye, Raaza, and Eigg, are attached to Invernesshire; and Canna, Rum, Muck, Coll, Tirey, Mull, Jura, Isla, and Sanda belong to Argyleshire.

HECLA, MOUNT, Island of Iceland, on the S. coast; a volcano 5,364 feet high, and conspicuous from the sea, in lat. 63° 59′ N., long. 19° 42′ W., according to Lieut. Raper.

HEDIC, or HŒDIC ISLAND, France, 3½ miles from Houat island, in Quiberon bay. It is surrounded by numerous rocks, both above and under water, which require a good berth in every direction. At 300 fathoms W. from the E. point of this island there is a fixed light, elevated 85 feet above high-water mark, visible 9 miles distant. S.E. of Hœdic lie the rocks called the Cardinals, the largest of which is always above water, and is about a mile distant from the island. The passage between the islands of Hœdic and Houat is foul ground. The tides are very strong near this island. The time of high water at full and change is 3^h 45^m.

HEDRON, Greek Archipelago, a small island W. of Hydra, between that and the mainland.

HEELA or HELA POINT, Gulf of Dantzick. From Rezerhooft towards Hela a narrow isthmus or neck of sandy land extends S.E.; and at its extremity stands the town and lighthouse of Heela, in lat. 54° 36′ N., long. 18° 49′ E., revolving light, 130 feet high. When abreast of Heela, a vessel, in clear weather, will perceive the city of Dantzick. The course from Rezerhooft towards Heela will be S.E. by S., the latter being distant from the former 7 leagues. From Heela, the course S.W. ¼ W., about 12 miles, the water shoaling all the way into the bight, from 30 fathoms near the Point of Heela, to 5 fathoms near Dantzick. The best anchorage is to bring Dantzick light-tower to bear S. or S. by W., and then run into 6, 5, or even 4 fathoms. The opening to the S.W. of Heela peninsula is called Putziger Wik; it is filled with sand, and has only from 4 to 2 feet water in it. The Gulf of Dantzick, from Rezerhooft to Brusterort lights, is 19 leagues across, and its depth about 11. See *Dantzick.*

HEIGHTS OF MECA, Spain, W. coast, to the N.E. of Cape Trafalgar. See *Altos.*

HELBRE SWASH, Cheshire coast, the E. passage into the River Dee, between the S. and E. Hoyle banks. Its entrance bears S. from the N.W. light-vessel in Liverpool bay.

HELDER, Holland, a town with a strong fort, commanding the entrance of the Texel, in the N. part of the province, 24 miles N. of Alkmaar.

HELEN'S REEF, E. of Rockal, which last is in lat. 57° 37′ N., long. 13° 41′ W. This reef was unknown until the wreck of the Helen upon it. It lies 1 league or somewhat less from Rockal, E.N.E. ½ E. by compass. Sixteen persons perished, and twelve escaped in one of the boats. The captain believed Rockal lay in lat. 13° 40′ N., and reported that the ship struck violently twice on a group of rocks not much bigger than a ship's length, on which the sea broke occasionally; no other breakers were in sight. Rockal bore by compass W.S.W. ½ W., he thought about 6 miles distant. Captain Vidal, R.N. has since surveyed the Rockal bank. See *Rockal.*

HELEN'S, ST., England, Hants. co., a village on the E. part of the Isle of Wight, opposite to Portsmouth, having a large bay. It is N. of Brading harbour. See *Portsmouth.*

HELEN'S, ST., ROAD. See *Portsmouth.*

HELFORD HAVEN, Cornwall. The entrance is from St. Anthony's point; it is 4 miles S.W. ¼ W. In going into or coming out of this place, it is necessary to keep in mid-channel or nearer to the S. shore, because the N. is foul, and the Gedge's shoal lies there, having only 6 feet water over it, one-third of a mile off shore. On the S. side there is a steep point, and on the N. one low and flat. Within Mawnan point a vessel may anchor, by bringing Mawnan church to bear E.N.E., in from 2 to 3½ fathoms. Helford town or village lies 1½ mile from the entrance to the harbour. A little to the E. of the town is a bar, on which there are only 12 feet. The best water over it is nearer to the S. shore. There is the same depth at Helford as at the bar. Small vessels only visit this port.

HELGOLAND ISLAND, North sea, a small elevated island, consisting of table land, in front of the rivers Jahde, Elbe, and Weser. It may be seen, in clear weather, 6 or 7 leagues off, and has a lighthouse upon it, 257 feet above the level of the sea, which is lighted constantly. This object vessels generally make, when bound for the Elbe or Weser; and pilots are to be found here for both rivers. To the E. of Helgoland is a small sandy island. Both these islands are surrounded with dangerous reefs, running chiefly to the N. S.E. from Helgoland is the Klip, or Steen rock, upon which a buoy is placed; but it is advisable to attend to the marks for it, which are, the two beacons on Sandy island in one, and the lighthouse in a line with a wooden beacon on Helgoland. Off the S. end of the island is a remarkably sharp broken point of land, called the Monk, Helgoland lighthouse is in lat. 54° 11′ 20″ N., and long.

7° 53′ 13″ E. It bears from Yarmouth E., distant 79 leagues; from the Spurn E.S.E. ¼ E., 92¾ leagues; from Flamborough head E.S.E., 92 leagues; from May island, off the Frith of Forth, S.E., 126 leagues. Vessels anchoring in the road of Helgoland, which is not to be recommended, from the rocky nature of the ground and the little security there, should, entering by the N. passage, bring Helgoland to bear S.E., and proceed in that direction until the island is about a mile distant; but not lose sight of the new lighthouse lantern, or observe the top of the old lighthouse over the cliff; in the night, keeping the light in sight just above the cliff. When the N. point of Helgoland bears S., or the beacon in the lower town is open of the cliff, a vessel may run in; or at night, when the light bears S. by E. ½ E., she may bear up for the anchorage, always taking care that the light be not obscured by the cliff. The lead must be constantly employed; and a ship must come no nearer the downs than 4 fathoms. In mid-channel there are 6 and 7 fathoms. The best anchorage for large vessels is to the E. of the downs. Over the bar, abreast of the town, there are only 2 to 2½ fathoms at low water. This island on all sides presents perpendicular cliffs of a reddish coloured stone, and is accessible only in one place, where 180 rude steps have been cut from the beach to the top of the cliff. The lighthouse is on the N. side. The pop. is about 2,000, principally dependent upon the fisheries. It has belonged to England since its surrender in 1807.

HELIER, ST., the chief town in Jersey island, on the E. side of a bay on the S. coast of the island, in which are the small islands of St. Aubin and Helier. On the latter is Elizabeth castle, commanding the entrance. It stands on the bay of St. Aubin, and has a harbour and stone pier, having the sea on the S.W. and hills on the N., which shelter it from the cold. This town is in lat. 49° 11′ N., and long. 2° 6′ W. Pop. 10,118.

HELIER'S, ST., ROCK, S. of St. Helier's island, Jersey, between which and the E. point of the bay is the passage in.

HELIGENHAFEN, Denmark, a small port of Femern or Femeren island.

HELLESPONT, the ancient name of the entrance from the Archipelago and Mediterranean into the Sea of Marmora and up to Constantinople, now known as the passage of the Dardanelles.

HELLEVOET SLUYS, Holland, on the S. side of Voorn island. There is a fixed light here, placed on a tower upon the W. pier of Hellevoet harbour. Lat. 51° 49′ N., long. 4° 8′ E. It is placed 46 feet about high water-mark, and is visible 8 miles from S.E. through the S. to N.W. There is a canal through the Island of Voorn to the Maas; and a vessel can go to sea from the Maas at any time by passing through this canal, for which purpose it is used by those not drawing more than 16½ feet of water, as they make the passage into the Maas, near Rotterdam, in about 4 hours. This town is strongly fortified, and has a good pier.

HELL GATE, America, United States, state of New York, a narrow part of the channel which communicates with and forms the passage to Long Island sound, N.E. from the city of New York.

HELLIS OE ISLAND, 4 leagues to the E. of the entrance of Mandal, Norway, and 4 miles W. of the island called Flekker Oe: it is distinguished by two towers or beacons painted white, with a high bar upon each, so that in clear weather they may be seen at the distance of 3 or 4 leagues. Within this island there is an excellent harbour for vessels of all sizes.

HELLY HUNTER, Ireland, E. coast, Loch Carlingford, a rocky shoal, the S.W. end of which lies 1½ mile S.E. by E. ¼ E. from the lighthouse on Haulbowline rock, and E. ¼ N. from Cooley point: this shoal then stretches a mile N.E. Its W. end dries with very low tides only, and in other parts the least water is 6 feet. There is a beacon buoy in 2½ fathoms a quarter of a mile S.S.W. from the largest rock.

HELSINGBORG, Sweden, W. coast, a town in the Sound, the nearest point to Cronborg on the Danish side. There is a harbour fixed light here, seen 8 miles, in long. 12° 42′ 12″ E., lat. 56° 3′ N.

HELSINGFORS, the capital of Finland, west of Borgo, and near the meridian of 25° E.; a populous town, considered the best port in Finland for large vessels. The entrance is defended by several forts; the principal is that of Sweaborg, extending its fortifications through seven islands, having barracks, magazines, and an arsenal, all bomb-proof. Since the Russians have had possession of Helsingfors, the town has been considerably improved, and the commerce extended: there are now two basins or docks for building and repairing ships. Its exports are corn, fish, logs, deals, and salted provisions.

HELSTON, Cornwall, on the N. part of the Lizard peninsula. The tower of the church serves as a sea-mark, but the town itself stands inland.

HELWICK HEAD, Ireland, S. coast; the S. point of Dungarvon bay, bearing from Hook point W. by N. ¼ N., distant 22 miles.

HELWICK SANDS, coast of South Wales; the E. end lies S.W. by W. 4 cables' length from Porth Einion head, thence extending W.N.W. nearly 6 miles, divided by passages of deeper water. The N. of these sands lies on what is called the Helwick channel. The W. Helwick stretches 2 miles from the W. side of the swashway between, and has from 7 feet to 2½ fathoms upon it. Off the S.W. end is a floating light, revolving every minute, in 13 fathoms. The W. end swashway and E. end are buoyed. The soundings to the W. of the Helwick shoal suddenly from 10 to 5 fathoms; along the S. side they rise from 18, 16, and 14 fathoms, coarse ground, to 6 and 3 fathoms, fine sand; on the N. side they shoal from 10, 9, and 8 fathoms, to 6 and 5, without any variety in the bottom. The sands are steep-to, and therefore dangerous. To sail to the W. of Rossily Parsonage must be open of Worms Table Land, E. by N. This will lead clear two-thirds of a mile in 5 fathoms. To sail to the S., bring the Mumbles light E. by S. open of Oxwich head, which will lead by a quarter of a mile, in 15 fathoms.

HEN AND CHICKEN ROCKS, Devonshire, N. coast, W. of the Constable rock, at the N.W. end of Lundy island. These rocks are the only ones detached from the island that require any precaution of moment to be taken regarding them. They do not extend more than 3 cables' length from the N.W. point, and some of their heads may be seen at low water. There being 25 fathoms close to them, a good berth must be given, as the flood tide from a mile to the W., has a tendency to set over them. The race off these rocks N.N.W. from them, presents an alarming appearance, although at the distance of half a mile from the island there is a depth of 25 fathoms. The day mark for passing clear of these rocks, is to keep the detached Black rock lying off the S.W. point of the island in sight, about a cable's length open.

HEN AND CHICKEN ROCKS, or the STIRKS, N. coast of Ireland, about a mile S.E. ¾ E. of the E. end of the Skerries, and N.N.W. ½ W. three-quarters of a mile from Dunluce castle, in the fairway of vessels passing between the Skerries and the main, and only covered at high water.

HENLOPEN, CAPE, America, United States, on the S. side of the entrance of the Delaware river, in lat., fixed, according to Lieut. Raper, at 38° 47′ N., and long. 75° 6′, but in long. 75° 5′ W. after the American statements. It has a fixed light, 180 feet high. See *Delaware*.

HENRIETTA MARIA, CAPE, America, the N.W. limit of James' bay, whence the coast trends one way almost due S. to form the W. side of James' bay and the other to the W. and N.W. towards the S.W. part of Hudson's bay. Lat. 55° 10′ N., long. 82° 30′ W.

HENRY, CAPE, America, United States, the S. cape of the entrance to the Bay of the Chesapeak, Cape Charles being the N. It lies in lat. 36° 58′ N., long. 75° 56′ W. There is a lighthouse here 120 feet high, carrying a fixed light, but one not seen at a great distance. In making Cape Henry from the S., a vessel should keep in 7 fathoms until she begins to draw up with False cape, which lies about 7 leagues from Cape Henry towards Currituck, then 9 and 10 fathoms are full near enough to False cape. After getting to the N. of False cape, a ship may keep again in 7, 8, and 9 fathoms until up with Cape Henry. The shore between Cape False and Cape Henry makes in like a bay, and in thick weather might be mistaken for Lynhaven bay, and False cape for Cape Henry, if so thick that the lighthouse on Cape Henry shall not be seen. The passage between Cape Charles and the Outer Middle is little known, and is only used by vessels drawing 8 or 10 feet water.

HERACLIA, or EREKLI, Sea of Marmora, on the N. side, E. from Rodosto 20 miles. This is a considerable town, the port spacious and circular in form. The land bends inward beyond Heraclia. Lat. 40° 58′ N., long. 27° 58′ E.

HERACLISTA, in the Sea of Marmora, a small town on the N. shore. Long. 27° 11′ E.

HERCULES, TOWER OF, Spain, N. coast, 5½ miles nearly S.W. by W. from Cape Priorino, a lofty square building 160 feet high so named, now turned into a lighthouse, 1 mile N.W. ¼ N. from the city of Coruña, having a revolving light 364 feet above the sea, in lat. 43° 23′ 45″ N., long. 8° 23′ 30″ W.; the period of revolution between the flashes is three minutes.

HERD SAND, Northumberland co., a large sand on the S. side of the inlet called Prior's Haven, between Sunderland and Tynemouth; opposite the inlet is the Sparrow Hawk, a dangerous rock, requiring a vessel to keep offshore when proceeding between these two ports.

HERIN POINT and BAY, West Indies, Island of Trinidad, S. coast, E. of Point Icaque.

HERM ISLAND, English channel, one of the smaller isles which shelter the anchorage of Guernsey roads, and belonging to the same group. It is very fertile, and has numerous houses, a mill, and two beacons upon it. The higher beacon is 214 feet above the level of high water at spring-tides. The base of the mill is 198 feet in elevation.

HERMITAGE COVE and BAY, Newfoundland, between Chapeau Rouge and Cape Ray, an extensive bay, bounded on the S.W. by Pass island, and on the N. by the islands that form the Bay of Bonne and Great Jervis harbour, the width being 2 leagues. The Fox islands, distant from Pass island 8 miles, are opposite to the entrance of Hermitage cove, about three-quarters of a mile from the land. Off the N. Fox island there are several rocks above water, and a sunken rock lies off the S. side of the island. To enter Hermitage cove a vessel should keep between the islands and the shore, borrowing somewhat towards the mainland in 30, 32, and 37 fathoms water. The cove will now be seen open, and a vessel may turn to the S., having deep water, and without the least danger. From hence Hermitage bay runs in nearly E. for 14 miles with very deep water, until near the head, where it lessens to 25 and 22 fathoms.

HERMITAGE OF SANTA CATALINA, Spain, N. coast, a mark for the anchorage of large vessels in the road of Gijon, bringing this Hermitage in one with Point Otero and the little Island Orrio in one with the town of Candas.

HERNE BAY, Kent co., situated S.E. by S. about 7 miles from the white beacon buoy on the W. end of the Spaniard shoal, and W.S.W. 2½ miles from the W. buoy of the Last. In this bay there is a town rising into some note as a watering place, and having a handsome pier running out in a N. ½ W. direction, full 3,000 feet, for the convenience of landing passengers. The distance from this place to Margate is 11½ miles E.

HERNOSAND, PORT, Sweden, Gulf of Bothnia, in Angermanland, at the mouth of the river. Pop. 2,500.

HERON ISLAND, N. America, Gulf of St. Lawrence, at the head of Chaleur bay, surrounded by a reef off Heron Island, a mile and a quarter offshore to the S., and half a mile offshore, cleared when the River Charlo bears to the S. of S.S.W.

HERRADURA BAY, Spain, S. coast, E. by S. 5 miles from Nerja. It has a watch-tower upon each point, distant from each other about 1½ mile, having the castle of Mora on the beach: it is a dangerous roadstead in winter, from vessels moored there not having room to turn, and consequently being certain of going on shore.

HERRING COVE, America, United States, 1 mile to the S. of Race point, in Massachusetts, near Cape Cod harbour.

HERRING, or SCHOONER'S, COVE, Nova Scotia, situated on the N.E. side of Liverpool bay, affording good shelter from sea winds, in 3 fathoms water, on a bottom of mud.

HERRING GUT HARBOUR, America, United States, Maine; Marshall's point is at the E. entrance of this harbour, and carries a fixed light, elevated 30 feet above the level of the sea. Old Cilly, a low black rock, always seen above the surface of the water, bears from the light on Marshall's point S., distant about 3 miles. A reef extends E. nearly a quarter of a mile, which must be avoided. Bradford island forms the W. side of this harbour, the directions for entering which must be particularly attended to, and it may be made in perfect safety.

HERVAGAULT'S BREAKERS, lat. 41° 2′ N., long. 49° 23′ W. These were originally stated as a danger upon the authority of M. Hervagault, commander of the Conquerant, of Nantes, as seen June 26, 1723. He described them as composed of two parts, between which he was obliged to pass, being, when he first saw them, a cable's length from one, and not more than the one-eighth of a league from the other. The sea was clear between, and broke heavily on these rocks. On the 12th of May, 1827, the commander of the ship Home, Mr. Maxwell, on his passage from Liverpool to New York, fell in with three sunken rocks, a tremendous sea breaking upon them. They were apparently from 4 to 6 feet under the surface, in lat. 41° 2′ N., long. 49° 23′ W., about 30 feet in circumference; the last of them tailed off to the N.E. with a long ledge. Soundings were not attempted, the ship being under a press of canvas, and it being dusk. The temperature of the air at the moment was 63°, and an hour before the danger was perceived it had been tried at a depth of 2 fathoms, and the water found to be 45°; an hour afterwards it was tried again, and the water found to be 60°. From the first to the last of the rocks was a mile. Wind at the time W.S.W; ship's head N.W.; going 7½ knots an hour. The chronometer on board asserted to be so correct as to be depended upon to a mile. In 1816 it is said that Captain Lourp, master of the brig Alexander Savage, saw it, and he placed it in lat. 41° 6′ 23″ N., and long. by dead reckoning 49° 57′ W.

HERWIT REEF, Scotland, off the Frith of Forth, to the S. of the Brigs of Inch Keith, 1 mile distant from Inch Keith. It is seen at spring ebbs, and stretches S.E. by S. nearly half a mile from the brigs. Between there is a channel of 3 fathoms at low water, and close to the outer point of the Herwit, there are 15 fathoms water which shoal off to the S.; a black buoy on the S.E. end.

HESSELOE ISLAND, in the Cattegat, separated from the Lyse ground by a channel having 10 and 13 fathoms midway, commonly hard sand. Between Hesselöe island and the shore of Zealand, having midway 10, 12, and 14 fathoms in a passage 11 and 12 fathoms wide. This island is small and low, with a reef running out at each end, that towards the coast of Zealand extending a full mile and a half, and sometimes becoming dry. The reef on the other end of the island runs out N.W. by N. nearly 2 miles. A lighthouse stands on Hesselöe; the light revolves in one minute in such a direction that for 19 seconds it appears a fixed light; then a strong glaring one for 11 seconds; then a fixed light for 19, and after that it disappears wholly for 11 seconds. This light is 85 feet above the level of the sea, and visible, in a clear atmosphere, 14 miles. The lighthouse is 32 feet high, and painted white.

HESTE HEAD, Denmark, on the E. coast of Falster island, and the most E. point.

HEVE, CAPE LA, France, W. coast, 10½ miles from Cape Antifer, S.W. ¼ W. At this cape the steep white cliffs along the coast terminate. Upon the cape are two lighthouses of freestone, placed on a level, and of equal height. Each tower is 56 feet high from the ground, and the lanterns 15 feet. They exhibit bright fixed lights, elevated 446 feet above the level of the sea, and may be seen 7 leagues off, so placed as not to be seen in a line by any vessels from the N., N.W., and W., in order that, when coming in for Havre, they may never be mistaken for the lighthouses of D'Ailly or Cape Barfleur, which stand singly.

HEVER, or HEEVER RIVER, Sleswick, to the N. of the Eyder about 8 miles N.N.E. The opening consists for the most part of inlets among sand-hills, which are intricate and deceptive to those not well acquainted with the coast.

HEYST, coast of Flanders, a town between Blankenburg and Cadsand. Here a fixed red light is established to the N. of the town, in lat. 51° 20′ 22″ N., long. 3° 14′ 7″ E. The lighthouse is 25 feet high, and the light elevated 48 feet above the level of the sea at spring tides, visible between the bearings of E. round to W. by S. by compass.

HICACOS, or YCACOS, POINT, KAYS, and ISLETS, West Indies. N. coast of Cuba; the latter, three in number, lie N.E. of the point on the edge of the bank.

HICKMAN'S HARBOUR, Newfoundland, between Bacalhao island and Cape Bonavista, 3 leagues from Random N. head, where there is good anchorage in 15 fathoms. Random N. head bears from Random S. head N.E. ½ E., distant 3 miles.

HIDE SAND, Holland and the Low Countries, on the coast of Zealand, near the Island of Schouwen. It is somewhat flat, and a ship may approach pretty near on the W. side.

HIELM ISLAND, Denmark, a small island in the Cattegat, E. of Hasselore head, on the E. coast of Jutland. There are several patches within a mile around this island, but a reef of 12 to 18 feet extends 4 miles to the S. The passage between this island and Jutland should never be attempted, for though there are 12 and 13 fathoms water, there are shoals which render it dangerous. To the E. of this island there is good anchorage in 10 and 9 fathoms on sand. This island, the Basser and Hatter reefs and Samsoe island, form the W. points of the entrance into the Great Belt. The Hatter is a low stony reef extending N.W. by W. and S.E. by E. about 2 miles, having shoal water around it, in some places not more than 4 feet, deepening to the N.E. and in the direction of Seyer Oe to 3 and 4 fathoms. To the N. this shoal and the Basser extend about 3½ miles, having 3 fathoms upon it in some places, and in others being foul and rocky. About 1½ mile S. of the Hatter reef is a small knoll of sand called Hatter Barn, having on its middle no more than 2½ fathoms water, its direction being E. and W. with 6 and 7 fathoms on its extremities; between this and the Hatter reef shoal are from 10 to 20 fathoms. The sea frequently breaks over these reefs. By bringing Reefness to bear S.W. by S., a ship will go to the E. of them.

HIGUERA, CAPE, Spain, N. coast, at the entrance of the Bidassoa river. It has a fixed light, in lat. 43° 23′ N., long. 1° 46′ W.

HIGUEROTA ISLAND, Spain, at the entrance of the River Guadiana, having the little town of Cannels on its W. side.

HIGUEZ BAY, West Indies, Island of Hayti, S.W. of Point Espada.

HILLSBOROUGH BAY, North America, Prince Edward's Island, Gulf of St. Lawrence. This bay contains the principal harbour and capital of the island, as well as being the largest bay on its shores. It has numerous dangers, which render its navigation difficult to strangers, but these may be obviated by the Admiralty chart and directions, it having been carefully surveyed. In Charlotte town harbour there are three rivers uniting, one of which is the Hillsborough, navigable for the largest ships to the distance of 7 or 8 miles, and for small vessels 14 miles above Charlotte town, where there is a bridge 2 miles from the head of the river. There is a portage of less than a mile across from Hillsborough near its head to Savage harbour on the N. coast of the island. The Bay of Hillsborough enumerates within its limits St. Peter's island, 3 miles in circumference; St. Peter's shoals, Spit, and Head; the Front rock, Squaw shoal, Governor island and shoals; Fitzroy rock, Huntly rock; Prim island and reef, Charlotte town harbour, Canseau shoal and Middle Ground. Charlotte town contains about 5,000 inhabitants; it stands on the N. bank of the Hillsborough river. It is high water there at full and change at $10^h 45^m$; rise at ordinary springs, 9½ feet; at neaps, 7 feet. The rise is much affected by the winds, N.E. gales raising them 11 feet, and neaps, during S.W. gales, only 6 feet. The E. part of the bay is little visited by shipping, and it has many dangers, to obviate which directions would avail little. Pownall bay, Gallows point, Orwell bay, Pinette harbour and shoals, Flat river, Rifleman reef, Ball point, Indian rocks, Wood islands, White sand, Black rock point, and Cape Bear, are the names of localities enumerated in this part of the bay. There is a lighthouse on Prim point, the S.E. point of the bay, a low point, with sandstone cliffs, 10 or 15 feet high. It is built of brick, and shows a fixed light 68 feet above the sea at high water, visible 4 or 5 leagues. It is of great use to vessels approaching from the E., as by its bearing it guides them clear of the Rifleman and Pinette shoals, and enables them to enter the bay in the darkest night. The W. extreme of St. Peter's island bears from it N.W. ½ W. 7¾ miles, and the W. end of Governor island N. ¾ E. 5 miles. This light is in lat. 46° 3′ 12″ N., long. 63° 6′ 30″ W., after Capt. Bayfield, R.N.

HILLSBOROUGH BAY and RIVER, West Indies, Island of Tobago, W. of Barbadoes bay.

HILLSBOROUGH, TOWN and BAY, West Indies, Island of Cariaco, one of the Grenadines.

HILS BANK, coast of France and Flanders; this bank is joined to the Braak, and limits the E. part of Dunkirk road. It is very dangerous, on account of its extent and shallowness, and the great depth of water close to its S. edge. Its general direction is parallel with the shore as far as its S.E. point, where there is a black buoy in 13 feet water, bearing E.N.E. from Dunkirk steeple, and N. from that of Zuydcoote.

HILTON ROCKS, W. of the Azores, lat. 39° 18′ N., long. 35° 50′ W. The bark Secret, Mr. R. Hilton, commander, from Valparaiso to Liverpool, May 25, 1845, had breakers reported as he happened to be taking a meridian altitude. They were of no great extent, but he saw some objects in the hollow of the waves which he felt certain were rocks. The swell was not very heavy, and in smooth water he thought they would be level with the surface of the sea. The breakers were about 1½ or 2 miles S.W. by compass from the vessel, and at the time she was running 7½ or 8 knots; there was not time for any particular remarks.

HINDER, WEST, a bank N.W. off Ostend; dangerous from its great distance off land, and being in the way of vessels bound for the Scheldt. It is long and narrow, lying in the direction of N.E. ½ E. about 13 miles in length. Its S. end, in 5 fathoms, lies N.W. by N., distant about 4 miles from the N.E. end of the Cliff bank or East Dyck, in lat. 51° 23′ N., and its N.E. in lat. 51° 34′ N. The depths very irregular, from 4½ to 3 fathoms, shoaling from 15 to 9 very suddenly, the latter depth being close to the edge of the bank. The S. end of this bank bears from the North Foreland lighthouse E.S.E., distant 36 miles; from the Galloper light-vessel S.S.E. ½ S., 29 miles; from Ostend N.W. ¼ N., 20 miles; and from West Kapelle W. ¾ N., 39 miles. The N. end bears from the Galloper light-vessel S.E. ¼ S. 26 miles; from Ostend N. by W., 24 miles; and from West Kapelle W.N.W. ½ N., 32 miles.

HINDER BANK, Holland; this bank lies directly before the entrance of Hellevoet-sluys, to the N. of the Island of Goeree, and the W. of Voorn island, forming on its S. side, with the strand of Goeree, the West Gat; and on its N.E. side, with the West Plaat, the North Gat. Near the middle of this bank is a patch which dries at half-ebb, about a mile in length, and nearly 1½ mile from Goeree island: this is called the Steele Hinder. On the other parts of the Hinder are from 2 feet to 2 fathoms; and to the westward of it the depths gradually increase to 4 or 5 fathoms. Along its S.W. and S. sides, is a long and shallow patch, called the Bol, or Hompel. The West Gat is marked by an outer black buoy, which lies N.W. by W. ¼ W. nearly, distant 2 miles from Houten Kaap, with the latter and Goeree church in a line. On the bar are 13 feet water. The latter mark will lead to the S. Bol or Hompel; and when within 2 cables' length of the shore, keeping at that distance until arriving at Kwade Hoek, leaving all the buoys on the port side. After passing the Kwade Hoek, off which a shallow spit extends, on which a white buoy is placed, the channel takes a S.S.E. and S.E. direction, and is called the S. deep. The Hinder Bank continues, mid-channel, to the S.E., and has over it some swashways, which are to be crossed in proceeding to Hellevoet-sluys, but these require a pilot to lead through. The entrance to the North Gat, between the Hinder Bank and the West Plaat, is about half a mile wide, with 2½ fathoms of water.

HINDER, NORTH and EAST. The S. part of the N. Hinder bank bears N. by E., distant 2 miles from the N. end of the West Hinder, having in the channel between them from 12 to 19 fathoms. It thence extends N.E. by N. 6 miles, to lat. 51° 42′ N., having from 4½ to 5 fathoms on its shoalest parts, which runs about 2 miles along the centre of the bank; from this to the extremes of the bank, are 6½ to 8 fathoms. The East Hinder has its S. end in lat. 51° 30′ N., and lies 1½ mile to the E. of the N. end of the West Hinder, having from 12 to 20 fathoms in the channel between them. This bank is about 11 miles in length, and three-quarters of a mile in breadth, with a narrow ridge of 4 fathoms, running nearly the whole length of the bank. Its N. end lies in lat. 51° 40′ N. Between this bank and the North Hinder, are from 14 to 21 fathoms; the channel is 3 miles wide.

HINLOPEN, or HINDELOPEN HARBOUR, Holland, on the E. coast of the Zuyder Zee, or shore of Friesland, nearly due E. from the Ton on the Bight, and N. of the Fries Plat buoy. This harbour is small, and stands in lat. 52° 52′ N., long. 5° 26′ E.

HINSHORT LOCH, Scotland, W. coast, to the N. of Loch Lusford 1 league. This loch is sometimes called Loch Inchard. It may be known at sea by Arkall hills, or Balag island, and may be considered the last bay before rounding Cape Wrath. There is depth of water, and the bottom is good in Ardicaar loch, a little further N.E., but with W. winds it is very hard riding there, and Hinshort loch is well sheltered.

HIRTA ISLAND, more correctly ST. KILDA. See *St. Kilda.*

HIRTSHOLM ISLAND, Denmark, off Falstrand, in the N.W. part of the Cattegat. It bears a light, revolving once in 1½ minute, on a quadrangular tower. The lantern is 42 feet above the level of the sea, and the light visible at the distance of 10 miles. The tower is white, and serves as a sea-mark. The anchorage for large vessels is in 5 and 6 fathoms. See *Falstrand.*

HISKERE ISLAND. See *Causamul.*

HISPANIOLA ISLAND, a former name of the Island of St. Domingo, now Hayti.

HITHE, or HYTHE, England, Kent co., one of the Cinque ports, with a harbour choked up; having also a small fort. Pop. 2,287. It was called East Hithe, and derived its consequence, before its harbour was in its own turn ruined, by the decay of the harbours of West Hithe and Lyme from a similar misfortune.

HITTARP REEF, coast of Sweden, at the entrance of the Sound from the Cattegat. The town of Viken on the Swedish shore has a little to the S. several houses near the land, and at the back two with red tops. When these bear N.N.E., a vessel will be abreast of this reef.

HITTERO ISLAND, Norway, from the Naze 5 leagues W.N.W. S. of which is the entrance into Witford, to the N.E. and N. of which is the passage to Berg sound, E.N.E.

HOBART'S SHOAL, West Indies, S.E. of the Mira-por-vos group of rocks, discovered by Mr.W. James, commander of the Lord Hobart, in 1821. When thus seen, Castle island bore N. by W. 14 or 15 miles; heavy breakers were observed ahead. The ship tacked to the E., the Mira-por-vos bearing N.W., and Castle island N.E., the breakers N.W. by N., distant half a mile. This shoal had been seen by the same commander three years before. Five miles S.E. of the Mira-por-vos is a spot of 3 fathoms in Admiralty Chart, 1847.

HOFT'S SHOAL, Gulf of Finland, 10 miles S. of Stonescar, a dangerous bank, formed of six or seven rocks above water, joined by a reef extending from the N. rock 2¼ miles towards the main. There are 13 fathoms between that and the shore, and from 25 to 30 between that and Stonescar. There is a pyramidical beacon stands upon these rocks, having a ball at the top.

HOG ISLES, African coast; four small islands, named in the country Rouban, Banak, Chiveya, and Corett. The latter is covered with large trees, and is the most N. They lie near Kanyabac island. They belong to the Bijooga group, the N.E. of them, and their E. point lying in lat. 11° 20′ N., long. 15° 40′42″ W., according to Capt. F. W. Owen, R.N.

HOGHALLA HEAD, Sweden, W. coast, at the bottom of Laholm bay, S.W. of Halmstadt, and N.E. of Koll, W. of the S. end of the Great Middle ground.

HOGNES or HOGANES HAVEN, Sweden, 6½ miles S. of Koll, the E. boundary of the entrance into the Sound. Here is a temporary anchorage, where vessels may ride in 3½ and 4 fathoms, and be tolerably sheltered from the seas caused by N.W. winds. A pilot for the Sound is always to be obtained.

HOGSTIES, West Indies. See *Cayman Islands.*

HOGSTY REEF, or LOS CORRALES, West Indies, three small low sandy islets or kays, having on the E. side a white shoal, surrounded by a reef

extending 6 miles. They lie at the distance of 11 leagues N. by W. ¾ W. from the N.W. point of Heneagua, or Inagua, in the Crooked Island passage. There are rocks and breakers and much broken ground about them, and in blowing weather the sea breaks over them all. The most S. is the smallest, and requires a good berth, it being shoal all round. It bears from the middle quay about S.S.E. The N.W. kay is in lat. 21° 40′ 30″ N., long. 73° 50′ W.

HOGUE, CAPE LA, France, the N.W. point. There is a fixed light here, 157 feet high; lat. 49° 43′ N., long. 1° 57′ W.

HOHER WEG SAND, in the River Weser, on the W. side, one of the shoals which border the entrance of that navigable river.

HOKOPING, Sweden, a town on the W. coast, N.E. of Falsterbo point.

HOLBECK, Denmark, a town at the bottom of the Jese Fiord, on the N. of Zealand island, nearly on the same parallel as Copenhagen, to the W.

HOLDEN'S BED, Ireland, 2 miles N.E. from Greenore point, E. coast, a shoal running E.N.E. and W.S.W., 4 miles long, and 1 broad. There are 6 feet water upon its S.W. end, and on its other parts 2 and 2½ fathoms. To sail clear of the S.W. end, keep the most S. of two high hills N.W., or the N. hill N.W. by N.

HOLDERNESS ISLAND, America, Nova Scotia, called also SOUTH-WEST ISLAND, between Halifax and Cape Sable. It is a very rocky island, 50 feet high, and steep on all sides. Lat. 44° 34′20″ N., long. 63° 58′48″ W., according to the Admiralty surveys.

HOLE IN THE ROCK, West Indies, at the S.E. end of Abaco island. See *Bahamas.*

HOLKHAM BAY, also called WELLS ROAD, Norfolk co., between Wells and Burnham. Here vessels may find anchorage in 3 fathoms, bringing Wells church to bear S. by E., Holkham church S.W. ¾ S., and the N. part of the Scald Head hills, W.N.W., distant 1⅜ mile from the shore.

HOLLAND HARBOUR, or CASCUMPE-QUE, America, Prince Edward's island, on the N.E. side, on which side it is the most W. harbour. The sands form a bar here running about a mile and a half off the shore. This harbour is easily known by the sand-hills which extend along the coast. Half way between the entrance of Richmond bay and Holland harbour there is one high sand-hill near the Conway inlet, much higher than the rest. Holland bay may be known by its being at the W. end of all the range of sand-hills. There is good anchorage close to the bar, in from 5 to 8 fathoms, and 18 feet of water upon the bar, and it is easily entered by strange vessels not drawing more than 12 feet. Surveyed by Captain Bayfield, R.N., 1839.

HOLLANDES POINT, or PUNTA DE OLANDES, West Indies, Island of Cuba, W. of Corrientes bay, and bearing W. ¼ N., distant 5¼ leagues from Cape Corrientes, terminating to the W. of the bay of that name. Near and E. of it commences a reef, which extends in that direction half a mile, and has no danger, being bold-to and close to the coast. Point Hollandes has a pleasing appearance, bearing some resemblance to a curtain on a wall for about two miles. The shore is woody after passing it. There are some sunken rocks off this point to the E. The coast here begins to bend to the N.E., forming a deep bay of 12 miles broad, and bold to the land; but there are no soundings in it.

HOLLEPOORT, coast N. of the Texel, the principal entrance to the Vlie Stroom, lying between a bank on the N. side, called the Noorden or North Buitengrund, and on the S., called the Wester Grund. The outer buoy of the Hollepoort is red, and lies with the beacon on the Strond Vlieland, in one with a mill on the S. side of the island, bearing S. ¼ W. and the S.W. end of the Ter Schelling S.E. by E. At the entrance to this channel the depth is 5 fathoms.

HOLLESLEY BAY, Suffolk co., England. The best anchorage in this bay is with Hollesley church bearing N. by W. ¾ W., or with the parsonage house in one with the Red Barn, in 4 or 4½ fathoms. There is also good anchorage in the N.E. part of the bay, between the middle or hook buoy, which is W. of the Whiting and the N.E. of the Middle Ground, in a depth of 6 or 7 fathoms. The tide flows here at full and change of the moon at 11 o'clock. A vessel should stand no nearer on either side than 6, nor to the beach than 7 or 8, fathoms.

HOLM HEAD, Baltic sea, the N.E. extremity of the Island of Gottland, in long. 19° 22' 30" E.

HOLM SAND, a large sandy flat joined to the Corton sand, between Orfordness and Yarmouth roads. See *Corton Sand.*

HOLMEN, or the HOLMS, Denmark, W. coast of Jutland, W. by S. from Robsnout 43 miles.

HOLMS, entrance of the Severn river, two small islands lying someway apart in the middle of the channel; that to the S. is called the Steepholm, a high island, the other to the N. by E. ½ E. about 2 miles, is called the Flatholm (See *Flatholm*). The tides run between these islands with great violence. At the Steepholm the tide flows on full and change 6h 37m. Spring tides rise 38 feet, neaps 21, and equinoctial springs 41 feet. Few vessels pass to the S. of Steepholm unless they are going to or returning from Bridgewater river. The Steepholm is of a narrow oblong shape, about half a mile long, in a S.E. by E. direction; the shore rises 220 feet above high water, or three times the height of Flatholm, and is a good mark to make out at night. The eastern end only is accessible, abreast of some fish stakes, where a spit of shingle dries out to the S.E., and there, as well as at the western end, a three-fathom shelf of rocks projects half a mile from the island; otherwise it is pretty bold-to. There are from 6 to 8 fathoms, sand and mud, between Steepholm and the Culver Sand, and similar depths towards Brean Down, but the bottom is foul; between it and Flatholm there are from 7 to 15 fathoms, the bottom mostly rocky. In fact so strong are the tides and so many are the changes in the banks, that there is no safe navigation in this vicinity, but for those well acquainted with the navigation. Of the necessity of this, a judgment may be formed from the fact of a communication made to the Admiralty in October last by Captain Beechey, R.N., and made public in the *Nautical Magazine.* Captain Beechey, then busily employed upon a survey of the noble estuary of the Severn, has discovered the most astonishing changes in the ground there, in a space of time so short as to render them still more extraordinary. The statement thus made is in substance, that between 1827 and 1839 great changes took place near this locality. One was the "fact of a shoal having risen on one side of the channel, nearly 80 feet in perpendicular height, and nearly a mile in extent, since the previous survey had been executed; and

that on the other side of the channel, on the English grounds, and about where a light-ship was to be placed, a shoal half a mile in length, that was before dry, was now not only annihilated, but there were thirty feet at low water spring tides where it had stood; and a bay had been scooped out of the English grounds a mile and a half in length, half a mile in breadth, and at an average increase of 15 feet in its depth; that there had from this spot alone been removed 337,500,000 cubic feet of earth ! and that on the northern side there had, on the contrary, accumulated 480,000,000 cubic feet of sand. That a channel had opened in the sands more than half a mile in width, in which there were generally 3 or 4 fathoms, where there had before existed a bank dry at half tide, 2 miles in length, by an average of one-third of a mile, or 2,000 feet in width, that there had, in short, been removed from here in addition to the above 840,000,000 cubic feet of sand." Such extraordinary changes show how much caution is required in this navigation, and how necessary it is that attention should be paid to the extracts from the Admiralty directions for the Bristol channel furnished by the able officer who compiled them.

HOLME'S HOLE, America, United States, Massachusetts, a harbour to which vessels resort during the winter season, there being a post office and mail made up there twice a week for Boston. There is a lighthouse having a fixed light, erected on the W. part of the harbour on the starboard hand going in, 60 feet high. This light is in lat. 41° 29' N., long. 70° 36' 42" W.

HOLTENAU, Baltic, at the N. entrance of the Holstecuke canal, at the port of Kiel, one of the three harbour lights there so denominated.

HOLY HEAD HARBOUR, North Wales, Caernarvon co. Upon entering this harbour, so well known as that used for the communication of the sister country with Ireland, there are two rocks to be avoided near Ynys Gybi point; the first is the Platters, lying nearly quarter of a mile N. ½ E. from Ynys Gybi, or Salt island, the least water on which is 4 feet; to avoid it on the N. side, the North Stack head must be brought out by the low point of Ynys Wellt or the west mill on with Ynys Gybi point; the E. windmill in a line with the lighthouse on Holy Head pier, will clear it on the E. side. The other rock is small, lies about half a cable's length N. from the middle of Ynys Gybi, and dries with spring-ebbs only. A vessel waiting for water to sail into the harbour must anchor at about two cables' length from the entrance in 5 fathoms, with the mouth of the harbour a little open. Ships drawing 15 or 16 feet may go into the harbour about high water, with spring-tides; and vessels drawing 11 feet may go in at high water, with neap-tides. A stone pier, 200 fathoms in length, extending true E. and W., projects from the S.E. end of Salt island, and forms the N. side of the harbour; on the pier head is a lighthouse, where, by night, a brilliant fixed gaslight is exhibited; there is also a red light shown from the same lighthouse, in a N.N.E. direction only 44 feet high; this is intended to lead you clear of the Platters, on which there is a black buoy, with bells attached to it. On the south side of the entrance is another pier, running out in a true N. direction; and to the eastward of this are some rocks called the Pibeo and Nimrod. About E. ¼ N., nearly a quarter of a mile from the lighthouse on the pier head, lies the dangerous rock which had so long

PART II.   NOVEMBER, 1847.

THE

NAUTICAL GAZETTEER;

OR,

DICTIONARY

OF

MARITIME GEOGRAPHY.

UNDER THE ESPECIAL SANCTION
OF THE
LORDS COMMISSIONERS OF THE ADMIRALTY.

LONDON:
HENRY HURST, KING WILLIAM STREET, CHARING CROSS.

SOLD ALSO BY
R. B. BATE, AGENT TO THE ADMIRALTY, 21 Poultry;
NORIE AND CO., CHART AND MAP SELLERS, 157 Minories; LETTS, 8 Royal Exchange;
AND BY ALL BOOKSELLERS IN TOWN AND COUNTRY.

Price 2s. 6d.

ADDRESS.

It is not the custom to regard any effort to be useful with indifference, and the attempt to concentrate information of the present character from sources widely scattered must be admitted to come within the category of utility. Our national greatness reposes almost exclusively upon a maritime superiority, in which scientific is united with mechanical proficiency, or, in other words, cultivated intellect with manual skill and experience. Neither the dauntless enterprise and courage struggling against physical obstacles which marked the period of early British navigation, nor its historical details down to the close of the last century, abounding as they do in pictures of matchless perseverance and triumphant result, exhibited that union of intellectual with physical effort which advanced the science of navigation so greatly during the half century that has just run its course. Within the term of human memory the subjugation of the ocean has gone rapidly forward. Great progress has been made in acquiring a knowledge of the phenomena of the winds and waves; although in exploring their causes we may as yet have done no more than pick up a pebble here and there from the shore of the great ocean of truth. We are advancing from secondaries to primaries. We are busy in penetrating into existing mysteries through the collection of facts for data, and in establishing their correctness. We are exploring those latent, perhaps simple because powerful agencies, which work out atmospherical and oceanic changes, elevating tides, directing the course of currents, and wielding the terrific energies of the tempest. More remarkable still, we are in the present day doing all this not with the exclusive view of avoiding what may be prejudicial to navigation, but with that of turning this very prejudicial action to account. This advancement of the intellectual part of the science has overcome many of its physical difficulties. The suffering experienced at sea by former navigators arising from disease, unwholesome sustenance, and ill-found vessels, has been exchanged for the healthiest position in which the human body can be placed, and in well-found vessels, proportional security. The modern seaman braves voyages of discovery with ease under which the amazing power of endurance of his forefathers must have sunk, and this facility he owes to scientific advancement. In voyages of exploration and in examining and surveying the most dangerous coasts, the ancient persevering spirit of investigation is eminently awakened. Even the course and action of hurricanes are submitting to a careful investigation. In regard to these the theory of their progression in vortices, advanced recently, may perhaps be found correct and made available for use or security. In the laborious and hazardous task of surveying, the names of Mudge, Owen, Smyth, Vidal, Bayfield, Stokes—but it is invidious to particularize,—have with others contributed largely to expand the sphere of knowledge, and widen the bounds of maritime geography, securing to themselves a boundless debt of national gratitude. The Government of late years has not been wanting in lending assistance to this most important branch of the science, and it has recorded the results under the revision and superintendence of one of the most distinguished hydrographical officers this or any other country ever had the advantage of possessing.

It is under such circumstances and at such a period, that the present Publication is submitted to general use. As yet but few points on the earth's surface are established in a perfect agreement of position, a disadvantage to which time and superior instruments will administer a remedy. Many, therefore, are drawn from former authorities which have alone recorded them, others differ, and some are a mean of two or three observations by different individuals, a circumstance inseparable from the vast mass of materials existing. This last fact has been the cause of a degree of compression in the different articles not as first contemplated. The object of the concentration of so many facilities, will, however, be attained, and a reference will always be at hand for the purpose of instruction and information, or to gratify curiosity in relation to what is at present scattered far and wide over the ocean surface. For the Australasian regions and shores in the Pacific Ocean, where a vast extent of coast has been explored, named, and colonized, within human memory, a work of reference of the present character does not exist in any shape, and was therefore imperiously required to carry onward the march of improvement and extend the limits of the world.

An acknowledgment of the obligations of this work to Lieut. Raper's *Nautical Positions* is a duty. It is to be regretted that the list of this distinguished scientific labour is so limited. The vast number of places required renders a reference here continual to sources, no two of which accord, an evil inseparable from the absence of such an authentic

PART III.   DECEMBER, 1847.

THE

NAUTICAL GAZETTEER;

OR

DICTIONARY

OF

MARITIME GEOGRAPHY:

UNDER THE   SPECIAL SANCTION

OF   THE

LORDS COMMISSIONERS   OF THE ADMIRALTY.

LONDON:

HENRY HURST, KING WILLIAM STREET, CHARING CROSS.

SOLD ALSO BY

R. B. BATE, Chart Agent to the Admiralty, Poultry;

NORIE and CO., Chart and Map Sellers to the Admiralty, Leadenhall Street;

AND BY ALL BOOKSELLERS IN TOWN AND COUNTRY.

Price 1s. 6d.

ASTRONOMICAL ALMANACK,

Just Published by the Stationers' Company, price 1s.

WHITE'S CELESTIAL ATLAS, or an Improved Knowledge for 1844, wherein are contained the Places of the Old and New Planets, the Eclipses, Occultations, and other Celestial Phenomena ... the year, adapted to the Meridian of Greenwich. By W. S. B. WOOLHOUSE, F.R.A.S. &c.

Published by GEORGE GREENHILL, Stationers' Hall, Ludgate-street.

Also, REQUISITE TABLES for the Use of Nautical Men, Astronomers, and others, price 4s. 6d. in cloth, or, bound with the Almanack, 5s. 6d.

THE NAUTICAL GAZETTEER.

IT is not the custom to regard any effort to be useful with indifference, and the attempt to concentrate information of the present character from sources widely scattered must be admitted to come within the category of utility. Our national greatness reposes almost exclusively upon a maritime superiority, in which scientific is united with mechanical proficiency; or, in other words, cultivated intellect with manual skill and experience. Neither powerless enterprise and courage struggling against physical obstacles which marked the period of early British navigation, nor its historical details down to the close of the last century abounding as they do in pictures of matchless perseverance and triumphant results, exhibit that union of intellectual with physical effort which advanced the empire of navigation so greatly during the half century that has just run its course. Within the term of human memory the subjugation of the ocean has gone rapidly forward. Great progress has been made in acquiring a knowledge of the phenomena of the winds and waves, though, in exploring their causes we may as yet have done no more than pick up a pebble here and there from the shore of the great ocean of truth. We are advancing from secondaries to primaries. We are busy in penetrating into existing mysteries through the collection of facts for data, and in establishing their correctness. We are exploring those laws, perhaps simple because powerful agencies, which work out atmospherical and oceanic changes, regulating tides, directing the course of currents, and wielding the terrific energies of the sea-god. More remarkable still, we are in the present day doing all this not with the exclusive view of ascertaining what may be prejudicial to navigation, but with that of turning this very judicial action to account. This advancement of the intellectual part of the science has overcome many of its physical difficulties. The suffering experiences at sea of former navigators arising from disease, unwholesome provisions, and ill-found vessels, has been exchanged for the healthiest position in which the human body can be placed, and in well-found vessels, proportional security. The modern seaman braves voyages of discovery with ease under which the crushing power of endurance of his forefathers must have sunk, and this facility he owes to scientific advancement. In voyages of exploration and in examining and surveying the most dangerous coasts, the patient persevering spirit of investigation is eminently awakened. Even the cause and union of hurricanes are submitting to a careful investigation. In regard to these the theory of their progression is perhaps advanced recently, may perhaps be found correct and made available for useful security. In the laborious and hazardous task of surveying, the names of Bayley, Owen, Smyth, Vidal, Bayfield, Stokes—but it is invidious to particularize, have with others contributed largely to expand the sphere of knowledge, and made the science of marine geography securing to themselves a boundless debt of national gratitude. The Government of late years has not been wanting in lending assistance to this most important branch of the science, and it has recorded the results under the revision and superintendence of one of the most distinguished hydrographical officers this or any other country ever had the advantage of possessing.

It is under such circumstances and at such a period, that the present Publication is submitted to general use. As yet but few points on the earth's surface are established to a perfect agreement of position, a disadvantage to which time and superior instruments will administer a remedy. Many, therefore, are drawn from former authorities which have alone recorded them; others differ, and some are a mean of two or three observations by different individuals; a circumstance inseparable from the vast mass of materials existing. This last fact has been the cause of a degree of compression in the different articles not at first contemplated. The object of the concentration of so many localities will, however, be attained, and a reference will always be at hand for the purpose of instruction and information, or to gratify curiosity in relation to what is at present scattered far and wide over the ocean surface. For the Australasian regions and shores in the Pacific Ocean, where a vast extent of coast has been explored, named, and colonised, within human memory, a work of reference of the present character does not exist in any shape, and was therefore imperatively required to carry onward the march of improvement and extend the limits of the useful.

An acknowledgment of the obligations of this work to Lieut. Raper's Marine Positions is a duty. It is to be regretted that the list of this distinguished scientific Officer is so limited. The vast number of places required renders a recurrence to a multitude of sources, no two of which accord, an evil inseparable from the absence of such an authentic reference.

Part IV.   January, 1848.

THE

NAUTICAL GAZETTEER;

OR,

DICTIONARY

OF

MARITIME GEOGRAPHY.

UNDER THE   ESPECIAL SANCTION

OF   THE

LORD COMMISSIONERS   OF THE ADMIRALTY.

LONDON.

HENRY HURST, KING WILLIAM STREET, CHARING CROSS.

SOLD ALSO BY

R. B. BATE, CHART AGENT TO THE ADMIRALTY, Poultry;

NORIE AND CO., CHART AND MAP SELLERS TO THE ADMIRALTY, Leadenhall Street;

AND BY ALL BOOKSELLERS IN TOWN AND COUNTRY.

Price 2s. 6d.

THE NAUTICAL GAZETTEER.

It is not the custom to regard any effort to be useful with indifference, and the attempt to concentrate information of the present character from sources widely scattered must be admitted to come within the category of utility. Our national greatness reposes almost exclusively upon a maritime superiority, in which scientific is united with mechanical proficiency, or, in other words, cultivated intellect with manual skill and experience. Neither the dauntless enterprise and courage struggling against physical obstacles which marked the period of early British navigation, nor its historical details down to the close of the last century; abounding as they do in pictures of matchless perseverance and triumphant result, exhibited that union of intellectual with physical effort which advanced the science of navigation so greatly during the half century that has just run its course. Within the term of human memory the subjugation of the ocean has gone rapidly forward. Great progress has been made in acquiring a knowledge of the phenomena of the winds and waves; although in exploring their causes we may as yet have done no more than pick up a pebble here and there from the shore of the great ocean of truth. We are advancing from secondaries to primaries. We are busy in penetrating into existing mysteries through the collection of facts for data, and in establishing their correctness. We are exploring those latent, perhaps simple because powerful agencies, which work out atmospherical and oceanic changes, elevating tides, directing the course of currents, and wielding the terrific energies of the tempest. More remarkable still, we are in the present day doing all this not with the exclusive view of avoiding what may be prejudicial to navigation, but with that of turning this very prejudicial action to account. This advancement of the intellectual part of the science has overcome many of its physical difficulties. The suffering experienced at sea by former navigators arising from disease, unwholesome sustenance, and ill-found vessels, has been exchanged for the healthiest position in which the human body can be placed, and in well-found vessels, proportional security. The modern seaman braves voyages of discovery with men under which the amazing power of endurance of his forefathers must have sunk, and this facility he owes to scientific advancement. In voyages of exploration and in examining and surveying the most dangerous coasts, the patient persevering spirit of investigation is eminently awakened. Even the course and action of hurricanes are submitting to a careful investigation. In regard to these the theory of their progression in vortices, advanced recently, may perhaps be found correct and made available for use or security. In the laborious and hazardous task of surveying, the names of Mudge, Owen, Smyth, Vidal, Bayfield, Stokes—but it is invidious to particularise,—have with others contributed largely to expand the sphere of knowledge, and widen the bounds of maritime geography, securing to themselves a boundless debt of national gratitude. The Government of late years has not been wanting in lending assistance to this most important branch of the science, and it has recorded the results under the revision and superintendence of one of the most distinguished hydrographical officers this or any other country ever had the advantage of possessing.

It is under such circumstances and at such a period, that the present Publication is submitted to general use. As yet but few points on the earth's surface are established in a perfect agreement of position, a disadvantage to which time and superior instruments will administer a remedy. Many, therefore, are drawn from former authorities which have alone recorded them, others differ, and some are a mean of two or three observations by different individuals; a circumstance inseparable from the vast mass of materials existing. This last fact has been the cause of a degree of compression in the different articles not at first contemplated. The object of the concentration of so many localities, will, however, be attained, and a reference will always be at hand for the purpose of instruction and information, or to gratify curiosity in relation to what is at present scattered far and wide over the oceanic surface. For the Australasiatic regions and shores in the Pacific Ocean, where a vast extent of coast has been explored, named, and extended within human memory, a work of reference of the present character does not exist in any shape, and was therefore imperiously required to carry onward the march of improvement and extend the limits of the useful.

An acknowledgment of the obligations of this work to Lieut. Raper's *Nautical Positions* is a duty. It is to be regretted that the list of this distinguished scientific Officer is so limited. The vast number of places required renders a recurrence here continual to sources, no two of which accord, an evil inseparable from the absence of such an authentic